THE PLATES OF EARTH'S LITHOSPHERE

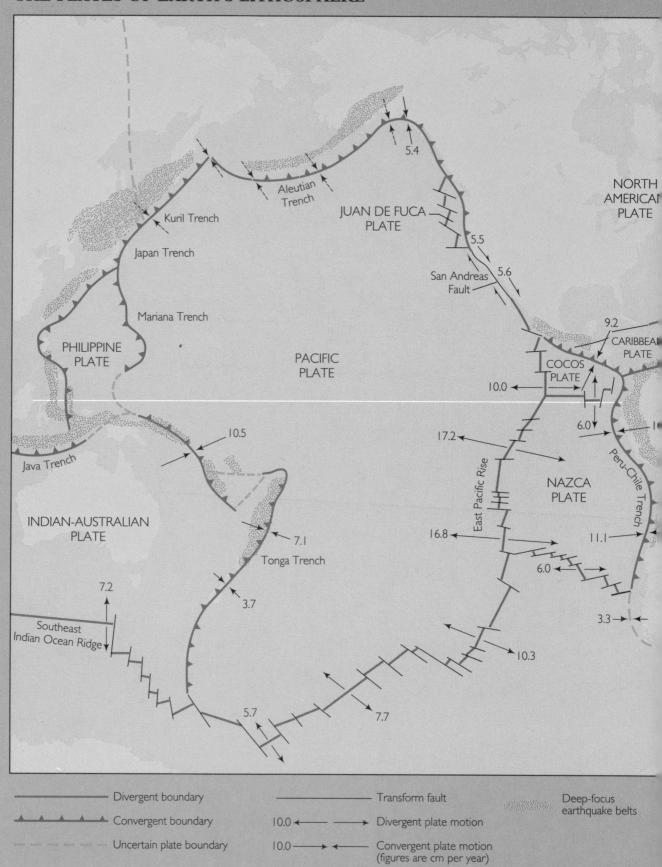

Divergent boundary

Convergent boundary

Uncertain plate boundary

Transform fault

10.0 ⟵ ⟶ Divergent plate motion

10.0 ⟶ ⟵ Convergent plate motion
(figures are cm per year)

Deep-focus
earthquake belts

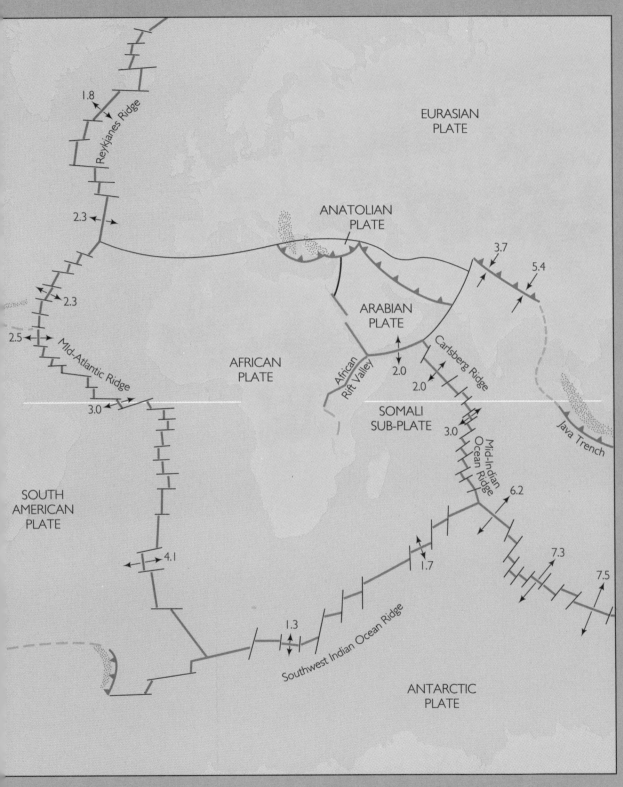

EURASIAN
PLATE

ANATOLIAN
PLATE

ARABIAN
PLATE

AFRICAN
PLATE

SOMALI
SUB-PLATE

SOUTH
AMERICAN
PLATE

ANTARCTIC
PLATE

Reykjanes Ridge

Mid-Atlantic Ridge

African Rift Valley

Carlsberg Ridge

Mid-Indian Ocean Ridge

Java Trench

Southwest Indian Ocean Ridge

1.8

2.3

2.3

2.5

3.0

4.1

1.3

3.7

5.4

2.0

2.0

3.0

1.7

6.2

7.3

7.5

Understanding Earth

Understanding Earth

SECOND EDITION

FRANK PRESS
Carnegie Institution of Washington

RAYMOND SIEVER
Harvard University

W. H. Freeman and Company
NEW YORK

TO OUR CHILDREN, AND OUR CHILDREN'S CHILDREN;
MAY THEY LIVE IN HARMONY WITH EARTH'S ENVIRONMENT.

ACQUISITIONS EDITOR: *Holly Hodder*
DEVELOPMENT EDITORS: *Susan Seuling; Nancy Fleming*
PROJECT EDITORS: *Mary Louise Byrd; Penelope Hull*
TEXT AND COVER DESIGNER: *Vertigo Design*
COVER ILLUSTRATION: *Tomo Narashima*
INTERIOR ILLUSTRATIONS: *Ian Warpole, Network Graphics, and Tomo Narashima*
PHOTO RESEARCH: *Alexandra Truitt; Jerry Marshall*
PRODUCTION COORDINATOR: *Paul W. Rohloff*
COMPOSITION: *York Graphic Services*
MANUFACTURING: *RR Donnelley & Sons Company*

LIBRARY OF CONGRESS CATALOGING-IN-PUBLICATION DATA

Press, Frank.
Understanding earth / Frank Press, Raymond Siever.—2nd ed.
p.   cm.
Includes index
ISBN 0-7167-2836-2   ISBN 0-7167-3099-5 (minibook)
1. Earth sciences.   I. Siever, Raymond.   II. Title.
QE28.P9   1997
550–dc21   96-43804
CIP
© 1998, 1994 by W. H. Freeman and Company

Printed in the United States of America.

Second printing, 1998

Brief Contents

Contents

Contents

Contents

Contents

Boxed Features

To the Instructor

When we wrote our first textbook, *Earth*, geology was a field flush with the excitement of new discovery. Recognition of continental drift only a decade earlier had triggered a revolution in our understanding of our planet. For the first time in the history of the discipline, an all-encompassing synthesis of geological knowledge was being advanced. Plate-tectonic theory gave us a framework for learning about the immense forces turning cyclically in the core, in the surrounding mantle, in the crust, and in the air, oceans, and biosphere, keeping our planet in a constant state of change. This new picture of Earth as a dynamic, coherent system was central to our writing of that book and its successor, *Understanding Earth*. We wanted to share with as many students as possible something of the exhilaration and intellectual excitement the profession was feeling. We have been enormously gratified by the warmth and loyalty with which each edition of *Earth,* and later *Understanding Earth,* was received.

More than thirty years into the plate-tectonics revolution, an understanding of the whole Earth system is more necessary than ever. The world in which we live has changed. With its growing population and increasing industrial societies, our world is using more resources, contributing more to pollution of air, water, and land, and becoming more vulnerable to environmental disaster than ever before. Human disturbances of the environment now equal or exceed natural disturbances in magnitude and rate. Our planet enjoys a unique and delicate balance of conditions that allowed life to begin and countless life forms to flourish. Because they affect multiple systems and ultimately the whole Earth, relentless abuses of the environment are a threat to life on the planet.

Students now entering college belong to the generation that will lead our world through the first decades of the next century. We believe the social, political, and economic issues they face will prove many times more challenging than those we have already encountered. To make wise decisions about such issues as resource development, waste management, environmental protection, and land use, they will have a tremendous need for scientific literacy in general and an understanding of geology in partic-

ular. We have brought this conviction to our textbook.

Understanding Earth is for today's students, especially beginning students whose one course in geology may be their sole college exposure to the physical sciences. Naturally, as geologists we want to share many of the fascinating aspects of our discipline with students. As teachers we have tried to do this in a way that is compelling, accurate, and above all, up to date. We do so with the hope that learning how scientists think and work, and understanding something of the geological systems and processes that underlie the past, present, and future of our planet, will help our readers to think more deeply and responsibly about the issues they will confront as citizens.

GOALS

Understanding Earth is designed for a one-term introductory course in physical geology for non-science majors. We took it as a challenge to present the essential material, both traditional and modern, that a good geology course should cover, but in terms accessible to a student who has had no previous college science. We have deliberately emphasized a broad view, one stressing concepts and trying to show by many examples what is meant by the "scientific method." We have tried to impart something of what motivates contemporary geologists. As much as possible, we have described contemporary technology and methods and discovery. We have tried to integrate in a natural way the newest discoveries and environmental concerns with the traditional discussions of such basic topics as geomorphology, sedimentation, petrology, volcanism, and structural geology.

PEDAGOGY

- **Illustrations** The explanatory drawings, diagrams, and photos in *Understanding Earth* are of considerable value in simplifying otherwise difficult concepts. These illustrations and their captions serve as alternative restatements of concepts presented in the text and as intermittent summaries. Much of our attention in this edition was devoted to making the proven illustrations even clearer and adding more color photos. We replaced more than half the photos in this edition and included an even broader range of photographic examples from all over the world.

- **Maps** Maps are essential to understanding plate tectonics and consequent phenomena such as earthquakes and volcanoes. In addition to the exceptionally clear schematic world maps in the first edition, the second edition of *Understanding Earth* incorporates global relief maps designed by Dr. Peter W. Sloss under the auspices of the National Geophysical Data Center.
- **Boxed Features** The program of boxed features has been expanded and refined to reflect the major themes of the book: Living on Earth, Interpreting the Earth, and Technology and Earth.

    Living on Earth boxed features carry forward the theme of environmental awareness by tackling such issues as mitigating natural catastrophes and using resources wisely in ways that preserve the Earth.

    Interpreting the Earth boxed features carry forward the theme of scientific discovery made through observation and logic. This category includes features on scientific method and the meaning of particular geologic formations.

    Technology and Earth boxed features look at how geologists use technology to gather information. This category includes features on seafloor and ice core drilling, monitoring volcanoes, and satellite altimetry mapping of the seafloor.
- **Study and Research Aids** Each chapter begins with an engaging story or piece of information followed by a statement of goals that offers a unifying view of the chapter in a manner that is conceptually accessible to the beginning student. **Key terms** are highlighted in bold type within each chapter and are listed at the end of the chapter for easy reference and review. Key terms and definitions are compiled in the **Glossary** at the back of the book. **Chapter Summaries** serve as systematic reviews of major concepts. **Exercises** help students test their comprehension of key chapter material. **Thought Questions** ask students to apply ideas and principles to situations not specifically covered in the text. **Suggested Readings** point curious students toward books and articles, both popular and technical, on subjects they wish to explore more deeply. Internet sources direct students to the latest research and a variety of related materials available on Web sites. **Team**

Projects, at the end of selected chapters, offer instructors a variety of inventive ideas for short-term, long-term, written, and oral group assignments.

ORGANIZATION

To accommodate the many ways that instructors may want to structure the course, we have made each chapter as self-sufficient as possible. Nevertheless, few geologic processes can be taught as wholly independent subjects; they must be seen in the context of a larger system. We have thus used the recurrence of many important topics as an opportunity for review and alternative restatement. This approach enhances learning and increases flexibility in the way the book can be used.

Relying on a consensus of views from colleagues and reviewers, we arrived at the following four-part organization for this text.

PART 1 UNDERSTANDING THE EARTH SYSTEM   Part 1 could almost be considered a basic text in the foundations of geology that evolved in the first half of the century, illuminated by the insights of plate-tectonic theory. An introductory chapter proceeds from a discussion of the origin of Earth to a first treatment of the geological cycle and elementary plate tectonics. This enables the instructor to relate plate tectonics to the many subjects covered before Chapter 20, the chapter devoted solely to plate tectonics.

A comprehensive chapter on mineralogy is followed by a short chapter introducing the rock cycle and the three major classes of rocks. The advantage of the rock cycle chapter is that it establishes the relationship of the rock cycle to other Earth systems and offers the instructor flexibility in assigning the more intensive chapters on each class of rock. The chapter on igneous rocks is followed by the closely related chapter on volcanism, so that knowledge of the origin of igneous rocks precedes the particulars of eruptive processes. In contrast, the weathering and erosion chapter is a logical prelude to the origin of sedimentary rocks in the following chapter. Instructors who choose to reverse the order of presentation of igneous and sedimentary rocks will find that the text is adaptable and flexible in this regard. After the major rock types are covered, the student is introduced to geologic time, stratigraphy, and structure.

PART 2 SURFACE PROCESSES   Part 2 covers the major topics of geomorphology: water, rivers, wind, and ice. These subjects are the most easily comprehended areas of geology because most students can draw on personal experience with Earth's landscapes. Completing this part is a chapter on the oceans, which continue to grow in importance in geology and in global environmental issues. Here we emphasize shorelines and shallow water processes, because they are the most readily observed, and we discuss those aspects of the deep sea that relate most closely to plate tectonics and to the formation of marine sediments. Part 2 examines the surface systems individually and in relation to each other and to the interior.

PART 3 INTERNAL PROCESSES, EXTERNAL EFFECTS

Part 3 is an exploration of Earth's interior and its dynamic interaction with the crust. A chapter on earthquakes leads into a chapter on what seismic waves, heat flow, and magnetism reveal about the structure and behavior of Earth's interior. This chapter leads into a full treatment of plate tectonics as it is understood today. Following is a chapter on the role of plate tectonics in the formation and deformation of the continents, a culmination that brings together all the elements of geologic history, which defied interpretation in the earlier decades of this century.

PART 4 CONSERVING EARTH'S BOUNTY

Part 4 begins with chapters on mineral and energy resources. These chapters weigh economic benefit against environmental degradation and highlight the advantages of policies that encourage economic development and use of nonpolluting energy sources such as solar and wind power.

The last chapter, "Earth Systems and Cycles," is new to this edition. This chapter develops the theme of interdependency among geosphere, atmosphere, hydrosphere, and biosphere, and ends with an eloquent argument for a responsible approach to using Earth's resources.

TRENDS AND THEMES

Plate tectonics continues to inspire a fast pace of activity in today's geology, geochemistry, and geophysics. To lend currency and excitement to the introductory course, we integrate, at an appropriate level, the flavor of today's research with the basic principles of physical geology.

We have introduced a number of topics that are the focus of current research in important areas of geology. Among these topics are the relationship of sedimentary sequences to sea level, the evolution of orogenic belts in relation to plate tectonics, current ideas on glaciation and climate, crust–mantle interactions, and the influence of comet and asteroid impacts on Earth's surface environments.

We call attention to newer approaches to some subjects, for example, correlating sea-level changes with climate and tectonics and interpreting continental geology in terms of microplates and ancient plate interactions.

We introduce students to newer technologies, such as side-scan acoustic radar used in oceanography and seismic wave tomography used in the study of the interior. We want to impart the idea that geology, like other sciences, is constantly changing, renewing itself as new ideas and technology come to the fore.

Geology has been and remains an eminently practical science. We discuss geologic hazards and hazard prevention in chapters on volcanism (dangerous eruptive behavior), weathering and erosion (soil loss), mass wasting (landslides), rivers (floods), water (groundwater contamination), and the interior (earthquakes). We also discuss the economic and social issues that hinge on geology.

All the common metals used in industry originate as ore deposits discovered by geologists. Sand, gravel, and limestone are used for construction and for many other industrial activities. Our major sources of energy—coal, oil, and gas—are geological deposits. Exploration for oil and gas, in fact, remains one of the major occupations of geologists. We discuss reasons for extracting and using these resources in ways that minimize damage to the Earth, and we look at the benefits and costs of alternative energy sources.

Global change is seen increasingly as a threat to life on Earth. We cover the connections among climate and weathering and erosion, landscape, and sedimentation throughout the book and devote the new, final chapter to the interaction of Earth's systems. We discuss climate in relation to natural and human-induced conditions that trigger changes in glaciation, the greenhouse effect, and the amount of carbon dioxide in the atmosphere.

One of the most important aspects of geology is its role in the evolution of life on Earth. Geology is a framework for understanding the physical conditions—surface changes and climate changes—underlying the myriad forms of life that evolved over billions of years. Global climate change is seen increasingly as a threat to life.

SUPPLEMENTS

We are again proud to announce the availability of a vast selection of print, visual, and electronic supplements for students and instructors using *Understanding Earth*. Many are completely new to this edition and take advantage of the best and latest technology.

- *CD-ROM* An innovative *CD-ROM* has been developed especially for this book to bring an added dimension to *Understanding Earth*. Among the features are **Q&A**—electronic self-quizzes with built-in feedback that students can use to practice for tests; **Added Views**—a wide selection of geological videos and original animations, as well as photographs that complement images found in the book and expanded topic coverage; **Illustrated Glossary**—an electronic device that defines, illustrates, and helps students recall the hundreds of terms used in introductory geology; **WebNotes™**—hundreds of geological URLs referenced to specific sections in the book; **Geographic Locator**—an electronic tool designed to show and help students learn the relative locations of map sites mentioned in the book; **Presentation Manager**—enables instructors to easily queue up a sequence of images, videos, and animations for display during lectures; **Tools and Resources**—a collection of components such as unit converter, calculator, interactive periodic table, and interactive geologic time scale to help the students interact with various geologic concepts.

- *Study Guide* David M. Best of Northern Arizona University has thoroughly updated his *Study Guide to Understanding Earth*. Many exercises use illustrations from the book and all were designed to focus on the key content in each chapter. Special "CD Lab" exercises help students see the most effective use of the CD-ROM. The *Study Guide* also contains detailed chapter summaries, complete chapter outlines, and practice multiple-choice tests.

- *Instructor's Resource Manual* Tying together the selection of supplements for instructors is the *Instructor's Resource Manual,* by Philip M. Astwood of the University of South Carolina, Columbia. In addition to the more traditional features, the *Manual* contains information on using the CD-ROM, the Web Site, and other items in the supplements package. It also offers ideas on using some of the newer electronic products available in the geology classroom and lab.

- *Slide Set with Lecture Notes* Peter L. Kresan of the University of Arizona has augmented the first-edition *Slide Set* with an additional 100 images and lecture notes carefully selected to enhance the images in the book. This brings the total number of slides offered to 220.

- *Overhead Transparency Set* About 110 vivid color diagrams are reproduced from the text in the *Overhead Transparency Set*. Some come with overlays of the labels. Oversized type is used for labels where appropriate.

- *Test Bank* Simon M. Peacock of Arizona State University has thoroughly revised the *Test Bank* for the new edition. A sizable number of the questions now contain geological diagrams, and the Windows and Macintosh software of the computerized versions allow instructors easily to change and add questions as well as to import their own electronic drawings.

- *Understanding Earth Videodisc* Produced by Videodiscovery, Inc., in cooperation with W. H. Freeman and Company, the *Understanding Earth Videodisc* is again available to qualified adopters.

- *Press Releases* This new supplement provides instructors with support materials on geologically significant events on an ongoing basis.

- *Understanding Earth Web Site* Also new to this edition, this electronic supplement provides continual support to the book and the CD-ROM. In addition to the information about important geological events, the site provides new images every month that can be exported into presentation programs, updated WebNotes™, another source of self-quizzes for students, and other features. Its address is: www.whfreeman.com/understandingearth

ACKNOWLEDGMENTS

It is a challenge both to teachers and to authors of geology texts to encompass the many important aspects of geology in a single course and to inspire interest and enthusiasm in the student. To meet this challenge, we have called on the advice of many colleagues teaching in all kinds of college and university settings. From the earliest planning stages of each edition of this book we relied on a consensus of views in deciding on an organization for the text and in choosing which topics to include. As we wrote and rewrote the chapters, we again relied on our colleagues to guide us in making the presentation pedagogically sound, accurate, and accessible and stimulating to students. To each one we are grateful.

Wayne M. Ahr
Texas A & M
Gary Allen
University of New Orleans
N. L. Archbold
Western Illinois University
Allen W. Archer
Kansas State University
Richard J. Arculus
University of Michigan, Ann Arbor
Philip M. Astwood
University of South Carolina
R. Scott Babcock
Western Washington University
Evelyn J. Baldwin
El Camino Community College
Lukas P. Baumgartner
University of Wisconsin-Madison
David M. Best
Northern Arizona University
Dennis K. Bird
Stanford University
Stuart Birnbaum
University of Texas, San Antonio
David L. S. Blackwell
University of Oregon
Arthur L. Bloom
Cornell University
Phillip D. Boger
State University of New York, Geneseo
Robert L. Brenner
University of Iowa
David S. Brumbaugh
Northern Arizona University
Edward Buchwald
Carleton College
Robert Burger
Smith College
Timothy Byrne
University of Connecticut
J. Allan Cain
University of Rhode Island
F. W. Cambray
Michigan State University
Ernest H. Carlson
Kent State University
Max F. Carman
University of Houston
Allen Cichanski
Eastern Michigan University
George R. Clark II
Kansas State University
G. S. Clark
University of Manitoba

Roger W. Cooper
Lamar University
Peter Dahl
Kent State University
Jon Davidson
University of California, Los Angeles
Larry E. Davis
Washington State University
Robert T. Dodd
State University of New York at Stony Brook
Bruce J. Douglas
Indiana University
Grenville Draper
Florida International University
William M. Dunne
University of Tennessee, Knoxville
R. Lawrence Edwards
University of Minnesota
C. Patrick Ervin
Northern Illinois University
Stanley Fagerlin
Southwest Missouri State University
Jack D. Farmer
University of California, Los Angeles
Stanley C. Finney
California State University, Long Beach
Tim Flood
Saint Norbert College
Richard M. Fluegeman, Jr.
Ball State University
Michael F. Follo
Colby College
Richard L. Ford
Weaver State University
Charles Frank
Southern Illinois University
William J. Frazier
Columbus College
Robert B. Furlong
Wayne State University
Sharon L. Gabel
State University of New York, Oswego
Alexander E. Gates
Rutgers University
Gary H. Girty
San Diego State University
William D. Gosnold
University of North Dakota
Richard H. Grant
University of New Brunswick
Bryan Gregor
Wright State University

G. C. Grender
Virginia Polytechnic Institute and State University
Mickey E. Gunter
University of Idaho
David A. Gust
University of New Hampshire
Kermit M. Gustafson
Fresno City College
Bryce M. Hand
Syracuse University
Ronald A. Harris
West Virginia University
Eric Hetherington
University of Minnesota
J. Hatten Howard III
University of Georgia
Herbert J. Hudgens
Tarrant County Junior College
Ian Hutcheon
University of Calgary
Ruth Kalamarides
Northern Illinois University
Frank R. Karner
University of North Dakota
Phillip Kehler
University of Arkansas, Little Rock
Peter L. Kresan
University of Arizona
Albert M. Kudo
University of New Mexico
Robert Lawrence
Oregon State University
Don Layton
Cerritos College
Peter Leavens
University of Delaware
Barbara J. Leitner
University of Montevallo
John D. Longshore
Humboldt State University
Stephen J. Mackwell
Pennsylvania State University
Peter Martini
University of Guelph
G. David Mattison
California State University, Chico
George Maxey
University of North Texas
Lawrence D. Meinert
Washington State University, Pullman
Robert D. Merrill
California State University, Fresno

Kula C. Misra
  University of Tennessee, Knoxville
Roger D. Morton
  University of Alberta
J. Nadeau
  Rider University
Peggy A. O'Day
  Arizona State University
Terrence M. Quinn
  University of South Florida
Simon M. Peacock
  Arizona State University
E. Kirsten Peters
  Washington State University, Pullman
Donald R. Prothero
  Occidental College
C. Nicholas Raphael
  Eastern Michigan University
J. H. Reynolds
  West Carolina University
Robert W. Ridkey
  University of Maryland
James Roche
  Louisiana State University

William F. Ruddiman
  University of Virginia
Charles K. Scharnberger
  Millersville University
James Schmitt
  Montana State University
Fred Schwab
  Washington and Lee University
Steven C. Semken
  Navajo Community College
D. W. Shakel
  Pima Community College
Charles R. Singler
  Youngstown State University
David B. Slavsky
  Loyola University of Chicago
Douglas L. Smith
  University of Florida
Richard Smosma
  West Virginia University
Donald K. Sprowl
  University of Kansas

Don Steeples
  University of Kansas
Randolph P. Steinen
  University of Connecticut
Bryan Tapp
  University of Tulsa
John F. Taylor
  Indiana University of Pennsylvania
Kenneth J. Terrell
  Georgia State University
Thomas M. Tharp
  Purdue University
Nicholas H. Tibbs
  Southeast Missouri State University
Herbert Tischler
  University of New Hampshire
Jan Tullis
  Brown University
Kenneth J. Van Dellen
  Macomb Community College
Lorraine W. Wolf
  Auburn University

Peter W. Sloss (National Oceanic and Atmospheric Administration–National Environmental Satellite, Data, and Information Service–National Geophysical Data Center) created the digital images of Earth's crustal topography used throughout the book.

We have also had the benefit of informal advice on content and checking for accuracy from many colleagues, including especially C. W. Burnham, S. B. Jacobsen, Jane Selverstone, J. B. Thompson, Jr., Sean Solomon, Peter Molnar, Tom Jordan, George Wetherill, Steve Shirey, and Paul Silver.

Michael F. Follo (Colby College) contributed to the second edition in countless ways and supplied the boxed feature in Chapter 24. Dallas Rhodes (Whittier College) contributed the Internet Sources. Jill S. Schneiderman (Vassar College) contributed the Team Projects.

Others have worked with us more directly in writing and preparing manuscript for publication. At our side always were the editors at W. H. Freeman and Company, with Susan Seuling and Nancy R. Fleming. The superb photographs that illuminate the text were obtained by Alexandra Truitt, who tirelessly combed many sources looking for the best possible choices.

The quality of the finished book would not have been possible without the final, and most skillful, efforts of Vertigo Design; Susan Wein, illustration coordinator; and Paul Rohloff, production coordinator. We are especially indebted to our illustrators—Ian Warpole, Tomo Narashima, and Network Graphics—for transforming our often very rough sketches into many outstanding drawings.

Geology fascinates and excites us. We wrote *Understanding Earth* to help you discover for yourselves how interesting geology is in its own right and how important an understanding of geology has become for making decisions of public policy. What can we do to protect people and property from natural disasters such as volcanoes, earthquakes, and landslides? How can we use the resources of Earth—coal and oil, minerals, water, and air—in ways that minimize damage to the environment? In the end, understanding Earth helps us understand how to preserve life on Earth.

People tend to enjoy what they do well, and we want you to enjoy your geology course. Here are some tips for using our book in ways that will help you do well.

- **Chapter goals** Pay particular attention to the paragraph in each chapter that begins "This chapter will." This statement of goals will give you a firm idea of what the chapter covers and why.
- **Photographs and diagrams** Much of the information in geology is conveyed through illustrations. Our photographs are carefully chosen and diagrams carefully drawn to convey essential ideas. Whenever you see a reference to a figure number, examine the illustration thoroughly and read the caption. Reviewing the figures is a good way to refresh your memory of the chapter before class and before exams.
- **Boxed features** These typically are self-contained stories on topics of interest that reinforce the major themes of the book.
- **Summaries and Key Terms** The summary at the end of each chapter provides a concise description of the topics covered. Key terms and concepts are highlighted in boldface within the chapter, where they are first explained, and are

listed with page references at the end of the chapter to help you review and test your understanding.

- **Exercises, Thought Questions, and Team Projects** These focus on important aspects of the chapter. Your instructor might select some of these for written homework or use them as a basis for test questions. Reviewing them quickly will help you test your memory and understanding of the chapter.
- **Suggested Readings** This list of books and journal articles will guide you to information on topics that interest you.
- **Internet Sources** Gain access to the latest research, visual materials, and more by trying the Web sites listed at the end of most chapters.
- **Study Guide** This learning aid features a wide variety of exercises to reinforce your knowledge of concepts and processes, self-tests, and suggestions for using the CD-ROM as a study tool.
- **CD-ROM** Your study and enjoyment of this book will be expanded to another dimension with this useful CD-ROM. Interactive exercises, practice quizzes, and a variety of electronic tools and resources have been designed to involve and engage you as you explore the key ideas of the textbook.
- *Understanding Earth* **Web Site** Updated regularly, this resource provides new information, new images, self-quizzes, and other helpful features. The address is: www.whfreeman.com/understandingearth

We hope you will find geology both intellectually satisfying and of practical value in preserving Earth for yourself and for future generations.

Understanding the Earth System

Our planet works as an interacting system of matter and energy that generates volcanoes, glaciers, mountains, lowlands, continents, and oceans. The energy that drives the system comes from the Earth's internal heat, which is responsible for plate tectonics, and solar radiation, which circulates the atmosphere and oceans and powers erosion. The matter of Earth—its rocks and minerals—and its structure are the relics of Earth system dynamics evolving over 4.6 billion years of geologic time. The three great clans of rocks and their geologic structures reflect geologic processes. Igneous rocks are linked to volcanism, sedimentary rocks to weathering and erosion, and metamorphic rocks to mountain building.

1

Aerial view of the Himalayas from Nepal. Mt. Everest, the world's highest peak, is to the rear on the left. (*Paul Steel/The Stock Market*)

Building a Planet

Earth is a unique place, home to more than a million life forms, including ourselves. No other planet yet discovered has the same delicate balance of conditions necessary to sustain life. **Geology** is the science that studies Earth—how it was born, how it evolved, how it works, and how we can help preserve it.

This chapter gives a broad picture of how geologists think. It starts with the scientific method, the objective approach to the physical universe on which all scientific inquiry is based. Throughout this book, you will see the scientific method in action as you discover how geologists gather and interpret information about our planet. The chapter then describes the most generally accepted scientific explanations for how Earth formed and why it continues to change.

You may find as you read these pages that your idea of time will start to change. A geologist's view of time must accommodate spans so large that our minds sometimes have trouble comprehending them. Geologists estimate that Earth is 4.5 billion years old. About 3.5 billion years ago living cells developed on Earth, but our human origins date back only a few million years—a mere few hundredths of 1 percent of Earth's existence. The scales that measure individual lives in decades and mark off periods of written human history in hundreds or thousands of years are inadequate as we study Earth. Geologists must explain features that evolve over tens of thousands, hundreds of thousands, or millions of years.

THE SCIENTIFIC METHOD

The goal of all science is to explain with increasing precision how the universe works. The **scientific method,** on which all scientists rely, is a general research strategy based on the principle that every physical event has a physical explanation, even if it may be beyond our present ability to discover.

When scientists propose a **hypothesis**—a tentative explanation based on data collected through observations and experiments—they present it to the community of scientists for criticism and repeated testing against new data. A hypothesis that is confirmed by other scientists gains credibility, particularly if it predicts the outcome of new experiments.

A hypothesis that has survived repeated challenges and accumulated a substantial body of support is elevated to the status of a **theory.** Although its explanatory and predictive power has been demonstrated, a theory can never be considered finally proved. The essence of science is that no explanation, no matter how believable or appealing, is immune to question. If convincing new evidence indicates that a theory is wrong, scientists may modify or discard it. The longer a theory holds up to all scientific challenges, however, the more confidently it is held.

To encourage the atmosphere of challenge, scientists share their ideas and data by presenting them at professional meetings, publishing them in professional journals, and discussing them in informal conversations with colleagues. Scientists learn from one another's work as well as from the discoveries of the past. Most of the great concepts of science, whether they emerge as a flash of insight or in the course of painstaking analysis, result from untold numbers of such interactions. Albert Einstein said it this way: "In science . . . the work of the individual is so bound up with that of his scientific predecessors and contem-

poraries that it appears almost as an impersonal product of his generation."

Because such free intellectual exchange is subject to abuses, a code of ethics has evolved among scientists. Scientists must acknowledge the contributions of all others on whose work they have drawn. They must not falsify data. And they must accept responsibility for training the next generation of researchers and teachers.

THE PRINCIPLE OF UNIFORMITARIANISM

Much of what we have come to understand about the geologic past is based on observation of the workings of our planet today. We can observe today the growth of continents, the erosion of mountains, the eruptions of volcanoes. A fundamental principle of geology, advanced in the eighteenth century by the Scottish physician and geologist James Hutton, is that "the present is the key to the past." This **principle of uniformitarianism,** as it is now known, holds that the geologic processes we see in operation as they modify Earth's crust today have worked in much the same way over geologic time.

The rates of geological processes vary over a wide range. Continents can drift apart and mountains can be raised over millions of years. A large meteorite or comet can strike Earth and gouge out a vast crater, volcanoes can blow their tops, and earthquakes can rupture Earth's surface in seconds. These are events in an ongoing system that has continued to shape and reshape Earth since its birth 4.5 billion years ago (Figure 1.1).

Uniformitarianism, together with the laws of physics and chemistry, provides the basis for the theory and practice of geology. We will call upon uniformitarianism frequently as we attempt to decipher Earth's geologic history.

THE ORIGIN OF OUR SYSTEM OF PLANETS

The search for the origins of the universe and our own small part of it goes back to the earliest recorded mythologies. Today the most generally accepted scientific explanation is the "Big Bang" theory, which holds that our universe began some 10 billion to 15 billion years ago with a cosmic "explo-

FIGURE 1.1 Geologic phenomena can stretch over thousands of centuries or can occur with dazzling speed. (left) Mount Kerkeslin in the Canadian Rocky Mountains, Alberta, Canada. This mountain range was raised and deformed over a period of tens of millions of years. *(Carr Clifton.)* (right) Meteor Crater, Arizona. The explosive impact of a meteorite about 25,000 years ago excavated this crater in a few seconds. *(John Sanford/Photo Researchers.)*

sion." Before that moment, all matter and energy were compacted into a single, inconceivably dense point. Although we know little of what happened in the first fraction of a second when time began, astronomers have acquired a general understanding of the billions of years that followed. During that time, in a process that still continues, the universe has expanded and thinned out to form the galaxies and stars. Geologists focus on the last 4.5 billion years of this vast expanse, when our solar system—the star we call the Sun and the planets revolving around it—formed and evolved. Specifically, geologists look to the formation of the solar system in order to understand the formation of Earth.

The Nebular Hypothesis

In 1755 the German philosopher Immanuel Kant suggested that the origin of the solar system could be traced to a rotating cloud of gas and fine dust. Discoveries made in the past few decades have led astronomers back to this old idea, now called the **nebular hypothesis.** Equipped with modern telescopes, they have found that outer space beyond our solar system is not as empty as it once was thought to be. Astronomers have recorded many clouds of the type that Kant surmised, and they have named them *nebulae.* They also have identified the materials that form these clouds. The gases are mostly hydrogen and helium, the two elements that make up all but a small fraction of our Sun. The dust-sized particles are chemically similar to materials found on Earth.

How could our solar system take form from such a cloud? Part of the answer lies in the force of gravity, the attraction between pieces of matter because of their mass. This diffuse, slowly rotating cloud contracted under the force of gravity (Figure 1.2[a]). Contraction in turn accelerated the rotation of the particles (just as ice skaters spin more rapidly when they pull in their arms), and the faster rotation flattened the cloud into a disk (Figure 1.2[b]).

THE SUN FORMS Under the pull of gravity, matter began to drift toward the center, accumulating into a proto-Sun, the precursor of our present Sun

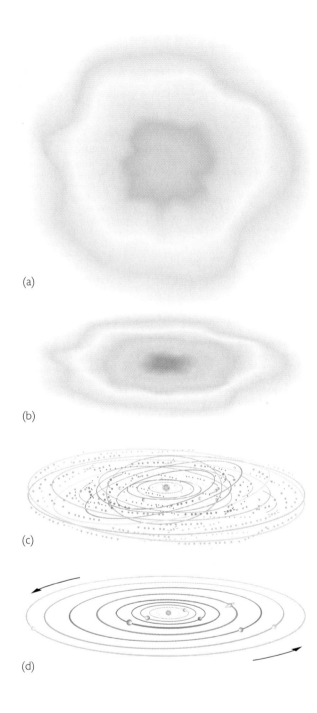

(a)

(b)

(c)

(d)

FIGURE 1.2 Evolution of the solar system. (a) A diffuse, roughly spherical, slowly rotating nebula begins to contract. (b) As a result of contraction and rotation, a flat, rapidly rotating disk forms with matter concentrated at the center that will become the proto-Sun. (c) The enveloping disk of gas and dust forms grains that collide and clump together into small chunks, or planetesimals. (d) The terrestrial planets build up by multiple collisions and accretion of planetesimals by gravitational attraction. Accretion of gas by the giant outer planets is not shown.

(Figure 1.2[c]). Compressed under its own weight, the material in the proto-Sun then became dense and hot. The internal temperature of the proto-Sun rose to millions of degrees, at which point nuclear fusion began. The Sun's nuclear fusion, which continues today, is the same nuclear reaction that occurs in a hydrogen bomb. In both cases hydrogen atoms, under intense pressure and at high temperature, combine (fuse) to form helium. Some mass is converted into energy in the process. This conversion is represented by Albert Einstein's famous equation $E = mc^2$, where E is the amount of energy released by conversion of mass (m) and c is the speed of light. Because c is a very large number (about 300,000 km per second) and c^2 is huge, a small amount of mass can yield an enormous amount of energy. Some of that energy is released as sunshine in the case of the Sun, and as a great explosion in the case of the H-bomb.

THE PLANETS FORM Although most of the matter in the original nebula was concentrated in the proto-Sun, a disk of gas and dust, called the solar nebula, remained to envelope it. The solar nebula grew hot as it flattened into a disk, hotter in the inner region, where more of the matter accumulated, than in the less dense outer regions. Once formed, the disk began to cool and many of the gases condensed. That is, they changed to their liquid or solid form, just as water vapor condenses into droplets on the outside of a cold glass and water solidifies into ice when it cools below the freezing point. Gravitational attraction caused the dust and condensing material to collide and accrete (clump together) as small chunks, or *planetesimals*. As the planetesimals collided and stuck together, larger Moon-size bodies formed (Figure 1.2[c]). In a final stage of cataclysmic impacts, a few of these larger bodies with their larger gravitational attraction swept up the others to form our nine planets in their present orbits (Figure 1.2[d]). Theoretical calculations indicate that all of this activity could have occurred in the remarkably short time of less than 100 million years. These rapid events occurred about 4.56 billion years ago, based on the age of meteorites that occasionally strike the Earth and that are believed to be remnants of that distant age.

As the planets formed, those in orbits close to the Sun and those in orbits farther from the Sun developed in markedly different ways.

◾ *The Inner Planets* The four inner planets, closest to the Sun, are Mercury, Venus, Earth, and Mars (Figure 1.3). They are also known as the terrestrial ("Earthlike") planets. In contrast to the outer planets, the four inner planets are small and rocky. They grew where conditions were so hot that volatile materials—those that become gases and boil away at relatively low temperatures—could not be retained in quantity. Radiation and matter streaming from the Sun blew away most of their hydrogen, helium, water, and other light gases and liquids, leaving behind dense metals such as iron and other heavy, rock-forming substances. By about 4.5 billion years ago, the inner planets had emerged as dense, rocky masses.

▪ *The Giant Outer Planets* According to this same scenario, most of the volatile materials swept from the region of the terrestrial planets were carried to the cold outer reaches of the solar system. Some accumulated on the giant *outer planets*—Jupiter, Saturn, Uranus, and Neptune—and their satellites. The rest were carried into outer space beyond. The giant planets were big enough and their gravitational attraction was strong enough to enable them to hold on to the lighter nebular constituents. Thus, although they have rocky cores, like the Sun, they are composed mostly of hydrogen and helium and the other light constituents of the original nebula. Tiny Pluto, orbiting farthest from the Sun, is a strange frozen mixture of gas, ice, and rock.

Testing the Nebular Hypothesis

This standard model of solar-system formation should be taken for only what it is—a tentative

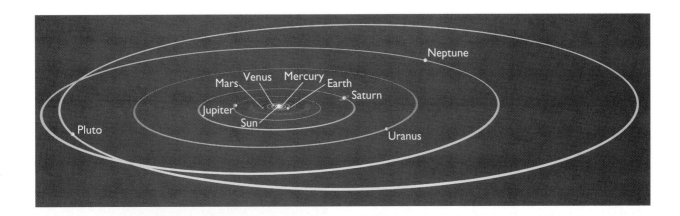

FIGURE 1.3 The solar system. The four inner planets—Mercury, Venus, Earth, and Mars—are closest to the Sun and are small and rocky. The four giant outer planets—Jupiter, Saturn, Uranus, and Neptune—and their satellites are mostly gaseous with rocky cores. The outermost planet, Pluto, is a snowball of methane, water, and rock. The upper panel shows the planetary orbits drawn roughly to scale; the distance from Pluto to the Sun averages about 5.9 billion km. The lower panel shows the relative sizes of the planets and the asteroid belt separating the inner and outer planets.

explanation that many scientists think best fits the known facts. Perhaps the model comes close to what actually happened. More important, however, is that this model gives us a way to think about the origin of the solar system. Scientists can examine its parts and modify it if necessary as more evidence comes in from new research. Such evidence has been gathered by American and Russian spacecraft carrying planetary probes. They have returned data on the nature and composition of the atmospheres and surfaces of Mercury, Venus, Mars, Jupiter, Saturn, Uranus, Neptune, and the Moon. A startling finding, which poses a problem for scientists who had hoped to learn how the planets evolved by comparing them, is that in our solar system—consisting of 9 planets and about 60 satellites—no two bodies are the same! Planetary scientists are searching for an answer.

Nebulae at various stages of development are being studied with powerful telescopes, and these observations are providing information about what goes on in the remote sections of the universe. A major discovery—made possible by the Hubble Space Telescope, which travels in orbit around Earth—was the first direct evidence of envelopes of gas, dust, and planets around several nearby stars. This finding strongly supports the idea that systems of planets in orbit around a sun also probably occur elsewhere in the universe. As the findings from these and other scientific investigations modify our working model, a clearer picture of the origin of our own solar system should emerge.

We have dwelt on the question of the origin of our solar system because the initial state of a planet helps determine its evolutionary course. Earth's current state is reasonably well known to us. The state of Earth at these two points in time, separated by some 4.5 billion years, must be fitted into any explanations we develop for the changes Earth has undergone throughout the course of its planetary evolution.

EARTH AS AN EVOLVING PLANET

How did Earth evolve from a rocky mass to a living planet with continents, oceans, and an atmosphere? The answer lies in **differentiation**—the transformation of a random mix of chunks of matter to a body in which the interior is divided into concentric layers that differ from one another both physically and chemically. Differentiation occurred early in Earth's history, when the planet got hot enough to melt.

The Earth Heats Up and Melts

To understand Earth's present layered structure, we must mentally return to the time when Earth was still subject to violent impacts by planetesimals and larger bodies. A moving body carries large amounts of kinetic energy, or energy of motion. (Think of how the energy of motion crushes a car in a collision.) A planetesimal colliding with Earth at a velocity of about 11 km per second would deliver as much energy as the same weight of TNT. When planetesimals and larger bodies crashed into the primitive Earth, most of this energy of motion was converted to heat, which is another form of energy. The impact energy of a Mars-sized body in collision with Earth would be equivalent to a trillion 1-megaton nuclear explosions, enough to eject into space a vast amount of debris and to deposit enough heat to melt most of what remained of Earth.

Many scientists now think such a cataclysm did indeed occur. Not only did it cause extensive melting, it also gave us the Moon. The giant impact propelled into space a shower of debris from both Earth and the impacting body, and the Moon aggregated from this debris (Figure 1.4). It would have re-formed as a largely molten body. This huge impact also knocked the spin axis of the Earth from vertical

FIGURE **1.4** Artist's rendering of the collision of a Mars-size body with Earth about 4.5 billion years ago. The impact energy would have caused extensive melting of Earth and would have ejected debris that aggregated to form the Moon. (Painting by Alfred T. Kamajian, *Scientific American,* July 1994, cover.)

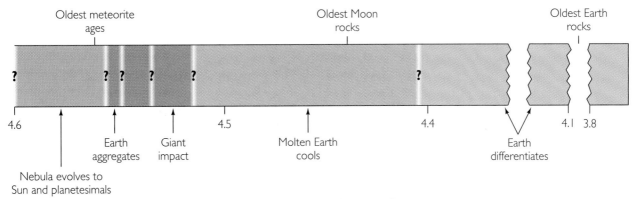

Age (billion years before present)

FIGURE 1.5 This timeline shows the origin of the Sun and Earth, the giant impact that melted much of the Earth and created the Moon, and the beginning of Earth's differentiation. The ages of the events listed above the bar are based on actual datings of meteorites and of lunar and Earth rocks. The question marks within the bar indicate a high degree of uncertainty about the timing of the events so marked. (Extensively modified from D. J. DePaolo, "Strange Bedfellows," *Nature,* Vol. 10, November 10, 1994, p. 131.)

with respect to the plane of the orbit to its present 23° inclination and sped up Earth's rotation. The oldest Moon rocks brought back by the *Apollo* astronauts are dated at 4.44 billion years ago, which should be close to the time of this violent event. If this hypothesis is correct, Earth's age can be bracketed between the age of meteorites, 4.56 billion years, and the age of the oldest Moon rocks.

In addition to the giant impact, another source of heat would have caused melting early in Earth's history. Several elements (uranium, for example) are radioactive. Although these elements occur in only tiny amounts, their radioactivity has had a profound effect on Earth's evolution. Atoms of radioactive elements spontaneously disintegrate by emitting subatomic particles. As these particles are absorbed by the surrounding matter, their energy of motion is transformed into heat. Radioactivity-generated heat, in addition to meteorite impacts, would have contributed to warming and melting in the young Earth. Radioactivity is a long-lived source of heat that continues to keep the interior warm.

Differentiation Begins

Early melting of Earth began the process of differentiation, perhaps the most significant event in Earth's history. It led to the formation of Earth's crust and eventually of the continents. It brought lighter materials to the outer layers of Earth and initiated the

escape of even lighter gases from the interior, which eventually led to the formation of the atmosphere and oceans. Even today, gases continue to escape in the emissions that accompany volcanic eruptions. All of this activity began during the early melting of Earth. Figure 1.5 depicts as a timeline the major events in Earth's early history.

EARLY MELTING   Although Earth probably began as an unsegregated mixture of the planetesimals and other remnants of the solar nebula, it did not retain this form long. Large-scale melting would have occurred as a result of the giant impact. Some workers in the field speculate that 30 to 65 percent of Earth melted, forming an outer layer hundreds of kilometers thick, which they call a "**magma** (molten rock) ocean." The interior, too, would have heated to a "soft" state in which its components could move around—heavy material sinking to the interior and lighter material floating toward the surface. The rising lighter matter would bring interior heat to the surface, where it could radiate into space. In this way Earth cooled and mostly solidified and was transformed into a differentiated or zoned planet with three main layers: a central core and an outer crust separated by a mantle (Figure 1.6).

EARTH'S CORE   Iron, which accounted for about a third of the primitive planet's material, is denser than the other elements and sank to form most of a central **core.** Scientists have found that the core,

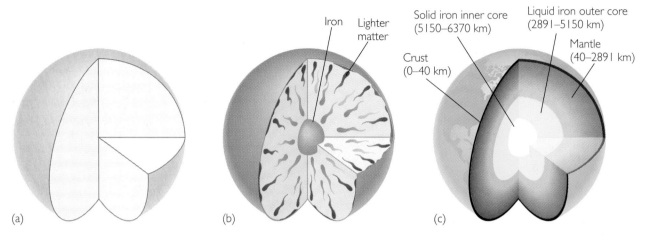

(a)   (b)   (c)

FIGURE **1.6** Early Earth (a) was probably a homogeneous mixture with no continents or oceans. In the process of differentiation, iron sank to the center and light material floated upward to form a crust (b). As a result, Earth is a zoned planet (c) with a dense iron core, a crust of light rock, and a residual mantle between them.

which exists at a depth beginning at 2900 km, is molten on the outside, but the inner core, from about 5200 to 6400 km, is solid. The reason is that the temperature at which any material melts increases with pressure. The iron core is solid nearest to Earth's center, where pressures are the highest.

EARTH'S CRUST Other molten materials were less dense than the parent substances from which they separated, so they floated toward the surface of the magma ocean. There they cooled to form Earth's solid **crust,** a thin outer layer ranging up to about 40 km in thickness. The crust contains relatively light materials with low melting temperatures. These are mostly compounds of the elements silicon, aluminum, iron, calcium, magnesium, sodium, and potassium, combined with oxygen. All of these materials, other than iron, are among the lightest of the solid elements. (Chapter 2 discusses chemical compounds and the elements from which they form.)

EARTH'S MANTLE Between the core and crust is the **mantle,** a region that is the bulk of the solid Earth. The mantle is the material left in the middle zone after most of the heavy matter sank and the light matter rose toward the surface. The mantle ranges from about 40 to 2900 km in depth. It consists of rock of intermediate density, mostly compounds of oxygen with magnesium, iron, and silicon.

Chemical analysis of rocks indicates that Earth contains about 90 naturally occurring chemical elements, but 99 percent of its mass is made up of only eight (see Figure 1.7). Furthermore, about 90 percent

of Earth consists of only four elements: iron, oxygen, silicon, and magnesium. When we compare the relative abundance of elements in the crust with their abun-

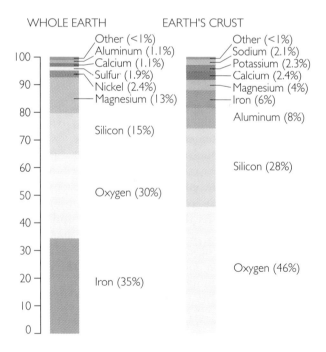

FIGURE **1.7** The relative abundance of elements in the whole Earth compared to that of elements in Earth's crust, given as percentages by weight. Differentiation has created a light crust, depleted of iron and rich in oxygen, silicon, aluminum, calcium, potassium, and sodium. Note that oxygen, silicon, and aluminum alone account for over 80 percent of the crust.

dance in the whole Earth, we find that iron accounts for a full 35 percent of Earth's mass, but because of differentiation there is little iron in the crust, where the light elements predominate. As you can see in Figure 1.7, the crustal rocks you stand on are almost 50 percent oxygen.

Earth's Continents, Oceans, and Atmosphere Form

Aside from the early, heat-producing impacts, from its very beginning Earth's history has been dominated by two constant heat engines, one internal, the other external. A heat engine—for example, the gasoline engine of an automobile—transforms heat energy released from fuel into mechanical motion or work. Earth's internal heat engine is powered by the heat generated by radioactivity. Earth's external heat engine is driven by solar energy—heat supplied to Earth's surface by the Sun. The internal heat melts rocks, forges volcanoes, and supplies the energy to build and move continents and to thrust mountains upward. The external heat is responsible for our climate and weather, and it drives the rain and wind that erode mountains and shape our landscape.

CONTINENTS   Continental growth began soon after differentiation, and it has continued throughout geologic time. We have only the most general notion of what caused the formation of continents. We think magma floated up from the molten interior of Earth to the surface, where it cooled and solidified to form a crust of rock. This primeval crust melted and solidified repeatedly, allowing the lighter materials to separate gradually from the heavier ones and float to the top to form the primitive nucleus of the continents. Rainwater and other components of the atmosphere caused rocks to decompose and disintegrate. Water, wind, and ice then loosened and moved rocky debris to low-lying places, where it accumulated in thick layers, forming beaches, deltas, and the floors of adjacent seas. As this process was repeated through countless cycles, continents formed.

OCEANS AND ATMOSPHERE   Most geologists believe that the origin of the oceans and atmosphere can be traced to Earth itself, that the oceans and atmosphere formed from water and gases that boiled off during heating and differentiation. A few other geologists propose an origin outside of Earth. Comets, they point out, are composed largely of water plus carbon dioxide, and other gases in frozen form. Countless comets may have bombarded Earth early

in its history, carrying in water and gases that formed the early oceans and atmosphere.

Geologists who believe in an internal origin argue this way: Originally the water was locked up; that is, chemically bound as oxygen and hydrogen in certain minerals. Similarly, nitrogen and carbon were chemically bound in minerals. As Earth heated and its materials partially melted, water vapor and other gases were freed and carried to the surface by magmas and released through volcanic activity.

The gases released from volcanoes some 4 billion years ago probably consisted of the same substances that are expelled from present-day volcanoes: water vapor, hydrogen, carbon dioxide, nitrogen, and a few other gases (Figure 1.8). The earliest atmosphere thus was entirely different from the one we live in now, which consists primarily of nitrogen and oxygen. How did the atmosphere change? The production of significant amounts of free oxygen and its persistence in the atmosphere probably came about only after life had evolved at least to the complexity of photosynthetic algae, simple one-celled forms of life. Like other organisms that employ photosynthesis, algae use carbon dioxide and water as raw materials and the energy of sunlight to manufacture organic matter, and they release oxygen as a waste product. This oxygen began to accumulate in the atmosphere and gradually built up to its present concentration.

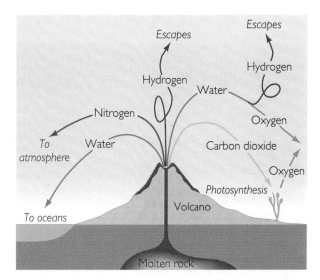

FIGURE 1.8 Volcanic activity has contributed enormous amounts of water, carbon dioxide, and other gases to the atmosphere and solid materials to the continents. Photosynthesis by plants removed carbon dioxide and added oxygen to the primitive atmosphere. Hydrogen, because it is light, easily escapes into space.

How the Other Planets and the Moon Evolved

By about 4 billion years ago Earth had become a fully differentiated planet. The core was still hot and mostly molten, but the mantle was fairly well solidified, and a primitive crust and continents had developed. Oceans and atmosphere had been produced, probably formed from substances released from Earth's interior, and the geologic processes we observe today were set in motion.

But what of the other planets? Did they go through the same early history? Information transmitted from our planetary spacecraft indicates that all the terrestrial planets have undergone differentiation, but their evolutionary paths have varied.

MERCURY   Mercury, for example, has an iron core even larger than Earth's and a heavily cratered lightweight crust. Mercury had no blanket of air to protect it from being riddled by the impacts of large meteorites over the past 4.5 billion years. It is now geologically dead—that is, without the ongoing processes of mountain making, volcanic activity, and earthquakes that we observe on our own planet. It has a trace of an atmosphere but no wind or water to erode and smooth its ancient, cratered surface.

VENUS   Venus is Earth's twin in mass and size, but it differs from Earth profoundly in the nature of its atmosphere and in its current geological behavior.

Somehow—perhaps because its distance from the Sun is three-quarters that of Earth's, or because its early history of large impacts differed from that of Earth, or for some other reason—Venus evolved into a bone-dry planet. It is wrapped in a heavy, poisonous, incredibly hot (475°C) atmosphere, composed mostly of carbon dioxide and clouds of corrosive sulfuric acid droplets. Venus surpasses most descriptions of Hell: a human standing on its surface would be crushed by the pressure, boiled by the heat, and eaten away by the sulfuric acid. Or, as one planetary scientist exclaimed, "Earth is Earth, Venus is Venus; vive la différence!"

Because dense clouds shroud the surface of Venus, we knew relatively little about this terrestrial planet until recently. Now, radar images from the *Magellan* spacecraft, which went into orbit around Venus in 1990, suggest that the planet was geologically active in the past. The images show volcanoes, mountains, plateaus, plains, and other evidence of a dynamic surface (Figure 1.9). Yet its crust, unlike Earth's, today is relatively immobile. Something happened on Venus about 300 to 500 million years ago—perhaps a global outpouring of lava—that paved over the surface, obliterating many of the features of the first 85 percent of the planet's history. After this catastrophic event, the pace of geologic change seems to have slowed.

MARS   Outermost of the terrestrial planets, Mars has a little over half the diameter of Earth. It has a

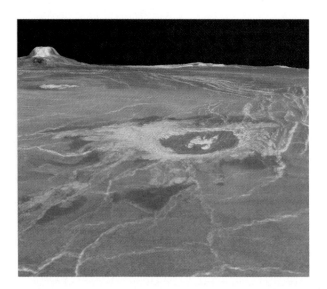 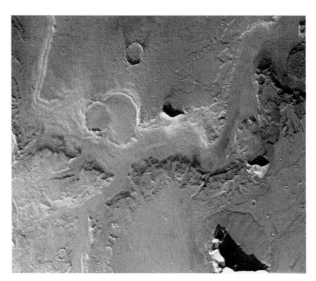

FIGURE 1.9 Evidence of geological activity on Venus and Mars. (left) Venusian crater and volcano revealed by radar on the *Magellan* spacecraft, 1992. Crater in foreground, formed by meteorite impact, is 48 km in diameter. Volcano near horizon is 3 km high. *(NASA.)* (right) *Viking Orbiter* picture of the surface of Mars. Flat-floored valley may have been caused by water run-off early in the history of the planet. Large crater resulted from a more recent meteorite impact. *(Astrogeology Team, USGS.)*

crust, a core, and a composition similar to Earth's, and it has experienced many of the same geological processes that have shaped the Earth. Mars's surface shows that volcanic activity has occurred as recently as 10 million years ago and may still be going on. Craters, evidence of ancient meteorite impacts, are still preserved, in greater abundance than on Earth, where they have been mostly obliterated by subsequent geological activity. Although no liquid water is present on the surface today, networks of valleys and dry river channels indicate that liquid water was abundant on Mars before 3.5 billion years ago. This water may still be present as ice stored below the surface or sequestered in polar icecaps, or it may have evaporated and been lost to space.

Mars's thin atmosphere is composed almost entirely of carbon dioxide. It is unlikely that life exists on Mars today, but it is a serious question whether it could have existed in the past.

THE MOON   The Moon is the best-known body in the solar system other than Earth because of its proximity and the programs of manned and unmanned exploration. As we discussed earlier, the favored theory of its formation proposes that it coalesced as a largely molten body after a giant impact ejected its matter from Earth. Analysis of lunar rocks brought back by the *Apollo* astronauts indicates that as this early magma ocean cooled and began to separate, lightweight, aluminum-rich compounds floated up to form a thick crust, and iron-rich materials sank to the interior. Thus the Moon underwent differentiation to form a small core and a thick crust, a process completed perhaps 4.35 billion years ago. In bulk, the Moon's materials are lighter than those of Earth because the heavier matter of the giant impactor and its primeval target remained embedded in the Earth.

The Moon has no atmosphere and, like Venus, is bone-dry, having lost its volatiles (substances that boil away at low temperatures) in the heat generated by the giant impact. The Moon cooled rapidly, and geological activity ceased some 3 billion years ago. The surface we see today is that of a very old, geologically dead body, with craters and mountains shaped almost entirely by the ancient impacts of large meteorites.

Evolution of the Outer Planets

The giant gaseous outer planets will remain a puzzle for a long time. They are so distinct chemically and so large that they must have followed an evolutionary course entirely different from that of the much smaller terrestrial planets. (It has even been proposed that Jupiter and its 15 moons are akin to a small solar system whose sun—Jupiter—never got quite massive enough for nuclear fusion to begin and then hot enough to shine.) Even less is understood about the evolution of Pluto.

Evidence of an Early Bombardment Period

The crater-pocked surface and the cratered surfaces of the Moon, Mars, Mercury, and other bodies are evidence of an important piece of the early history of the solar system—the **Heavy Bombardment period.** During this period, which may have lasted for 600 million years after the planets formed, the planets swept up and collided with the residual matter that had been left behind when they were assembled.

Every few million years a large chunk of matter still collides with Earth, sometimes with devastating effect. Most scientists believe that the impact of a large meteorite about 65 million years ago caused the extinction of many species, including the dinosaurs. In 1994, a once-in-a-millennium event occurred, when the comet Shoemaker-Levy crashed into Jupiter, exploding a plume of Jupiter's matter more than 1000 km above its atmosphere (Figure 1.10).

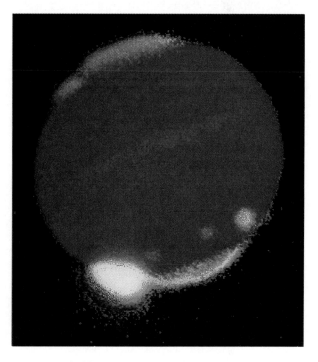

FIGURE **1.10** Infrared image showing the fireball created by the collision of the Comet Shoemaker-Levy with Jupiter on July 18, 1994. *(MSSSO, ANU/Photo Researchers.)*

Although such collisions have become extremely rare, some scientists have proposed that a number of telescopes should be assigned to search space and warn us months to years in advance of sizable bodies that might slam into Earth and devastate life over a large area.

PLATE TECTONICS: A UNIFYING THEORY FOR GEOLOGICAL SCIENCE

In the 1960s a great revolution in thinking shook the world of geologists. Physics had a comparable revolution at the beginning of the twentieth century, when the theory of relativity unified the physical laws that govern space, time, mass, and motion. Biology had a comparable revolution in the middle of this century when the discovery of DNA allowed biologists to explain how organisms transmit the information that controls their growth, development, and functioning from generation to generation. For almost 200 years geologists supported various theories of mountain building, volcanism, and other major phenomena of Earth, but no theory was general enough to explain well the whole range of geologic processes. We now have a single, all-encompassing concept that explains many of Earth's major geologic features. Furthermore, such topics as the classification and distribution of rocks and the positions and characteristics of volcanoes, earthquake belts, mountain systems, and ocean basins were formerly described more or less in iso-

lation. Today we can treat these and other topics in the context of a unifying theory, **plate tectonics.** In the history of science, simple theories that explain many observations, as this one does, are the most enduring.

Plate tectonics is the idea that Earth's behavior is caused largely by the formation, movement, interactions, and destruction of the large rigid plates found at the surface of our planet. To understand what these plates are and how they were formed, we should return to our earlier discussion of Earth's crust and mantle and core.

We described the three main zones of Earth as chemically distinct layers that took their forms during differentiation. We could, however, also describe zones of Earth in terms of their physical properties. For example, you will see in Chapter 20 that Earth's zones can be characterized as strong and weak. We speak of "strong" here in the sense that a ceramic material is strong, and "weak" in the sense that modeling clay or wax is ductile, or plastic. The one is rigid and not easily deformed but can crack; the other is easily molded, like a tube of toothpaste. The **lithosphere** (from the Greek *lithos*, meaning "stone"), which includes the crust and the top part of the mantle, is depicted in Figure 1.11 as the strong, solid outermost shell, 50 to 100 km thick. The continents are raftlike inclusions embedded in the lithosphere. The lithosphere rides on the solid but weak **asthenosphere** (from the Greek *asthenes*, meaning "weak"). The lithosphere is strong because it is relatively cool, being so close to the surface. The asthenosphere, at greater depths, is a weak solid because it is hot, almost at the melting point.

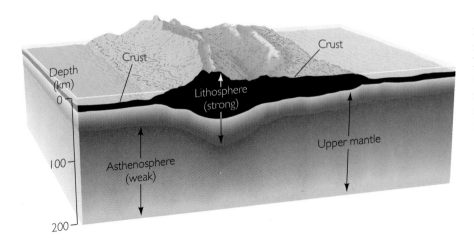

FIGURE 1.11 Earth's outermost shell is the strong, solid lithosphere, composed of the crust and top of the mantle. It rides on the weak, partially molten region of the mantle called the asthenosphere.

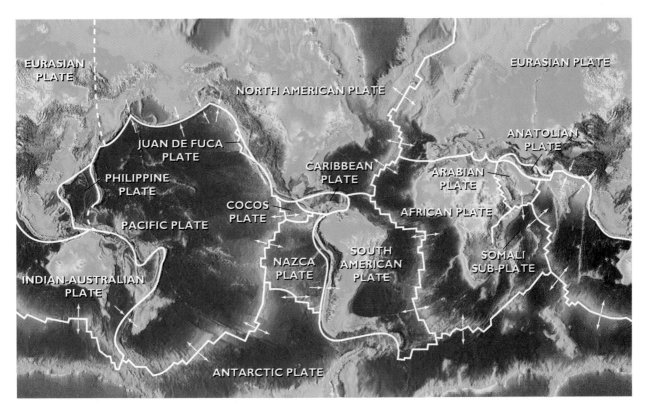

FIGURE 1.12 Earth's plates today. This flattened view of Earth's land and undersea topography shows plate boundaries, where plates separate (← →), collide (→ ←), or slide past each other (⇌). Note that plates and continents are not identical—the North American Plate, for example, is more extensive than the landmass that is the North American continent. *(Digital image by Peter W. Sloss, NOAA-NESDIS-NGDC.)*

The fact that the zones have different strengths determines that the lithosphere tends to behave as a rigid and brittle shell and the underlying asthenosphere "flows" as a ductile solid when it is subjected to forces.

Plates and Their Movements

According to the theory of plate tectonics, the lithosphere is not a continuous shell; it is broken into about a dozen large rigid plates that are in motion over the Earth's surface. Each plate moves as a distinct rigid unit, riding on the asthenosphere, which is also in motion. The major plates and the directions in which they move are shown in Figure 1.12. The North American Plate, for example, extends from the Pacific coast of the continent of North America to the middle of the Atlantic Ocean, where it meets the Eurasian and African plates.

Why should the plates move? Because Earth's interior is still hot. And even though the mantle beneath the lithosphere is mostly solid, it is hot and ductile or moldable. It can flow or "creep" if driven by forces. **Convection** supplies the forces. Convection is a mechanism of heat transfer that allows hot, less dense material to rise and dense, cool surface material to sink.

We tend to think of convection as a property of fluids and gases. We are all familiar with the circulating currents of boiling water in a pot, smoke rising from a chimney, heated air floating up to the ceiling, and cooled air sinking to the floor. Convection motion occurs in a flowing material, either a liquid or a moldable solid that is heated from below and cooled from above. The motion of solid flow is much slower than that of fluid flow. In either case, the heated matter rises under the forces of buoyancy because it has become less dense than the matter above it. It gives up heat and cools as it moves along the surface,

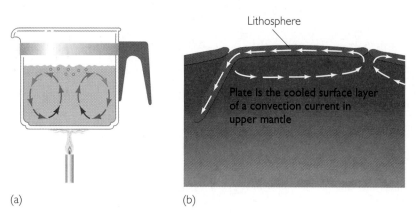

(a)

(b)

Lithosphere

Plate is the cooled surface layer of a convection current in upper mantle

FIGURE 1.13 (a) In this familiar instance of convection, water rises as it heats and then falls again as it cools near the surface. (b) A simplified view of the way convection currents in the interior of Earth may be the reason why plates form and move. As hot matter rises under some plate boundaries, the plates form and diverge. At other boundaries where plates converge, a cooled plate sinks and is dragged under a neighboring plate. Note that continents are embedded in plates and move with them.

as depicted in Figure 1.13, becoming more dense. When it becomes "heavier" than the underlying material, it sinks under the pull of gravity. The circulation continues as long as heat remains to be transferred from the hot interior to the cool surface.

On Earth, the cooled surficial layer of the convective flow system becomes the rigid lithospheric plate. (Although the chemistry and convection system are entirely different, you might think of an ice sheet on a lake as a cooled surficial layer.) In this way the lithosphere forms from rising hot mantle where plates separate, cools as it moves away from this boundary, and sinks back into the asthenosphere, dragging the plate along. This process occurs at boundaries where plates converge.

The movement of plates at a rate of a few centimeters a year is the surface manifestation of Earth's

convective system, still being driven by internal heat generated more than 4 billion years ago. This is a very simplified description of the mechanism of plate tectonics. Many details have been left out, such as the fate of the continents that are carried by the plates. We will be discussing these matters in due course. Although geologists generally agree on the "big picture," however, the details are still subjects of research and controversy.

Boundaries Between Plates

Many large-scale geologic features occur at the boundaries of plates, where the plates interact. These boundaries are of three types, all shown in Figure 1.14 and discussed in the following pages.

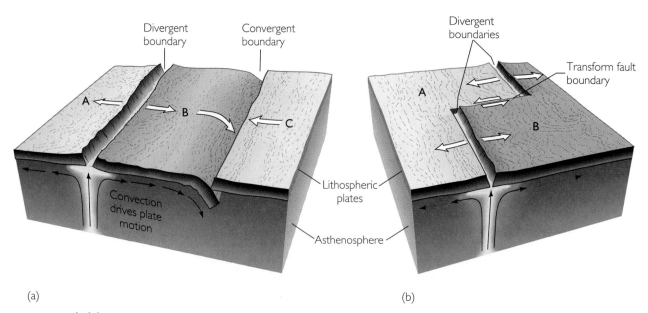

(a)

(b)

FIGURE 1.14 The three types of plate boundaries: (a) a divergent boundary, where plates A and B separate, and a convergent boundary, where plates B and C collide; (b) a transform fault boundary, where plates A and B slip past each other.

- **Divergent boundaries,** where plates separate and move in opposite directions, allowing new lithosphere to form from upwelling magma.

- **Convergent boundaries,** where plates collide and one sinks beneath the other, returning existing lithosphere to the interior.

- **Transform fault boundaries,** where plates slide past each other, approximately at right angles to their divergent boundaries.

DIVERGENT BOUNDARIES   A divergent boundary is typified by a rift, or cracklike valley, at the crest of a **mid-ocean ridge,** a chain of volcanic mountains that winds along the bottom of the world's oceans. The Mid-Atlantic Ridge, for example, runs up the middle of the Atlantic Ocean and surfaces in several places, most extensively in Iceland (Figure 1.15). Other major divergent boundaries are the East Pacific Rise and the Indian Ocean ridges, which you can locate on the front endpapers.

FIGURE 1.15 The Mid-Atlantic Ridge, a plate divergence boundary, surfaces above sea level in Iceland. The cracklike valley indicates that plates are being pulled apart. *(Gudmundur E. Sigvaldason, Nordic Volcanological Institute.)*

Divergent boundaries are characterized by earthquakes and volcanoes as the plates move apart and magma rises in the space between them. The magma solidifies as rock in the crack between the plates, and the plates grow by the gradual accretion of this fresh rock. If plates continue to separate, new seafloor is created as ocean basins widen—a process called **seafloor spreading** (Figure 1.16).

CONVERGENT BOUNDARIES   Because the plates cover the globe, if they separate in one place, they must converge somewhere else; and they do. Plates collide head-on along their convergent boundaries.

A profusion of geologic activities is associated with a plate collision. One plate sinks beneath the other, a process called **subduction** (Figure 1.16). Ocean lithosphere thus descends into the asthenosphere. This downbuckling produces a long, narrow deep-sea trench (about 100 km wide), where the ocean floor reaches its greatest depths (about 10 km below sea level). The edge of the overriding plate (in Figure 1.16, a plate with a continent on its edge) is crumpled and uplifted to form a mountain chain roughly parallel to the trench. The enormous forces of collision and subduction produce great earthquakes. Materials may be scraped off the descending slab and incorporated into the adjacent mountains. Imagine yourself as a geologist attempting to figure out the meaning of such tangled evidence. Furthermore, during subduction, parts of the descending plate may begin to melt. Magma formed where plates sink into the mantle floats upward, and can reach the surface and erupt from volcanoes.

Recall that divergent zones are sources of new lithosphere. **Subduction zones** at boundaries of convergence are sinks in which lithosphere is consumed by being returned to the mantle.

The west coast of South America, where the South American Plate collides with the oceanic Nazca Plate, is a subduction zone at a convergent boundary. (This and other convergent boundaries can be located on the front endpapers.) The Andes Mountains rise on the continental side of this boundary, and the Chilean deep-sea trench lies just off the coast. In this locale, volcanoes are active and deadly. One of them, Nevado del Ruiz in Colombia, was responsible for the deaths of 25,000 people when it erupted in 1985. Some of the world's greatest earthquakes have also been recorded along this boundary.

Another subduction zone is the boundary between the small Juan de Fuca Plate and the North American Plate, just off the coasts of British Columbia, Washington, and Oregon This convergent boundary gives rise to the volcanoes of the Cascade Range, including the dangerously active Mount St. Helens. As

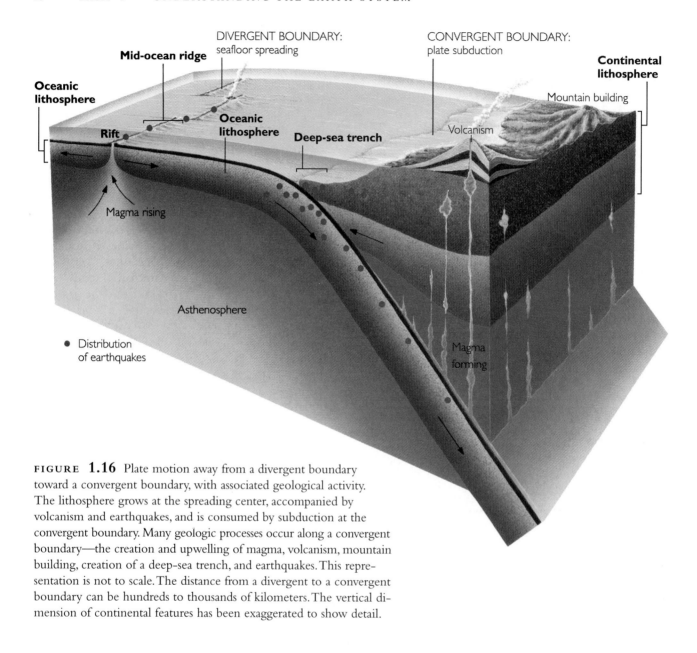

FIGURE 1.16 Plate motion away from a divergent boundary toward a convergent boundary, with associated geological activity. The lithosphere grows at the spreading center, accompanied by volcanism and earthquakes, and is consumed by subduction at the convergent boundary. Many geologic processes occur along a convergent boundary—the creation and upwelling of magma, volcanism, mountain building, creation of a deep-sea trench, and earthquakes. This representation is not to scale. The distance from a divergent to a convergent boundary can be hundreds to thousands of kilometers. The vertical dimension of continental features has been exaggerated to show detail.

our understanding of this subduction zone grows, scientists are increasingly worried that it will be the site of a future great earthquake affecting Oregon, Washington, and British Columbia.

TRANSFORM FAULT BOUNDARIES Some plates do not collide; they slip past each other horizontally along a transform fault. The famed San Andreas fault of California is such a boundary. There the Pacific Plate slides past the North American Plate in a northwesterly direction (Figure 1.17). Because the plates have been sliding past each other for millions of years, rocks facing each other on the two sides of the fault are of different types and ages. The sliding is not smooth, but more of a stick-slip process, in which sudden slips produce damaging earthquakes. One such earthquake destroyed San Francisco in 1906. There is much concern that a sudden slip along the San Andreas fault or related faults near Los Angeles or San Francisco may be extremely destructive within the next 25 years or so.

FIGURE 1.17 The view southeast along the San Andreas fault in the Carrizo Plain of central California. The San Andreas is a transform fault, forming a portion of the sliding boundary between the Pacific plate on the right and the North American plate on the left. *(Michael Collier.)*

Plate Tectonics and Planetary History

If continents are embedded in plates, it follows that the geography of the world must have been different in the remote past. The movement of continents over geologic time, termed **continental drift**, is an important part of modern geologic analysis. Some 200 million years ago all of the continents were assembled together into the supercontinent of Pangaea, as shown in Figure 1.18. (We discuss the geologic evidence for continental drift and Pangaea in Chapter 20.)

The plate motions we have just described represent the general pattern of the work output of Earth's internal heat engine as we can see it today. Because of Earth's internal heat, plates have emerged, moved, and been destroyed since the large-scale differentiation of Earth ended some 4 billion years ago. Geologists ponder why plate tectonics is not active today on the other terrestrial planets, but may have been in the past. In later chapters we will consider in more detail how a living planet generates and gets rid of its heat so that we can understand how that planet works.

The Theory of Plate Tectonics and the Scientific Method

Earlier we discussed the scientific method and the ways in which it guides the work of geologists. In the context of the scientific method, plate tectonics is not a dogma but a confirmed theory whose strength lies in its simplicity and generality. Theories can be overturned, but the theory of plate tectonics—like the theories of the age of Earth, the evolution of life,

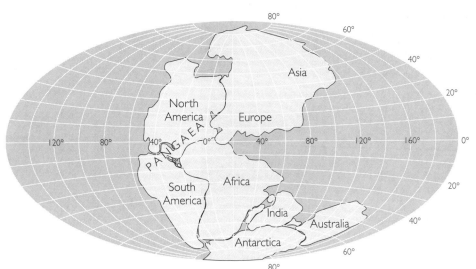

FIGURE 1.18 The supercontinent of Pangaea ("all lands"). Our current understanding of plate motions on the Earth leads us to believe that the geography of the world may have been different 200 million years ago. At that time, all of the continents were probably joined together as a single supercontinent, as shown here.

and genetics—explains so much so well and has survived so many efforts to prove it false that most geologists treat it as fact (see Feature 1.1).

Nevertheless, competing hypotheses have been advanced to explain the forces that cause plates to move, and others will probably be advanced in the future. Our story of the origin and early evolution of Earth is a hypothesis, and we can expect it to change many times because of the difficulty of recovering the information contained in the oldest rocks—information largely destroyed during the violent process of differentiation.

Because available information is incomplete and scientists' ability to observe nature is limited, some questions can be answered no better by scientists than by philosophers or poets. What, for instance, existed before the Big Bang? Or, as Walt Whitman asked:

> Great is the Earth, and the way it became what
> it is,
> Do you imagine it is stopped at this?

In this chapter we have made many statements without providing supporting observations or an underlying rationale. We have done so in order to preview the big picture, reserving for later chapters all of the substantiation called for by the scientific method. We began with the origin of the universe, but most of our discussion has dealt with the events that led to the formation and evolution of our planet to the point where continents, oceans, and an atmosphere developed. Now that we have outlined plate tectonics, the paradigm that has shaped geologic thought in the second half of the twentieth century, we can consider continents and ocean basins and interpret information contained in the various kinds of rocks within that conceptual framework. Although plate tectonics does not explain everything, it appears to be the best foundation on which to structure Earth's story.

GEOLOGISTS AT WORK

Although all geologists approach their work as scientists, they vary in the aspects of geology they embrace most avidly. Some are interested in geology as a pure science, a fascinating intellectual pursuit in its own right. Other geologists are more interested in using their knowledge in practical ways (Figure 1.19). They may join in the search for essential resources—the coal, oil, natural gas, metals, chemicals, and other materials on which modern civilization depends. They may warn of impending volcanic eruptions or help in the design of river flood control systems.

Scholarly Geology

Like most scientists, scholarly geologists seem to have an unbounded curiosity about the world around them, and they tend to feel uneasy when important natural phenomena remain unexplained. With the impulse of explorers, geologists will be hammering at rock outcrops, drawing up geologic maps, probing the seafloor, and scrutinizing Moon rocks as long as

FIGURE **1.19** Geologist collecting samples from a lava flow on Kilauea volcano, Hawaii. (*David R. Frazier/Photo Researchers.*)

the workings of our planet remain incompletely explained. They may serve as teachers and researchers in universities, or work for government agencies or for private firms that engage in construction or draw materials from the Earth.

Mitigating Nature's Threats

Some geologists are primarily interested in advancing our understanding of processes that can result in natural disasters. As human populations multiply and become increasingly concentrated, our vulnerability to natural disasters also increases. With greater urgency than ever before we must seek safeguards against nature's threats: earthquakes and the destructive sea waves that often accompany them, volcanic eruptions, landslides, floods, and droughts.

Geologists have had some success in this ongoing search. We have learned, for example, that certain earthquakes below the seafloor can trigger giant waves with the potential to destroy communities on distant coasts. A warning system now provides people along the Pacific coast with a few hours' notice of the arrival of such waves. Systems are now in place in other countries to monitor volcanoes and predict eruptions so that authorities can evacuate endangered populations. Japan, China, Russia, and the United States are testing methods to predict impending earthquakes. And in many countries around the globe, geologists play a role in identifying areas threatened by landslides and floods so that they can be rezoned to prohibit construction and thus preclude later devastation (Figure 1.20).

PROTECTING THE ENVIRONMENT Of all nature's threats to our planet, humans may be the greatest. Our species has gained the power to foul the land, seas, and sky. Urbanization, mining, agricultural operations, and warfare now rival nature in their ability to modify Earth's surface.

Groundwaters, which are the source of much of our drinking water, are beginning to show worrisome levels of toxic chemicals. These substances originate from agricultural and industrial wastes that seep through the soil and rock to the water stored below.

Acid rain, in part a product of coal combustion and automobile exhaust, threatens not only our lakes but also our forests.

Earth's atmosphere has functioned like a greenhouse, raising Earth's temperatures to levels that make life possible. The combustion of coal and oil, along with other industrial activities and the destruction of forests, releases gases that may be intensifying the greenhouse effect. It now seems quite possible that our climate will grow warmer in 50 to 100 years as a result. Unchecked, this trend could in time convert some fertile agricultural regions to semiarid lands, and melting glaciers could raise the level of the seas until it floods some low-lying coastal cities. By studying the causes and effects of past natural climatic changes, geologists are providing important knowledge about the possible impact of human activities on the environment.

SEARCHING FOR UNDISCOVERED RESOURCES From early Stone Age times, humans have been

FIGURE **1.20** Some natural disasters can be avoided. Many buildings were destroyed when heavy rains caused this landslide in Hong Kong. It is too late to relocate this area of downtown Hong Kong, but in other areas geologists can help determine whether environmental risks to a proposed building site dictate special construction or no construction at all.

1.1   INTERPRETING THE EARTH

Continental Drift: A Case History of the Scientific Method at Work

The idea that continents can move over Earth's surface through geologic time has a 140-year history of hypotheses successively advanced, criticized, rejected, and modified and improved with new data.

Francis Bacon first noticed in 1660 the jigsaw-puzzle fit of Africa and South America. Not until the mid-nineteenth century, however, was the conformity of the two coastlines proposed as evidence of the breakup and separation of the two continents. Few supported this farfetched idea. There was evidence that continents had moved up and down through geologic time but none, as yet, that they could move across Earth's surface. Uniformitarianism's tenet that the present is the key to the past had become firmly implanted in geologic thought. The conventional view was that the continents were and therefore always would be anchored in their present spots.

Although several scientists in the following years offered additional observations in support of continental drift, it was the German meteorologist Alfred Wegener who in 1912 ventured outside his own specialty and proposed a hypothesis that could not be ignored.

From paleontology, Wegener found evidence that many fossil and living organisms are remarkably similar between North America and Europe, South America and Africa, Australia and India, South Africa and southernmost South America. From geology, he found that different types of rock formations matched up if the continents were reassembled. Wegener explained his data by postulating a supercontinent, "Pangaea," that broke apart about 200 million years ago (see Figure 1.18). The fragments, which became the continents we know today, moved apart, and oceans filled the growing voids between them. Wegener wrote: "It is just as if we were to refit the torn pieces of a newspaper by matching their edges and then check whether the lines of print run smoothly across. If they do, there is nothing left but to conclude that the pieces were in fact joined this way" (Hallam 1973).

Wegener's hypothesis found few supporters. Some scientists pointed out that the coastlines did not fit together precisely. They also questioned whether the rock formations on both sides of the Atlantic matched as Wegener claimed, and whether this necessarily implied that the continents once were joined.

Paleontologists offered an alternative hypothesis for the similarities in fossil plants and animals. A few land bridges across the ocean could have provided a path for the migration of plants and animals. If the land bridges subsided and disappeared beneath the sea 200 million years ago, there was no need to postulate the breakup of a supercontinent.

Others argued against Wegener's hypothesis because they thought that the rocks of Earth's outer layers were too stiff for continents to plow through them. Moreover, Wegener had not provided a believable physical

searching Earth's crust for sources of energy and materials to support their current level of culture and society. That search has depleted some of the most accessible supplies of Earth's resources. Gone are the days when prospectors could easily find deposits of oil, iron, copper, tin, uranium, and other materials important to the world's economies. But if we are to see life improve for people in less developed countries, let alone maintain our own present standard of living, new mineral deposits must be found. The challenge for economic geologists is to reexplore the world, using new tools and techniques to search out the remaining undiscovered deposits of coal, oil, natural gas, and the ores from which our metals and industrial chemicals are derived.

The exploitation of Earth's mineral wealth has raised new and pressing concerns for geologists, principally protection of the environment and conservation of resources. How can we most efficiently exploit nature's wealth without waste and without devastating the landscape? Somehow we must find answers to this question.

We hope that some students taking this geology course will choose careers in Earth sciences or environmental sciences and that all readers will develop a lifelong interest in understanding Earth.

mechanism to explain why continents should be forced to move at all.

Wegener himself admitted that his hypothesis was based on circumstantial evidence and incomplete data. His opppenents accused him of ignoring contradictory data. However, his ideas found some support from a few important geologists, who corrected some of Wegener's mistakes, found additional evidence, and began to suggest believable mechanisms to move continents. Still, it was not enough to elevate Wegener's hypothesis to a respectable theory. An important reason was that almost all their data came from the continents, with little from the unexplored seafloor which covers 80 percent of Earth's surface.

World War II saw the development of new types of scientific instruments for military purposes. When the wartime geologists returned to their civilian life they began to use many of these powerful tools in their peacetime oceanographic research. They were able to map the topography of the seafloor rapidly and study its rocks with remote sensors. They discovered the mid-ocean ridges and showed that the seafloor is progressively older away from the ridges, as should occur with seafloor spreading. They found no evidence of submerged land bridges across the Atlantic—a death knell for that hypothesis. With this and other evidence they could provide overwhelming support for seafloor spreading, continental drift, and plate tectonics.

Wegener's hypothesis now has a basis in firm and abundant data, and supporting evidence continues to accumulate. Global Positioning Satellites, which can locate any position on Earth to a centimeter or so, can show on a yearly basis that the Atlantic Ocean is opening at a rate as fast as your fingernails grow.

Wegener died in 1930 on an expedition on the Greenland icecap, never knowing that he would be regarded as the grandfather of the plate tectonics revolution. Geologists are still researching the mechanism that causes plates to move and continents to drift with them.

Globe generated from digital data bases of land and seafloor elevations clearly shows the mid-Atlantic Ridge, where seafloor spreading is moving the South American and African plates apart.
(Peter W. Sloss, NOAA-NESDIS-NGDC.)

Summary

What is geology? Geology is the science that deals with Earth—its history, its composition and internal structure, and its surface features.

How do geologists study Earth? Geologists, like other scientists, use the scientific method. They share the data they develop and check one another's work. A hypothesis is a tentative explanation of a body of data. If it is confirmed repeatedly by other scientists' experiments, it may be elevated to a theory. Many theories are abandoned when subsequent experimental work shows them to be false. Confidence grows in those theories that withstand repeated tests and are able to predict the results of new experiments. Geologists have confidence in the principle of uniformitarianism. They believe that the processes that have shaped Earth have not changed over geologic time, and that therefore the key to understanding the past lies in observing how those processes work today.

How did our solar system originate? The Sun and its family of planets probably formed when a primeval cloud of gas and dust condensed about 4.5 billion years ago. The planets vary in chemical composition in accordance with their distance from the Sun and with their size.

How did Earth form and evolve over time? Earth probably grew by accretion of colliding chunks of matter. Soon after it was formed, it was struck by a giant meteorite. The impact had a profound effect on our planet. Matter ejected into space from both Earth and the impactor reassembled to form the Moon. The impact also melted much of the Earth. Radioactivity also contributed to early heating and melting. Heavy matter, rich in iron, sank toward Earth's center, and lighter matter floated up to form the outer layers that became the crust and continents. Outgassing gave rise to the oceans and a primitive atmosphere. In this way Earth was transformed to a differentiated planet with chemically distinct zones: an iron core; a mantle that is mostly magnesium, iron, silicon, and oxygen; and a crust rich in the light elements oxygen, silicon, aluminum, calcium, potassium, and sodium and in radioactive elements.

What are the basic elements of plate tectonics? The lithosphere, Earth's outermost shell, is broken into about a dozen large, rigid plates. These plates and their origin and movement are a manifestation of solid-flow convection currents in the mantle. The plates jostle each other as they move in their individual courses. The boundaries between divergent and convergent plates are zones of intense activity: mountain building, volcanism, seafloor creation and destruction, and earthquakes. Major earthquakes occur on transform fault boundaries.

How is geology both a basic and an applied science? Geology is a basic science in that it creates new understanding and knowledge of the planet. It is an applied science in its involvement with discovering Earth's mineral wealth and mitigating hazards such as earthquakes, volcanoes, floods, and environmental damage.

KEY TERMS AND CONCEPTS

geology (p. 3)
scientific method (p. 4)
hypothesis (p. 4)
theory (p. 4)
principle of uniformitarianism (p. 4)
nebular hypothesis (p. 5)
differentiation (p. 8)
magma (p. 9)

core (p. 9)
crust (p. 10)
mantle (p. 10)
Heavy Bombardment period (p. 13)
plate tectonics (p. 14)
lithosphere (p. 14)
asthenosphere (p. 14)
convection (p. 15)

divergent boundaries (p. 17)
convergent boundaries (p. 17)
transform fault boundaries (p. 17)
mid-ocean ridge (p. 17)
seafloor spreading (p. 17)
subduction (p. 17)
subduction zones (p. 17)
continental drift (p. 19)

EXERCISES

1. What is the difference between an experiment, a hypothesis, a theory, and a fact?

2. What factors made Earth a particularly congenial place for life to develop?

3. How and why do the inner planets differ from the giant outer planets?

4. What caused Earth to differentiate, and what was the result?

5. How does the chemical composition of Earth's crust differ from that of its deeper interior? From that of its core?

6. Describe the central idea of the theory of plate tectonics.

7. Describe and explain the large-scale geologic activities associated with the three types of plate boundaries.

THOUGHT QUESTIONS

1. How does the discovery of solid matter around other stars contribute to the debate about the possibility of life elsewhere in the cosmos? What are the implications of the existence of life on the planets of other stars?

2. If you were an astronaut exploring another planet, how would you decide whether the planet was differentiated and whether it was still geologically active?

3. What are the advantages and disadvantages of living on a differentiated planet? On a geologically active planet?

4. If a giant impact occurred after life had formed on Earth, what would have been the consequences?

5. Speculate on what life would be like today if the ancient continent of Pangaea had remained a single landmass instead of breaking up into Eurasia, Africa, and the Americas.

6. According to biblical interpretation, Earth is some 5000 years old. Is that figure a hypothesis, a theory, a fact, or an article of faith? What is the difference between a statement based on faith and a scientific theory?

SUGGESTED READINGS

Ahrens, Thomas J. 1994. The origin of the Earth. *Physics Today* (August): 38–35.

Allegre, Claude. 1992. *From Stone to Star*. Cambridge, Mass.: Harvard University Press.

Brandt, J. C., and S. P. Maran. 1979. *New Horizons in Astronomy*, 2nd ed. San Francisco: W. H. Freeman.

Hallam, A. 1973. *A Revolution in the Earth Sciences: From Continental Drift to Plate Tectonics*. Oxford: Clarendon Press.

Hurley, P. M. 1968. The confirmation of continental drift. In *Continents Adrift, Readings from Scientific American*, 56-67. San Francisco: W. H. Freeman.

Exploring Space. 1990. Special issue of *Scientific American*.

Managing Planet Earth. 1989. Special issue of *Scientific American* (September).

May, Robert H. 1992. How many species inhabit Earth? *Scientific American* (April): 42–48.

National Academy of Sciences. 1992. *Science and Creationism*. Washington, D.C.: National Academy Press.

National Research Council. 1990. *The Search for Life's Origins*. Washington, D.C.: National Academy Press.

National Research Council. 1993. *Solid-Earth Sciences and Society*. Washington, D.C.: National Academy Press.

Press, Frank, and Raymond Siever. 1986. The planets: A summary of current knowledge. Chapter 22 in *Earth*, 4th ed. New York: W. H. Freeman.

Stanley, Steven M. 1993. *Exploring Earth and Life Through Time*. New York: W. H. Freeman.

Takeuchi, H., S. Uyeda, and H. Kanamori. 1970. *Debate About the Earth*. San Francisco: W. H. Freeman, Cooper & Co.

Taylor, G. Jeffrey. 1994. The scientific legacy of *Apollo. Scientific American* (July): 40–47.

Understanding Earth 2.0 CD-ROM. 1997. New York: W. H. Freeman.

Walter, William J. 1992. *Space Age*. New York: Random House.

Weiner, Jonathan. 1986. *Planet Earth*. New York: Bantam Books.

Westbroek, Peter. 1991. *Life as a Geologic Force*. New York: W. W. Norton.

Wetherill, George W. 1990. Formation of the Earth. *Ann. Rev. Earth Planet. Sci.* 18: 205–256.

INTERNET SOURCES

Ask-A-Geologist
ⓘ **http://walrus.wr.usgs.gov/docs/ask-a-ge.html**
Questions addressed to this site will be answered by a geologist with the U.S. Geological Survey. The site also includes a list of Frequently Asked Questions (FAQ).

The Geologist's Lifetime Field List
ⓘ **http://www.uc.edu/~ACOMBTY/geologylist.html**
This site provides a list of the features and events that most geologists would like to experience firsthand during their careers. Links are provided to more information about the items on the list.

Earth Science Site of the Week
ⓘ **http://agcwww.bio.ns.ca/misc/geores/sotw/ sotw.html**
The Atlantic Division of the Canadian Geological Survey features an outstanding site dealing with a different aspect of earth science each week. A list of previous selections is available with links to the sites.

The Nine Planets
ⓘ **http://www.seds.org/billa/tnp/**
More than 60 "pages" of text and images dealing with the solar system are available here. You will find multiple images of the features comprising the solar system, video of a few features, a glossary, and a list of "Open Issues" about each member of the solar system. Take the "Quick Tour" for an overview of the site.

Views of the Solar System
ⓘ **http://bang.lanl.gov/solarsys/homepage.htm**
This site includes an image gallery from NASA and other sources. With links to pages for each component of the solar system, this site provides basic data, images, video, and more. The text is available in English and Spanish.

Welcome to the Planets
ⓘ **http://pds.jpl.nasa.gov/planets/**
This NASA site provides data and images for each of the planets, a glossary, and information on the spacecraft that produced the images.

Web Site for "Understanding Earth"
ⓘ **www.whfreeman.com/understandingearth/**
The publisher of *Understanding Earth* updates this Web site periodically to provide current information and URLs on topics relevant to all chapters.

2

Crystals of amethyst, a variety of quartz. The planar surfaces are crystal faces, which reflect the underlying arrangement of the atoms that make up the crystals. (*Chip Clark.*)

Minerals: Building Blocks of Rocks

In Chapter 1 we saw how plate tectonics describes Earth's large-scale structure and dynamics, but we touched only briefly on the wide variety of materials that appear in plate-tectonic settings. In this chapter and the next, we focus on rocks, the records of geologic processes, and on minerals, the building blocks of rocks.

To tell Earth's story accurately, geologists often adopt a Sherlock Holmesian perspective, using current evidence to deduce the processes and events that occurred in the past at some particular place on Earth. The kinds of minerals found in some volcanic rocks, for example, give evidence of the eruptions that brought molten rock, at temperatures perhaps as high as 1000°C, to Earth's surface. The minerals of a granite give evidence that it crystallized deep in the Earth's crust under conditions that formed mountains such as the Himalayas. Such conditions, which arise when two continental plates collide, produce temperatures as high as 700°C and pressures more than 10,000 times higher than at Earth's surface.

This process of deduction is essential as we attempt to understand the geology of a region and make informed guesses about the location of as yet undiscovered deposits of economically important resources such as metal ores. One of the richest sources of evidence is the focus of this chapter: **mineralogy**—the branch of geology that studies the composition, structure, appearance, stability, occurrence, and associations of minerals.

WHAT ARE MINERALS?

Geologists define a **mineral** as a *naturally occurring, solid crystalline substance, generally inorganic, with a specific chemical composition.* Minerals are the building blocks of rocks, which can be made up of varying assemblages of minerals. Minerals are homogeneous: they cannot be divided by mechanical means into smaller components. With the proper tools and effort, most rocks can be separated into their constituent minerals. A few kinds of rocks, such as limestone, contain only a single kind of mineral (calcite). Others, such as granite, are made of several kinds of minerals. To identify and classify the many kinds of rocks found in the Earth and understand how they are formed, we must be informed about minerals.

We begin by examining each part of our definition of a mineral in a little more detail.

NATURALLY OCCURRING . . .   To qualify as a mineral, a substance must be found in nature. Diamonds mined from the Earth in South Africa are minerals. Synthetic versions produced in industrial laboratories are not considered to be true minerals. Nor are the thousands of laboratory products invented by chemists.

SOLID CRYSTALLINE SUBSTANCE . . .   Minerals are solid substances—they are neither liquids nor gases. When we say that a mineral is *crystalline,* we mean that the tiny particles of matter—the atoms—that compose it are arranged in an orderly, repeating, three-dimensional array. Solid materials that have no such orderly arrangement are referred to as *glassy* or *amorphous* (without form). Window glass is amorphous, as are some natural glasses formed during volcanic eruptions. Later in this chapter, we will explore in detail the process by which crystalline materials form.

GENERALLY INORGANIC . . .   The stipulation that minerals are inorganic substances follows historical usage and excludes the organic substances that make up plant and animal bodies. These organic substances are made of organic carbon, the form of carbon found in all biological materials. Decaying vegetation in a swamp may be geologically transformed into coal, which is also made of organic carbon, but though it is found as a natural deposit, coal is not traditionally considered a mineral. Many minerals are, however, secreted by organisms. One such mineral, calcite, forms the shells of oysters and many other organisms, and it contains inorganic carbon. The calcite of these shells, which constitute the bulk of many limestones, fits the definition of a mineral because it is inorganic and crystalline.

. . . WITH A SPECIFIC CHEMICAL COMPOSITION
The key to understanding the materials and composition of the Earth lies in understanding how the chemical elements are organized into minerals. What makes each mineral unique is the combination of its chemical composition and the arrangement of its atoms in an internal structure. A mineral's chemical composition either is fixed or varies within defined limits. The mineral quartz, for example, has a fixed ratio of two atoms of oxygen to one of silicon. This ratio never varies, although quartz is found in many different kinds of rock. The components of olivine, a slightly more complex mineral, always have a fixed ratio. Iron, magnesium, and silicon make up olivine. Although the ratio of iron to magnesium atoms may vary, the sum of those atoms in relation to the number of silicon atoms always forms a fixed ratio. Figure 2.1 summarizes the definition of a mineral.

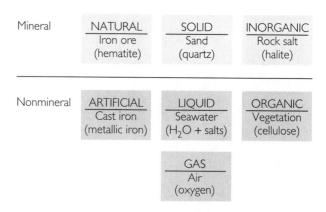

FIGURE 2.1 Minerals are distinguished from other materials in being naturally occurring, solid, inorganic substances with specific chemical compositions.

THE ATOMIC STRUCTURE OF MATTER

A modern dictionary lists many meanings for the word *atom* and its derivatives. One of the first is "anything considered as the smallest possible unit of any material." To the ancient Greeks, *atomos* meant "indivisible." John Dalton (1766–1844), an English chemist and the father of modern atomic theory, theorized that atoms were particles of matter of several kinds that were so small that they could not be seen with any microscope and so universal that they composed all substances. In 1805 Dalton hypothesized that the various chemical elements consist of different kinds of atoms, that all atoms of any given element are identical, and that chemical compounds are formed by various combinations of atoms of different elements in definite proportions.

By the early twentieth century, physicists, chemists, and mineralogists, building on Dalton's ideas, had reached an understanding of the structure of matter, much as we know it today. We now know that an **atom** is the smallest unit of an element that retains the physical and chemical properties of that element. We also know that atoms are the small units of matter that combine in chemical reactions, but that atoms themselves are divisible into even smaller units.

The Structure of Atoms

Understanding the structure of atoms allows us to predict how chemical elements will react with one another and form new crystal structures.

THE NUCLEUS: PROTONS AND NEUTRONS   At the center of every atom is a dense **nucleus** containing practically all the mass of the atom in two kinds of particles, protons and neutrons (Figure 2.2). For convenience, each of these particles is taken to have a mass of 1 atomic mass unit.[1] Protons carry an electrical charge; neutrons do not. Thus a **proton** has a positive electrical charge of +1. A **neutron** is electrically neutral—that is, uncharged. Atoms of the same chemical element may have different numbers of neutrons, but the number of their protons does not vary.

---

[1]The atomic mass unit is equal to 1/12 of the actual mass of a carbon atom with mass number 12, approximately 1.6604×10^{-24} grams.

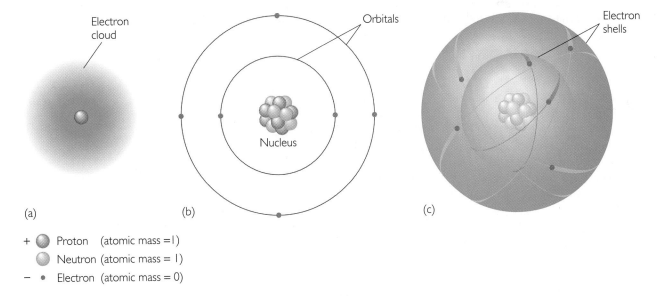

FIGURE 2.2 Electron structure of hydrogen and carbon atoms. (a) The position of the single electron of the simplest element, hydrogen, shown as an electron cloud surrounding the nucleus. (b) Electrons of a carbon atom are represented as lying in definite orbits around the nucleus, which contains six protons, each with a charge of +1, and six neutrons, each with zero charge. Six electrons are found in two concentric shells, an inner one with two electrons and an outer one with four. (c) A somewhat more realistic representation of the electron shells of a carbon atom. The size of the nucleus is greatly exaggerated; it is much too small to show on a true scale.

ELECTRONS   Surrounding the nucleus is a cloud of moving **electrons,** each with a mass so small that it is conventionally taken to have no mass. Each electron carries an electrical charge of −1. The number of protons in the nucleus of any atom is balanced by the same number of electrons in the cloud outside, so an atom is electrically neutral. Modern models of atomic structure give the locations of electrons around the nucleus as *orbitals* (Figure 2.2b). They can be thought of as spherical **shells,** or regions, around the nucleus where an electron is most likely to be found, not as fixed orbits or paths of travel. For convenience in diagrams, however, we usually depict orbitals as concentric spherical shells around a nucleus (see Figure 2.2c).

Atomic Number and Atomic Mass

The number of protons in the nucleus of an atom is called its **atomic number.** Because all atoms of the same element have the same number of protons, they also have the same atomic number. All atoms with six protons, for example, are carbon atoms (atomic number 6). Since all the protons are balanced by an equal number of electrons, each element has a distinctive number of electrons, too. The atomic number of an element determines how it will react chemically with other elements.

The **atomic mass** of an element is the sum of the masses of its protons and neutrons. (Electrons, because they have so little mass, are not included in this sum.) Although the number of protons is constant, atoms of the same chemical element may have different numbers of neutrons, and therefore different atomic masses. These various kinds of atoms are called **isotopes.** Isotopes of the element carbon, for example, all with six protons, exist with six, seven, and eight neutrons, giving atomic masses of 12, 13, and 14 (see Figure 2.3). In nature, the chemical elements occur as mixtures of isotopes, so their atomic masses are never whole numbers. Carbon's atomic mass, for example, is 12.011. It is close to 12 because the isotope carbon-12 is overwhelmingly abundant. The relative abundance of the different isotopes of an element on Earth is determined by geological processes that enhance the abundance of some isotopes over others. Carbon-12, for example, is favored by some reactions, such as photosynthesis, in which organic carbon compounds are produced from inorganic carbon compounds.

CHEMICAL REACTIONS

The structure of any particular kind of atom determines its chemical reactions with other atoms. **Chemical reactions** are interactions of the atoms of two or more chemical elements in certain fixed proportions that produce new chemical substances—

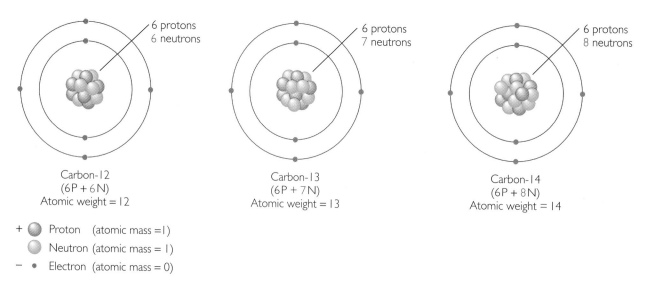

Carbon-12
(6P + 6N)
Atomic weight = 12

Carbon-13
(6P + 7N)
Atomic weight = 13

Carbon-14
(6P + 8N)
Atomic weight = 14

+ ◯ Proton   (atomic mass =1)

◯ Neutron (atomic mass = 1)

− • Electron (atomic mass = 0)

FIGURE 2.3 These three carbon isotopes all have the same number of protons and thus the same atomic number, 6. Their atomic masses differ, however, because they have slightly different numbers of neutrons. The atomic mass of any element is the average of the weighted sum of the atomic masses of its various isotopes. One isotope of an element—for example, carbon-12—is far more abundant than the others because natural processes favor that particular isotope.

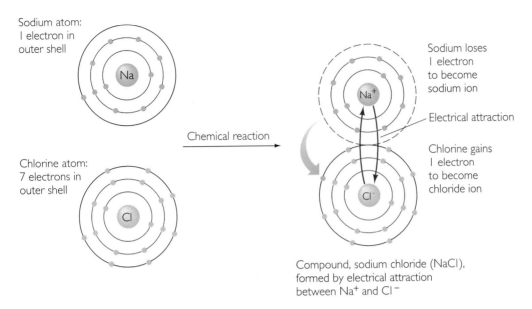

Sodium atom:
1 electron in
outer shell

Chlorine atom:
7 electrons in
outer shell

Chemical reaction

Sodium loses
1 electron
to become
sodium ion

Electrical attraction

Chlorine gains
1 electron
to become
chloride ion

Compound, sodium chloride (NaCl),
formed by electrical attraction
between Na$^+$ and Cl$^-$

FIGURE 2.4 Formation of a chemical compound by reaction of chlorine and sodium atoms. When sodium (Na) and chlorine (Cl) react to form table salt (sodium chloride, NaCl), the sodium atom loses one electron from its outer shell and the chlorine atom acquires that electron. During this reaction, an orderly array of ions is formed.

chemical compounds. When two hydrogen atoms combine with one oxygen atom, they form a new chemical compound that we call water (H_2O). The properties of a chemical compound formed in the course of a reaction may be entirely different from those of its constituent elements. For example, when an atom of sodium, a metal, combines with an atom of chlorine, a noxious gas, they form the chemical compound sodium chloride, better known as table salt. We represent this compound by the chemical formula *NaCl,* the symbol *Na* standing for the element sodium and the symbol *Cl* for the element chlorine. (Every chemical element has been assigned its own symbol, which we use as a kind of shorthand for writing chemical formulas and equations.)

Chemical reactions take place primarily through the interactions of electrons. To understand those interactions we need to know the number of electrons in an atom and how they are arranged in electron shells.

Gaining and Losing Electrons

Electrons surround the nucleus of an atom in a unique set of concentric spheres called electron shells, each of which can hold up to a certain number of electrons. In the chemical reactions of most elements, only the electrons in the outermost shells interact. In the reaction between sodium (Na) and chlorine (Cl) that forms sodium chloride (NaCl), the sodium atom loses an electron from its outer shell of electrons and the chlorine atom gains an electron in its outer shell (Figure 2.4).

IONS After *gain or loss of an electron,* the atoms in the new chemical compound are no longer electrically neutral. When the sodium atom loses an electron, it becomes a sodium **ion,** with an electrical charge of $+1$, because it still has the same number of protons but one less electron. It is now represented by the symbol Na$^+$. The chlorine atom, in gaining an electron, has become a *chloride* ion with an electrical charge of -1, written as Cl$^-$. Positive ions, such as sodium, are called **cations;** negative ions, such as chloride, are called **anions.** The compound NaCl itself remains electrically neutral because the positive charge on Na$^+$ is exactly balanced by the negative charge on Cl$^-$.

Groups of ions may join to form *complex ions,* such as the common sulfate ion (SO_4^{2-}), a component of the mineral anhydrite ($CaSO_4$) and an abundant constituent of seawater. The sulfate ion is a unit made up of one sulfur ion with a $+6$ charge and four oxygen ions, each with a -2 charge, the net charge adding to -2.

ELECTRON SHELLS AND ION STABILITY Before reacting with chlorine, the sodium atom has one electron in its outer shell. When it loses that electron, its outer shell is eliminated and the next shell inward, which has eight electrons (the maximum this shell can hold), becomes the outer shell. The original chlorine atom had seven electrons in its outer shell, with room for a total of eight. By gaining an electron, its outer shell is filled. Many elements have a strong tendency to acquire a full outer electron shell, some by gaining electrons and some by losing them in the course of a chemical reaction. The stability of ions with fully occupied outer shells is related to the interactions of electrons in various orbitals around the nucleus.

Many chemical reactions involve gains and losses of several electrons as two or more elements combine. The element calcium (Ca), for example, becomes a doubly charged cation, Ca^{2+}, as it reacts with two chlorine atoms to form calcium chloride. (In the chemical formula for calcium chloride, $CaCl_2$, the presence of two chloride ions is symbolized by the subscript 2. Chemical formulas thus show the relative proportion of atoms or ions in a compound. Common practice is to omit the subscript 1 next to single ions in a formula. For the single Ca in $CaCl_2$, therefore, the subscript 1 is omitted.)

FIGURE 2.5 Electron sharing in diamond. The mineral diamond is composed of the single element carbon. Each carbon atom has four electrons in its outer shell, and each acquires four more by sharing its electrons with four adjacent carbon atoms.

Electron Sharing

Not all chemical elements react by gaining or losing electrons. Some have a strong tendency to combine chemically by engaging in **electron sharing** with atoms of the same or a different element to achieve a stable configuration of electrons. Carbon and silicon, two of the abundant elements of Earth's crust, are elements that tend toward electron sharing.

The mineral diamond is composed of the single element carbon. Each carbon atom has four electrons in its outer shell, and each acquires four more by sharing its electrons with four adjacent carbon atoms (see Figure 2.5). When these electrons are shared, all of the atoms act as if each had a full complement of eight electrons in its outer shell. The shared electrons cannot be considered to have been gained or lost. In a sense, the nuclei of the sharing atoms have "gained" the electron for whatever part of the time it can be visualized as belonging to the outer shell of one or the other atom. Nevertheless, because the atoms still have their original number of electrons, we do not ordinarily refer to them as ions.

Periodic Table of Elements

Chemists have long known that some groups of elements have similar chemical properties, such as boiling and melting points and tendencies to react chemically with other elements. These groups differ markedly from one another. As the atomic structure of elements became known, these properties proved to correspond to the electron shell patterns of the elements.

The periodic table (Figure 2.6) organizes the elements (from left to right along a row) in order of atomic number (number of protons), which also means increasing numbers of electrons in the outer shell. The third row from the top, for example, starts at the left with sodium (atomic number 11), which has one electron in its outer shell. The next is magnesium (atomic number 12), which has two electrons in its outer shell, followed by aluminum (atomic number 13), with three, and silicon (atomic number 14), with four. Then come phosphorus (atomic number 15), with five; sulfur (atomic number 16), with six; and chlorine (atomic number 17), with seven. The last element in this row is argon (atomic number 18), with eight electrons, the maximum possible, in its outer shell. Each column in the table forms a vertical grouping of elements with similar electron shell patterns.

ELEMENTS THAT TEND TO LOSE ELECTRONS

The elements in the leftmost column all have a single electron in their outer shells and have a strong tendency to lose that electron in chemical reactions.

Of this group, hydrogen (H), sodium (Na), and potassium (K) are found in major abundance at Earth's surface and in its crust.

The second column from the left includes two more elements of major abundance, magnesium (Mg) and calcium (Ca). Elements in this column have two electrons in their outer shells and a strong tendency to lose them both in chemical reactions.

ELEMENTS THAT TEND TO GAIN ELECTRONS

Toward the right side of the table, the two columns headed by oxygen (O), the most abundant element in the Earth, and fluorine (F), a highly reactive toxic gas, group the elements that tend to gain electrons in their outer shells. The elements in the column headed by oxygen have six of the possible eight electrons in their outer shells and tend to gain two electrons. Those in the column headed by fluorine have seven electrons in their outer shells and tend to gain one.

OTHER ELEMENTS

The columns between the two on the left and the two headed by oxygen and fluorine have varying tendencies to gain, lose, or share electrons. The column toward the right side of the table headed by carbon (C) includes silicon (Si), of major abundance in the Earth. As we noted earlier, both silicon and carbon tend to share electrons.

The elements in the last column on the right, headed by helium (He), have full outer shells and thus no tendency either to gain or to lose electrons. As a result, these elements, in contrast to those in other columns, do not react chemically with other elements, except under very special conditions.

We can predict a great many chemical reactions from the information conveyed by the columns and rows in the periodic table. Figure 2.7 illustrates diverse patterns of gaining, losing, or sharing electrons in a comparison of five common elements: hydrogen, sodium, magnesium, oxygen, and chlorine.

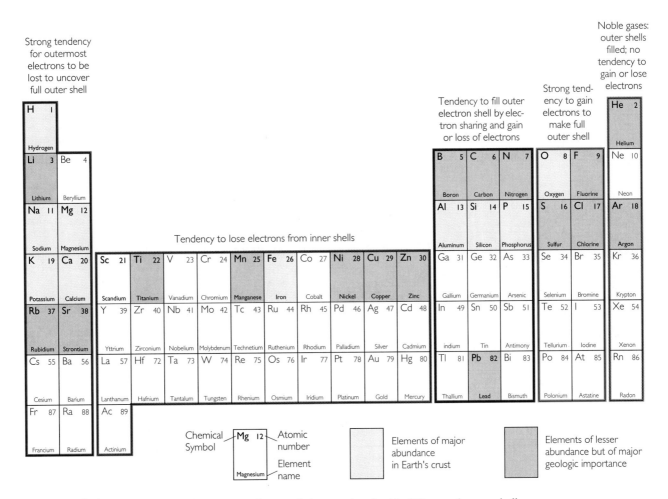

FIGURE 2.6 Periodic table of elements. Characteristics associated with different electron shell patterns are indicated at the top of the table. (Two special groups of rare elements—atomic numbers 58–71 and 90–103—have been omitted.)

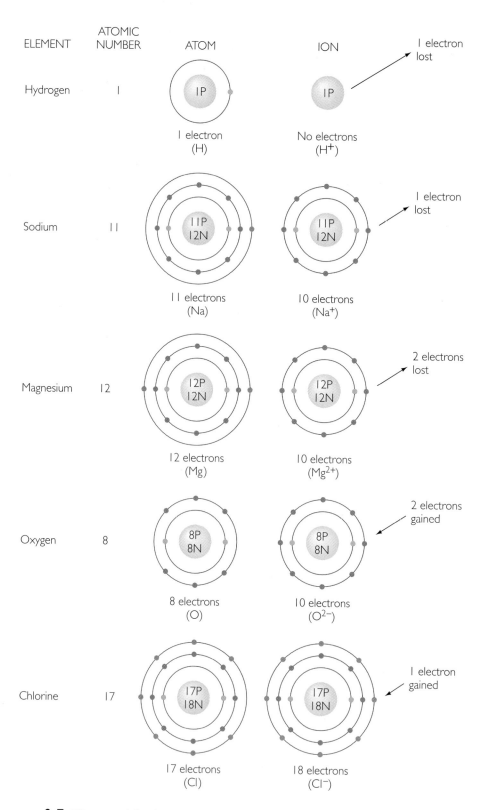

FIGURE 2.7 These models of the atoms and ions of five common elements illustrate the diversity of electron shells. Ions are formed as electrons are gained or lost from outer shells. (P stands for proton, N for neutron.)

CHEMICAL BONDS

The ions or atoms of elements that make up compounds are held together by electrical forces of attraction between electrons and protons, which we call chemical bonds. The electrical attractions either of shared electrons or of gained or lost electrons may be strong or weak, and the bonds created by these attractions are correspondingly strong or weak. Strong chemical bonds keep a substance from chemically decomposing into its elements or into other compounds. They also make minerals hard and keep them from cracking or splitting. Two major types of bonds are found in most rock-forming minerals: ionic bonds and covalent bonds.

Ionic Bonds

The simplest form of chemical bond is the **ionic bond.** Bonds of this type are formed by electrical attraction between ions of the opposite charge, such as Na^+ and Cl^- in sodium chloride (see Figure 2.4). This attraction is of exactly the same nature as the static electricity that can make clothing of nylon or silk cling to the body. The strength of an ionic bond decreases greatly as the distance between ions increases. Bond strength increases as the electrical charges of the ions increase. Ionic bonds are the dominant type of chemical bonds in mineral structures; *about 90 percent of all minerals are essentially ionic compounds.*

Covalent Bonds

Elements that do not readily gain or lose electrons to form ions and instead form compounds by sharing electrons are held together by **covalent bonds.**

Covalent bonds are generally stronger than ionic bonds. One mineral with a covalently bound crystal structure is diamond, consisting of the single element carbon. As we saw in the case of diamond, carbon has four electrons in its outer shell and acquires four more by electron sharing to achieve a full outer shell with eight electrons. In diamond, every carbon atom (not an ion) is surrounded by four others arranged in a regular *tetrahedron,* a four-sided pyramidal form, each side a triangle (Figure 2.8a). In this configuration, each carbon atom shares an electron with each of its four neighbors and thus achieves a stable set of eight electrons in its outer shell.

Atoms of metallic elements, which have strong tendencies to lose electrons, pack together as cations, while freely mobile electrons are shared and dispersed among the ions. This free electron sharing results in a kind of covalent bond that we call a **metallic bond.** It is found in a small number of minerals, among them the metal copper and some sulfides.

The chemical bonds of some minerals are intermediate between pure ionic and pure covalent bonds because some electrons are exchanged and others are shared.

ATOMIC STRUCTURE OF MINERALS

Minerals can be looked at in two complementary ways: as crystals (or grains) we can see with the naked eye, and as assemblages of submicroscopic atoms organized in an ordered three-dimensional array (Figure 2.8). The preceding discussion of how substances are formed by chemical bonding between atoms and ions prepares us for a closer view of the

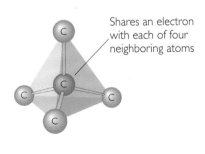

Shares an electron with each of four neighboring atoms

(a)

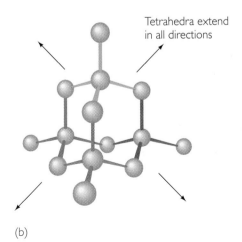

Tetrahedra extend in all directions

(b)

FIGURE 2.8 The carbon tetrahedron of diamond. (a) A single tetrahedron formed by a carbon atom bonded to four other carbon atoms. (b) A network of carbon tetrahedra linked to one another.

orderly forms that characterize minerals' structure and of the conditions under which minerals form. Later in this chapter, you will see that the crystal structures of minerals are reflected in their physical properties. First, however, we turn to the question of how minerals form.

How Do Minerals Form?

Minerals are formed by the process of **crystallization,** the growth of a solid from a material whose constituent atoms can come together in the proper chemical proportions and crystalline arrangement. (Remember that the atoms in a mineral are arranged in an ordered, three-dimensional array.) The bonding of carbon atoms in diamond, a covalently bonded mineral, is one example of crystallization and crystal structure. In satisfying the requirements of electron sharing, carbon atoms come together in tetrahedra, each tetrahedron attaching to another and building up a regular three-dimensional structure from a great many atoms (see Figure 2.8b). As a diamond crystal grows, it extends its tetrahedral structure in all directions, always adding new atoms in the proper geometric arrangement. Diamonds can be synthesized under very high pressures and temperatures that mimic conditions in Earth's mantle.

The sodium and chloride ions that make up sodium chloride, an ionically bonded mineral, also crystallize in an orderly three-dimensional array. In Figure 2.9 we can see the geometry of their arrangement, with each ion of one kind surrounded by six ions of the other in a series of cubes extend-

(a)

(b)

FIGURE **2.10** (a) Ultra-high vacuum scanning tunneling microscope image of a galena (PbS) surface with atomic resolution. The scale of the image is approximately $26 \times 10^{-8} \times 26 \times 10^{-8}$ cm. The cubic structure of galena is reflected in the arrangement of both the bright and dark spots. Theoretical calculations suggest that the bright spots correspond to lead atoms and the dark spots correspond to sulfur sites. *(Kevin M. Rosso and Michael F. Hochella, Jr., Virginia Polytechnic and State University.)* (b) Galena crystals. *(Chip Clark.)*

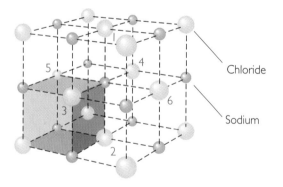

FIGURE **2.9** Structure of halite (sodium chloride). The dotted lines between the ions show the cubic geometry of this mineral; they do not represent bonds. In this example it is easy to see one sodium ion surrounded by six chloride ions. (Ions are not drawn to scale.)

Chloride

Sodium

ing in three directions. This arrangement is called a cubic structure.

Atoms and ions are so small—most of them only a few ten-millionths of a centimeter—that we cannot see the crystalline arrangement of a mineral directly even with the most powerful ordinary microscope. With especially high-powered electron and other, newer kinds of microscopes, however, we can now image the atomic arrangements of crystals (Figure 2.10).

Crystallization starts with the formation of microscopic single **crystals,** bodies whose boundaries are natural flat (plane) surfaces. These surfaces, called *crystal faces,* are the defining external characteristic of a crystal. The crystal faces of a mineral are the external expression of the mineral's internal atomic structure. Figure 2.11 pairs drawings of perfect crystals (which are very rare in nature) and photos of two actual minerals. The simple geometric cubes of sodium chloride crystals (the mineral halite, or rock salt) correspond to the cubic arrangement of its ions. The six-sided (hexagonal) shape of the quartz crystal corresponds to its hexagonal internal atomic structure.

During crystallization, the initially microscopic crystals grow larger, maintaining their crystal faces as long as they are free to grow. Large crystals with well-defined faces form when growth is slow and steady and space is adequate to allow growth without interference from other crystals nearby. For this reason, most large mineral crystals form in open spaces in rocks, such as open fractures or cavities (Figure 2.12).

Often, however, the spaces between growing crystals fill in or crystallization proceeds too rapidly. Crystal faces then grow over one another, and the former crystals coalesce to become a solid mass of crystalline particles, or *grains*. In the crystallized mass, few or no grains show crystal faces (see Figure 2.12). Large crystals that can be seen with the naked eye are relatively unusual, but many microscopic minerals in rocks display crystal faces.

Glassy materials, which solidify from liquids so quickly that they lack any internal atomic order, do not form crystals with plane faces. Instead they are found as masses with curved, irregular surfaces. The most common glass is volcanic glass.

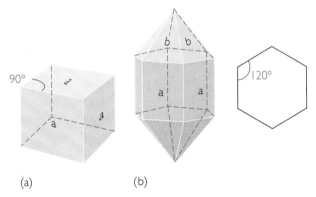

(a)   (b)

FIGURE 2.11 Perfect crystals. A perfect crystal is rare, but no matter how irregular the shapes of the faces may be, the angles are always exactly the same. (a) Halite, a cubic crystal: structural diagram and sample of crystals *(Ed Degginger/Bruce Coleman.)* (b) Quartz, a hexagonal crystal: structural diagrams and sample of a nearly perfect crystal. *(Breck P. Kent.)* The cross section at right angles to the long dimension of the crystal shows a regular hexagon with "A" faces at 120° angles.

FIGURE 2.12 This sample of quartz crystals displays both macro- and microcrystallinity. Large, well-formed crystals were able to form in the central cavity where adequate space could accommodate their slow growth. The exterior of this mass is rimmed by a whitish layer of smaller crystal grains displaying few crystal faces, as would be expected in a space confined by other, similar masses of crystals. *(Chip Clark.)*

When Do Minerals Form?

Lowering the temperature of a liquid below its freezing point is one way to start the process of crystallization. In the case of water, 0°C is the temperature below which crystals of ice, a mineral, start to form. Similarly, magma, a hot, molten liquid rock, crystallizes solid minerals when it cools. As a magma falls below its melting point, which may be over 1000°C, crystals of silicate minerals such as olivine or feldspar begin to form. (Geologists usually refer to melting points of magmas rather than freezing points, since freezing normally implies cold.)

Another set of conditions that can produce crystallization occurs during precipitation, as liquids evaporate from a solution. A solution is formed when one chemical substance is dissolved in another, such as salt in water. As the water evaporates from a salt solution, the concentration of salt eventually gets so high that the solution is said to be saturated—it can hold no more salt. If evaporation continues, the salt starts to **precipitate,** or drop out of solution as crystals. Deposits of halite or table salt form under just these conditions when seawater evaporates to the point of saturation in some hot, arid bays or arms of the ocean.

Crystals also form when atoms and ions in solids become mobile and rearrange themselves at high temperatures. For most minerals, temperatures must reach at least 250°C before this kind of rearrangement forms new minerals with different crystal structures. The mineral mica forms this way.

Two major factors control the arrangement of atoms and ions in a crystal structure: the number of neighboring atoms or ions, and their size.

Sizes of Ions

We can think of ions as if they were solid spheres, packed together in close-fitting structural units. Figure 2.13 shows the relative sizes of the ions in NaCl. There are six neighboring ions in NaCl's basic structural unit. The relative sizes of the sodium (smaller) and chloride ions allow them to fit together in a closely packed arrangement.

Ion size is related to the atomic structures of the elements (Figure 2.14). The sizes of ions increase with the number of electrons and electron shells. An ion's charge also affects its size. The more electrons an element loses to become a cation, the stronger its positive charge and the stronger the electrical attraction of its nucleus for the remaining electrons. Many of the cations of abundant minerals are relatively small; most anions are large. This is the case with the most common Earth anion, oxygen. Because anions

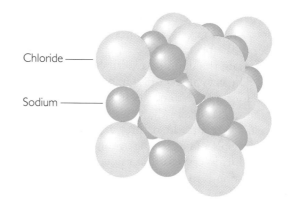

Chloride ——
Sodium ——

FIGURE 2.13 The relative sizes of sodium and chloride ions allow them to pack together in a cubic structure. Ions here are shown in their correct relative sizes.

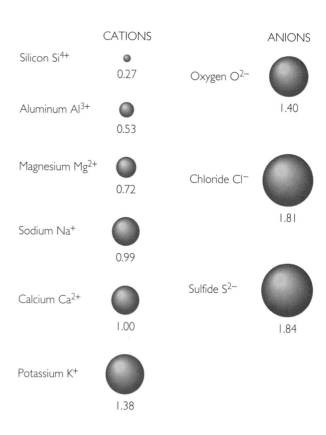

CATIONS   ANIONS

Silicon Si^{4+} •
                    0.27

Aluminum Al^{3+} •   Oxygen O^{2-}
                    0.53   1.40

Magnesium Mg^{2+} •
                    0.72   Chloride Cl$^-$

Sodium Na$^+$ •
                    0.99   1.81

Calcium Ca^{2+} •   Sulfide S^{2-}
                    1.00   1.84

Potassium K$^+$ •
                    1.38

FIGURE 2.14 Sizes of ions as they are commonly found in rock-forming minerals. Ionic radii are given in 10^{-8} cm. (After L. G. Berry, B. Mason, and R. V. Dietrich, *Mineralogy* [San Francisco: W. H. Freeman, 1983].)

tend to be larger than cations, it is apparent that most of the space of a crystal is occupied by the anions and that cations fit into the spaces between them. As a result, crystal structures are determined largely by the way the anions are arranged and the way the cations fit between them.

CATION SUBSTITUTION: SAME CRYSTAL STRUCTURE, DIFFERENT CHEMICAL COMPOSITIONS

Cations of similar sizes and charges tend to substitute for one another and to form compounds with the same crystal structure but differing chemical composition. Cation substitution is common in silicate minerals, in which cations combine with the silicate ion $(SiO_4)^{4-}$. This process is illustrated by olivine, a mineral abundant in many volcanic rocks.

Iron (Fe) and magnesium (Mg) ions are similar in size, and both have two positive charges, so they easily substitute for each other in the structure of olivine. The composition of pure magnesium olivine is Mg_2SiO_4; the pure iron olivine is Fe_2SiO_4. The composition of olivine with both iron and magnesium is given by the formula $(Mg,Fe)_2SiO_4$, which simply means that the number of iron and magnesium cations may vary, but their combined total (expressed as a subscript 2) does not vary in relation to each $(SiO_4)^{4-}$ ion. The proportion of iron to magnesium is determined by the relative abundance of the two elements in the molten material from which the olivine crystallized. In many silicate minerals, aluminum (Al) substitutes for silicon (Si). Aluminum and silicon ions are so similar in size that aluminum can take the place of silicon in many crystal structures. The difference in charge between aluminum (3+) and silicon (4+) ions is balanced by an increase in one of the other cations, such as sodium (1+).

POLYMORPHS: DIFFERENT CRYSTAL STRUCTURE, SAME CHEMICAL COMPOSITION

The same combinations of elements in the same proportions can sometimes form more than one kind of crystal structure, and therefore more than one kind of mineral. These alternative possible structures for a single chemical compound are called **polymorphs** ("many forms"). The structure that actually forms depends on the conditions of pressure and temperature, and therefore on the depth within the Earth at the time and place crystallization occurs.

Diamond and graphite (the material that is used as the "lead" in pencils) are polymorphs. These two minerals, both formed from carbon, have different crystal structures and very different appearances (Figure 2.15). From experimentation and geological observation, we know that diamond forms and remains stable at the very high pressures and temperatures of Earth's mantle. The high pressure in the mantle forces the atoms in diamond to be closely packed. Diamond therefore has a higher density (mass per unit volume), 3.5 g/cm³, than graphite, which is less closely packed and has a density of only 2.1 g/cm³. Graphite forms and is stable at relatively moderate pressures and temperatures, such as those in Earth's crust.

Low temperatures can also produce closer packing. Quartz and cristobalite are polymorphs of silica (SiO_2). Quartz is formed at low temperatures and is relatively dense (2.7 g/cm³). Cristobalite, formed at a higher temperature, has a more open structure and is therefore less dense (2.3 g/cm³).

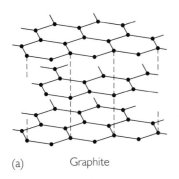

(a)   Graphite

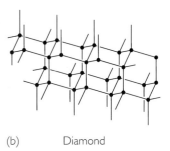

(b)   Diamond

FIGURE 2.15 (a) In graphite, sheets of carbon atoms arranged in hexagons are stacked above one another with weak bonds (dashed lines) between the sheets. (b) In diamond, carbon atoms are arranged in a tetrahedral network. Graphite (at the left in photo) forms flat, platelike masses of crystals. Diamond (at the right in photo), when well crystallized, typically forms octahedral (eight-faced) crystals. *(Chip Clark.)*

ROCK-FORMING MINERALS

All minerals have been grouped into eight classes according to their chemical composition; six of those classes of minerals are shown in Table 2.1. Some minerals, such as copper, occur naturally as unionized pure elements, and they are classified as *native elements*. Most others are classified by their anions. Olivine, for example, is classed as a silicate by its silicate anion, $(SiO_4)^{4-}$. Another mineral we have discussed, halite (NaCl), is classed as a halide from its chloride anion, Cl^-. So is its close relative sylvite, potassium chloride (KCl). The anion of the carbonate minerals is $(CO_3)^{2-}$.

Although many thousands of minerals are known, geologists commonly encounter only about thirty of them. These are the minerals that are the building blocks of most crustal rocks, and thus they are called *rock-forming minerals*. Their relatively small number reflects the small number of elements that are found in major abundance in Earth's crust: as we saw in Chapter 1, 99 percent of Earth's crust is made up of only nine elements. Figure 1.7 shows the relative abundance, by weight, of these elements.

In the following pages, we discuss the most common rock-forming minerals:

- *Silicates,* which are the most abundant minerals in Earth's crust and are composed of oxygen (O) and silicon (Si)—the two most abundant elements in Earth's crust—mostly in combination with the cations of other elements.

- *Carbonates,* minerals made of carbon and oxygen in the form of the carbonate anion $(CO_3)^{2-}$, in combination with calcium and magnesium; calcite ($CaCO_3$) is one such mineral.

- *Oxides,* a group of compounds of oxygen and metallic cations, such as the mineral hematite (Fe_2O_3).

- *Sulfides,* compounds of the sulfide anion S^{2-}, and metallic cations, a group that includes the mineral pyrite (FeS_2).

- *Sulfates,* compounds of the sulfate anion $(SO_4)^-$ and metallic cations and including the mineral anhydrite ($CaSO_4$).

The other chemical classes of minerals, including native elements and halides, are not found as commonly as rock-forming minerals.

Silicates

The basic building block of all silicate mineral structures is the *silicate ion,* formed by four oxygen ions (O^{2-}) surrounding and sharing electrons with a silicon ion (Si^{4+}), giving the formula $(SiO_4)^{4-}$ (Figure 2.16a). This configuration results in a four-sided pyramidal form, called a *tetrahedron,* in which each side forms a triangle (Figure 2.16b). Each silicon-oxygen tetrahedron is an anion with four negative charges, which must be balanced by four positive charges to make an electrically neutral mineral in one of two ways: The ion can bond with cations such as sodium (Na^+), potassium (K^+), calcium (Ca^2), magnesium (Mg^{2+}), and iron (Fe^{2+}). Alternatively, the ion can share oxygens with other silicon-oxygen tetrahedra. All silicate minerals are made up

TABLE **2.1**

SOME CHEMICAL CLASSES OF MINERALS

CLASS	DEFINING ANIONS	EXAMPLE
Native elements	None: no charged ions	Copper metal (Cu)
Oxides and hydroxides	Oxygen ion (O^{2-}) Hydroxyl ion (OH^-)	Hematite (Fe_2O_3) Brucite ($Mg[OH]_2$)
Halides	Chloride (Cl^-), fluoride (F^-), bromide (Br^-), iodide (I^-)	Halite (NaCl)
Carbonates	Carbonate ion (CO_3^{2-})	Calcite ($CaCO_3$)
Sulfates	Sulfate ion (SO_4^{2-})	Anhydrite ($CaSO_4$)
Silicates	Silicate ion (SiO_4^{4-})	Olivine (Mg_2SiO_4)

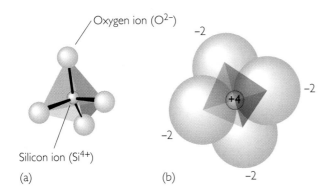

Oxygen ion (O^{2-})

Silicon ion (Si^{4+})

(a)   (b)

FIGURE 2.16 Silicate ion. (a) A model showing the structure of the silicon-oxygen tetrahedron. In the center of this figure is one silicon ion carrying a positive charge of 4. It is surrounded by four oxygen ions, each carrying a minus charge of 2. After bonding, the resulting silicate ion carries a negative charge of 4 (4 minus 8 equals -4). (b) A more realistic model of the silicate ion with the atoms drawn to scale and filling space.

of silicon-oxygen tetrahedra as basic units, linked in combinations of these two ways. Tetrahedra may be isolated, or they may be linked in rings, single chains, double chains, sheets, or frameworks, as shown in Figure 2.17.

ISOLATED TETRAHEDRA   Isolated tetrahedra are linked by the bonding of each oxygen ion of the tetrahedron to a cation (Table 2.2; Figure 2.17a); the cations in turn bond to the oxygens of other tetrahedra. The tetrahedra are thus isolated from one an-

other by cations on all sides. Olivine is a rock-forming mineral with this structure.

RING LINKAGES   Rings of tetrahedra form as two oxygens of each tetrahedron bond to adjacent tetrahedra in closed rings (Table 2.2; Figure 2.17b). In these rings, each tetrahedron shares two of its oxygens with other tetrahedra, one on each side. Rings may link three, four, or six tetrahedra. Cordierite, common in metamorphic rocks, is a rock-forming mineral with this structure.

SINGLE-CHAIN LINKAGES   Single chains also form by sharing oxygens. Two oxygens of each tetrahedron bond to adjacent tetrahedra, but in an open-ended chain instead of a closed ring (Figure 2.17c). Single chains are linked to other chains by cations. Minerals of the pyroxene group are single-chain silicate minerals. Enstatite, a pyroxene, is composed of iron and/or magnesium ions limited to a chain of tetrahedra in which the two cations may substitute for each other, as in olivine. The formula $(Mg,Fe)SiO_3$ reflects this structure.

DOUBLE-CHAIN LINKAGES   Two single chains may combine to form double chains linked to each other by shared oxygens (Table 2.2; Figure 2.17d). Adjacent double chains linked by cations form the structure of the amphibole group of minerals. Hornblende, a member of this group, is an extremely common mineral in both igneous and metamorphic rocks. It has a complex composition including calcium (Ca^{2+}), sodium (Na^+), magnesium (Mg^{2+}), iron (Fe^{2+}), and aluminum (Al^{3+}).

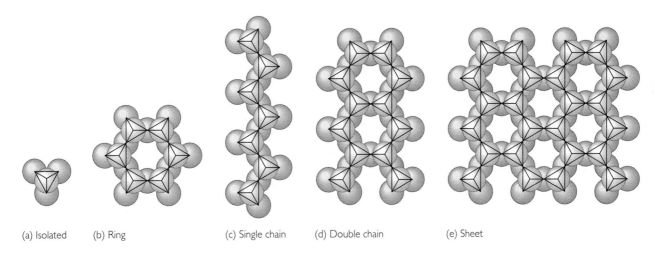

(a) Isolated   (b) Ring   (c) Single chain   (d) Double chain   (e) Sheet

FIGURE 2.17 The crystal structures of silicate minerals, which are classified according to the different ways in which silica tetrahedra can be linked.

TABLE 2.2

MAJOR SILICATE STRUCTURES

GEOMETRY OF LINKAGE OF SiO$_4$ TETRAHEDRA		EXAMPLE MINERAL	CHEMICAL COMPOSITION
Isolated tetrahedra: No sharing of oxygens between tetrahedra; individual tetrahedra linked to each other by bonding to cation between them		Olivine	Magnesium-iron silicate (Mg,Fe)$_2$SiO$_4$
Rings of tetrahedra: Joined by shared oxygens in three-, four-, or six-membered rings		Cordierite	Magnesium-iron-aluminum silicate Al$_3$(Mg,Fe)$_2$Si$_5$AlO$_{18}$
Single chains: Each tetrahedron linked to two others by shared oxygens; chains bonded by cations		Pyroxene (enstatite)	Magnesium-iron silicate (Fe,Mg)SiO$_3$
Double chains: Two parallel chains joined by shared oxygens between every other pair of tetrahedra; to cations that lie between the chains		Amphibole (hornblende)	Calcium-magnesium-iron silicate Ca(Mg,Fe)$_4$Al(Si$_7$Al)O$_{22}$(OH,F)
Sheets: Each tetrahedron linked to three others by shared oxygens; sheets bonded by cations		Kaolinite	Aluminum silicate Al$_2$Si$_2$O$_5$(OH)$_4$
		Mica (muscovite)	Potassium-aluminum silicate KAl$_3$Si$_3$O$_{10}$(OH)$_2$
Frameworks: Each tetrahedron shares all its oxygens with other SiO$_4$ tetrahedra (in quartz) or AlO$_4$ tetrahedra		Feldspar (orthoclase)	Potassium-aluminum silicate KAlSi$_3$O$_8$
		Quartz	Silicon dioxide SiO$_2$

SHEET LINKAGES   Sheets are structures in which each tetrahedron shares three of its oxygens with adjacent tetrahedra to build stacked sheets of tetrahedra (Table 2.2; Figure 2.17e). Cations may be interlayered with tetrahedral sheets. The micas (Figure 2.18) and clay minerals are the most abundant sheet silicates. Muscovite (KAl$_3$Si$_3$O$_{10}$[OH]$_2$) is one of the commonest of sheet silicates, found in many types of rocks. It can be separated into extremely thin, transparent sheets. Kaolinite (Al$_2$Si$_2$O$_5$[OH]$_4$), which also has this structure, is a common clay mineral found in sediments and is the basic raw material for pottery and china.

FRAMEWORKS   Three-dimensional frameworks form as each tetrahedron shares all its oxygens with other tetrahedra (see Table 2.2). Members of the feldspar group, the most abundant minerals in Earth's crust, are framework silicates, as is another of the most common minerals, quartz (SiO$_2$) (Figure 2.18).

<figure>**FIGURE 2.18** Silicate minerals (clockwise from upper left): feldspar, a framework structure; mica, a sheet structure; pyroxene, a single-chain structure; quartz, a framework structure; and olivine, an isolated tetrahedral structure. *(Chip Clark.)*</figure>

COMPOSITION OF SILICATES Chemically, the simplest silicate is silicon dioxide, also called silica (SiO_2), which is found most frequently as the mineral quartz. The tendency of silicon to bond with oxygen is so strong that silicon is never found in nature as the pure element; it is always combined with oxygen. (The pure silicon used in computer chips is artificially prepared by advanced chemical techniques.) When the silicate tetrahedra of quartz are linked, sharing two O's for each Si, the total formula adds up to SiO_2.

In other silicate minerals, the basic units—rings, chains, sheets, and frameworks—are bonded to such cations as sodium (Na^+), potassium (K^+), calcium (Ca^{2+}), magnesium (Mg^{2+}), and iron (Fe^{2+}). As we noted in our discussion of cation substitution, aluminum (Al^{3+}) substitutes for silicon in many silicate minerals. All of the several varieties of feldspar, for example, contain aluminum, with various combinations of potassium, sodium, and calcium.

Carbonates

Of the nonsilicate minerals (Figure 2.19), the mineral calcite (calcium carbonate, $CaCO_3$) is one of

<figure>**FIGURE 2.19** Nonsilicate minerals (clockwise from upper left): halite, spinel, gypsum, hematite, calcite, pyrite, and galena. *(Chip Clark.)*</figure>

the abundant minerals of Earth's crust and is the chief constituent of a group of rocks called limestones. Its basic building block, the carbonate ion, $(CO_3)^{2-}$, consists of a carbon atom surrounded by three oxygen atoms in a triangle, as in Figure 2.20a. The carbon shares electrons with oxygens. Groups of carbonate ions are arranged in sheets somewhat like the sheet silicates and are bonded by layers of cations (Figure 2.20b). The sheets of carbonate ions in calcite are separated by layers of calcium ions. The mineral dolomite, $CaMg(CO_3)_2$, another major mineral of crustal rocks, is made up of the same carbonate sheets separated by alternating layers of calcium ions and magnesium ions.

Oxides

Oxide minerals are compounds in which oxygen is bonded to atoms or cations of other elements, usually metallic ions such as iron (Fe^{2+} or Fe^{3+}). Most oxide minerals are ionically bonded, their structures varying with the size of the metallic cations. This group is of great economic importance because it includes the ores of most of the metals, such as chromium and titanium, used in industrial and technological manufacture of metallic materials and devices. Hematite (Fe_2O_3) is a chief ore of iron.

Another of the abundant minerals in this group, spinel, is an oxide of two metals, magnesium and aluminum ($MgAl_2O_4$). Spinel has a closely packed cubic structure and high density (3.6 g/cm^3), reflecting the conditions of high pressure and temperature under which it is formed. Transparent, gem-quality spinel resembles ruby and sapphire and is found in the crown jewels of England and Russia.

Sulfides

The chief ores of many valuable minerals, such as copper, zinc, and nickel, are members of the sulfide group of minerals. This group includes compounds of the sulfide ion, S^{2-}, with metallic cations. The sulfide ion is one in which a sulfur atom has gained two electrons in its outer shell. Most sulfide minerals look like metals, and almost all are opaque. The structures of these minerals are diverse, depending on the way the sulfide anions combine with the metallic cations. The most common sulfide mineral is pyrite (FeS_2), frequently called "fool's gold" because of its yellowish metallic appearance.

Sulfates

In sulfates, sulfur is present as the sulfate ion, a tetrahedron made up of one sulfur atom that has lost six electrons from its outer shell, joined with four oxygen ions (O^{2-}), giving it the formula SO_4^{2-}. The sulfate ion is the basis for a variety of structures. One of the most abundant minerals of this group is gypsum, the primary component of plaster. Gypsum is formed when seawater evaporates. During evaporation, Ca^{2+} and SO_4^{2-}, two ions abundant in seawater, combine and precipitate as layers of sediment, forming calcium sulfate ($CaSO_4 \cdot 2H_2O$). (The dot in this formula signifies that two water molecules are bonded to the calcium and sulfate ions.)

Another calcium sulfate, anhydrite ($CaSO_4$), differs from gypsum in containing no water. Its name is derived from the word *anhydrous*, meaning "free from water." Gypsum is stable at the low temperatures and pressures found at Earth's surface, whereas

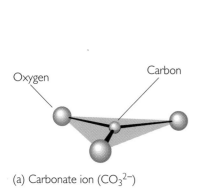

(a) Carbonate ion (CO_3^{2-})

Oxygen   Carbon

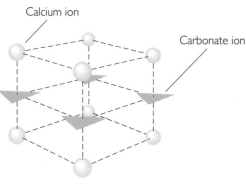

Calcium ion

Carbonate ion

(b) Calcium carbonate structure

FIGURE 2.20 Carbonate minerals, such as calcite (calcium carbonate, $CaCO_3$), have a layered structure. (a) Top view of the carbonate building block, a carbon atom surrounded in a triangle by three oxygen ions, with a net charge of -2. (b) View of the alternating layers of calcium and carbonate ions.

FIGURE 2.21 Acid test. One easy but effective way to identify certain minerals is to drop diluted hydrochloric acid (HCl) on the substance. If it fizzes, indicating the escape of carbon dioxide, the mineral is likely to be calcite. *(Chip Clark.)*

anhydrite is stable at the higher temperatures and pressures of buried sedimentary rocks.

Crystal structure and composition are not useful solely for organizing our knowledge of the minerals of the world. They also explain the physical properties of minerals that make them useful or decorative.

PHYSICAL PROPERTIES OF MINERALS

Geologists use the compositions and structures of minerals to understand the origin of the rocks they make up and thus the nature of the geological processes that operate on and in the Earth. This understanding often begins in the field with attempts to identify and classify unknown minerals. At such times, geologists place great reliance on chemical and physical properties that can be observed with relative ease. Most of what we know about the chemical composition of minerals was originally learned through the use of ordinary chemical methods to dissolve minerals and rocks, separate them into their constituent elements, and then measure their weights or volumes. In the nineteenth and early twentieth centuries, geologists used to carry field kits for the rough partial chemical analysis of minerals that would help in identification. One such test is the origin of the phrase "the acid test." It consists of dropping diluted hydrochloric acid (HCl) on a mineral to see if it fizzes. The fizzing indicates that carbon dioxide (CO_2) is escaping, and that means the mineral is likely to be calcite, a carbonate mineral (Figure 2.21).

In the remainder of this chapter we review the physical properties of minerals, many of which contribute to their practical and decorative value. (See Feature 2.1, "What Makes Gems So Special?")

Hardness

Hardness is a measure of the ease with which the surface of a mineral can be scratched. Just as a diamond, the hardest mineral known, scratches glass, so a quartz crystal, which is harder than feldspar, scratches a feldspar crystal. In 1822 Friedrich Mohs, an Austrian mineralogist, devised a scale, now known as the **Mohs scale of hardness,** based on the ability of one mineral to scratch another. At one extreme is the softest mineral (talc); on the other, the

What Makes Gems So Special?

No one can be sure when the first human picked up a crystal of a mineral and kept it for its rare beauty, but we do know that gems were being worn as necklaces and other adornments at the dawn of civilization in Egypt, at least 4000 years ago. These early Egyptians were undoubtedly attracted to the color and play of light on the polished surfaces of such minerals as carnelian, lapis lazuli, and turquoise. Color and luster, or the ability to reflect light, are two of the qualities that still serve to define gemstones. Although the value placed on a gemstone varies from one culture and historical period to another, other required qualities seem to be beauty, transparency, brilliance, durability, and rarity.

Most minerals have some of these remarkable qualities, but the stones considered most precious are ruby, sapphire, emerald, and, of course, diamond. A diamond—geologically speaking, at least—may not be forever, as the advertisers claim, but it is special. Its glitter is unique, as are the play of colors and the sparkle it emits. The source of these qualities is the way diamond refracts, or bends, light. They are enhanced by diamond's remarkable ability to split perfectly along certain directions of the crystal, which diamond cutters use to advantage in carefully cutting gem-quality stones. Diamond's multiple facets (faces superficially similar to crystal faces) can be polished to enhance this sparkle. These facets can be ground only by other diamonds, for this is the hardest mineral known, so hard it can scratch any other mineral and remain undamaged. This mineral's tightly packed crystal structure and strong covalent bonds between carbon atoms give it these characteristics that allow it to be identified with certainty by mineralogists and jewelers.

TABLE 2.3

MOHS SCALE OF HARDNESS

MINERAL	SCALE NUMBER	COMMON OBJECTS
Talc	1	
Gypsum	2	
		Fingernail
Calcite	3	Copper coin
Fluorite	4	
Apatite	5	Knife blade
Orthoclase	6	Window glass
		Steel file
Quartz	7	
Topaz	8	
Corundum	9	
Diamond	10	

hardest (diamond) (Table 2.3). The Mohs scale is still one of the best practical means to identify an unknown mineral. With a knife blade and a few of the minerals on the hardness scale, a field geologist can gauge an unknown mineral's position on the scale. If the unknown mineral is scratched by a piece of quartz but not by the knife, for example, it lies between 5 and 7 on the scale.

Recall from our earlier discussion that covalent bonds are generally stronger than ionic bonds. The hardness of any mineral depends on the strength of its chemical bonds; the stronger the bonds, the harder the mineral. Crystal structure varies in the silicate group of minerals, and so does hardness. For example, within the silicates, hardness varies from 1 in talc, a sheet silicate, to 8 in topaz, a silicate with isolated tetrahedra. Most silicates fall in the 5 to 7 range on the Mohs scale. Only sheet silicates are relatively soft, with hardnesses between 1 and 3.

Rubies and sapphires are gem-quality varieties of the common mineral corundum (aluminum oxide), which is widespread and abundant in a number of rock types. Although not as hard as diamond, corundum is extremely hard. Small amounts of impurities produce the intense colors we value. Ruby, for example, is red because of small amounts of chromium, the same substance that gives emeralds their green color.

Less valuable, sometimes called semiprecious, gemstones are topaz, garnet, tourmaline, jade, turquoise, and zircon. Most, like garnet, are common constituents of rocks, occurring mostly as small imperfect crystals with many impurities and poor transparency. But under special conditions, gem-quality garnets form. From time to time, some minerals that are not ordinarily considered gems may enjoy sudden—perhaps temporary—popularity. Hematite (iron oxide) currently enjoys this status, appearing in necklaces and bracelets.

Sapphire (blue) and diamond (colorless) brooch by Fortunato Pio Castellani, Smithsonian Institution, nineteenth century. *(Aldo Tutino/Art Resource.)*

Within groups of minerals having similar crystal structures, increasing hardness is related to factors that also increase bond strength:

- *Size*   The smaller the atom or ion, the smaller the distance between the atoms or ions, and thus the stronger the electrical attraction.

- *Charge*   The larger the charge of ions, the greater the attraction between ions, and thus the stronger the bond.

- *Packing of atoms or ions*   The closer the packing of atoms or ions, the smaller the distance between atoms or ions, and the stronger the bond.

Size is an especially important factor for most metallic oxides and sulfides of metals with high atomic numbers—such as those of gold, silver, copper, and lead. Members of this group are soft, with hardnesses of less than 3, because their metallic cations are so large. Carbonates and sulfates, groups in which the structures are packed less densely, are also soft, with hardnesses of less than 5. In each of these groups, their chemical bonds are reflected in their hardness.

Cleavage

Cleavage is the tendency of a crystal to break along flat planar surfaces. The term is also used to describe the geometric pattern produced by such breakage. Cleavage varies inversely with bond strength—if bond strength is high, cleavage is poor; if bond strength is low, cleavage is good. Because of their strength, covalent bonds generally give poor or no cleavage. Ionic bonds are relatively weak, so they give excellent cleavage.

If the bonds between some of the planes of atoms or ions in a crystal are weak, the mineral can

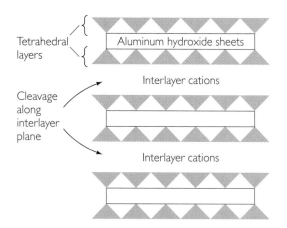

FIGURE 2.22 Cleavage of mica. The diagram shows the cleavage planes in the mineral structure, oriented perpendicular to the plane of the page. Horizontal lines mark the interfaces of silica-oxygen tetrahedral sheets and sheets of aluminum hydroxide bonding the two tetrahedral layers into a "sandwich." Cleavage takes place between composite tetrahedral-aluminum hydroxide sandwiches. The photograph shows thin sheets separating along the cleavage planes. *(Chip Clark.)*

FIGURE 2.23 Example of rhomboidal cleavage in calcite. Calcite can be cleaved by a light hammer blow on a chisel oriented parallel to one of its planes. *(Chip Clark.)*

be made to split along those planes. Muscovite, a mica and a sheet silicate, breaks along smooth, lustrous, flat, parallel surfaces, forming thin transparent sheets less than a millimeter thick. Mica's excellent cleavage is the result of weakness of the bonds between the sandwiched layers of cations and tetrahedral silica sheets (Figure 2.22).

Cleavage is classified according to two primary sets of characteristics: the number of planes and pattern of cleavage, and the quality of surfaces and ease of cleaving.

NUMBER OF PLANES; PATTERN OF CLEAVAGE
The number of planes and patterns of cleavage are identifying hallmarks of many rock-forming minerals. Muscovite has only one plane of cleavage, but calcite and dolomite have three excellent cleavage directions that give them a rhomboidal shape (Figure 2.23).

A crystal's structure determines its cleavage planes and its crystal faces. Crystals have fewer cleavage planes than possible crystal faces. Faces may be formed along any of numerous planes defined by rows of atoms or ions. Cleavage occurs along any of those planes across which the bonding is weak. All crystals of a mineral exhibit its characteristic cleavage, whereas only some crystals display particular faces.

Galena (lead sulfide, PbS) and halite (sodium chloride, NaCl) cleave along three planes, forming perfect cubes. Distinctive angles of cleavage help identify two important groups of silicates, the pyrox-

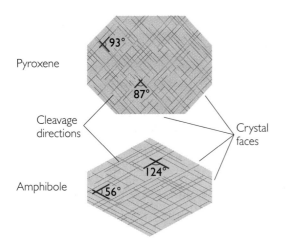

FIGURE 2.24 Comparison of cleavage directions and typical crystal faces in pyroxene and amphibole. These two minerals often look very much alike; but their angles of cleavage differ. These angles are frequently used to identify and classify them.

enes and amphiboles, that otherwise often look alike (Figure 2.24). Pyroxenes have a single-chain linkage, and they are bonded so that their cleavages are almost at right angles (93°) to each other. In cross section, the cleavage pattern of pyroxene is nearly a square. In contrast, amphiboles, the double chains, bond to give two cleavage directions, at 56° and 124° to each other. They produce a diamond-shaped cross section.

QUALITY OF SURFACES; EASE OF CLEAVING   A mineral's cleavage is assessed as perfect, good, or fair, according to the quality of surfaces produced and the ease of cleaving. Muscovite can be cleaved easily, producing extremely high-quality, smooth surfaces; its cleavage is *perfect.* The single- and double-chain minerals (pyroxenes and amphiboles, respectively) show *good* cleavage. Although these minerals break easily along the cleavage plane, they also break across it, producing cleavage surfaces that are not as smooth as those of mica. *Fair* cleavage is shown by the ring silicate beryl. Beryl's cleavage is more irregular, and the mineral breaks relatively easily along directions other than cleavage planes.

Many minerals are so strongly bonded that they lack even fair cleavage. Quartz, one of the commonest, is a framework silicate; it is so strongly bonded in all directions that it breaks only along irregular surfaces. Garnet, an isolated tetrahedral silicate, is also bonded strongly in all directions and so has no

cleavage. This absence of a tendency to cleave is found in most framework silicates and in silicates with isolated tetrahedra.

Fracture

Fracture is the tendency of a crystal to break along irregular surfaces other than cleavage planes. All minerals show fracture, either across cleavage planes or in any direction in minerals, such as quartz, with no cleavage. Fracture is related to the way bond strengths are distributed in directions that cut across crystal planes. Breakage of these bonds results in irregular fractures. Fractures may be *conchoidal,* showing smooth, curved surfaces like those of a thick piece of broken glass. A common fracture surface with an appearance like split wood is described as *fibrous* or *splintery.* The shape and appearance of many kinds of irregular fractures depend on the particular structure and composition of the mineral.

Luster

The way the surface of a mineral reflects light gives it a characteristic **luster.** Mineral lusters are described by the terms listed in Table 2.4. Luster quality is controlled by the kinds of atoms present and their bonding, both of which affect the way light passes

TABLE 2.4

MINERAL LUSTER	
Metallic	Strong reflections produced by opaque substances
Vitreous	Bright, as in glass
Resinous	Characteristic of resins, such as amber
Greasy	The appearance of being coated with an oily substance
Pearly	The whitish iridescence of such materials as pearl
Silky	The sheen of fibrous materials such as silk
Adamantine	The brilliant luster of diamond and similar minerals

through or is reflected by the mineral. Ionically bonded crystals tend to be glassy, or vitreous, but covalently bonded materials are more variable. Many tend to have an adamantine luster, like that of diamond. Metallic luster is shown by pure metals, such as gold, and sulfides, such as galena (lead sulfide, PbS). Pearly luster is the result of multiple reflections of light from planes beneath the surfaces of translucent minerals, such as the mother-of-pearl inner surfaces of many clam shells, which are made of the mineral aragonite. Luster quality, although an important criterion for field classification, depends heavily on visual perception of reflected light. Textbook descriptions fall short of the actual experience of holding the mineral in your hand.

Color

The **color** of a mineral is imparted by light—either transmitted through or reflected by crystals, irregular masses, or a streak. **Streak** is the name given to the color of the fine deposit of mineral dust left on an abrasive surface, such as a tile of unglazed porcelain, when a mineral is scraped across it. Such tiles are called *streak plates* (see Figure 2.25). A streak plate is a good diagnostic tool because the uniform small grains of mineral that occur in the powder on the ceramic tile permit a better analysis of color than

does a mass of the mineral. A mass formed of hematite (Fe_2O_3), for example, may be black, red, or brown, but this mineral will always leave a trail of reddish-brown dust on a streak plate.

Color is determined both by the kinds of ions found in the pure mineral and by trace impurities. Color is a complex and as yet not fully understood property of minerals; its details, other than those given below, are beyond the scope of this book.

IONS AND COLOR OF MINERALS   The color of pure substances depends on the presence of certain ions, such as iron or chromium, which strongly absorb portions of the light spectrum. Olivine containing iron, for example, absorbs all colors except green, which it reflects, so we see this type of olivine as green. We see pure magnesium olivine as white (transparent and colorless). Most ionically bonded pure minerals whose ions have full, stable outer electron shells, such as halite, are colorless.

TRACE IMPURITIES AND MINERAL COLOR   All natural minerals contain impurities. In the past few decades, new instruments have made it possible to measure even very small quantities of some elements—as little as a billionth of a gram in some cases. Elements that make up much less than 0.1 percent of a mineral are reported as "traces," and many of these are called trace elements.

Some trace elements are useful in interpreting the origins of the minerals in which they are found. Others, such as the trace amounts of uranium in some granites, contribute to local natural radioactivity. Still others, such as small dispersed flakes of hematite that color a feldspar crystal brownish or reddish, are notable because they give a general color to an otherwise colorless mineral. Many of the gem varieties of minerals, such as emerald (green beryl) and sapphire (blue corundum), get their color from trace impurities dissolved in the solid crystal (see Feature 2.1). Emerald derives its color from chromium; the source of sapphire's blue is iron and titanium.

The color of a mineral may be distinctive, but it is not the most reliable clue to its identity. Some minerals always show the same color; others may have a range of colors. Many minerals show a characteristic color only on freshly broken surfaces or only on weathered surfaces. Some—precious opals, for example—show a stunning display of colors on reflecting surfaces. Others change color slightly with a change in the angle of the light shining on their surface.

FIGURE 2.25 Hematite may be black, red, or brown, but it always leaves a reddish brown streak when scratched along a ceramic plate. *(Breck P. Kent.)*

TABLE **2.5**

PHYSICAL PROPERTIES OF MINERALS

PROPERTY	RELATION TO COMPOSITION AND CRYSTAL STRUCTURE
Hardness	Strong chemical bonds give high hardness. Covalently bonded minerals are generally harder than ionically bonded minerals.
Cleavage	Cleavage is poor if bond strength in crystal structure is high and is good if bond strength is low. Covalent bonds generally give poor or no cleavage; ionic bonds are weak, so give excellent cleavage.
Fracture	Type is related to distribution of bond strengths across irregular surfaces other than cleavage planes.
Luster	Tends to be glassy for ionically bonded crystals, more variable for covalently bonded crystals.
Color	Determined by kinds of atoms and trace impurities. Many ionically bonded crystals are colorless. Iron tends to color strongly.
Streak	Color of fine powder is more characteristic than that of massive mineral because of uniform small size of grains.
Density	Depends on atomic weight of atoms and their closeness of packing in crystal structure. Iron minerals and metals have high density; covalently bonded minerals have more open packing and so have lower density.

Specific Gravity and Density

The difference in weight between a piece of hematite iron ore and a piece of sulfur of the same size is easily felt when the two pieces are hefted. A great many common rock-forming minerals, however, are too similar in density for such simple tests as hefting. **Density** is mass per unit volume (usually expressed in grams per cubic centimeter—g/cm^3). Scientists therefore needed some method that would make it easy to measure this property of minerals. A standard measure of density is **specific gravity,** which is the weight of a mineral in air divided by the weight of an equal volume of pure water at 4°C.

Density depends on the atomic weight of the mineral's ions and the closeness with which they are packed in a mineral's crystal structure. Consider the iron oxide magnetite, with a density of 5.2 g/cm^3. This high density results partly from the high atomic weight of iron and partly from the closely packed structure magnetite shares with the other members of the spinel group of minerals (see page 44). The density of the iron silicate olivine, 4.4 g/cm^3, is

lower than that of magnetite for two reasons. First, the atomic weight of silicon, one of the elements from which olivine is formed, is lower than that of iron. Second, this olivine has a more openly packed structure than that of the spinel group. The density of the magnesium olivine is even lower, 3.32 g/cm^3, because magnesium's atomic weight is much lower than that of iron. Table 2.5 summarizes the physical properties of minerals.

Increases of density caused by pressure affect the way minerals transmit light, heat, and earthquake waves. Experiments at extremely high pressures have shown that olivine converts to the denser structure of the spinel group at pressures corresponding to a depth of 400 km. At a greater depth, 670 km, mantle materials are further transformed to silicate minerals with the even more densely packed structure of the mineral perovskite (calcium titanate, $CaTiO_3$). Because of the huge volume of the lower mantle, silicate with the perovskite structure is probably the most abundant mineral in the Earth as a whole (see page 39). Some perovskite minerals have been synthesized to be high-temperature semiconductors,

2.2 LIVING ON EARTH

Asbestos: Health Hazard, Overreaction, or Both?

Mention of asbestos, once used extensively as a fireproof insulator and flame retardant in plaster, ceiling and floor tile, and automobile insulation, has come to provoke fear in the last two decades. During that time, asbestos has been linked to several fatal lung diseases—such as asbestosis (characterized by progressive lung stiffening and difficulty in breathing) and lung cancer, including a specific cancer called mesothelioma, which attacks the lining of the lung. The specific link seems to be heavy exposure to certain minerals lumped under the commercial name *asbestos* or exposure to them over a prolonged period of time. The exact way these substances cause the diseases is still not well understood, but the sharp fibrous crystal habit of some of the minerals has been implicated.

The health problems associated with exposure to asbestos came to public attention when lawyers representing workers and their families brought class-action lawsuits against some of the major companies that fabricated asbestos products. The lawsuits accused the companies of responsibility for the deaths and disabilities of large numbers of people formerly employed in asbestos factories. In the wake of settlements awarded as a result of these lawsuits, some of the companies went bankrupt. They left behind them a public intensely concerned over asbestos-containing materials in place in schools, hospitals, and other public and private buildings. Many states now require disclosure of such materials during negotiations for the sale of private homes, for example.

Although research specialists who identify and treat lung diseases have been vocal in their warnings to the public, other responsible scientists think the frenzied concern about all forms of asbestos is an overreaction. At the center of this debate lies mineralogy.

Six distinct minerals are lumped under the commercial term *asbestos:* chrysotile, a sheet silicate member of the serpentine group;

Asbestos (chrysotile). Fibers are readily combed from the solid mineral. (*Runk/Schoenberger/Grant Heilman Photography.*)

crocidolite, a double-chain silicate member of the amphibole group; and four other double-chain silicates in the serpentine group. Although crocidolite does form sharp fibers, many of the other minerals do not. Crocidolite has been heavily implicated in lung diseases. U.S. government regulations nevertheless apply to all of these minerals.

Physicians and mineralogists vary widely in their opinions on whether all forms of asbestos should be eliminated from buildings and factories. Any decision must consider several facts. First, heavy exposure to crocidolite is dangerous, especially for smokers. Smokers may develop lung diseases without any exposure to asbestos, but smokers are much more prone to asbestos-related lung diseases than are nonsmokers. Prolonged occupational exposure to some other forms of asbestos, as by workers in asbestos factories, may also create a danger.

Abundant evidence exists, however, to indicate that people exposed to moderate amounts of chrysotile, the most commonly used asbestos in North America, for long periods of time show no asbestos-related lung disease. The lack of a correlation between exposure to specific asbestos minerals, verified by laboratory analysis, and the occurrence of specific lung diseases also contributes to confusion about the danger of exposure to asbestos. The situation is particularly ambiguous in respect to the danger posed to the general population by asbestos in large public buildings.

Many medical scientists and mineralogists doubt that we need to spend the $50 billion to $150 billion it would cost to clean up relatively harmless chrysotile and the four forms of amphibole that do not form sharp fibers. Before any decision can be made with certainty, we need more mineralogical evaluation and additional medical studies to determine the nature of the problem confronting us.

which conduct electricity without loss and may have large commercial potential. Mineralogists experienced with the natural material helped unravel the structure of these newly created materials. Temperature also affects density: the higher the temperature, the more open and expanded the structure and thus the lower the density.

Crystal Habit

A mineral's **crystal habit** is the shape in which its individual crystals or aggregates of crystals grow. Crystal habits are often named after common geometric shapes, such as blades, plates, and needles. Some minerals have such a distinct and characteristic crystal habit that they are easily recognizable, such as quartz's six-sided column topped by a pyramidlike set of faces. These shapes reflect not only the planes of atoms or ions in the mineral's crystal structure but also the typical speed and direction of crystal growth. Thus a needlelike crystal is one that grows very quickly in one direction and very slowly in all other directions. In contrast, a plate-shaped crystal (often referred to as *platy*) grows fast in all directions that lie perpendicular to its single direction of slow growth. Fibrous crystals take shape as multiple long, narrow fibers, essentially aggregates of long needles. *Asbestos* is a generic name for a group of silicates with a more or less fibrous habit that allows the crystals to become embedded after having been inhaled into the lungs (see Feature 2.2, "Asbestos: Health Hazard, Overreaction, or Both?").

In summary, minerals exhibit a variety of physical and chemical properties that result from their chemical compositions and atomic structures. Many of these properties are useful to the mineralogist or geologist for purposes of identification or classification. Geologists study the compositions and structures of minerals to understand the origin of the rocks in which they are found, and thus the nature of the geological processes that operate on and in the Earth.

In the chapters that follow, we will refer to various minerals in one geological context or another. Appendix 3, just before the glossary at the end of this book, lists the properties of the most common minerals in Earth's crust. We urge you to consult this appendix when you want more information about these minerals.

SUMMARY

What is a mineral? Minerals, the building blocks of rocks, are naturally occurring inorganic solids with specific crystal structures and chemical compositions that are fixed or vary within a defined range. A mineral is constructed of atoms, the small units of matter that combine in chemical reactions. An atom is composed of a nucleus of protons and neutrons, surrounded by electrons traveling in orbitals around the nucleus. The atomic number of an element is the number of protons in its nucleus, and its atomic weight is the sum of the masses of its protons and neutrons.

How are atoms combined to form the crystal structures of minerals? Chemical substances react with each other to form compounds either by gaining or losing electrons to become ions or by sharing electrons. Either way, the atoms achieve stable configurations of electron shells. The atoms or ions of a substance are held together by ionic and covalent bonds formed by electrostatic attraction between nuclei and electrons of the constituent elements. When a mineral crystallizes, atoms or ions come together in the proper proportions to form a crystal structure, which is an orderly, three-dimensional geometric array in which the basic arrangement is repeated in all directions.

What are the major rock-forming minerals? Silicates, the most abundant minerals in Earth's crust, are crystal structures built of silicate tetrahedra linked in various ways. Tetrahedra may be isolated (olivines) or in rings (cordierite), single chains (pyroxenes), double chains (amphiboles), sheets (micas), or frameworks (feldspars). Carbonate minerals are made of carbonate ions bonded to calcium and/or magnesium. Oxide minerals are compounds of oxygen and metallic elements. Sulfide and sulfate minerals are structures made up of sulfur atoms in combination with metallic elements.

What are the physical properties of minerals? Physical properties, which reflect the compositions and structures of minerals, include hardness, or the ease with which a mineral surface is scratched; cleavage, or the ability of a mineral to split or break along flat surfaces; fracture, or the way in which minerals break along irregular surfaces; luster, or the nature of a mineral's reflection of light; color, imparted by either transmitted or reflected light to crystals, irregular masses, or a streak (the color of a fine powder); density, or the mass per unit volume; and crystal habit, or the shapes of individual crystals or aggregates.

KEY TERMS AND CONCEPTS

mineralogy (p. 28)

mineral (p. 28)

atom (p. 29)

nucleus (p. 29)

proton (p. 29)

neutron (p. 29)

electron (p. 30)

shell (p. 30)

atomic number (p. 30)

atomic mass (p. 30)

isotope (p. 30)

chemical reaction (p. 30)

ion (p. 31)

cation (p. 31)

anion (p. 31)

electron sharing (p. 32)

ionic bond (p. 35)

covalent bond (p. 35)

metallic bond (p. 35)

crystallization (p. 36)

crystal (p. 37)

precipitate (p. 38)

polymorph (p. 39)

hardness (p. 45)

Mohs scale of hardness (p. 45)

cleavage (p. 47)

fracture (p. 49)

luster (p. 49)

color (p. 50)

streak (p. 50)

density (p. 51)

specific gravity (p. 51)

crystal habit (p. 53)

MINERAL NAMES TO REMEMBER

(See Appendix 3 for specific mineral properties.)

amphibole	garnet	olivine
anhydrite	gold	opal
aragonite	graphite	perovskite
calcite	gypsum	pyrite
clay mineral	halite	pyroxene
corundum	hematite	quartz
diamond	hornblende	spinel
dolomite	kaolinite	talc
enstatite	magnetite	
feldspars	mica	
galena	muscovite	

EXERCISES

1. Define a mineral.

2. What is the difference between an atom and an ion?

3. Draw the atomic structure of sodium chloride.

4. What are two types of chemical bonds?

5. What are the two polymorphs of carbon?

6. List the basic structure of silicate minerals.

7. How does the cleavage of mica reflect its atomic structure?

8. Name three groups of minerals, other than silicates, based on their chemical composition.

9. What two factors account for the densities of mantle minerals?

THOUGHT QUESTIONS

(Consult Appendix 3 for specific mineral properties.)

1. Joan and Alex are comparing rubies. Joan's is natural and Alex's is synthetic. Are both of them looking at minerals? Why, or why not?

2. Hydrogen (H), the lightest element, has an atomic number of 1 and an atomic weight of 1.008 in nature. What does this information tell you about possible isotopes of hydrogen?

3. Draw a simple diagram to show how silicon and oxygen in silicate minerals share electrons. Model your diagram on Figure 2.5.

4. An isotope of the element aluminum has 13 protons and an atomic weight of 27. Use the periodic table of elements to find its atomic number and then draw its electron shells and its nucleus, using Figure 2.3 as a model.

5. Use iron and magnesium in silicate minerals to illustrate cation substitution.

6. Diopside, a pyroxene, has the formula $(Ca,Mg)_2Si_2O_6$. What does this tell you about its crystal structure and cation substitution?

7. Oxygen exists as three isotopes with atomic weights of exactly 16, 17, and 18. The atomic weight of oxygen found in nature is approximately 16. What does this information tell you about the relative abundance of the three isotopes in nature?

8. In some places in bodies of granite, we can find very large crystals, some as much as a meter across, yet these crystals tend to have few crystal faces. What can you deduce about the conditions under which these large crystals grew?

9. What physical properties of sheet silicates are related to their crystal structure and bond strength?

10. How might you identify and differentiate between a single- and a double-chain silicate?

11. What physical properties would make calcite a poor choice for a good gemstone?

12. Choose two minerals from Appendix 3 that you think might make good abrasive or grinding stones for sharpening steel, and describe the physical property that causes you to believe they would be suitable for this purpose.

13. Aragonite, with a density of 2.9 g/cm^3, has exactly the same chemical composition as calcite, with a density of 2.7 g/cm^3. Other things being equal, which of these two minerals is more likely to have formed under high pressure?

14. What properties of talc make it suitable for face and body powder?

SHORT-TERM TEAM PROJECT: ASBESTOS

Children were in danger. News reports of asbestos in the New York City schools touched off a wave of public outrage, and the Board of Education postponed the start of the 1993 school year for three weeks to complete an asbestos-removal program.

Reports of asbestos in public buildings almost always focus on the risk of lung disease. Seldom, however, do these reports include interviews with mineralogists. As a result, the public knows too little about asbestos to ask the right questions and make an informed judgment.

You now have the chance to educate the public. Working in a team of four students over the next two weeks, prepare a half-hour radio program representing a variety of perspectives on the asbestos issue. The program will show how useful a basic knowledge of mineralogy can be in deciding important policy questions. Assemble a cast of experts with opposing positions on the issue. From what disciplines would they come? With what arguments and facts would they support their positions? Write a script for the program and deliver an oral summary of your list of experts and their positions.

SUGGESTED READINGS

Berry, L. G., B. Mason, and R. V. Dietrich. 1983. *Mineralogy*, 2nd ed. San Francisco: W. H. Freeman.

Dietrich, R. V., and B. J. Skinner. 1990. *Gems, Granites, and Gravels.* Cambridge: Cambridge University Press.

Keller, P. C. 1990. *Gemstones and Their Origins.* New York: Chapman & Hall.

Klein, C., and C. S. Hurlbut, Jr. 1993. *Manual of Mineralogy*, 21st ed. New York: Wiley.

McQuarrie, D. A., and P. A. Rock. 1991. *General Chemistry*, 3rd ed. New York: W. H. Freeman.

Prinz, M., G. Harlow, and J. Peters. 1978. *Simon & Schuster's Guide to Rocks and Minerals.* New York: Simon & Schuster.

INTERNET SOURCES

Mineral Gallery

ⓘ **http://mineral.galleries.com/**

This commercial site provides a data base to search for minerals by name, chemical composition, class (oxides, silicates, etc.), and groupings (birthstones, gemstones, etc.). A search produces an image of the mineral, data on composition, characteristics, properties, and uses. The physical properties of minerals are also explained.

Mineralogy

ⓘ **http://un2sg1.unige.ch/www/athena/mineral/mineral.html**

This site, based in Switzerland, provides images and a data base that can be searched for minerals by name or chemical composition.

Smithsonian Gem & Mineral Collection

🛈 **http://galaxy.einet.net/images/gems/gems-icons.html**

This site features images and descriptions of outstanding mineral specimens in the Smithsonian's collection, including the Hope Diamond and the Star of Bombay sapphire.

Minerals in Thin Sections

🛈 **http://www.science.ubc.ca/~geol202/s/cgi-bin/mineral.cgi**

Maintained by the University of British Columbia, this site includes photomicrographs of minerals in plane and polarized light and lists their optical properties.

USGS Minerals Information

🛈 **http://minerals.er.usgs.gov/minerals/**

The U.S. Geological Survey provides information under such headings as What's New, Publications and Information Products (Mineral Year Book, etc.), Gemstone Production, World Gold, and more.

3

Sandstone formations, painted cliffs, Tasmania, Australia. These sandstones represent a former cycle of uplift, erosion, and sedimentation. Erosion of the sandstone cliffs is part of the current cycle. *(John Cancalosi/DRK.)*

Rocks: Records of Geologic Processes

What determines the appearance of the rocks we encounter? They vary in color, in the sizes of their crystals or grains, and in the kinds of minerals that make them up. Along a road cut, for example, we may find a black, homogeneous rock, its constituent particles—volcanic glass and crystals of pyroxene and feldspar—too small to be seen with the naked eye. Near it may be a brownish rock; this one, transformed by heat and pressure deep in the Earth, has abundant large glittering crystals of mica and some grains of quartz and feldspar. Overlying both the black rock and the brown one may be the remains of a former beach, horizontal layers of light-brown rock that appear to be made up of sand grains cemented together.

The appearance of these rocks is determined partly by their mineralogy and partly by their texture. Mineralogy, or the relative proportions of a rock's constituent minerals, helps determine its appearance and other properties, as you will recall from Chapter 2. So does the rock's **texture,** or the sizes and shapes of its mineral grains and crystals and the way they are put together. These grains or crystals, only a few millimeters in diameter in most rocks, are categorized as *coarse* (if they are large enough to be seen with the naked eye) or *fine* (if they are not). Mineral grains or crystals also vary in shape: they may be needle-shaped, flat, platy, or equant (about the same dimension in all directions, like a sphere or a cube). These variations in mineralogy and texture combine to produce larger features that define individual rocks.

The mineralogy and texture that determine a rock's appearance are themselves determined by the rock's geologic origin—where and how it was made (Figure 3.1). The dark rock in our road cut, called *basalt,* was formed by a volcanic eruption; its mineralogy and texture depend on the chemical composition of rocks that were melted deep in Earth's interior and on the nature of the eruption—whether it was explosive or a quieter lava flow. All rocks that were formed by the solidification of molten rock are called **igneous rocks.**

The light-brown layered rock of the road cut, a *sandstone,* was formed as sand particles accumulated, perhaps on a beach, and eventually were covered over, buried, and cemented together to form a rock. All rocks that were formed as the burial products of layers of sediments—such as sand, mud, and calcium carbonate shells—whether they were laid down on the land or under the sea, are called **sedimentary rocks.**

The brownish rock of our road cut, a *schist,* contains crystals of mica, quartz, and feldspar and was formed deep in Earth's crust as high temperatures and pressures transformed the mineralogy and texture of a buried sedimentary rock. All rocks that are formed by transformations of preexisting rocks in the solid state under the influence of high pressure and temperature are called **metamorphic rocks.**

Understanding rock properties and reasoning from them to deduce their geologic origins is the primary aim of a geologist. Such deductions, as you learned earlier, not only are essential to the ongoing process of understanding the planet we live on but also are important sources of information about fuel reserves and solutions to environmental problems.

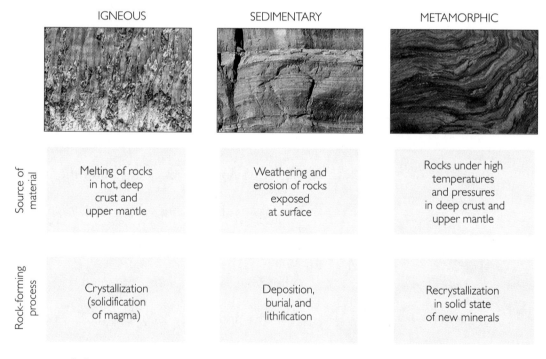

IGNEOUS	SEDIMENTARY	METAMORPHIC
Source of material: Melting of rocks in hot, deep crust and upper mantle	Weathering and erosion of rocks exposed at surface	Rocks under high temperatures and pressures in deep crust and upper mantle
Rock-forming process: Crystallization (solidification of magma)	Deposition, burial, and lithification	Recrystallization in solid state of new minerals

FIGURE 3.1 The minerals and textures of the three great rock groups are formed in different places in the Earth by different geologic processes. As a result, geologists use mineralogical and chemical analyses of rocks to determine the origins of rocks and the processes that formed them. The igneous rock shown here is basalt, formed by volcanism, in Yellowstone National Park. *(Willard Clay.)* The sedimentary rock is sandstone, formed from sand particles, in Sedona, Arizona. *(Michael Long/Visuals Unlimited.)* The metamorphic rock is schist, produced by very high temperatures and pressures acting on accumulated crystals of mica and other minerals, on the Blue Ridge Parkway in North Carolina. *(Gary Meszaros.)*

Knowing that oil, for example, is formed in certain kinds of sedimentary rocks that are rich in organic remains of biological origin, we can explore for new oil reserves more intelligently. Similarly, our knowledge of the properties of rocks is necessary for the discovery of other useful and economically valuable mineral and energy resources, such as gas, coal, and metal ores.

Understanding how rocks form also guides us to the solution of environmental problems. Will this rock be prone to earthquake-triggered landslides? How may it transmit polluted waters in the ground? The underground storage of radioactive and other wastes depends on the analysis of the rock to be used as a repository.

If rocks are clues to many of the things we want to know about the Earth, how do we go about interpreting them? We need a key, just as historians needed the Rosetta stone to crack the "code" of Egyptian hieroglyphics before they could read the inscriptions on temples and tombs. The first step in finding this key is to recognize the various kinds of rocks. The second step is to understand what their characteristics tell us about the surface and subsurface conditions under which they formed.

This chapter gives an overview of how geologists interpret the three great families of rock—igneous, sedimentary, and metamorphic. We see what the appearance, texture, mineralogy, and chemical composition of a rock reveal about how and where it formed. We look at how rock patterns found in subsurface drilling and in outcrops can help us reconstruct geologic history. Finally, we trace the rock cycle—the set of processes that convert each type of rock into the other two—and see how these processes are all driven by plate tectonics.

IGNEOUS ROCKS

Igneous rocks (from Latin *ignis,* fire) form by crystallization from a magma, a mass of melted rock that originates deep in the crust or upper mantle, where temperatures reach the 700°C or more needed to melt most rocks. When magmas cool slowly in the interior, microscopic crystals start to form. As the magma cools below the melting point, some of these crystals have time to grow to several millimeters or larger before the whole mass is crystallized as a coarse-grained igneous rock. But when a magma erupts from a volcano onto Earth's surface, it cools and solidifies so rapidly that individual crystals have no time for gradual growth. In that case, many tiny crystals form simultaneously, and the result is a fine-grained igneous rock. Geologists distinguish two major types of igneous rocks—intrusive and extrusive—on the basis of the sizes of their crystals.

Intrusive Igneous Rocks

Intrusive igneous rocks are formed by slowly crystallizing magmas that have intruded rock masses deep in the interior of Earth. They can be recognized by their interlocking large crystals, which grew slowly as the magma gradually cooled (Figure 3.2). Magmas cool slowly in Earth's interior because they

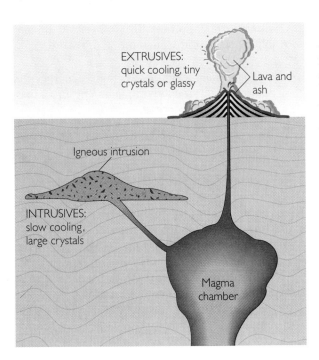

Basalt: an igneous extrusive

Granite: an igneous intrusive

FIGURE **3.2** *Extrusive igneous rocks* are formed when magma erupts at the surface, rapidly cooling to fine ash or lava and forming tiny crystals. The resulting rock, such as the basalt sample here, is finely grained or has a glassy texture. *Intrusive igneous rocks* crystallize when molten rock intrudes into unmelted rock masses deep in Earth's crust. Large crystals grow during the slow cooling process, producing coarsely grained rocks such as the granite sample shown here.

invade rock masses that conduct heat slowly; in addition, the temperatures of some of the rock masses may not be a great deal cooler than the magma itself. *Granite* is an intrusive igneous rock.

Extrusive Igneous Rocks

Rocks such as *basalt,* which form from rapidly cooled magmas that erupt at the surface, are called **extrusive igneous rocks.** They are easily recognized by their glassy or fine-grained texture (see Figure 3.2). Extrusive igneous rocks are formed by **volcanism**—the processes that form volcanoes. These rocks may consist of almost instantaneously crystallized ash particles that were blown high into the atmosphere as a volcano erupted, or of lavas, which flow as liquids for some distance on Earth's surface before they solidify.

Common Minerals

Most of the minerals of igneous rocks are silicates, partly because silicon is so abundant in the Earth and partly because many silicate minerals, but few oxide minerals, melt at the temperatures and pressures reached in lower parts of the crust and the mantle. As Table 3.1 indicates, the common silicate minerals found in igneous rocks include quartz, feldspar, mica, pyroxene, amphibole, and olivine, the minerals that typify the various crystal structures we described in Table 2.2.

TABLE 3.1

SOME COMMON MINERALS OF IGNEOUS, SEDIMENTARY, AND METAMORPHIC ROCKS

IGNEOUS ROCKS	SEDIMENTARY ROCKS	METAMORPHIC ROCKS
Quartz★	Quartz★	Quartz★
Feldspar★	Clay minerals★	Feldspar★
Mica★	Feldspar★	Mica★
Pyroxene★	Calcite	Garnet★
Amphibole★	Dolomite	Pyroxene★
Olivine★	Gypsum	Staurolite★
	Halite	Kyanite★

Asterisks indicate that a mineral is a silicate.

SEDIMENTARY ROCKS

Sediments, the precursors of sedimentary rocks, are found on Earth's surface as layers of loose particles, such as sand, silt, and shells of organisms. Particles such as sand grains and pebbles form at the surface of the Earth as rocks undergo **weathering**—that is, they are broken up into fragments of various sizes. The fragmented rock particles created by weathering are then transported by **erosion**—the set of processes that loosen soil and rock and move them downhill or downstream, where they are laid down as layers of sediment (see Figure 3.3). Weathering and erosion produce two types of sediments:

- **Clastic sediments** are physically deposited sedimentary particles, such as grains of quartz and feldspar derived from a weathered granite. (*Clastic* is derived from the Greek word *klastos,* meaning "broken.") These sediments are laid down by running water, wind, and ice, in the process forming layers of sand, silt, and gravel.

- **Chemical and biochemical sediments** are new chemical substances that form by precipitation when some of a rock's components dissolve during weathering and are carried in river waters to the sea. These sediments include layers of minerals such as halite (sodium chloride) and calcite (calcium carbonate, most frequently found in the form of shells).

From Sediment to Solid Rock

Lithification is the process that converts sediments into solid rock, and it occurs in one of two ways:

- By *compaction,* as grains are squeezed together by the weight of overlying sediment into a mass denser than the original.

- By *cementation,* as minerals precipitate around deposited particles and bind them together.

Sediments are compacted and cemented after burial under additional layers of sediment. Thus sandstone forms by the lithification of sand particles, and limestone forms by the lithification of shells and other particles of calcium carbonate.

Sediments and sedimentary rocks are characterized by **bedding,** the formation of parallel layers by the settling of particles to the bottom of the sea, a river, or a land surface. Bedding may reflect varia-

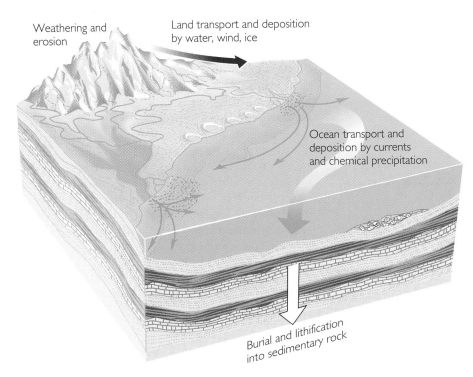

FIGURE 3.3 Weathering breaks down rock into smaller particles that are then carried downhill and downstream by erosion to be deposited as layers of sediment. Other sediment is produced by chemical precipitation. As layers accumulate and are buried deeper and deeper, they lithify, hardening into sedimentary rock.

tions in mineralogy, as when sandstone is interbedded with limestone, or differences in texture, as when a coarse-grained sandstone becomes interbedded with a fine-grained one.

Formed by surface processes, sedimentary rocks cover much of Earth's land surface and seafloor. Although most rocks found at the Earth's surface are sedimentary, they form only a thin layer atop the igneous and metamorphic rocks that make up the main volume of the crust (Figure 3.4).

Common Minerals

The common minerals of clastic sediments are silicates, reflecting the dominance of silicate minerals in rocks that weather to form sedimentary particles (see Table 3.1). Thus the most abundant minerals in clastic sedimentary rocks are quartz, feldspar, and clay minerals.

The most abundant minerals of chemically or biochemically precipitated sediments are carbonates.

FIGURE 3.4 Sediments and sedimentary rocks cover much of the land surface and the seafloor. Nevertheless, they are only a thin layer distributed over the igneous and metamorphic rocks that account for most of the crust's volume. The pie charts at the top of this figure compare igneous and sedimentary rocks as percentages of crustal volume and as percentages of surface area. In the pie charts, the area and volume of metamorphic rocks are divided between igneous and sedimentary, reflecting the type of parent rock that was metamorphosed.

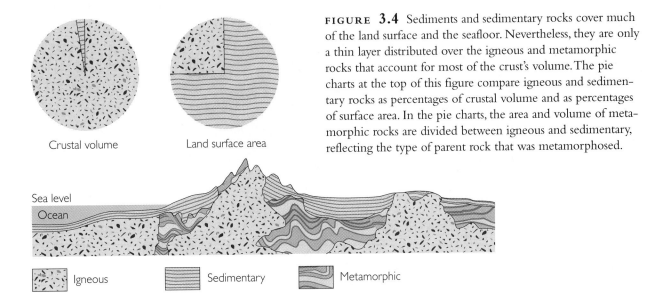

One is calcite, the main constituent of limestone. Another is dolomite, which is also found in limestone and is a calcium-magnesium carbonate formed by precipitation during lithification. Two others—gypsum and halite—form by chemical precipitation as seawater evaporates.

METAMORPHIC ROCKS

Metamorphic rocks take their name from the Greek words for "change" (*meta*) and "form" (*morphe*). These rocks are produced when high temperatures and pressures deep in the Earth cause any kind of rock—igneous, sedimentary, or other metamorphic rock—to change its mineralogy, texture, or chemical composition while maintaining its solid form. The temperatures are below the melting points of the rocks (about 700°C), but they are high enough (above 250°C) for the rocks to change by recrystallization and chemical reactions.

Regional and Contact Metamorphism

Metamorphism may take place over a widespread or a limited area. Where high pressures and temperatures extend over large regions, rocks are subject to **regional metamorphism.** Regional metamor-phism accompanies plate collisions that result in mountain building and the folding and breaking of sedimentary layers that once were horizontal (Figure 3.5a). Where high temperatures are restricted to smaller areas, such as the rocks near and in contact with an intrusion, rocks are transformed by **contact metamorphism** (Figure 3.5b).

Many regionally metamorphosed rocks, such as schists, have characteristic **foliation,** wavy or flat planes produced when the rock was structurally deformed into folds. Granular textures are more typical of most contact metamorphic rocks, which contain minerals with equant-shaped crystals, and of some regional metamorphic rocks formed by very high pressure and temperature.

Common Minerals

Silicates are the most abundant minerals of metamorphic rocks because these rocks are transformations of other rocks that are rich in silicates (see Table 3.1). Typical minerals of metamorphic rocks are quartz, feldspar, mica, pyroxene, and amphibole, the same kinds of silicates characteristic of igneous rocks. Several other silicates—kyanite, staurolite, and some varieties of garnet—are characteristic of metamorphic rocks alone. They form under conditions of high pressure and temperature in a crustal setting

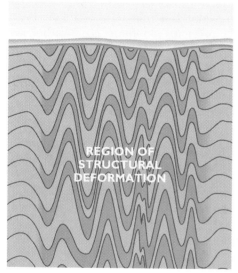

(a) Regional metamorphism

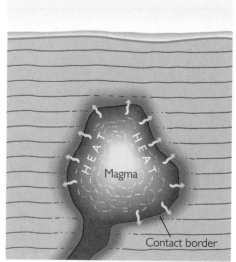

(b) Contact metamorphism

FIGURE 3.5 Metamorphism may take place over a widespread or limited area. (a) Regional metamorphism occurs where high pressures and temperatures extend over large areas, as happens when heat and pressure accompanying plate collisions strongly deform rocks. (b) Contact metamorphism occurs in more limited areas, when rocks in border zones of an intrusion of magma are metamorphosed by heat emanating from the magma.

and are not characteristic of igneous rocks. They are therefore good indicators of metamorphism. Calcite is the main mineral of marbles, which are metamorphosed limestones.

THE CHEMICAL COMPOSITION OF ROCKS

When geologists try to infer the geologic origin of rocks, they frequently make chemical analyses to determine the relative proportions of the rocks' chemical elements. Chemical analyses complement mineralogical studies, particularly for the very fine-grained or glassy rocks, such as volcanic lavas, in which few individual minerals can be observed, even with a microscope. The chemical composition of one of the abundant igneous rocks, basalt, for example, reveals the kinds of rock that melted to form the magma from which the basalt crystallized. By convention the elements are stated in terms of oxides, the compounds they form by combining with oxygen, the most abundant element in the Earth. This convention is followed even though the oxides are not actually present as such but are mostly in the form of silicates. Table 3.2 shows the major elements, as percentages by weight, for a sample of basalt. In that table you will see that the element silicon, for example, is given as the oxide silica (SiO_2) and that it makes up 48 percent of the basalt. In an ideal analysis, the percentages of all the elements would total 100, but small errors in analysis or the omission of minor elements, as in Table 3.2, may produce a smaller total.

Basalt is a typical igneous rock; about two-thirds of its weight consists of silica and alumina, the oxide of aluminum. Iron oxides are present in smaller amounts (14.7 percent), followed by calcium, magnesium, sodium, and potassium oxides (see Table 3.2). These seven *major elements,* along with oxygen, make up the great bulk of all rocks in the Earth.

So how can such an analysis help us understand basalt's geologic origin? Differences of a few percentage points or less in the proportions of some elements—such as the major cations—may indicate, for example, whether a basalt was formed at a mid-ocean ridge, where plates diverge, or at a subduction zone, where plates converge.

One difference geologists watch for in distinguishing among different kinds of rocks is the amount of water (the oxide of hydrogen) chemically bound in minerals. Water is a constituent of many rocks, but rarely in abundance. Many igneous rocks, for example, are only about 1 percent water, whereas sedimentary rocks may contain as much as 5 percent water, not counting the water present as liquid in their pore spaces. The larger amount of water in sedimentary rocks is traceable to the abundance of clay minerals in them. Finding a rock with such a large water content, a geologist would expect it to be a sedimentary rock, not an igneous rock.

In Chapter 4 we will look more closely at chemical differences among igneous rocks, and at the way these differences function as clues to the origin of the magmas from which the rocks came and the conditions under which they crystallized. In Chapter 7 we examine in more detail the chemical compositions of sedimentary rocks and the way they indicate the kinds of rocks that weathered and provided the sediment, as well as the chemical conditions under which precipitated minerals were formed. Similarly, in Chapter 8 we will see that chemical analyses of metamorphic rocks are guides to the preexisting rocks that were transformed by heat and pressure.

TABLE 3.2

CHEMICAL ANALYSIS OF MAJOR ELEMENTS OF BASALT, AN IGNEOUS ROCK		
ELEMENT	FORMULA[1]	PERCENT OF WEIGHT
Silicon	SiO_2	48.0
Aluminum	Al_2O_3	16.0
Iron	Fe_2O_3, FeO	14.7
Calcium	CaO	10.0
Magnesium	MgO	3.9
Sodium	Na_2O	3.5
Potassium	K_2O	1.5
All major elements		97.6[2]

[1]Formulas are conventionally shown as oxides, though the elements are not actually present in the rock in this form.

[2]Percentages do not add up to 100 because of the omission of minor elements and small errors in analysis.

WHERE WE SEE ROCKS

Rocks are not found in nature conveniently divided into separate bodies—igneous here, sedimentary there, metamorphic in another place. Instead, they are found jumbled together in patterns determined by the geologic history of the region. Geologists map those patterns both at the surface and projected into the interior, and try to deduce the geologic past from the present variety and distribution of the rocks.

If we were to drill a hole into any spot on Earth, we would find rocks that reflect the geologic history of that region. In the top few kilometers of most regions we would probably find sedimentary rock. Drilling deeper, perhaps 6 to 10 km down, we would eventually penetrate an underlying area of older igneous and metamorphic rock.

In fact, thousands of relatively shallow holes have been drilled on the continents in the search for oil, water, and mineral resources, and these holes are major sources of information, mainly about sedimentary rocks and their history. In the quest for more data on the deep continental crust, the governments of several countries, including the United States, Germany, and Russia, have drilled to great depths on the continents. The deepest hole, in Russia, measures more than 12 km, exceeding the depth of any commercial drilling.

A large part of our knowledge about the rocks of the ocean floor comes from the hundreds of holes punched down by the Deep-Sea Drilling Program, an ongoing project to drill the world's seafloor for geological information. Started by the United States in the late 1960s, at the same time that plate tectonics swept the geological community, it is now an international program (the Ocean Drilling Program) carried on with the cooperation of the major maritime countries of the world.

Even with all these sources of information on what lies beneath Earth's surface, geologists continue to rely on the rocks exposed in **outcrops,** places where bedrock—the underlying rock beneath the loose surface materials—is laid bare (Figure 3.6). Outcrops vary from region to region because they reflect the geologic structure of the Earth at a particular spot. On a trip across North America, we might run across many kinds of outcrops. Starting from the Pacific, we would encounter sea cliffs from Mexico to Canada. From the West Coast to the Rocky Mountain front, which stretches from New Mexico in the south to Alberta in Canada, outcrops of all kinds of rock are abundant in the canyons, mountainsides, and cliffs of the relatively dry mountainous regions of the western third of the continent (Figure 3.7).

From the Rockies eastward to the Appalachian Mountains, the landscape is dominated by the plains and prairies of the American Midwest and the Canadian Plains provinces (Figure 3.8). In this region, outcrops are scarce because most of the sedi-

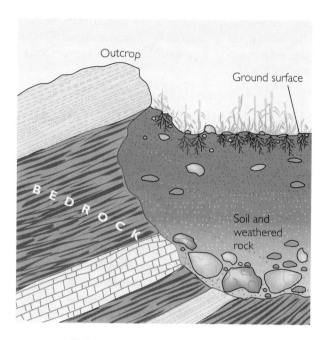

FIGURE 3.6 Outcrops are places where bedrock—the underlying rock beneath loose surface materials such as soil and boulders—is laid bare.

mentary bedrock is covered by soils and the sediments deposited by such rivers as the Missouri and their tributaries. Here geologists search the low hills and gentle valleys, looking for evidence of geologic history along dry creeks and interstate highway cuts.

Outcrops become more numerous once we reach the Appalachians. In this more humid climate, most of the rocks of the low ridges are covered by abundant vegetation and soil, but there are many outcrops along rocky cliffs and ledges, especially on the higher ridges and mountains.

Low coastal plains cover the region from southeastern New Jersey to the Carolinas and Georgia in the east and Texas, Louisiana, Mississippi, and Alabama in the south. Here barely lithified, relatively soft sedimentary rocks are exposed in outcrops similar to those of the Great Plains. Good exposures can be found in the occasional bluff along the shoreline. To the north, in the hilly, rugged landscape of New England and the Maritime provinces of Canada, we can find good outcrops, with the best exposures displayed along rocky coastlines (Figure 3.9). Other good outcrops are displayed in creek bottoms.

As this travelogue indicates, the presence and type of outcrops depend on the nature of the landscape, which in turn depends on the geologic structure of the region, its history, and its present climate. In later chapters, we explore in more detail the way rock types relate to geologic structures (Chapter 10)

FIGURE **3.7** Rocky cliffs on the Pacific coast at Cape Kiwanda, Oregon. Shoreline cliffs such as these provide ready accessibility to bedrock for the geologist. *(Fred Hirschmann.)*

FIGURE **3.8** Granite outcrops on Bonaventure Island, Quebec, Canada, form low ledges of bedrock, often the only exposure of bedrock in gentle topography. In some cases this type of outcrop extends to the Atlantic seaboard of Canada. *(Russ Kinne/Comstock.)*

FIGURE 3.9 Shawangunk Mountains, New York. Even though these are ancient mountains, part of the Appalachian Mountain chain, excellent outcrops of bedrock are formed along steep slopes. *(Carr Clifton.)*

and to landscape (Chapter 16). Now, however, we turn to the rock cycle, which—in combination with plate tectonics—reveals the interrelationships among the three groups of rocks and thus geologic structure and history.

THE ROCK CYCLE

The **rock cycle** is a set of geologic processes by which each of the three great groups of rocks is formed from the other two. The Scotsman James Hutton described this cycle in an oral presentation in 1785 before the Royal Society of Edinburgh; ten years later he presented it in more detail in his book *Theory of the Earth with Proof and Illustrations.* As is often the case in the history of science, other scientists, both in England and on the European continent,

had also recognized elements of the cyclic nature of geological change. Hutton's role was that of synthesizer—he presented the larger picture that has enabled us to understand the process.

We present an account of one particular cycle here, recognizing that such cycles vary with time and place. We can start our account with a magma deep in Earth's interior, where temperatures and pressures are high enough to melt any kind of preexisting rock: igneous, metamorphic, or sedimentary (see Figure 3.10). This activity deep in Earth's crust Hutton called the *plutonic episode,* for Pluto, the Roman god of the underworld. We now refer to all igneous intrusives as **plutonic rocks,** whereas the extrusives are known as **volcanic rocks.** As preexisting rocks melt, all their component minerals are destroyed and their chemical elements are homogenized in the resulting hot liquids. As the magma cools, crystals of new minerals grow and form new

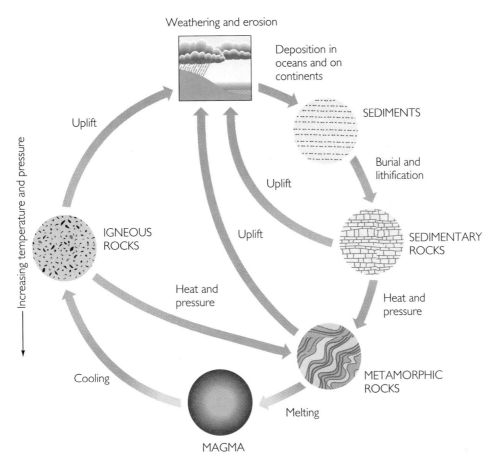

FIGURE 3.10 The rock cycle, as proposed by James Hutton 200 years ago. Subjected to weathering and erosion, rocks form sediments, which are deposited, buried, and lithified. After deep burial, the rocks undergo metamorphism, melting, or both. Through orogeny and volcanic processes, rocks are uplifted, only to be recycled again.

igneous rock. Melting and formation of igneous rock take place mainly along the boundaries of colliding or diverging tectonic plates, as well as in mantle plumes, as you will see in later chapters.

The igneous rocks that form at the boundaries where plates collide, together with associated sedimentary and metamorphic rocks, are then uplifted as a high mountain chain as a section of Earth's crust becomes crumpled and deformed. Geologists call this process, which begins with plate collision and ends in mountain building, **orogeny.** After uplift, the rocks of the crust overlying the uplifted igneous rock gradually weather, creating loose material that erosion then strips away, exposing the igneous rock at the surface.

The igneous rock, now in cool, wet surroundings far from its birthplace in the hot interior, also weathers, and some of its minerals undergo chemical changes. Iron minerals, for example, may "rust" to

form iron oxides. High-temperature minerals such as feldspars may become low-temperature clay minerals. Such substances as pyroxene may dissolve completely as rain pours over them. The weathering of the igneous rock again produces various sizes and kinds of rock debris and dissolved material, which are carried away by erosion. Some are transported by water and wind on land. Much of the debris is transported by streams to rivers and ultimately to the ocean, where it is deposited as layers of sand, silt, and other sediments formed from dissolved material, such as the calcium carbonate from shells.

These sediments laid down in the sea, as well as those deposited by water or wind on the land, are buried under successive layers of sediment, where they gradually lithify into sedimentary rock. Burial is accompanied by **subsidence;** that is, a depression or sinking of the Earth's crust. As subsidence continues, additional layers of sediment can accumulate.

FIGURE 3.11 Many processes combine to produce the rock cycle. These processes are all driven by plate tectonics. The five parts of this figure illustrate those relationships in various tectonic settings.

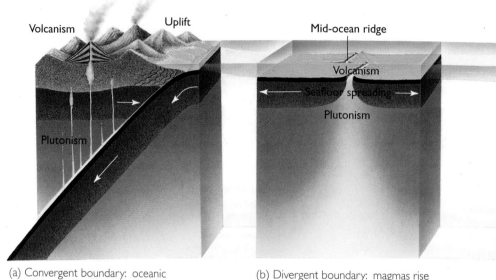

(a) Convergent boundary: oceanic plates subduct and melt.

(b) Divergent boundary: magmas rise and seafloors spread.

As the lithified sedimentary rock is buried more deeply in the crust, it gets hotter. As the depth of burial exceeds 10 km and temperatures climb to over 300°C, the minerals in the still-solid rock start to change to new minerals that are more stable at the higher temperatures and pressures of the deeper parts of the crust. This is the process of metamorphism, which transforms the previously sedimentary rocks into metamorphic rocks. With further heating, the rocks may melt and form a new magma from which igneous rocks will crystallize, starting the cycle all over again.

As we noted earlier, this cycle is only one variation among the many that may occur in the rock cycle. Any type of rock—metamorphic, sedimentary, or igneous—can be uplifted during an orogeny and weathered and eroded to form new sediments. Some stages may be omitted; as a sedimentary rock is uplifted and eroded, for example, metamorphism and melting are skipped. And stages may occur out of sequence, as when an igneous rock formed in the interior is metamorphosed before it is uplifted. And, as we know from deep drilling, some igneous rocks many kilometers down in the crust may never have been uplifted or exposed to weathering and erosion.

The rock cycle never ends; it is always operating at different stages in different parts of the world, forming and eroding mountains in one place and laying down and burying sediments in another. The rocks that make up the solid Earth are recycled continuously, but we can see only the surface parts of the cycle and must deduce the recycling of the deep crust and the mantle from indirect evidence.

PLATE TECTONICS AND THE ROCK CYCLE

Plutonism, volcanism, tectonic uplift, metamorphism, weathering, sedimentation, transportation, deposition, and burial are the geologic processes that combine in the rock cycle to convert the three groups of rocks to one another. These processes are all driven by plate tectonics.

Plutonism and volcanism are the result of the interior heat of the Earth, and we identify them primarily with three plate-tectonic settings:

- Convergent boundaries, where oceanic plates descend into the mantle and melt, and igneous rocks eventually form (Figure 3.11a)

- Divergent boundaries at mid-ocean ridges, where seafloor spreading allows basaltic magmas to rise from Earth's mantle to form oceanic crust (Figure 3.11b)

- Mantle plumes or hot spots, where magmas rise through the mantle and pour out at the surface (Figure 3.11d)

Sediment is carried away from mountains and deposited on continents and ocean floors as plates slowly subside while they move away from mid-ocean ridges. Multiple layers of sediment accumulate, burying lower layers (Figure 3.11c).

In contrast to the Earth's interior, at the surface the Sun's heat drives the circulation of the oceans

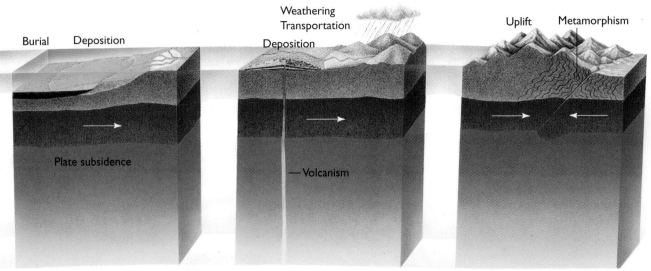

(c) Ocean floor and shoreline: subsidence of an oceanic plate causes deposition of sediment, burial, and lithification.

(d) Continental interior: stable plates are dominated by weathering, deposition, and transportation. Volcanism may be generated by mantle plumes or "hot spots."

(e) Convergent boundaries: plates carrying continents collide, uplifting mountains and metamorphosing large regions.

and atmosphere, producing weathering and transportation of sediment by wind, water, and ice (Figure 3.11d).

Metamorphism and uplift occur when continental plates collide at convergent boundaries. These collisions uplift mountains and create the great pressures and temperatures that metamorphose rocks (Figure 3.11e).

With this introduction to the rock world, we are ready to begin the study of rocks. In Chapters 4 and 5 we look at the geologic origin of magmas, the types of igneous rocks that form by their crystallization, the larger picture of plate-tectonic control of igneous processes, and the dynamics of volcanoes and their eruptions. In Chapters 6 and 7 we explore weathering, the nature of sedimentary particles, and the ways various sediments and sedimentary rocks are produced. We complete our discussion of rocks in Chapter 8 by seeing how high heat and pressure affect preexisting rocks, transforming them to metamorphic rocks, and how metamorphism relates to plate tectonics and orogeny.

SUMMARY

What determines the properties of the various kinds of rocks that are formed in and on the surface of the Earth? Mineralogy (the kinds and proportions of minerals that make up a rock) and texture (the sizes, shapes, and spatial arrangement of its crystals or grains) define a rock. The mineralogy and texture of a rock are determined by the geologic conditions, including chemical composition, under which it formed, either in the interior under various conditions of high temperature and pressure or at the surface, where temperatures and pressures are low.

What are the three types of rock and how are they formed? Igneous rocks are formed by the crystallization of magmas as they cool. Intrusive igneous rocks form in Earth's interior and have large crystals. Extrusive igneous rocks, which form at the surface of the Earth where lavas and ash erupt from volcanoes, have a glassy or fine-grained texture. Sedimentary rocks are formed by the lithification of sediments after burial. Sediments are derived from the weathering and erosion of rocks exposed at the Earth's surface. Metamorphic rocks are formed by alteration in the solid state of igneous, sedimentary, or other metamorphic rocks as they are subjected to high temperatures and pressures in the interior.

How does the rock cycle describe the formation of rocks as the products of geologic processes? The rock cycle relates rock-forming geologic processes to the formation of the three types of rocks from one another. One can view the processes by starting at any point in the cycle. We began with the formation of igneous rocks by crystallization of a magma in the interior of the Earth. The igneous rock then is uplifted to the surface in the mountain-building process. There it is exposed to weathering and erosion, which produce sediment. The sediment is cycled back to the interior by burial and lithification to sedimentary rock. Deep burial leads to metamorphism or melting, at which point the cycle begins again. Plate tectonics is the mechanism by which the cycle operates.

KEY TERMS AND CONCEPTS

texture (p. 60)
igneous rocks (p. 60)
sedimentary rocks (p. 60)
metamorphic rocks (p. 60)
intrusive igneous rocks (p. 61)
extrusive igneous rocks (p. 62)
volcanism (p. 62)
sediments (p. 62)

weathering (p. 62)
erosion (p. 62)
clastic sediments (p. 62)
chemical and biochemical
    sediments (p. 62)
lithification (p. 62)
bedding (p. 62)
regional metamorphism (p. 64)

contact metamorphism (p. 64)
foliation (p. 64)
outcrops (p. 66)
rock cycle (p. 68)
plutonic rocks (p. 68)
volcanic rocks (p. 68)
orogeny (p. 69)
subsidence (p. 69)

EXERCISES

1. What are the differences between extrusive and intrusive igneous rocks?

2. What are the differences between regional and contact metamorphism?

3. What are the differences between clastic and chemical or biochemical sedimentary rocks?

4. List three common silicate minerals found in each group of rocks: igneous, sedimentary, and metamorphic.

5. Of the three groups of rocks, which are formed at Earth's surface and which in the interior of the crust?

6. Where on the continents can you see bare rock?

THOUGHT QUESTIONS

1. What geologic processes transform a sedimentary rock to an igneous rock?

2. Name a mineral found only in sedimentary rocks that you might use to distinguish between a fine-grained sedimentary rock formed from lithified mud and an extrusive igneous rock.

3. As a magma cools, what might cause differences in the sizes of the crystals of two intrusive igneous rocks, one with crystals about 1 cm in diameter, the other with crystals about 2 mm in diameter?

4. Which igneous intrusion would you expect to have a wider contact metamorphic zone: one intruded by a very hot magma or one intruded by a magma of a more moderate temperature?

5. Describe the geologic processes by which an igneous rock is transformed to a metamorphic rock and then exposed to erosion.

6. Describe the kinds of outcrop that are found in various places in your hometown. If none, explain how you would determine the nature of the buried bedrock.

7. How does plate tectonics explain plutonism?

SHORT-TERM TEAM PROJECT: IDENTIFYING BUILDING STONES

Building stones often are a clue to local geology. Local stone is both cost-effective and a source of community pride. With a partner, examine the building stones you find on campus or in your community. Choose between four and six different-looking types of stone and note their locations on a local map. Draw each stone on a separate piece of paper and note such features as color, grain size, presence or absence of layering, and whether the stone appears to contain one mineral or more. Also describe any evidence of chemical or physical weathering and judge how good the stone is for building. Then decide whether the stone is most likely to be igneous, metamorphic, or sedimentary, and explain why. Finally, compare a geologic map of the area with your findings in the field and explain why the stones you described do or do not reflect local geology. Submit an organized folder containing your drawings, observations, and inferences.

SUGGESTED READINGS

Blatt, Harvey, and Robert J. Tracy. 1996. *Petrology: Igneous, Sedimentary, and Metamorphic,* 2nd ed. New York: W. H. Freeman.

Dietrich, Robert V., and Brian J. Skinner. 1980. *Rocks and Rock Minerals.* New York: Wiley.

Ernst, W. Gary. 1969. *Earth Materials.* Englewood Cliffs, N.J.: Prentice Hall.

Prinz, Martin, George Harlow, and J. Peters. 1978. *Simon & Schuster's Guide to Rocks and Minerals.* New York: Simon & Schuster.

INTERNET SOURCES

Rock Cycle

ⓘ **http://www.science.ubc.ca/~geol202/s/rock_cycle/rockcycle.html**

The University of British Columbia established a Web site for its Introduction to Petrology course (Geology 202) with a variety of URLs. This URL will take you directly to the Rock Cycle.

Rock Families

ⓘ **http://www.science.ubc.ca/~geol202/**

This URL for the University of British Columbia's Introduction to Petrology course (Geology 202) provides links to a variety of data and images concerning igneous, sedimentary, and metamorphic rocks. There are links to additional pages on terminology and classification of rocks and an Image Gallery that can be searched.

4

Columnar basalts, North Iceland. Masses of this kind of extrusive igneous rock fracture along more or less symmetrical columnar joints when they cool. (*Thomas Dressler/DRK.*)

Igneous Rocks: Solids from Melts

Igneous rocks are formed by the solidification of magmas. Chapter 3 examined igneous rocks in the context of the rock cycle; Chapter 5 discusses them in the context of the magmas that feed volcanoes. In this chapter, we take a broader and yet more detailed view of the wide range of igneous rocks, both intrusive and extrusive, and the processes by which they form.

Geologists' understanding of the variety and formation of igneous rocks is a rich chapter in the history of science. It began more than 2000 years ago, when the Greek scientist and geographer Strabo traveled to Sicily to observe the volcanic eruptions of Mount Etna. Strabo observed that the hot liquid lava spilling down from this active volcano onto Earth's surface cooled and solidified into solid rock within a few hours. Two millennia later, eighteenth-century geologists made another connection, and it would be a milestone in the development of geology into a modern observational and experimental science. They began to understand that some of the bands or sheets of rock that cut across other formations had also been formed by the cooling and solidifying of magma. In this case, however, the magma had cooled and solidified slowly because it had remained buried in the Earth's crust. Today we know that at places deep in the hot crust and mantle of the Earth, rocks melt and rise toward the surface. Some solidify before they reach the surface, and some break through and solidify on the surface. Both groups are igneous rocks.

As we saw in Chapter 3, much of Earth's crust is composed of igneous rock, some metamorphosed and some not. Since igneous rocks are the products of the cooling and solidification of liquid rock, it follows that understanding the processes that melt and resolidify rocks is a key to understanding how Earth's crust is formed.

As we also saw in Chapter 3, modern geologists have learned that the origins of a wide variety of igneous rocks are connected with plate tectonics, especially with the spreading apart of two divergent plates and the sinking of one convergent plate below another. Although much is still to be learned about the exact *mechanisms* of melting and solidification, geologists now have good answers to questions that long perplexed them: How do igneous rocks differ from one another? Where do igneous rocks form? How do rocks solidify from a melt? Where do melts form?

HOW DO IGNEOUS ROCKS DIFFER FROM ONE ANOTHER?

Late eighteenth-century geologists had sought an understanding of igneous rocks in their field observations. By the late nineteenth century, some geologists had moved their inquiries to the laboratory, running experiments to determine the mineral and chemical composition of igneous rocks collected in the field. They classified their rock samples in the same general way we do today:

- By texture.
- By mineral and chemical composition.

Texture

Two hundred years ago, the first division of igneous rocks was made on the basis of texture, as geologists classified rocks as either coarsely or finely crystalline. Texture is a simple and practical distinction that a geologist can easily see in the field. A coarse-grained rock such as granite has separate crystals quite visible to the naked eye. In contrast, the crystals of fine-grained rocks such as basalt are too small to be seen, even with the aid of the small magnifying lens that no field geologist would be without. (Figure 4.1 shows two samples of granite and basalt, accompanied by photographs, called photomicrographs, of very thin transparent rock slices. Photomicrographs, which are taken through a microscope, give an enlarged view of minerals and their textures.) The textural difference was clear to early geologists; its meaning would be unraveled only through further work.

FIRST CLUE: VOLCANIC ROCKS   The first clue to the meaning of texture came from the study of volcanic rocks. Early geologists could observe rocks forming from lava during volcanic eruptions. (*Lava*, you may recall from Chapter 3, is the term we apply to magma flowing out on the surface.) Geologists noted that when lava cooled rapidly, it formed either a finely crystalline rock or a glassy one in which no crystals could be distinguished. Where lava cooled more slowly, as in the middle of a thick flow many meters high, somewhat larger crystals were present.

SECOND CLUE: LABORATORY STUDIES OF CRYSTALLIZATION   The second clue to the implications of texture came in the nineteenth century, as experimental scientists studying familiar liquids came to understand the nature of crystallization. Anyone who has frozen an ice cube knows that water solidifies to ice in a few hours as its temperature drops below the freezing point. If you have ever attempted to retrieve your ice cubes before they were completely solid, you may have seen thin ice crystals forming at the surface and along the sides of the tray. During crystallization, the water molecules take up fixed positions in the solidifying crystal structure, and they are no longer able to move freely, as they did when the water was liquid. All other liquids, including magmas, crystallize in this way.

The first tiny crystals form a pattern. Other atoms or ions in the crystallizing liquid then attach themselves in such a way that the tiny crystals grow larger. It takes some time for the atoms or ions to "find" their correct places on a growing crystal, and large crystals form only if they have time to grow slowly. Recall from Chapter 3 that if a liquid solidifies very quickly, as a magma does when it erupts onto the cool surface of the Earth, the crystals have no time to grow larger. Instead, a large number of tiny crystals form simultaneously as the liquid cools and solidifies.

FIGURE 4.1 Texture was the first characteristic by which igneous rocks were classified. Early geologists assessed texture with a small hand-held magnifying lens. Modern geologists have access to high-powered polarizing microscopes, which produce photomicrographs of thin, transparent rock slices like those shown here. The coarsely crystalline rock on the left is granite; the finely crystalline rock on the right is basalt. *(Hand-sample photos by Chip Clark. Photomicrographs by Raymond Siever.)*

THIRD CLUE: GRANITE—EVIDENCE OF SLOW COOLING The study of volcanoes allowed early geologists to link finely crystalline textures with quick cooling at Earth's surface and to see finely crystalline igneous rocks as evidence of former volcanism. But in the absence of direct observation, what geological evidence could allow them to deduce that coarse-grained rocks formed by slow cooling deep in the interior? Granite—one of the commonest rocks of the continents—turned out to be the crucial clue (Figure 4.2). James Hutton, one of the founders of modern geology, saw granites cutting across and disrupting the bedding, or layers, of sedimentary rocks as he worked in the field in Scotland near the end of the eighteenth century. He noticed that the granite had somehow fractured and invaded the sedimentary rocks, as though the granite had been forced into the fractures as a liquid.

As Hutton looked at more and more granites, he began to focus on the sedimentary rocks bordering them. He observed that the minerals of the sedimentary rocks in contact with the granite were

FIGURE 4.2 Granite exposed in Joshua Tree National Monument, California. Weathering has brought out the coarsely crystalline texture of this intrusive igneous rock, which is made up of large crystals of quartz, feldspar, and other silicates. *(David Muench.)*

different from those found in sedimentary rocks at some distance from the granite. He concluded that the changes in the sedimentary rocks must have resulted from great heat and that the heat must have emanated from the granite. Hutton also noted that granite was composed of crystals interlocked like the pieces of a jigsaw puzzle (see Figure 4.1). By this time, chemists had established that a slow crystallization process produces this pattern.

Assessing these three lines of evidence, Hutton proposed that granite had formed from a hot molten material that solidified deep in the Earth. The evidence was conclusive, as no other explanation could accommodate all the facts. Other geologists, seeing the same characteristics of granites in widely separated places in the world, came to recognize that granite and many similar coarsely crystalline rocks were the products of magma that had crystallized slowly in the interior of the Earth.

INTRUSIVE IGNEOUS ROCKS   The full significance of textural distinction in igneous rocks is now clear. As we have seen, texture is linked to the rapidity, and therefore the place, of cooling. Slow cooling of magma in Earth's interior allows adequate time for the growth of the interlocking large coarse crystals that characterize intrusive igneous rocks, also known as plutonic rocks. An **intrusive igneous rock** is igneous rock that has forced its way into surrounding rock. This surrounding rock is called **country rock.** (Later in this chapter we examine some special forms of intrusive igneous rocks.)

EXTRUSIVE IGNEOUS ROCKS   Rapid cooling at Earth's surface produces the finely grained texture or glassy appearance of the **extrusive igneous rocks.**

These rocks form when lava or other volcanic material erupts from volcanoes; therefore, they are also known as volcanic rocks. They are of two major categories:

- *Lavas* Volcanic rocks formed from lavas range in appearance from smooth and ropy to sharp, spiky, and jagged. We now know that many of these special textural properties are clues to the conditions under which these rocks formed, the composition of their parent magma, and the way they were ejected from the volcano.

- *Pyroclastic rocks* In more violent eruptions, pyroclasts are formed when broken pieces of lava and glass are thrown high in the air (Figure 4.3). **Pyroclasts** include crystals that started to form before the explosion, fragments of previously solidified lava, and pieces of glass that cooled and then fractured during the eruption. The finest pyroclasts are **volcanic ash,** extremely small fragments, usually of glass, that form when escaping gases force a fine spray of magma from a volcano. Volcanic ash accumulates as layers of loose and uncemented material. All volcanic rocks lithified from these pyroclastic materials are called **tuff.**

Volcanic glass comes in a variety of forms when it is the sole constituent of an igneous rock. One common glassy rock type is **pumice,** a frothy mass with a great number of *vesicles,* holes that remained after trapped gas escaped from the solidifying melt. Another wholly glassy volcanic rock is **obsidian,** which, unlike pumice, contained no trapped gases and so is solid and dense. Chipped or fragmented obsidian produces very sharp edges, and Native

FIGURE 4.3 Pyroclasts form when violent volcanic eruptions hurl broken pieces of lava and glass high in the air. Obsidian, shown here on the left, is a solid glassy pyroclast. Volcanic ash, in the center, forms when a fine spray of magma erupts. The multiple small holes in pumice, on the right, are evidence of the gas formerly trapped in this pyroclast. *(Chip Clark.)*

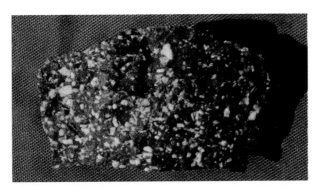

FIGURE 4.4 A quartz-rich felsic porphyry. The white larger crystals are phenocrysts of feldspar. *(A. J. Copley / Visuals Unlimited.)*

Americans and many other hunting groups used it for arrowheads and a variety of cutting tools.

A **porphyry** has a mixed texture in which large crystals "float" in a predominantly fine crystalline matrix (Figure 4.4) The large crystals, called *phenocrysts*, formed while the magma was still below Earth's surface. Then, before other crystals could grow, a volcanic eruption brought the magma to

Earth's surface, where as lava it quickly cooled to a finely crystalline mass. In Chapter 5 we will look more closely at how these and other volcanic rocks form during volcanism. Now, however, we turn to the second way igneous rocks are classified.

Chemical and Mineral Composition

Igneous rocks can be subdivided on the basis of their chemical and mineral compositions as well as according to texture. Volcanic glass, which is formless even under the microscope, is frequently classified by chemical analysis, as are some very finely grained rocks (see Chapter 3). One of the earliest classifications of igneous rocks was based on a simple chemical analysis of their silica (SiO_2) content. Silica, as noted in Chapter 3, is abundant in most igneous rocks and accounts for 40 to 70 percent of their total weight. We still refer to rocks rich in silica, such as granite, as silicic.

Modern classifications now group igneous rocks according to their relative proportions of silicate minerals (Table 4.1). These proportions were

TABLE 4.1

COMMON MINERALS OF IGNEOUS ROCKS

COMPOSITIONAL GROUP	MINERAL	CHEMICAL COMPOSITION	SILICATE STRUCTURE
FELSIC	Quartz	SiO_2	Frameworks
	Potassium feldspar	$KAlSi_3O_8$	
	Plagioclase feldspar	$\begin{cases} NaAlSi_3O_8 \\ CaAl_2Si_2O_8 \end{cases}$	
	Muscovite (mica)	$KAl_3Si_3O_{10}(OH)_2$	Sheets
MAFIC	Biotite (mica)	$\left.\begin{array}{c} K \\ Mg \\ Fe \\ Al \end{array}\right\} Si_3O_{10}(OH)_2$	
	Amphibole group	$\left.\begin{array}{c} Mg \\ Fe \\ Ca \\ Na \end{array}\right\} Si_8O_{22}(OH)_2$	Double chains
	Pyroxene group	$\left.\begin{array}{c} Mg \\ Fe \\ Ca \\ Al \end{array}\right\} SiO_3$	Single chains
	Olivine	$(Mg,Fe)_2SiO_4$	Isolated tetrahedra

determined by thousands of mineralogical analyses performed by geologists all over the world. The silicate minerals—quartz, feldspar (both orthoclase and plagioclase), muscovite and biotite micas, the amphibole and pyroxene groups, and olivine—form a systematic series. Felsic minerals are high in silica; mafic minerals are low in silica. The adjectives *felsic* (from *fel*dspar and *si*lica) and *mafic* (from *ma*gnesium and *fer*ric, from Latin *ferrum*, iron) are applied to both the minerals and the rocks that are high in these minerals. Mafic minerals crystallize at higher temperatures—that is, earlier in the cooling of a magma—than those at which felsic minerals crystallize.

As the mineral and chemical compositions of igneous rocks became known, geologists soon noticed that some extrusive and intrusive rocks were identical in composition and differed only in texture. Basalt, for example, is an extrusive rock formed from lava. Gabbro has exactly the same mineral composition as basalt but is formed deep in the Earth's crust (Figure 4.5). Similarly, rhyolite and granite are identical in composition and different in texture. Thus extrusive and intrusive rocks form two chemically and mineralogically parallel sets of igneous rocks. Conversely, most chemical and mineral compositions can appear as either extrusive or intrusive rocks. The sole exceptions are very highly mafic rocks that rarely or never appear as extrusive igneous rocks.

Figure 4.6 is a model that portrays the relationships we have been discussing. Notice that the horizontal axis plots silica content as a percentage of a given rock's weight. The percentages given—from high-silica content at 70 percent to low-silica content at 40 percent—cover the range found in igneous rocks. The vertical axis displays a scale measuring the mineral content of a given rock. If, for example, you had established that the silica content of a coarsely grained rock sample was about 70 percent, you could run a vertical line up from the 70 percent marker on the horizontal axis, which in fact corresponds to the left border of Figure 4.6. As the vertical line moves upward, it intersects the minerals amphibole, biotite, muscovite, plagioclase feldspar, quartz, and orthoclase feldspar, but it excludes the minerals pyroxene and olivine. Using the vertical axis to estimate the percentages of minerals at the points where the line intersects, you could then assert that the sample was probably granite, which has a mineral content like the one mapped by your vertical line. It contains about 6 percent amphibole, 3 percent biotite, 5 percent muscovite, 14 percent plagioclase feldspar, 22 percent quartz, and 50 percent orthoclase feldspar. Rhyolite, which has a mineral composition identical to granite, has the same mineral profile, but its texture would eliminate it—as a volcanic rock, rhyolite has a fine texture, not a coarse one.

We can use Figure 4.6 to help with the discussion of the intrusive and extrusive igneous rocks. We begin with the felsic rocks, on the extreme left of the model.

FIGURE 4.5 Two parallel sets of extrusive and intrusive igneous rocks. The finely grained basalt *(upper left)* is identical in composition to the coarsely grained gabbro *(upper right)*. Similarly, the finely grained rhyolite *(lower left)* is identical in composition to the coarsely grained granite *(lower right)*. *(Chip Clark.)*

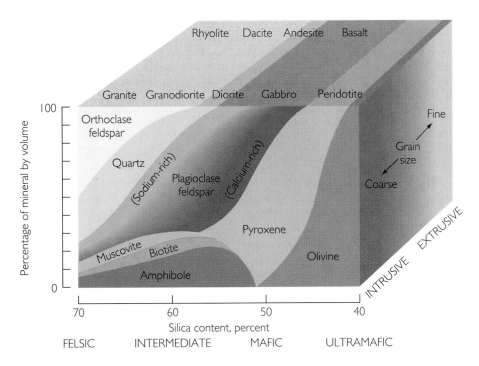

FIGURE 4.6 Classification model of igneous rocks. The vertical axis measures the mineral composition of a given rock as a percentage of its volume. The horizontal axis is a scale of silica content by weight. Thus if you knew by chemical analysis that a coarsely textured rock sample contained about 70 percent silica, you could determine that it also contained about 6 percent amphibole, 3 percent biotite, 5 percent muscovite, 14 percent plagioclase feldspar, 22 percent quartz, and 50 percent orthoclase feldspar. Your rock would be granite. Although rhyolite has the same mineral profile, its fine texture would eliminate it.

FELSIC ROCKS Felsic rocks are light-colored igneous rocks that are poor in iron and magnesium and rich in minerals that are high in silica, such as quartz, orthoclase feldspar, and plagioclase feldspar. The plagioclase feldspars contain both calcium and sodium. As Figure 4.6 indicates, they are richer in sodium near the felsic end and richer in calcium near the mafic end. In accord with the point made earlier, that mafic minerals crystallize at temperatures higher than those at which felsic minerals crystallize, calcium-rich plagioclases crystallize at higher temperatures than the sodium-rich plagioclases.

Felsic minerals and rocks tend to be light in color. **Granite,** the best known and one of the most abundant intrusive igneous rocks, is a felsic rock, with about 70 percent silica. Granite contains abundant quartz and orthoclase feldspar and a lesser amount of plagioclase feldspar (see the left side of Figure 4.6). These light-colored felsic minerals give granite its pink or gray color. It also contains small amounts of muscovite and biotite micas and amphibole.

Rhyolite is the extrusive equivalent of granite. This light-brown to gray igneous rock shares granite's felsic composition and light coloration, but it is much more finely grained. Many rhyolites are composed largely or entirely of volcanic glass.

INTERMEDIATE IGNEOUS ROCKS Midway between the felsic and mafic ends of the series are the **intermediate igneous rocks;** as their name indicates, these rocks are neither as rich in silica as the felsic rocks nor as poor in it as the mafic rocks. We find the intrusive intermediate rocks to the right of granite in Figure 4.6. The first is **granodiorite,** a light-colored felsic rock that looks something like granite. It is similar to granite in having abundant quartz, but its predominant feldspar is plagioclase, not orthoclase. To its right is **diorite,** still lower in silica and dominated by plagioclase feldspar, with little or no quartz. Diorites contain a moderate amount of the mafic minerals biotite, amphibole, and pyroxene, and they tend to be darker than granite or granodiorite.

The volcanic equivalent of granodiorite is **dacite.** To its right in the extrusive series is **andesite,** the volcanic equivalent of diorite. Andesite derives its name from the Andes, the volcanic mountain chain in South America.

MAFIC ROCKS Mafic rocks are high in pyroxenes and olivines, minerals relatively poor in silica but rich in magnesium and iron, from which these minerals get their characteristic dark colors. Here, at even lower levels of silica than those found in the intermediate range, we find **gabbro,** a coarsely grained, dark-gray intrusive igneous rock. Gabbro has an abundance of mafic minerals, especially pyroxene; it contains no quartz and only moderate amounts of calcium-rich plagioclase feldspar.

Basalt is dark gray to black and is the fine-grained extrusive equivalent of gabbro. Basalt is the most abundant igneous rock of the crust and underlies virtually all the floors of the oceans. On the continents, thick and extensive sheets of basalt make up large plateaus in some places, such as the Columbia River plateau of Washington State and the remarkable formation known as the Giant's Causeway in Northern Ireland.

ULTRAMAFIC ROCKS **Ultramafic rocks** consist primarily of mafic minerals and contain less than 10 percent feldspar. Here, at the very low silica level of only about 45 percent, we find **peridotite,** a coarsely grained, dark greenish-gray rock made up primarily of olivine with small amounts of pyroxene and amphibole. Ultramafic rocks are rarely found as extrusives. They form by the accumulation of crystals from a magma and have never constituted a liquid, hence do not form lavas.

Thus igneous rocks can be classified on the basis of their composition as well as by their texture. The compositional groups can be explained geologically by a strong correlation between mineralogy and the temperatures of crystallization. As Table 4.2 indicates, mafic minerals crystallize at temperatures higher than those at which felsic minerals crystallize. This increase in crystallization temperatures is a mirror image of the temperatures at which rocks melt. As we move from the mafic group toward the felsic rocks, silica content also increases. Increasing silica content is expressed as an increasingly complex silicate structure (see Table 4.1). The increasingly complex silicate structure forms an inverse correlation with a melted rock's ability to flow: as structure grows more complex, ability to flow decreases. Thus **viscosity**—the measure of a liquid's *resistance* to flow—increases as silica content increases. Viscosity is an important factor in the behavior of lavas, as you will see in Chapter 5.

It is clear that knowing a rock's minerals can yield important information about the conditions under which the rock's parent magma formed and crystallized. To interpret this information accurately, however, we must understand more about igneous processes, the topic we turn to next.

TABLE 4.2

CHANGES IN SOME MAJOR CHEMICAL ELEMENTS FROM FELSIC TO MAFIC ROCKS

	FELSIC	INTERMEDIATE		MAFIC
COARSE-GRAINED (INTRUSIVE)	Granite	Granodiorite	Diorite	Gabbro
FINE-GRAINED (EXTRUSIVE)	Rhyolite	Dacite	Andesite	Basalt

SILICA INCREASING

SODIUM INCREASING

POTASSIUM INCREASING

CALCIUM INCREASING

MAGNESIUM INCREASING

IRON INCREASING

(Viscosity increasing)

(Melting temperature increasing)

How Do Magmas Form?

We know from the way the Earth transmits earthquake waves that the bulk of the Earth is solid for thousands of kilometers down to the boundary of the core. The evidence of volcanic eruptions tells us, however, that there must also be liquid regions where magmas originate. How do we resolve this apparent contradiction? The answer lies in the processes that melt rocks and create magmas.

Although geologists can observe volcanic eruptions of lavas and study pyroclasts in their laboratories, most igneous processes cannot be directly observed. The study of magmas and igneous processes depends primarily on geological inferences and experimental simulations. To know where rocks melt in the Earth, for example, we have to know both the conditions under which various rocks melt and the regions of the Earth in which those conditions are found.

How Do Rocks Melt?

Although we do not yet understand the exact mechanisms of melting and solidification, geologists have learned a great deal from laboratory experiments on how rocks melt. From these experiments, we know that a rock's melting point depends on the rock's composition and on conditions of temperature and pressure.

TEMPERATURE AND MELTING As early twentieth-century geologists ran experiments on rocks, they discovered that a rock of varied composition does not melt completely at any given temperature. The **partial melting** that these early geologists had discovered occurs because the minerals that compose a rock melt at different temperatures. As temperatures rise, some minerals melt and others remain solid. If melting is halted and conditions are maintained at any given temperature, melting ceases and the current mixture of solid rock and melt is maintained. The fraction of rock that has melted at a given temperature is a **partial melt.** To visualize a partial melt, you might think of how a chocolate chip cookie would look if you heated it to the point where the chocolate chips had melted while the main part of the cookie stayed solid.

The ratio of liquid to solid in a partial melt depends on the composition and melting temperatures of the original rocks and on the temperature at the depth in the crust or mantle where melting takes place. At the lower end of its melting range, a partial melt might be less than 1 percent of the volume of the original rock. Much of the hot rock would still be solid, but appreciable amounts of liquid would be present as small droplets in the tiny spaces between crystals throughout the mass. Many partial melts of basaltic magma in the upper mantle, for example, are estimated to be only 1 percent to 2 percent melt. At the high end of the melting temperature range, much of the rock would be liquid, with lesser amounts of unmelted crystals in it. This would be the case for a reservoir of a granitic magma and crystals just below a volcano.

Geologists seized on the new knowledge of partial melts to help them determine how different kinds of magma form at different temperatures and in different regions in Earth's interior. As you can imagine, the composition of a partial melt where only the minerals with the lowest melting points have melted may be significantly different from the composition of a completely melted rock. Thus basaltic magmas that formed in different regions in the mantle may have somewhat different compositions. From this observation we could deduce that the different magmas come from different proportions of partial melt.

PRESSURE AND MELTING To get the whole story on melting, we have to consider pressure, which increases with depth in the Earth as a result of the increased weight of overlying rock. Geologists found that as they melted rocks under various pressures, increasing pressure led to increasing melting temperatures. Thus rocks at melting temperatures at Earth's surface would remain solid at the same temperature in the interior, where pressures are high. For example, a rock that melted at 1000°C at Earth's surface might have a melting temperature much higher, perhaps 1300°C, at depths in the interior where pressures are many thousands of times greater than those at the surface. It is the pressure effect that explains why rocks in most of the crust and mantle do not melt. Only where composition and both temperature and pressure are right can rock melt. You will learn more about pressure and its effects on rocks in the Earth's interior in Chapter 8, "Metamorphic Rocks."

WATER AND MELTING The many experiments on melting temperatures and partial melting paid other dividends as well. One of these was a better understanding of the role of water in rock melting. Geologists knew from analyses of natural lavas that there was water in some magmas, so they added small amounts of water to the rocks they were melting. In doing so, they discovered that the compositions of partial and complete melts vary not only with temperature and pressure but also with the amount of water present. Consider, for example, the effect of water content on pure albite, the high-sodium plagioclase feldspar, at the low pressures of the Earth's surface.

If only a small amount of water is present, pure albite will remain solid at temperatures just over 1000°C, hundreds of degrees above the boiling point of water. At these high temperatures, the water in the albite is present as a vapor (gas). If large amounts of water are present, the melting temperature of the albite will be lower, dropping to as low as 800°C (see Figure 4.7). This behavior follows the general rule that dissolving some of one substance in another lowers the melting point of the solution. If you live in a cold climate, you are probably familiar with this process because you know that towns and municipalities sprinkle salt on icy roads to lower the melting point of the ice.

In the same way, the melting temperature of the albite—and of all the feldspars and other silicate minerals—drops considerably in the presence of large amounts of water. In this case, the melting points of various silicates are lowered in proportion to the amount of water dissolved in the molten silicate. This is an important point in our knowledge of melting rocks. Water content is an important factor in lowering the melting temperatures of mixtures of sedimentary and other rocks. Sedimentary rocks contain an especially high volume of water in their pore spaces, higher than that found in igneous or metamorphic rocks. As you will see later in this chapter, this water plays an important role in melting in Earth's interior.

The Formation of Magma Chambers

Most substances have a lower density in the liquid form than in the solid form. The density of a melted rock is lower than the density of a solid rock of the same composition—that is, a given volume of melt would weigh less than the same volume of solid rock. Geologists reasoned that large bodies of magma could form in the following way. If the less dense melt were given a chance to move, it would move upward, just as oil, which is less dense than water, rises to the surface of a mixture of oil and water. Being liquid, the partial melt could move slowly upward through pores and along the boundaries between crystals of the overlying rocks. As the hot drops moved upward, they would coalesce with other drops, gradually forming larger pools of molten rock within Earth's solid interior.

We now know that the large pools of molten rock envisioned by early geologists form **magma chambers**—magma-filled cavities in the lithosphere that form as ascending drops of melted rock push aside surrounding solid rock. Magma chambers may encompass a volume as large as several cubic kilometers. The exact manner in which they form is still the subject of ongoing research, and we cannot yet say exactly what magma chambers look like in three dimensions. We think of them as large liquid-filled cavities in solid rock, which expand as more of the surrounding rock melts or as liquid migrates in through cracks and other small openings between crystals. Magma chambers contract as they expel magma to the surface in eruptions. We know magma chambers exist because earthquake waves can show us the depth, size, and general outlines of magma chambers underlying some active volcanoes.

With this knowledge of how rocks could melt to form magma, we can better consider where different kinds of magma form at different temperatures and regions in Earth's interior.

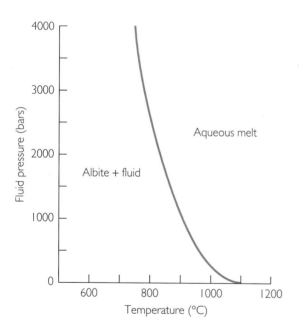

FIGURE 4.7 The melting curve for albite in the presence of water. Albite is a silicate mineral, and the melting temperature of silicate minerals drops considerably in the presence of large amounts of water. When only a small amount of water is present, pure albite remains solid at temperatures just over 1000°C; when large amounts of water are present, albite's melting point may drop as low as 800°C.

WHERE DOES MAGMA FORM?

We noted earlier in this chapter that our understanding of igneous processes stems from geological inferences as well as laboratory experimentation. Our inferences are based mainly on data from two sources. The first is volcanoes on land and under the sea—everywhere that molten rock erupts—which give us information about where magmas are located. The second source of data is temperatures recorded at deep drill holes and mine shafts in various locations, which show that the temperature of the interior of the Earth increases with depth. Using such figures, scientists have been able to estimate the rate at which temperature rises as depth increases.

Some temperatures recorded at various locations are much hotter than others at the same depth, an indication that some parts of Earth's mantle and crust are hotter than others. In tectonically and volcanically active areas, for example, temperature increases at an exceptionally rapid rate, reaching 1000°C at a depth of 40 km, not far below the base of the crust. This temperature is almost high enough to melt basalt. In tectonically stable regions, temperature rises much more slowly, reaching only 500°C at the same depth.

We now know that different kinds of rock can solidify from a magma through the process of partial melting. And we know that increasing temperatures in Earth's interior could cause magmas to form. But our picture is not complete; for more information, we turn again to the theory of plate tectonics.

Tectonic Activity, Rock Composition, and Types of Magmas

Laboratory experiments have established the temperatures and pressures at which different kinds of rock melt, and this information gives us some idea of where melting may take place. Mixtures of sedimentary rocks, for example, melt at temperatures several hundred degrees lower than the melting point of basalt. This information leads us to expect that basalt may start to melt near the base of the crust in tectonically active regions of the upper mantle and that sedimentary rocks melt at shallower depths than basalt. The geometry of plate motions is the link we need to tie tectonic activity and rock compositions to melting (Figure 4.8). Two types of plate boundaries are associated with magma formation: mid-ocean ridges, where the divergence of two

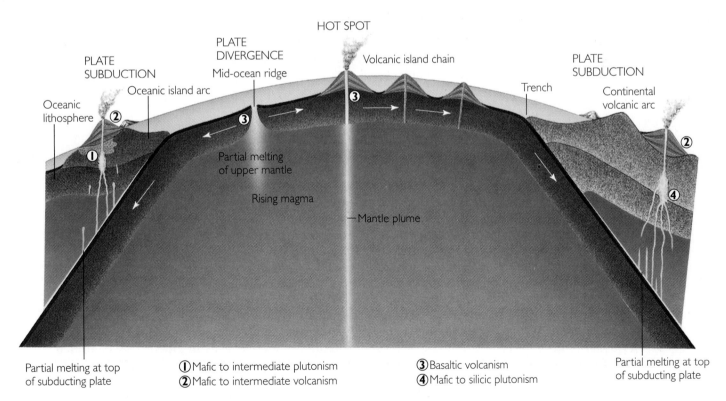

FIGURE 4.8 The three main types of magma—basaltic, andesitic, and rhyolitic—form under conditions that are strongly connected to movements of lithospheric plates. These movements control where rocks of the crust and upper mantle melt and whether they will be intruded or extruded.

plates causes the seafloor to spread, and subduction zones, where one plate dives beneath another.

Before proceeding, recall the terminology used to describe igneous rocks and remember that rocks can have identical compositions but different textures, depending on their intrusive (plutonic) or extrusive (volcanic) form. Geologists label magmas in the same way, using the names of corresponding igneous rock groups. The names most commonly used are those of the volcanic rocks, such as rhyolitic (the felsic group), andesitic (the intermediate group), and basaltic (the mafic group). We will use this terminology not only in this chapter but in Chapter 5 and throughout the text.

MID-OCEAN RIDGES   At mid-ocean ridges, heat in the form of rising convection currents in the mantle causes the formation of basaltic magma. Basaltic magma forms in the hot upper mantle below mid-ocean ridges; it then rises to collect in shallow, narrow, wedge-shaped magma chambers near the crest of the ridge. Tremendous quantities of basaltic magma flow intermittently from the rifts and fissures of mid-ocean ridges, giving rise to the abundant lavas of the seafloor.

SUBDUCTION ZONES   Other kinds of magmas underlie regions in which volcanoes are highly concentrated, such as the Andes Mountains of South America and the Aleutian Islands of Alaska. Both of these regions were generated by the subduction of one plate under another. The magmas of subduction zones form partly from a mixture of seafloor sediments and partly from basaltic and felsic crust. Sediments have some water remaining in pore spaces. In addition, shales, the most abundant sedimentary rocks, are very high in clay minerals, which contain much water chemically bound in their crystal structure. Sediments become very deeply buried as the subducting lithospheric plate moves down into the lower crust. At moderate depths of about 5 km, much of this water is released by chemical reactions as the temperature increases to about 150°C. Almost all of the remaining water is released at greater depths, from 10 to 20 km or deeper. As this water moves up from the top of the subducting slab, it promotes the melting of the mantle wedge of the plate overlying the subducting plate, and magmas of varying composition are formed.

The compositions of the sedimentary and basaltic and felsic materials that become part of the magma determine the types of igneous rock that can form. The igneous rocks of these subduction zones are generally more silicic than the basalts of mid-ocean ridges. They include much andesite and lesser amounts of more felsic volcanic rocks.

Deep in the crust, beneath the volcanoes, intrusive rocks of intermediate to silicic compositions— from diorite to granite—are formed at the same time that magmas erupt at Earth's surface. These intrusives are added to the base of the crust, thickening it by a process called *underplating*.

MANTLE PLUMES   Basalts similar to those produced at mid-ocean ridges are found in thick accumulations over some parts of continents distant from plate boundaries. In the states of Washington, Oregon, and Idaho, the Columbia and Snake rivers flow over a great area covered by this kind of basalt, which solidified from lavas that flowed out millions of years ago. Large quantities of basalt are also erupted in isolated volcanic islands far from plate boundaries, such as the Hawaiian Islands. In such places, slender, pencillike *plumes* of hot basaltic magma rise from deep in the mantle, perhaps as deep as the boundary of the core and the mantle. Mantle plumes, most of them far from plate boundaries, are the "hot spots" of the Earth and are responsible for the outpouring of huge quantities of basalt.

To sum up, basaltic magmas form in the upper mantle beneath mid-ocean ridges and in the lower mantle beneath intraplate hot spots. Magmas of varying composition form at subduction zones, depending on how much felsic material and water the rocks overlying the subduction zone contribute to the melt.

MAGMATIC DIFFERENTIATION

The processes we have been discussing all demonstrate how rocks melt to form magmas. But what accounts for the variety of igneous rocks? Do they arise from magmas of different chemical compositions made by the melting of different kinds of rocks? Or do some processes produce variety from an originally uniform parent material? By the early twentieth century, accumulating geological data on igneous rocks were leading to these questions. Again, the answers came from experiments, as geologists mixed chemical elements in proportions simulating those of natural igneous rocks and melted them in high-temperature furnaces. As the melts subsequently cooled and solidified, geologists carefully observed the temperatures at which crystals formed and they recorded the chemical compositions of those crystals. This research gave rise to the theory of **magmatic differentiation,** a process by which a uniform parent magma may lead to rocks of a variety of compositions. Magmatic differentiation occurs because different minerals crystallize at different temperatures. During the crystallization process the composition of the magma changes as it is depleted of the chemical elements withdrawn to make the crystallized minerals.

In a mirror image of partial melting, the first minerals to crystallize from the cooling molten rock were the last to melt in partial-melting experiments. As this initial crystallization withdrew chemical elements from the melt, the magma's composition changed. As cooling continued, the next minerals to crystallize were those that had melted in the same temperature range in melting experiments. Again the magma's chemical composition changed as different elements were withdrawn. Finally, at the temperature at which the magma solidified completely, the last minerals to crystallize were the ones that melted first when the rock was heated to the beginning of its melting temperature range.

In the course of many experiments, two patterns of crystallization emerged:

▪ *Continuous and gradual change* In this pattern, illustrated by the plagioclase feldspars, the composition of the successively formed feldspars changed continuously and gradually as crystallization proceeded.

▪ *Abrupt and discrete change* In the other pattern, characteristic of the mafic minerals such as olivine and pyroxene, the composition of the crystals changed discontinuously during cooling, one mineral abruptly changing to another at a particular temperature.

Because these crystallization patterns are so basic to an understanding of magmatic differentiation, we will examine them in more detail.

The Continuous Reaction Series

When melts of various plagioclase feldspar compositions were cooled, the first crystals to form were always richer in calcium than the melt. Their formation partially depleted the melt of calcium, so that the remaining melt became richer in sodium. As a result, when the melt continued to cool, the next crystals to form were more sodium-rich. At the same time, the calcium-rich crystals formed earlier reacted chemically with the now more sodium-rich

melt. In this reaction, calcium ions in the crystal were replaced by sodium ions from the melt so that the calcium-rich crystals formed earlier became richer in sodium. All crystals, both earlier and new, now had the same composition (Figure 4.9). As the process continued, both melt and crystals gradually became richer in sodium and poorer in calcium, until, when crystallization was complete, the final

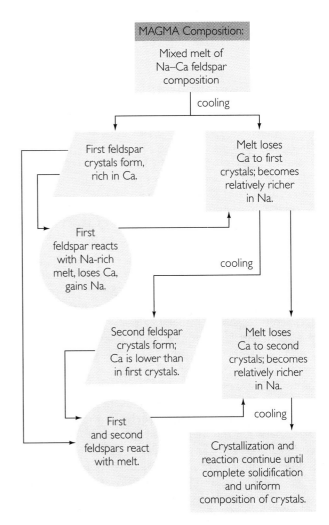

FIGURE 4.9 A continuous reaction series in plagioclase feldspar. As molten plagioclase feldspar cools, crystallization occurs. The crystals that form are always richer in calcium than the liquid they leave behind, which is correspondingly richer in sodium. In successive stages, liquid and crystals interact. Newly formed crystals and earlier crystals react with the liquid, so that at any given time all existing crystals, old and new, have the same composition. At the point of complete solidification, all crystals have reacted to attain a uniform composition identical to that of the original liquid.

homogeneous solid mass of crystals had arrived at the same composition as the original melt. At all times the mineral being crystallized was a plagioclase feldspar.

The key to this process is the continuous reaction of crystals with the melt. Each continuously changes by small amounts, so that at any point in the course of crystallization all crystals have the same composition. Crystals and melt move through a series of compositions, which in the earlier stages are richer in calcium and in later stages are richer in sodium. With continued cooling this **continuous reaction series** proceeds until crystallization is complete.

The Discontinuous Reaction Series

A somewhat different process is involved in the crystallization of mafic minerals, such as olivine, pyroxene, amphibole, and biotite mica. Experiments like those on plagioclase feldspars, in which melts made up of the components of mafic minerals were allowed to cool slowly so that crystals could react with the liquid, also showed a systematic order of crystallization. At 1800°C, olivine crystallized first; it continued to crystallize until the melt cooled to 1557°C. Below that temperature, pyroxene, a completely different mineral, abruptly formed, and all of the earlier olivine crystals were converted to pyroxene (Figure 4.10). At 1543°C, cristobalite, a high-temperature silica mineral, began to form, and pyroxene crystallization continued until solidification was complete. In other experiments with melts of different compositions, first amphibole and then biotite mica crystallized at temperatures successively lower than the olivine-pyroxene series. In this **discontinuous reaction series,** reactions take place between the melt and minerals of two definite compositions only at particular temperatures. This is a different process from the gradual evolution of plagioclase feldspars and parent melt over a continuous range of compositions and temperatures.

The crystal structures of the minerals of the two reaction series are part of the differences in crystallization patterns (see Table 4.1). Throughout the changes in the continuous reaction series, the basic feldspar crystal structure remains constant, although the proportions of calcium and sodium change. In contrast, the crystal structures of the discontinuous reaction series change as temperatures fall, forming increasingly complex arrangements of silica tetrahedra. At the highest temperatures, the crystal structure of olivine is that of isolated silica tetrahedra, the basic building blocks of silicate minerals (see Chapter 2).

In the next stage, the pyroxenes are single chains of tetrahedra. Then come the amphiboles, double chains of tetrahedra, followed by the micas, sheets of tetrahedra. At the end stages of both the continuous and discontinuous series, quartz and the feldspars are three-dimensional frameworks of silica tetrahedra.

In the cooling of a natural magma, which normally contains the chemical elements of both plagioclase feldspars and mafic minerals, both patterns of crystallization go on simultaneously. As the temperature of such a magma drops below 1550°C, for example, pyroxene crystals form through discontinuous reaction and a pure calcium plagioclase feldspar crystallizes through continuous reaction.

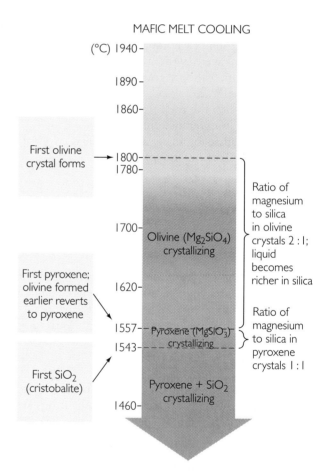

FIGURE 4.10 Part of a discontinuous reaction series. The sequence illustrated here occurs during the crystallization of a cooling liquid of magnesium and silica, in which silica is about 50 percent by weight. This is one part of a discontinuous reaction series by which different minerals successively crystallize from a melt in a series of abrupt changes. Olivine does not first crystallize at exactly 1800° in every melt; the exact temperature depends on the composition of the cooling melt.

Although these two reaction series explain the composition of many igneous rocks, they fall short of accounting for many others. Consider a natural magma in which all crystals have reacted completely with the liquid rock at all stages of crystallization. Given these conditions, we would expect to find only the final products of a crystallization—a single plagioclase feldspar corresponding to the composition of the original melt and a pyroxene. No traces of the first crystallization products—the original calcium-rich feldspar and the original olivine—would be left. Geologists examining volcanic rocks knew some part of the theory was missing when they found many such rocks with calcium-rich plagioclases and with olivine.

Fractional Crystallization

The theory of magmatic differentiation needed one more essential piece: a way to account for the preservation of minerals formed earlier as the composition of the melt changed. N. L. Bowen, a Canadian geologist, proposed such a mechanism about 75 years ago. Even as an undergraduate, Bowen had been interested in the chemical basis for igneous rock formation. Later he focused on the continuous and discontinuous crystallization series. He was especially interested in the course of crystallization in situations in which plagioclase feldspars—or mafic minerals—failed to change composition through reaction with the remaining liquid. Such might be the case, for example, if a magma cooled more rapidly than usual. In such a magma, plagioclase feldspar crystals would have time to grow, but only the outer surfaces of existing crystals would have enough time to react with the changing liquid. As a result, only the outer layer of each crystal would change composition. As crystallization proceeds, each successive layer of feldspar is covered by plagioclase that is progressively richer in sodium.

The end product would be what we now call a **zoned crystal,** a single crystal of one mineral that has a different chemical composition in its inner and outer parts (Figure 4.11). In Bowen's example, the crystals' compositions would change in a series of gradual steps from calcium-rich interiors to sodium-rich exteriors. There is more to the problem of limited reaction than the effects of rapid crystallization, however. If the calcium-rich centers of the growing crystals were encapsulated, the liquid would not reach a state of equilibrium with crystals, as it would during a slow continuous reaction. The liquid would remain rich in sodium because the calcium from the

FIGURE 4.11 A zoned crystal. The colors in this crystal correspond to differences in chemical composition in various zones within the crystal. Each zone represents a fractional crystallization stage, during which the earlier-formed inner zones did not react with the liquid. *(Chip Clark.)*

crystal interiors would not be available to replace the sodium in the melt.

Relying on both experimentation and field observation, Bowen proposed a new theory of magmatic differentiation. Though the mechanisms he suggested are no longer accepted for the differentiation of most igneous rocks, his ideas served as a foundation for most later work and still teach us much today. Bowen proposed a process by which the first-formed crystals would be segregated from the remaining melt. This segregation could happen in several ways. For example, crystal settling might cause early crystals formed in a magma chamber to settle to the chamber's floor and thus be removed from further reaction with the remaining liquid. Another possibility is that structural deformation during the crystallization process might squeeze the remaining liquid from the chamber, segregating and compressing the crystals as a distinct intrusive body. The magma would then migrate to new locations, forming new chambers.

By either scenario, settling of crystals or structural deformation, crystals that had formed early would be segregated from the remaining melt, which would then behave as though it had just begun to crystallize. In the continuous reaction series, for example, the melt, already enriched in sodium at that point, would start to crystallize a feldspar much richer in sodium than any that would have crystallized from an unsegregated magma. Continued crystallization would produce a mass of such feldspars

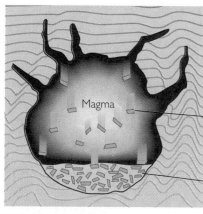

Crystals form from magma cooling and settle to floor of chamber

Crystals from early cooling accumulate

(a) Early crystallization

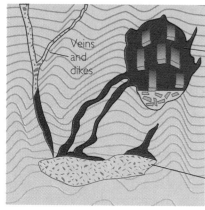

Magma migrates to secondary chamber, where it continues to crystallize

Mass of crystals formed early are segregated and compressed to form separate intrusive body

(b)   Later deformation squeezes remaining liquid from crystal mush

FIGURE 4.12 Two stages in the evolution of a magma differentiated by fractional crystallization. In the first stage (a), early crystals settle to the floor of the magma chamber. As cooling proceeds, structural deformation may squeeze the remaining liquid from the chamber, segregating and compressing those crystals to form a separate intrusive body (b). The separated liquid migrates to form veins, dikes, and other magma chambers in new locations, where it continues to crystallize.

that would be much richer in sodium than the rock that the original melt would have produced. Meanwhile, the segregated calcium-rich crystals that had formed first would form a mass of feldspar much richer in calcium than the original melt. **Fractional crystallization** is the term used for this separation and removal of successive fractions of crystals formed from a cooling magma (Figure 4.12). Bowen believed this process could account for the preservation of early-formed calcium-rich feldspars and the crystallization of sodium-rich plagioclases from an originally calcium-rich magma.

Bowen went on to show that fractional crystallization could work with the discontinuous mafic mineral series as well. Just as the first plagioclase feldspars to crystallize may be removed, the first-formed crystals of olivine in a discontinuous reaction series may settle out and so be removed from further reaction. These mafic minerals would be found with their corresponding plagioclase feldspars. With the first-formed olivine removed, the magma would then crystallize pyroxene. Thus both continuous and discontinuous crystallization paths might produce a range of products similar to those found in natural igneous rocks. But like any scientific theory, the theory of fractional crystallization had to be tested before it could be accepted.

From Laboratory to Field Observation: The Palisades Intrusion

A perfect test case for the theory of fractional crystallization was the Palisades, a massive cliff facing the city of New York on the west bank of the Hudson River. The Palisades is an igneous formation about 80 km long and in places more than 300 m from top to bottom. It was intruded as a melt of basaltic composition into almost horizontal sedimentary rocks, and it contains abundant olivine near the bottom, pyroxene and plagioclase feldspar in the middle, and mostly plagioclase feldspar near the top (Figure 4.13a). The variation in mineral composition from top to bottom made this formation a perfect site for testing Bowen's theory and showed how laboratory experiments could help explain field observations.

From experiments on the melting of rocks with about the same proportions of the various minerals found in the Palisades intrusion, geologists knew that the temperature of the melt had to have been about 1200°C. The parts of the magma within a few meters of the relatively cold upper and lower contacts of the surrounding sedimentary rocks cooled quickly, forming a fine-grained basalt and preserving the chemical composition of the original melt. But the hot interior of the intrusion cooled more slowly, as the slightly larger crystals testify.

Bowen's experiments on fractional crystallization lead us to think that the first mineral to crystallize from the slowly cooling interior would be olivine, a heavy mineral that would sink through the melt to the bottom of the intrusion. There it can be found today in a coarse-grained, olivine-rich layer just above the chilled, fine-grained basaltic layer along the bottom contact in the Palisades (Figure 4.13b). Continued cooling would produce pyroxene

(a)

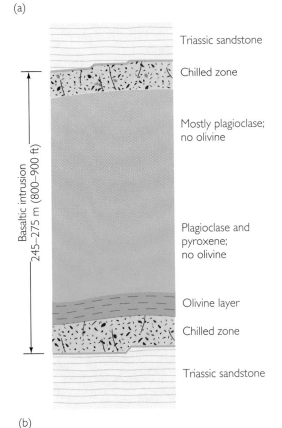

Triassic sandstone

Chilled zone

Mostly plagioclase;
no olivine

Plagioclase and
pyroxene;
no olivine

Olivine layer

Chilled zone

Triassic sandstone

Basaltic intrusion
245–275 m (800–900 ft)

(b)

FIGURE 4.13 The Palisades demonstrates fractional crystallization. (a) The Palisades is the result of an intrusion of basaltic magma into sedimentary rocks some 200 million years ago along what is now the Hudson River. *(Breck P. Kent.)* (b) In a classic application of laboratory science to field observation, geologists used the concept of fractional crystallization to interpret the Palisades' vertical variation in texture and mineral composition.

position until successive layers of settled crystals would be topped off by a layer of mostly plagioclase, sodium-rich feldspar crystals.

The explanation of the layering of the Palisades intrusion as the result of fractional crystallization was an early success of the first version of the theory of magmatic differentiation. It firmly tied field observations to laboratory experiments and was solidly based on chemical knowledge. More than two-thirds of a century of geological research has passed since the Palisades was first seen as a test case, and we now know that the Palisades has a more complex history, including several injections of magma and a more complicated story of olivine settling. Nevertheless, the general picture remains a valid instance of fractional crystallization.

Bowen's Theory of Magmatic Differentiation

If fractional crystallization and magmatic differentiation were to account for the differences among igneous rocks, they would have to explain two seemingly contradictory facts:

crystals, followed almost immediately by calcium-rich plagioclase feldspar. These minerals, too, had settled out through the magma to accumulate in the lower third of the Palisades intrusion. The abundance of plagioclase feldspar in the upper parts of the intrusion corresponded with the expectation that the melt would then continue to change com-

- The widespread distribution and abundance of granite. This intrusive rock is at the silicic end of the igneous rocks, and it contains abundant sodium-rich plagioclases and other minerals with low melting temperatures.

- The equally widespread distribution and abundance of basalt. Basalt, an extrusive rock, is at the mafic, or low silica, end of the spectrum; it contains calcium-rich plagioclase feldspars and other minerals with high melting temperatures.

Studies of the lavas of volcanoes showed that basaltic magmas are common, far more common than the rhyolitic magmas that correspond in composition to granites. How could the abundant granites have been derived from basaltic magmas?

Bowen's idea was that an originally basaltic magma would gradually cool and differentiate to a more silicic, lower-temperature melt by fractional crystallization. Early stages of differentiation of a basaltic magma by fractional crystallization would produce andesitic magma, which might erupt to form andesite lavas or solidify by slow crystallization to form diorite intrusives. Intermediate stages would make magmas of granodiorite composition. If this process were carried far enough, its late stages would form rhyolite lavas and granite intrusions.

Bowen's Reaction Series

In 1928 Bowen capped more than 10 years of active experimentation by proposing a simplified general scheme for magmatic differentiation that combined the continuous and discontinuous fractional crystallizations of the major minerals of igneous rocks. The **Bowen reaction series,** as it is called, starts with the cooling of a high-temperature basaltic magma that gradually differentiates by fractional crystallization along two paths simultaneously (Figure 4.14):

1. The continuous path of the plagioclase feldspars, starting at high temperatures with calcium-rich feldspar and proceeding to the lower-temperature sodium-rich feldspar.

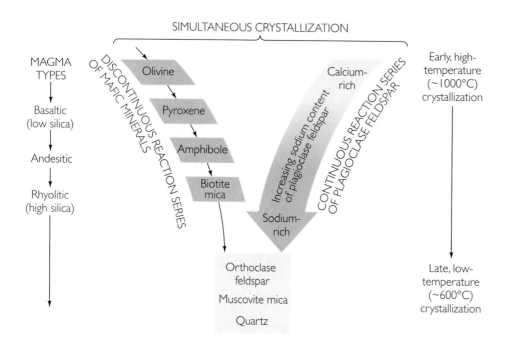

FIGURE 4.14 Bowen's reaction series. Bowen proposed this simplified general scheme to show how the sequence of fractional crystallization of a melt could lead to the formation of differentiated magmas. Two different but simultaneous paths would allow a variety of minerals to crystallize during the cooling of a high-temperature basaltic magma. Though this series does describe what happens to a hypothetical magma, it does not adequately explain many examples of igneous rock origin, such as the widespread intrusion of granite. Nevertheless, it remains a valuable chart for understanding the role of fractional crystallization.

2. The discontinuous path of the mafic minerals, starting at the high-temperature end with olivine, then progressing to pyroxene, amphibole, and biotite mica as the magma cools.

The paths of the two series converge at a final, low-temperature (about 600°C) magma crystallizing the minerals of granite: albite (sodium-rich plagioclase feldspar) and orthoclase (potassium feldspar), muscovite mica, and quartz.

To sum up the story thus far, a train of observations and experiments led to the Bowen reaction series as a possible explanation for the great variety of igneous rocks:

■ Melting and crystallization behavior of rocks and melts revealed the continuous and discontinuous reaction series.

■ Fractional crystallization explained how magmas could evolve through a series of compositions by the separation of the crystals formed earlier from the magma.

■ The continuous and discontinuous series were put together as an explanation for the evolution of granites and other intrusive and extrusive rocks from an originally basaltic magma.

Modern Theories since Bowen

At first Bowen's theory of magmatic differentiation seemed to be a great success. It explained well how different types of igneous rocks could form by fractional crystallization, and it provided an understanding of the kinds of rocks seen in the field, such as the Palisades mafic intrusion. Bowen's theory also seemed to explain how rhyolite, the extrusive equivalent of granite, could form toward the end of a series of eruptions that started with basaltic lavas.

As often happens when a scientific theory quickly dominates an area of science, further work showed the need to modify and add to Bowen's original theory of magmatic differentiation. One line of research showed that such great lengths of time would be needed for small crystals of olivine to settle through a dense, viscous magma that they might never reach the bottom of a magma chamber. Other researchers demonstrated that many layered intrusions similar to but much larger than the Palisades do not show the simple progression of layers predicted by Bowen's theory.

The biggest problem, however, was the source of granite. The first sticking point is that the great volume of granite found on Earth could not have been formed as Bowen's reaction series suggests. Large quantities of liquid volume are lost by crystallization during successive stages of differentiation. To produce the existing amount of granite, an initial volume of basaltic magma 10 times the size of a granitic intrusion would be required. That abundance would imply the crystallization of huge quantities of basalt underlying granite intrusions. But geologists could not find anything like that amount of basalt. Even where great volumes of basalts are found—at mid-ocean ridges—there is no wholesale conversion to granite through magmatic differentiation.

Most in question is Bowen's original idea that all granitic rocks evolve from the differentiation of a single type of magma, a basaltic melt. Instead, geologists discovered that melting of varied source rocks of the upper mantle and crust is responsible for much variation in magma composition, as follows:

1. Rocks in the upper mantle might partially melt to produce basaltic magma.

2. A mixture of sedimentary rocks and basaltic oceanic rocks such as those found in subduction zones might melt to form andesitic magma.

3. A melt of sedimentary, igneous, and metamorphic continental crustal rocks might produce granitic magma.

Magmatic differentiation does operate, but its mechanisms are more complex than Bowen recognized. Here are just a few of the points that have altered Bowen's original theory:

■ Partial melting is of great importance in the production of magmas of varying composition. Magmatic differentiation can be achieved by partial melting of mantle and crustal rocks over a range of temperatures and water contents. Thus basaltic magma can be formed by a 10 to 15 percent partial melting of rocks of the upper mantle at depths of around 100 km. An andesitic magma can be formed by partial melting of a water-rich basaltic oceanic crust that heats up as it descends along a subduction zone. A rhyolitic magma may be formed by partial melting in the lower crust of a mixture of continental crustal rocks or, alternatively, of andesite. In all of these cases one can apply

Bowen's reaction series in reverse to predict the composition of the magma as it is formed from partial melting.

- Magmas do not cool uniformly; they may exist at a wide range of temperatures within a magma chamber.

- The differences in temperature in magma chambers may cause the chemical composition of the magma to vary from one region to another.

- A few melt compositions are immiscible—they do not mix with one another, just as oil and water do not mix. When such magmas coexist in one magma chamber, each forms its own crystallization products.

- Some magmas that are miscible may give rise to a crystallization path different from that followed by any one magma alone.

We also now know more about the physical processes that interact with crystallization within magma chambers (Figure 4.15). Magma at various

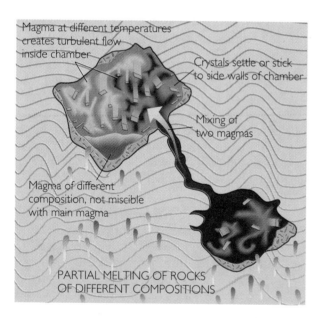

FIGURE **4.15** Modern ideas of magmatic differentiation. Geologists now recognize that magmatic differentiation does operate, but its mechanisms are more complex than Bowen recognized. Melting is usually partial. Some magmas derived from rocks of varying compositions may mix, while other magmas are immiscible. Crystals may be transported to various parts of the magma chamber by turbulent currents in the liquid.

temperatures in different parts of a magma chamber may flow turbulently, crystallizing as it circulates. Crystals may settle, then again be caught up in currents, and eventually be deposited on the chamber's walls. The margins of such a magma chamber may be a "mushy" zone of crystals and melt lying between the solid rock border of the chamber and the completely liquid magma within the main part of the chamber. And at some mid-ocean ridges, such as the East Pacific Rise, a mushroom-shaped magma chamber may be surrounded by hot basaltic rock with only small amounts (1 to 3 percent) of partial melt.

Bowen's original theory of magmatic differentiation has been supplanted since he proposed it many decades ago. Nevertheless, as we noted earlier, most later work on the differentiation of igneous rocks was built on the foundation of Bowen's ideas.

FORMS OF MAGMATIC INTRUSIONS

As we noted earlier, geologists cannot directly observe the forms that intrusive igneous rocks take when magmas intrude the crust. We can only deduce their shapes and distributions from evidence gained in geological fieldwork done millions of years after the rocks were formed, long after the magma cooled and the rocks were uplifted and exposed to erosion.

To be sure, we do have indirect evidence of current magmatic activity. Earthquake waves, for example, show us the general outlines of magma chambers that underlie some active volcanoes, but they cannot tell us the shape or size of intrusions supplied from those magma chambers. In some nonvolcanic but tectonically active regions, such as an area near the Salton Sea in southern California, high temperatures in deep drill holes reveal a crust much hotter than normal, which may be evidence of an intrusion at depth.

But in the end, most of what we know about intrusive igneous rock is based on the work of field geologists who have mapped and compared a wide variety of outcrops and have reconstructed their history. Their studies have resulted in the description and classification of the many irregular and variable forms of intrusive bodies. In the following pages, we consider some of these bodies—plutons, sills and dikes, and veins. Figure 4.16 illustrates a variety of extrusive and intrusive structures.

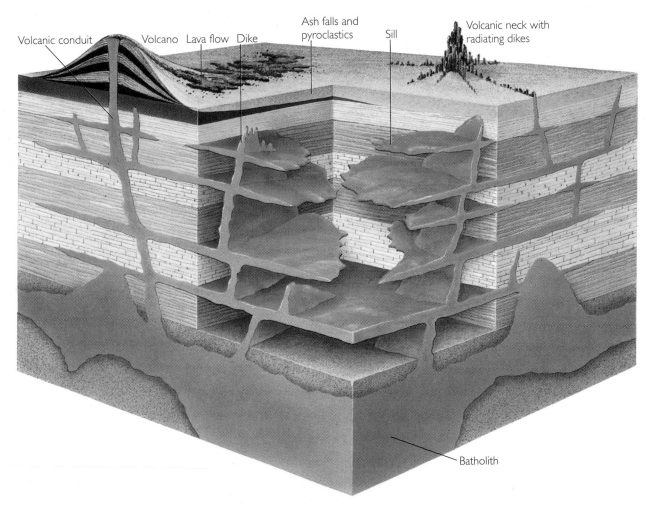

FIGURE **4.16** Basic extrusive and intrusive igneous structures. Notice that dikes cut across layers of country rock, but sills run parallel to them. Batholiths are the largest forms of plutons.

Plutons

Plutons are large igneous bodies that formed at depth in Earth's crust; they range in size from 1 km³ to hundreds of cubic kilometers. These large bodies become accessible to study when uplift and erosion uncover them or when mines or drill holes cut into them. Plutons are highly variable, not only in size but also in shape and in their relation to the surrounding country rock.

This wide variability in part reflects the different ways magma makes space for itself as it rises through the crust. Most magmas intrude at great depths—deeper than 8 to 10 km. At these depths, few holes or openings exist because the great pressure of the overlying rock would close them. But even that high pressure is overcome by the presence of the upwelling magma.

Magma rising through the crust makes space for itself in three ways:

■ *Wedging open the overlying rock* As the magma lifts that great weight, it fractures the rock, penetrates the cracks, wedges them open, and so flows into the rock. Overlying rocks may bow up during this process.

■ *Breaking off large blocks of rock* Magma can push its way upward by breaking off blocks of the invaded crust. These blocks sink into the magma, melt, and blend into the liquid, in some places changing the composition of the magma.

■ *Melting surrounding rock* Magma also makes its way along by melting walls of country rock.

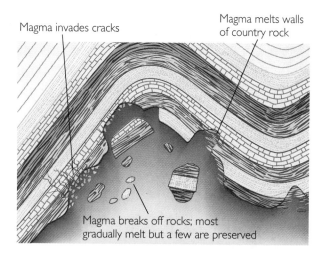

Magma invades cracks

Magma melts walls of country rock

Magma breaks off rocks; most gradually melt but a few are preserved

FIGURE 4.17 Magmas make their way into country rock in three basic ways: by invading cracks and wedging open overlying rock, by breaking off rock, and by melting surrounding rock. Here we see a magma intruding an area of surrounding folded rock.

Figure 4.17 illustrates these three methods of magmatic intrusion.

Most plutons show sharp contacts with country rock and other evidence of intrusion of a liquid magma into solid rock. Other plutons grade into country rock and show structures vaguely resembling those of sedimentary rocks. This latter set of features suggests that these plutons formed from preexisting sedimentary rocks that underwent granitization. **Granitization** is the process by which granite is formed from other rocks by recrystallization, with or without complete melting. (We discuss granitization in more detail in Chapter 8, "Metamorphic Rocks.") This type of pluton was converted to granite by partial melting and invasions by hot solutions and gases percolating up from great depths with the magma.

Batholiths, the largest plutons, are great irregular masses of coarse-grained igneous rock that by definition cover at least 100 km² (see Figure 4.16). Similar but smaller plutons are called **stocks.** Both batholiths and stocks are **discordant intrusions;** that is, they cut across the layers of the country rock they intrude.

Batholiths are found in the cores of tectonically deformed mountain belts. Geological field evidence is accumulating to show that batholiths are horizontal sheetlike or lobate thick bodies extending from a funnel-shaped central region. Their bottoms may extend 10 to 15 km deep, and a few are estimated to go even deeper. Batholiths' coarse grain is a consequence of slow cooling at great depths.

Sills and Dikes

Sills and dikes are similar to plutons in many ways, but they are smaller and they have a different relation to the layering of the surrounding intruded rock. A **sill** is a tabular sheetlike body that was formed by injection of magma between parallel layers of preexisting bedded rock (see Figure 4.16). Sills are **concordant intrusions**—that is, their boundaries lie parallel to these layers, whether or not the layers are horizontal. Sills range in thickness from a single centimeter to hundreds of meters, and they can extend over considerable areas. Figure 4.18

FIGURE 4.18 Dark sill formed by magma intruded between layers of sedimentary rock (light-colored bedded rock at top) in Big Bend National Park, Texas. The sill shows columnar jointing. *(Tom Bean.)*

shows a large sill at Big Bend National Park in Texas. The 300-meter-thick Palisades intrusion (see Figure 4.13) is another large sill.

Sills may superficially resemble layers of lava flows and pyroclastic material, but they differ from these layers in four ways:

■ They lack the ropy, blocky, and vesicle-filled structures that characterize many volcanic rocks.

■ They are more coarsely grained than volcanics because the sills have cooled more slowly.

■ Rocks above and below sills show the effects of heating; their color may have been changed, or they may have been mineralogically altered by contact metamorphism.

■ Many lava flows overlie weathered older flows or soils formed between successive flows; sills do not.

Dikes are the major route of magma transport in the crust. They are like sills in being tabular igneous bodies, but dikes cut across layers of bedding in country rock, rather than run parallel to them (see Figure 4.16). Dikes sometimes form by forcing open fractures that formed earlier, but they more often create channels through new cracks opened by the pressure of magmatic injection. Some individual dikes can be followed across country for tens of kilometers. Their widths vary from many meters to a few centimeters. In some dikes, one can see *xenoliths*—fragments of country rock completely surrounded by the intrusive material. These fragments, which once floated in the intruding magma, are good evidence of disruption of the surrounding rock during the intrusion process (Figure 4.19). Dikes rarely occur alone; more typically large numbers, or swarms, of hundreds or thousands of dikes are found in a region that has been deformed by a large igneous intrusion.

The texture of dikes and sills varies. Many are coarsely crystalline, with an appearance typical of

FIGURE 4.19 Dike of igneous rock (dark) intruded into shaly sedimentary rock (reddish brown) in Grand Canyon National Park, Arizona. *(Tom Bean/DRK.)*

intrusive rocks. Many others are finely grained and look much more like volcanic rocks. Because texture reflects rate of cooling, we know that the fine-grained dikes and sills invaded country rock nearer the Earth's surface, where rocks are cold in comparison with intrusions. Their fine texture is a result of fast cooling. The coarse-grained ones formed at depths of many kilometers and invaded warmer rocks whose temperatures were much closer to their own.

Veins

Veins are deposits of minerals found within a rock fracture that are foreign to the host rock. Irregular pencil- or sheet-shaped veins branch off from the tops and sides of many intrusive bodies. They may be a few millimeters to several meters across, and they tend to be tens of meters to kilometers long or wide. The famous Mother Lode of the Gold Rush of 1849 in California is a vein of quartz bearing crystals of gold. Veins of extremely coarse-grained granite cutting across a much finer-grained country rock are called **pegmatites** (Figure 4.20). They are crystallized from a water-rich magma in late stages of solidification. Pegmatites provide ores of many rare elements, such as lithium and beryllium.

Some veins are filled with minerals that contain large amounts of chemically bound water and that are known to crystallize from hot-water solutions.

FIGURE 4.20 A granite pegmatite dike. The center of the dike displays the coarse crystallinity associated with slow cooling. The finer crystals along the boundaries of the dike cooled more rapidly. *(Martin Miller.)*

From laboratory experiments, we know that these minerals crystallize at high temperatures—typically from 250 to 350°C—but not nearly as high as the temperatures of magmas. The solubility and composition of the minerals in these **hydrothermal** (from Greek *hydōr*, water, and *thermē*, heat) **veins** indicate that abundant water was present as the veins formed. Some of the water may have come from the magma itself, but some may be from underground water in the cracks and pore spaces of the intruded rocks. Groundwaters originate as rainwater seeps into the soil and surface rocks. Hydrothermal veins are abundant along mid-ocean ridges. In these areas, seawater infiltrates cracks in basalt and circulates down into hotter regions of the basalt ridge, emerging at hot vents on the seafloor in the rift valley between the spreading plates. (The geology of hydrothermal veins and the valuable ores they contain are discussed in detail in Chapter 23.)

IGNEOUS ACTIVITY AND PLATE TECTONICS

Since the advent of the theory of plate tectonics in the 1960s, geologists have been trying to fit the facts and theories of igneous rock formation into its framework (see Figure 4.8). We noted that batholiths, for example, are found in the cores of many mountain ranges. These mountain ranges were formed by the convergence of two plates. This observation implies a connection between plutonism and the mountain-making process and between both of them and the forces responsible for plate movements—plate tectonics.

The major sites of igneous rock formation are divergence zones—mid-ocean ridges. At these locations, basaltic magmas derived from partial melting of the mantle wells up along rising convection currents. Magma is extruded as lavas, fed by magma chambers below the ridge axis. At the same time, gabbroic intrusions are emplaced at depth (Figure 4.21).

Subduction zones, where one plate dives below another, are major sites of rock melting. The top of a subducting lithospheric plate includes oceanic crust, which is largely basalt formed originally at a mid-ocean ridge. The plate also carries water and still-soft oceanic sediment, which it accumulated during its travels from mid-ocean ridge to subduction zone. As the plate moves downward, it encounters increasing temperature and pressure, which convert the sediments first to sedimentary rocks and then at greater

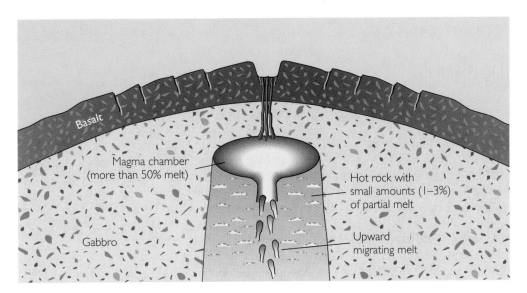

FIGURE 4.21 This diagrammatic cross section of a mid-ocean ridge shows a mushroom-shaped magma chamber below the ridge's central axis. Surrounding the chamber is a thick zone of hot rock containing small amounts of partial melt.

depths to metamorphic rocks. Because of the relatively large amounts of water they contain, these materials have lower melting temperatures than dry crustal or mantle rock. As the lithospheric plate moves deeper, temperatures rise, reaching the melting point of the sedimentary or metamorphic rocks. Continuing downward, the plate finally encounters temperatures that melt the top parts of the basalt. Subduction thus creates magma, or perhaps magmas of several kinds.

As the magmas and water from dehydration reactions work their way upward from the top of the melting subducting slab, they may melt portions of the overriding plate and change their composition. At the same time, the magmas may differentiate by fractional crystallization. The result is a range of igneous rocks, both intrusives and extrusives. Volcanoes over the deeper parts of the subduction zone, where melting is going on, extrude basaltic, andesitic, and rhyolitic lavas and pyroclasts, forming a wide range of volcanic rocks. These volcanoes and the volcanics they expel form the islands of oceanic volcanic arcs, such as the Aleutian Islands of Alaska.

Where subduction takes place beneath a continent, the many volcanoes and the volcanic rocks coalesce to form a mountainous arc on land. Subduction of an offshore oceanic plate has generated one such arc—the Cascade Range with its active volcanoes, such as Mount St. Helens—in northern California, Oregon, and Washington.

As mountains are forming above, intrusive magma bodies are crystallizing deep below, forming igneous rocks that vary from mafic to felsic, depending on the magma's composition and degree of differentiation. Working backward from the patterns and compositions of igneous rocks, we can estimate the composition of the parent magma and the depth of the descending slab. In so doing, geologists can reconstruct events that occurred millions of years ago in the subduction zone (see Feature 4.1).

The terrain of the Japanese Islands is a prime example of the complex of intrusives and extrusives that evolves over many millions of years at a subduction zone. Everywhere in this small country are all kinds of extrusive igneous rocks of various ages, intercalated with mafic and intermediate intrusives, metamorphosed volcanic rocks, and sedimentary rocks derived from erosion of the igneous rocks. The erosion of these kinds of rocks has contributed to the distinctive landscapes portrayed in so many classical and modern Japanese paintings.

In all of these ways, igneous rocks reflect the major forces shaping the Earth. Each plate-tectonic setting produces its own pattern of igneous rocks: the lava flows and pyroclastics extruded from volcanoes; the batholiths, dikes, and sills intruded at depth; and the wide variety of rocks that come from magmas of distinctive compositions following their own routes of differentiation.

4.1 INTERPRETING THE EARTH

Japan, a Growing Island Arc

Japan lies at the intersection of three converging oceanic plates, the giant Pacific and Eurasian plates and the small Philippine Plate. Just east of the Japanese Islands are deep trenches marking the lines along which the Pacific and Philippine plates subduct beneath the Eurasian Plate. As these plates slip downward, they provoke the earthquakes that are so prevalent throughout the islands. Japan is dotted with active and dormant volcanoes, the most famous of which, Fujiyama, is a traditional object of reverence.

Japanese geologists have worked out the history of this subduction complex by mapping the intrusive and extrusive rocks of various ages to show that originally the archipel-

ago was a narrow arc of small islands more like the Marianas of the western Pacific. As subduction continued, the volcanism, accompanied by intrusions at depth, built up a widening belt of land, while some of the igneous rocks emplaced earlier underwent structural deformation. In the course of this igneous and tectonic activity, mountains were elevated, one chain of them spectacular enough to be called the Japanese Alps. Thus the consequence of continued subduction has been the growth of sizable islands that through magmatic differentiation, structural deformation, and sedimentation have come to resemble tiny continents.

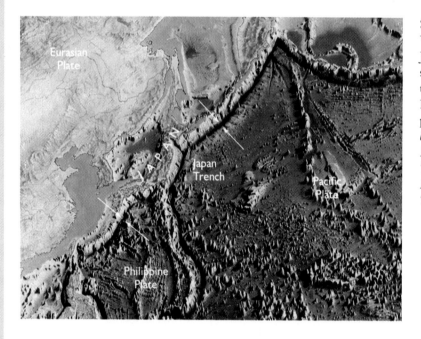

Seafloor topography of the Western Pacific shows the Japan Trench, part of the system of subduction zones that bound the Pacific, Philippine, and Eurasian plates. (*World Ocean Floor, based on bathymetric studies by Bruce C. Heezen and Marie Tharp. Painting by Heinrich C. Berann. Copyright © Marie Tharp, 1977.*)

SUMMARY

How are igneous rocks classified? All igneous rocks can be divided into two broad textural classes: (1) the coarsely crystalline rocks, which are intrusive and therefore cooled slowly, and (2) the finely crystalline ones, which are extrusive and cooled rapidly.

Within each of these broad categories, the rocks are classified on the chemical basis of their silica content or by the mineralogical equivalent, the proportions of lighter, felsic minerals and darker, mafic minerals. Felsic rocks, such as granite and its corresponding extrusive, rhyolite, are rich in silica and dominated by quartz, potassium feldspar, and sodium-rich plagioclase feldspar. Mafic rocks, such as gabbro and its corresponding extrusive, basalt, are poor in silica and consist primarily of pyroxene, olivine, and calcium-rich feldspar. Intermediate rocks are granodiorite and diorite and their corresponding extrusives, dacite and andesite.

How and where do magmas form? Magmas form at places in the lower crust and mantle where temperatures and pressures are high enough for at least partial melting of water-containing rock. Basalt can partially melt in the upper mantle where convection currents bring hot rock upward at mid-ocean ridges. Mixtures of basalt and other igneous rocks with sedimentary rocks, which contain significant quantities of water, have lower melting points than dry igneous rocks. These mixtures therefore melt when they are heated during subduction into the mantle.

How does magmatic differentiation account for the variety of igneous rocks? Minerals crystallize from magmas along two paths: (1) a continuous reaction series of the plagioclase feldspars and (2) a discontinuous reaction series of the mafic minerals. In these series, crystals continuously react with the melt through successive stages of crystallization and magma composition until they solidify completely, at which point the final product has the same composition as the original magma. If there is fractional crystallization, so that the crystals do not react with the melt, either because they grow very rapidly or because they are separated from the liquid, the final product may be more silicic than the earlier, more mafic crystals.

Bowen's continuous and discontinuous reaction series explain how fractional crystallization can produce mafic igneous rocks from earlier stages of crystallization and differentiation and felsic rocks from later stages, but Bowen's theory does not adequately explain the abundance of granite. Magmatic differentiation of basalt does not explain the composition and abundance of igneous rocks. Different kinds of igneous rocks may be produced by variations in the compositions of magmas caused by the melting of different mixtures of sedimentary and other rocks and by mixing of magmas.

What are the forms of intrusive and extrusive igneous rocks? Igneous bodies of large size are plutons. The largest plutons are batholiths, which are thick tabular masses with a central funnel. Stocks are smaller plutons. Less massive than plutons are sills, which are concordant, with the intruded rock, following its layering, and dikes, which are discordant with the layering, cutting across it. Hydrothermal veins are formed where water is abundant, either in the magma or in surrounding country rock.

How are igneous rocks related to plate tectonics? The two major sites of magmatic activity are mid-ocean ridges, where basalt wells up from the upper mantle, and subduction zones, where a series of differentiated magmas produces both extrusives and intrusives in island or continental volcanic arcs as the subducting oceanic lithosphere moves down into the deep crust and upper mantle. Large volumes of basalt are produced at oceanic islands and on landmasses that overlie mantle plumes.

KEY TERMS AND CONCEPTS

igneous rocks (p. 75)
intrusive igneous rock (p. 78)
country rock (p. 78)
extrusive igneous rock (p. 78)
pyroclast (p. 78)
volcanic ash (p. 78)
tuff (p. 78)
pumice (p. 78)
obsidian (p. 78)
porphyry (p. 79)
felsic rocks (p. 81)
granite (p. 81)
rhyolite (p. 81)
intermediate igneous rocks (p. 81)
granodiorite (p. 81)

diorite (p. 81)
dacite (p. 81)
andesite (p. 81)
mafic rocks (p. 81)
gabbro (p. 82)
basalt (p. 82)
ultramafic rocks (p. 82)
peridotite (p. 82)
viscosity (p. 82)
partial melting (p. 83)
partial melt (p. 83)
magma chamber (p. 84)
magmatic differentiation (p. 87)
continuous reaction series (p. 88)
discontinuous reaction series (p. 88)

zoned crystal (p. 89)
fractional crystallization (p. 90)
Bowen reaction series (p. 92)
pluton (p. 95)
granitization (p. 96)
batholith (p. 96)
stock (p. 96)
discordant intrusion (p. 96)
sill (p. 96)
concordant intrusion (p. 96)
dike (p. 96)
vein (p. 98)
pegmatite (p. 98)
hydrothermal vein (p. 98)

EXERCISES

1. Why are intrusive igneous rocks coarsely crystalline and extrusive rocks finely crystalline?

2. What kinds of minerals would you find in a mafic igneous rock?

3. What kinds of igneous rock contain quartz?

4. Name two intrusive igneous rocks with a higher silica content than that of gabbro.

5. What is the difference between the continuous and discontinuous reaction series?

6. How does fractional crystallization lead to magmatic differentiation?

7. Where in the crust, mantle, or core would you find a partial melt of basaltic composition?

8. How do you distinguish a sill from a dike?

9. What field geological evidence could you use to tell whether a basalt formation was a dike or a lava flow?

10. In which plate-tectonic settings would you expect magmas to form?

11. Why do melts migrate upward?

12. Where on the ocean floor would you find basaltic magmas being extruded?

THOUGHT QUESTIONS

1. How would you classify a coarse-grained igneous rock that contained about 12 percent quartz, 10 percent potassium feldspar, 35 percent plagioclase feldspar, and small amounts of biotite and amphibole?

2. What kind of rock would contain some plagioclase feldspar crystals about 5 mm long "floating" in a dark-gray matrix of crystals of less than 1 mm?

3. What differences in crystal size might you expect to find between two sills, one intruded at a depth of about 12 km, where the country rock was very hot, and the other at a depth of 0.5 km, where the country rock was moderately warm?

4. If you were to drill a hole through the crust of a volcanic island arc, what intrusive or extrusive igneous rocks might you expect to encounter at or near the surface? What intrusive or extrusive igneous rocks might you expect at the base of the crust?

5. Assume that a magma with a certain ratio of calcium to sodium starts to crystallize. If fractional crystallization occurs during the solidification process, will the plagioclase feldspars formed after complete crystallization have the same ratio of calcium to sodium that characterized the magma?

6. Would you expect to find a magma crystallizing olivine at the same stage of crystallization as a sodium-rich plagioclase feldspar? Why, or why not?

7. In what way is a zoned crystal evidence of fractional crystallization?

8. Why are plutons more likely than dikes to show the effects of magmatic crystallization?

9. What might be the origin of a rock composed almost entirely of olivine?

10. Are porphyries more likely to occur in rapidly cooled extrusive rocks or in very slowly cooled intrusive rocks?

11. What kinds of rock would you expect to find if you drilled a hole in a mid-ocean ridge?

12. Water is abundant in the sedimentary rocks and oceanic crust of subduction zones. How would the water affect melting in these zones?

SUGGESTED READINGS

Barker, D. S. 1983. *Igneous Rocks.* Englewood Cliffs, N.J.: Prentice Hall.

Best, M. G. 1982. *Igneous and Metamorphic Petrology.* San Francisco: W. H. Freeman.

Blatt, Harvey, and Robert J. Tracy. 1996. *Petrology: Igneous, Sedimentary, and Metamorphic,* 2nd ed. New York: W. H. Freeman.

Coffin, Millard F., and Olav Eldholm. 1993. Large igneous provinces. *Scientific American* (June), pp. 42–49.

Philpotts, Anthony R. 1990. *Igneous and Metamorphic Petrology.* Englewood Cliffs, N.J.: Prentice Hall.

Raymond, L. A. 1995. *Petrology.* Dubuque, Iowa: Wm. C. Brown.

INTERNET SOURCES

Igneous Rocks

ⓘ **http://www.science.ubc.ca/~geol202/igneous/igneous.html**

This URL for the University of British Columbia's Introduction to Petrology course (Geology 202) goes directly to the section on igneous rocks.

Granite

ⓘ **http://uts.cc.utexas.edu/~rmr/general.html#links**

Established by a graduate student at the University of Texas, this site includes abundant information about granite (especially in Texas) and links to a variety of other sites.

5

Lava stream flowing from a vent near the summit of Mount Etna, Sicily, in the eruption of January 1992. *(Roger Ressmeyer/Starlight.)*

Volcanism

I magine a volcanic eruption in which an area of ground the size of New York City collapses, a region larger than Vermont is buried under hot ash that snuffs out all life, and fields 1000 or 2000 km away are blanketed by 20 cm of ash and rendered infertile. Imagine that the volcanic dust thrown high into the stratosphere dims the Sun for a year or two, so that there are no summers. Unbelievable? Yet it has happened at least twice in what is now the United States: at Yellowstone in Wyoming 600,000 years ago and at Long Valley, California, 700,000 years ago. This was long before humans first reached North America, only 30,000 years ago, but not very long ago on the 4.5-billion-year geologic time scale. We know that these events took place because volcanic rocks formed by the eruptions have been identified.

A large portion of Earth's crust, oceanic as well as continental, is made up of volcanic rock.

Beginning as magma deep inside the Earth, volcanic rock serves, in a sense, as a window through which we can dimly perceive Earth's interior. This chapter examines **volcanism,** the process by which magma from the interior of the Earth rises through the crust, emerges onto the surface as lava, and cools into hard, volcanic rock. We will look at the major types of lava, eruptive styles, the landforms they create, and the kinds of environmental disruption that volcanoes can cause. We will see how plate tectonics can explain why the majority of volcanoes occur at plate boundaries but a few occur at "hot spots" within plates. Finally, we will discuss ways to control the destructive potential of volcanoes and benefit from their chemical riches and heat energy.

Ancient philosophers were awed by volcanoes and their fearsome eruptions of molten rock. In their efforts to explain volcanoes, they spun myths about a hellish, hot underworld below Earth's surface. The ancients had the right idea. Modern scientists, also seeking an explanation, see in volcanoes evidence of Earth's internal heat.

Temperature readings of rocks as far down as humans have drilled (about 10 km) show that the Earth does get hotter with depth. Inferring trends to even greater depths, geologists now believe that temperatures at the depths of the asthenosphere, which extends from about 75 to 250 km, reach 1100 to 1200°C—high enough for the rocks there to begin to melt. This is why geologists identify the asthenosphere as a main source of magma, the molten rock below Earth's surface that we call **lava** after it

erupts. Melting of sections of the solid lithosphere that rides above the asthenosphere may be another source. The magma rises buoyantly, that is, it floats up because the fraction that melts at these temperatures is less dense than the residual surrounding rocks. In effect, the denser surrounding rock exerts pressure on the melt squeezing it upward. In some places the melt may find a path to the surface through fractures in the lithosphere. In other places geologists believe the rising magma melts a path to the surface. Some of the magma eventually reaches the surface and erupts as lava. A **volcano** is a hill or mountain that forms from the accumulation of matter that erupts at the surface. Figure 5.1 is a simplified, generic diagram of the plumbing system of a volcano, which taps a pool of molten rock at depth and gives vent to it at the surface. Note the pipelike conduit through which the magma rises from the magma chamber. This shallow reservoir in the crust below the summit periodically fills with magma rising from below and empties to the surface in cycles of eruptions. Lava can also erupt from cracks on the flanks of the volcano.

Because lava is a sample of the Earth's interior, it is interesting to geologists. Unfortunately, it is not a perfect sample. Lava differs from magma at depth. It will have lost some gaseous constituents to the atmosphere or ocean as it erupted and may have gained or lost other chemical components on its way to the surface. Despite these differences, lava and other eruptive materials still provide us with important information clues to the chemical com-

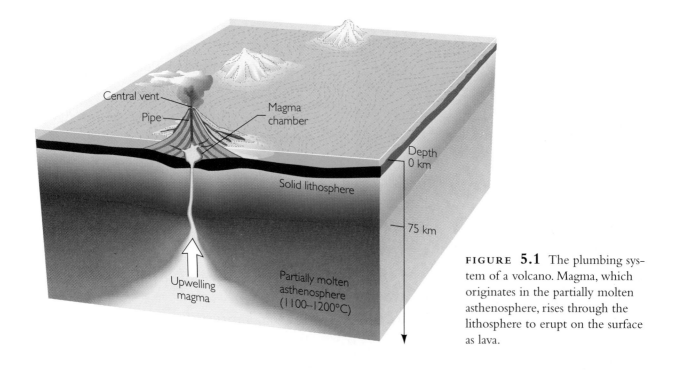

FIGURE 5.1 The plumbing system of a volcano. Magma, which originates in the partially molten asthenosphere, rises through the lithosphere to erupt on the surface as lava.

position and physical state of the upper mantle. Hardened into volcanic rock, these materials can also tell us something about eruptions that created them many thousands or millions of years ago. The chemical and mineralogical compositions of lava have much to do with the way it erupts and the kind of landform it will create when it solidifies.

VOLCANIC DEPOSITS

The major types of lavas and the rocks they form differ according to the magmas from which they were derived. As we saw in Chapter 4, igneous rocks and their precursor magmas are divided into three major groups—felsic, intermediate, and mafic—based on their chemical composition (see Table 4.2). The rocks are further classified as intrusive (they cooled below the surface and are coarse-grained as a result) or extrusive (they cooled on the surface and are finer-grained). The major intrusive rocks are granite (felsic), diorite (intermediate), and gabbro (mafic). The major extrusive counterparts are rhyolite (felsic) and the more common andesite (intermediate) and basalt (mafic). These classifications are summarized in Figure 4.6. With this framework in mind, let us examine the types of lava and how they flow and solidify.

Types of Lava

The several types of lavas leave behind different landforms—volcanic mountains that vary in shape and solidified lava flows that vary in character. These variations reflect differences in the chemical composition, gas content, and temperature of lavas. The higher the silica content and the lower the temperature, for example, the more viscous the lava is and the more slowly it flows. The more gas a lava contains, the more violent its eruption is likely to be.

BASALTIC LAVAS   Basaltic lava, dark in color, erupts at temperatures of 1000 to 1200°C—close to the temperature of the upper mantle. Because of its high temperature and low silica content, basaltic lava is extremely fluid and can flow downhill fast and far. Streams flowing as fast as 100 km per hour have been observed, although velocities of a few kilometers per hour are more common. In 1938 two daring Russian volcanologists measured temperatures and collected gas samples while they floated down a river of molten basalt on a raft of colder solidified lava. The surface temperature of their raft was 300°C, and the river had a temperature of 870°C. Lava streams that traveled more than 50 km from their source have been witnessed in historical times.

Basaltic lava flows vary according to the conditions under which they erupt. Important examples include:

- *Flood Basalts* Highly fluid basaltic lava that erupts on flat terrain, however, can spread out in thin sheets as a flood of lava. Successive flows often pile up into immense basaltic lava plateaus, called flood basalts, as at the great Columbia Plateau of Oregon and Washington (Figure 5.2).

- *Pahoehoe and Aa* Cooling basaltic lava flowing downhill falls into one of two categories,

FIGURE 5.2 View of Columbia Plateau, Washington. Successive flows of flood basalts piled up to build this immense plateau, which covers a large area of Washington and Oregon. *(Martin G. Miller.)*

FIGURE 5.3 Two types of lava, ropy pahoehoe (*bottom*) and jagged blocks of aa (*top*). Mauna Loa volcano, Hawaii. (*Kim Heacox/DRK.*)

face, having been exposed to cool air for more time, has developed a thicker outer layer.

▪ *Pillow Lavas* A geologist who comes across **pillow lavas**—piles of ellipsoidal, sacklike blocks of basalt about a meter wide—knows they formed in an underwater eruption (Figure 5.4) even if they are now on dry land. In fact, pillow lavas are an important indicator that a region was once under water. Geologist-divers have actually observed pillow lavas forming on the ocean floor off Hawaii. Tongues of molten basaltic lava develop a tough, plastic skin on contact with the cold ocean water. Because lava inside the skin cools more slowly, the pillow's interior develops a crystalline texture, while the quickly chilled skin solidifies to a crystalless glass.

RHYOLITIC LAVAS   Rhyolite, light in color and the most felsic lava, has a lower melting point than basalt and erupts at temperatures of 800 to 1000°C. It is much more viscous because of its lower temperature and higher silica content. Rhyolite moves 10 or more times more slowly than basalt, and because it resists flow, it tends to pile up in thick, bulbous deposits.

ANDESITE LAVAS   Andesite, with an intermediate silica content, has properties that fall between those of basalts and rhyolites.

according to its surface form: **pahoehoe** (pronounced pa-ho-ee-ho-ee) or **aa** (ah-ah). Figure 5.3 shows examples of both.

Pahoehoe (the word is Hawaiian for "ropy") forms when a highly fluid lava spreads in sheets and a thin, glassy elastic skin congeals on its surface as it cools. The skin is dragged and twisted into coiled folds that resemble rope as the molten liquid continues to flow below the surface (see Figure 5.3).

"Aa" is what the unwary exclaim after venturing barefoot onto lava that looks like clumps of moist, freshly plowed earth. Aa is lava that has lost its gases and consequently become more viscous than pahoehoe. It moves more slowly, allowing a thick skin to form. As the flow continues to move, the thick skin breaks into rough, jagged blocks (see Figure 5.3). The blocks ride on the viscous, massive interior, piling up a steep front of angular boulders that advances like a tractor tread. Aa is truly treacherous to cross. A good pair of boots may last about a week on it, and the traveler or geologist can count on cut knees and elbows.

A single downhill basaltic flow will commonly have the features of pahoehoe near its source, where the lava is still fluid and hot, and of aa farther downstream, where the flow's sur-

FIGURE 5.4 Pillow lava, characteristic of underwater volcanic eruptions, on the seafloor near the Galápagos Islands. Pillow lava found on land indicates earlier period of submergence. (*Woods Hole Oceanographic Institute.*)

Textures of Lavas

Lavas have other features that reflect the temperature and pressure conditions under which they formed. They can have a glassy or fine-grained texture if they cool quickly or a coarsely crystalline texture if they cool slowly beneath the surface. They can also have little bubbles, created when pressure falls suddenly as the lava rises and cools. Lava is typically charged with gas, like soda in an unopened bottle. When lava rises, the pressure on it decreases, just as the pressure on the soda drops when the bottle cap is removed. And just as the soda's carbon dioxide creates bubbles as it is released, water vapor and other dissolved gases escaping from lava create gas cavities, or *vesicles* (Figure 5.5). A frothy texture in solidified lava provides geologists with details of the rock's volcanic origins. One extremely vesicular, generally rhyolitic volcanic rock is pumice. Some pumice has so much void space that it is light enough to float.

Pyroclastic Deposits

Water and gases in magmas can have even more dramatic effects on eruptive styles. Before eruption, the confining pressure of the overlying rock keeps these volatiles from escaping. When the magma rises close to the surface and the pressure drops, the volatiles may be released with explosive force, shattering the lava and any overlying solidified rock into fragments of various sizes, shapes, and textures (Figure 5.6). Explosive eruptions are particularly likely with gas-rich, viscous rhyolitic and andesitic lavas.

VOLCANIC EJECTA Pyroclasts, as described in Chapter 4, are any fragmentary volcanic rock materials that are ejected into the air. These rocks, miner-

FIGURE 5.6 Pyroclastic eruption at Arenal volcano, Costa Rica. *(Gregory G. Dimijian/Photo Researchers.)*

als, and glasses are classified according to size. The finest fragments, less than 2 mm in diameter, are called **ash**. Fragments ejected as blobs of lava that become rounded and cooled in flight or chunks torn loose from previously solidified volcanic rock can be much larger (Figure 5.7). Volcanic ejecta as big as a house have been thrown more than 10 km in violent eruptions. Volcanic ash fine enough to stay

FIGURE 5.5 Vesicular basalt sample, approximately 1 ft across. *(Glenn Oliver/Visuals Unlimited.)*

FIGURE 5.7 Volcanologist Katia Krafft examines a volcanic bomb ejected from Asama volcano, Japan. *(Science Source/Photo Researchers.)*

FIGURE 5.8 Volcanic breccia. The view covers about 1 ft across. *(Doug Sokell/Visuals Unlimited.)*

aloft can be carried great distances. Within two weeks of the 1991 eruption of Mount Pinatubo in the Philippines, its volcanic dust was traced all the way around the world by orbiting Earth satellites.

Sooner or later pyroclasts fall, usually building deposits near their source. As they cool, hot sticky fragments become welded together (or lithified). The rocks created from smaller fragments are called **volcanic tuffs;** those formed from larger fragments are called **volcanic breccias** (Figure 5.8).

PYROCLASTIC FLOWS   One particularly spectacular and often devastating form of eruption occurs when hot ash, dust, and gases are ejected in a glowing cloud that rolls downhill at speeds of up to 200 km per hour. The solid particles are actually buoyed up by hot gases, so there is little frictional resistance to this incandescent **pyroclastic flow** (Figure 5.9).

In 1902 a pyroclastic flow with an internal temperature of 800°C exploded from the side of Mont Pelée, on the Caribbean island of Martinique, with very little warning. The avalanche of choking hot gas and glowing volcanic ash plunged down the slopes at a hurricane speed of 160 km per hour. In one minute and with hardly a sound, the searing emulsion of gas, ash, and dust enveloped the town of St. Pierre and killed 29,000 people. It is sobering to scientists who render advice to others to recall the statement of one Professor Landes, issued the day before the cataclysm: "The Montagne Pelée presents no more danger to the inhabitants of St. Pierre than does Vesuvius to those of Naples." Professor Landes perished with the others. French volcanologists Maurice and Katia Krafft, whose photographs appear in this chapter, were killed by a pyroclastic flow at Mount Unzen, Japan, in 1991.

FIGURE 5.9 A pyroclastic flow plunges down the slopes of Mount Unzen, in Japan, in June 1991. Note the fireman and fire engine in the foreground, trying to outrun the hot ash cloud descending on them. Three scientists who were studying this volcano died when they were engulfed by a similar flow. *(AP/Wide World Photos.)*

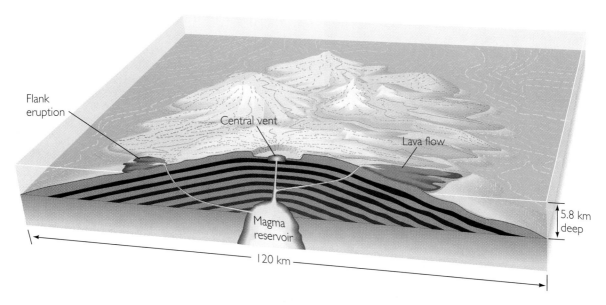

Flank
eruption

Central vent

Lava flow

Magma
reservoir

5.8 km
deep

120 km

FIGURE 5.10 A shield volcano is built up by the accumulation of thousands of thin basaltic lava flows that spread widely and cool as gently sloping sheets. Each layer shown in this schematic diagram represents an accumulation of many hundreds of thin lava flows. Magma can find alternate routes to the surface and erupt on the flanks of a volcano. Modeled after Mauna Loa, Hawaii.

ERUPTIVE STYLES AND LANDFORMS

Now that we have examined various types of volcanic materials that pour or blow out of Earth's interior, we can look more closely at styles of eruptions and the characteristic formations they leave behind. Eruptions do not always create a majestically symmetrical cone. The hundreds of thousands of square kilometers of monotonous layers of basalt that make up the Columbia Plateau of Washington and Oregon present another variant. Volcanic landforms vary in shape with the properties of the lava and the conditions under which it erupts.

Central Eruptions

Central eruptions create the most familiar of all volcanic features—the volcanic mountain shaped like a cone. These eruptions discharge lava or pyroclastic materials from a **central vent,** an opening atop a pipelike feeder channel rising from the magma chamber through which the material rises to erupt at the Earth's surface.

SHIELD VOLCANOES   A lava cone is built by successive flows of lava from a central vent. If the lava is

basaltic, it flows easily and spreads widely. If flows are copious and frequent, they create a broad, shield-shaped volcano many tens of kilometers in circumference and more than 2 km high. The slopes are relatively gentle. Mauna Loa, on Hawaii, is the classic example of a **shield volcano** (Figure 5.10). Although it rises only 4 km above sea level, it is actually the world's tallest structure: measured from its base on the seafloor, the volcano is 10 km high. It has a base diameter of 120 km—an area roughly three times that of Rhode Island. It grew to this enormous size by the accumulation of thousands of lava flows, each only a few meters thick, over a period of a few million years. In fact, the island of Hawaii actually consists of the tops of a series of overlapping active shield volcanoes emerging through the ocean surface.

VOLCANIC DOMES   In contrast to basaltic lavas, felsic lavas are so viscous that they can just barely flow. They usually produce a **volcanic dome,** a rounded, steep-sided mass of rock. Domes look as though lava had been squeezed out of a vent like toothpaste, with very little lateral spreading. Domes often plug vents, trapping gases. Pressures increase until an explosion occurs, blasting the dome into fragments. This occurred in the eruptions of Mount St. Helens in 1980 (Figure 5.11).

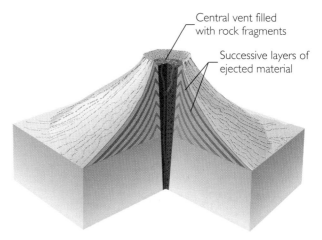

FIGURE 5.12 In a cinder cone, ejected material is deposited as layers that dip away from the crater at the summit. The vent beneath the crater is filled with fragmental debris.

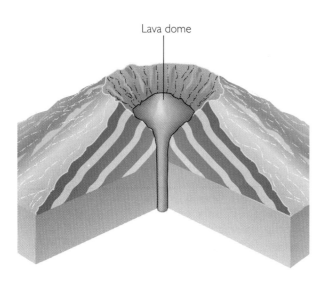

FIGURE 5.11 Volcanic domes are bulbous masses of felsic lava, which are so viscous that instead of flowing they pile up over the vent. Shown here is a growing dome within the crater of Mount St. Helens after the eruption. (*Lyn Topinka/USGS Cascades Volcano Observatory.*)

FIGURE 5.13 Cerro Negro in 1968. This volcano, near Managua, Nicaragua, is a cinder cone built on an older terrain of lava flows. (*Mark Hurd Aerial Surveys.*)

CINDER-CONE VOLCANOES When volcanic vents discharge pyroclasts, the solid fragments build up and form **cinder cones** (Figure 5.12). The profile of a cone is determined by the maximum angle at which the debris remains stable instead of sliding downhill (see Figure 11.1). The larger fragments, which fall near the summit, can form very steep but stable slopes. Finer particles are carried farther from the vent and form gentle slopes at the base of the cone. The classic concave-shaped volcanic cone with its summit vent reflects this variation in slope (Figure 5.13).

COMPOSITE VOLCANOES When a volcano emits lava as well as pyroclasts, alternating lava flows and beds of pyroclasts build a concave-shaped **composite volcano** or **stratovolcano** (Figure 5.14). This is the

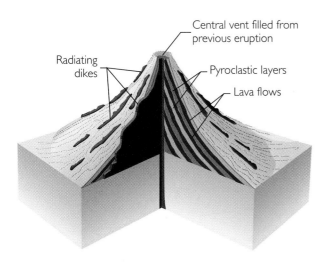

FIGURE 5.14 A composite volcano is built up of alternating layers of pyroclastic material and lava flows. Lava that has solidified in fissures forms riblike dikes that strengthen the cone. *(After R. G. Schmidt, USGS.)*

most common form of such large volcanoes as Fujiyama in Japan (Figure 5.15), Mounts Vesuvius and Etna in Italy, and Mount St. Helens in Washington.

CRATERS A bowl-shaped pit or **crater** is found at the summit of most volcanoes, centered on the vent. During the eruption of a lava volcano, the upwelling lava overflows the crater walls. When eruption ceases, the lava that remains in the crater often sinks back into the vent and solidifies. When the next eruption occurs, the material is literally blasted out of the crater in a pyroclastic explosion. The crater later becomes partially filled by the debris that falls back into it. Because a crater's walls are steep, they may cave in or become eroded in time. In this way the diameter of a crater can grow to several times that of the vent and hundreds of meters deep. The crater of Mount Etna in Sicily, for example, is presently 300 m (more than three football fields) in diameter and at least 850 m deep.

CALDERAS After a violent eruption in which large volumes of magma are discharged from a magma chamber a few kilometers below the vent, the empty chamber may no longer be able to support its roof. In such cases the overlying volcanic structure can

FIGURE 5.15 The composite volcano Fujiyama, Japan. *(Raga/The Stock Market.)*

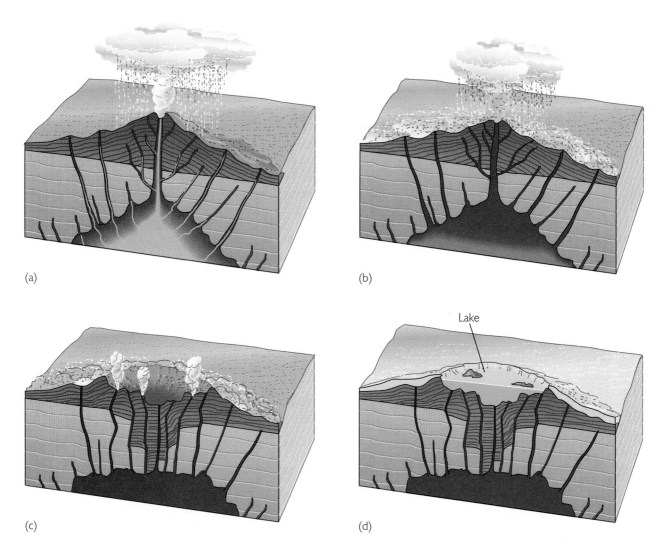

(a)

(b)

(c)

(d)

Lake

FIGURE 5.16 Stages in the evolution of a caldera. (a) Fresh magma fills a magma chamber and triggers a volcanic eruption of lava and columns of incandescent ash. (b) Eruption of lava and pyroclastic flows continue and the magma chamber becomes partially depleted. (c) Caldera results from the collapse of the mountain summit into empty chamber. Large pyroclastic flows ejected from fractures accompany the collapse, blanketing the caldera and the surrounding area. (d) A lake can form in the caldera. As the residual magma in the chamber cools, minor eruptive activity continues, with hot springs and gas emissions, and a small volcanic cone forms in the caldera.

collapse catastrophically, leaving a large, steep-walled, basin-shaped depression, much larger than the crater, called a **caldera** (Figure 5.16). Calderas are impressive features, ranging in size from a few kilometers to as much as 50 km or more in diameter, or about the area of greater New York City (Figure 5.17).

After some hundreds of thousands of years, fresh magma can reenter the collapsed magma chamber and reinflate it, forcing the caldera floor to dome upward again—perhaps to repeat the cycle of eruption, collapse, resurgence, eruption, and so on. This phenomenon is known as a *resurgent caldera*. The collapse

of large resurgent calderas is one of the most destructive natural phenomena on Earth. Yellowstone Caldera in Wyoming, marked today by a leftover relic, Old Faithful geyser, ejected some 1000 km^3 of pyroclastic debris during its eruptive stages about 600,000 years ago, more than a thousand times the amount of material ejected by Mount St. Helens in 1980. Ash deposits fell over much of what is now the United States. Other resurgent calderas are Valles Caldera in New Mexico, Long Valley Caldera in California, the still-active Kilauea and Rabaul (New Guinea) calderas, and the dormant Crater Lake in Oregon.

FIGURE **5.17** Crater Lake, Oregon, fills a caldera 8 km in diameter. The caldera is all that remains of an earlier composite volcano that was destroyed in the collapse that formed the caldera. *(Greg Vaughn / Tom Stack.)*

Monitoring caldera unrest is very important to geologists today because of the potential for large destruction. Fortunately, no catastrophic collapses with global consequences have occurred during human history, but geologists are wary of an increasing occurrence of small earthquakes in Yellowstone and Long Valley calderas and other indications of activity in the magma chambers in the crust below. For example, carbon dioxide leaking into soil from magma deeper in the crust has been killing trees since 1992 on Mammoth Mountain, a volcano on the boundary of Long Valley Caldera. A forest service ranger was almost asphyxiated by volcanic carbon dioxide leaking into a cabin on the flanks of Mammoth Mountain. Other indications of resurgence of the Long Valley Caldera are the occurrence of nearly continuous swarms of small earthquakes and uplift of the center of the caldera by about half a meter over the past 15 years. Rabaul Caldera showed uplift of about 6 meters in the 20 years before violent eruptions occurred in 1994. Following the eruption the caldera subsided by about 2 meters.

At one time it was thought that a caldera was formed by a huge explosion in which a volcano literally blew its top. However, geologic mapping of the debris and the pattern of faulting around calderas produces a picture more consistent with the collapse of the roof than with its ejection upward.

PHREATIC EXPLOSIONS When hot, gas-charged magma encounters groundwater or seawater, the vast quantities of superheated steam generated cause **phreatic,** or steam, **explosions** (Figure 5.18). One of the most destructive volcanic eruptions in history, that of Krakatoa in Indonesia, was a phreatic explosion (see Feature 5.1).

FIGURE **5.18** Phreatic explosion on Nisino-sima, a new volcano that rose above the sea in 1973 following a submarine eruption in the Pacific Ocean about 900 km south of Tokyo. *(Maritime Safety Agency, Japan.)*

5.1 INTERPRETING THE EARTH

The Explosion of Krakatoa

The 1883 explosion of the volcano Krakatoa, in the strait between Java and Sumatra, was one of the greatest ever witnessed. Now almost completely submerged, Krakatoa was then a small island formed from a group of volcanic cones in an ancient caldera. The caldera, 6 km across, was a remnant of a collapsed prehistoric andesitic stratovolcano. On August 27, after many smaller explosions, Krakatoa blew its top in a phreatic explosion with the energy of 100 million tons of TNT (5000 times greater than the nuclear explosion that destroyed Hiroshima). It is believed that much of the energy was provided by the violent expansion of hot steam after the walls of the volcano first ruptured, letting seawater into the magma chamber. The result can be viewed as the biggest steam-boiler explosion and the loudest noise in recorded history.

The explosion was heard in Australia, nearly 2000 km away. Volcanic ash fell over an area of some 700,000 km². Almost total darkness settled on Jakarta, 150 km away, when the dust blotted out the Sun's rays. Fine dust rose to the stratosphere and drifted around the Earth, lowering Earth's mean annual temperature a few degrees for the next year or so by blocking 13 percent of the Sun's light from reaching Earth. The explosion also generated a *tsunami*, or giant sea wave, that reached a height of almost 40 m, destroying 295 coastal towns as far as 80 km away and drowning 36,000 people. The tsunami was recorded on tide gauges as far away as the English Channel. After the eruption, most of Krakatoa disappeared, leaving in its place the current 300-m-deep water-covered basin.

Vent of Anak Krakatoa

Outline of 1883 caldera

KRAKATOA TODAY

5 km

Anak Krakatoa ("Child of Krakatoa") is a new volcano that is building up in the caldera left by the 1883 eruption. The new cone rose above sea level in 1928. Since then, successive eruptions have largely filled in the 1883 caldera and may eventually build a new island in roughly the same position as the original Krakatoa. (*Katia Krafft/Explorer.*)

DIATREMES Sometimes when hot matter from the deep interior escapes explosively, the vent and feeder channel below are left filled with breccia as the eruption wanes. The resulting structure is called a **diatreme.** Shiprock, which towers over the surrounding plain in New Mexico, is a diatreme exposed by the erosion of the sedimentary rocks through which it originally burst. To transcontinental air travelers, Shiprock looks like a gigantic black skyscraper in the red desert (Figure 5.19).

The eruptive mechanism that produces diatremes has been pieced together in great detail from the geologic record. The kinds of minerals and rocks found in some diatremes could have been formed only at great depths—100 km or so, well within the upper mantle. This observation indicates that diatremes are formed when gas-charged magmas melt their way upward, finally ejecting gases, lava fragments from the vent walls, and fragments from the deep crust and mantle, all with explosive energy and sometimes at supersonic speed. Such an eruption would probably look like the exhaust jet of a giant rocket upside down in the ground blowing rocks and gases into the air.

Another diatreme is encountered in the underground workings of the fabled Kimberly mines of South Africa, one of the world's richest sources of diamonds. This diatreme is a peridotite, an ultramafic rock composed mostly of the mineral olivine. It also contains diamonds, which form from carbon under the great pressures found in the mantle, and other scrambled fragments of mantle rock picked up by the magma en route to the surface. Geologists view this diatreme much as they would a 300-km drill core into the mantle. Its fragments provide our only direct evidence of mantle materials and so have been studied extensively. This is an example of the evidence that enables geologists to conclude that peridotite is a major constituent of the upper mantle.

Fissure Eruptions

Imagine basaltic lava flowing out of a crack in the Earth's surface tens of kilometers long, flooding vast areas. Such **fissure eruptions** have occurred on Earth countless times in the last four billion years (Figure 5.20). Among the geologically important fissure eruptions are those that occur along mid-ocean ridges. In recorded history humans have witnessed such an eruption only once, in 1783 on Iceland, which is an exposed segment of the Mid-Atlantic Ridge. One-fifth of the Icelandic population perished as a result. A fissure 32 km long opened and

FIGURE **5.19** Shiprock, towering 515 m above the surrounding flat-lying sediments of New Mexico, is a diatreme, or volcanic pipe, exposed by erosion of its enclosing sedimentary rock. *(Fred Padula.)*

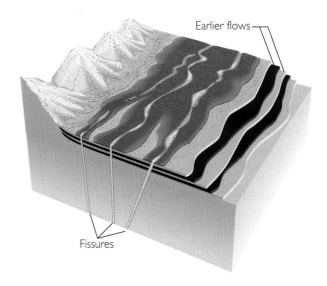

FIGURE **5.20** In a fissure eruption of highly fluid basalt, lava rapidly flows away from fissures and forms widespread layers, rather than building up into a volcanic mountain. *(After R. S. Fiske, USGS.)*

FIGURE 5.21 Volcanic cones along the Laki fissure (Iceland) that opened in 1783 and erupted the largest flow of lava on land in the course of human history. *(Tony Waltham.)*

spewed out some 12 km³ of basalt, enough to cover Manhattan about halfway up the Empire State Building (Figure 5.21). Fissure eruptions continue on Iceland, although on a smaller scale than the 1783 catastrophe.

FLOOD BASALTS (BASALTIC LAVA PLATEAUS)
The geologic record contains ample evidence of prehistoric basaltic flooding from great fissures. When **flood basalts** erupt from fissures, the lavas build a plain or accumulate as a plateau, rather than piling up as a volcanic mountain as they do when they erupt from a central vent. The flood basalts that made the Columbia Plateau (see Figure 5.2) buried 200,000 km² of preexisting topography (Figure 5.22). Some individual flows were more than 100 m thick, and some were so fluid that they spread more than 60 km from their source. An entirely new landscape

FIGURE 5.22 The area covered by the Columbia River flood basalts. *(After R. S. Fiske, USGS.)*

with new river valleys has since evolved atop the lava that buried the old surface. Plateaus made by flood basalts are found on every continent.

ASH-FLOW DEPOSITS Fissure eruptions of pyroclastic materials have produced extensive sheets of hard volcanic tuffs called **ash-flow deposits** (Figures 5.23 and 5.24). As far as is known, humans have never witnessed one of these spectacular events. The early Tertiary ash-flow deposits of the Great Basin in Nevada and adjacent states, formed in this way, cover an area of about 200,000 km² and are as much as 2500 m thick in some places. Yellowstone National

FIGURE 5.23 Welded tuff from an ash-flow deposit in the Sierra Nevada of California. Image covers an area of 2 ft × 3 ft. *(Gerald and Buff Corsi/Visuals Unlimited.)*

FIGURE **5.24** These ash-flow sheets on O-shima Island, Japan, were formed when intensely hot gas-charged volcanic dust, ash, and pumice spread swiftly over the surface to settle, cool, and harden. *(S. Aramaki.)*

Park, in Wyoming, has been covered by a number of ash-flow sheets (see Figure 5.24). A succession of forests there were buried by these deposits.

Other Volcanic Phenomena

LAHARS Among the most dangerous volcanic events are the torrential mudflows of wet volcanic debris that are called **lahars.** They can form when a pyroclastic flow meets a river or a snowbank; when the wall of a crater lake breaks, suddenly releasing water; when glacial ice is melted by a lava flow; or when heavy rainfall transforms new ash deposits into mudflows. One extensive layer of volcanic debris in the Sierra Nevada of California contains 8000 km³ of material of lahar origin, enough to cover all of Delaware with a deposit more than a kilometer thick. Lahars have been known to carry huge boulders for tens of kilometers. When Nevado del Ruiz in the Colombian Andes erupted in 1985, lahars triggered by the melting of glacial ice near the summit plunged down the slopes and buried the town of Armero 50 km away, killing more than 25,000 people.

EDIFICE COLLAPSE The cone or edifice of a volcano is constructed from thousands of deposits of lava or ash or both. This is not the way to build a solid structure. In recent years volcanologists have discovered many examples of catastrophic structural failures in which a big piece of the summit breaks

off, perhaps precipitated by an earthquake, and slides downhill in a massive, destructive landslide. Edifice failure occurs about four times a century. The collapse of one side of Mount St. Helens was the most damaging part of its 1980 eruption (look ahead to the photos in Feature 5.3, page 128). Surveys of the seafloor of Hawaii have discovered many giant landslides on the underwater flanks of the Hawaiian ridge. When they occurred, these massive movements would have triggered huge tsunami. Actually coral-bearing marine sediments have been found on one of the Hawaiian Islands some 300 m above sea level. These sediments were probably deposited by giant sea waves that were excited by a prehistoric volcanic collapse. It is worrisome that the southern flank of Kilauea volcano is advancing toward the sea at a rate of 25 cm per year. Should it break off and slide catastrophically into the ocean, it would prove disastrous for Hawaii, California, and all Pacific coastal areas.

VOLCANIC GASES The nature and origin of volcanic gases are of considerable interest and importance. It is thought that over geologic time these gases have created the oceans and the atmosphere and may even affect today's climate. Volcanic gases have been collected by courageous volcanologists and analyzed to determine their composition. Water vapor is the main constituent of volcanic gas (70 to 95 percent), followed by carbon dioxide, sulfur dioxide, and traces of nitrogen, hydrogen, carbon

FIGURE 5.25
Volcanologist Katia Krafft in a heatproof suit examining a lava flow on Kilauea volcano, Hawaii. *(Maurice Krafft/ Photo Researchers.)*

monoxide, sulfur, and chlorine. Every eruption releases enormous amounts of these gases (Figure 5.25). Some volcanic gas may come from deep within the Earth, making its way to the surface for the first time. Some may be recycled groundwater and ocean water, recycled atmospheric gas, or gas that has been trapped in earlier generations of rocks.

The relation between volcanic eruptions and changes in weather and climate is receiving increasing attention. For instance, the 1982 eruption of El Chichón in southern Mexico and the 1991 eruption of Mount Pinatubo injected sulfurous gases into the atmosphere, 10 km above the Earth. Through various chemical reactions the gases formed an aerosol (a fine airborne mist) representing tens of millions of metric tons of sulfuric acid. This aerosol partially blocked enough of the Sun's radiation from reaching Earth's surface to lower global temperatures for a year or two. The eruption of Mount Pinatubo, one of the largest explosive eruptions of the century, led to a global cooling of at least 0.5°C in 1992. For similar reasons, the debris lofted into the atmosphere during the 1815 eruption of Mount Tambora in Indonesia caused even greater cooling. The Northern Hemisphere suffered a very cold summer with snow storms in 1816. The drop in temperature and ash fallout caused widespread crop failures. More than 90,000 people perished as a result in that "year without a summer." This terrible year inspired Byron's gloomy poem "Darkness":

I had a dream, which was not all a dream.
The bright sun was extinguish'd, and the stars
Did wander darkling in the eternal space,
Rayless, and pathless, and the icy earth
Swung blind and blackening in the moonless air;
Morn came and went—and came, and brought
*   no day.*
And men forgot their passions in the dread
Of this their desolation; and all hearts
Were chill'd into a selfish prayer for light . . .

Chlorine emissions from Pinatubo have also hastened the loss of ozone in the atmosphere, nature's shield that protects humankind from the Sun's ultraviolet radiation.

FUMAROLES, HOT SPRINGS, AND GEYSERS

Volcanic activity does not cease when lava or pyroclastic materials cease to flow. For decades and in some cases centuries after a major eruption, volcanoes continue to emit gas fumes and steam through small vents called **fumaroles.** All of these emanations contain dissolved materials that precipitate on surrounding surfaces as the water evaporates or cools. Various sorts of encrusting deposits (such as travertine) are formed, including some that contain valuable minerals (Figure 5.26).

Circulating groundwater that reaches buried magma (which retains heat for hundreds of thousands of years) is heated and returned to the surface

FIGURE 5.26 Fumarole becoming encrusted with sulfur deposits on Sierra Negra volcano, Galápagos Islands. *(Christian Grzimek/Photo Researchers.)*

FIGURE 5.27 Strokkur geyser, in Iceland, throws a column of steam and superheated water 20 to 30 m into the air every few minutes. *(Simon Fraser/Photo Researchers.)*

as hot springs and geysers (Figure 5.27). A geyser is a hot-water fountain that spouts intermittently with great force, frequently accompanied by a thunderous roar. The best known geyser in the United States is Old Faithful in Yellowstone Park, which erupts about every 65 minutes, sending a jet of hot water as high as 60 m.

THE GLOBAL PATTERN OF VOLCANISM

Before the advent of plate-tectonics theory, geologists noted a concentration of volcanoes around the rim of the Pacific Ocean and nicknamed it the "Ring of Fire." We now know that the Ring of Fire coincides with plate boundaries. This correlation has been an important clue to the origin of volcanoes.

The 500 to 600 active volcanoes of the world are not randomly distributed, but show a definite pattern. About 80 percent are found at boundaries where plates converge, 15 percent where plates separate, and the remaining few within plates (Figure 5.28). In addition, Chapter 4 shows that lava compositions vary with plate-tectonic setting (see Figure 4.8). Can we weave these observations into a hypothesis that not only describes events but explains them as well? Thanks to the theory of plate tectonics, we can.

Spreading-Zone Volcanism

As we saw in Chapters 1 and 4, the seafloor is broken up by a worldwide system of rifts, along which plates separate and basalt erupts. The fissure between the separating plates extends down to the asthenosphere. Basaltic magmas, partial melts of the hot, ultramafic rock in the asthenosphere (see pages 83–86 and Figure 4.8), rise buoyantly in the gap between the separating plates and overflow the fissure to form ocean ridges, volcanoes, basaltic seafloor crust. Enormous amounts of basalt have poured out of this world-encircling system of cracks. In the past 200 million years, enough magma has been released to lay down the crust of all the present seafloor.

Much of the volcanic heat on the seafloor is removed when cold seawater circulates in the fissures of the ocean-ridge volcanic system. Seawater that has been heated and enriched in dissolved minerals by its contact with magmas forms extremely hot

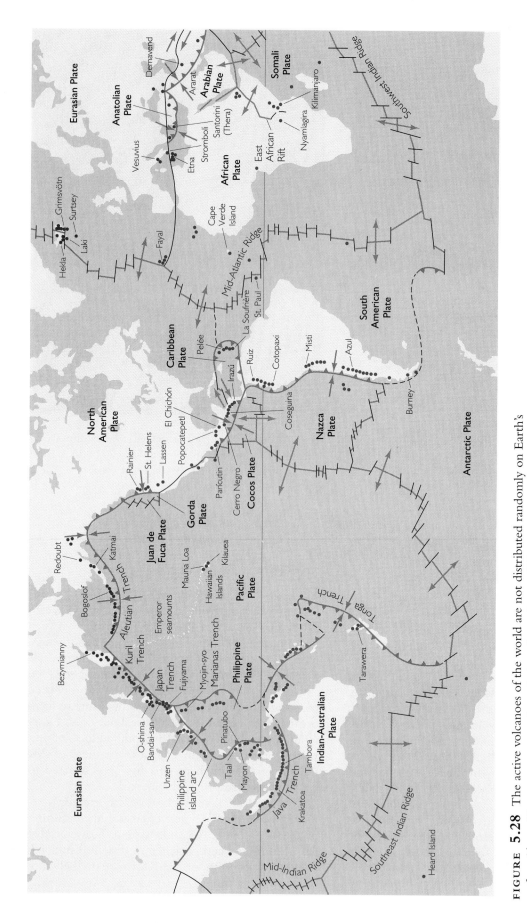

FIGURE 5.28 The active volcanoes of the world are not distributed randomly on Earth's surface. About 80 percent occur at boundaries where plates collide, 15 percent where plates separate, and the remaining few at intraplate hot spots. Convergent boundaries are shown in blue, divergent boundaries in orange. Black lines are transform faults. Active volcanoes are marked by red dots.

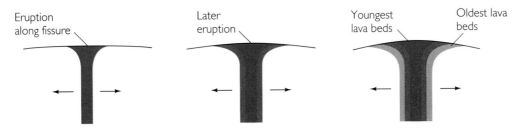

Eruption along fissure   Later eruption   Youngest lava beds   Oldest lava beds

FIGURE 5.29 Iceland is an exposed part of the Mid-Atlantic Ridge. Repeated fissure eruptions and lateral spreading are the mechanisms of its growth.

(350°C) springs and smoking vents along the cracks. These sites are a major source of minerals including ores of zinc, copper, and iron (see Feature 17.3).

Iceland, one of the few exposed segments of the Mid-Atlantic Ridge, provides an unmatched opportunity to view the process of fissure eruption and seafloor spreading directly (see Figure 1.15). The island is composed mostly of basalt. Iceland is in a state of tension, literally being pulled apart—one half is moving eastward in the Eurasian Plate, the other westward in the North American Plate. Tensional forces cause fissures to develop, and magma flows in from below and overflows onto the surface. At the conclusion of each episode, the lava solidifies to form a nearly vertical dike in the fissure and nearly horizontal beds on the adjacent surfaces. With each new episode of lateral spreading, a new crack opens and another flow pours out over the old one (Figure 5.29). Iceland thus grows by repeated eruptions, primarily from long fissures but also from localized vents. Although the details may differ under water, it is likely that the seafloor crust grows in a similar fashion.

Convergence-Zone Volcanism

Scientists are now sorting out the many phenomena that occur where plates converge and subduction occurs. One of the most striking features is the chain of volcanoes that parallels convergent boundaries, whether their type is ocean-ocean or ocean-continent (Figure 5.30). Magmas that feed convergence-zone volcanoes are more varied than the basalts of mid-ocean ridge volcanism and range from mafic to felsic, that is, from basalts to andesites to rhyolites. Subduction provides several mechanisms to explain these observations.

PLATE SUBDUCTION AND CONVERGENCE-ZONE MAGMAS   Because water lowers the melting temperature of rock (Chapter 4), water from the veneer of seafloor sediments atop the subducted slab can induce melting in the hot mantle above it. This process provides the basaltic magmas that feed convergence-zone volcanoes (see pages 85–86). In addition to basaltic magmas that derive from partial melting of the mantle, intermediate and more felsic magmas would form if sources were available to contribute silica and the other elements to the melt (Chapter 4; see Table 4.2 and Figure 4.8). In subduction zones, where the initially cool subducted slab heats up as it plunges into the hot mantle, such materials would derive from melting of the seafloor sediments and ocean crust atop the subducted ocean slab. Also, magmas rising from the mantle can invade and partially melt the felsic crust of an overriding continental plate. This process accounts for the more felsic kinds of magmas that feed volcanoes of an overriding continental plate margin. Thus the association of volcanism with subduction and the generation of magmas of different types is predicted by plate-tectonics theory.

VOLCANISM AT OCEAN–OCEAN CONVERGENCE   In the case of ocean–ocean convergence, an arc of volcanic islands builds up from the seafloor of the overriding plate, mostly by the extrusion of basalts, occasionally andesites, and rarely rhyolites. The basalts probably derive from the asthenosphere above the descending plate. The more silicic andesites would occur when elements are added, derived from varying degrees of partial melting of the basaltic crust and the ocean-bottom sediments attached to the descending plate (as described in Chapter 4). The creation of the Aleutian and Mariana island arcs are prototypes of this process. The collision of two

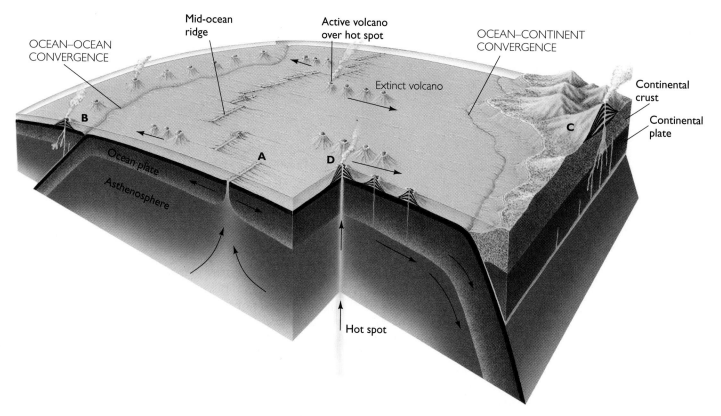

FIGURE 5.30 Volcanism is associated with plate tectonics. Plate separation at a mid-ocean ridge and partial melting of upwelling ultramafic mantle result in basaltic volcanism (A). At ocean–ocean convergent boundaries, magmas originating from partial melting of the mantle give rise to volcanic island arcs erupting mostly basaltic lavas (B). Magmas formed at ocean–continent convergences are mixtures of basalts from the mantle, remelted felsic continental crust, and materials melted off the top of the subducted plate. They give rise to volcanoes erupting andesitic and rhyolitic lavas (C). Plate motion over hot spots accounts for the creation of mid-plate chains of basaltic volcanic islands (D).

ocean plates is also responsible for Fujiyama, an andesitic volcanic cone, and Mount Pelée on the island of Martinique, from which viscous felsic lavas and explosive pyroclastic flows erupt.

VOLCANISM AT OCEAN–CONTINENT CONVERGENCE

When a plate carrying a continent on its leading edge overrides an ocean plate, an arcuate volcanic mountain chain grows in the zone of collision near the continental margin (see Figure 5.30). The Andes mark the convergence boundary between South America and the Nazca Plate. Farther north, the subduction of the small Juan de Fuca Plate under the North American Plate gives rise to the volcanoes of the Cascade Range, which stretch from northern

California to British Columbia. Mount St. Helens (see Feature 5.3, page 128) is among them. During a typical volcanic eruption, large quantities of ash and of andesitic lavas, and more rarely rhyolite, are ejected. The source is probably a mixture of basaltic magma rising from the mantle and remelted felsic continental crust through which it passes. Materials melted off the subducted slab could also contribute to the mixture.

Intraplate Volcanism

For many years, volcanism far from plate boundaries posed a problem for the theory of plate tectonics: it seemed to be an exception to the neat correlation of volcanism and plate boundaries. Take the Hawaiian Islands, in the middle of the Pacific Plate. This island chain begins with the active volcanoes on Hawaii and continues as a string of progressively older, extinct, eroded, and submerged volcanic ridges and mountains. Frequent large earthquakes do not occur along the Hawaiian chain—that is, it is essentially aseismic (without earthquakes)—so it is called an **aseismic ridge.**

Aseismic ridges of volcanic origin, which also occur elsewhere in the Pacific and in the other large oceans, were difficult to incorporate into the plate-tectonics framework until the concept of **hot spots** was introduced. Hot spots have also been proposed to explain some forms of volcanism within continents, far from plate boundaries. Yellowstone is an example. According to this hypothesis, illustrated in Figure 5.30, hot spots are the volcanic manifestations of jets or plumes of hot solid material that rise from deep within the mantle (perhaps even from the core-mantle boundary). When the plume reaches the lower pressures of shallow depths it begins to melt. The magma penetrates the lithosphere, and erupts at the surface. These columnar currents are thought to be fixed in the mantle and not to move with the lithospheric plates. As a result, the hot spot leaves a trail of extinct, progressively older volcanoes as the plate moves over it. The tracks of the extinct volcanoes that constitute the Hawaiian Islands and the Emperor seamount chain (shown in Figure 5.28) trace the motion of the Pacific Plate over a hot spot marked by the active volcanoes on Hawaii. The bend in the chain records a change in the direction of plate motion. (The satellite map of the seafloor in Feature 17.3 shows this hot spot trail clearly.)

If hot spots are indeed fixed in the mantle, the trail of volcanoes carried away from the hot spot provides a powerful method of measuring the velocity of plate motion. We could calculate the velocity, for example, by dividing the distance an extinct volcano has traveled from its hot spot by the age of that volcano's youngest volcanic rock. Data from deep-sea drilling along the Hawaiian and Emperor seamount chains have provided evidence consistent with the hypothesis that they originated over hot spots. Drill core samples confirm that the farther islands in these chains are from an active hot spot, the older they are. Calculations tell us that the plates are moving several centimeters a year and that the change in direction of plate motion occurred 40 million years ago.

The origin of fissure eruptions of basalt on continents, such as those that formed the Columbia River Plateau, and even larger lava plateaus in Brazil–Paraguay, India, and Siberia, is a subject of debate. Some geologists suggest that because of the immense amount of lava released (well over a million cubic kilometers in some cases) superplumes rising from deep within the mantle are the sources of the fissure eruptions that build the great continental lava plateaus. Others propose that fractures (of unknown origin) penetrated the continental lithosphere and that basaltic lavas, which represent partial melts of the underlying mantle, spurted rapidly to the surface without much contamination from the felsic crust. The volcanism that covered much of Siberia with lava is of special interest because it occurred at the end of the Permian some 250 million years ago and may have caused the greatest mass extinction of species in the geological record (see Chapter 24).

Fissure eruptions that mark the initial stages of continental rifting and the opening of a new ocean can be documented in several parts of the world. For example, basalt is found in the rift valleys of East Africa (see Figure 5.28)—a feature that some geologists interpret as signs of a breakup of Africa that was never completed.

VOLCANISM AND HUMAN AFFAIRS

Of the many volcanoes that have affected communities worldwide, one made a particularly powerful impact on Western civilization. Teams of archeologists and marine geologists have pieced together the story of the demise of Thera (formerly Santorini), a volcanic island in the Aegean Sea. The eruption of Thera in about 1623 B.C. appears to have been far more violent than that of Krakatoa. The center of the island collapsed, forming a caldera visible today as a lagoon about 60 km in circumference and as much as 500 m deep, with two small active volcanoes in the center. The lagoon is rimmed by two crescent-shaped islands known for their wine exports and scenic beauty and still subject to destructive earthquakes. The resultant volcanic debris and tsunami from this ancient catastrophe destroyed dozens of coastal settlements over a large part of the eastern Mediterranean. Some scientists have attributed the mysterious disappearance of the Minoan civilization to this cataclysm. And it is thought that the legend of the lost continent of Atlantis may have its origin in the land collapse that accompanied the eruption. The course of history was probably changed by this one volcanic event. It could happen again.

The date is known from a layer of volcanic ash transported to Greenland from Santorini. The ash layer was found in 1994 in an ice core extracted from deep within the Greenland ice sheet. Annual cycles of snow deposition can be counted in a core as rings in a tree are counted to determine its age. The ice core goes back 7000 years and provides evidence of some 400 ancient volcanic eruptions.

Can volcanic eruptions be predicted? To some extent they can (Feature 5.2)—fortunately for us all, because there are about 100 high-risk volcanoes in the world, and some 50 erupt each year. Certainly with our growing understanding of volcanism we can improve the terrible record of the past. Over the last 500 years alone some 200,000 people have been killed by volcanic eruptions.

Reducing the Risks of Hazardous Volcanoes

Of Earth's 500 to 600 active volcanoes, one out of six has claimed human lives. Volcanoes kill people and damage property by edifice collapse, explosive blasts, ash falls, lethal gas release, lava flows, and mudflows or lahars.

Scientists monitoring Mount St. Helens (Feature 5.3) and Mount Pinatubo were able to issue warnings of imminent major eruptions (Figure 5.31).

Government infrastructures were in place to evaluate the warnings and to issue and enforce evacuation orders. In the case of Pinatubo, the warning was issued a few days before the cataclysmic eruption on May 17, 1991. A quarter of a million people were evacuated, including some 16,000 residents of the nearby U.S. Clark Air Force Base (since permanently abandoned). Tens of thousands of lives were saved from the lahars that destroyed everything in their path. Casualties were limited to the few who disregarded the order. In 1994, 30,000 residents of Rabaul, Papua New Guinea, were successfully evacuated by land and sea hours before two volcanoes on either side of the town erupted, destroying or damaging most of it. Many owe their lives to the government for conducting evacuation drills and to scientists at the local volcano observatory who issued a warning when their seismographs recorded the ground tremor that signaled magma moving toward the surface.

Contrast these success stories with the tragedy at Nevado del Ruiz in Colombia in 1985. Scientists knew this volcano to be dangerous and were prepared to issue warnings, but no evacuation procedure was in place. As a result, 25,000 lives were lost in the lahars triggered by a minor eruption.

Volcanology has progressed to the point that we can identify the world's dangerous volcanoes and characterize their potential hazards from deposits laid down in earlier eruptions. These hazard assess-

FIGURE 5.31 Scientists carried by helicopter into the caldera of Mount Pinatubo, Philippines, collecting samples of gas, water, and volcanic debris. *(Roger Ressmeyer/ Corbis.)*

5.2 TECHNOLOGY AND EARTH

Kilauea: Monitoring a Volcano

The giant shield volcano Mauna Loa and the smaller Kilauea on its eastern flank make up the southern half of the island of Hawaii. Because the U.S. Geological Survey operates a volcano observatory on the rim of Kilauea Caldera, this volcano, which is actively erupting, is perhaps the best studied in the world. What has been learned from it has profoundly influenced our notions of volcanic processes.

A modern network of instruments and laboratory facilities is used to track the movement of magma within the volcano and the changing chemistry of the erupting lavas and gases. Seismographs, which measure movements within the Earth, detect and locate the small earthquakes that are often correlated with movements of magma. Seismographs can locate earthquakes as deep as 55 km beneath Kilauea. Such quakes often mark the entrance of magma into the channels leading from the asthenosphere through the lithosphere to the Earth's surface. The upward migration of the magma can be followed over a period of months because seismic disturbances occur progressively nearer the surface as the magma

rises. Tiltmeters, which measure tilting of the ground, indicate when the volcano begins to swell as the rising magma fills a magma chamber not far below the summit.

The first sign that an outbreak of lava is imminent is a swarm of small earthquakes, thousands of them, signifying that the magma is splitting rock as it forces its way to the surface. Very often, geologists know where the eruption will occur from the location of the earthquakes and changes in their pattern. In January 1960, for example, U.S. Geological Survey scientists detected a swarm of earthquakes not far from the village of Kapoho, on the flank of Kilauea. As they expected, an eruption broke out, destroying Kapoho but causing no casualties because the village had been evacuated. A new landscape was created as the lava flowed to the sea. Twenty-foot walls were built in a futile attempt to divert the lava and save a seashore community. When it was all over, the tiltmeters showed that the volcano had deflated, signifying that the magma chamber below had been drained in the Kapoho eruption. The cycle is repeated every few years.

Destruction of property engulfed by lava flow in the May 1990 eruption of Kilauea volcano, Hawaii. (*James Cachero/Sygma.*)

5.3 INTERPRETING THE EARTH

Mount St. Helens: Dangerous but Predictable

Long before Mount St. Helens erupted in the spring of 1980, geologists knew it to be the most active and explosive volcano in the contiguous United States. They could piece together a 4500-year history of destructive lava flows, hot pyroclastic flows, lahars, and distant ash falls by examining the geologic record. Monitoring efforts were intense. Beginning on March 20, 1980, a series of small to moderate earthquakes under the volcano signaled the start of a new eruptive phase after 123 years of dormancy. The earthquakes moved the U.S. Geological Survey (USGS) to issue a formal hazard alert. The first outburst of ash and steam erupted from a newly opened crater on the summit one week later. In April the seismic tremors increased, indicating that magma was moving beneath the summit, and an ominous swelling of the northeastern flank was noticed. The USGS issued a more serious warning, and people were ordered out of the vicinity.

On May 18 the climactic eruption began abruptly. A large earthquake apparently triggered the collapse of the north side of the mountain, loosening a massive landslide, the largest in recorded world history. As a huge flow of debris plummeted down the mountain, gas and steam under high pressure were released in a tremendous lateral blast that blew out the northern flank of the mountain. USGS geologist David A. Johnston was monitoring the volcano from his observation post 8 km to the north. He must have seen the advancing blast wave before he radioed his last message: "Vancouver, Vancouver, this is it!" A northward-directed jet of superheated (500°C) ash, gas, and steam roared out of the breach with hurricane force, devastating a zone 20 km outward from the volcano and 30 km wide. A vertical eruption sent an ash plume 25 km into the sky, twice as high as a commercial jet flies. The ash cloud drifted to

Mount St. Helens before and after the cataclysmic eruptions of May 1980.
(*Before: Emil Muench/Photo Researchers. After: David Weintraub/Photo Researchers.*)

ments can be used to guide zoning regulations to restrict land use—the most effective measure to reduce casualties. Instrumented monitoring (as described in Feature 5.2) can detect signals such as earthquakes, swelling of the volcano, and gas emissions that warn of impending eruptions. People at risk can be evacuated if the authorities are organized

and prepared. Volcanic eruptions cannot be prevented, but their catastrophic effects can be significantly reduced by a combination of science and enlightened public policy.

As a part of a United Nations program—the International Decade of Natural Disaster Reduction—leading volcanologists have designated 15 active

the east and northeast with the prevailing winds, bringing darkness at noon to an area 250 km to the east and depositing ash up to 10 cm deep on much of Washington, northern Idaho, and western Montana. The energy of the blast was equivalent to about 25 million tons of TNT. The volcano's summit was destroyed, its elevation reduced by 400 m, its northern flank disappeared. In effect, the mountain was "hollowed out."

Local devastation was spectacular. Within an inner blast zone extending 10 km, the thick forest was denuded and buried under several meters of pyroclastic debris. Beyond this zone, out to 20 km, trees were stripped of their branches and blown over like broken matchsticks aligned radially away from the volcano. As far as 26 km away, the hot blast was so intense that it overturned a truck and melted its plastic parts. Some fishermen were severely burned and survived only by jumping into a river. More than 60 other people were killed by the blast and its effects.

A lahar formed when the landslide and pyroclastic debris—fluidized by groundwater, melted snow, and glacial ice—flowed 28 km down the valley of the Toutle River. The valley bottom was filled to a depth of 60 m. Beyond this debris pile, muddy water flowed into the Columbia River, where sediments clogged the ship channel and stranded many vessels in Portland. Mount St. Helens may go on erupting for 20 years or more until its present episode of activity comes to an end.

Although much of the devastated area is still barren, after a decade almost 20 percent of the surface showed evidence of revegetation: native species of grass, legumes, and young trees are beginning to come back.

volcanoes for special study for the period 1995–2000 (Table 5.1). Many of these are among the most dangerous in the world. The goals of the project are to understand all volcanoes better, to evaluate the hazards posed by these particular volcanoes, to improve the prediction of explosive eruptions, and to prepare the public for crises that may develop.

TABLE **5.1**

15 VOLCANOES CHOSEN BY THE UNITED NATIONS FOR STUDY, 1995–2000

VOLCANO	COUNTRY
Colima	Mexico
Etna	Italy
Galeras	Colombia
Mauna Loa	United States
Merapi	Indonesia
Niragongo	Zaire
Rainier	United States
Sakurajima	Japan
Santa Maria/Santiaguito	Guatemala
Santorini	Greece
Taal	Philippines
Teide	Spain (Canary Islands)
Ulawun	Papua New Guinea
Unzen	Japan
Vesuvius	Italy

Volcanologists who study active volcanoes are motivated to understand them so that warnings of eruptions can be made more reliable. Their work is hazardous—nine have lost their lives since 1991. In one case a team of seven volcanologists gathering data in the crater of Galeras volcano in the Colombian Andes was caught in a pyroclastic eruption. Six of the seven members lost their lives in the ash and incandescent boulders that exploded from the crater. The lone survivor, Professor Stanley Williams of Arizona State University, is now at work developing an instrument that can analyze volcanic gases in the crater of an active volcano and transmit this information to scientists at a safe distance. An impending eruption might be predicted from the kinds of gases that are emitted.

Can volcanic eruptions be controlled? Not likely, although in special circumstances and on a small scale the damage can be reduced. Perhaps the most successful attempt to control volcanic activity was made on the Icelandic island of Heimaey in January 1973. By spraying advancing lava with seawater, Icelanders cooled and slowed the flow, preventing the lava from blocking the port entrance and saving some homes from destruction.

In the years ahead, the best policy for protecting the public will be the establishment of more warning and evacuation systems and more rigorous restriction of settlements in potentially dangerous locations. But even these precautions may not help. Dormant or long-extinct volcanoes can come to life suddenly—as Vesuvius and St. Helens did after hundreds of years. (Some potentially dangerous volcanoes in the United States and Canada are identified in Figure 5.32.) An even more difficult problem of prediction is posed by eruptions like that of Paricutín, which rose up with little warning from a small hole in a Mexican cornfield in 1943. Whole towns were quickly buried by ash and lava as the new volcano grew by repeated eruptions. Learning how to sense the movements of deep lava in relation to possible new outlets to the surface is a real challenge for geologists.

Reaping the Benefits of Volcanoes

We have seen something of the beauty of volcanoes and also something of their destructiveness. Volcanoes contribute to our well-being in many ways. In Chapter 1 we mentioned that the atmosphere and the oceans may have originated in volcanic episodes of the distant past. Soils derived from volcanic materials are exceptionally fertile because of the mineral nutrients they contain. Volcanic rock, gases, and steam are also sources of important industrial materials and chemicals, such as pumice, boric acid, ammonia, sulfur, carbon dioxide, and some metals. Seawater circulating through fissures in the ocean-ridge volcanic system is a major factor in the formation of ores.

Thermal energy from volcanism is being harnessed in more and more places. Most of the houses

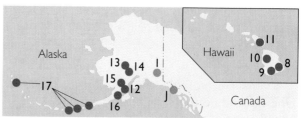

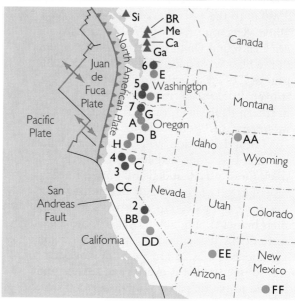

● U.S. volcanoes that have short-term eruption periodicities (100–200 years or less), or have erupted in the past 200–300 years, or both:

Cascades	Hawaii	Alaska
1 Mount St. Helens	8 Kilauea	12 Augustine volcano
2 Mono-Inyo craters	9 Mauna Loa	13 Redoubt volcano
3 Lassen Peak	10 Hualalai	14 Mount Spurr
4 Mount Shasta	11 Haleakala	15 Iliamna volcano
5 Mount Rainier		16 Katmai volcano
6 Mount Baker		17 Aleutian volcanoes
7 Mount Hood		

● U.S. volcanoes that appear to have eruption periodicities of 1000 years or greater and last erupted 1000 years or more ago:

Cascades	Alaska
A Three Sisters	I Mount Wrangell
B Newberry volcano	J Mount Edgecumbe
C Medicine Lake volcano	
D Crater Lake (Mount Mazama)	
E Glacier Peak	
F Mount Adams	
G Mount Jefferson	
H Mount McLoughlin	

● U.S. volcanoes that last erupted more than 10,000 years ago, but beneath which exist large, shallow bodies of magma that are capable of producing exceedingly destructive eruptions:

AA Yellowstone Caldera	DD Coso volcanoes
BB Long Valley Caldera	EE San Francisco Peak
CC Clear Lake volcanoes	FF Socorro

▲ Danger classifications are not available for Canadian volcanoes:

Si Silverthrone	BR Bridge River	Me Meagher Mountain
Ca Mount Cayley	Ga Mount Garibaldi	

FIGURE 5.32 Locations of potentially hazardous volcanoes in the United States and Canada. Volcanoes within each U.S. group are listed in the order of declining probable cause for concern, subject to revision as studies progress (danger classifications are not available for Canadian volcanoes). Note the relationship between the Cascade volcanoes, which extend from northern California to British Columbia, and the subduction plate boundary between the North American Plate and the Juan de Fuca Plate. (After R. A. Bailey, P. R. Beauchemin, F. P. Kapinos, and D. W. Klick, USGS.)

FIGURE 5.33
Crops growing on fertile volcanic soils in the Canary Islands. *(Cesar Lucas/ The Image Bank.)*

in Reykjavík, Iceland, are heated by hot water tapped from volcanic springs. Geothermal steam, originating in water heated by contact with hot volcanic rocks below the surface, is exploited as a source of energy for the production of electricity in Italy, New Zealand, the United States, Mexico, Japan, and the former Soviet Union. Figure 5.33 shows the agricultural benefits of volcanic soils.

SUMMARY

Why does volcanism occur? Volcanism occurs when molten rock inside the Earth rises buoyantly to the surface because it is less dense than surrounding rock. In effect, the melt is squeezed up by the weight of the overlying layers.

What are the three major categories of lava? Lavas are classified as felsic (rhyolite), intermediate (andesite), or mafic (basalt), on the basis of the decreasing amounts of silica and the increasing amounts of magnesium and iron they contain. The chemical composition and gas content of lava are important factors in the form an eruption takes.

How are the structure and terrain of a volcano related to the kind of lava it emits and the style of its eruption? Basalt can be highly fluid. On continents it can erupt from fissures and flow out in thin sheets to build a lava plateau. A shield volcano grows from repeated eruptions of basalt from vents. Silicic magma is more viscous and, when charged with gas, tends to erupt explosively. The resultant pyroclastic debris may pile up into a cinder cone or cover an extensive area with ash-flow sheets. A stratovolcano is built of alternating layers of lava flows and pyroclastic deposits. The rapid ejection of magma from a magma chamber a few kilometers below the surface, followed by collapse of the chamber's roof, results in a large surface depression, or caldera. Giant resurgent calderas are among the most destructive natural cataclysms.

How is volcanism related to plate tectonics? The ocean crust forms from basaltic magma that rises from the asthenosphere into fissures of the ocean ridge–rift system where plates separate. Basaltic, andesitic (intermediate), and rhyolitic (felsic) lavas tend to erupt in convergent zones. The basalts derive from partial melting of the mantle above the subducted plate, induced by water streaming off subducted seafloor sediments. Basalts are typical of volcanic islands found at ocean–ocean plate convergences. Andesites and rhyolites are more commonly found in the volcanic belts of ocean–continent convergent plate boundaries. The addition to basaltic magma of silica and other elements derived from remelting of felsic continental crust or from melting of seafloor sediments and crust atop the downgoing slab can produce andesites and rhyolites. Within plates volcanism may occur above hot spots, which are manifestations of plumes of hot material that rise from deep in the mantle.

What are some beneficial effects of volcanism? Over the course of Earth's evolution, volcanic eruptions released the water and gases that formed the oceans and atmosphere. Geothermal heat drawn from areas of recent volcanism is of growing importance as a source of energy. An important ore-forming process occurs when groundwater circulates around buried magma or seawater circulates through ocean floor rifts.

KEY TERMS AND CONCEPTS

volcanism (p. 106)
volcano (p. 106)
lava (p. 106)
pahoehoe (p. 108)
aa (p. 108)
pillow lava (p. 108)
ash (p. 109)
volcanic tuff (p. 110)
volcanic breccia (p. 110)
pyroclastic flow (p. 110)

central vent (p. 111)
shield volcano (p. 111)
volcanic dome (p. 111)
cinder cone (p. 112)
composite volcano/
  stratovolcano (p. 112)
crater (p. 113)
caldera (p. 114)
phreatic explosion (p. 115)
diatreme (p. 117)

fissure eruption (p. 117)
flood basalts (p. 118)
ash-flow deposit (p. 118)
lahar (p. 119)
fumarole (p. 120)
aseismic ridge (p. 125)
hot spot (p. 125)

EXERCISES

1. The asthenosphere has been identified as a major source of magma. Why? What forces magma to rise to the surface?

2. What is the difference between magma and lava? Give examples of types of volcanic rocks and their coarse-grained, intrusive counterparts.

3. Describe the principal styles of eruptions and the deposits and landforms each style produces.

4. The accompanying photograph shows the remains of a building in a village 5 km south of El Chichón volcano, in southeastern Mexico. The village was destroyed when El Chichón erupted in April 1982. From the debris and the bent reinforcing rods evident in the photograph, what can you conclude about the nature of the flow, its force, and its direction?

5. What is the association between plate boundaries and volcanism? Can the eruptive style and composition of volcanic deposits be correlated with plate boundaries?

6. Under what circumstances do lahars occur? Hot springs? Ash-flow deposits?

7. Name the most dangerous features of volcanoes.

8. What signals an impending eruption?

THOUGHT QUESTIONS

1. What public policy initiatives do the eruptions of Mount St. Helens, Mount Pinutabo, and Nevado del Ruiz suggest should be undertaken in such areas as zoning, land use, insurance, warning systems, and public education?

2. Do a risk-benefit analysis of volcanoes—that is, tabulate their dangers and their contributions to humankind—and decide whether you would prefer an Earth with or without them.

3. What have we learned about the Earth's interior from volcanoes?

SUGGESTED READINGS

A.G.U. Special Report. 1992. *Volcanism and Climatic Change.* Washington, D.C.: American Geophysical Union.

Decker, R. W., and B. Decker. 1989. *Volcanoes,* rev. and updated ed. New York: W. H. Freeman.

Dvorak, John J., Carl Johnson, and Robert I. Tilling. 1982. Dynamics of Kilauea volcano. *Scientific American* (August):46–53.

Edmond, John M., and Karen L. Von Damm. 1992. Hydrothermal activity in the deep sea. *Oceanus* (Spring):74–81.

Francis, Peter. 1983. Giant volcanic calderas. *Scientific American* (June):60–70.

Heiken, G. 1979. Pyroclastic flow deposits. *American Scientist* 67:564–571.

Krakaner, Jon. 1996. Geologists worry about dangers of living "under the volcano." *Smithsonian,* 33–125.

McPhee, John. 1990. Cooling the lava. In *The Control of Nature.* New York: Farrar, Straus & Giroux.

National Research Council. 1994. *Mount Rainier, Active Cascade Volcano.* Washington, D.C.: National Academy Press.

Simkin, T., L. Siebert, L. McClelland, D. Bridge, C. Newhall, and J. H. Latter. 1981. *Volcanoes of the World.* New York: Academic Press.

Tilling, Robert I. 1989. Volcanic hazards and their mitigation: Progress and problems. *Reviews of Geophysics* 27 (no. 2):237–269.

Vink, Gregory E., and W. Jason Morgan. 1985. The Earth's hot spots. *Scientific American* (April):50–57.

White, Robert S., and Dan P. McKenzie. 1989. Volcanism at rifts. *Scientific American* (July):62–72.

Wright, Thomas L., and Thomas C. Pierson. 1992. *Living with Volcanoes.* U.S. Geological Survey Circular 1073. Washington, D.C.: U.S. Government Printing Office.

INTERNET SOURCES

Cascades Volcano Observatory
❶ **http://vulcan.wr.usgs.gov/home.html**
This U.S. Geological Survey laboratory is the major source of information about the Mount St. Helens eruption and monitoring of the Cascade volcanoes. The site provides links to images, a data archive, and information on preparing for a Cascade volcano eruption.

The Electronic Volcano
❶ **http://www.dartmouth.edu/pages/rox/volcanoes/elecvolc.html**
This site is a "front door" for links to data sets and images of volcanoes around the world.

Hawaii's Center for Volcanology
❶ **http://www.soest.hawaii.edu/GG/hcv.html**
Maintained by the School of Ocean and Earth Science and Technology of the University of Hawaii (Manoa), this site includes images and data for past and current eruptions in the Hawaiian Islands.

Volcanoes
❶ **http://www.geo.mtu.edu/volcanoes/**
Michigan Technological University maintains this extensive site with reference maps, recent and ongoing activity, information about volcanic hazards mitigation, a glossary, and even volcano humor.

Volcano World
❶ **http://volcano.und.nodak.edu/vw.html**
This site, at the University of North Dakota, is a "front door" for almost everything available on volcanoes. Features include Today in Volcano History, What's Erupting Now? (providing a map and links to each eruption with images and a glossary), Ask a Volcanologist, Volcanic Peaks and Monuments, and Volcanoes of the World.

Japan's Volcano Research Center
❶ **http://hakone.eri.u-tokyo.ac.jp/vrc/VRC.html**
Current volcanic activity in Japan is the focus of this site at the University of Tokyo. Current reports are maintained on the major Japanese volcanoes, including Fuji and Unzen.

6

Bryce Canyon, Utah.
Weathering has sculpted
these sandstones into a
wonderful array of strange
forms.
(Zandria Muench Beraldo.)

Weathering and Erosion

Solid as the hardest rocks may seem, they—like rusting old automobiles and yellowed old newspapers—eventually weaken and crumble when they are exposed to water and the gases of the atmosphere. Unlike those cars and newspapers, however, rocks may take thousands of years to disintegrate. In this chapter, we look closely at two processes, weathering and erosion. **Weathering** is the general process by which rocks are broken down at Earth's surface. Weathering produces all the clays of the world, all soils, and the dissolved substances that are carried by rivers to the ocean. Weathering takes place in two ways:

- **Chemical weathering** occurs when the minerals in a rock are chemically altered or dissolved. The blurring or disappearance of lettering on old gravestones and monuments is attributable mainly to chemical weathering.

■ **Physical weathering** occurs when solid rock becomes fragmented by physical processes that do not change its chemical composition. The rubble of broken stone blocks and columns that were once stately temples in ancient Greece and the cracks and breaks in the ancient tombs and monuments of Egypt are primarily the result of physical weathering.

Chemical and physical weathering help and reinforce each other. The faster the decay, the weaker the pieces and the more susceptible to breakage; the smaller the pieces, the greater the surface area available for chemical attack and the faster the decay.

WEATHERING, EROSION, AND THE ROCK CYCLE

After tectonics and volcanism have made mountains, chemical decay and physical breakup join with rainfall, wind, ice, snow, and downward movements of material over Earth's surface to wear away those mountains. This is the course of **erosion,** which we defined in Chapter 3 as the set of processes that loosen and move soil and rock downhill or downwind. Erosion moves weathered material from Earth's surface, carrying it away and depositing it elsewhere. And as erosion carries away weathered solid material, it exposes fresh, unaltered rock to weathering.

Weathering and erosion not only are closely intertwined, they are major geological processes in the rock cycle, as we also saw in Chapter 3. With tectonics and volcanism, other elements of the rock cycle, weathering and erosion change the form of Earth's surface and alter rock materials, converting igneous and other rocks into sediment and forming soil. In some instances, weathering and erosion are inseparable. When a rock such as pure limestone or rock salt weathers by dissolving in rainwater, for example, all of the material is completely dissolved and carried away in the water as ions in solution. Material dissolved during chemical weathering contributes most of the dissolved material in the oceans.

The early sections of this chapter emphasize the chemical aspect of weathering, for it is in some ways the more fundamental driving force of the whole process. The effect of physical weathering, always important, is largely dependent on chemical decay, but it reinforces chemical action and is itself promoted by decay. First, however, we examine the factors that control weathering.

WHY DO SOME ROCKS WEATHER MORE RAPIDLY THAN OTHERS?

All rocks weather, but the manner and rate of their weathering vary. The four key factors that control the fragmentation and decay of rocks are the properties of the parent rock, the climate, the presence or absence of soil, and the length of time they are exposed to atmospheric conditions.

Properties of the Parent Rock

The nature of a parent rock affects weathering because (1) different minerals weather at different rates and (2) the structure of rocks affects their susceptibility to cracking and fragmentation. Century-old inscriptions on gravestones offer clear evidence of the varying rates at which rocks weather. The carved letters on a recently erected gravestone stand out in sharp relief from the stone's polished surface. After a hundred years in a moderately rainy climate, limestone's surface will be dull and the letters inscribed on it will almost have melted away, much as the name on a bar of soap disappears after a few washes (Figure 6.1). Granite will show only minor changes. The differences in the weathering of granite and limestone reflect the different **solubilities** of their constituent minerals in water; that is, the extent to which they can dissolve in water. In a later section of this chapter, we will define solubility more precisely. Given enough time, however, even a resistant rock will ultimately decay. After several hundred years, the granite monument will also have weathered appreciably, and its surface and letters will be somewhat dulled and blurred. Looking more closely at the weathered granite, perhaps with a hand lens, we would see different patterns of weathering in its various mineral grains. The feldspar crystals would show signs of corrosion, and their surfaces would be chalky and covered with a thin layer of soft clay. The outer layers of the grains of feldspar will have undergone a change in chemical composition and turned into a new mineral. The quartz crystals will appear fresh—clear and unaltered.

A rock's structure also affects physical weathering. Granite monuments may remain unbroken and uncracked even after centuries, though they may show evidence of some chemical weathering. Other intrusive igneous rocks, like many granites, may be massive, showing no planes of weakness that contribute to cracking or fragmentation. In contrast, shale, a sedimentary rock that splits easily across thin

FIGURE 6.1 Early nineteenth-century gravestones at Wellfleet, Massachusetts. The stone on the right is limestone and is so weathered as to be unreadable. The stone on the left is slate, which retains its legibility under the same conditions. *(R. Siever.)*

bedding planes, breaks into small pieces so quickly that only a few years after a new road is cut through a shale, the rock will become rubble.

Climate: Rainfall and Temperature

A tour of graveyards across the continent, from the lower United States to northern Canada and Alaska, would reveal that the rate of weathering—chemical and physical—varies not only with the properties of the rock but also with the **climate**—the amount of rainfall and temperature. Old gravestones in hot, humid Florida are badly chemically weathered, but those of the same age in the equally hot but arid Southwest are hardly touched. And gravestones in cold, dry arctic regions show even less chemical weathering than those found in the Southwest. Climate exerts strong control over the rate of chemical weathering. High temperatures and heavy rainfall speed chemical weathering; cold and dryness impede the process. In cold climates, water is chemically unreactive because it is frozen. In arid regions,

water is relatively unavailable. In both cases, chemical weathering proceeds slowly.

In climates in which chemical weathering is minimal, physical weathering may be active. Freezing water may act as a wedge, widening cracks and pushing the rock apart.

Presence or Absence of Soil

Soil, one of any country's most valuable natural resources, is composed of fragments of bedrock, clay minerals formed by the alteration of bedrock minerals, and organic matter produced by the organisms that live in it. Although soil is itself a product of weathering, its presence or absence affects the chemical and physical weathering of other substances. An old nail that has been buried in soil usually will be so badly rusted that you can snap it like a matchstick. Yet a nail pried from the wood of a centuries-old house may still be strong, covered with only a thin layer of rust. Similarly, a mineral in soil in a lowland valley may be badly altered and corroded, while the same mineral exposed in a nearby cliff of bedrock will be much less affected by weathering. Although the cliff is exposed to occasional rain, the bare rock is usually dry, and weathering proceeds very slowly. No soil forms on the cliff because rain quickly carries loosened particles down to lower areas, where they can accumulate.

Soil production is a *positive-feedback process;* that is, the product of the process works to advance the process itself. Once soil starts to form, it works as a geological agent to weather rock more rapidly. The soil retains rainwater, and it plays host to a variety of vegetation, bacteria, and organisms. These life forms create an acidic environment that, in combination with moisture, promotes chemical weathering, which alters or dissolves minerals. The plant roots and organisms tunneling through the soil aid physical weathering by helping create fractures. Chemical and physical weathering in turn lead to the production of more soil.

Length of Exposure

The longer a rock weathers, the greater its alteration, dissolution, and physical breakup. Rocks that have been exposed at Earth's surface for many thousands of years form a rind—an external layer of weathered material several millimeters to several centimeters thick, which surrounds the fresh, unaltered rock. In dry climates some rinds have grown as slowly as 0.006 mm per 1000 years.

The lavas and ash deposits newly extruded from volcanoes have had a very short period of exposure at Earth's surface, and they are relatively unweathered. Because we know the dates of modern eruptions, such as those of Mount St. Helens in 1980, we can measure the times required for various degrees of weathering to occur. In the years since those eruptions, the volcanic ash deposits have weathered appreciably and have altered to form other minerals. After the same length of time, masses of solidified lava are still relatively fresh. The difference in the extent of weathering occurs mainly because the ash is made up of very small particles, which weather faster than the more massive volcanic rocks.

Four factors, then (summarized in Table 6.1), control rates of chemical and physical weathering:

1. Minerals and rocks of different compositions weather at different rates.

2. Climate, which includes both rainfall and temperature, strongly affects chemical weathering: warmth and heavy rainfall accelerate chemical weathering, whereas cold and aridity hinder it.

TABLE **6.1**

MAJOR FACTORS CONTROLLING RATES OF WEATHERING

	SLOW	← WEATHERING RATE →	FAST
PROPERTIES OF PARENT ROCK			
Mineral solubility	Low (e.g., quartz)	Moderate (e.g., pyroxene, feldspar)	High (e.g., calcite)
Rock structure	Massive	Some zones of weakness	Very fractured or thinly bedded
CLIMATE			
Rainfall	Low	Moderate	Heavy
Temperature	Cold	Temperate	Hot
PRESENCE OR ABSENCE OF SOIL			
Thickness of soil layer	None—bare rock	Thin to moderate	Thick
Organic activity	Sparse	Moderate	Abundant
LENGTH OF EXPOSURE	Short	Moderate	Long

Physical weathering, which is also aided by water, may be appreciable in cold or arid climates, even though chemical weathering proceeds very slowly.

3. The presence of soil, itself a product of weathering, promotes weathering.

4. The longer a rock is exposed at the surface, the more weathered it becomes.

We can now consider each type of weathering in more detail. Although chemical and physical weathering are not completely separable, it is easier to show their relationships after we have discussed their separate characteristics.

CHEMICAL WEATHERING

Chemical weathering results from chemical reactions between minerals in rocks and air and water. In the following pages we have eliminated some details of these reactions, which are the subject of extensive research by geologists and soil chemists. The essential ideas are that during chemical reactions, some minerals dissolve and others combine with water and components of the atmosphere, such as oxygen and carbon dioxide, to form new chemical compounds. These are the new minerals formed by weathering. As you saw in the preceding discussion of factors that control rates of weathering, we can deduce some chemical reactions from observations in the field. We can get a better picture of the mechanisms of chemical weathering if we combine field observations with laboratory experiments that simulate natural processes. We begin our investigation by examining the chemical weathering of feldspar, the most abundant mineral in Earth's crust.

From Feldspar to Kaolinite Clay

Feldspar is a key mineral in a great many igneous, sedimentary, and metamorphic rocks. Feldspar is also representative of the many other kinds of rock-forming silicate minerals. Feldspar is one of many silicates that are altered, by chemical reactions, to form the water-containing minerals known as clay minerals. An understanding of feldspar's behavior during weathering contributes much to our grasp of the weathering process in general, for two reasons:

■ There is an overwhelming abundance of silicate minerals in the Earth.

■ The chemical processes of dissolution and alteration that characterize feldspar weathering are also characteristic of weathering in other kinds of minerals.

Recall that when we described the minerals in the weathered granite gravestone earlier in this chapter, we noted that the feldspar crystals were corroded and altered. A more extreme example of feldspar weathering can be found in granite boulders in soils of the humid tropics. Here many of the factors that promote weathering—heavy rainfall, high temperature, the presence of soil, and abundant organic activity—are at a maximum. Granite boulders found in the tropics are so weakened that they can easily be kicked or pounded into a heap of loose mineral grains. Most of the feldspar particles in these boulders have been altered to clay. Greatly magnified under an electron microscope, any remaining feldspar grains can be seen to be corroded and coated with a clay rind (Figure 6.2). Quartz crystals, in contrast, would be relatively intact and unaltered.

In a sample of unweathered granite, the rock is hard and solid because the interlocking network of quartz, feldspar, and other crystals holds it tightly together with strong cohesive forces. But when the feldspar is altered to a loosely adhering clay, the

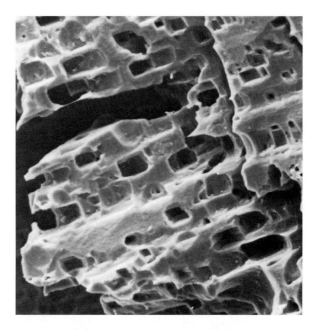

FIGURE 6.2 A scanning electron micrograph of feldspar etched and corroded by chemical weathering in soil. (From R. A. Berner and G. R. Holden, Jr., "Mechanism of Feldspar Weathering: Some Observational Evidence," *Geology,* vol. 5 [1977], p. 369.)

Incipient cracks form along crystal boundaries, and feldspar, biotite, and magnetite start to decay

Decay of feldspar, biotite, and magnetite progresses: as cracks open, rock weakens and is easily disintegrated

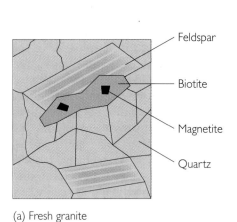

(a) Fresh granite

Feldspar

Biotite

Magnetite

Quartz

(b) Granite starting to weather

(c) Granite extensively weathered

FIGURE 6.3 Microscopic views of stages in the disintegration of granite.

network is weakened and the mineral grains are separate (Figure 6.3; see also Figure 4.1). In this instance, chemical weathering, by producing the clay, also promotes physical weathering because the rock now fragments easily along widening cracks at mineral boundaries.

The white to cream-colored clay produced by the weathering of feldspar is **kaolinite,** named for Gaoling, a hill in southwestern China where it was first obtained. Chinese artisans had used pure kaolinite as the raw material of pottery and china for centuries before Europeans borrowed the idea in the eighteenth century.

Only in the severely arid climates of some deserts and polar regions does feldspar remain relatively unweathered. This observation points to water

as an essential component of the chemical reaction by which feldspar turns to kaolinite. Kaolinite is a hydrous (that is, containing water in the crystal structure) aluminum silicate. In the reaction that produces kaolinite, the solid feldspar undergoes the process of **hydration;** that is, it gains water. The feldspar also loses several chemical components. The alteration is analogous to the chemical reaction that takes place when we make coffee. Solid coffee reacts chemically with hot water to make a solution—the liquid coffee. Caffeine and other components of the bean are extracted from the solid by the reaction, leaving behind spent coffee grounds. Similarly, rainwater filters into the ground, altering feldspar in rock particles and leaving behind the kaolinite as a residue (Figure 6.4).

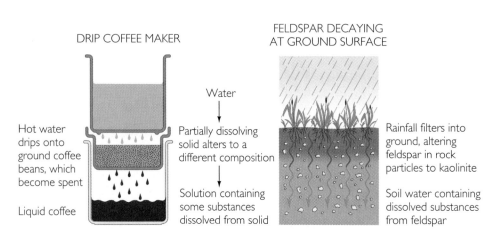

DRIP COFFEE MAKER

Water

Hot water drips onto ground coffee beans, which become spent

Liquid coffee

Partially dissolving solid alters to a different composition

Solution containing some substances dissolved from solid

FELDSPAR DECAYING AT GROUND SURFACE

Rainfall filters into ground, altering feldspar in rock particles to kaolinite

Soil water containing dissolved substances from feldspar

FIGURE 6.4 The process by which feldspar decays is analogous to the brewing of coffee. In both processes, water dissolves some of the solid, leaves behind an altered material, and produces a solution containing substances drawn from the original solid.

The finer the coffee beans are ground, the more coffee that can be extracted from them and the stronger the brew can become. The rate of a chemical reaction increases with an increase in the surface area of the solid (the total of the surface areas of all the grains used) that is exposed to the fluid reacting with it. Because the only part of a solid available for reaction with a fluid is its surface, the more we increase the surface area, the more we speed up the reaction. As we grind coffee beans into finer and finer particles, we increase the ratio of their surface area to their volume. As with coffee beans, so with minerals and rocks: the smaller the fragments, the greater the surface area. The ratio of surface area to volume increases greatly as the average particle size decreases, as shown in Figure 6.5.

We can now write a partial chemical reaction for the weathering of the common feldspar of granite, orthoclase ($KAlSi_3O_8$), which is made up of potassium (K), aluminum (Al), silicon (Si), and oxygen (O):

$$feldspar \quad + \quad water \quad \longrightarrow \quad kaolinite$$
$$KAlSi_3O_8 \qquad H_2O \qquad Al_2Si_2O_5(OH)_4$$

EXPERIMENT 1: DISSOLVING FELDSPAR IN PURE WATER To learn more about the loss of components as feldspar weathers, we can perform a simple laboratory experiment: immerse feldspar in pure water and analyze the solution for the kinds of material that have dissolved. First we grind the feldspar to a powder, to speed the reaction by exposing more surface area to the water. The feldspar dissolves very slowly, and after some time we take samples from the solution and analyze them. We find small amounts of potassium and silica (SiO_2) dissolved in the water,

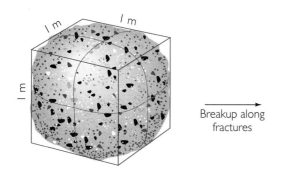

Single boulder, approximately 1 m on a side
Volume = 1 m³
Surface area = 6 m²

Breakup along fractures

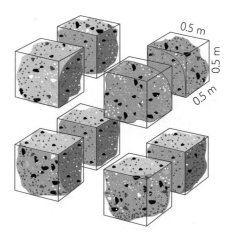

0.5 m
0.5 m
0.5 m

8 fragments, each approximately 0.5 m on a side
Volume = (0.5)³ × 8 = 1 m³
Surface area = 12 m²

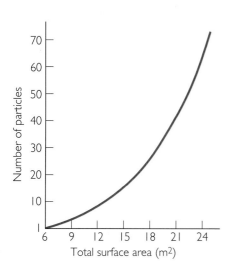

FIGURE **6.5** As a rock mass breaks into smaller pieces, much more surface becomes available for the chemical reactions of weathering. The graph shows the increase of total surface area as the number of particles increases with no change in the total volume (1 m³).

which has acted as a *solvent;* that is, a liquid that is able to dissolve a substance. We can now write a qualitatively more complete but still partial chemical reaction:

feldspar + water ⟶
$KAlSi_3O_8$ H_2O

kaolinite   + dissolved + dissolved
                    silica   potassium ion
$Al_2Si_2O_5(OH)_4$ SiO_2 K^+

Two major points about this reaction give us information on the gains and losses of material that occur as feldspar weathers:

- The potassium and silica dissolved from the feldspar appear as dissolved materials in the water solution.

- Water is used up in the reaction; it is absorbed into the kaolinite crystal structure. This absorption of water is hydration, one of the major processes of weathering.

EXPERIMENT 2: ADDING CARBON DIOXIDE TO SPEED THE REACTION   The reaction of feldspar with pure water in a laboratory is a very, very slow process. It would take thousands of years to weather even a small amount of feldspar under such conditions, so this reaction cannot account for the more rapid weathering we observe widely in nature. If we wanted to, we could speed up weathering by adding a strong acid, such as hydrochloric acid, and dissolve the feldspar in a few days. An acid is a substance that releases hydrogen ions (H^+) to a solution. A strong acid produces abundant hydrogen ions; a weak one, relatively few. Since hydrogen ions have a strong tendency to combine chemically with other substances, acids make excellent solvents.

On the earth's surface, the most common natural acid—and the one responsible for increasing weathering rates—is carbonic acid (H_2CO_3). This weak acid is formed by the solution in rainwater of a small amount of carbon dioxide (CO_2) gas from the atmosphere:

carbon dioxide + water ⟶ carbonic acid
      CO_2 H_2O H_2CO_3

We are familiar with everyday solutions of carbon dioxide in water in the form of carbonated soft drinks. The bottler carbonates the liquid by pump-

ing carbon dioxide into it under pressure. A large quantity of carbon dioxide becomes dissolved in the beverage, making it acidic. As you open the bottle, pressure drops and the dissolved gas bubbles out of solution, making the solution less acidic. The amount of carbon dioxide dissolved in the liquid decreases as the amount of gaseous carbon dioxide in contact with the water decreases. When the amount of carbon dioxide in the beverage reaches the amount in the atmosphere, no more carbon dioxide bubbles out and the beverage is "flat" and only mildly acidic, like rainwater.

The amount of carbon dioxide dissolved in rainwater is small because the amount of carbon dioxide gas in the atmosphere is small. About 0.03 percent of the molecules in Earth's atmosphere are carbon dioxide. Small as that number is, it makes carbon dioxide the fourth most abundant gas, just behind argon (0.9 percent), oxygen (21 percent), and nitrogen (78 percent). The amount of carbonic acid formed in rainwater is very small, only about 0.0006 gram per liter. As carbon dioxide from the burning of oil, gas, and coal increases in the atmosphere, the amount of carbonic acid in rain increases slightly. Most of the acidity of acid rain, however, comes from sulfur dioxide and nitrogen gases, which react with water to form strong sulfuric and nitric acids. Volcanoes and coastal marshes emit gases of carbon, sulfur, and nitrogen into the atmosphere, but by far the largest source is industrial pollution. (See Chapter 24 for more information on acid rain.)

Although rainwater contains only a relatively small amount of carbonic acid, that amount is enough to weather feldspars and dissolve great quantities of rock over a long time. Adding carbon dioxide, in the form of carbonic acid, to our equation, we can now write the full balanced form of the weathering reaction. A *balanced* reaction is one in which the numbers of atoms or ions on one side of the equation exactly equal those on the other side.

feldspar   + carbonic + water ⟶
                    acid
$2KAlSi_3O_8$ $2H_2CO_3$ H_2O

                                                dissolved
kaolinite   + dissolved + dissolved + bicarbonate
                    silica   potassium   ion
$Al_2Si_2O_5(OH)_4$ $4SiO_2$ $2K^+$ $2HCO_3^-$

The simple experiment of adding hydrochloric acid to pure water to speed the dissolution of feldspar includes the three main chemical effects of chemical weathering on silicates: It *leaches,* or dissolves away, cations and silica. It *hydrates,* or adds water to, the minerals. And it makes the solutions less acidic. Specifically, the carbonic acid in rainwater helps to weather feldspar in the following way (Figure 6.6):

- A small proportion of carbonic acid molecules ionizes, forming hydrogen ions (H^+) and bicarbonate ions (HCO_3), making the water droplets slightly acidic.

- The slightly acidic water dissolves potassium ions and silica from feldspar, leaving a residue of kaolinite, a solid clay. The hydrogen ions from the acid combine with the oxygens of the feldspar to form the water in the kaolinite structure. The kaolinite becomes part of the soil or is carried away as sediment.

- The solution becomes less acidic as the reaction goes on.

- The dissolved silica, potassium ions (K^+), and bicarbonate ions are carried away by rain and river waters and ultimately are transported to the ocean.

IN NATURE: FELDSPAR IN OUTCROPS AND IN MOIST SOIL   Now that we understand the chemical reaction by which rainwater weathers feldspar, we can better understand why feldspars on bare rock surfaces are much better preserved than those buried in damp soils. The equation for feldspar weathering gives us two separate but related clues: amount of water and amount of acid available for the chemical reaction. The feldspars on a bare rock weather only while it is moist with rainwater; during all the dry periods, the only moisture that touches the bare rock is dew. The feldspar in moist soil is constantly in contact with the small amounts of water retained in spaces between grains in the soil. Thus feldspar weathers continuously in moist soil.

Second, there is more acid in the water in the soil than there is in rainwater. Rainwater carries its original carbonic acid into the soil. As it filters through the soil, it picks up additional carbonic acid and other acids produced by the roots of plants, by the many insects and other animals that live in the soil, and by the bacteria that degrade plant and animal remains. Recently it has been discovered that bacteria release organic acids, even in waters hundreds of meters deep in the ground. These organic

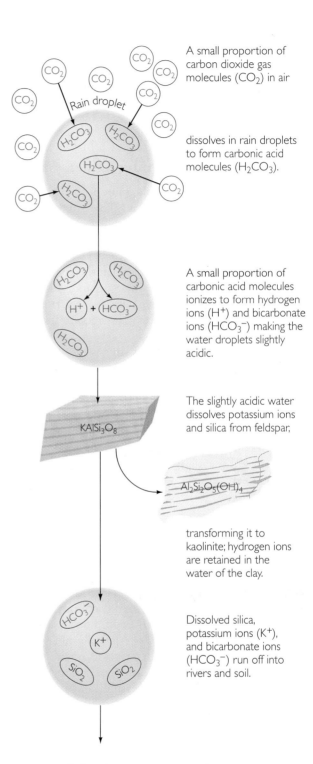

A small proportion of carbon dioxide gas molecules (CO_2) in air

dissolves in rain droplets to form carbonic acid molecules (H_2CO_3).

A small proportion of carbonic acid molecules ionizes to form hydrogen ions (H^+) and bicarbonate ions (HCO_3^-) making the water droplets slightly acidic.

The slightly acidic water dissolves potassium ions and silica from feldspar,

transforming it to kaolinite; hydrogen ions are retained in the water of the clay.

Dissolved silica, potassium ions (K^+), and bicarbonate ions (HCO_3^-) run off into rivers and soil.

FIGURE 6.6 Feldspar weathering when it is in contact with carbonic acid from rainwater containing carbon dioxide. Two products are formed: kaolinite clay and a solution containing dissolved silica, potassium ions, and bicarbonate ions.

acids then weather feldspar and other minerals in rocks below the surface.

Rock weathers more rapidly in the tropics than in temperate and cold climates mainly because plants and bacteria grow quickly in warm, humid climates, contributing the acids that promote weathering. Also, most chemical reactions, weathering included, speed up with an increase in temperature.

Other Silicates Forming Other Clays

Clay minerals are a principal component of soils and sediments all over Earth's surface. This is understandable because the rock-forming silicates constitute so large a fraction of the Earth's crust and weathering at the surface is so widespread. Deposits pure enough to be used as the raw material for pottery, chinaware, and industrial ceramics are found in some uncommon soils and sedimentary rocks. Brick, clay tiles, and other structural ceramics require less pure material.

These clays form through the weathering of a variety of silicate minerals, not just from feldspar. Amphibole, mica, and granite also weather to form clays. The chemical reactions by which these silicates produce clays follow the general course of feldspar weathering. As the mineral weathers, it absorbs water and loses silica and ions such as sodium, potassium, calcium, and magnesium to the solution. The kinds of clays formed depend on two factors:

- The composition of the parent silicates
- The climate

The clay mineral montmorillonite, for example, which swells when it absorbs large quantities of water, typically forms from the weathering of volcanic ash. Montmorillonite is a common weathering product of other silicates, too, especially in semiarid environments, such as the high plains of the southwestern United States. Because there are many parent silicates, many climatic regimes and many clay minerals, predicting the course of silicate weathering to specific clay minerals is a task for trained geologists and soil scientists.

Not all silicates weather to form clay minerals. Rapidly weathering silicates, such as some pyroxenes and olivines, may dissolve completely in humid climates, leaving no clay residue. Quartz, one of the slowest of the abundant silicate minerals to weather, also dissolves without forming any clay mineral.

Silicate weathering can also form materials other than clay minerals. **Bauxite**—an ore composed of aluminum hydroxide and the major source of aluminum metal—is one such product. Bauxite forms when clay minerals derived from the weathered silicates continue to weather until they have lost all their silica and ions other than aluminum. Bauxite is found in tropical regions where rainfall is heavy and weathering is intense.

From Iron Silicates to Iron Oxides

Iron is one of the eight most abundant elements in the Earth's crust, but iron metal, the chemical element in its pure form, is rarely found in nature. It occurs only in certain kinds of meteorites that fall to Earth from other places in the solar system. Weathering is responsible for the formation of most of the iron ores used for the production of iron and steel. These ores are composed of iron oxide minerals that originally formed during the weathering of iron-rich silicate minerals, such as pyroxene and olivine. The iron released by dissolution of these minerals combines with oxygen from the atmosphere to form iron oxide minerals by a chemical reaction called **oxidation**—a chemical combination of an element with oxygen. Oxidation is defined more generally as a chemical reaction in which an ion or element loses one or more electrons (see Chapter 2). Like hydration, oxidation is one of the important chemical weathering processes.

The iron in minerals may be present in one of several forms: metallic iron, ferrous iron, and ferric iron. In the metallic iron of meteorites, iron's atoms are uncharged; that is, they have neither gained nor lost electrons by reaction with another element. In the **ferrous iron** (Fe^{2+}) in silicate minerals such as pyroxene, the iron atoms have lost two of the electrons they would have in the metallic form and thus have become ions. The iron in the most abundant iron oxide at the Earth's surface, **hematite** (Fe_2O_3), is **ferric** (Fe^{3+}); that is, these iron atoms have lost three electrons of the number in the metallic form. Iron ions oxidize by losing an additional electron, going from 2+ (ferrous) to 3+ (ferric). All of the several kinds of iron oxides formed at the surface of the Earth are ferric. The electrons lost by the iron are gained by oxygen atoms as they become oxygen ions (O^{2-}). Thus oxygen atoms from the atmosphere oxidize ferrous iron to ferric iron.

When an iron-rich mineral such as pyroxene dissolves in water, its silicate structure dissolves, releasing silica and ferrous iron to solution, and its ferrous iron is oxidized by oxygen to the ferric form (Figure 6.7). Because of the strength of the chemical bonds between ferric iron and oxygen, ferric iron is very insoluble in most natural surface waters. It therefore precipitates from the solution, forming a solid ferric iron oxide. We are familiar with ferric iron oxide in another form—rusting iron, which has formed from iron metal exposed to the atmosphere.

We can show this overall weathering reaction of iron-rich minerals by the equation

iron pyroxene + oxygen → hematite + dissolved silica
$$4FeSiO_3 \qquad O_2 \qquad 2Fe_2O_3 \qquad 4SiO_2$$

Although the equation does not show it explicitly, water is required for this reaction to proceed.

Iron minerals, which are widespread, weather to the characteristic red and brown colors of oxidized iron (Figure 6.8). Iron oxides are found as coatings and encrustations that color soils and weathered surfaces of iron-containing rocks. The red soils of Georgia and other warm, humid regions are colored by iron oxides. Iron minerals weather so slowly in frigid regions that iron meteorites frozen in the ice of Antarctica are almost entirely unweathered.

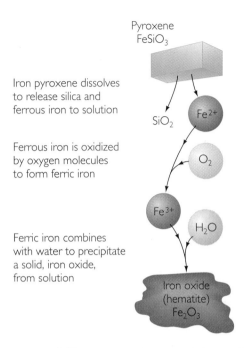

Iron pyroxene dissolves to release silica and ferrous iron to solution

Ferrous iron is oxidized by oxygen molecules to form ferric iron

Ferric iron combines with water to precipitate a solid, iron oxide, from solution

FIGURE 6.7 The general course of chemical reactions by which an iron-rich mineral, such as pyroxene, weathers in the presence of oxygen and water.

FIGURE 6.8 Red and brown iron oxides color weathering rocks in Monument Valley, Arizona. *(Betty Crowell.)*

Fast Weathering: Dissolution of Carbonates

Even olivine, the silicate mineral that weathers most rapidly, is relatively slow to dissolve in comparison with some other rock-forming minerals. Limestone, made of the calcium and magnesium carbonate minerals calcite and dolomite, is one of the rocks that weather most quickly in humid regions. Old limestone buildings show the effects of dissolution by rainwater (Figure 6.9). Underground water dissolves away great quantities of carbonate minerals, hollowing out caves in limestone formations. Farmers and gardeners use ground limestone to counterbalance acidity in soil because of the stone's ability to dissolve rapidly over the growing season. When limestone dissolves, no clay minerals are formed. The solid dissolves completely, and its components are carried off in solution.

Carbonic acid promotes the dissolution of limestone as well as the weathering of silicates. The overall reaction by which calcite, the major mineral of limestones, dissolves in rain or other water containing carbon dioxide is

$$
\begin{array}{cccc}
\text{calcite} + & \text{carbonic} \rightarrow & \text{calcium} + & \text{bicarbonate} \\
 & \text{acid} & \text{ion} & \text{ion} \\
CaCO_3 & H_2CO_3 & Ca^{2+} & 2HCO_3^-
\end{array}
$$

The reaction proceeds only in the presence of water, which contains the carbonic acid and dissolved ions. When calcite dissolves, the calcium and bicarbonate ions are carried away in solution. Dolomite ($CaMg[CO_3]_2$), another abundant carbonate mineral, dissolves in the same way, producing equal amounts of magnesium and calcium ions.

Chemical Stability: A Speed Control for Weathering

Much larger areas of Earth's surface are covered by silicate rocks than by carbonate rocks. But because carbonate minerals dissolve faster and in greater amounts than any silicates, the weathering of limestone accounts for more of the total chemical weathering of the land surface each year than that of any other rock.

The weathering rates of minerals cover a great range, from the rapid rates of carbonates to the slow rate of quartz. The varying rates at which minerals weather reflect their chemical stability under weathering conditions—that is, in the presence of water at given surface temperatures.

Chemical stability is a measure of the tendency for a chemical substance to remain in a given chemical form rather than to react spontaneously to become a different chemical form. We can think of this stability as somewhat similar to mechanical stability. A book lying flat on a table is stable; it will remain in the same position unless it is moved. A book balanced on its edge is unstable; a bump or a push will cause it to fall to the flat, stable position. Just as the flat book is mechanically stable, iron metal in a meteorite in outer space is chemically stable; exposed to no oxygen or water, it remains unaltered for billions of years. If that meteorite were to fall to Earth, where it would be exposed to oxygen and water, it would be chemically unstable and would spontaneously react to form iron oxide.

Mechanical and chemical stability differ in an important way, however, and for that reason we must abandon our analogy. Unlike the book on the table, which is said to be simply stable or unstable, chemical substances are stable or unstable in relation to a specific environment or set of conditions. Feldspar, for example, is stable at the conditions found deep in Earth's crust (high temperatures and small amounts of water), but it is unstable at the conditions of Earth's surface (lower temperatures and abundant water). Two characteristics of a mineral—its solubility and its rate of dissolution—help determine its chemical stability.

SOLUBILITY   The solubility of a specific mineral is measured by the amount of the mineral dissolved in water when the solution reaches the point of satura-

FIGURE 6.9 Weathered limestone blocks and columns of 2500-year-old Greek ruins at Segesta, Italy, show pitted, etched surfaces caused by chemical solution. *(Ric Ergenbright.)*

tion—the point at which the water cannot hold any more of the dissolved substance. The higher a mineral's solubility, the lower its stability under weathering. Rock salt, for example, is unstable under weathering conditions; it is highly soluble in water and is leached from a soil by even small amounts of water. Quartz, in contrast, is fairly stable under most weathering conditions; its solubility in water is very low (only about 0.008 gram per liter of water), and it is not easily leached from a soil.

RATE OF DISSOLUTION

A mineral's rate of dissolution is measured by the amount of the mineral that dissolves in an unsaturated solution in a given length of time. The faster the mineral dissolves, the less stable it is. Thus feldspar dissolves at a much faster rate than quartz, and primarily for that rea-

son, it is less stable than quartz under weathering conditions.

RELATIVE STABILITY OF COMMON ROCK-FORMING MINERALS

In a tropical rain forest, where chemical weathering is intense, only the most stable minerals will be left on an outcrop or in the soil. In a region such as the desert of North Africa, where weathering is at a minimum, alabaster (gypsum) monuments remain intact, as do many unstable minerals. Knowing the relative chemical stabilities of different minerals enables us to predict the intensity of weathering in a given area. Geologists have compared the stabilities of all the common rock-forming minerals and compiled the series shown in Table 6.2. It ranges from salt and carbonate minerals at the least stable end to iron oxides at the most stable end. The

TABLE **6.2**

STABILITY OF COMMON MINERALS UNDER WEATHERING CONDITIONS COMPARED WITH BOWEN'S REACTION SERIES

STABILITY OF MINERALS	RATE OF WEATHERING	BOWEN'S REACTION SERIES
Most stable	**Slowest**	
Iron oxides (hematite)		
Aluminum hydroxides (gibbsite)		**Last to crystallize**
Quartz		Quartz
Clay minerals		
Muscovite mica		Muscovite
Potassium feldspar (orthoclase)		Orthoclase Albite
Biotite mica		Biotite
Sodium-rich feldspar (albite)		
Amphiboles		Amphibole
Pyroxene		Pyroxene
Calcium-rich feldspar (anorthite)		Anorthite
Olivine		Olivine
Calcite		
Halite		
Least stable	**Fastest**	**First to crystallize**

Mafic minerals (discontinuous crystallization series)

Plagioclase feldspars (continuous crystallization series)

positions of the silicate minerals in this series are approximately the reverse of their positions in Bowen's reaction series (discussed in Chapter 4), which lists silicate minerals in the order in which they crystallize from a basalt magma. Bowen's series, too, is shown in Table 6.2. Note that olivine and calcium plagioclase are the first minerals to crystallize as the melt cools, indicating their stability under high temperatures and pressures. They also are the least stable and the first to disappear when they are exposed at Earth's surface, under conditions of greatly reduced temperature and pressure.

The nature of the chemical bonds that characterize the crystal structures of the silicate minerals determines the relative stability revealed by both weathering order and reaction series (see Chapter 2). The silicate minerals that are least stable under weathering conditions are the isolated tetrahedra. The mineral olivine, which appears near the bottom of the first list in Table 6.2, has this structure. Somewhat more stable are the single-chain silicates, the pyroxenes, and the double chains, the amphiboles. Next in order of stability are the sheet silicates; the micas and clay minerals; the framework silicate, quartz; and the aluminum and iron oxides.

PHYSICAL WEATHERING

Now that we have surveyed chemical weathering alone, we can turn to its partner, physical weathering. We can see the working of physical weathering most clearly by examining its role in arid regions, where chemical weathering is minimal.

Physical Weathering in Arid Regions

Weathered outcrops in arid regions are covered by a rubble of various-sized fragments, from individual mineral grains only a few millimeters in diameter to boulders more than a meter across. The differences in size reflect varying degrees of physical weathering and patterns of breakage of the parent rock. As physical weathering continues, the larger particles are cracked and broken into smaller ones. Some of these fragments are broken along planes of weakness in the parent rock (Figure 6.10). Grains of sand are formed when individual crystals of various minerals, such as quartz, break apart from one another, or when fine-grained rocks such as basalt become fragmented. The finest materials are clay particles, formed from the chemical weathering of silicates.

Arid regions show some signs of chemical

FIGURE 6.10 Weathered, enlarged joint patterns developed in two directions in rocks at Point Lobos State Reserve, California. *(Jeff Foott/DRK.)*

weathering, such as clays and altered feldspars, but the dominant form of weathering in this climate is physical. Though physical weathering is the most common form in dry regions, even here chemical weathering has prepared the way for it. Slight alterations of feldspar and other minerals weaken the cohesive forces that hold the crystals in a rock together. As small cracks form and widen, individual quartz or feldspar crystals are freed by a combination of physical and chemical weathering and fall to the ground. Fractures enlarge, and large blocks of rock are separated from the outcrop.

Physical Weathering in All Regions

The chemical weathering that promotes physical weathering is itself promoted by fragmentation, which opens channels where water and air can penetrate and react with minerals inside the rock. The breaking up of the rock into smaller pieces exposes more surface area to weathering and so speeds the chemical reactions.

Physical weathering is not always so dependent on chemical weathering. There are processes by which unweathered rock masses are broken up, such as the freezing of water in cracks. And some rocks are made particularly susceptible to physical weathering by fracturing produced by tectonic forces as rocks are bent and broken during mountain building. On the Moon, physical fragmentation works alone, for there is no water to make chemical weathering possible. On that lifeless terrain, rocks are broken into boulders and fine dust by the impacts of large and small meteorites.

What Determines How Rocks Break?

Rocks can break for a variety of reasons, including stress along natural zones of weakness and biological and chemical activity.

NATURAL ZONES OF WEAKNESS Rocks have natural zones of weakness along which they tend to crack. In sedimentary rocks such as sandstone and shale, these zones are the bedding planes formed by the successive layers of solidified sediment. Metamorphic rocks such as slate form parallel planes of fractures that enable them to be split easily to form roofing tiles. Granites and other rocks are massive— that is, large masses that show no changes in rock type or structure. Massive rocks tend to crack along regular fractures at intervals of one to several meters called **joints** (see Chapter 10 for further discussion of joints). These and more irregular fractures form while rocks are still deeply buried in Earth's crust. Through uplift and erosion, the rocks rise gradually to Earth's surface. There, freed from the weight of tons of overlying rock, the fractures open slightly. Once the fractures open a little, both chemical and physical weathering work to widen the crack.

ACTIVITY BY ORGANISMS Both chemical and physical weathering are affected by the activity of organisms, from bacteria to tree roots, all working in ways that destroy the rock. Bacteria and algae invade cracks, as you learned earlier in this chapter. The acidity these organisms produce then promotes chemical weathering. Many of us have seen a crack

FIGURE 6.11 Organisms such as these tree roots invade fractured rock, widening cracks and promoting further chemical and physical weathering. *(Peter Kresan.)*

in a rock that has been widened when a tree took root in it. The physical force of the growing root system helps pry cracks apart (Figure 6.11).

FROST WEDGING One of the most efficient mechanisms for widening cracks is **frost wedging**— breakage resulting from the expansion of freezing water. As the water freezes, it exerts an outward force strong enough to wedge open the crack and split the rock (Figure 6.12). This is the same process that can crack open the engine block of a car that is not protected by antifreeze.

FIGURE 6.12 Gneiss boulder, 3 m high, fractured by frost action. Taylor Valley, Victoria Land, Antarctica. *(Michael Hambrey.)*

FIGURE 6.13 Exfoliation on Half Dome, Yosemite National Park, California. *(Tony Waltham.)*

MINERAL CRYSTALLIZATION Other expansive forces that can split rocks are generated when minerals crystallize from solutions in rock fractures. This phenomenon is most common in arid regions, where dissolved substances derived from chemical weathering may crystallize as a solution evaporates. The minerals are commonly calcium carbonate, occasionally gypsum, and rarely rock salt.

ALTERNATING HEAT AND COLD A recurring idea among geologists who research weathering is that rocks can break as a result of the daily alternation of hot days and cold nights in a desert, where temperatures may drop from 43°C to 15°C in an hour at twilight. Part of the breakage process may be weakening of the rock caused by its expansion in the heat and contraction during the cold. Camping and forest fires have taught us that fires built over a rock surface can crack rocks. Various attempts at laboratory simulations of natural breakage caused by temperature extremes have failed to confirm the hypothesis. But supporters of the idea point out that a few months of laboratory experiments are no match for the thousands of years of expansion in the heat and contraction in the cold that might weaken a rock and cause it to fracture. The matter remains unsettled.

EXFOLIATION AND SPHEROIDAL WEATHERING Two forms of rock breakage not directly related to earlier, preweathering fractures or joints are exfoliation and spheroidal weathering. **Exfoliation** is a physical weathering process in which large flat or curved sheets of rock are fractured and detached from an outcrop. These sheets may look like the layers peeled from a large onion (Figure 6.13). **Spheroidal weathering** is also a cracking and splitting off of curved layers from a generally spherical boulder, but usually on a much smaller scale (Figure 6.14). As common as exfoliation and spheroidal weathering are, no generally accepted explanation of their origin has yet emerged. It is apparent that exfoliation and spheroidal weathering are other instances in which chemical weathering induces cracks that follow the shape of the surface in some way, particularly in strongly jointed rock. Some geologists have suggested that both exfoliation and spheroidal weathering result from uneven distribution of expansion and contraction derived from chemical weathering and temperature changes.

OTHER FORCES Rivers excavate bedrock valleys by beating on the bedrock of the channel with transported rocks and by hurling their own force against the bedrock at waterfalls and rapids. Rock masses are also broken up by the scouring and plucking action of glaciers, as you will see in Chapter 15. And in Chapter 17 you will see how waves, pounding on rocky shores with a force equal to hundreds of tons per square meter, fracture exposed bedrock.

FIGURE 6.14 This remarkable structure is a product of spheroidal weathering. As cabbage-leaf layers fall away, a rock core remains, retaining the original shape of the rock on a smaller scale. *(Michael Follo.)*

Physical Weathering and Erosion

As we noted earlier, weathering and erosion are closely related, interacting processes. Physical weathering and erosion are closely tied to how wind, water, and ice work to transport weathered material.

Consider the following sequence. Physical weathering fractures a large rock into smaller pieces, more easily transported and therefore more easily eroded than the larger mass was. The first steps in the erosion process are downhill movements of masses of weathered rock, such as landslides, and transportation of individual particles by flows of rainwater down slopes. The steepness of slopes affects both physical and chemical weathering, which in turn affect erosion. Weathering and erosion are more intensive on steep slopes and by their action make slopes gentler. Wind may blow away the finer particles and glacial ice can carry away large blocks torn from bedrock.

As you can see, the sizes of materials formed by physical weathering are closely related to various erosional processes. As weathered material is transported, it may again change in size and shape, and its composition may change as a result of chemical weathering. When transportation stops, deposition of the sediment formed by weathering begins.

Figure 6.15 is a flow chart summarizing the processes of physical and chemical weathering. All of these processes, which are detailed in Chapters 11 through 17, contribute to the formation of different kinds of landscape.

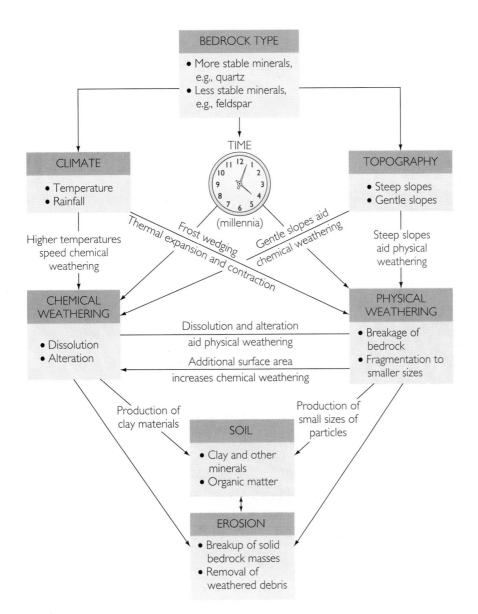

FIGURE **6.15** Summary flow chart of weathering, soil formation, and erosion. Starting at the top of the flow chart, bedrock type denotes the proportion of stable and unstable minerals in a rock. Rocks with a high proportion of unstable minerals will weather more than rocks with a high proportion of stable minerals. Climate, including temperature and rainfall; topography, including steep and gentle slopes; and time have strong effects on chemical and physical weathering. Soil is the product of weathering in place, while erosion removes weathering products.

SOIL: THE RESIDUE OF WEATHERING

Not all weathering products are eroded and immediately carried away by streams or other transport agents. On moderate and gentle slopes, plains, and lowlands, a layer of loose, heterogeneous weathered material remains overlying the bedrock. It may include particles of weathered and unweathered parent rock, clay minerals, iron and other metal oxides, and other products of weathering. Engineers and construction workers refer to this entire layer as "soil." Geologists, however, prefer to call this material **regolith,** reserving the term *soil* for the topmost layers, which contain organic matter and can support plant life. We can easily see the difference between regolith and soil if we consider the regolith found on the Moon. The Moon's regolith is a loose layer of fragmented rocks and dust, but it is quite dead. It contains little or no organic matter and can-

6.1 LIVING ON EARTH

Soil Erosion

Before settlers began to farm the prairies of the United States and Canada in the nineteenth century, the soils were moderately thick. Soil formed relatively quickly in this area because the glaciation that ended about 10,000 years ago had deposited an abundance of easily weathered ground-up bedrock material. But soils form very slowly; even in rapid weathering environments they may gain as little as 2 mm per year.

Because soils take such a long time to form, they cannot be renewed quickly after

Contour plowing, as on these farms in Pennsylvania, minimizes soil erosion.
(Larry Lefever/Grant Heilman.)

not support life. The organic matter in Earth's soil is **humus,** the remains and waste products of the many plants, animals, and bacteria living in it. Leaf litter contributes significantly to the soil of forests.

Soils vary in color, from the brilliant reds and browns of iron-rich soils to the black of soils rich in organic matter. Soils also vary in texture. Some soils are full of pebbles and sand; others are composed entirely of clay. Because soils are easily eroded, they do not form on very steep slopes or where high altitude or frigid climate prevents plant growth.

Soil is such an essential part of our environment and economy that a separate science, soil science, has developed in the twentieth century. Soil scientists, agronomists, geologists, and engineers study the composition and origin of soils, their suitability for agriculture and construction, and their value as a guide to climatic conditions in the past. Many scientists are focusing special attention on ways to combat the serious threat of soil erosion (see Feature 6.1).

they have become eroded. There is some balance between the moderate natural erosion of soils by streams and the wind and the slow formation of new soil. If soil is formed and eroded at roughly the same rate, its thickness remains constant. If soil is eroded more slowly than it is formed, it grows thick. If it erodes much more rapidly than it forms, new soil has no opportunity to form and the existing soil is quickly lost.

Agriculture accelerates erosion because plowing breaks up the soil and eliminates the erosion-resistant natural plant cover. Soil erosion has been especially severe in many places on the globe. One such place is the North American prairies, where plows have bitten deep and conservation practices were long ignored. The loosened soils have thinned and been carried away by the region's rivers. The prairies are now vulnerable to further erosion as dust storms strike during long periods of drought (see Chapter 14).

Contour plowing like that shown in the accompanying photo can lessen the damage caused by agriculture. This practice covers the field with curved furrows that follow the contours of the natural slopes, rather than running parallel to property lines that intersect slopes. Contour plowing inhibits erosion because much of the rainwater is channeled along the contoured furrows instead of running off downhill.

Despite widespread promotion of such agricultural techniques, however, soils are still eroding much more rapidly than they can be replaced. In parts of the United States and Canada, over 10 tons of topsoil per acre of cropland are lost annually. In the United States alone, 2 billion tons of topsoil are lost to erosion each year—twice the amount of soil formed during the same period. A loss on that scale is equivalent to destroying 780,000 acres of cropland. Comparable losses are hitting other agricultural regions of the world, such as the steppes of Russia, parts of Africa, and deforested areas of the Amazon Basin in Brazil. It has been estimated that all the direct and indirect costs of soil erosion add up to $44 billion in the United States and $400 billion over the entire globe. If such losses persist, the next century will see decreasing agricultural yields from thinned soils and the inevitable abandonment of agriculture in the regions most seriously affected. And once soil is gone, it takes thousands of years to form again. The estimation and prediction of soil erosion are a backdrop to policy decisions on optimal agricultural methods and ways of maintaining agriculture at a sustainable level while preserving Earth's soil cover.

Soil Profiles

A road or trench cut through a soil reveals its vertical structure, the soil profile (Figure 6.16). Its topmost layer, usually not much more than a meter or two thick, is usually the darkest, containing the highest concentration of organic matter. The topmost layer of soil is called the **A-horizon** (a particular level in a rock section is commonly called a "horizon"). In a thick soil that has formed over a long period of time, the inorganic components of this top layer are mostly clay and insoluble minerals such as quartz. Soluble minerals have been leached

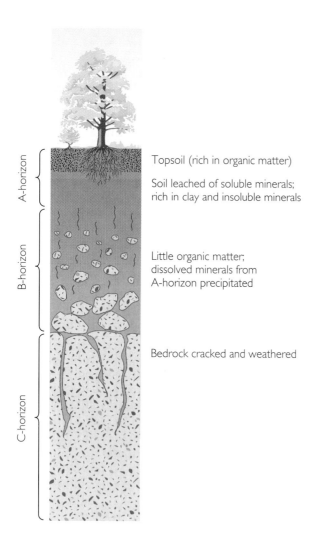

Topsoil (rich in organic matter)

Soil leached of soluble minerals; rich in clay and insoluble minerals

Little organic matter; dissolved minerals from A-horizon precipitated

Bedrock cracked and weathered

FIGURE 6.16 Soil profile. The thickness of the soil profile depends on the climate, the length of time the soil has been forming, and the composition of the parent rock. The transition from one horizon to another is normally indistinct.

from this layer. Beneath the topmost section is the **B-horizon,** where organic matter is sparse. In this layer soluble minerals and iron oxides have accumulated in small pods, lenses, and coatings. The lowest layer, the **C-horizon,** is slightly altered bedrock, broken and decayed, mixed with clay from chemical weathering. The transition from one horizon to another is normally indistinct.

Soils also are described as either residual or transported. Residual soils evolve in one place from bedrock to well-developed soil horizons. Most soils are residual, and they form faster and become thicker where weathering is intense. Even under intense weathering, the A-horizon may take thousands of years to develop to the point where it can support crops. Soil forms so slowly because the most active chemical weathering occurs only during short intervals of fresh rain. During dry periods the reactions continue, but very slowly and only when some moisture remains in the soil. When soil dries out completely between rains, chemical weathering stops almost completely.

Transported soils may accumulate in some limited areas of lowlands after being eroded from surrounding slopes and carried downhill. They can be recognized because in texture and composition they are similar to soils rather than to ordinary sediments. In some cases, parts of the original upland soil profile are preserved. These soils owe their thickness to deposition rather than to weathering in place. Transported soils are common and may be confused with ordinary sediment laid down by rivers, wind, and ice.

Climate, Time, and Soil Groups

Because climate so strongly affects weathering, it has a great influence on the characteristics of the soil formed on any given parent rock. The characteristics of a soil in a warm and humid region, for instance, differ from those of soils in arid and temperate regions. Because soil is so important to agriculture, soil scientists and other researchers set out to map soil characteristics over much of the world. This research has led to a very detailed level of mapping for use in preventing soil erosion and fostering efficient agricultural practices (see Feature 6.2). For our purposes, we can distinguish three major soil groups on the basis of their mineralogy and chemical composition, which also correlate with climate (Figure 6.17).

WET CLIMATES: LATERITES In warm, humid climates, weathering is fast and intense and soils become thick. The higher the temperature and humidity, the lusher the vegetation. Abundant vegetation and moisture and warm temperatures so greatly speed up chemical weathering that the uppermost layer of the soil is leached of all soluble and easily weatherable minerals. The residue of this fast weathering is **laterite,** a deep-red soil in which feldspar and other silicates have been completely altered, leaving behind mostly aluminum and iron oxides and hydroxides (Figure 6.17a). Silica as well as calcium carbonate has been leached from laterite. Although these soils may support lush vegetation in equatorial jungles, they are not very productive for

crop plants. Most of the organic matter is constantly recycled from the surface to the vegetation, with at best a very thin layer of humus at the top of the soil. The clearing of trees and tilling of the soil allow its superficial layer of rich humus to oxidize quickly and disappear, uncovering the infertile layer below.

For this reason, many laterites can be farmed intensely for only a few years after they have been cleared before they become barren and have to be abandoned. Large regions of India are now in this condition. As sections of the Amazon rain forest of Brazil are cleared, they too become barren after only a few years. Under natural conditions it takes a very

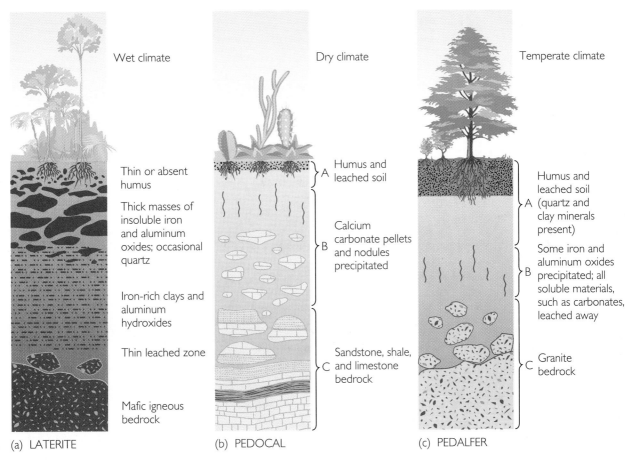

Wet climate

Thin or absent humus

Thick masses of insoluble iron and aluminum oxides; occasional quartz

Iron-rich clays and aluminum hydroxides

Thin leached zone

Mafic igneous bedrock

(a) LATERITE

Dry climate

A — Humus and leached soil

B — Calcium carbonate pellets and nodules precipitated

C — Sandstone, shale, and limestone bedrock

(b) PEDOCAL

Temperate climate

A — Humus and leached soil (quartz and clay minerals present)

B — Some iron and aluminum oxides precipitated; all soluble materials, such as carbonates, leached away

C — Granite bedrock

(c) PEDALFER

FIGURE 6.17 Major soil types. (a) Laterite soil profile developed on a mafic igneous rock in a tropical region. In the upper zone, only the most insoluble precipitated iron and similar oxides remain, plus occasional quartz. All soluble materials, including even relatively insoluble silica, are leached; thus the whole soil profile may be considered to be an A-horizon directly overlying a C-horizon. (b) Pedocal soil profile developed on sedimentary bedrock in a region of low rainfall. The A-horizon is leached; the B-horizon is enriched in calcium carbonate precipitated by evaporating soil waters. (c) Pedalfer soil profile developed on granite in a region of high rainfall. The only mineral materials in the upper parts of the soil profile are iron and aluminum oxides and silicates such as quartz and clay minerals, all of which are very insoluble. Calcium carbonate is absent.

Mapping Soil Types with a Newer Classification

Mapping of soils by the U.S. Soil Conservation Service and similar government agencies in other countries is a necessity for soil scientists. These scientists need accurate maps to understand soil formation better in order to reduce soil erosion and make optimum agricultural use of soils. In order to map soils, they need a good working classification. While the terms we have used, *pedalfer, pedocal,* and *laterite,* are useful and widely recognized as basic soil types, they are not easy to use in practical mapping, partly because they do not take sufficient account of the differences in bedrocks. Below is a list of the ten orders, as they are called in this classification, which is in official use in the United States and other countries.

SOIL ORDER	PROPERTIES	CORRELATION WITH MAJOR SOIL TYPES
Entisol	Soil formation only slightly advanced, soil profile rudimentary	—
Inceptisol	Weakly developed soils with definite A-horizon and poorly developed B-horizon	Pedalfer
Mollisol	High organic matter content in A-horizon	Pedalfer Pedocal
Alfisol	Some organic matter in A-horizon; A is thin and overlies clay-enriched B-horizon	Pedalfer
Spodosol	Organic-rich A-horizon overlies iron-enriched B-horizon	—
Ultisol	A-horizon overlies highly weathered B-horizon	Pedalfer Laterite
Oxisol	A-horizon overlies extremely weathered B-horizon	Laterite
Aridosol	Little organic matter in A-horizon and salt or silica accumulations at depth	Pedocal
Histosol	Peaty (abundant in preserved dead vegetation) soils	—
Vertisol	High content of clay minerals, which expand under wet conditions and contract in dry seasons	—

Soil types of North and South America. Alfisols and mollisols make the best farmland. (From William P. Cunningham and Barbara Saigo, *Environmental Science: A Global Concern* [Dubuque, Iowa: Wm. C. Brown, 1990], p. 188.)

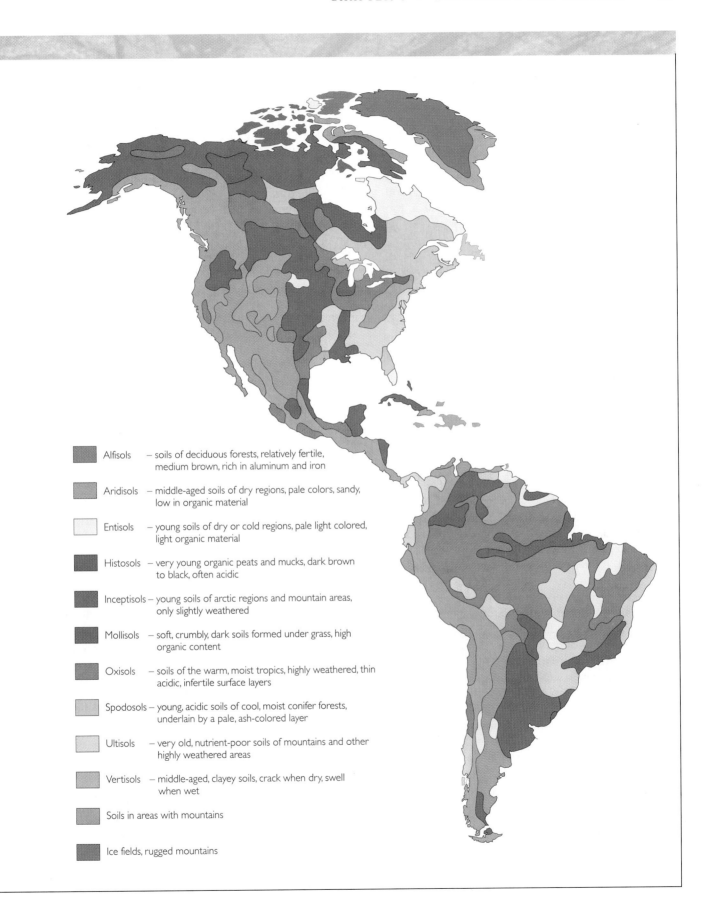

Alfisols — soils of deciduous forests, relatively fertile, medium brown, rich in aluminum and iron

Aridisols — middle-aged soils of dry regions, pale colors, sandy, low in organic material

Entisols — young soils of dry or cold regions, pale light colored, light organic material

Histosols — very young organic peats and mucks, dark brown to black, often acidic

Inceptisols — young soils of arctic regions and mountain areas, only slightly weathered

Mollisols — soft, crumbly, dark soils formed under grass, high organic content

Oxisols — soils of the warm, moist tropics, highly weathered, thin acidic, infertile surface layers

Spodosols — young, acidic soils of cool, moist conifer forests, underlain by a pale, ash-colored layer

Ultisols — very old, nutrient-poor soils of mountains and other highly weathered areas

Vertisols — middle-aged, clayey soils, crack when dry, swell when wet

Soils in areas with mountains

Ice fields, rugged mountains

long time, many thousands of years, to restore a forest on laterite soil.

DRY CLIMATES: PEDOCALS   Lack of water and absence of vegetation impede weathering, so soils are thin in arid regions. In cold arid regions, where chemical weathering is very slow, the parent rock's influence is dominant, even where soils have been forming over long periods. As a result, A-horizons contain much unweathered parent rock mineral and fragments of the parent rock. When rainfall is too sparse to dissolve significant amounts of soluble minerals, these minerals may remain in the A-horizon.

Pedocals are the dominant soils in arid regions (Figure 6.17b). These soils are rich in calcium from the calcium carbonate and other soluble minerals they contain and low in organic matter. Pedocals are found in the southwestern United States and in similar areas. In such climates, much of the soil water is drawn up near the surface and evaporates between rainfalls, leaving precipitated nodules and pellets of calcium carbonate, mostly in the middle layer of the soil. Because the combination of mineralogy and dryness is less favorable for a large population of organisms in the soil, pedocals are not as fertile as the pedalfers, which we consider next.

TEMPERATE CLIMATES: PEDALFERS   The characteristics of soils in regions with moderate rainfall and temperatures depend on the climate, the type of parent rock, and the length of time the soil has had to develop and thicken. Intense weathering decreases the influence of the parent rock, as does length of exposure to weathering. Thus soil developed after a relatively short time on a granite bedrock in a climate with moderate temperatures and humidity may differ greatly from the soil formed on limestone under the same conditions. The soil on the granite may still contain remnants of the silicate minerals and be dominated by the clay minerals forming from feldspar, a chief constituent of the parent rock. The soil on the limestone may still have a few remnants of calcium carbonate, but most of the limestone fragments will be dissolved. The clay minerals will be mainly those found as impurities in the parent limestone. After many thousands of years, though, the differences between the two soils may dwindle or even disappear. Both soils may develop the same clay minerals, depending on the exact nature of the climate, and both will have lost all soluble minerals in the upper layers.

Areas of moderate to high rainfall in the eastern United States, most of Canada, and much of Europe are home to the **pedalfer** group of soils (Figure 6.17c). Pedalfers take their name from *pedon,* the Greek word for "ground" or "soil," and *al* and *fer,* from the chemical symbols for aluminum and iron. The upper and middle layers of pedalfers contain abundant insoluble minerals such as quartz, clay minerals, and iron alteration products. Carbonates and other more soluble minerals are absent. Pedalfers make good agricultural soils.

PALEOSOLS: WORKING BACKWARD FROM SOIL TO CLIMATE   Recently there has been much interest in ancient soils that have been preserved as rock in the geological record, some of them over a billion years old. These *paleosols,* as they are called, are being studied as guides to ancient climates and even to the amounts of carbon dioxide and oxygen in the atmosphere in former times. Paleosols billions of year old provide evidence in their mineralogy that there was no oxidation of soils at that early stage of Earth's history and therefore that oxygen had not evolved to become a major fraction of the atmosphere.

HUMANS AS WEATHERING AGENTS

People tend not to think of themselves as part of the natural landscape, but they are. Humans are responsible for acid rains, which enhance chemical weathering perceptibly in the case of monuments but imperceptibly break down rock outcrops as well. Because chemical weathering is promoted by physical weathering, humans speed up both processes by their innumerable activities that break up rocks, from digging foundations and constructing highways to massive mining operations. It has been estimated that road building alone moves 3000 trillion tons of rock and soil throughout the world each year.

Over the last decade, researchers have discovered that soils may also be long-term sources of pollution after spills of toxic materials contaminate the soil. Contaminants then leak slowly from the soil into ground and surface waters. Humans have also loaded the ground with salt, pesticides, and petroleum products, all of which may significantly interfere with plant growth, and thus render soils vulnerable to accelerated erosion. Humans have significantly affected the surface of Earth in the few thousands of years of civilization.

FIGURE 6.18 Sand is a sediment composed of grains of various minerals broken down from various parent rocks in the course of chemical and physical weathering. *(Rex Elliott.)*

Weathering Makes the Raw Material of Sediment

Soil is only one of many products of weathering. Processes that weather rocks and break them apart produce fragments that vary greatly in size and shape, from huge boulders 5 m across to small particles too fine to see without a microscope. Pieces larger than a large grain of sand (2 mm in diameter) tend to be rock fragments containing mineral grains of the parent rock. Sand and silt grains are usually individual crystalline grains of any of the various minerals that make up the rock (Figure 6.18). The smallest pieces of weathered material are fine clay minerals, which are the products of the chemical alteration of silicate minerals.

Collectively, all the atoms and ions that are products of chemical and physical weathering are equal to all the atoms and ions of the original rock that weathered. A certain mass of granite, for example, is broken down by weathering into the following classes:

- Mineral and rock fragments of the parent rock that are still identifiable as such
- Solid products of chemical alteration, such as clay minerals and iron oxides
- Ions dissolved in rainwater and soil water

The mass of all these products, minus the water and carbon dioxide derived from the atmosphere, equals the original mass of the granite that was weathered.

In this way, a granite rock is transformed into the raw material of sediment. The weathered products are eventually carried away from the weathering site by wind, water, and ice. Ultimately they are deposited as various kinds of sediment, such as the sand, silt, and mud found in river valleys. As the rock cycle proceeds, this sediment is buried by additional deposits, and gradually it turns into sedimentary rock, the subject of Chapter 7.

Summary

What is weathering and how is it geologically controlled? Rocks are broken down at the surface of the Earth by chemical weathering, the chemical alteration or dissolution of a mineral, and by physical weathering, the fragmentation of rocks by physical processes. Erosion, the wearing away of the land, carries away the products of weathering, the raw material of sediment. The nature of the parent rock affects weathering because different minerals weather at different rates. Climate affects weathering: warmth and heavy rainfall speed weathering; cold and dryness slow it down. The presence of soil promotes weathering by providing moisture and an acidic environment, which promote chemical weathering and the growth of plant roots that aid physical weathering. All else being equal, the longer a rock weathers, the more completely it breaks down.

How does weathering work? Potassium feldspar ($KAlSi_3O_8$), in a way typical of silicate minerals, weathers in the presence of water through hydration. In a chemical reaction, potassium (K) and silica (SiO_2) are lost to the water solution and the solid feldspar alters to a clay mineral, kaolinite ($Al_2Si_2O_5[OH]_4$). Carbon dioxide dissolved in water promotes chemical weathering by providing acid, in the form of carbonic acid. Dissolved ions and silica are carried away by water in the ground and by streams on the surface. Iron (Fe), which is found in ferrous form in many silicates, weathers by oxidation, producing ferric oxides in the process. Carbonates weather by dissolving completely, leaving no residue. These processes operate at varying rates, reflecting the degrees of chemical stability of minerals under weathering conditions.

What are the processes of physical weathering? Physical weathering breaks rocks into fragments, either along crystal boundaries or along joints in rock masses. Physical weathering is promoted by chemical weathering, which weakens grain boundaries; by frost wedging, crystallization of minerals, and burrowing and tunneling by organisms and tree roots, all of which expand cracks; by fires; and perhaps by alternating extremes of heat and cold. Patterns of breakage such as exfoliation and spheroidal weathering result from interactions between chemical and physical weathering processes.

How do soils form as products of weathering? Soil is a mixture of clay minerals, weathered rock particles, and organic matter that forms as organisms interact with weathering rock and water. Because weathering is controlled by climate and the activity of organisms, soils form faster in warm, humid climates than in cold, dry ones. Young soils reflect the composition of the parent rock, but older soils primarily reflect climate. The three major soil groups are pedocals, which form in warm, dry climates; pedalfers, found in temperate climates; and laterites, found in the humid tropics.

KEY TERMS AND CONCEPTS

weathering (p. 135)
chemical weathering (p. 135)
physical weathering (p. 136)
erosion (p. 136)
solubility (p. 136)
climate (p. 137)
soil (p. 137)
kaolinite (p. 140)
hydration (p. 140)

bauxite (p. 144)
oxidation (p. 144)
ferrous iron (p. 144)
hematite (p. 144)
ferric iron (p. 144)
chemical stability (p. 146)
joints (p. 149)
frost wedging (p. 149)
exfoliation (p. 150)

spheroidal weathering (p. 150)
regolith (p. 152)
humus (p. 153)
A-horizon (p. 154)
B-horizon (p. 154)
C-horizon (p. 154)
laterite (p. 155)
pedocal (p. 158)
pedalfer (p. 158)

EXERCISES

1. What do the various kinds of rocks used for monuments tell us about weathering?

2. What rock-forming minerals found in igneous rocks weather to clay minerals?

3. How does abundant rainfall affect weathering?

4. Which weathers faster, a granite or a limestone?

5. How does physical weathering affect chemical weathering?

6. How do climates affect chemical weathering?

7. What are the main factors that control the development of different soil types?

8. What accelerates soil erosion?

THOUGHT QUESTIONS

1. You are planning to use polished decorative limestone facings for monuments in Tucson, Arizona (a hot, arid region), and Seattle, Washington (a cool, rainy region). How do you think each of these monuments might look after a hundred years?

2. In northern Illinois you can find two soils developed on the same kind of bedrock; one is 10,000 years old and the other is 40,000 years old. What differences would you expect to find in their composition or profile?

3. Which igneous rock would you expect to weather faster, a granite or a basalt? What factors influenced your choice?

4. Assume that a granite with crystals about 4 mm across and a rectangular system of joints spaced about 0.5 to 1 m apart is weathering at Earth's surface. What size would you ordinarily expect the largest weathered particle to be?

5. You have been comparing weathering in two rocks: (a) a basalt rich in volcanic glass, with very small crystals of pyroxene and calcium-rich feldspar that are about 0.5 mm across; and (b) a gabbro of exactly the same mineral composition but with large crystals about 3 mm across. What can you say about the speed of weathering in these two rocks?

6. Why do you think a road built of concrete, an artificial rock, tends to crack and develop a rough, uneven surface in a cold, wet region even when it is not subjected to heavy traffic?

7. Pyrite is a mineral in which ferrous iron is combined with sulfide ion. What major chemical process weathers pyrite?

8. Rank the following rocks in the order of the rapidity with which they weather in a warm, humid climate: granite; a sandstone made of pure quartz; a limestone made of pure calcite; a deposit of rock salt (halite, $NaCl$).

9. What differences do you expect to find in the weathering of a pure magnesium olivine (Mg_2SiO_4) and a pure iron olivine (Fe_2SiO_4)?

10. What would a world look like if there were no weathering at the surface?

11. Go to a local graveyard and compile a one-page description of visible evidence of weathering of different kinds of rock used for gravestones. If you find no such evidence, explain why.

SUGGESTED READINGS

Blatt, Harvey. 1992. *Sedimentary Petrology*, 2nd ed. New York: W. H. Freeman.

Carroll, Dorothy. 1970. *Rock Weathering*. New York: Plenum.

Colman, S. M., and D. P. Dethier. 1986. *Rates of Chemical Weathering of Rocks and Minerals*. New York: Academic Press.

Gauri, K. L. June 1978. The preservation of stone. *Scientific American*, p. 126.

Loughnan, F. C. 1969. *Chemical Weathering of the Silicate Minerals*. New York: Elsevier.

Martini, I. P., and W. Chesworth, eds. 1992. *Weathering, Soils, and Paleosols*. New York: Elsevier.

Nahon, D. B. 1991. *Introduction to the Petrology of Soils and Chemical Weathering*. New York: Wiley.

Retallack, G. J. 1990. *Soils of the Past*. Boston: Unwin Hyman.

INTERNET SOURCES

Acid Rain Resources

🛈 **http://www.econet.apc.org/acidrain/**
The EcoNet organization provides this gateway to a variety of Web resources: conferences, publications, information sites for places and problems around the world, educational resources, and the "strange and unusual."

U.S. Soil Survey

🛈 **http://www.nhq.nrcs.usda.gov/BCS/soil/survey.html**
Operated by the U.S. Department of Agriculture's Ecological Sciences Division, this site is a gateway to a variety of soils information and resources, including the National Soil Survey Information System, the National Soil Survey Center, and the National Soil Data Access Facility.

7

Layers of sedimentary rock in Marble Canyon, part of the Grand Canyon, which was cut by the Colorado River through what is now northern Arizona. These layers record millions of years of geologic history. (*Tom Bean / DRK.*)

162

Sediments and Sedimentary Rocks

uch of Earth's surface is covered with sediments, layers of loose particles that originated from the weathering of the continents as solid debris from rocks or as dissolved substances. Because sedimentary rocks were once sediments, they are records of the conditions at the surface when and where they were deposited. Using evidence provided by a sedimentary rock's texture, physical structure, mineral content, and environment, geologists can work backward to infer the sources of the sediments from which the rocks were formed and the kinds of places in which the sediments were originally deposited. The Grand Canyon, for example, contains beds of limestone in which fossils of marine organisms are embedded. This evidence allows us to infer that long before the Grand Canyon was formed, this region, now high on the continent, must have been the floor of an ocean.

Even a single quartz grain in a sandstone has a story to tell. It may originally have been a crystal in a granite. Perhaps weathering freed it from the other minerals of the granite, and it became a loose sand grain. It was then carried by a stream to the ocean, where currents transported it to its ultimate resting place. There it joined other grains to form a layer of sediment. Finally, it may have become cemented to its neighboring grains by the precipitation of additional quartz around the original grain.

The kind of analysis that allows geologists to make inferences about rock formations in the Grand Canyon also enables them to deduce ancient shorelines, mountains, plains, deserts, and swamps in other regions. In one part of a region, for example, sandstone may record an earlier time when sedimentation deposited sands under a sea that no longer exists.

In a bordering area, carbonate reefs may have been laid down along the edge of a continental shelf. Beyond, there may have been a nearshore area in which the sediments were shallow ocean carbonate muds that later became thin-bedded limestone. In reconstructing such environments, we can map the continents and oceans of long ago.

Further inferences are also possible. We can work backward to understand former plate-tectonic settings and movements using the evidence of clastic sedimentary rocks, which reflect their origin in volcanic arcs, rift valleys, or collisional mountains. And because the components of sediments and sedimentary rocks are derived from the weathering of pre-existing rocks, we can form hypotheses about the ancient climate and weathering regime. We can also use marine sediments, the products of ocean

---

7.1   INTERPRETING THE EARTH

Sedimentary Rocks Illuminate the Principle of Uniformitarianism

In Chapter 1, we introduced the principle of uniformitarianism—the idea that geological processes have operated in the past much as they do today—as one of the great guiding generalizations of geology. The principle applies even though the rates of some processes may have changed and catastrophes such as cometary impacts have occurred in the past. Since the beginning of the nineteenth century, sediments and sedimentary rocks have been central to explaining uniformitarianism. The reason is that sedimentary processes are directly observable at the surface of the Earth, whereas igneous and metamorphic processes are largely hidden deep in the Earth (although we do see volcanic eruptions in action).

If the present is the key to the past, then our best way to discover, for example, how river sediments are laid down is to observe a river today. In the middle of the nineteenth century, Edward Hitchcock, a pioneer of American geology, did just that when he observed from his home in Northampton, Massachusetts, a flood of the Connecticut River. He noted how the river gradually rose until it spilled over its banks and covered the entire width of the valley. And he watched as

the river receded to its channel, leaving river-laid deposits of sands, silts, and muds. Those sediments could be matched with ancient sandstones and shales that had all the same characteristics, such as cross-bedding and ripples. Hitchcock and other geologists all over the world have left us a legacy of that kind of description of geological processes, which enables us to use the principle of uniformitarianism to work back from observations of ancient sedimentary (or other) rocks to the nature of the environment in which the sediments were deposited.

Have rivers always behaved exactly like the Connecticut? Perhaps not exactly. Some geologists now believe that before land plants, such as grasses and trees, had evolved, about 450 to 500 million years ago, river banks, lacking plant root systems, were much more easily eroded than those of today; the greater erodibility led to different patterns of channels and flooding. In addition, carbonate sedimentation in the deep ocean before about 100 million years ago was very different from that of today, because the unicellular organisms whose calcium carbonate shells make up the bulk of deep-sea carbonate today had not yet evolved.

water chemistry, to read the history of Earth's oceans.

Not only is the study of sediments and sedimentary rocks enormously interesting in its own right, it also has great practical value. Oil, gas, and coal, our most valuable sources of energy, are found in these rocks. So is much of the uranium used for nuclear power. Phosphate rock used for fertilizer is sedimentary, as is much of the world's iron ore. Knowing how these kinds of sediments are formed helps us to explore for additional resources more intelligently.

Many geologists are interested in sedimentary processes because they provide vital clues to understanding the environment. Because virtually all sedimentary processes take place on Earth's surface where we humans live, they are background to our understanding of environmental problems. Although sedimentary geology has been actively studied for

hundreds of years, the environmental aspects of this area of geology have been recognized as a critical interest only relatively recently, having come to the fore in the 1960s.

In this chapter, we will see how geological processes such as weathering, transportation, sedimentation, and diagenesis produce sediments and sedimentary rocks. We will describe the compositions, textures, and structures of sediments and sedimentary rocks and examine how they correlate with the kinds of environments in which the sediments and rocks are laid down. We will see how the interpretation of the origin of sediments and sedimentary rocks is heavily based on the doctrine of uniformitarianism (Feature 7.1). Throughout the chapter, we will apply our understanding of sediment origins to the study of human environmental problems and to the exploration for energy and mineral resources.

Yet the fundamental nature of river and deep-sea carbonate sedimentation has not changed, just some of the details.

Uniformitarianism plays a role in understanding planetary geology, too. Our satellite and ground observations of Mars lead us to

Itkillik River, Brooks Range, Alaska, carries a high sediment load. *(Tom Bean.)*

believe that rivers once ran there. Planetary geologists used their knowledge of how Earth's rivers work to infer that valleys and sediments on Mars more than a billion years ago behaved much as they do here on Earth today, eroding valleys and depositing sands, silts, and muds in the same way. Broadly interpreted, the principle of uniformitarianism can be used to infer the geological behavior of strange planetary bodies such as Titan, Saturn's largest moon, which has an atmosphere of nitrogen, ethane, methane, and other organic compounds and has frozen lakes of some of these gases.

In some ways, the principle reduces to a simple statement: the laws of chemistry and physics do not change, only specific conditions at one place or another. Yet in a complex environment, such as a river channel and valley, that is not much help, for geological processes cannot be deduced from the basic laws of physics and chemistry. Geological processes are studied in their full detail to reveal their complexity. But because these processes always remain completely compatible with physics and chemistry, uniformitarianism remains a powerful tool for interpreting the past.

SEDIMENTARY ROCKS AND THE ROCK CYCLE

Sediments and the sedimentary rocks formed from them are important elements of the rock cycle, as described in Chapter 3. Sedimentary rocks occur in the surface part of the cycle, between rocks brought up from the interior by tectonics and rocks returned to the interior by burial. Figure 7.1 reviews the several overlapping processes that make up the sedimentary stages of the rock cycle:

- *Weathering*—Physical weathering fractures rocks; chemical weathering converts minerals and rocks to altered solids, solutions, and sometimes precipitates.

- *Erosion*—Erosion carries away the particles produced by weathering.

- *Transportation*—Currents of wind and water, and the movement of glaciers, transport particles to new locations downhill or downstream.

- *Deposition* (also called *sedimentation*)—Detrital particles settle out as wind dies down, water currents slow, or glacier edges melt; these particles form layers of sediment on land or under the sea. In the ocean or in land aquatic environments, chemical or biochemical precipitates are deposited.

- *Burial*—As layers of sediment accumulate, older deposited material is compacted and eventually buried in Earth's crust.

- *Diagenesis*—The term **diagenesis** refers to the physical and chemical changes—including pressure, heat, and chemical reactions—by which buried sediments undergo alterations and are lithified, assuming a new identity as sedimentary rocks. Diagenesis includes all postdepositional alterations to sediment or sedimentary rock, such as changes in mineral composition and the amount of space between the grains, but does not include erosion or metamorphism. Geological processes that result in diagenesis are termed *diagenetic.*

Weathering and Erosion Yield the Raw Materials: Particles and Dissolved Substances

As we saw in Chapter 6, chemical weathering and mechanical fragmentation of rock at the surface make both solid and dissolved products, and erosion carries away these materials. The end products are known as clastic sediments and chemical and biochemical sediments.

CLASTIC SEDIMENTS   Clastic particles are the physically transported solid fragments produced by weathering of preexisting rocks. These particles range in size from boulders to pebbles to particles of sand, silt, and clay. They also vary widely in shape. The shapes of boulders, cobbles, and pebbles are determined by natural breakage along joints, bedding planes, and other fractures in the parent rock. Sand

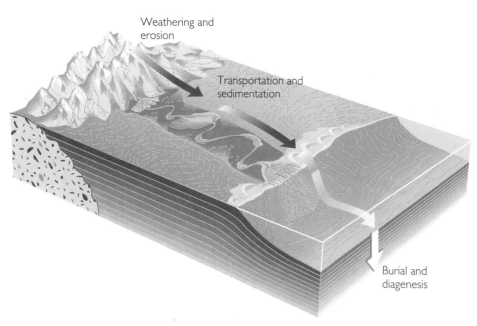

FIGURE **7.1** The sedimentary stages of the rock cycle embrace several overlapping processes: *physical and chemical weathering; erosion; transportation and sedimentation;* and *burial.* Burial converts the sediment to rock through *diagenesis.*

grains tend to inherit their shapes from the individual crystals formerly interlocked in the parent rock.

Clastic sediments are accumulations of clastic particles. These sediments are also often called *siliciclastic* because they are produced by the weathering of rocks composed largely of silicate minerals. The mixture of minerals in clastic sediment varies. Minerals such as quartz are resistant to weathering and thus are stable. There may be partially altered fragments of minerals that are less resistant to weathering and so less stable, such as feldspar. Still other minerals may be newly formed, such as clay minerals. Varying intensities of weathering can produce different sets of minerals in sediments derived from the same parent rock. Where weathering is intense, the sediment will contain only clastic particles made of chemically stable minerals, mixed with clay minerals. Where weathering is slight, many minerals that are unstable under surface conditions will survive as clastic particles. Table 7.1 shows three possible sets of minerals for a typical granite outcrop.

On average, clastic particles form and accumulate much more rapidly than chemical or biochemical precipitates. In most of the world, much more rock is fragmented by physical weathering than is dissolved by chemical weathering. As a result, clastic sediments are about 10 times more abundant in the Earth's crust than are chemical and biochemical sediments.

CHEMICAL AND BIOCHEMICAL SEDIMENTS

The dissolved products of weathering are ions or molecules in the waters of soils, rivers, lakes, and oceans. These dissolved substances are precipitated from water by chemical and biochemical reactions. **Chemical sediments** are formed at or near their place of deposition, usually from seawater. **Biochemical sediments** contain the mineral remains of organisms or minerals precipitated as a result of biological processes. We distinguish these two sediment types here for convenience only; in practice, many chemical and biochemical sediments overlap.

Many layers of biochemical sediment consist of biologically precipitated sedimentary particles, such as whole or fragmented shells. Other layers are transported and deposited as **bioclastic particles.** These layers are composed of detritus of previously deposited and broken up calcium carbonate sediments. In the deep sea, for example, calcium carbonate sediments are made of the shells of only a few kinds of organisms and contain few (if any) bioclastic particles. They are composed of only one calcium carbonate mineral, calcite. In contrast, calcium carbonate sediments found in shallow waters may be mixtures of the shells of many different kinds of organisms as well as bioclastic detritus. These shallow-water sediments also comprise two calcium carbonate minerals, calcite and aragonite.

Other chemical sediments are formed inorganically. For example, the evaporation of seawater commonly leads to the precipitation of layers composed of gypsum or halite.

Transportation and Deposition: The Downhill Journey to Sedimentation Sites

Once clastic particles and dissolved ions have been formed by weathering, they start a journey to a sedimentation area. Chemically or biochemically

---

TABLE **7.1**

MINERALS IN CLASTIC SEDIMENTS DERIVED FROM AN AVERAGE GRANITE OUTCROP UNDER VARYING INTENSITIES OF WEATHERING			
	INTENSITY OF WEATHERING		
	LOW	**MEDIUM**	**HIGH**
Minerals remaining in sediment	Quartz	Quartz	Quartz
	Feldspar	Feldspar	Clay minerals
	Mica	Mica	
	Pyroxene	Clay minerals	
	Amphibole		

produced particles may also be transported from one area to another, usually nearby.

Almost all agents of transportation carry material downhill. Rocks falling from a cliff, sand carried by a river flowing to the sea, and glacial ice slowly creeping downhill are all responses to gravity. Although winds may blow material from a low elevation to a height and back again, in the long run gravity is relentless, and windblown sand and dust settle in response to its pull. Once a windblown particle drops to the ocean and settles through the water, it is trapped. It can be picked up again only by an ocean current, which can transport it only to another depositional site on the seafloor.

CURRENTS AS TRANSPORT AGENTS FOR CLASTIC PARTICLES   Much of the transportation of sediment is accomplished by currents, which are movements of fluids such as air and water. (Scientists include air and other gases among fluids because, like liquids, they take the shapes of their containers, whereas solids maintain their own shapes.) Rivers, like ocean currents, are currents of water. The enormous quantities of all kinds of sediment found in the oceans indicate the transporting capabilities primarily of rivers, which annually carry a solid and dissolved sediment load of about 25 billion tons (250×10^{14} g).

Currents in air move material too, but in far smaller quantities than rivers or ocean currents. As the particles are lifted into the fluid, the current carries them downriver or downwind. The stronger the current—that is, the faster it flows—the larger the particles it transports.

CURRENT STRENGTH, PARTICLE SIZE, AND SORTING   Sedimentation starts where transportation stops, and for clastic particles, the driving force is in large part the effect of gravity. Working against the ability of a current to carry a particle is the particle's tendency to settle to the bottom of the flow in response to the pull of gravity. Although it is a basic law of physics that particles of all sizes fall to Earth at the same speed in a *vacuum,* large grains settle faster than small ones in a *fluid.* The settling velocity is proportional to the density of the particle as well as to its size. But the most common minerals in sediments have roughly the same density (about 2.6 to 2.9 g/cm^2). We therefore use size, which is more conveniently measured than density, as an indicator of the settling velocity of minerals in sediments.

As a current carrying particles of various sizes begins to slow, it can no longer keep the largest par-

ticles suspended, and they settle. As the current slows even more, smaller particles settle. When the current stops completely, even the smallest particles settle (see Feature 7.1). In more specific terms, currents sort particles in the following way:

- *Strong currents* (over 50 cm/s) carry gravels, together with an abundant supply of coarse and fine detritus. Such currents are common in rapidly flowing streams in mountainous terrains, where erosion is rapid. Beach gravels are deposited where ocean waves actively erode rocky shores. Glaciers also produce and deposit clastic debris of all sizes as they move forward.

- *Moderately strong currents* (20–50 cm/s) lay down sand beds. Currents of moderate strength are common to most rivers, which carry and deposit sand in their channels. Rapidly flowing floodwaters may spread sands over the width of a river valley. Sands are also blown and deposited by winds, especially in deserts, and by waves and currents on beaches and in the ocean.

- *Weak currents* (less than 20 cm/s) carry muds composed of the finest clastic particles. These weakest currents are found on the floor of a river valley when floodwaters recede slowly or stop flowing entirely. Muds are deposited in the ocean generally some distance from shore, where currents are too slow to keep even fine particles in suspension. Much of the floor of the open ocean is covered by mud particles originally transported by surface waves and currents or by the wind. These particles gradually settle to depths where currents and waves are stilled and, ultimately, all the way to the bottom.

As you can see, currents may begin by carrying particles of widely varying size, which then become separated as the velocity of the current varies. A strong, fast current may lay down a bed of gravel, while keeping sands and muds in suspension. If the current weakens and slows, it will lay down a bed of sand on top of the gravel. If the current then stops altogether, it will deposit a layer of mud on top of the sand bed. This tendency for variations in current velocity to segregate sediments according to size is called **sorting.** A well-sorted sediment is homogeneous—it consists mostly of particles of a uniform size. A poorly sorted sediment contains particles of many sizes (Figure 7.2).

FIGURE 7.2 As currents decrease in velocity, sediment is segregated according to size of particles. The relatively homogeneous group of sand grains on the left is well sorted; the group on the right is poorly sorted. *(Rex Elliott.)*

Particles are generally transported intermittently rather than steadily. Fast currents give way to weak flows or stop entirely. A river may transport large quantities of sand and gravel when it floods, but it will drop them as the flood recedes, only to pick them up again and carry them even farther in the next flood. Strong winds may carry large amounts of dust for a few days, then die down and deposit the dust as a layer of sediment. Similarly, strong tidal or other shallow-water currents along some ocean margins may transport eroded particles of previously deposited calcium carbonate sediments to places farther offshore and drop them there. Both chemical and physical weathering processes continue during transportation. Like transportation agents, the slower weathering processes tend to operate intermittently.

CHEMICAL WEATHERING IS INTERMITTENT Although clastic material is still in contact with the chief agents of chemical weathering—water and the oxygen and carbon dioxide of the atmosphere— slow weathering reactions do not have much effect during the brief periods in which material is actually being transported by a current. Most weathering is accomplished during the long intermittent periods when the sediment is temporarily deposited before being picked up again by the current. For ex-

ample, when a river floods its valley for a few days, it deposits sand, silt, and clay. After the flood recedes, weathering of the deposits resumes and continues until the next flood. That flood may pick up the sediment of the earlier flood and redeposit it farther downstream, where the sediment begins to weather again.

In this way, episodes of transportation and deposition can alternate with episodes of chemical weathering. Because deposition may be intermittent, the total time between the formation of clastic debris and final deposition may be many hundreds or thousands of years, depending on the distance to the final depositional area and the number of stop-offs along the way. Clastic particles eroded by the headwaters of the Missouri River in the mountains of western Montana, for example, take hundreds of years to travel the 2000 miles down the Missouri and Mississippi rivers to the Gulf of Mexico. The particles carried by these rivers may be affected by physical weathering during the long transportation process as well as by intermittent chemical weathering.

PHYSICAL WEATHERING REDUCES SIZE, ROUNDS ANGLES Transportation by water and wind currents continues physical weathering

Distance of transport

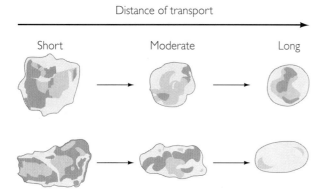

Short   Moderate   Long

FIGURE 7.3 Transportation reduces the size and angularity of clastic particles. Grains become rounded and slightly smaller as they are transported, although the general shape of the grain may not change significantly.

processes, which affect particles in two ways: they reduce particle size, and they round the originally angular fragments (Figure 7.3).

As particles are transported, they tumble and strike one another or rub against bedrock. Pebbles or large grains that collide forcefully may break into two or more smaller pieces. Weaker hits may chip small pieces from edges and corners. Abrasion by bedrock, together with grain impacts, also rounds the particles, wearing off and smoothing sharp edges and corners.

Larger particles such as gravels lose a greater proportion of their size by breakage and become more rounded for a given distance of transport than smaller ones such as sand grains. Boulders break into smaller pieces after they have traveled short distances; cobbles and pebbles travel somewhat farther before they break up and become rounded. Sand grains are much less likely to break into several pieces and tend to become rounded over long distances.

GLACIERS AS TRANSPORT AGENTS FOR CLASTIC PARTICLES Glaciers also carry clastic particles. As these rivers of ice move downhill in response to gravity, they incorporate large quantities of solid particles that have been eroded from soil and bedrock and carry this material away. Because the particles cannot settle through the solid ice, they do not become sorted, and the ice carries a heterogeneous mixture of sizes. As a glacier melts, however, it deposits a large quantity of clastic debris ranging from gravels to clays at the edge of the melting ice. The debris is then carried away by rivers of meltwater and becomes subject to the same sorting process that applies to other clastic particles.

During glacial transport, particles become smaller, but they are not rounded. This kind of transport slowly presses particles against one another and splits them. Glaciers also break up bedrock at their bottom and sides. (Transportation and sedimentation by glaciers are discussed in detail in Chapter 15.)

SOLUTIONS: VEHICLES FOR TRANSPORTING DISSOLVED MATERIALS The driving force of chemical and biochemical sedimentation is chemistry rather than gravity. Chemical substances dissolved in water during weathering are carried along with the water as a homogeneous solution. Because materials such as dissolved calcium ions are part of the water solution itself, they can never settle out of it. As dissolved materials flow down rivers, they ultimately enter lake waters or the ocean.

OCEANS: CHEMICAL MIXING VATS The ocean may be thought of as a huge chemical mixing tank. At any given time, dissolved materials are constantly being brought in by rivers, rain, wind, and glaciers. Smaller quantities of dissolved materials enter the ocean by hydrothermal chemical reactions between seawater and hot basalt at mid-ocean ridges. Currents and waves mix these materials with ocean water. The ocean continuously loses water by evaporation at the surface. The inflow and outflow of water to and from the oceans are so exactly balanced that the amount of water in the oceans remains constant over geologically short times such as years, decades, or even centuries. Over millions of years, the balance may shift somewhat.

The entry and exit of dissolved materials, too, are balanced. Each of the many dissolved components of seawater participates in some chemical or biochemical reaction that precipitates it out of the water and onto the seafloor. For example, although small quantities of the chemical substances transported by streams are precipitated in saline or alkaline lakes, far huger quantities enter the saline surroundings of the sea, an environment that is chemically very different from the more or less fresh waters of soils and rivers. In the chemical reactions that ensue, these dissolved substances exit the ocean waters as chemical and biochemical precipitates. As a result, the ocean's salinity—the total amount of dissolved substances in a given volume of seawater—remains constant. Totaled over all the oceans of the world, precipitation balances the total inflow of dissolved material from continental weathering and hydrothermal activity at mid-ocean ridges.

We can see some of the mechanisms by which this chemical balance is maintained by focusing on the element calcium. Calcium is an important component of the most abundant biochemical precipitate formed in the oceans, calcium carbonate ($CaCO_3$). Calcium is dissolved when limestone and silicates containing calcium, such as some feldspars and pyroxenes, are weathered on land and brought as ions (Ca^{2+}) to the oceans. There, a wide variety of marine organisms biochemically combine the calcium ions with bicarbonate ions (HCO_3^-), also present in seawater, to form their calcium carbonate shells. The calcium that entered the ocean as dissolved ions leaves it as a solid sediment when the organisms die and their shells settle and accumulate as calcium carbonate sediment on the seafloor, ultimately to be transformed by postdepositional processes into limestone. The chemical balance that keeps the levels of calcium dissolved in the ocean constant is thus regulated in part by the activities of organisms.

Chemical balance is also maintained by nonbiological mechanisms. For example, sodium ions (Na^+) brought into the oceans react chemically with chloride ions (Cl^-) to form the precipitate sodium chloride ($NaCl$) when evaporation raises the amounts of sodium and chloride ions to the point of supersaturation. As we saw in Chapter 2, supersaturated solutions crystallize minerals when they become so rich in dissolved materials that they react spontaneously to form precipitates. The intense evaporation required to crystallize salt takes place in warm, shallow arms of the sea.

Organic sedimentation is yet another type of biochemical precipitation. Vegetation may be preserved from decay in swamps and accumulate as a rich organic material, **peat,** which contains more than 50 percent carbon. Peat ultimately is buried and transformed by diagenesis to coal. In both lake and ocean waters, the remains of algae, bacteria, and other microscopic organisms may accumulate in sediments as organic matter that is transformed to oil and gas.

SEDIMENTARY ENVIRONMENTS

Of the many ways in which sedimentation can be classified, geologists have found the concept of sedimentary environments most useful. A **sedimentary environment** is a geographic location characterized by a particular combination of geological processes and environmental conditions (Figure 7.4). Environmental conditions include

- The kind and amounts of water (ocean, lake, river, arid land)

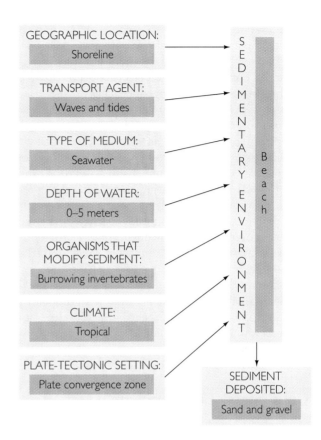

FIGURE 7.4 A sedimentary environment is characterized by a particular set of environmental conditions and geological processes. Properties of one specific example—a Caribbean island beach—are highlighted in the green boxes.

- The topography (lowland, mountain, coastal plain, shallow ocean, deep ocean)

- Biological activity

Geological processes include the nature of currents that transport and deposit sediments (water, wind, ice), the tectonic settings that may affect sedimentation and burial of sediments, and the presence of volcanic activity. Thus, a beach environment brings together the dynamics of waves approaching and breaking on the shore, the resultant currents, and the distribution of sediments on the beach.

Sedimentary environments are related to plate-tectonic settings. For example, the deep trenches of the oceans are found at subduction zones, whereas thick alluvial deposits are typically associated with mountains formed by the collision of continents. Alluvial environments are also found along the borders of continental rift valleys. Sedimentary environments may be affected or determined by climate as well as tectonics. For example, a desert environment implies an arid climate and a glacial environment a cold one.

Humans have extensively modified some environments. For example, the construction of seawalls and breakwaters has changed beaches greatly (see Chapter 17). Arid environments have been changed to humid ones in and around major cities such as Phoenix, Arizona, and at sites in Saudi Arabia, by long-distance water transport used for "greening" of the desert. Desert environments may expand and take over less arid neighboring environments as a result of overgrazing by cattle and deleterious agricultural practices. Stream environments may be altered by damming or artificial channelization. The chemical compositions of the waters of lakes and the coastal oceans may be changed by the introduction of polluted waters.

Sedimentary environments are often grouped by their locations on the continents, near shorelines, or in the ocean. Figure 7.5 and the following text summarize these groupings.

Continental Environments

Sedimentary environments on continents are diverse, reflecting the wide range in temperature and rainfall on the surface of the land. These environments are built around rivers, deserts, lakes, and glaciers:

- An *alluvial environment* encompasses a river channel, the borders of the channel, and the flat valley floor on either side of the channel that is covered by water when the river floods. Because rivers are ever-present on the continents, alluvial deposits are widespread. Organisms are abundant in the muddy flood deposits and are responsible for organic sediments. Climates vary from arid to humid.

- A *desert environment* is arid. Sediment there is formed by a combination of wind action and the work of rivers that flow (mostly intermittently) through it. Because the aridity inhibits organic growth, organisms have little effect on the sediment. Desert sand dunes provide a special sandy environment.

- A *lake environment* is controlled by the relatively small waves and moderate currents of inland bodies of fresh or saline water. Freshwater lakes may be the sites of chemical sedimentation of organic matter and carbonates. Saline lakes such as those found in deserts evaporate and precipi-

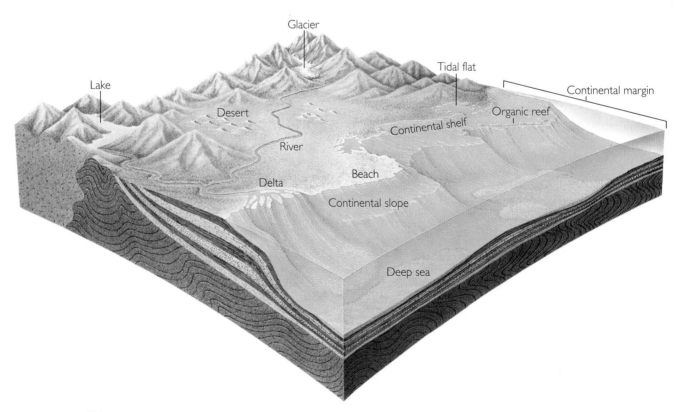

FIGURE 7.5 Common sedimentary environments.

tate a variety of evaporite minerals such as halite. The Great Salt Lake is an example.

▪ A *glacial environment* is controlled by the dynamics of moving masses of ice and is characterized by a cold climate. Vegetation is present but has small effects on sediment. At the melting border of a glacier, meltwater streams form a transitional alluvial environment.

Shoreline Environments

Shoreline environments are dominated by the dynamics of waves, tides, and currents on sandy shores. Organisms may be abundant in these shallow waters, but they do not influence clastic sedimentation much except where carbonate sediments are abundant. Shoreline environments include

▪ *Deltaic environments,* where rivers enter lakes or the sea.

▪ *Tidal flat environments,* where extensive areas exposed at low tide are dominated by tidal currents.

▪ *Beach environments,* which bring together the geological processes of the strong waves approaching and breaking on the shore and the distribution of sediments on the beach. In these environments, strips of sand or gravel are laid down by wave action.

Marine Environments

Marine environments encompass a wide variety of subenvironments, ranging from those affected by the salinity of the water, such as evaporites, to physically deposited sediments affected by currents. Because depth of water determines the kinds of currents, marine environments are usually subdivided on the basis of water depth. Alternatively, they can be classified on the basis of distance from land.

▪ *Continental shelf environments* are located in the shallow waters off continental shores, where sedimentation is controlled by relatively gentle currents. Sedimentation may be either clastic or chemical, depending on the source of clastics and the extent of carbonate-producing organisms or evaporite conditions.

▪ *Continental margin environments* are found in the deeper waters at the edges of the continents, where sediment is deposited by turbidity currents.

▪ *Organic reefs* are composed of carbonate structures formed by carbonate-secreting organisms built up on continental shelves or on oceanic volcanic islands.

▪ *Deep-sea environments* include all the floors of the deep ocean, far from the continents, where the quiet waters are disturbed only occasionally by ocean currents. These environments include the *deep trenches* of the oceans found at subduction zones, the *mid-ocean ridges* that lie over divergent plate boundaries, and *abyssal plains* built up by turbidity currents traveling far from continental margins.

Facies: Coexisting Sets of Sedimentary Environments

Different sedimentary environments exist at the same time in different parts of a region. Thus, along a continental margin, we might find beach, tidal flat, and deeper continental shelf environments, each with its characteristic kind of sediment. We call such sets of sediments *facies* to describe the differing mineralogical, textural, and structural aspects of sediments deposited simultaneously in the various environments. For example, we can speak of an alluvial facies found in a stream environment and a muddy carbonate facies found at the same time in a continental shelf carbonate environment offshore. (We will see in the next chapter, on metamorphic rocks, that *facies* is used in a different sense to describe the formation of metamorphic rocks in terms of pressure and temperature.)

Sedimentary environments are distinguished by the kinds of sedimentary facies found in them. For instance, the sediments in almost all alluvial environments are clastic, whereas chemical and biochemical carbonate sediments are dominant in coral reef or shallow carbonate platform environments. Another way geologists group sedimentary environments is according to the dominant category of sedimentation. Grouping in this manner produces two broad classes: clastic sedimentary environments and chemical and biochemical sedimentary environments.

Clastic versus Chemical and Biochemical Environments

Clastic sedimentary environments are those dominated by clastic sediments. They include the continental alluvial (stream), desert, lake, and glacial environments, as well as the shoreline environments

TABLE 7.2

CLASTIC SEDIMENTARY ENVIRONMENTS

ENVIRONMENT	AGENT OF TRANSPORTATION, DEPOSITION	SEDIMENTS
CONTINENTAL		
Alluvial	Rivers	Sand, gravel, mud
Desert	Wind	Sand, dust
Lake	Lake currents, waves	Sand, mud
Glacial	Ice	Sand, gravel, mud
SHORELINE		
Delta	River + waves, tides	Sand, mud
Beach	Waves, tides	Sand, gravel
Tidal flats	Currents	Sand, mud
MARINE		
Continental shelf	Waves, tides	Sand, mud
Continental margin	Ocean currents	Mud, sand
Deep sea	Ocean currents, settling	Mud

transitional between continental and marine: deltas, beaches, and tidal flats. They also include those oceanic environments of the continental shelf, continental margin, and deep ocean floor where clastic sands and muds are deposited (Table 7.2). The sediments of these clastic environments are often called **terrigenous,** to indicate their origin on land.

Chemical and biochemical sedimentary environments are those characterized principally by chemical and biochemical precipitation (Table 7.3). In most places, they contain little clastic sediment; if present in abundance, clastic sediment may dilute chemical sediment or modify chemical sedimentary processes. By far the most abundant are

TABLE 7.3

MAJOR CHEMICAL AND BIOCHEMICAL SEDIMENTARY ENVIRONMENTS

ENVIRONMENT	AGENT OF PRECIPITATION	SEDIMENTS
SHORELINE AND MARINE		
Carbonate (includes reef, bank, deep sea, etc.)	Shelled organisms, some algae; inorganic precipitation from seawater	Carbonate sands and muds, reefs
Evaporite	Evaporation of seawater	Gypsum, halite, other salts
Siliceous: deep sea	Shelled organisms	Silica
CONTINENTAL		
Evaporite	Evaporation of lake water	Halite, borates, nitrates, other salts
Swamp	Vegetation	Peat

carbonate environments, marine settings where calcium carbonate, principally of biochemical origin, is the main sediment. Carbonate shell materials are secreted by literally hundreds of species of mollusks and other invertebrate organisms as well as by calcareous (calcium-containing) algae. Various populations of these organisms live at different depths of water, both in quiet areas and in places where waves and currents are strong. As they die, their shells accumulate to form sediment.

Except for those of the deep sea, carbonate environments are mostly found in the warmer tropical or subtropical regions of the oceans, where chemical conditions are favorable for the precipitation of calcium carbonate. These regions include organic reefs, carbonate sand beaches, tidal flats, and shallow carbonate banks. In a few places, carbonate sediments may form from cooler waters that are supersaturated with respect to carbonate.

An **evaporite environment** is created when the warm seawater of an arid inlet or arm of the sea evaporates more rapidly than it can mix with the connected open marine seawater. The degree of evaporation and the length of time it has proceeded control the salinity of the evaporating seawater and thus the kinds of sediments formed. Evaporite environments also form in lakes with no outlet rivers. Such lakes may produce halite, borate, nitrates, and other salts as sediments.

Siliceous environments are special deep-sea environments named for the remains of silica shells deposited in them. The organisms that secrete silica grow in surface waters where nutrients are abundant. Their shells settle to the ocean floor and accumulate as layers of siliceous sediment.

SEDIMENTARY STRUCTURES

All kinds of bedding and a variety of other structures formed at the time of deposition are called **sedimentary structures. Bedding** or **stratification** is a hallmark of sediments and sedimentary rocks. The parallel layers of different grain sizes or compositions indicate successive depositional surfaces that were formed at the time of sedimentation. Bedding may be thin, on the order of centimeters or even millimeters. At the opposite extreme, bedding may be meters or even many meters thick. Bed thickness is a guide to depositional processes, as we note shortly. Although we commonly refer to sediments as horizontally bedded, sedimentary rocks display several types of bedding, not all of which are horizontal.

FIGURE 7.6 Cross-bedding in a desert environment. The varying directions of cross-bedding in this sandstone reflect changes in wind direction at the time the sand dunes were deposited. Navajo sandstone, Zion National Park, southwestern Utah. *(Peter Kresan.)*

Sedimentary structures are strongly linked to the sedimentary environment in which they form. Cross-bedding in desert sandstone (Figure 7.6), for example, has a geometry that reflects the varying directions of the wind during deposition. Graded beds are found almost exclusively in continental slope and deep-sea sediments deposited by a special variety of bottom current called a turbidity current (see Chapter 17) and are almost never found in shallow shoreline environments.

Cross-Bedding

Cross-bedding consists of sets of bedded material deposited by wind or water and inclined at angles up to 35° from the horizontal (Figure 7.7). Cross-beds form when grains are deposited on the steeper, downcurrent (lee) slopes of sand dunes on land or of sandbars in rivers and under the sea. Cross-bedding of wind-deposited sand dunes may be complex,

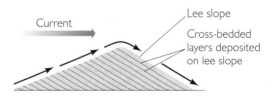

FIGURE 7.7 Cross-beds form when grains are deposited on the steeper lee or downcurrent slope of a dune or ripple.

FIGURE **7.8** Ripples in modern sand on a beach *(left)* and ancient ripple-marked sandstone *(right)*. *(Left: Raymond Siever. Right: Reg Morrison/Auscape.)*

reflecting rapidly changing wind directions. Cross-bedding is common in sandstones, and it also occurs in gravels and some carbonate sediments. Cross-bedding is easier to see in sandstones than in sands, which must be trenched or otherwise excavated to see a cross section.

Graded Bedding

In **graded bedding,** beds progress from coarse grains at the base to fine grains at the top. The grading reflects a waning of the current that deposited the grains. A graded bed, comprising one set of coarse to fine beds, normally ranges from a few centimeters to several meters thick. Accumulations of many graded beds, the total thickness reaching many hundreds of meters, are deposited in deep ocean waters by turbidity currents flowing along the ocean bottom. (We return to this topic in Chapter 17.)

Ripples

Ripples are very small dunes of sand or silt whose long dimension is at right angles to the current. They form low, narrow ridges or corrugations, most only a centimeter or two high, separated by wider troughs. These sedimentary structures are common in both modern sands and ancient sandstones (Figure 7.8). Ripples can be seen on the surfaces of windswept dunes, on underwater sandbars in shallow streams, and under the waves at beaches. Geologists can distinguish the symmetrical ripples made by waves moving back and forth on a beach from the asymmetrical ripples formed by currents moving in a single direction over river sandbars or windswept dunes (Figure 7.9).

Bioturbation Structures

Bedding in many sedimentary rocks is broken or disrupted, sometimes crossed by roughly cylindrical tubes a few centimeters in diameter that extend vertically through several beds. These sedimentary structures are remnants of burrows and tunnels excavated by clams, worms, and many other marine organisms that live on the bottom of the sea. In a process called **bioturbation,** these organisms rework existing sediments, burrowing through muds and sands. They ingest sediment for the bits of organic matter it contains and leave behind the reworked sediment, which fills the burrow (Figure 7.10). From bioturbation structures, geologists can

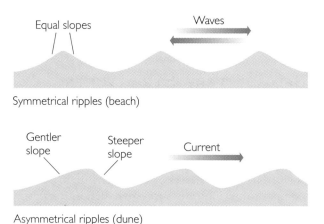

FIGURE **7.9** The shapes of ripples on beach sand, produced by the back-and-forth movements of waves, are symmetrical. Ripples on dunes and river sandbars, produced by the movement of a current in one direction, are asymmetrical.

FIGURE 7.10 Bioturbation structures. These tracks are thought to have been made by trilobites living in middle Cambrian muddy sediment in Montana about 500 million years ago. This rock is crisscrossed with fossilized tracks and tunnels originally made as the organisms burrowed through the mud. *(Chip Clark.)*

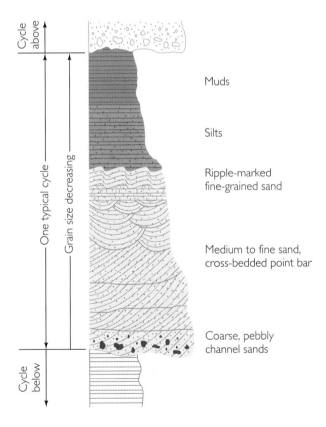

Muds

Silts

Ripple-marked
fine-grained sand

Medium to fine sand,
cross-bedded point bar

Coarse, pebbly
channel sands

FIGURE 7.11 A typical alluvial cycle. The widths of the sections drawn are proportional to the sizes of the grains in the sediment. The thickness of a cycle can range from a few meters for small streams, up to 20 m or more for large ones.

deduce the behaviors of organisms that burrowed the sediment and thus reconstruct the sedimentary environment.

Bedding Sequences

Bedding sequences are patterns of interbedding of sandstone, shale, and other sedimentary rock types. The nature of these bedding sequences helps geologists reconstruct how all of the sediments were deposited. Figure 7.11 shows an alluvial cycle, a cycle of bedding sequences typically formed by rivers. A river lays down repeated cycles of bedding sequences, in which sediments grade from coarse, with large-scale cross-bedded layers at the base, to fine, with small-scale cross-bedding, and then to horizontal bedding at the top. These sequences are formed as a river migrates laterally over its valley bottom. Other characteristic bedding sequences can be used to recognize deposition at the shoreline and in the deep sea. (These sequences are discussed further in Chapter 9.)

BURIAL AND DIAGENESIS: FROM SEDIMENT TO ROCK

Most of the clastic particles produced by weathering and erosion of the land end up on the bottom of the ocean, which is blanketed by many kinds of sediment brought there by rivers, the wind, and glaciers. A relatively smaller amount of clastic sediment is laid down on land, and of this, only a fraction will escape erosion and be buried and preserved.

As is the case with the clastic sediments, the larger proportion of chemical and biochemical sediments is deposited on the floor of the ocean. But, as we have seen, some chemical and biochemical sediments are deposited in lakes or swamps. Compared with sediments laid down on land, a larger fraction of sediment deposited on the ocean floor is buried and preserved for a long time. Biological and nonbiological precipitates that settle through the ocean's waters are trapped there. They can be picked up again by an ocean current, which may transport them to a different depositional site on the seafloor. But in most parts of the deep ocean, bottom currents are not strong enough to erode the sediment once it has settled.

Although most sediments are deposited in the oceans, a large proportion of the sedimentary rocks that geologists observe on the continents have been deposited in alluvial and shallow marine environments. Many of these rocks are or were formerly deeply buried in the continental crust. How does sedimentation lead to burial?

Accumulated Sediment Leads to Burial

As sedimentation continues in various environments over time, thick piles of sediment of various facies may accumulate, partly in response to subsidence of the crust. **Subsidence** is a gentle movement by which a broad area of the crust sinks relative to the elevation of the crust of surrounding areas in the region. Subsidence is induced in part by the additional weight of sediments on the crust and in part by tectonic mechanisms, such as regional downfaulting. **Sedimentary basins** are regions of considerable extent (at least 10,000 km^2) where the combination of deposition and subsidence has formed thick accu-

---

7.2   INTERPRETING THE EARTH

Rifting Can Create a Sedimentary Basin

Sedimentary basins have been known since the middle of the nineteenth century, but only in the past two decades have we come to a fuller understanding of their formation. Studies have been stimulated primarily by exploration for oil and gas, which are abundant in sedimentary basins. Our growing knowledge has enabled us to infer the deep structure of basins and thus to learn more about the continental lithosphere. We start with the role of continental rift valleys in initiating basins and then give a brief evolution of one basin.

Just as a mid-ocean ridge is a divergent plate boundary, so is a rift on a continent. Continental rifting creates valleys as the crust tries to pull apart and the bottom of the valley drops down (see the figure). The rift valleys subside and receive much sediment from the eroding borders. In Chapter 20, which discusses plate tectonics more fully, we show how continental rift valleys are incipient plate divergences and either evolve to a mid-ocean ridge or fail to evolve that way and stay as long, narrow, downdropped continental valleys bounded on either side by steep faults—cracks in the Earth's crust along which movement occurs. The rift valleys of East Africa, the rifted valley of the Rio Grande, and the Jordan Valley in the Middle East are current examples of this kind of valley. As these valleys tectonically subside, they receive large volumes of sediment. Their weight loads the crust and induces further subsidence.

One rift valley lies deeply buried beneath the Illinois Basin, a spoon-shaped sedimentary basin beneath the plains of southern Illinois. About a billion years ago, the proto–North American continent was split by a rift valley running north-northeast from an area to the south of the present confluence of the Ohio and Mississippi rivers to the northeastern part of the lower peninsula of Michigan. This valley subsided as it rifted apart and received river-borne sediments that accumulated to at least 3 km in thickness. Gradually, over millions of years as the rift became inactive, the original tectonic valley was buried deeply beneath sediment. Subsidence continued from the sediment load as the formerly hot crust of the rift zone slowly cooled and contracted. By about 500 million years ago, subsidence of the crust reached the point where it was below sea level, and shallow oceans spread over this region of the continent. Sediments of the subsiding basin became marine sands and then shallow marine carbonate sediments, including reefs at times. Sediments were comparatively thin, because at this stage the basin subsidence was slight.

Carbonate sedimentation was brought to a close about 330 million years ago as large rivers carrying clastics eroded from the rising Appalachian Mountains transported sand, gravel, and mud into the now spoon-shaped basin. This ushered in a period of alluvial sedimentation that accumulated another kilometer

mulations of sediment and sedimentary rock. The accumulations of various facies found in these basins form geometrical shapes ranging from narrow troughs to circular or oval spoon-shaped depressions. Most of the sedimentary rock of the world is found in these basins, and they are the reservoir of most of the oil and gas accumulations that have been drilled or will be drilled in the future. The mechanisms by which basins subside and accumulate sediment are being actively researched by geologists (Feature 7.2).

Diagenesis: Heat, Pressure, and Chemistry Transform Sediment to Rock

After sediments are deposited and buried, they are subject to diagenesis—the many physical and chemical changes that continue until the sediment or sedimentary rock is either exposed to weathering or metamorphosed by heat and pressure. Burial promotes these changes because buried sediments are subjected to increasingly high temperatures and

of sediment in the basin, this time with abundant organic sediment that gradually converted to the coal beds that are so prevalent in the region. The basin was finally immobilized about 250 million years ago when it stopped subsiding as the flow of debris from the Appalachian Mountains stopped or was detoured to other basins. Since then, the Illinois Basin has been stable, the top layers undergoing slight erosion and showing neither appreciable subsidence nor tectonic uplift. Yet the underlying rift still moves along the buried major faults and causes earthquakes, such as the severe earthquake at New Madrid, Missouri, in 1812.

The history of the Illinois Basin is similar to that of a great many sedimentary basins. Others are formed in lowlands adjacent to mountain uplifts, where an abundance of debris from erosion of the mountains loads the crust and depresses it, initiating a subsiding basin. Once the mountains become stable and are worn down by erosion, the rate of loading of sediment in the basin drops and the basin stops subsiding and accumulating more sediment. No doubt other mechanisms for the subsidence of basins will be found, for there are many hundreds of sedimentary basins in the world and geologists and geophysicists are studying their origins intensively.

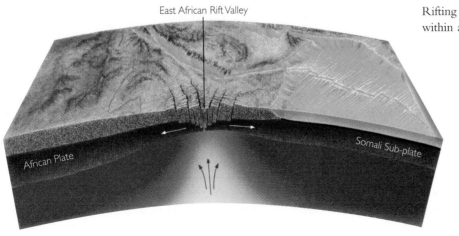

Rifting and plate separation within a continent.

pressures in Earth's interior. These diagenetic processes are illustrated in Figure 7.12.

Earth's temperature increases with depth, but the rate of increase varies among the Earth's interior zones. The most rapid rate of temperature increase occurs in the crust, where the temperature rises at an average rate of 30°C for each kilometer of depth. At a depth of 4 km, buried sediments may reach temperatures of 120°C or more. (The rate at which temperature increases with depth is known as the *geotherm* or geothermal gradient and is discussed more fully in Chapter 19.) Many chemical reactions among the minerals and pore waters of sedimentary rocks take place only at these higher temperatures.

The second factor causing diagenesis is the increase in pressure with burial—on average, about 1 atmosphere for each 4.4 m of depth. This pressure is responsible for the compaction of sediments.

When temperatures become even higher, diagenesis gives way to metamorphism. Those temperatures typically are around 300 to 350°C, corresponding to a depth of about 10 to 12 km.

CHEMICAL ALTERATION: CEMENTATION   A major *chemical* diagenetic change is **cementation,** in which minerals are precipitated in the pores of sediments, forming cements that bind clastic sediments and rocks. Cementation results in a decrease in **porosity,** the percentage of a rock's volume occupied by open pores between grains. Cementation also results in **lithification,** the diagenetic processes by which soft sediment is hardened into rock. In some sands, for example, calcium carbonate is precipitated as calcite, which acts as a cement that binds the grains and hardens the resulting mass into sandstone (Figure 7.13). Other minerals such as quartz may cement sands, muds, and gravels into sandstone, mudstone, and conglomerate.

Another chemical diagenetic change in clastic sediments and rocks is the chemical alteration of clay minerals originally deposited as clastic particles. Kaolinite, for example, may be transformed into illite, a clay mineral similar to muscovite mica.

PHYSICAL ALTERATION: COMPACTION   The major *physical* diagenetic change is **compaction,** a decrease in the volume and porosity of sediment, which results when the grains are squeezed closer together by the weight of overlying sediment. Sands are fairly well packed during deposition, so they do not compact much. Newly deposited muds, however, are highly porous; often, more than 60 percent of the sediment is water in pore space. As a result, muds compact greatly after burial, losing more than half of their water.

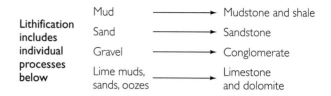

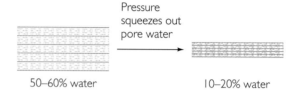

Compaction (primarily of muds)

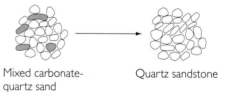

50–60% water   10–20% water

Precipitation of new minerals or additions to existing ones

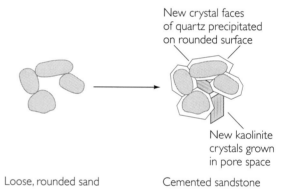

New crystal faces of quartz precipitated on rounded surface

New kaolinite crystals grown in pore space

Loose, rounded sand   Cemented sandstone

Dissolution of more soluble minerals

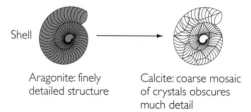

Mixed carbonate-quartz sand   Quartz sandstone

Recrystallization of unstable minerals

Shell

Aragonite: finely detailed structure

Calcite: coarse mosaic of crystals obscures much detail

FIGURE 7.12 Diagenetic processes produce changes in composition and texture. Most of the changes tend to transform a loose, soft sediment into a hard, lithified sedimentary rock.

FIGURE 7.13 Photomicrograph of sandstone showing quartz grains (white and gray) cemented by calcite cement (brightly colored and variegated) introduced after deposition. *(Peter Kresan.)*

Carbonate sediments are strongly prone to chemical and textural alteration. For example, aragonite, the less stable form of calcium carbonate, is the main constituent of many of the shell materials that make up carbonate sediments. During diagenesis soon after the beginning of burial, the aragonite tends to recrystallize to the more stable form of calcium carbonate, calcite, which is the most common mineral of limestones. Carbonate sediments are commonly lithified to limestone soon after deposition and burial.

Compaction is often, but not always, accompanied by the precipitation of cementing materials. Many of the muddy sediments of the deep sea, even after burial to hundreds of meters, are compacted but uncemented and therefore are not lithified into hard rock.

CLASSIFICATION OF CLASTIC SEDIMENTS AND SEDIMENTARY ROCKS

We can now use our knowledge of sedimentation to classify sediments and their lithified counterparts, sedimentary rocks. The major divisions are the clastic and the chemical and biochemical. Clastic sediments and sedimentary rocks—muds and shales, sands and sandstones, gravels and conglomerates—constitute more than three-quarters of the total mass of all types of sedimentary rocks in the Earth's crust. We therefore begin with them.

Classification by Particle Size

Clastic sediments and rocks are classified by their textures, primarily the sizes of the grains, yielding the following three broad categories (Table 7.4).

TABLE 7.4

MAJOR CLASSES OF CLASTIC SEDIMENTS AND SEDIMENTARY ROCKS

PARTICLE SIZE	SEDIMENT	ROCK
Coarse	*Gravel*	
Larger than 256 mm (about 10 in.)	Boulder	
256–64 mm (about 2.5 in.)	Cobble	Conglomerate
64–2 mm (about 0.08 in.; actual size about ●)	Pebble	
Medium		
2–0.062 mm	*Sand*	Sandstone
Fine	*Mud*	
0.062–0.0039 mm	Silt	Siltstone
Finer than 0.0039 mm	Clay	Mudstone (blocky fracture) Shale (breaks along bedding) Claystone

- *Coarse*—gravels and conglomerates

- *Medium*—sand and sandstone

- *Fine*—silt and siltstone; mud, mudstone, and shale; clay and claystone

Classifying the various clastic sediments and rocks on the basis of the sizes of their particles in effect also distinguishes them by the conditions of sedimentation. As we have seen, the larger the particle, the stronger the current needed to move and deposit it. This relationship between current strength and particle size is the reason like-sized particles tend to accumulate in sorted beds. That is, most sand beds do not contain pebbles or mud, and most muds consist only of finer particles.

Within each textural category, clastics can be further subdivided by mineralogy, which reflects the parent rocks. Thus, there are quartz-rich and feldspar-rich sandstones and calcareous, siliceous, and organic-rich shales. Some sediments are bioclastic; they were formed when material such as carbonate originally precipitated as a shell but then was broken and transported by currents.

Of the various types of clastic sediments and sedimentary rocks, silt, siltstone, mud, mudstone, and shale are by far the most abundant—about three times more common than the coarser clastics (Figure 7.14). The abundance of the fine-grained clastics, which contain large amounts of clay minerals, reflects the chemical weathering to clay minerals of the large quantities of feldspar and other silicate minerals in the Earth's crust. The following discussion considers each of the three groups of clastic sediments and sedimentary rocks in more detail.

Coarse-Grained Clastics: Gravel and Conglomerates

Gravel is the coarsest clastic sediment, consisting of particles larger than 2 mm in diameter and including pebbles, cobbles, and boulders (see Table 7.4). **Conglomerates** are the lithified equivalents of gravel (Figure 7.15a). Because of their large size, pebbles are easy to study and identify. Granite pebbles in a river-deposited conglomerate provide firm evidence of a mass of exposed granite in the drainage areas feeding the rivers that transported the gravel.

There are relatively few environments in which currents are strong enough to transport pebbles: mountain streams, rocky beaches with high waves, and the meltwaters of glaciers. Strong currents also carry sand, and we almost always find sand between the pebbles, some of it deposited with the gravel and some of it infiltrated into pores between fragments after the larger clastic sediments have been deposited. Pebbles and cobbles abrade and become rounded very quickly in the course of transport on land or in water. Travel of 100 km can make pebbles smooth and round. Beach gravels constantly moved back and forth by strong waves also become rounded.

Some coarse-grained clastic rocks escape rounding and show little or no signs of abrasion. These are not conglomerates; they are **sedimentary breccias,** characterized by the sharp angularity of their component rock fragments (Figure 7.15b). Sedimentary breccias are found in deposits close to their source, where sediments were deposited before they had been transported very far. Not all breccias are of sedimentary origin; some form by the breaking up of volcanic materials during eruptions or by the fragmentation of rocks along faults.

Medium-Grained Clastics: Sand and Sandstone

Sand is formed from medium-sized particles, 0.062 to 2 mm in diameter (see Table 7.4). These sediments are moved by moderate currents, such as those of rivers, waves at shorelines, and the winds that blow sand into dunes. Sand particles are large enough to be seen with the naked eye, and many of their features are easily discerned with a low-power magnifying glass.

The lithified equivalent of sand is **sandstone** (Figure 7.15c). Of all the clastic groups, sandstone

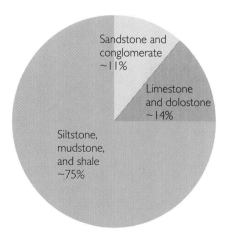

FIGURE 7.14 The relative abundance of the major sedimentary rock types. In comparison with these three, all other rock types—including evaporites, cherts, and other chemical sediments—occur in only minor amounts.

FIGURE 7.15
Clastic sedimentary rocks:
(a) conglomerate, (b) breccia,
(c) sandstone, (d) shale.
(Breck P. Kent.)

(a) (b) (c) (d)

has received the most attention, both because it is abundant and widespread and because the information it contains about its origin is easy to read. For example, many sandstones are cross-bedded. Since cross-bedding indicates the direction of the transporting current, sandstones are useful for mapping paleocurrents that reveal the directions of former stream, wind, or shallow marine water flows.

SIZES OF SAND GRAINS Sand particles may be fine, medium, or coarse. The average size of the grains in any one sandstone reflects both the strength of the current that carried them and the sizes of the crystals eroded from the parent rock. But size does not tell the whole story; the range and relative abundance of the various sizes are also significant. If all the grains are close to the average size, the sand is well sorted; if many grains are much larger or smaller than the average, the sand is poorly sorted. The degree of sorting can help distinguish, for example, between sands of beaches (well sorted) and muddy sands deposited by glaciers (poorly sorted).

SHAPES OF SAND GRAINS The shapes of sand grains can also be important clues to their origin. Sand grains, like other clastic particles, are rounded

by abrasion as grains are knocked together during transport. Angular grains imply short distances of transportation; rounded ones indicate long journeys down a large river system. Sand grains also become rounded as they are moved constantly back and forth by the waves on beaches. Most sand grains, however, inherit their spheroidal, elongate, or disc shapes from the shapes of the original crystals in the parent rock.

MINERALOGY From the mineralogy of sands and sandstones, geologists can deduce the nature of the source areas that were eroded to produce the sand grains. The presence of sodium- and potassium-rich feldspars with abundant quartz, for example, might indicate that the sediments were eroded from a granitic terrain. Other minerals, as we will see in Chapter 8, would be indicative of metamorphic parent rocks.

The mineralogy of a sand or sandstone may not match the mineralogy of the parent rocks exactly. Chemical weathering at the source area can alter and dissolve—and thus remove—much of the feldspar from a granite. Under these conditions, what remains might be grains mainly of quartz. Thus we can infer, from an analysis of the mineralogy of a sand or sandstone, its parent rocks and something about the factors, such as climate, that influence weathering in the source area. The mineralogical composition of parent rocks can also be correlated with plate-tectonic settings. Sandstones containing abundant fragments of mafic volcanic rocks, for example, are derived from the volcanic arcs of subduction zones.

MAJOR KINDS OF SANDSTONES   Sandstones fall into several major groups on the basis of their mineralogy and texture.

- *Quartz arenites* are made up almost entirely of quartz grains, usually well sorted and rounded (Figure 7.16a). These pure quartz sands result from extensive weathering that occurred before and during transport and removed everything but quartz, the most stable mineral.

- *Arkoses* contain more than 25 percent feldspar; the grains tend to be poorly rounded and less well sorted than those of pure quartz sandstones (Figure 7.16b). These feldspar-rich sandstones come from rapidly eroding granitic and metamorphic terrains where chemical weathering is subordinate to physical weathering.

- *Lithic sandstones* contain many fragments derived from fine-grained rocks, mostly shales, volcanic rocks, and fine-grained metamorphic rocks (Figure 7.16c).

- *Graywacke* is a heterogeneous mixture of rock fragments and angular grains of quartz and feldspar, the sand grains being surrounded by a fine-grained clay matrix (Figure 7.16d). Much of this matrix is formed by the chemical alteration and mechanical compaction and deformation of relatively soft rock fragments, such as those of shale and some volcanic rocks, after deep burial of the sandstone formation.

Both groundwater geologists and petroleum geologists have a special interest in sandstones. Groundwater geologists rely on their understanding of the origins of sandstones to predict possible supplies of water in areas of porous sandstone such as found in the western plains of North America. Petroleum geologists must know about the porosity and cementation of sandstones because much of the oil and gas discovered in the last 150 years has been found in buried sandstones. In addition, a good deal of the uranium used for nuclear power plants and weapons has come from diagenetic uranium in sandstones.

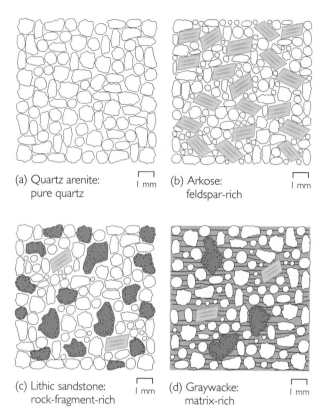

(a) Quartz arenite: pure quartz | 1 mm

(b) Arkose: feldspar-rich | 1 mm

(c) Lithic sandstone: rock-fragment-rich | 1 mm

(d) Graywacke: matrix-rich | 1 mm

FIGURE 7.16 The mineralogy of four major groups of sandstones.

Fine-Grained Clastics: Silt and Siltstone; Mud, Mudstone, and Shale

The finest grained clastic sediments and sedimentary rocks are the silts and siltstones, the muds and mudstones, and the shales. The particles comprising these sediments are all less than 0.062 mm in diameter, but they vary widely in their range of grain sizes and mineral compositions. Fine-grained sediments are deposited by the gentlest currents, which allow the finest particles to settle slowly to the bottom in quiet waves.

SILT AND SILTSTONE   Siltstone is the lithified equivalent of **silt,** a clastic sediment in which most of the grains are between 0.0039 and 0.062 mm in diameter. Siltstone looks similar to mudstone or very fine-grained sandstone.

MUD, MUDSTONE, AND SHALE   Mud is any clastic sediment in which most of the particles are less than 0.062 mm in diameter (see Table 7.4). Muds are deposited by rivers and tides. After a river has flooded its lowlands and the flood recedes, the current slows and mud settles; this mud contributes to the fertility of river bottomlands. Muds are left behind by ebbing tides along many tidal flats where wave action is mild. Much of the deep ocean floor, where currents are weak or absent, is blanketed by mud.

The fine-grained rock equivalents of muds are mudstones and shales (see Figure 7.15d). **Mudstones** are blocky and show poor or no bedding. Bedding may have been well marked in many mudstones when the sediments were deposited, but it may have been lost by bioturbation (see Figure 7.10).

Shales are composed of silt and clay, and they tend to break readily along bedding planes. Many muds, mudstones, and shales contain more than 10 percent carbonate, forming deposits of calcareous shales. Black, or organic, shales contain abundant diagenetically altered organic matter; some, called oil shales, contain large quantities of oily organic material, which makes them a potentially important source of oil. (We discuss the oil shales in more detail in Chapter 22.)

CLAY AND CLAYSTONE The most abundant component of fine-grained sediment and sedimentary rock is **clay**-sized material, which consists largely of clay minerals. Clay-sized particles (we are referring only to the sizes of the particles, not to clay minerals) are less than 0.0039 mm in diameter (see Table 7.4). Rocks made up exclusively of clay-sized particles are called *claystones*. Dust, which includes both clay- and silt-sized particles, is deposited by the wind after dust storms on arid plains (see Chapter 14).

Certain clay minerals in fine-grained sediments, especially kaolinite, are economically valuable for making ceramic products. These and other clays can make up *paleosols*, which are the lithified equivalent of soils. Some of these ancient paleosols, although they do not look like a modern soil now, can be shown to have a mineralogical profile that is clearly derived from a soil profile (see Chapter 6).

CLASSIFICATION OF CHEMICAL AND BIOCHEMICAL SEDIMENTS AND SEDIMENTARY ROCKS

Whereas clastic sediments and sedimentary rocks give us information about continental parent rocks and weathering, the chemical and biochemical sediments tell us about chemical conditions in the environment of sedimentation, which is predominantly the ocean. Although chemical sedimentation takes place in some lakes, particularly those of arid regions where evaporation is intense, such as the Great Salt Lake of Utah, such sediments account for only a very small fraction relative to the amounts deposited along the ocean's shorelines, on continental shelves, and in the deep ocean.

Classification by Chemical Composition

Chemical and biochemical sediments and sedimentary rocks are classified by their chemical composition (Table 7.5). In the case of marine sediments, this classification reflects the chemical elements dissolved in seawater. The ions of the most abundant such elements are as follows:

- Chloride (Cl^-)
- Magnesium (Mg^{2+})
- Sodium (Na^+)
- Sulfur (as sulfate, SO_4^{2-})
- Potassium (K^+)
- Carbonate (CO_3^{2-})
- Calcium (Ca^{2+})

Two other major components of some sedimentary rocks—silica (SiO_2) and phosphorus—are found in minor quantities in seawater.

We break nonclastic sediments into chemical and biochemical to emphasize the importance of organisms as the chief mediator of this kind of sedimentation. The shells of organisms, biochemically precipitated, account for much of the carbonate sediment of the world, and carbonate is by far the most abundant nonclastic sediment. The chemical sediments are precipitated by inorganic processes alone and are relatively less abundant.

Carbonate Sediments and Sedimentary Rocks: Limestone and Dolostone

Carbonate sediments and sedimentary rocks are formed from the accumulation of carbonate minerals precipitated organically or inorganically. The precipitation may occur during sedimentation or diagenesis. The minerals are either calcium or calcium-magnesium carbonates. Carbonate rocks are abundant because of the large amounts of calcium and carbonate present in seawater—carbonate derived from the carbon dioxide in the atmosphere

TABLE **7.5**

CLASSIFICATION OF BIOCHEMICAL AND CHEMICAL SEDIMENTS AND SEDIMENTARY ROCKS

SEDIMENT	ROCK	CHEMICAL COMPOSITION	MINERALS
BIOCHEMICAL			
Sand and mud (primarily bioclastic)	Limestone	Calcium carbonate $CaCO_3$	Calcite (aragonite)
Siliceous sediment	Chert	Silica SiO_2	Opal Chalcedony Quartz
Peat, organic matter	Organics	Carbon compounds Carbon compounded with oxygen and hydrogen	(Coal) (Oil) (Gas)
CHEMICAL			
No primary sediment (formed by diagenesis)	Dolostone	Calcium-magnesium carbonate $CaMg(CO_3)_2$	Dolomite
Iron oxide sediment	Iron formation	Iron silicate; oxide; carbonate Fe_2O_3	Hematite Limonite Siderite
Evaporite sediment	Evaporite	Sodium chloride; calcium sulfate $NaCl; CaSO_4$	Gypsum Anhydrite Halite Other salts
No primary sediment (formed by diagenesis)	Phosphorite	Calcium phosphate $Ca_3(PO_4)_2$	Aspatite

and calcium and carbonate from the easily weathered limestone on the continents.

Most carbonate sediments of shallow marine environments are bioclastic. They were originally secreted biochemically as shells by organisms living near the surface or on the bottom of the oceans and then broken and transported by currents. They are found from the coral reefs of the Pacific and Caribbean to the shallow banks of the Bahama Islands. Less accessible for study than these spectacular vacation spots is the deep ocean floor, the site of most of the carbonate deposited today.

BIOLOGICAL SOURCES OF CARBONATE SEDIMENTS Most of the carbonate sediments of the ocean are derived from the calcite shells and skeletons of **foraminifera,** tiny single-celled organisms that live in surface waters, and from other organisms that secrete calcium carbonate extracted from seawater. When the organisms die, the shells and skeletons settle to the seafloor and accumulate there as sediment (Figure 7.17). In addition to calcite, most carbonate sediments contain aragonite, a less stable form of calcium carbonate. As noted earlier in this chapter, some organisms precipitate calcite, others aragonite, and some both.

Reefs are mounds or ridge-shaped organic structures constructed of the carbonate skeletons of millions of organisms. In the warm seas of today, most reefs are built by coral. The calcium carbonate of the coral and other organisms, in contrast to the soft, loose sediment produced in other environments, forms a rigid, wave-resistant structure—a cemented buttress of solid limestone—that is built up to and slightly above sea level (Figure 7.18). The solid limestone of the reef is produced directly by the action of organisms; there is no soft sediment stage. On and around these reefs live hundreds of species of other carbonate-precipitating organisms, many of them similar to the familiar clams, oysters, and mussels of

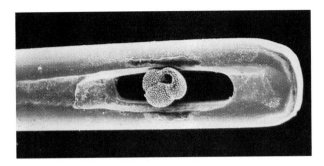

FIGURE 7.17 Scanning electron micrograph of a foraminifer in the eye of a needle. These single-celled organisms secrete shells of calcium carbonate, which they extract from seawater. *(Chevron Corporation.)*

bonate and precipitate carbonate by the following chemical reaction:

$$\begin{array}{cc} \text{calcium ion} & \text{bicarbonate ion} \\ & \text{(dissolved)} \\ + & \\ Ca^{2+} & 2HCO_3^- \end{array} \longrightarrow$$

$$\begin{array}{cc} \text{calcium carbonate} & \text{carbonic acid} \\ \text{(precipitated)} & + & \text{(dissolved)} \\ CaCO_3 & H_2CO_3 \end{array}$$

When living organisms secrete their carbonate shells, they are effectively producing the same chemical reaction by biochemical means. The chemical reaction that precipitates calcium carbonate is the reverse of the reaction that takes place during the chemical weathering of limestone (see Chapter 6):

$$CaCO_3 + H_2CO_3 \longrightarrow Ca^{2+} + 2HCO_3$$

our shorelines. Carbonate is also precipitated by marine algae, single-celled organisms similar to primitive plants, which grow on reef tracts and in other carbonate environments.

INORGANIC PRECIPITATION OF CARBONATE SEDIMENTS

Until recently, many geologists thought there was no such thing as direct inorganic precipitation of calcium carbonate from seawater, but research in lagoons and banks in the Bahama Islands has shown that a significant fraction of the carbonate mud on shallow banks is inorganically precipitated directly from seawater. The chemical basis for inorganic carbonate precipitation is the relative abundance of calcium (Ca^{2+}) and bicarbonate (HCO_3^-) ions in seawater. The warm, tropical parts of the ocean are supersaturated with calcium car-

CARBONATE SEDIMENTS OF MIXED ORIGIN

Carbonate sediments contain some fine-grained carbonate muds of mixed origin, consisting partly of microscopic fragments of shells and calcareous algae and partly of inorganic precipitates. These muds form in the lagoons behind reefs and on extensive shallow banks such as those of the Bahamas. Reefal and carbonate bank deposits are found on passive continental margins and fringing volcanic island arcs in the warmer parts of the oceans where they are not flooded with clastic sediment.

Carbonate platforms, both in past geological ages and at present, are another major carbonate environment. Like the Bahama banks, these platforms

FIGURE 7.18 A coral reef surrounding a central volcanic island. In the foreground is the reef, which shelters the shallow lagoon behind it. *(Jean-Marc Truchet/Tony Stone Worldwide.)*

are shallow, extensive flat areas where both biological and nonbiological carbonates are deposited. Below the level of the platform are carbonate ramps, gentle slopes to deeper waters that also accumulate carbonate sediment, much of it fine-grained. In other times and places, platforms are rimmed carbonate shelves in which there is a clear demarcation of the shelf margin by reefs of various organisms and buildups of shoals of bioclastic and other material. Below the rims are steep slopes that are covered with detritus derived from the rim materials.

LIMESTONE   The dominant biochemical sedimentary rock lithified from carbonate sediments is limestone. **Limestone** is composed mainly of calcium carbonate ($CaCO_3$) in the form of the mineral calcite (Figure 7.19a; see Table 7.5). Limestone is formed from carbonate sand and muds. Carbonate sands are abundant in many environments; mud dominates in others. Carbonate muds are laid down in deeper waters or in protected lagoons or bays, where waves and currents do not have the power to carry away the fine particles. These muds become compacted and cemented after burial, forming dense, fine-grained limestone.

FORMED BY DIAGENESIS: DOLOSTONE   Another abundant carbonate rock is **dolostone,** made up of the mineral dolomite, which is composed of calcium-magnesium carbonate, $CaMg(CO_3)_2$ (see Table 7.5). Dolostones are diagenetically altered carbonate sediments and limestones. The mineral dolomite does not form as a primary precipitate from ordinary seawater, and no organisms secrete shells of dolomite. Instead, the original calcite or aragonite of a carbonate sediment is converted to dolomite soon after deposition by the addition of magnesium ions from seawater slowly passing through the pores of the sediment.

Many dolostones form in shallow bays and flats where seawater is concentrated. As the seawater becomes concentrated, the proportion of magnesium ions increases in relation to calcium ions. This increase occurs as the precipitation of calcite and aragonite and gypsum or anhydrite causes a local depletion of calcium ions in the seawater. As seawater and sediment then react chemically, some calcium ions in the calcite or aragonite are exchanged for magnesium ions, converting these calcium carbonate mineral, $CaCO_3$, to dolomite, $CaMg(CO_3)_2$.

Other dolostones form long after deposition and after moderate or deep burial of limestones.

(a)

(b)

(c)

(d)

FIGURE 7.19 Chemical and biochemical sedimentary rocks: (a) limestone, lithified from carbonate sediments; (b) gypsum and (c) halite, marine evaporites that crystallize out of shallow seawater basins; and (d) chert, made up of silica sediment. (*Breck P. Kent.*)

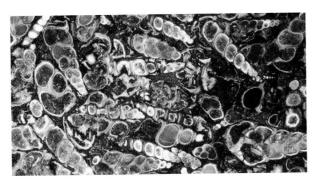

FIGURE 7.20 Fossiliferous limestone made up of fossil shells of gastropods (snails) in the Eocene Green River formation of Wyoming. *(Peter Kresan.)*

These extensive dolomitized zones are built up by a reaction of the buried limestone with groundwaters containing a large proportion of magnesium ions relative to calcium ions.

By comparing the mineral composition and textures of carbonate sediments being deposited today with those of limestones and dolostones of the past, geologists can tell how the older rocks were formed. Sedimentary structures of sand-sized carbonate sediments, some of them cross-bedded, can be found in the carbonate sands of today's beaches and underwater banks and bars in the Caribbean and the Bahamas. These structures allow us to view today the sedimentation patterns that long ago led to the formation of many ancient limestones. Other limestones and dolostones were originally deposited as limestone reefs, as evidenced by the fossilized remains of the organisms that made them (Figure 7.20). Today, reefs are constructed mainly by corals; but at earlier times in Earth's history, other organisms—such as a now-extinct variety of oyster—built

wave-resistant structures, some of them cemented buttresses of solid limestone.

The marine environments where carbonate sedimentation produces rigid limestone structures—including reefs, carbonate banks, and deepwater deposits in the open ocean—are discussed further in Chapter 17.

Evaporite Sediments: Sources of Halite, Gypsum, and Other Salts

Evaporite sediments and sedimentary rocks are precipitated inorganically from evaporating seawater and from water in arid-region lakes that have no river outlet.

MARINE EVAPORITES   **Marine evaporites** are the chemical sediments and sedimentary rocks formed by the evaporation of seawater. These sediments and rocks contain minerals formed by the crystallization of sodium chloride (halite), calcium sulfate (gypsum and anhydrite), and other combinations of the ions commonly found in seawater. As evaporation proceeds, seawater becomes more concentrated and minerals crystallize in sequence. Some are primary precipitates; others are formed by diagenetic reaction. As dissolved ions precipitate to form each of these minerals, the evaporating seawater changes composition.

The great volume of many marine evaporites, some hundreds of meters thick, shows that they could not have formed from the small amount of water that could be held in a shallow bay or pond. A huge amount of seawater had to have evaporated. The way such large quantities of seawater are evaporated is very clear in bays or arms of the sea that meet three conditions (Figure 7.21):

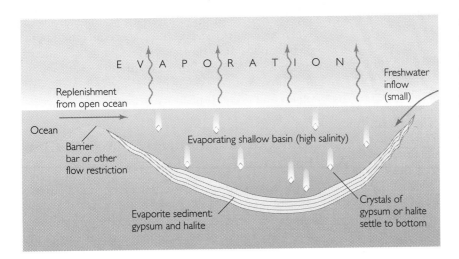

FIGURE 7.21 When seawater evaporates in a shallow basin with a restricted connection to the open ocean, gypsum and halite form as evaporite sediments. Evaporation removes much more water than the fresh water flowing in can replace. As the evaporating basin gets appreciably more saline than the water of the open sea, gypsum is precipitated. A further increase in salinity leads to the crystallization of halite.

- The freshwater supply from rivers is small.
- Connections to the open sea are constricted.
- The climate is arid.

In such locations, water evaporates steadily, but the openings allow seawater to flow in to replenish the evaporating waters of the bay. As a result, those waters stay at constant volume, but they are more saline than the open ocean. The evaporating bay waters stay more or less constantly supersaturated and steadily deposit evaporite minerals on the floor of the evaporite basin.

The first precipitates to form as seawater starts to evaporate are the carbonates, first calcite alone, then dolomite by diagenetic reaction. Continued evaporation leads to the precipitation of **gypsum,** calcium sulfate ($CaSO_4 \cdot 2H_2O$) (see Figures 7.19b and 7.21). Gypsum is the principal component of plaster. By the time gypsum precipitates, no carbonate ions are left in the water.

After still further evaporation, the mineral halite (NaCl)—one of the commonest chemical sediments formed from evaporating seawater—starts to form (see Figures 7.19c and 7.21). Halite, you may remember from Chapter 2, is table salt. Deep under the city of Detroit, Michigan, for example, beds of salt laid down by an evaporating arm of an ancient ocean are mined in extensive workings.

In the final stages of evaporation, after the sodium chloride is gone, magnesium and potassium chlorides and sulfates are precipitated.

This sequence of precipitation has been studied in the laboratory and is matched by the bedding sequences found in certain natural salt formations. Most of the evaporites of the world consist of thick sequences of dolomite, gypsum, and halite and do not contain the final-stage precipitates. Many do not even go so far as halite. The absence of the final stages indicates that the water did not evaporate completely but was replenished by normal seawater as evaporation continued.

NONMARINE EVAPORITES Evaporite sediments also form in arid-region lakes that typically have little or no river outlet. In such lakes, evaporation controls the lake level, and incoming salts derived from chemical weathering accumulate. The Great Salt Lake of Utah is one of the best known of these lakes. River waters enter the lake, bringing salts dissolved in the course of weathering. In the dry climate of Utah, evaporation has more than balanced the inflow of fresh water from rivers and rain. As a result, concentrated dissolved ions in the lake make it one of the saltiest bodies of water in the world—eight times saltier than seawater.

In arid regions, small lakes may collect unusual salts, such as borates (compounds of the element boron), and some become alkaline. The water in this kind of lake is poisonous. Economically valuable resources of borates and nitrates (minerals containing the element nitrogen) are found in the sediments beneath some of these lakes.

Siliceous Sediment: Source of Chert

One of the first sedimentary rocks to be used for practical purposes by our prehistoric ancestors was **chert,** a sedimentary rock made up of chemically or biochemically precipitated silica (SiO_2) (Figure 7.19d). Early hunters used it for arrowheads and other tools because it could be chipped and shaped to form hard, sharp implements. A common name for chert is **flint,** and the terms are practically interchangeable. The silica in most cherts is in the form of extremely finely crystalline quartz; some geologically young cherts consist of a less well crystallized form of silica, opal.

Like calcium carbonate, much silica sediment is precipitated biochemically, secreted as shells by ocean-dwelling organisms. When these organisms die, they sink to the deep ocean floor, where their shells accumulate as layers of silica sediment. After these silica sediments are buried by later sediments, they are diagenetically cemented into chert. Chert may also form as diagenetic nodules and irregular masses replacing carbonate in limestones and dolomites.

Formed by Diagenesis: Phosphorite

Among the many other kinds of chemical and biochemical sediments that are deposited in the sea are phosphorites. **Phosphorite,** sometimes called phosphate rock, is composed of calcium phosphate precipitated from phosphate-rich seawater in places where currents of deep, cold water containing phosphate and other nutrients rise along continental margins. The phosphorite forms diagenetically by the interaction between muddy or carbonate sediments and the phosphate-rich water.

Iron Oxide Sediment: Source of Iron Formation

Iron formations are sedimentary rocks, usually containing more than 15 percent iron in the form of iron oxides and some iron silicates and iron carbonates. Most of these rocks formed early in Earth's history, when there was less oxygen in the atmosphere, and, as a result, iron was more easily soluble. In soluble form, iron was transported to the sea and precipitated there.

Organic Matter: Source of Coal, Oil, and Gas

Coal is a biochemically produced sedimentary rock composed almost entirely of organic carbon formed by the diagenesis of swamp vegetation. Coal is classified as an **organic sedimentary rock,** a group of sedimentary rocks that consist entirely or partly of organic carbon-rich deposits formed by the decay of once-living material that has been buried.

Oil and **gas** are fluids that are not normally classed with sedimentary rocks. But because they are formed by the diagenesis of organic material in the pores of sedimentary rocks, they can be considered organic sediments. Deep burial alters organic matter originally deposited along with inorganic sediments to a fluid that then escapes to other, porous formations and becomes trapped there. As noted earlier in this chapter, oil and gas are found mainly in sandstones and limestones (see Chapter 22).

SUMMARY

What are the major processes involved in the formation of sedimentary rock? Weathering and erosion produce the clastic particles and dissolved ions that compose sediment. Currents of water, wind, and ice transport the sediment to its ultimate resting place, the site of sedimentation. Sedimentation (also called deposition), a settling of particles from the transporting agent, produces bedded sediments in river channels and valleys, on sand dunes, and at the edges and floors of the oceans. Lithification and diagenesis harden the sediment to sedimentary rock after it has been buried by additional sediment.

What are the two major divisions of sediments and sedimentary rocks? The two major divisions are *clastic* and *chemical and biochemical.* Clastic sediments are formed from the fragments of parent rock produced by physical weathering and the clay minerals produced by chemical weathering. These solid products are carried by water and wind currents and ice to the oceans, sometimes being deposited along the way. Chemical and biochemical sediments originate from the ions dissolved in water during chemical weathering. These ions are transported in solution to the oceans, where they are mixed into seawater. Through chemical and biochemical reactions, the ions are precipitated from solution and the precipitated particles settle to the ocean floor.

How do we classify the major kinds of clastic and chemical and biochemical sedimentary rocks? Clastic sediments and sedimentary rocks are classified in accordance with the sizes of their particles: as gravels and conglomerates; sands and sandstones; and muds, mudstones, and shales. This method of classifying sediments reflects the importance of the strength of the current as it transports and deposits solid materials. The chemical and biochemical sediments and sedimentary rocks are classified on the basis of their chemical compositions. The most abundant of these rocks are the carbonate rocks, limestone and dolostone. Limestone is made up largely of biochemically precipitated shells. Dolostone is formed by the diagenetic alteration of limestones. Other chemical and biochemical sediments include evaporites, siliceous sediments such as chert; phosphorites; iron formations; and peat and other organic matter that is transformed to coal, oil, and gas.

KEY TERMS AND CONCEPTS

diagenesis (p. 166)
clastic particles (p. 166)
clastic sediments (p. 167)
chemical sediments (p. 167)
biochemical sediments (p. 167)
bioclastic particles (p. 167)
sorting (p. 168)
peat (p. 171)
sedimentary environment (p. 171)
clastic sedimentary environment
 (p. 173)

terrigenous sediment (p. 174)
chemical and biochemical
 sedimentary environments (p. 174)
carbonate environment (p. 175)
evaporite environment (p. 175)
siliceous environment (p. 175)
sedimentary structures (p. 175)
bedding (p. 175)
stratification (p. 175)
cross-bedding (p. 175)
graded bedding (p. 176)

ripples (p. 176)
bioturbation (p. 176)
bedding sequence (p. 177)
subsidence (p. 178)
sedimentary basins (p. 178)
cementation (p. 180)
porosity (p. 180)
lithification (p. 180)
compaction (p. 180)
gravel (p. 182)
conglomerate (p. 182)

EXERCISES

1. What processes change sediment into sedimentary rock?

2. How do clastic sedimentary rocks differ from chemical and biochemical sedimentary rocks?

3. How and on what basis are the clastic sedimentary rocks subdivided?

4. What kind of sedimentary rocks were originally formed by the evaporation of seawater?

5. Define *sedimentary environment,* and name three clastic environments.

6. Name three kinds of sandstone.

7. Name two kinds of carbonate rocks and explain how they differ.

8. How do organisms produce or modify sediments?

9. Name two ions that are involved in the precipitation of calcium carbonate in a sedimentary environment.

10. In what two kinds of sedimentary rocks are oil and gas found?

THOUGHT QUESTIONS

1. Weathering of the continents has been much more widespread and intense in the past 10 million years than it was in earlier times. How might this observation be reflected in the sediments that now cover the Earth's surface?

2. In what respects might you consider a volcanic ash fall a sediment?

3. A geologist is heard to say that a particular sandstone was derived from a granite. What information could she have gleaned from the sandstone to lead her to that conclusion?

4. You are looking at a cross section of a rippled sandstone. What sedimentary structure would tell you the direction of the current that deposited the sand?

5. You discover a bedding sequence that has a conglomerate at the base; grades upward to a sandstone and then to a shale; and finally, at the top, grades to a limestone of cemented carbonate sand. What changes in the area of the sediment's source or in the sedimentary environment would have been responsible for this sequence?

6. From the base upward, a bedding sequence begins with a bioclastic limestone; passes upward into a dense carbonate rock made of carbonate-cementing organisms, including algae normally found with coral; and ends with beds of dolomite. Deduce the possible sedimentary environments represented by this sequence.

7. In what kinds of sedimentary environment would you expect to find carbonate muds?

8. How can you use the sizes and sorting of sediments to distinguish between sediments deposited in a glacial environment and those deposited on a desert?

9. Describe the beach sands that you would expect to be produced by the beating of waves on a coastal mountain range consisting largely of basalt.

10. What role do transporting currents play in the origin of some kinds of limestone?

11. Name a sedimentary rock that is essentially the product of diagenesis and that has no exact equivalent as a sediment.

12. Where are reefs likely to be found?

13. An ocean bay is separated from the open ocean by a narrow, shallow inlet. What kind of sediment would you expect to find on the floor of the bay if the climate were warm and arid as opposed to the kind of sediment you would find if the climate were cool and humid?

14. How are chert and limestone similar in origin?

SHORT-TERM TEAM PROJECTS: CARBONATE SEDIMENTS

The three main minerals of carbonate rocks are calcite, $CaCO_3$; aragonite, also $CaCO_3$; and dolomite, $CaMg(CO_3)_2$. Calcite and aragonite are polymorphs; they have the same chemical composition but different crystal structures. Some organisms make their shells and skeletons of aragonite, while others make them of calcite. Interestingly, fossils of very old organisms are never made of aragonite. The explanation is that aragonite is less stable than calcite and eventually breaks down to form calcite.

The origin of dolomite, in contrast, is unresolved. Dolomite is a mineral found in many sedimentary sequences, but it is not a constituent of shells or newly deposited carbonate sediment. Many sedimentologists regard it as a secondary mineral, forming when calcite or aragonite (primary minerals) combine with magnesium. Others think that dolomite sometimes forms as a primary mineral.

Maybe you can help to resolve this question. For this short-term project, you and a partner should propose and explain a hypothesis for the formation of dolomite. Describe how you could test your hypothesis, perhaps using standard equipment in a college laboratory. Ask your professor to help you set up such an experiment in the laboratory and follow through with the test for a few weeks.

INFERRING THE ORIGIN OF INTERESTING FORMATIONS

The Shawangunk conglomerate is a well-known sedimentary unit that runs through New Jersey and southern New York. Parts of it are exposed in outcrops. Resting unconformably on Ordovician shales, this Silurian-age unit varies in thickness, then thins and disappears from the Paleozoic stratigraphic section. The Shawangunk conglomerate consists mostly of quartz pebbles, so it is very resistant to erosion and forms spectacular cliffs at the eastern margins of the Catskill Mountains. Only in a few locations does it display pebbles of different rock types.

What type of source area would supply the coarse-grained sediments for the Shawangunk conglomerate? How could you determine the direction of the source area at the time the conglomerate was deposited? What might the few diverse pebble types tell you about the source? Using a geologic map of your own region, locate a coarse conglomerate unit and determine the source of its constituent particles. If there are no coarse conglomerates in your area, determine why this is the case.

SUGGESTED READINGS

Blatt, Harvey. 1992. *Sedimentary Petrology,* 2nd ed. New York: W. H. Freeman.

Goreau, T. F., N. I. Goreau, and T. J. Goreau. August 1979. Corals and coral reefs. *Scientific American,* 124–136.

Leeder, M. R. 1982. *Sedimentology.* London: Allen and Unwin.

McLane, Michael. 1995. *Sedimentology.* New York: Oxford University Press.

Prothero, Donald R., and Fred Schwab. 1996. *Sedimentary Geology.* New York: W. H. Freeman.

Siever, Raymond. 1988. *Sand.* New York: Scientific American Library.

INTERNET SOURCE

Sedimentary Rocks
ⓘ **http://www.science.ubc.ca/~geol202/s/sed/ sedimentary.html**
The University of British Columbia established a Web site for its Introduction to Petrology course. This URL links directly to the sedimentary rocks section of the course.

8

Metamorphic rock showing bands of different compositions produced by heat and pressure. These bands are not to be confused with the layering of sedimentary rocks. Pemaquid Point lighthouse, Maine. *(Larry Ulrich.)*

Metamorphic Rocks

W e are all familiar with the ways in which heat can transform material. Frying raw ground meat changes it into a hamburger composed of chemical compounds very different from those in the raw meat. Cooking batter in a waffle iron not only heats up the batter but also puts pressure on it, transforming it into a rigid solid. In similar ways, rocks change as they encounter high temperatures and pressures. Deep in Earth's crust, tens of kilometers below the surface, the temperatures and pressures are high enough to metamorphose rock without being high enough to melt it. Increases in heat and pressure and changes in the chemical environment can alter the mineral compositions and crystalline textures of sedimentary and igneous rocks, *even though they remain solid all the while.* The result is the third large class of rocks, the **metamorphic** or "changed form" **rocks,** which have undergone changes in mineralogy, texture, chemical composition, or all three.

A limestone filled with fossils, for example, may be transformed to a white marble in which no trace of fossils remains. The rock, originally made of small crystals of calcite, may be unchanged in mineral and chemical composition, but its texture may have changed drastically to one of large, intergrown crystals. Shale, a well-bedded rock so fine-grained that no individual mineral grain can be seen by the naked eye, may be changed to a form in which bedding is obscured and large crystals of mica glitter in the sun. In this metamorphic transformation, both mineral composition and texture have changed, but the overall chemical composition of the rock remains the same. Other rocks, such as those altered by heat and fluids derived from igneous activity, change in mineralogy, texture, *and* chemical composition. Silicate minerals that are diagnostic of metamorphism—their presence indicates that a rock is metamorphic—include the three polymorphs of aluminum silicate (Al_2SiO_5): kyanite, andalusite, and sillimanite; staurolite; and epidote. Some other minerals are common in metamorphic rocks but are also found in some igneous rocks. These minerals include garnet, quartz, muscovite, amphibole, and feldspar.

This chapter examines the causes of metamorphism, the types of metamorphism that occur under certain sets of conditions, and the origins of the various textures that characterize metamorphic rocks.

CAUSES OF METAMORPHISM

Sediments and sedimentary rocks belong to Earth's surface environments and igneous rocks to the melts of the lower crust and mantle. Metamorphic rocks now exposed at the surface are mainly the products of processes acting on rocks at depths ranging from the upper to the lower crust; most have formed at depths of 10 to 30 km, the middle to lower half of the crust. Large parts of the mantle may be metamorphic, but we see such rocks only in the exposed cores of some deeply eroded mountain belts.

Although most metamorphism occurs at depth in the crust and mantle, metamorphism can take place at the Earth's surface. We can see metamorphic changes in the baked surfaces of soils and sediments just beneath volcanic lava flows.

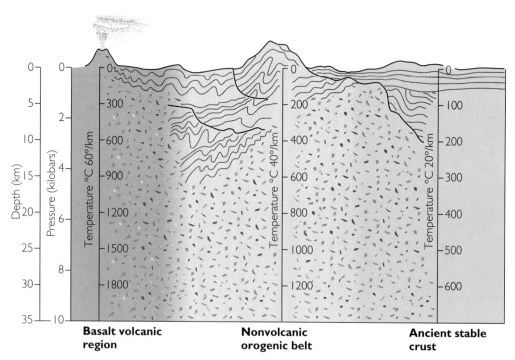

FIGURE 8.1 Pressure and temperature increase with depth in all regions, as shown in this cross section of a volcanic, a nonvolcanic orogenic, and an ancient stable crustal region. Pressure increases with depth at more or less the same rate everywhere, but temperature increases at different rates in different regions. (Pressure is measured in kilobars; one kilobar is approximately equal to 1000 times the atmospheric pressure at Earth's surface.)

The pressure and heat that drive metamorphism are consequences of three forces:

■ The internal heat of the Earth

■ The weight of overlying rock

■ The horizontal pressures developed as rocks become deformed

Long ago, miners and geologists learned through experience that the deeper we go into the Earth, the hotter it gets and the heavier are the supports needed to keep mining tunnels from collapsing under the pressure. Temperatures increase with depth at different rates in different regions of the Earth, ranging from 20 to 60°C/km (Figure 8.1). In much of Earth's crust, temperatures increase at a rate of 30°C/km. For the 30°C/km rate, for example, the temperature will be about 450°C at a depth of 15 km, much higher than the average temperature of the surface, which ranges from 10 to 20°C in most regions. (See also the discussion of geotherm in Chapter 19.) The pressure at a depth of 15 km would be equivalent to the weight of all the overlying rock, or about 4000 times the pressure at the surface.

High as these temperatures and pressures may seem, they are only in the middle range of metamorphism, as Figure 8.2 shows. We refer to the metamorphic rocks formed under the lower temperatures and pressures of shallower crustal regions as *low-grade rocks* and the ones formed at deeper zones of higher temperature and pressure as *high-grade rocks.*

Some metamorphic rocks may have been subjected early in their history to high pressures and temperatures, which produced a high-grade rock, then much later to lower temperatures and pressures. In that case, the new conditions remetamorphose the high-grade rock to a low-grade rock in a process called **retrograde metamorphism.**

PHYSICAL AND CHEMICAL FACTORS CONTROLLING METAMORPHISM

Metamorphic changes bring a preexisting rock into equilibrium with new surroundings. A sedimentary rock formed by diagenesis, for example, is in equilibrium with the moderate pressures and temperatures corresponding to burial a few kilometers deep. Later, this rock may be caught up in an orogeny that buries it much deeper and subjects it to a temperature of more than 500°C. Given enough time—short by geological standards but usually a million years or more—the rock is changed mineralogically and texturally so that it is brought into equilibrium with the new temperatures and pressures. The deeper (and thus hotter) the crust, the faster the metamorphic changes take place.

Temperature

Heat has a profound effect on a rock's mineralogy and texture. As we saw in Chapter 4, heat can break chemical bonds and alter the existing crystal structures of igneous rocks. As the rock adjusts to its new temperature, its atoms and ions recrystallize, linking up in new arrangements and creating new mineral assemblages. Many new crystals will grow larger than they were in the original rock, and the rock may become banded as minerals of different compositions are segregated into separate planes. (See the photograph that appears on the opening page of this chapter.) Because different minerals are known to crystallize and remain stable at different temperatures, the metamorphic geologist, like the igneous

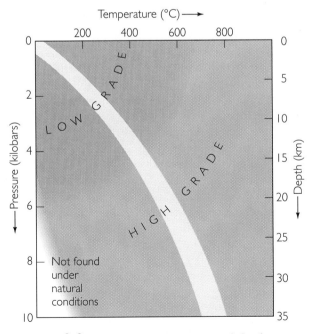

FIGURE 8.2 Temperatures, pressures, and depths at which low- and high-grade metamorphic rocks are formed. The wide band shows common rates at which temperature and pressure increase with depth over much of the continents.

geologist, uses a rock's composition to gauge the temperature at which the rock formed. Thus, the mineral assemblage of a metamorphic rock may constitute a "geothermometer."

Pressure

Pressure changes a rock's texture as well as its mineralogy. Solid rock is subjected to two basic kinds of pressure, also called **stress:**

- *Confining pressure,* a general pressure in all directions, like the pressure the atmosphere exerts at the surface of the Earth. High levels of confining pressure alter mineralogy by squeezing atoms together to form new minerals with denser crystal structures.

- *Directed pressure,* which is exerted in a particular direction, as when a ball of clay is squeezed between thumb and forefinger. The compressive action of converging plates is a form of directed pressure, which results in deformation of the rock. Because heat reduces the strength of rocks, directed pressure is likely to cause severe folding and deformation of metamorphic rocks in orogenic belts where temperatures are high.

Depending on the kind of stress applied to the rocks, metamorphic minerals may be compressed, elongated, or rotated to line up in a particular direction. Thus, directed pressure guides the shape and orientation of the new metamorphic crystals formed as the minerals recrystallize under the influence of both heat and pressure. During recrystallization of micas, for example, the crystals grow with the planes of their sheet-silicate structures aligned perpendicular to the directed stress. Crystals of elongate minerals, such as the amphiboles, also line up in planes perpendicular to the directed stress. Knowledge from experimental laboratory studies of the pressures required for these changes allows us to infer from the mineralogy and texture what the pressures were in a given area. Thus, metamorphic mineral assemblages can be used as pressure gauges or *geobarometers.*

Chemical Metamorphic Changes

A rock's chemical composition can be altered significantly during metamorphism by the introduction or removal of chemical components. Chemical changes in surrounding rocks, such as shale or lime- stone, commonly follow the intrusion of a magma. Hydrothermal fluids rise from the magma, carrying dissolved sodium, potassium, silica, copper, zinc, and other chemical elements soluble in hot water under pressure. These elements may be derived from both the magma and the intruded rock. As hydrothermal solutions percolate up to the shallower parts of the crust, they react with the rocks they penetrate, changing their chemical and mineral compositions and sometimes completely replacing one mineral by another without changing the rock's texture. This kind of change in a rock's bulk composition by fluid transport of chemical elements into or out of the rock is called **metasomatism.** Many valuable deposits of copper, zinc, lead, and other metal ores are formed by this kind of chemical substitution.

Fluids in Metamorphism

Many of the chemical and mineralogical changes that occur during metamorphism take place in fluids that permeate the solid rock. Although metamorphic rocks as we see them in outcrops appear to be completely dry and of extremely low porosity, most contain fluid in their minute, thin pores (the spaces between grains). The fluid is typically water containing carbon dioxide and minor amounts of other dissolved gases, salts, and traces of the components of the rock's minerals. This intergranular fluid acts as a medium that accelerates metamorphic chemical reactions. As changes in temperature and pressure break up crystal structures, atoms and ions move back and forth between the rock and its fluid. When in the fluid, the atoms and ions are able to migrate through the rock more rapidly and react with the solids to form new minerals.

The carbon dioxide in the fluids in metamorphosing rocks is derived largely from sedimentary carbonate—limestones and dolostones. The water is derived from chemically bound water in clay and other hydrous minerals, not from sedimentary pore waters, which are largely expelled during diagenesis.

Fluids move actively during metamorphism, at first along grain boundaries, then along opening channels as the rock cracks from the fluid pressure. As the fluids rise in the crust, they encounter cooler surroundings and often precipitate quartz in the channels, producing the quartz veins that are so abundant in low-grade metamorphic rocks.

As metamorphism proceeds, the water itself reacts with the rock as chemical bonds between minerals and water molecules form or break. The minerals of mafic volcanic rocks, for example, which contain almost no water in their crystal structures, take up some water molecules from pore fluids during early stages of metamorphism. In this reaction, these largely anhydrous (water-free) minerals form newly crystallized micas, chlorite, and other minerals that are hydrous—that is, with chemical bonds between water and other chemical components. The clay minerals in sedimentary rocks, by contrast, initially contain much chemically bound water, and the rocks contain additional water in the pores. Both chemically bound water and pore water are largely lost from the rock during metamorphosis and work their way upward to shallower crustal regions. The higher the metamorphic grade, the lower the water content of the rock.

KINDS OF METAMORPHISM

Thanks to modern technology, geologists working in laboratories can now duplicate metamorphic conditions and determine the precise combinations of pressure, temperature, and chemical composition under which transformations might take place. But to understand how any particular combination is related to the geology of metamorphism—that is, when, where, and how these conditions came about in the Earth—geologists must go into the field. Their observations in the field have allowed them to group metamorphic rocks into several categories on the basis of the geological circumstances of the rocks' origin. These categories are described below; Figure 8.3 locates them in relation to major plate-tectonic settings.

Regional Metamorphism

Regional metamorphism, the most widespread type of metamorphism, occurs where both high temperature and high pressure are imposed over large belts of the crust. We use this term to distinguish this type of metamorphism from more localized changes near igneous intrusions or faults. Regional

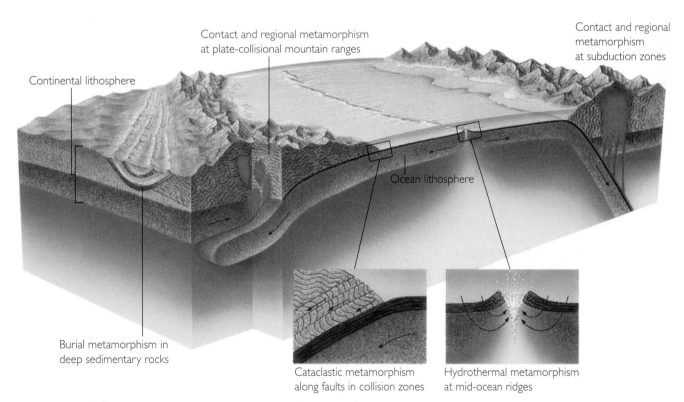

FIGURE 8.3 Metamorphic rocks are formed in four main plate-tectonic settings: subduction zones, continental collisions, mid-ocean ridges, and deeply subsiding regions on continents. Contact metamorphism by extrusives is not shown because it is quantitatively unimportant.

metamorphism destroys some or all of the original igneous or sedimentary textures as it causes the growth of new minerals. Some regional metamorphic belts are created by high temperatures and moderate to high pressures near the volcanic arcs formed where subducted plates dive deep into the mantle (see Figure 8.3). Other belts are formed by high pressures and moderate temperatures near oceanic trenches, where subduction drags down relatively cold oceanic crust. Regional metamorphism under very high pressures and temperatures occurs in deeper levels of the crust along boundaries where converging continental tectonic plates deform rock and raise high mountain belts.

Contact Metamorphism

Igneous intrusions metamorphose the immediately surrounding rock by their heat and pressure, subjecting the minerals of the preexisting rock to new conditions. This type of localized transformation, called **contact metamorphism,** normally affects only a thin region of intruded rock along the contact. In many contact metamorphic rocks, especially along shallow intrusions, the mineral transformations are largely related to the high temperature of the magma. Pressure effects are important where the magma was intruded at great depths. Contact metamorphism by extrusives is limited to very thin zones because lavas cool quickly at the surface and their heat has little time to penetrate deep into the surrounding rocks and cause metamorphic changes. Contact metamorphism, like the igneous activity with which it is linked, is found along plate convergences and oceanic and continental hot spots. Because igneous activity may also occur in regional metamorphic belts, contact metamorphism is also found in deformed mountain belts.

Cataclastic Metamorphism

Metamorphism may be found along faults, where tectonic movements cause the crust to crack and slip. As the rocks along the fault plane shear past each other, they grind and mechanically fragment solid rock into a pasty mass. This is **cataclastic metamorphism**—high-pressure metamorphism occurring primarily by the crushing and shearing of rock during tectonic movements. It produces the broken, pulverized texture found in strongly deformed mountain belts where faulting is extensive. Thus, cataclastic rocks are often found together with regionally metamorphosed rocks.

Hydrothermal Metamorphism

Another form of metamorphism, called **hydrothermal metamorphism,** is often associated with mid-ocean ridges, where plates spread apart and rising basalt magmas form new oceanic crust. Seawater percolating through the hot, fractured basalts of the ridge flanks becomes heated. The increase in temperature promotes chemical reactions between the seawater and the rock, forming altered basalts whose chemical compositions differ distinctively from that of the original basalt. Hydrothermal metamorphism also occurs on continents when fluids rising from igneous intrusions metamorphose both the overlying rocks and the deeply buried rocks that are undergoing regional metamorphism.

Burial Metamorphism

Recall from Chapter 7 that as sedimentary rocks are gradually buried by the sinking of the crust, they slowly heat up as they come into equilibrium with the crustal temperatures surrounding them. In this process, diagenesis alters the mineralogy and texture of the sedimentary rock. Diagenesis grades into **burial metamorphism,** a low grade of metamorphism caused by the heat and pressure exerted by overlying sediments and sedimentary rocks. Although temperatures and pressures are not so great as those that accompany the higher grades of regional metamorphism, they are high enough to produce partial alteration of the mineralogy and texture of the sedimentary rock. Bedding and other sedimentary structures are preserved. In general, burial metamorphic rocks are not strongly deformed but show broad, open folds. With the increasing temperatures and pressures that accompany orogenies, normally in association with volcanism, burial metamorphism grades into regional metamorphism.

METAMORPHIC TEXTURES

All the various kinds of metamorphism imprint new textures on the rocks they alter. The texture of a metamorphic rock is determined by the sizes, shapes, and arrangement of its constituent crystals. Some metamorphic textures are dependent on the particular kinds of minerals formed, such as the micas, which have a platy habit. Metamorphic textures may be inherited in part from the parent; thus, the sizes of grains in a sedimentary rock may be reflected in the sizes of crystals formed during metamorphism.

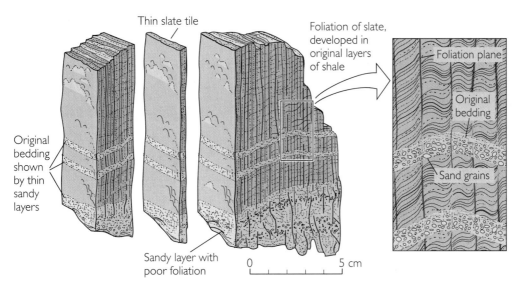

Thin slate tile

Foliation of slate,
developed in
original layers
of shale

Foliation plane

Original
bedding

Sand grains

Original
bedding
shown
by thin
sandy
layers

Sandy layer with
poor foliation

0   5 cm

FIGURE 8.4 Fragments of slate *(left)* show foliation (vertical lines) and relics of original bedding. The enlargement shows small faultlike offsets of the bedding along the surfaces of foliation. (After J. Gilluly, A. C. Waters, and A. O. Woodford, *Principles of Geology,* 4th ed., San Francisco, W. H. Freeman, 1975.)

Each textural variety tells us something about the metamorphic process that created it.

Foliation and Cleavage

The most prominent textural feature of regionally metamorphosed rocks is **foliation,** a set of flat or wavy parallel planes produced by deformation. The planes generally cut the rocks at an angle to the bedding of the original sediment, although they may coincide with the bedding in some places (Figure 8.4; also see metamorphic rock in Figure 3.1).

One of the main causes of foliation is the presence of platy minerals, chiefly the micas and chlorite. Platy minerals tend to crystallize as thin platelike or sheety crystals. The planes of all the platy crystals are aligned parallel to the foliation. The parallel planes are called the *preferred orientation* of the minerals (Figure 8.5). As platy minerals crystallize, their planes take a preferred orientation that is usually perpendicular to the main direction of the forces squeezing the rock during the deformation that accompanies metamorphism. Preexisting minerals may acquire a preferred orientation and thus produce foliation by

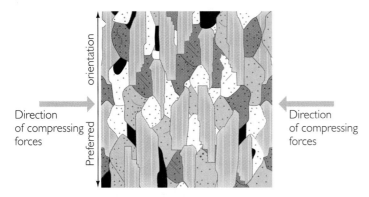

Preferred orientation

Direction
of compressing
forces

Direction
of compressing
forces

FIGURE 8.5 Preferred orientation of platy and elongate crystals in foliated rocks *(left).* The preferred orientation of crystallization of platy minerals is usually perpendicular to the forces that compress the rock during deformation. A photomicrograph of schist *(right)* shows the preferred orientation of mica (blue, green, purple) and staurolite (yellow) crystals. *(S. Dobos.)*

FIGURE 8.6 Slaty cleavage, the most familiar form of foliation, develops along thin, regular intervals. This outcrop in southwestern Montana shows slaty cleavage (nearly vertical fractures) and bedding (indistinct layers at an angle of about 45°). *(Martin Miller.)*

rotating crystals until they lie parallel to the developing plane. Plastic deformation, or the softening and bending of the hot rock without breakage, also may produce crystals with a preferred orientation.

Minerals whose crystals have an elongate, pencil-like shape, such as the amphiboles, also tend to assume a preferred orientation during metamorphism; the crystals normally line up parallel to the foliation plane. Rocks that contain abundant amphiboles, typically metamorphosed mafic volcanics, show this kind of texture.

The most familiar form of foliation is seen in slate, a common metamorphic rock, which is easily split along smooth, parallel surfaces into thin sheets (Figure 8.6). This *slaty cleavage* (not to be confused with the cleavage of a mineral such as muscovite) develops along moderately thin, regular intervals in the rock. Slate splitters learned long ago to recognize this property and use it to make thick or thin slates for roofing tiles and blackboards. We still use flat slabs of slate for flagstone walks in parts of the country where slate is abundant.

Foliated Rocks

The foliated rocks are classified according to four main criteria (Figure 8.7):

- The nature of their foliation
- The size of their crystals
- The degree to which their minerals are segregated into lighter and darker bands
- Their metamorphic grade

Examples of the major types of foliated rocks are shown in Figure 8.8.

SLATE   Slates are the lowest grade of foliated rocks. These rocks, with their excellent planar partings, are so fine-grained that their individual minerals cannot be seen easily without a microscope. They are commonly produced by the metamorphism of shales or, less frequently, of volcanic ash deposits. Slates are usually dark gray to black, colored by small amounts of organic material originally present in the parent shale (Figure 8.8a). Slates tinged red and purple get their color from iron oxide minerals, and greenish slates are colored by chlorite, a sheety iron silicate mineral closely related to the micas. As we saw in Chapter 7, shales may split along bedding planes. In contrast, the foliation along which slates split is usually at an angle to the bedding. This is good evidence that foliation is not necessarily related to the texture of the parent rock. Foliation is

INCREASING INTENSITY OF METAMORPHISM

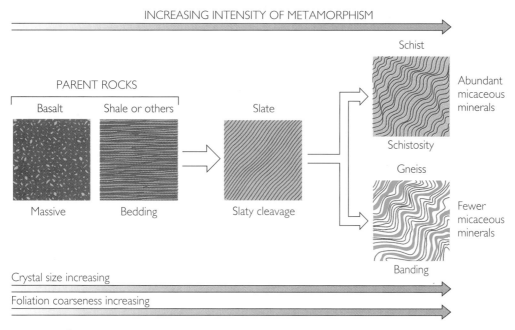

PARENT ROCKS

Basalt

Massive

Shale or others

Bedding

Slate

Slaty cleavage

Schist

Schistosity

Abundant micaceous minerals

Gneiss

Banding

Fewer micaceous minerals

Crystal size increasing

Foliation coarseness increasing

FIGURE **8.7** Classification of foliated rocks. Cleavage and schistosity do not generally correspond to the original bedding direction of sedimentary rocks.

the product of metamorphic rather than sedimentary processes.

PHYLLITE   The **phyllites** are of a slightly higher grade than the slates but are of similar character and origin. Phyllites tend to have a more or less glossy sheen from crystals of mica and chlorite that have grown a little larger than those of slates. Phyllites, like slates, tend to split into thin sheets but less perfectly than slates.

SCHIST   At low grades of metamorphism, crystals of platy minerals are generally too small to be seen,

foliation is closely spaced, and layers are very thin. As metamorphic rocks are more intensely metamorphosed to higher grades, the foliation becomes more conspicuous and pervasive throughout the rock. At the same time, the platy crystals grow to sizes visible to the naked eye, and the minerals may tend to segregate in lighter and darker bands. This parallel arrangement of sheet minerals produces the coarse, wavy, pervasive foliation that is known as schistosity and that characterizes **schists.** Schists are among the most abundant metamorphic rock types. They contain more than 50 percent platy minerals, mainly the micas muscovite and biotite (Figure 8.8b). Depend-

(a)   (b)   (c)

FIGURE **8.8** Foliated metamorphic rocks. (a) Slate. (b) Schist. (c) Gneiss. *(Breck P. Kent.)*

ing on the quartz content of the original shale, schists may contain thin layers of quartz, feldspar, or both. Schists are often named for their most abundant mineral constituent. Thus, there are mica schists, chlorite schists, and quartz schists.

GNEISS   Even coarser foliation is shown by high-grade **gneisses,** light-colored rocks with coarse bands of segregated light and dark minerals throughout the rock (Figure 8.8c). Gneisses do not split along the foliation, and there are few sheetlike, or micaceous, minerals along the foliation planes. Gneisses are coarse-grained, and the ratio of granular to platy minerals is higher than in slate or schist. The result is poor foliation and thus little tendency to split. Under conditions of high pressure and temperature, mineral assemblages of the lower grade rocks containing micas and chlorite alter to new assemblages dominated by quartz and feldspars, with lesser amounts of micas and amphiboles.

The banding of gneisses into light and dark layers results from the segregation of lighter colored quartz and feldspar and darker amphiboles and other mafic minerals. Some gneisses may be the metamorphic equivalents of sandstones, others of granitic rocks.

Nonfoliated Rocks

Not all metamorphic rocks are foliated. Some show a very weak preferred orientation of crystals, which results in little or no foliation. Other rocks show no preferred orientation and therefore no foliation. **Nonfoliated rocks** are composed mainly of crystals that grow in equant (equidimensional) shapes, such as cubes and spheres, rather than in platy or elongate shapes (Figure 8.9). These rocks may result from contact, regional, hydrothermal, or burial metamor-phism. Nonfoliated rocks include hornfels, quartzite, marble, argillite, greenstone, amphibolite, and granulite. All of them, except for hornfels, are defined by their mineral composition rather than their texture.

A **hornfels** is a high-temperature contact metamorphic rock of uniform grain size that has undergone little or no deformation. Its platy or elongate crystals are oriented randomly, and foliated texture is absent. Hornfels have a granular texture overall, even though they commonly contain pyroxene, which makes elongate crystals, and some micas.

Quartzites are very hard, nonfoliated white rocks derived from quartz-rich sandstones. Some quartzites are massive—that is, unbroken by preserved bedding or foliation. Other quartzites contain thin bands of slate or schist, relics of former interbedded layers of clay or shale (Figure 8.9a). Quartzites are found in both contact and regionally metamorphosed areas.

Marbles are the metamorphic products of heat and pressure acting on limestones and dolomites. They may result from either contact or regional metamorphism. Some white, pure marbles, such as the famous Italian Carrara marbles prized by sculptors, show an even, smooth texture of intergrown calcite crystals of uniform size. Other marbles show irregular banding or mottling from silicate and other mineral impurities in the original limestone (Figure 8.9b).

Argillite is a low-grade, nonfoliated metamorphic rock made from a mudstone or other clay-rich sedimentary rock. Argillite breaks with an irregular or conchoidal fracture. Its lack of foliation may be attributable partly to a lower grade of deformation and partly to an abundance in the parent mudstone of quartz silt or other minerals that are neither platy nor elongate.

Greenstones are metamorphosed mafic volcanic rocks. Many of these low-grade rocks form

(a)

(b)

FIGURE 8.9 Nonfoliated metamorphic rocks.
(a) Quartzite. (b) Marble.
(Breck P. Kent.)

when mafic lavas and ash deposits react with percolating seawater or other solutions. Large areas of the seafloor are covered with basalts slightly or extensively altered in this way at mid-ocean ridges. On the continents, buried volcanic and plutonic mafic igneous rocks react with groundwaters at temperatures of 150 to 300°C and form similar greenstones. An abundance of chlorite gives these rocks their greenish cast.

Amphibolite is a mostly nonfoliated rock made up of amphibole and plagioclase feldspar. It is typically the product of medium- to high-grade metamorphism of mafic volcanics.

The metamorphic rock type **granulite,** though it usually has a granular texture, is defined by its mineral composition, which indicates high or very high grade. Lower grade granular rocks, which are not all granulite, include quartzite and hornfels. The minerals found in granulites typically include quartz, plagioclase feldspar, pyroxenes, garnets, and sillimanite. Like other granular metamorphic rocks, granulites are medium- to coarse-grained rocks in which the crystals are equant and show only faint or no foliation. They are formed by the metamorphism of shale, impure sandstone, and many kinds of igneous rock.

FIGURE 8.10 Garnet porphyroblasts in schist matrix. *(Chip Clark.)*

Large Crystal Textures

New metamorphic minerals may grow into large crystals surrounded by a much finer grained matrix of other minerals. They superficially resemble phenocrysts of igneous rocks. These large crystals are **porphyroblasts;** they are found in both contact and regionally metamorphosed rocks (Figure 8.10). They grow by reorganization of the chemical components of the matrix and thus replace portions of the matrix, in contrast to phenocrysts, which are the earliest minerals to crystallize as an igneous rock is formed. Porphyroblasts form under conditions of a strong contrast between the chemical and crystallographic properties of the matrix and those of the porphyroblast minerals. This contrast causes the porphyroblast crystals to grow faster than the slow-growing minerals of the matrix, at the expense of the matrix. Porphyroblasts vary in size, ranging from a few millimeters to several centimeters in diameter. Their composition also varies. Garnet and staurolite are two common minerals that form porphyroblasts, but many others are also found. The precise composition and distribution of porphyroblasts of these two minerals can be used to infer the pressures and temperatures of metamorphism. Pure garnets, trans-

parent and beautifully colored in shades of red, green, and black, are valued as semiprecious gems.

Deformational Textures

Structural deformation accompanies most metamorphism. We have already seen that mechanical deformation along fault planes produces cataclastic metamorphism. As the movement of two rock surfaces against each other pulverizes minerals and strings them out in bands or streaks, metamorphic rocks called **mylonites** are formed. These fine-grained rocks can be foliated when they are formed very deep in Earth's crust, where rocks under very high pressures have been plastically deformed.

The textural effects of deformation are most obvious in mylonites, but they are also prominent in the foliated rocks. Metamorphism may precede the deformation, be contemporaneous with it, or follow it. Many metamorphic rocks have complex geological histories. A sedimentary rock may become a low-grade metamorphic greenish schist as a result of deep burial without significant deformation. Later it may be caught up in deformation associated with mountain building and become strongly foliated during a higher grade of metamorphism. Finally, when subjected to contact metamorphism as igneous intrusions are injected into the orogenic belt,

TABLE **8.1**

CLASSIFICATION OF METAMORPHIC ROCKS BASED ON TEXTURE

CLASSIFICATION	CHARACTERISTICS	ROCK NAME	TYPICAL PARENT ROCK
Foliated	Distinguished by slaty cleavage, schistosity, or gneissic foliation; mineral grains show preferred orientation	Slate Phyllite Schist Gneiss	Shale, sandstone
Granular (nonfoliated)	Granular, characterized by coarse or fine interlocking grains; little or no preferred orientation	Hornfels Quartzite Marble Argillite Greenstone Amphibolite[1] Granulite[2]	Shale, volcanics Quartz-rich sandstone Limestone, dolomite Shale Basalt Shale, basalt Shale, basalt
Porphyroblastic	Large crystals set in fine matrix	Slate to gneiss	Shale
Crushed	Grains pulverized and streaked by cataclastic deformation	Mylonite	Shale, sandstone

[1]Typically contains much amphibole, which may show alignment of long, narrow crystals.
[2]High-temperature, high-pressure rock.

the rock may be altered to an even higher grade nonfoliated metamorphic rock.

Table 8.1 summarizes the textural classes of metamorphic rocks and their main characteristics.

REGIONAL METAMORPHISM AND METAMORPHIC GRADE

Because metamorphic rocks form under a wide range of conditions, their minerals and textures are clues to the pressures and temperatures in the crust where and when they were formed. Geologists who study the geologic formation of metamorphic rocks constantly seek to determine the intensity and character of metamorphism more precisely than is indicated by a designation of "low grade" or "high grade." To make these finer distinctions, geologists read minerals as though they were pressure gauges and thermometers. The techniques are best illustrated by their application to regional metamorphism.

Mineral Isograds: Mapping Zones of Change

When geologists study broad belts of regionally metamorphosed rocks, they can see many outcrops,

some showing one set of minerals, some showing others. Different parts of these belts may be distinguished by their *index minerals,* or the characteristic minerals that define metamorphic zones formed under a restricted range of pressures and temperatures. For example, one may cross from a region of unmetamorphosed shales to a belt of weakly metamorphosed slates and then to a belt of high-grade schists (Figure 8.11a). Original sedimentary minerals—typically quartz, feldspar, calcite, and clay minerals—are found in the shales, but at the slate belt, a new mineral—chlorite—appears. Moving in the direction of increasing metamorphism, the geologist may successively encounter biotite, then garnet, then staurolite, then other metamorphic mineral zones, the schists becoming progressively more foliated.

We can show these zones, where one metamorphic grade changes to another, on a map. To do this, geologists define the zones by drawing lines called *isograds,* which connect the places where index minerals first appear. Figure 8.11b uses isograds to show a series of regionally metamorphosed rocks produced by the metamorphism of a shale. A pattern of isograds tends to follow the structural grain of a region as folds and faults reveal it. An isograd based on a single index mineral, such as the biotite isograd (see Figure 8.11b), is a good approximate measure of metamorphic pressure and temperature.

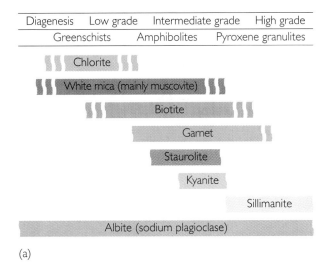

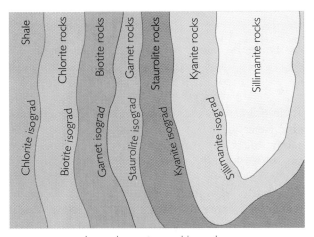

— Increasing metamorphic grade →

(b)

muscovite + quartz ⟶
    K-feldspar + sillimanite + water

$$KAl_3Si_3O_{10}(OH)_2 + SiO_2 \longrightarrow$$
$$KAlSi_3O_8 + Al_2SiO_5 + H_2O$$

Many groups of minerals have been carefully studied in the laboratory to determine more exactly the pressures and temperatures at which they formed. The results are used to calibrate the field mapping of isograds.

Because isograds reflect pressures and temperatures at which minerals form, the isograd sequence in one metamorphic belt may differ from that in another belt. This is true because pressure and temperature do not increase at the same rate in all geologic settings. Pressure increases more rapidly than temperature in some places, more slowly in others (Figure 8.12).

Metamorphic Grade and Parent Rock Composition

The kind of metamorphic rock that results from a given grade of metamorphism depends partly on the mineral composition of the parent rock. The shale metamorphism shown in Figure 8.11 reveals the effects of metamorphic conditions on rocks rich in

FIGURE 8.11 (a) Changes in the mineral composition of shales metamorphosed under low-grade to high-grade conditions. (b) A map view of a regionally metamorphosed terrain in which shales have been metamorphosed under the same conditions of pressure and temperatures as those shown in (a). The isograd lines mark the first appearance of the index minerals displayed as colored bars in diagram (a).

To determine pressure and temperature more precisely, geologists examine a group of two or three minerals whose textures indicate that they crystallized together. For example, a sillimanite isograd would be represented by the chemical reaction of muscovite and quartz to produce K-feldspar and sillimanite, liberating water (as water vapor) in the process:

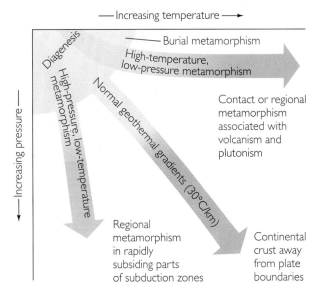

FIGURE 8.12 The routes taken by various combinations of increases in pressure and temperature in relation to depth increases depend on tectonic and igneous activity. The various routes give different sequences of metamorphic rock types.

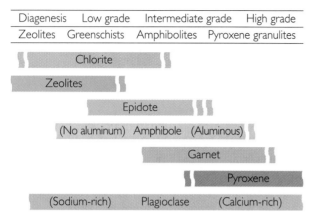

FIGURE 8.13 Changes in the mineral composition of basalts and other mafic rocks metamorphosed under conditions of low-grade to high-grade pressure and temperature. Compare the mineral assemblages of shales metamorphosed under the same conditions in Figure 8.11 to see the effect of original composition on metamorphic mineralogy.

clay minerals, quartz, and perhaps some carbonate minerals. Metamorphism of mafic volcanic rocks, composed predominantly of feldspars and pyroxene, follows a different course (Figure 8.13).

In the regional metamorphism of a basalt, for example, the lowest grade rocks characteristically contain various **zeolite** minerals. This class of silicate minerals contains water in cavities within the crystal structure. They are formed by alteration at very low temperatures and pressures. Rocks that include this group of minerals are thus identified as zeolite grade.

Overlapping with the zeolite grade is a higher grade of metamorphosed mafic volcanic rocks, the **greenschists,** whose abundant minerals include chlorite and epidote (an aluminosilicate). Next are the **amphibolites,** which contain large amounts of the key minerals hornblende (an amphibole mineral), plagioclase feldspar, and garnet. The highest grade of metamorphosed mafic volcanics are the **pyroxene granulites,** coarse-grained rocks containing pyroxene and calcium plagioclase.

Pyroxene granulites are the products of high-grade metamorphism in which the temperature is high and the pressure moderate. In the opposite situation, where the pressure is high and the temperature moderate, rocks of a variety of compositions—from mafic volcanic rocks to shaly sedimentary rocks—form rocks of **blueschist** grade. The name comes from the abundance in these rocks of glaucophane, a blue amphibole. Still another metamorphic rock, formed at extremely high pressures and moderate to high temperatures, is **eclogite,** a rock rich in garnet and pyroxene.

Metamorphic Facies

We can put all this information on metamorphic grades derived from parent rocks of many different chemical compositions on a graph of temperature and pressure (Figure 8.14). **Metamorphic facies** are groupings of rocks of various mineral compositions formed under different grades of metamorphism from different parent rocks. The essential points of the concept of metamorphic facies are that

- Different kinds of metamorphic rocks are formed from parent rocks of different composition at the same grades of metamorphism.

- Different kinds of metamorphic rocks are formed under different grades of metamorphism from parent rocks of the same composition.

Table 8.2 lists the major minerals of the metamorphic facies produced from shale and basalt. Because parent rocks vary so greatly in composition, there are no sharp boundaries between metamorphic facies (see Figure 8.14).

At the high-grade end of the series shown in Figure 8.14, metamorphic rocks become partly melted and thus are transitional to igneous rocks.

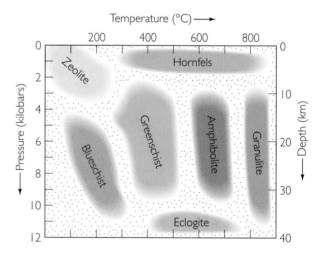

FIGURE 8.14 The various types of metamorphic rocks may be grouped in accordance with the temperature and pressure fields in which they were formed. There are no sharp boundaries between any of these groupings, or facies.

TABLE **8.2**

MAJOR MINERALS OF METAMORPHIC FACIES PRODUCED FROM PARENT ROCKS OF DIFFERENT COMPOSITION

FACIES	MINERALS PRODUCED FROM SHALE PARENT	MINERALS PRODUCED FROM BASALT PARENT
Greenschist	Muscovite, chlorite, quartz, sodium-rich plagioclase feldspar	Albite, epidote, chlorite
Amphibolite	Muscovite, biotite, garnet, quartz, plagioclase feldspar	Amphibole, plagioclase feldspar
Granulite	Garnet, sillimanite, plagioclase feldspar, quartz	Calcium-rich pyroxene, calcium-rich plagioclase feldspar
Eclogite	Garnet, sodium-rich pyroxene, quartz	Sodium-rich pyroxene, garnet

These rocks are badly deformed and contorted, and they are penetrated by many veins, small pods, and lenses of melted rock. This kind of very high grade veined gneiss is called **migmatite,** a term that is applied to rocks of mixed igneous and metamorphic origin. Some migmatites are mainly metamorphic, with only a small proportion of igneous material. Others have been so affected by melting that they are considered almost entirely igneous.

CONTACT METAMORPHIC ZONES

The best place to see the metamorphic effects of an intruding igneous body is at an outcrop where a shale has been intruded by a dike or sill. At the border, where the shale is in contact with the dike, the shale has lost all of its original texture: bedding is gone, fossils are obliterated, and the shale's mineralogy has completely changed. Instead of the fine-grained clay minerals of the shale, the rock immediately adjacent to the dike now consists of large crystals of pyroxenes and aluminosilicate minerals such as andalusite, which are not found in sedimentary rocks. Farther from the contact, from several centimeters to a meter away, faint outlines of the shale's bedding can still be seen, but the clay minerals have been altered to crystalline micas. At some greater distance from the contact, the shale is unaltered. Thus, contact metamorphic zones, like regional metamorphic ones, are characterized by index minerals that reflect different grades of metamorphism.

Contact Aureoles

The rim of metamorphically altered rock around an igneous intrusion is called the *contact aureole.* Its thickness and character depend on the temperature of the magma and the depth in the crust where the intrusion took place. As one might expect, the contact aureole is most prominent where a mafic intrusion at about 1000°C invaded shallow crustal rocks only a few kilometers deep, where normal temperatures are only about 60 to 90°C. In this kind of intrusion, the temperature is very high at the contact but drops rapidly a short distance away from it. Intrusions at lower temperatures, such as granites at about 600°C, invade deeper parts of the crust where temperatures are already high. Because they do not heat the surrounding rocks so much, these lower temperature intrusions cause less metamorphic change.

Metamorphic Grade and Parent Rock Composition

Contact metamorphic zones vary with the kinds of parent rock that are in contact with the hot intrusive. Although contact metamorphic rocks are not ordinarily included in metamorphic facies diagrams such as Figure 8.14, these rocks show the same relationships between grade and composition of parent rock. For example, the patterns of mineral grades shown by impure limestones, which are composed mostly of carbonate minerals, differ from those shown by shales, which are made almost entirely of silicate minerals. In a typical contact metamorphic change in limestones, the carbonate minerals react with silica impurities in the rock to form wollastonite, a light-colored calcium pyroxene-like mineral:

calcite + silica ⟶ wollastonite + carbon dioxide

$$CaCO_3 + SiO_2 \longrightarrow CaSiO_3 + CO_2$$

The carbon dioxide produced by this reaction generally escapes as a gas through fractures and pores in the rock. This reaction takes place at temperatures of about 500°C at near-surface pressures, or at higher temperatures as the pressure increases. Thus, wollastonite is a guide to the metamorphic grades of rocks whose parent rocks were carbonates. Near the contact is the mineral produced at the highest temperature, wollastonite, together with garnet (Figure 8.15) and diopside, a calcium magnesium pyroxene. Farther from the contact, the zone contains serpentine—a

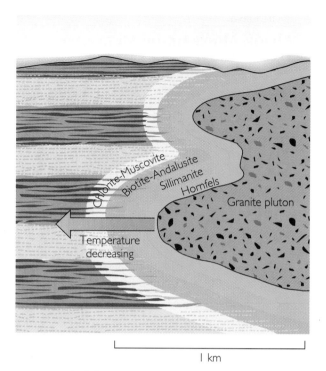

FIGURE 8.16 A contact aureole in sandstones and shales intruded by a granite pluton. This contact aureole is a series of zones characterized by minerals ranging from the unaltered quartz and clay minerals of the sandstones and shales to bands of aluminum, iron, and magnesium silicates. The rock in contact with the pluton is a pyroxene hornfels.

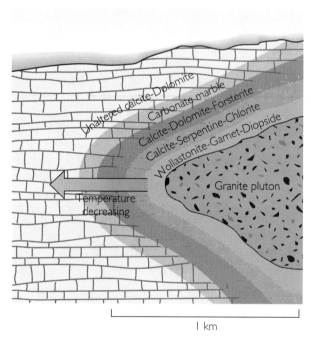

FIGURE 8.15 The contact metamorphism of a limestone composed of calcite and dolomite with quartz and clay mineral impurities produces a contact aureole of several zones of minerals. Grading from unaltered rock to the contact, they change from carbonate marble to bands of various calcium-magnesium silicate minerals and finally to a carbonate-free silicate rock.

magnesium silicate containing chemically bound water—together with chlorite and calcite. Farthest from the intrusion, the zones of lowest temperature contain a magnesium olivine, forsterite, and calcite and dolomite. Beyond this zone, the limestone mineralogy is unaffected. The whole aureole may be hundreds of meters wide. Because there can be appreciable chemical exchange among the aureole, the pluton, and the surrounding rock, this process is a variety of metasomatism, which you will recall leads to a chemical change in a rock's bulk composition.

Contact metamorphism of silicate rocks such as shale also produces progressive mineral zones, but the groups of minerals in them differ from those seen in metamorphosed limestones (Figure 8.16). Hornfels, containing pyroxene and mica, is at the contact. In inner zones, closer to the heat of the intrusion, pure aluminum silicates, such as sillimanite, are characteristic. In cooler outer zones, the quartz, clay, and carbonate of the shale are transformed to biotite micas, andalusite amphiboles, and calcite. In the outermost zones, chlorite and muscovite form. These kinds of aureoles may involve little, if any, chemical exchange among the aureole, the pluton, and the surrounding rock. Hence, there is no overall change in chemical composition.

PLATE TECTONICS AND METAMORPHISM

Soon after the theory of plate tectonics was proposed, geologists started to see how patterns of metamorphism fit into the larger framework of plate-tectonic movements that cause volcanism and orogeny. At the beginning of this chapter, we explored the link between plate-tectonic settings and the geologic processes that cause the various types of metamorphism (see Figure 8.3). We can also often deduce a rock's site of metamorphism on the basis of its grade and composition.

Greenstones—metamorphosed basalts—are associated with metamorphism at mid-ocean ridges, where the seafloor spreads and hot basaltic magma wells up from the mantle. The heat of the magma transforms newly extruded basalts into low-grade metamorphic rocks of the greenschist facies. Hydrothermal circulation through the basalt also plays a role in the alteration of basalts.

Blueschists—the metamorphosed volcanic and sedimentary rocks whose minerals indicate that they were formed under very high pressures but at relatively low temperatures—form in the forearc region of a subduction zone, the area between the seafloor trench and the volcanic arc. There sediments are carried down the subduction zone along the surface of a cool subducting lithospheric slab. The subducted plate moves down at such a high rate of speed that it heats up very slowly, while the pressure increases rapidly.

The opposite conditions of high temperature and low pressure are found along the volcanic island arcs on the overriding plate at a subduction zone. Here the magmas formed by melting of the deeply subducted plate rise to shallow depths in the overriding plate and by their heat transform shallowly buried volcanics and sediments into greenschist and higher grade rocks. Both kinds of metamorphism, high pressure–low temperature and low pressure–high temperature, are found as paired belts along plate convergences. On the oceanic side, close to the trench, where the descending slab is still cool, we find the high-pressure, low-temperature metamorphics. On the landward or volcanic arc side, we find the high-temperature, low-pressure rocks. Figure 8.17 shows these pressure-temperature relationships superimposed on the metamorphic facies diagram.

Regional metamorphic belts are associated with continental collisions that cause orogenies. In the cores of the major mountain belts of the world, from the Appalachians to the Alps, we find long belts of regionally metamorphosed and deformed sedimentary and volcanic rocks that parallel the lines of folds and faults of the mountains (see Feature 8.1).

As continents collide and the lithosphere thickens, the deep parts of the continental crust heat up and metamorphose to different grades, while in deeper zones melting may begin. In this way, the complex mixture of metamorphic and igneous rocks forms the cores of orogenic belts that evolve during mountain building. Millions of years afterward, when erosion has stripped off the surface layers, the cores are exposed at the surface, providing the geologist with a rock record of the metamorphic processes that formed the schists, gneisses, and other metamorphic rocks.

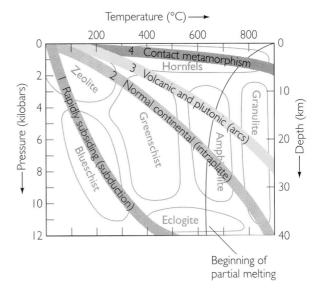

FIGURE **8.17** The routes taken by increases in pressure and temperature in various tectonic settings superimposed on metamorphic facies. Curve 1 shows rapidly increasing pressure with a slow rise in temperature, characteristic of the forearcs of subduction zones. Curve 2 is the path followed in continental regions away from plate boundaries, places where burial metamorphism occurs. Curve 3 is the higher temperature route followed in regions of volcanic arcs, typical of regional metamorphism. Curve 4 is the high-temperature, low-pressure path of contact metamorphism at shallow and moderate depths.

8.1   INTERPRETING THE EARTH

The New England Metamorphic Terrane

The rocky, mountainous terrane of New England is a historically important regionally metamorphosed area in North America that has been mapped by a great many geologists from the mid-nineteenth century to the present day. While the plains were still being explored, early geologists in New England, such as Edward and Charles Hitchcock and J. D. Dana, were struggling to understand the complex arrays of minerals in this perplexing terrane. They built on a foundation of understanding that came from studies of similar metamorphic terranes in the Scottish Highlands and the Alps. Today, the geology of New England still serves as a useful model of the relationships between metamorphism and deformation in an orogenic belt.

Knowledge of metamorphic processes in a terrane such as New England has enabled geologists to "look through" the metamorphic mineral assemblages to reconstruct the mineral and textural characteristics of the original sedimentary rocks. These rocks were deposited in the elongate sedimentary basin that occupied what is now the northern part of the Appalachian mountain chain. In researching this area, metamorphic geologists learned to deduce the intrusive and extrusive igneous rocks that were emplaced before, during, and after the main metamorphic event.

A map of the present metamorphic belt in New England shows how the terrane runs north-northeast to south-southwest from northern Maine and the Canadian Maritime provinces to the southern shore along Long Island Sound. In this terrane, the metamorphic facies parallel lines of folding and faulting.

Northern New England is dominated by low-grade greenschist facies rocks formed at low temperatures and pressures, typically containing chlorite, biotite, and garnet. Farther south, the grade increases to the amphibolite facies of medium grade, containing staurolite and kyanite. In eastern Vermont and western New Hampshire, there is an elongate region of high-grade granulite facies with rocks containing sillimanite and K-feldspar.

The main period of metamorphism extended over two major tectonic deformations, one about 460 million years ago and one about 400 million years ago. According to a plate-tectonic reconstruction of these times in New England, the first of these tectonic episodes was the result of active subduction along the eastern continental margin of North America. Partial evidence for this episode comes from the presence of metamorphosed volcanics along the eastern side of the terrane. The second deformation and continued metamorphism accompanied the convergence of two plates carrying what we now know as North America and an eastern continent covering parts of western Europe and northwest Africa. This continental collision continued for many more millions of years as the supercontinent Pangaea was assembled (see Chapter 20).

Note: Terrane is a term used for a distinctive block of the Earth's crust, usually delimited by major faults or displacements from surrounding regions (see Chapter 20).

Right: A map of the metamorphic belt of New England showing grades of metamorphism. (Modified from P. Robinson, *Styles of Metamorphism with Depth in the Central Acadian High, New England: A Field Trip Honoring J. B. Thompson, Jr.,* Contribution No. 63, Department of Geology and Geography, University of Massachusetts, Amherst, 1989.)

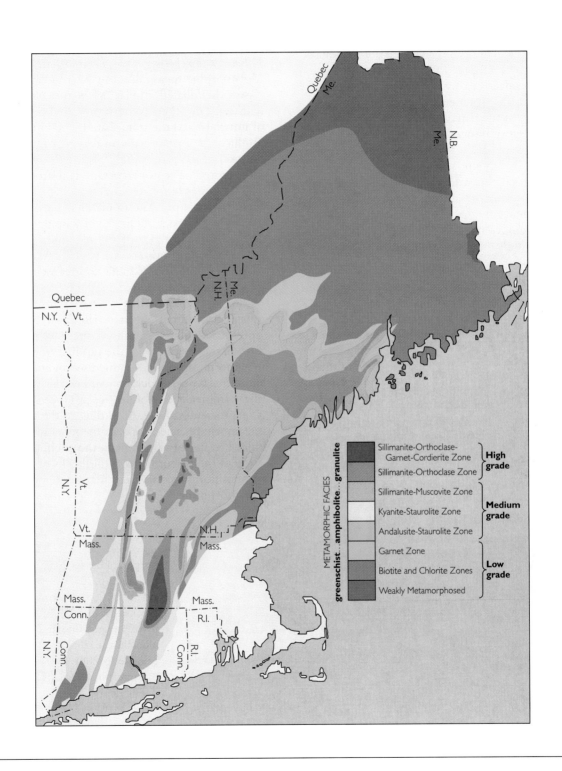

Sillimanite-Orthoclase-
Garnet-Cordierite Zone   **High
grade**
Sillimanite-Orthoclase Zone

Sillimanite-Muscovite Zone

Kyanite-Staurolite Zone   **Medium
grade**

Andalusite-Staurolite Zone

Garnet Zone

Biotite and Chlorite Zones   **Low
grade**

Weakly Metamorphosed

METAMORPHIC FACIES
greenschist...amphibolite...granulite

SUMMARY

What factors cause metamorphism? Metamorphism—alteration in the solid state of preexisting rocks—is caused by increases in pressure and temperature and reaction with chemical components introduced by migrating fluids. As pressures and temperatures deep within the crust increase as a result of tectonic or igneous activity, the chemical components of the parent rock rearrange themselves into a new set of minerals that are stable under the new conditions. Rocks metamorphosed at relatively low pressures and temperatures are referred to as low-grade rock, those metamorphosed at high temperatures and pressures as high-grade rocks. Chemical components of a rock may be added or removed during metamorphism, most commonly by the influence of fluids migrating from nearby intrusions.

What are the various kinds of metamorphism? The two major types of metamorphism are (1) regional metamorphism, during which large areas are metamorphosed by high pressures and temperatures generated during orogenies; and (2) contact metamorphism, during which rocks surrounding magmas are metamorphosed primarily by the heat of the igneous body. Three additional metamorphic types are (3) cataclastic metamorphism, during which rocks along fault planes are pulverized and mineralogically altered; (4) hydrothermal metamorphism, during which hot fluids percolate through and metamorphose various crustal rocks; and (5) burial metamorphism, a variety of regional metamorphism, during which deeply buried sedimentary rocks are altered by the more or less normal increase of pressure and temperature in the crust.

What are the chief types of metamorphic rocks? Metamorphic rocks fall into two major textural classes: the foliated (displaying fracture cleavage, schistosity, or other forms of preferred orientation of minerals) and the nonfoliated. The kinds of rocks produced by metamorphism depend on the composition of the parent rock and the grade of metamorphism. The regional metamorphism of a shale leads to zones of foliated rocks of progressively higher grade, from slate to phyllite, schist, and gneiss. These zones are marked by isograds, defined by the first appearance of an index mineral. Regional metamorphism of mafic volcanic rocks progresses from zeolite grade to greenschist, then to amphibolite and pyroxene granulite. Among nonfoliated rocks, marble is derived from the metamorphism of limestone, quartzite from quartz-rich sandstone, argillite from mudstone, and greenstone from basalt. Hornfels are the product of contact metamorphism of fine-grained sedimentary rocks and other types of rock containing an abundance of silicate minerals. Mylonites are produced by cataclastic metamorphism. According to the concept of metamorphic facies, rocks of the same grade may differ because of variations in the chemical composition of the parent rocks, while rocks of the same composition may vary because of different grades of metamorphism.

KEY TERMS AND CONCEPTS

metamorphic rocks (p. 195)

retrograde metamorphism (p. 197)

stress (p. 198)

metasomatism (p. 198)

regional metamorphism (p. 199)

contact metamorphism (p. 200)

cataclastic metamorphism (p. 200)

hydrothermal metamorphism (p. 200)

burial metamorphism (p. 200)

foliation (p. 201)

phyllite (p. 203)

schist (p. 203)

gneiss (p. 204)

nonfoliated rocks (p. 204)

hornfels (p. 204)

quartzite (p. 204)

marble (p. 204)

argillite (p. 204)

greenstone (p. 204)

amphibolite (p. 205)

granulite (p. 205)

porphyroblast (p. 205)

mylonite (p. 205)

zeolite (p. 208)

greenschist (p. 208)

pyroxene granulite (p. 208)

blueschist (p. 208)

eclogite (p. 208)

metamorphic facies (p. 208)

migmatite (p. 209)

EXERCISES

1. What kinds of metamorphism are related to igneous intrusions?

2. What does preferred orientation refer to in a metamorphic rock?

3. Name a mineral that is commonly found in a schist and that shows preferred orientation.

4. Name two nonfoliated metamorphic rocks.

5. What is a porphyroblast?

6. Contrast a schist and a gneiss.

7. What is an isograd?

8. What is the difference between a granite and a slate?

9. How are metamorphic facies related to temperatures and pressures?

10. In which plate-tectonic settings would you expect to find regional metamorphism?

THOUGHT QUESTIONS

1. What kinds of preferred orientation of minerals would you expect to find in an amphibolite?

2. Why are there no metamorphic rocks formed under natural conditions of very low pressure and temperature, as shown in Figure 8.2?

3. How is slaty cleavage related to deformation?

4. Are cataclastic rocks more likely to be found in a continental rift valley or in a volcanic arc?

5. Would you choose to rely on chemical composition or type of foliation to determine metamorphic grade?

6. You have mapped metamorphic rocks and have observed north-south isograd lines running from kyanite in the east to chlorite in the west. Were metamorphic temperatures higher in the east or in the west?

7. What determines which polymorph of Al_2SiO_5 is formed in a metamorphic rock?

8. Contrast the minerals found in the contact metamorphism of a pure limestone and a limestone containing appreciable layers of shale.

9. Which kind of pluton would produce the highest grade metamorphism, a granite intrusion 20 km deep or a gabbro intrusion at a depth of 5 km?

10. Why would you not expect to find burial metamorphosed rocks at a mid-ocean ridge?

SUGGESTED READINGS

Best, Myron G. 1982. *Igneous and Metamorphic Petrology.* New York: W. H. Freeman.

Blatt, Harvey, and Tracy, Robert. 1996. *Petrology: Igneous, Sedimentary, and Metamorphic,* 2nd ed. New York: W. H. Freeman.

Hyndman, Donald W. 1985. *Petrology of Igneous and Metamorphic Rocks.* New York: John Wiley and Sons.

Raymond, L. A. 1995. *Petrology.* Dubuque, Iowa: Wm. C. Brown.

Yardley, Bruce W. D. 1989. *An Introduction to Metamorphic Petrology.* Essex, England: Longmans Scientific and Technical; co-published in the United States with John Wiley and Sons, New York.

INTERNET SOURCES

Metamorphic Rocks
🛈 **http://www.science.ubc.ca/~geol202/s/meta/metamorphic.html**
The University of British Columbia established a Web site for its Introduction to Petrology course. This URL links directly to the section on metamorphic rocks.

Skarn Page
🛈 **http://www.wsu.edu:8080/~meinert/skarnHP.html**
This site at Washington State University is a resource for this interesting metamorphic alteration of carbonate rocks. Features include a general overview, photos of skarns that change periodically, skarn mineralogy, and skarn trivia.

9

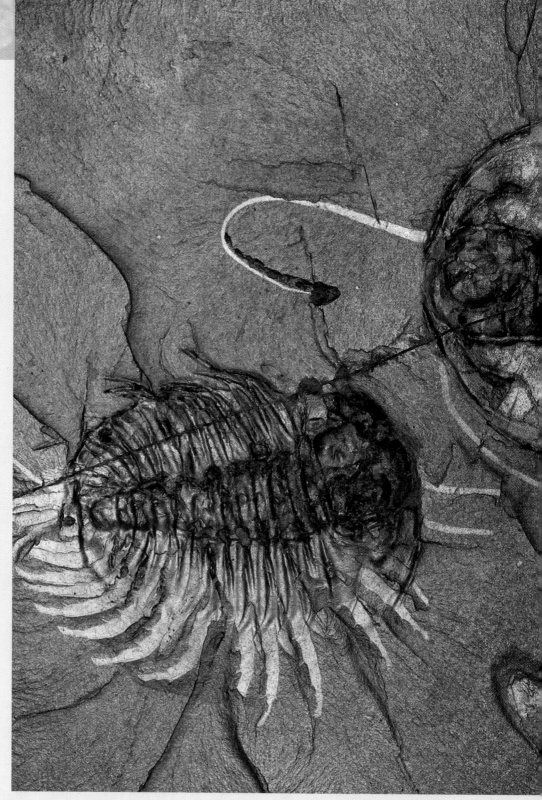

Trilobites in Burgess Shale from Canadian Rockies. These particular distant relatives of modern crayfish are now extinct. They are found only in rocks about 530 million years old. The relative ages of sedimentary rock layers can be surmised from the characteristic fossils in each layer. (*Chip Clark.*)

The Rock Record and the Geologic Time Scale

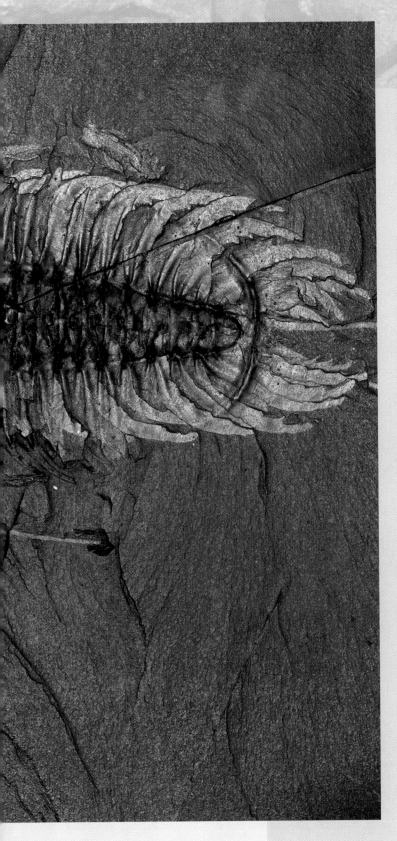

Despite an occasional earthquake or volcanic eruption, the Earth seems to provide a reasonably stable foundation on which to build a civilization. But organized human society is only a few thousand years old. Over the past millions of years, Earth has been far less stable than it presently seems to be: continents, oceans, and mountains have moved great distances. Even at this instant, there is hardly a place on Earth that is not moving vertically and horizontally, however slowly. Learning the patterns and rates of these movements is a key job of the geologist.

The geological processes that shape Earth's surface and give structure to its interior work over millions and billions of years. In this chapter, we examine some of the ways geologists have learned to deal with these extraordinarily long intervals, both to understand the nature of such slow processes and to reconstruct the geologic history of the planet.

One of the most important reasons for placing geological processes in their proper sequence is to learn about the evolution of the planet we see today. When were the Rocky Mountains formed? What was happening in East Africa when early humans were evolving? To answer these kinds of questions, we need a tool for organizing and dating the rock record: a geologic calendar to determine the sequence in which rock layers formed and their ages, and a way to compare the ages of rocks situated continents apart. Two centuries of modern geological research have resulted in such a tool: the geologic time scale. It enables scientists to determine the age of the Earth, to unravel complex geologic histories, and even to study the origin and evolution of life.

TIMING THE EARTH

Geologists (and astronomers) differ from most other scientists in their attitude toward time. Physicists and chemists study processes that last only fractions of a second, such as the splitting of an atomic nucleus or a fast chemical reaction. Other scientists perform experiments that last from minutes to hours. Geologists, by contrast, deal with Earth processes that unfold over a great range of time periods (Figure 9.1). Earthquake tremors last seconds or minutes, whereas the building of a mountain chain takes many millions of years.

We can measure some geological processes directly. A river's floodwaters, for example, rise and fall over a few days. We can even measure the relatively slow movement of glaciers, which may take a year to move 50 m. Other geological processes, however, such as the erosion of a hillside, are too slow to be measured directly. We may rely on historical records to determine the amounts of time required for some of these processes (Figure 9.2). But even the oldest historical records extend back only a few thousand years, far short of the time span required to measure the very slow geological processes that shape the planet. Our only resource for timing such processes is the rock record. Rocks formed in the past and preserved from erosion serve as Earth's memory, recording geological events, such as glaciations, that lasted many thousands or millions of years.

As you will see, geologists of the nineteenth century used their understanding of rock strata and fossils to determine the **relative ages** of sedimen-

tary rock layers—how old they are in relation to one another. These early researchers could then put the geological events that created these rock formations into chronological order. Today, geologists use the physics of radioactive decay to pinpoint a rock's **radiometric age,** often called its "absolute" age—

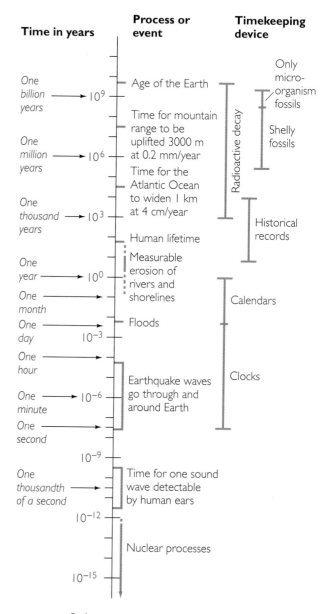

FIGURE 9.1 The amount of time required for some common processes and events. Times are given as orders of magnitude. The scale is logarithmic; that is, it has equal divisions between successive powers of 10.

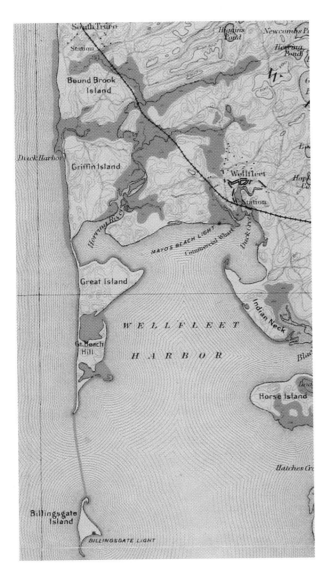

FIGURE 9.2 Historical records such as old maps and land surveys are useful for measurement of some geological processes. Since 1887, when this map was being prepared, sand, silt, and mud (brown areas added to original in 1988) brought in by waves and tides have filled in areas of tideland near Wellfleet Bay on Cape Cod in Massachusetts.

the actual number of years that have passed since the rock formed.

The geologists who worked out the geologic time scale did more than merely date rocks. They precipitated a revolution in the way we think about time, our planet, and even ourselves. They discovered that Earth is far older than anyone had previously imagined and, contrary to the common beliefs of earlier times, that its surface and interior have been changed and shaped repeatedly by the same geological processes that operate now. They found, too, that not only the planet but its inhabitants have evolved over time. And humans, they discovered, account for only the briefest moment of Earth's long history.

RECONSTRUCTING GEOLOGIC HISTORY THROUGH RELATIVE DATING

Geologists of the nineteenth century built a geologic time scale from the time and space relationships of rocks exposed at the surface or in drill holes. They began, as we do here, with evidence from stratigraphy—the description, correlation, and classification of strata in sedimentary rocks.

The Stratigraphic Record

Stratification—the layering that is the hallmark of sedimentary rocks—is basic to two simple principles used to interpret geologic events from the sedimentary rock record:

1. The **principle of original horizontality** states that sediments are deposited as essentially horizontal beds. Observation of modern marine and nonmarine sediments in a wide variety of environments supports this generalization. (Although cross-bedding, discussed in Chapter 7, is inclined, the overall orientation of cross-bedded units is horizontal.) If we find a sequence of sedimentary rock layers that is folded or tilted, we know that the rocks were deformed by tectonic stresses after their sediments were deposited.

2. The **principle of superposition** states that each layer of sedimentary rock in a tectonically undisturbed sequence is younger than the one beneath it and older than the one above it. Geological common sense tells us that a younger layer cannot slip beneath a layer that has already been deposited. This principle enables us to view a series of layers as a kind of vertical time line—that is, a partial or complete record of the

time elapsed from the deposition of the lowest bed to the deposition of the uppermost bed (Figure 9.3).

These two principles allow us to view a vertical set of strata, called a **stratigraphic sequence,** as a chronological record of the geologic history of a region. The corresponding time line based on the sequence is called the **geologic time** spanned by the sequence. (Geologic time is also used to refer to the entire span of time represented by Earth.) Stratigraphic sequences differ from sedimentary sequences, which were discussed in Chapter 7. Sedimentary sequences are vertical changes of lithology in sediments deposited in one environment of sedimentation. A stratigraphic sequence is broader in definition and includes a wide variety of beds of different origins. Whereas the emphasis in sedimentary sequences is the nature of successive sediment types, the emphasis in a stratigraphic sequence is the chronology of the constituent beds and the sedimentary conditions they imply.

With a geologic timekeeper, or "stratigraphic clock," geologists can tell whether one rock layer is older than another, although they cannot necessarily tell how much older in years. Ideally, a stratigraphic sequence might provide a measure of time in actual years if sediments accumulated continuously at a steady rate, compacted a constant amount as they lithified, and did not erode. If we knew that muddy sediments accumulated at a rate of 10 m per million years, for instance, then 100 m of mudstone would represent 10 million years of deposition.

In practice, however, we cannot accurately gauge time in years from stratigraphy, for several reasons. First, as you learned in Chapter 7, sediments do not accumulate at a constant rate in any sedimentary environment. During a flood, a river may deposit several meters of sand in its channel in just a few days, whereas in the years between floods it will deposit only a few centimeters of sand. Even in the deep ocean, where it may take 1000 years to deposit 1 mm of mud, sedimentation is unsteady, and the thickness of sediment cannot be used for precise timekeeping. In addition, the rate at which sediment is deposited varies widely in different sedimentary environments.

The second reason stratigraphy is an imprecise timekeeper is that the rock record does not tell us how many years have passed between periods of deposition. Many places on the floor of a river valley receive sediment only during times of flood. The times between floods are not represented by any sediment. Over the course of Earth's history in various places, there have been long intervals, some lasting *millions* of years, in which no sediments were deposited at all. In other places and at other times, sedimentary rocks may have been removed by erosion. Although we often can tell where a gap in the record occurs, we rarely can say exactly how long an interval it represents.

The final reason—and the one most important to geologists who wish to compare the geologic histories of different parts of the Earth—is that stratigraphy alone cannot be used to determine the relative ages of two widely separated beds. A geologist might

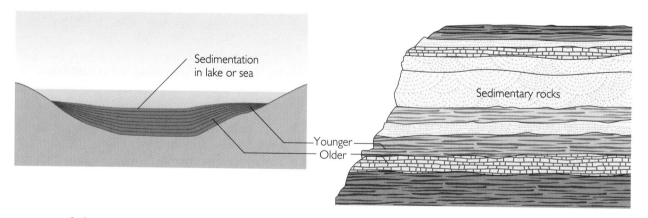

FIGURE 9.3 The principles of original horizontality and superposition. Sediments are deposited in horizontal layers, and the sediments gradually change into sedimentary rocks. If the rocks are undisturbed by tectonic processes, they remain horizontal, and their relative ages remain the same as those of their predecessor sediments: oldest on the bottom, youngest on the top.

be able to establish the relative age of one bed or a series of beds by following an outcrop over a limited distance, but there is no way of knowing whether a rock layer in Arizona, say, is older or younger than one in northern Canada.

Those early geologists found that the key to detecting missing time intervals and correlating the relative ages of rocks at different geographic locations lay in the discovery of fossils. Fossils became the single most important tool for constructing an accurate geologic time scale for the entire planet.

Fossils as Timepieces

To a great many students today, it must seem obvious that fossils are the remains of ancient organisms. Some look very much like animals now living; although others, such as the trilobite in the photo that opens this chapter, are the remains of extinct life forms. Fossils can be shells, teeth, bones, impressions of plants, or tracks of animals. The most common fossils in rocks of the last half-billion years are shells of invertebrates, such as clams, oysters, and the ammonite group shown in Figure 9.4a. Much less common are the bones of vertebrates, such as mammals, reptiles, and dinosaurs. It is only through the study of fossil dinosaur bones that we have gained an understanding of the nature of the immense beasts that have long been extinct. Plant fossils are abundant in some rocks, particularly those associated with coal beds, where leaves, twigs, branches, and even whole tree trunks can be recognized (Figure 9.4b). Fossils are not found in intrusive igneous rocks because the original biological material is lost in a hot melt. Fossils are rarely found in metamorphic rocks because any remains of organisms are usually too changed and distorted to recognize.

How did we get to this understanding of fossils? The ancient Greeks were probably the first to surmise that fossils are records of ancient life, but it was

(a)

(b)

FIGURE 9.4 Animal and plant fossils. (a) Ammonite fossils, ancient examples of a large group of invertebrate organisms that is now largely extinct. Their sole representative in the modern world is a single species, the chambered nautilus. *(Chip Clark.)* (b) Petrified Forest, Arizona. These ancient logs are millions of years old. Their substance was completely replaced by silica, which preserved all the original details of form. *(Tom Bean.)*

not until modern times that the concept took hold and its consequences were explored. One of the first modern thinkers to make the connection between fossils and once-living organisms was Leonardo da Vinci, in the fifteenth century. In the seventeenth century, Nicolaus Steno compared what he called "tongue-stones" found in the Mediterranean region with similarly shaped teeth of modern sharks and concluded that the stones were the remains of ancient life. Many people scoffed at Steno's conclusions; the idea seemed preposterous. To others it was blasphemous, for it ran counter to the belief that the stones were divine creations. But by the end of the eighteenth century, after hundreds of fossils and their relationships to modern organisms had been described and cataloged, the evidence that fossils are the remains of formerly living creatures had become overwhelming. Thus **paleontology,** the study of the history of ancient life from the fossil record, took its place beside geology, the study of Earth's history from the rock record.

The dividends from the study of fossils did not accrue to geology alone. Young Charles Darwin's famous voyage as naturalist on the *Beagle* (1831–1836) vastly extended his knowledge of the great variety of fossil organisms and what their presence in rocks portended. He also had an opportunity to see a host of unfamiliar animal and plant species in their native habitats. In 1859, putting his and others' knowledge of both ancient and modern organisms together with an understanding of Earth's history that came from the geologic time scale, Darwin proposed the theory of evolution. It revolutionized scientific thinking about the origins of the millions of species of animal and plant life and provided a sound theoretical framework for paleontology.

Well before Darwin, in 1793, William Smith, a surveyor working in southern England, had recognized that fossils could be used to date the relative ages of sedimentary rocks. Smith was fascinated by the variety of fossils, and he collected them from the sequences of rock strata he saw exposed along canals and outcrops. Smith knew nothing of the idea of organic evolution that Charles Darwin was to enunciate more than half a century later. He did note, however, that different layers contained different kinds of fossils, and he was able to tell one layer from another by the characteristic fossils in each. He established a general order for the sequence of fossils and strata, from lowermost (oldest) to uppermost (youngest) rock layers. Regardless of the location, Smith could predict the stratigraphic position of any particular layer or set of layers in any new outcrop in southern England, basing his prediction on its fossil assemblages. This stratigraphic ordering of the fossils is known as a *faunal succession.*

Smith was the first person to use faunal succession to correlate rocks from different outcrops. In each outcrop, he identified distinct formations. A **formation** is a series of rock layers that everywhere has about the same physical properties and contains the same assemblage of fossils. Some formations may consist of a single rock type, such as limestone. Others may be made up of thin, interlayered beds of different kinds of rocks, such as sandstone and shale. However they vary, each formation comprises a distinctive set of rock layers that can be recognized and mapped as a unit.

Using his knowledge of faunal successions, Smith matched up the similarly aged formations found in different outcrops. By noting the vertical order in which the formations were found in each

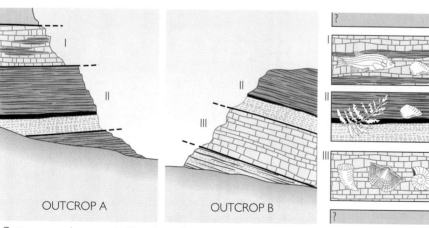

OUTCROP A   OUTCROP B

Outcrops may be separated by a long distance

FIGURE 9.5 William Smith could piece together the sequence of rock layers of different ages containing different fossils by correlating outcrops found in southern England. In this example, Formations I and II were exposed at outcrop A, and Formations II and III at outcrop B. A composite of the two would show Formations I and II both overlying Formation III and therefore younger than Formation III.

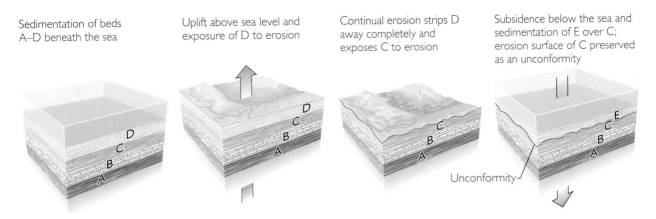

Sedimentation of beds A–D beneath the sea

Uplift above sea level and exposure of D to erosion

Continual erosion strips D away completely and exposes C to erosion

Subsidence below the sea and sedimentation of E over C; erosion surface of C preserved as an unconformity

Unconformity

FIGURE 9.6 An unconformity is a surface between two layers that were not laid down in an unbroken sequence. In the series of events outlined here, an unconformity forms through uplift and erosion, followed by subsidence and another round of sedimentation.

place, he compiled a composite stratigraphic sequence for the entire region. His composite series showed how the complete sequence would have looked if the formations at different levels in all the various outcrops could have been brought together in a single spot. Figure 9.5 shows such a composite for a series of three formations.

Relying on this approach of combining faunal successions with stratigraphic sequences, geologists of the last two centuries have carefully correlated formations around the world. The result, as we shall see, is a geologic time scale for the entire Earth.

Unconformities: Markers of Missing Time

In putting together sequences of formations, geologists often find places in which a formation is missing: either it was never deposited, or it was eroded away before the next strata were laid down. The boundary along which the two existing formations meet is called an **unconformity**—a surface between two layers that were not laid down in an unbroken sequence (Figure 9.6). An unconformity represents time, just as sedimentary rocks do.

Unconformities not only represent time but also imply tectonic forces that were responsible for the erosional interval by raising the land above sea level, where it became eroded. Alternatively, unconformities may represent times in which the land was eroded as sea level was lowered globally. Such lowering of sea level could be caused by, for example, the withdrawal of water from the oceans to form polar, glacial ice caps.

Geologists classify different kinds of unconformities in terms of the relationships between the upper and lower sets of layers. An unconformity in which the upper set of layers overlies an erosional surface developed on an undeformed, still horizontal lower set of beds is a **disconformity.** An unconformity in which the upper beds overlie metamorphic or igneous rock is a **nonconformity.** In many sequences, an unconformity forms after the lower beds have been folded by tectonic processes and then eroded to a more or less even plane. The next series is then laid down in horizontal layers. The resulting **angular unconformity** is an erosion surface that separates two sets of layers having bedding planes that are not parallel. Figure 9.7 is a dramatic angular unconformity found in the Grand Canyon.

FIGURE 9.7 The Great Unconformity in the Grand Canyon, Colorado, is an angular unconformity between the horizontal Tapeats sandstones above and the steeply folded Precambrian Wapatai shales below.
(GeoScience Features Picture Library.)

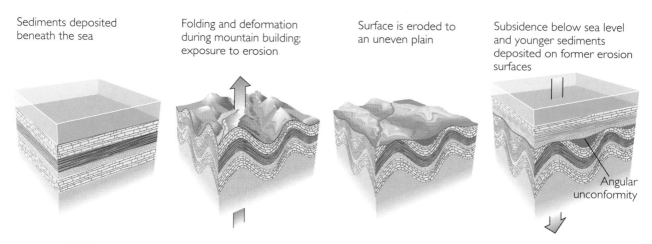

Sediments deposited beneath the sea

Folding and deformation during mountain building; exposure to erosion

Surface is eroded to an uneven plain

Subsidence below sea level and younger sediments deposited on former erosion surfaces

Angular unconformity

FIGURE 9.8 An angular unconformity is an erosion surface that separates two sets of layers having bedding planes that are not parallel. This sequence shows how such a surface can form.

Figure 9.8 illustrates the process by which angular unconformities may form.

Cross-Cutting Relationships

Other disturbances of the layering of sedimentary rocks also provide clues for dating. Recall from Chapter 4 that discordant dikes or other magmatic intrusions can cut through sedimentary layers, interrupting them. Faults displace bedding planes as they break apart blocks of rock (see Chapter 10). Faults can also displace dikes and sills. Because intrusions and faults can be fitted into stratigraphic sequences, they help us place geological events within a relative time frame. Since we know that deformational or intrusive events must have occurred after the affected sedimentary layers were deposited, such deformations or intrusions must be younger than the rocks they cut (Figure 9.9). If the intrusions or fault displacements are eroded and planed off at an unconformity, then overlain by a younger series of formations, we know that the intrusions or faults are older than the younger series of beds.

Sequence Stratigraphy

In the last two decades, a new form of stratigraphy—**sequence stratigraphy**—has arisen. This kind of stratigraphy was originally known as *seismic stratigraphy* because it took advantage of great im-

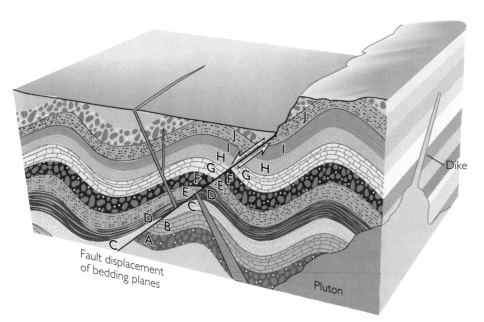

Fault displacement of bedding planes

Pluton

Dike

FIGURE 9.9 Cross-cutting relationships of igneous intrusions and deformation allow us to place geological events within the relative time frames given by the stratigraphic sequence. The faulting shown here had to have occurred after the folding of the bedding planes that were displaced.

provements in exploration seismology (see Chapter 19) that allowed geologists to see individual beds and packets of beds on a seismic profile, or cross section. Since then, the ideas generated by seismic stratigraphy have been worked out in outcrop sections as well. The basic unit used for sequence stratigraphy is the *sequence,* a series of sedimentary beds bounded above and below by unconformities. The alluvial sedimentary sequence discussed in Chapter 7 is one such sequence, representing just one cycle of stream deposits (see Figure 7.12). The sequences used for sequence stratigraphy are generally larger packets of beds that might extend over many such cycles. The unconformities that define sequences represent fluctuations in sea level that allowed land erosion to take place. The beds that make up the sequences show internal patterns that are diagnostic of changes in sedimentation.

For example, in a large river delta, sediment is laid down as the river enters the sea. This sediment gradually builds up the seafloor to sea level, thereby creating new land. Alluvial sediments then accumulate on this new surface, and—given millions of years—the delta may advance many miles into the sea (Figure 9.10a). If sea level were to rise as a result of either global events or tectonic subsidence, the shoreline would be shifted many miles inland. A new delta would start to be constructed overlying the former deltaic sequence, as illustrated in Figure 9.10b.

Using their knowledge of bedding patterns, sequence stratigraphers can match sequences of the same geologic age over wide areas. From such information, we can reconstruct a region's geologic history, including changes in sea level, in terms of successions of sequences.

The Geologic Time Scale

We have now seen several ways of ordering rock strata and correlating them with a time sequence of geologic events:

- We can determine the relative ages of sedimentary rocks both by the simple rule of superposition and by the local and global fossil record.

- We can use deformation and angular unconformities to date tectonic episodes in relation to the stratigraphic sequence.

- We can use cross-cutting relationships to establish the relative ages of igneous bodies or faults cutting through sedimentary rocks.

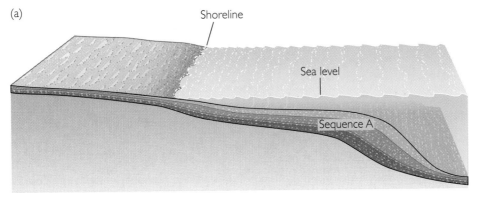

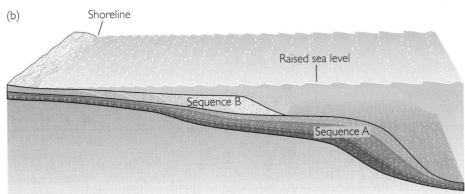

FIGURE 9.10 When tectonic subsidence or other global events have caused the sea level to rise, two deltaic sequences are found. (a) The lower sea level is accompanied by the deposition of sequence A. (b) The later raised sea level causes deposition of sequence B, which lies over sequence A. One result of the sea level changes is a displacement of the shoreline landward as sequence B is deposited.

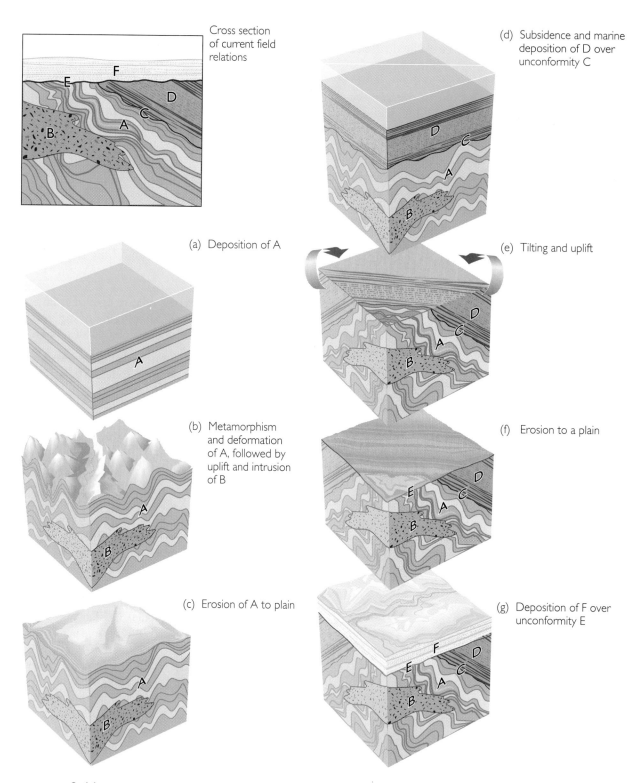

Cross section of current field relations

(a) Deposition of A

(b) Metamorphism and deformation of A, followed by uplift and intrusion of B

(c) Erosion of A to plain

(d) Subsidence and marine deposition of D over unconformity C

(e) Tilting and uplift

(f) Erosion to a plain

(g) Deposition of F over unconformity E

FIGURE **9.11** Here we see how geologists use field relationships for relative time dating. From field mapping, a geologist makes a cross section of four formations: A, deformed metamorphic rocks; B, a granite pluton; D, sandstones, limestones, and shales containing marine fossils; F, sandstones containing land fossils. A and D are separated by an unconformity (C). D and F are separated by an angular unconformity (E). The details of these formations enable a geologist to reconstruct the stages of this area's geologic history, as shown in diagrams (a) through (g).

Combining all three methods, we can decipher the history of geologically complicated regions (Figure 9.11).

During the nineteenth and twentieth centuries, geologists used these relative dating principles and pieced together information from outcrops all over the world to work out the entire **geologic time scale,** a relative-age calendar of Earth's geologic history. Each time interval on this scale is correlated with a corresponding set of rocks and fossils. Although the geologic time scale is still being refined here and there, its main divisions have remained constant for the past century.

As Figure 9.12 indicates, the geologic time scale is divided, in order of decreasing length, into four major time units: eons, eras, periods, and epochs. An **eon** is the largest division of history.

ARCHEAN EON   The oldest eon is the Archean (from Greek *archaios,* "ancient"). Archean rocks range from the earliest rocks known, about 4 billion years old, to rocks 2.5 billion years old. During the Archean eon, the Earth's basic structure and dynamics—from the core, mantle, and crust to the surface—were still being formed. Fossils of primitive unicellular microorganisms are found in some sedimentary rocks of this age.

PROTEROZOIC EON   The next younger set of rocks was formed during the Proterozoic eon (2.5 billion to 544 million years ago). During Proterozoic times, the surface and interior were close to the state they attained during later geologic times, with some significant exceptions. One exception was the level of oxygen in the atmosphere, which did not approach present levels until later in the Proterozoic. Most life forms were still unicellular; but in the later Proterozoic, more advanced forms started to evolve and to be preserved as fossils.

PHANEROZOIC EON   The most recent and best understood eon, covering the last 544 million years, is known as the Phanerozoic; many rock formations of this age contain an abundance of shells and other fossils, such as vertebrate bones. The Phanerozoic is subdivided into three **eras:**

- Paleozoic (544 million to 245 million years ago)

- Mesozoic (245 million to 65 million years ago)

- Cenozoic (65 million years ago to the present)

The eras are then subdivided into **periods,** most of which are named either for the geographic locality in which the formations are best displayed or were first described or from some distinguishing characteristic of the formations. The Jurassic period, for

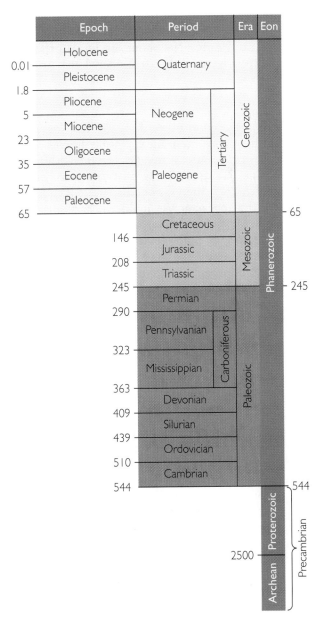

FIGURE 9.12 The geologic time scale. Numbers at the sides of the columns represent millions of years before the present. (Modified from W. B. Harland et al., *Geologic Timescale, 1989,* Cambridge: Cambridge University Press, 1990.)

example, is named for the Jura Mountains of France and Switzerland, and the Carboniferous period is named for the coal-bearing sedimentary rocks of Europe and North America.

Periods are further subdivided into **epochs;** the geologically best known are the subdivisions of the Tertiary period, such as the Pliocene, the youngest (5 million to 1.8 million years ago). In Feature 9.1, the geologic time scale is used to interpret one of the world's most spectacular outcrops, the Grand Canyon.

The names of the eons and eras originally were derived from early ideas about life forms preserved as fossils. Early researchers considered the Archean and the Proterozoic eons to be devoid of fossils and so called them *Azoic,* meaning "without life." When microscopic fossils were discovered in these rocks in the twentieth century, they were renamed *Proterozoic* (earlier life). *Phanerozoic* means visible life; *Paleozoic,* ancient life; *Mesozoic,* middle life; and *Cenozoic,* recent life. We now know that the start of the Paleozoic era (the Cambrian period) roughly marked the appearance of multicellular animals and that other major evolutionary developments happened at various times during the Proterozoic and Phanerozoic eons.

In working out this geologic time scale, geologists had to change their way of thinking about the Earth. This process of reconceptualization was led by James Hutton in the late eighteenth century and, in the nineteenth century, by Charles Lyell, author of one of the first and most influential geology textbooks *(Principles of Geology).* Following the lead of these two scholars, geologists came to understand that the planet was not shaped by a series of catastrophic events over a mere few thousand years, as many people believed, especially those who accepted the first chapter of Genesis as a literal account of Earth's creation. Rather, Earth was the product of ordinary geological processes operating steadily over much longer time intervals. As we noted in Chapter 3, Hutton was among the first to grasp the cyclical nature of the geologic changes that result from erosion, weathering, sedimentation, burial, igneous and tectonic activity, and finally the renewal of the cycle by mountain building and erosion again.

Also inherent in Hutton's thinking and enunciated most forcefully in Lyell's textbook was the principle of uniformitarianism—which, you may recall from Chapter 1, states that the processes we see shaping the Earth today are the same as those that have been operating during all of Earth's history. Although different kinds of sediments may have been deposited at different rates in different places throughout Earth's history, we can be sure that the depositional processes that laid down sediments millions and billions of years ago were working then in the same way they do today.

RADIOMETRIC TIME: ADDING DATES TO THE TIME SCALE

The geologic time scale based on studies of stratigraphy and fossils is a relative one; with it, geologists can say whether one formation is older than another, but they cannot pinpoint precisely when a rock formed. It's like knowing that World War I preceded World War II but not knowing the specific years in which each conflict began and ended. Nineteenth-century geologists could only estimate that it might take millions of years for one fossil assemblage to change to another. Although they had estimated times for various geological processes, such as the laying down of sediments on lake bottoms or the erosion of a river valley, they could not devise a way to measure the duration of those processes accurately. Some nineteenth-century physicists had estimated the age of the Earth and the solar system from astronomical and physical principles and calculated it to be many millions of years old. But these educated estimates were based on the physical principles of the day (some now outmoded), and they varied a good deal, from 25 million to 75 million years. Then, in 1896, an advance in modern physics paved the way for reliable and accurate measurements of geologic time in actual years. In that year, Henri Becquerel, a French physicist, discovered radioactivity in uranium. Within less than a year of Becquerel's find, the French chemist Marie Sklodowska-Curie discovered and isolated a different and highly radioactive element, radium. It soon became clear that rays emitted from radioactive materials are indications that these materials transform at constant rates from one chemical element to another.

In 1905, a New Zealander working in England, the physicist Ernest Rutherford, suggested that radioactivity could be used to measure the exact age of a rock. He was able for the first time to tell the age in years of a uranium mineral from measurements in his laboratory. In the next few years, the ages of many more rocks were determined as dating methods were refined and more radioactive elements were found. This was the start of **radiometric**

dating, the use of naturally occurring radioactive elements to determine the ages of rocks. When Rutherford announced the results of his first measurements, it became clear that the Earth is billions of years old and that the Phanerozoic eon alone was a little more than half a billion years long.

Radioactive Atoms: The Clocks in Rocks

How do geologists use radioactivity to determine the ages of rocks? What the pioneers of nuclear physics discovered at the turn of the century was that atoms of uranium, radium, and several other elements that exhibit **radioactivity** are unstable. The nucleus of a radioactive atom spontaneously disintegrates, forming an atom of a different element and emitting radiation, a form of energy, in the process. We call the original atom the parent; its decay product is known as the daughter. The parent isotope rubidium-87, for instance, forms the stable daughter isotope strontium-87 through radioactive decay. The rubidium-87 atom ejects an electron from one of the neutrons in its nucleus, leaving an additional proton. The former rubidium atom, which had 37 protons, thus becomes a strontium atom, with 38 protons (Figure 9.13). (Recall from Chapter 2 that an atomic nucleus consists of protons and neutrons and that the isotopes of a given element contain the same number of protons but different numbers of neutrons.)

When a given amount of a radioactive substance decays, all parents do not alter at the same time but at random times. Thus, a given mass of radioactive parent atoms is constantly firing off electrons to form daughters. The reason radioactive decay offers a dependable means of keeping time is that the *average* rate of nuclear disintegration is fixed. The decay rate does not vary with the changes in temperature, chemistry, or pressure that typically accompany geological processes in the Earth or other planets. This means that once atoms of a radioactive isotope are created anywhere in the universe, they start to act like a ticking clock, steadily altering from one type of atom to another at a fixed rate.

Radioactive decay rates are commonly stated in terms of an element's **half-life**—the time required for one-half of the original number of radioactive atoms to decay (Figure 9.14). The half-lives of geologically useful radioactive elements range from thousands to billions of years. At the end of the first half-life after a radioactive isotope is incorporated into a new mineral, half the number of parent atoms

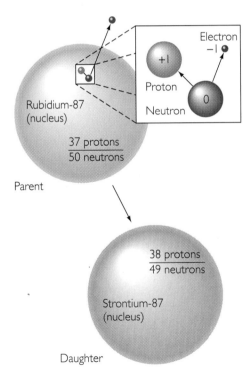

FIGURE 9.13 The radioactive decay of rubidium to strontium. A neutron in a rubidium-87 atom ejects an electron, leaving an additional proton, which changes the atom to strontium-87.

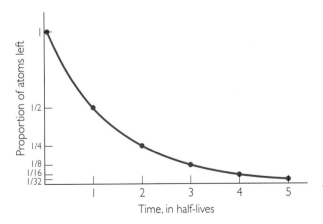

FIGURE 9.14 The number of radioactive atoms in any mineral declines at a fixed rate over time. This rate of decay is stated as a series of half-lives.

Interpreting the Grand Canyon Sequence

The rocks of the Grand Canyon have many stories to tell. They record a long history of sedimentation in a variety of environments, sometimes on land and sometimes under the sea. Unconformities mark erosion intervals, and angular unconformities also record ancient tectonic movements that built mountains. The rocks contain a succession of fossils, which reveals the evolution of new organisms and the extinction of old ones. From the rock sequences exposed along the canyon wall, geologists can reconstruct a geologic history billions of years long.

The lowermost—and therefore oldest—rocks exposed at the Grand Canyon are dark igneous and metamorphic rocks, mostly schists. These make up a complex Precambrian basement formation called the Vishnu group (after Vishnu's Temple, an erosion-sculpted mass on the Colorado River named for a Hindu god). Because no fossils are preserved in metamorphic rocks, there is no quick way to tell the Vishnu's geologic age.

Above the Vishnu are the younger Precambrian Grand Canyon beds—two groups, the Chuar and Unkar. These are made up of interlayered sandstones, shales, and limestones originally deposited as sands and muds along rivers, in lakes, and in shallow seas. An angular unconformity separates the Vishnu and Grand Canyon, signifying a period of structural deformation accompanying metamorphism of the Vishnu and erosion before the deposition of the Grand Canyon. The tilting of the Grand Canyon beds at an angle from their originally horizontal position shows that they, too, were folded after deposition and burial. None of the rocks in the Grand Canyon series contains fossils of shelled organisms, but some do include fossils of single-celled microorganisms, an indication that these rocks are about a billion years old.

Yet another angular unconformity divides the Grand Canyon beds from the overlying horizontal Tapeats sandstone. This unconformity indicates a long period of erosion after the lower rocks had been tilted. The Tapeats sandstone contains no fossils, but it can be dated by reference to the overlying Bright Angel shale. The Bright Angel shale can be dated as Cambrian by its fossils, many of which are trilobites, extinct relatives of modern crayfish. (The fossils shown in the photograph at the beginning of this chapter are trilobites.) These fossils also indicate that the sequence was laid down under a sea.

Above the Bright Angel shale is a group of horizontal limestone and shale formations (Muav limestone, Temple Butte limestone, Redwall limestone) that represent about 200 million years from the late Cambrian to the end of the Mississippian period. We know that there are unconformities between these formations because there are gaps in the faunal succession. In fact, there are so many time gaps that less than 40 percent of the Paleozoic is actually represented by rock strata.

The next formation, high up on the canyon wall, is the Supai group of formations (Pennsylvanian and Permian), which contains fossils of land plants like those found in coal beds of North America and other continents. Of even greater interest are fossil footprints of primitive land reptiles. Overlying the Supai is the Hermit, a sandy red shale.

Continuing up the canyon wall, we find another continental deposit, the Coconino sandstone, which contains more vertebrate animal tracks and is extensively cross-bedded. The animal tracks and cross-bedding suggest that the Coconino was formed by windblown sand in an arid environment during Permian times. All of these features not only help geologists to date the formation as Permian and understand its origin but also tell something of the paleoclimate, the climate of the past. Information about past climates might help us to understand the scientific, economic, and social impact of future climate changes.

At the top of the cliffs at the canyon rim are two more formations of Permian age: the Toroweap, made mostly of limestone, overlain by the Kaibab, a massive layer of sandy and cherty limestone. These two formations record subsidence below sea level and the deposition of marine sediments. Still higher, younger formations are exposed at some distance from the canyon rim. Some of these formations contain dinosaur bones and fossilized tree trunks of the famous Petrified Forest (see Figure 9.4b).

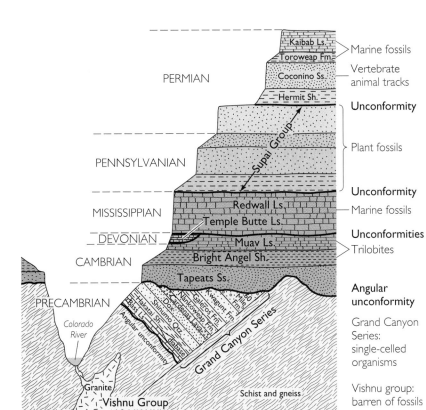

PERMIAN

Kaibab Ls. — Marine fossils
Toroweap Fm.
Coconino Ss. — Vertebrate animal tracks
Hermit Sh. — **Unconformity**

PENNSYLVANIAN

Supai Group — Plant fossils

Unconformity

MISSISSIPPIAN

Redwall Ls. — Marine fossils

DEVONIAN

Temple Butte Ls. — **Unconformities**
Muav Ls.
Bright Angel Sh. — Trilobites

CAMBRIAN

Tapeats Ss.

PRECAMBRIAN

Colorado River

Dox Ss.
Shinumo Qtz.
Hakatai Sh.
Bass Ls.
Cardenas Lavas
Nankoweap Fm.
Galeros Fm.
Kwagunt Fm.
Sixty Mile Fm.

Angular unconformity

diabase

Grand Canyon Series

Granite
Vishnu Group

Schist and gneiss

Angular unconformity

Grand Canyon Series: single-celled organisms

Vishnu group: barren of fossils

Left: Generalized stratigraphic section of the rock units in the Grand Canyon Sequence. (After S. S. Beus and M. Moral, eds., *Grand Canyon Geology,* New York, Oxford University Press/Museum of Northern Arizona Press, 1990, p. 463.) *Below:* View of the Grand Canyon from Lipan Point, with formations labeled. *(Michael Collier.)*

remain. At the end of a second half-life, half of that half, or one-quarter of the original number, are left. At the end of a third half-life, an eighth remains, and so on.

If we know the decay rate and can count the number of newly formed daughter isotope atoms as well as the remaining parent atoms, then we can calculate the time that has elapsed since the radioactive clock began to tick. In effect, we can work back to the time when there were no daughter isotopes, only those of the undecayed parent element.

Geologists count the number of parent and daughter isotopes with a mass spectrometer, a very precise and sensitive instrument that can detect even minute quantities of isotopes. Suppose we counted 950 rubidium-87 atoms and 50 strontium-87 atoms in a rock. Using the known rubidium-to-strontium decay rate, we could then calculate that 4 billion years had passed since the rubidium in our rock had begun to disintegrate. (Four billion years is the age of the oldest rocks on Earth.)

To geologists, this is the age of the rock, or more exactly, the time since rubidium was first trapped in newly formed minerals. Rubidium, like other elements, is incorporated into minerals as they crystallize from a magma or recrystallize in a metamorphic rock. During crystallization, rubidium and strontium are incorporated in different kinds of minerals. This means that rubidium is separated chemically from any strontium daughters produced before the magma solidified or the parent rock metamorphosed. This separation sets the radioactive clock back to zero. Inside the newly formed mineral, the radioactive decay of rubidium-87 continues, and new strontium atoms start to accumulate. These strontium atoms cannot escape unless recrystallization occurs again. Thus, rubidium-87 and other radioactive isotopes in igneous rocks provide a way of measuring when a magma was emplaced and cooled, and those in metamorphic rocks enable us to measure the time that has elapsed since they were metamorphosed. Rubidium decay is not used for dating sedimentary rocks because the minerals in clastic sediments are usually derived from older, preexisting rocks. Chemically and biochemically precipitated minerals in sediments, such as carbonates, usually contain too little newly precipitated rubidium for accurate isotopic analysis.

Geologists use a number of other naturally occurring radioactive elements, in addition to rubidium-87, to determine the ages of rocks. Table 9.1 lists the major radioactive elements, their effective dating ranges, and some of the minerals that can be dated. Among these elements are uranium-238, which decays to lead-206 by a complex series of transformations; potassium-40, which decays to argon-40; and carbon-14, which decays to nitrogen-14. One intermediate daughter product of the uranium decay series is radon, a radioactive gas that poses a hazard to health in some places (see Feature 9.2).

Each radioactive element has its own decay rate. Those that decay slowly over billions of years, like rubidium-87, are used to measure the ages of old rocks. Uranium-238 and potassium-40, like rubidium-87, have half-lives of billions of years and therefore are also excellent choices for dating old rocks (see Table 9.1). Radioactive elements that decay rapidly, so that most of the parent isotope disappears over only a few tens of thousands of years—carbon-14 is one—are useful for determining the ages of very young rocks. Radiometric dating is possible only if a measurable number of parent and daughter atoms remain in the rock. For example, if the rock is very old and the decay rate is fast, almost all the parent atoms may already have been transformed. In that case, we could determine that the radiometric clock has run down, but we would have no way of knowing how long ago it had stopped.

Carbon-14: Clocking "Recent" Activity

Carbon-14, which rapidly decays to nitrogen-14, has a half-life of 5730 years. In a rock that is 30,000 years old, then, more than five half-lives have passed, and only a little less than $\frac{1}{32}$ of the original amount of carbon-14 remains. By the time 70,000 years have passed, too few carbon-14 atoms are left to count accurately. Carbon-14 is thus most suitable for measuring times in the relatively recent geologic past.

Carbon-14 is an especially important tool for dating fossil, bone, shell, wood, and other organic materials in very young sediments because all of these materials contain carbon. Unlike radioactive elements such as rubidium and uranium, which are incorporated into rocks during metamorphism or during crystallization from a magma, carbon, including a small amount of carbon-14, is an essential element of the living cells of all organisms. As plants grow, they continuously incorporate into their tissues a small amount of carbon-14, along with other (stable) carbon isotopes, from carbon dioxide in the atmosphere. When a plant dies, it stops absorbing carbon dioxide, and so no new carbon of any kind is

TABLE 9.1

MAJOR RADIOACTIVE ELEMENTS USED IN RADIOMETRIC DATING

ISOTOPES		HALF-LIFE OF PARENT (YEARS)	EFFECTIVE DATING RANGE (YEARS)	MINERALS AND OTHER MATERIALS THAT CAN BE DATED
PARENT	**DAUGHTER**			
Uranium-238	Lead-206	4.5 billion	10 million–4.6 billion	Zircon Uraninite
Potassium-40	Argon-40	1.3 billion	50,000–4.6 billion	Muscovite Biotite Hornblende Whole volcanic rock
Rubidium-87	Strontium-87	47 billion	10 million–4.6 billion	Muscovite Biotite Potassium feldspar Whole metamorphic or igneous rock
Carbon-14	Nitrogen-14	5730	100–70,000	Wood, charcoal, peat Bone and tissue Shell and other calcium carbonate Groundwater, ocean water, and glacier ice containing dissolved carbon dioxide

added. At that moment, the amount of carbon-14 in relation to the stable carbon isotopes is identical to that in the atmosphere. But the amount of carbon-14 in the dead plant tissue steadily decreases as radioactive atoms decay. Because the nitrogen-14 daughter atoms of carbon-14 are gaseous and thus leak from the sediment, we cannot measure the daughters accurately. We can, however, compare the amount of carbon-14 left in the plant material with the original amount that would have been in equilibrium with the atmosphere, which is assumed to be approximately constant over the periods of time involved. This comparison will yield the time that has elapsed since the plant died. Similarly, we can use carbon-14 to date bones and other once-living organic material.

Limits and Uses of Radiometric Dating

Radiometric dating cannot be used for every rock a geologist samples. If a rock containing uranium were to lose some of its lead through weathering, for example, we would get an erroneously young age. Or

if an igneous rock were metamorphosed, the daughter isotopes that had accumulated since the crystallization of the magma might be lost, resetting the clock to the time of metamorphism rather than the time of initial formation. In addition to these factors, the accuracy and precision of radiometric dating depends on the accurate measurement of the often minute amounts of daughter atoms found in rocks. Techniques have now advanced to the point where lead from a single zircon crystal can be used to date a rock. Accuracy has improved so much in recent years that Precambrian rocks can be dated with an error of no more than 1 million years, a far cry from errors of more than 50 million years just a few decades ago.

One of the boundaries of the geologic time scale that has shifted a good deal as a result of both better radiometric dating and the acquisition of new samples from hitherto unexplored territory is the boundary between the Cambrian period and the Precambrian eon. In the past few years, this boundary has changed several times, from the former date of 570 million years to the current, more firmly placed value of 544 million years. This date is

9.2 LIVING ON EARTH

Radon: An Environmental Threat

Radon is an invisible, odorless, radioactive gas, possibly the most lethal of natural hazards. It is one of the products of the radioactive decay of uranium. Minerals containing uranium are unevenly distributed in rocks and soils and are usually present in amounts so small as to be negligible, although concentrations occur in certain rocks, such as shales formed from organic-rich muds and some granites. Under certain conditions, uranium is easily dissolved by groundwater, transported over considerable distances, and reprecipitated in other rocks and soils.

As a gas, radon can move freely from its source in rock or soil into the atmosphere, where it can be inhaled. In the open air, it is dispersed by winds and presents no danger. If it seeps into homes, however, it can build up to hazardous levels. Radon works its damage by releasing its own radioactive decay products, which can lodge in lung tissues. (Radon itself is inhaled and exhaled without significant buildup.) In a small percentage of individuals exposed to radon for many years, the result is lung cancer; smokers are particularly vulnerable to this disease. Some scientists estimate that there may be as many as 6000 to 25,000 radon-induced cancer deaths each year in the United States. This is the same order of risk as dying by an accidental fall or a fire in the home.

A survey of several thousand homes in seven states by the Environmental Protection Agency (EPA) revealed that nearly one in three homes exceeds federal health guidelines for indoor levels of radon. In most cases, the radon originates in the underlying soil or rocks and seeps into homes through cracks and openings in basement walls and floors. Fortunately, it is possible to identify homes at risk and take preventive measures. Elevated concentrations of radon often occur in "hot belts" corresponding to specific geological formations. Because radon escapes into the atmosphere so readily, it is hard to measure directly over a region. A map of the abundance of its parent, uranium, however, reveals the geographic distribution of areas in which the levels of uranium in soil and rock are high enough to produce radon exceeding safe levels.

There are some scientific critics of the radon "scare." Much of the information on the biological effects of radon comes from data on the incidence of lung cancer among uranium miners, who encounter high levels of radon in the mines. The critics do not believe that the incidence of disease at these high levels can be simply extrapolated to the low levels found in homes. Some critics say the data are not clear about the combined effects of cigarette smoking and exposure to radon. There is also continued argument about the health risks to people exposed to radon in drinking water, although estimates by the EPA and others agree that fewer than 200 people die each year from this cause.

Nevertheless, it is wise to follow the ad-

---

particularly important, for it relates to some of the most significant developments in the evolution of multicellular organisms.

The geologic time scale has many uses, aside from its value to geologists. Anthropologists reconstructing the path of human evolution and archeologists dating various human settlements use the most recent parts of the time scale, covering the last 5 million to 10 million years. Seismologists who study earthquake-prone regions use the time scale and stratigraphy of the recent past to map movements on faults that gave rise to earthquake shocks.

Other Geologic Clocks

Because time is so central to the study of the Earth, geologists continue to search for additional ways to gauge geologic time. Paleomagnetic stratigraphy, for

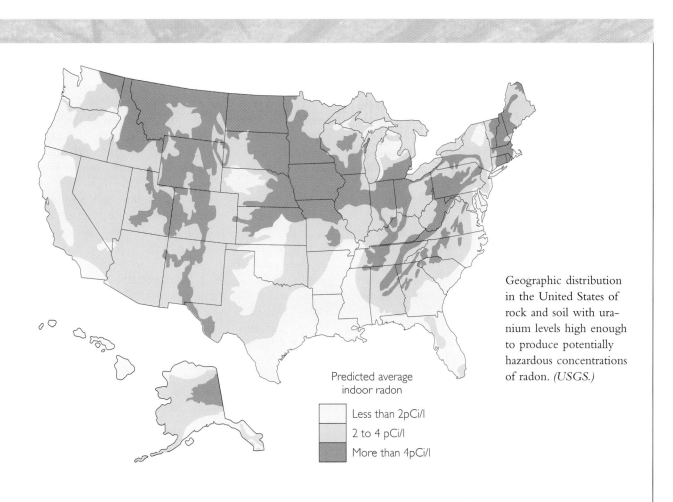

Geographic distribution in the United States of rock and soil with uranium levels high enough to produce potentially hazardous concentrations of radon. *(USGS.)*

Predicted average indoor radon

Less than 2pCi/l
2 to 4 pCi/l
More than 4pCi/l

vice of the U.S. Health Service and invest in an inexpensive device (costing $25 or less) available in hardware stores for measuring the level of radiation from radon in the home. This preliminary screening is good enough either to eliminate concern if its results are negative or to warrant more definitive tests if the results are positive. If further tests substantiate hazardous levels of radon, the home-owner can usually reduce them to safe levels by one or more of the following measures: seal the basement from the subsoil; increase the exchange of air with the outside; increase the air pressure in the house slightly so that the radon is not sucked in from below; install equipment to pump the air out of the soil or rock beneath the basement floor. These efforts typically cost a few thousand dollars.

example, is under steady development as an adjunct to radiometric dating. As you will see in Chapter 19, Earth's magnetic field reverses about every half-million years. These periodic reversals are recorded in the orientation of magnetic minerals in rocks, especially those on the seafloor. The magnetic time scale has been calibrated both by radioactive age determinations and by the stratigraphic ages of overlying and underlying fossiliferous formations.

FROM THREE LINES OF EVIDENCE: A RELIABLE DATING TOOL

Once geologists had determined radiometric ages and linked them to their earlier studies of fossils and stratigraphy, they could add absolute dates to the geologic time scale (see Figure 9.12). With these three sources of information, geologists can deduce approximate ages of rock formations, even those that contain no materials amenable to radiometric analysis. For example, if we know from radiometric dating that an igneous intrusion is 500 million years old, then sedimentary layers cross-cut by that intrusion must be older than 500 million years. And if these same layers overlie metamorphic rocks radiometrically dated at 550 million years, then we know that the sedimentary layers were formed between 500 and 550 million years ago. If these sedimentary rocks contain fossils indicating Cambrian and late Ordovician stratigraphic ages, we know the absolute ages of parts of these geological periods. With this kind of "bracketing" of ages, geologists worked out the entire geologic time scale. After almost a century of radioactive dating and continued work on the stratigraphy of the world, this time scale remains undisputed in all major respects.

Estimating the Rates of Very Slow Earth Processes

Now that we have a way of dating rocks, let us see what the time scale can tell us about the rates of some slow geological processes. Consider the opening of an ocean, in which seafloor plates spread away from each other at mid-ocean ridges. The southern Atlantic Ocean from South America to Africa is a little over 5000 km wide; that is how far the two continents have spread from each other. The seafloor at the edges of these continents is the oldest, having formed at the inception of spreading. From fossils, we know that these sediments are about 100 million years old (middle Cretaceous age). Thus, the average rate of spreading of this part of the ocean is 5000 km every 100 million years, or about 5 cm per year. In other parts of the oceans, spreading rates may be as low as 1 or 2 cm per year or as high as 10 or 12 cm per year.

This kind of rough calculation gives a surprisingly good estimate of the rate of seafloor spreading. It was put to the test in 1987, when scientists were first able to measure the spreading rate of the Atlantic seafloor directly, using the technology of long-ranging laser beams and satellites. Their results agreed with the spreading rates that marine geologists had calculated from the age and position of the seafloor.

The dating methods we have described in this chapter also help us understand another slow process, and one of direct concern to the people of California—the movement of crustal blocks along the San Andreas and adjacent faults, which have been responsible for many major earthquakes. Figure 9.15 shows how the Pacific Plate slides past the North American Plate along this transform boundary. One way to determine the rate of this movement is by measuring the offsets of distinctive geologic formations of various ages split by the plate boundary faults. Dividing the distance that now separates these formations by the time since they were formed and split gives the rate of movement. Seafloor spreading rates and satellite measurements provide other estimates of plate motions along the boundary. Using these observations we can estimate that the average movement over the past several million years was about 5 to 6 cm per year. If that rate continues, Los Angeles, on the Pacific Plate, will be contiguous to San Francisco, on the North American Plate, in about 10 million years.

We can also measure the rates of vertical movement by dating marine deposits that are now above sea level. Parts of the Alpine range, for instance, contain marine fossils known to be about 15 million years old. These fossils, now elevated 3000 m above sea level, were originally deposited on the shallow seafloor close to sea level. The sedimentary rocks containing the fossils must have been uplifted an average of about 2 mm per decade, although the rates may have been higher or lower for shorter time intervals or in different parts of the mountain range.

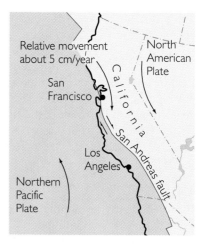

FIGURE 9.15 We can calculate the relative movement of about 5 cm per year between the North American and Pacific plates by the offset of the geological formations split by the fault, by satellite measurements, and by seafloor spreading rates.

The final case we consider here is erosional processes, which are continuously wearing down the land surface. These processes operate so slowly that two photographs of a river valley taken a hundred years apart show little difference (Figure 9.16). We can estimate erosion rates by adding up all the disintegrated and dissolved products of erosion being carried away from a land area by rivers and wind. The rate at which the North American continent erodes has been estimated to be about 3 mm per century.

FIGURE 9.16 Two photos of Bowknot Bend on the Green River in Utah taken nearly 100 years apart show that little has changed in the configuration of rocks and formations in that time interval.

At this rate, it would take 100 million years to reduce a 3000-m-high mountain to sea level.

Thus, in these particular cases, it took about 100 million years to open an ocean, 15 million years to raise a mountain range, and 100 million years to erode it. But as we will see, these time intervals are relatively short compared with the entire history of the planet. During that history, Earth has experienced many cycles of mountain building and erosion.

In addition to providing a way of measuring the rates of geological processes and reconstructing Earth's history, the time scale and the ideas on which it is based offer a tool for validating plate tectonics. Without a way to measure the ages of rocks on the seafloor and the continents, geologists would not have been able to deduce the spreading of the seafloor and other plate motions.

An Overview of Geologic Time

We can now combine the geologic time scale with absolute dates from the analysis of radioactive decay and the evolution of organisms to construct a time line for the whole history of the Earth, beginning at its birth 4.6 billion years ago (see the back endpapers). Figure 9.17 shows the whole sweep of geologic time as a spiral pathway, each revolution of the spiral corresponding to one billion years. From this illustration, we can see how short a proportion of the total of Earth's history is taken up by the eras of the Phanerozoic eon and what a tiny amount of time has elapsed since humans evolved.

As another way of comprehending this extraordinarily long period of time, think of Earth's age as being only one calendar year. On January 1, the Earth was formed. During January and part of early February, the Earth became organized into core, mantle, and crust. About February 21, life evolved. During all of spring, summer, and early fall, the Earth evolved to continents and ocean basins something like those of today, and plate tectonics became active. On October 25, at the beginning of the Cambrian period, complex organisms, including those with shells, arrived. On December 7 reptiles evolved, and on Christmas Day the dinosaurs became extinct. Modern humans, *Homo sapiens,* appeared on the scene at 11 P.M. on New Year's Eve, and the last glacial age ended at 11:58:45 P.M. Three-hundredths of a second before midnight, Columbus landed on a West Indian island. And a few thousandths of a second ago, you were born.

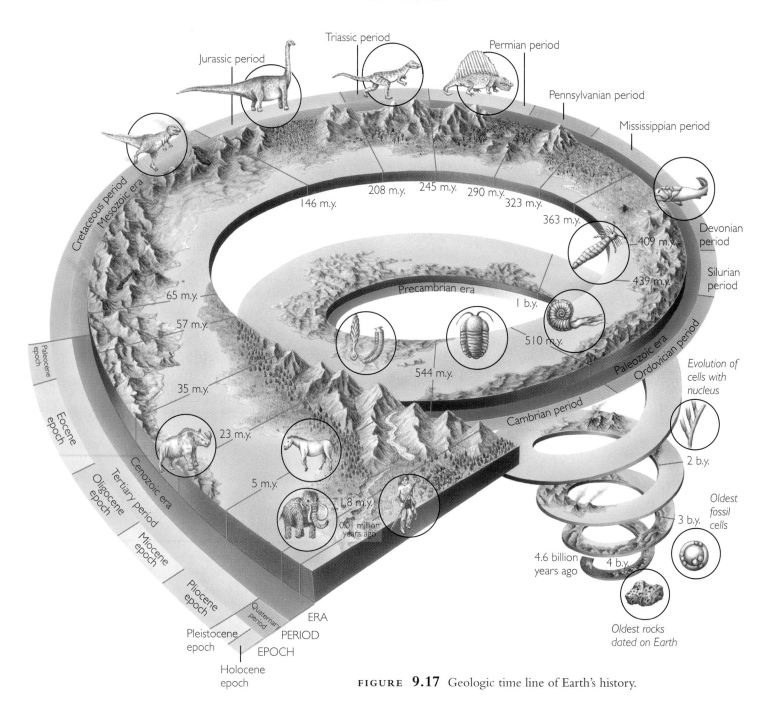

FIGURE 9.17 Geologic time line of Earth's history.

SUMMARY

How do geologists know how old a rock is and whether one rock is older than another? Geologists determine the order in which rocks were formed by studying their stratigraphy, fossils, and arrangement in the field. An undeformed sequence of sedimentary rock layers will be horizontal, with each layer younger than the layers beneath it and older than the ones above it. In addition, since animals and plants have evolved progressively over time, their fossil remains change in known ways in the stratigraphic sequence. Knowing the faunal succession makes it easier for geologists to spot missing sedimentary layers, known as unconformities. Most important, fossils enable geologists to correlate rocks located all over the world.

How did geologists create a geologic time scale that is applicable around the world? Using fossils to correlate rocks of the same geological age and piecing together the sequences exposed in hundreds of thousands of outcrops around the world, geologists compiled a stratigraphic sequence applicable everywhere in the world. The composite sequence represents the geologic time scale. The use of radiometric dating allowed scientists to assign absolute dates to the units of the time scale. Radiometric dating is based on the behavior of radioactive elements, in which unstable parent atoms are transformed into daughter isotopes at a constant rate. When radioactive elements are locked into minerals as rocks are formed, their daughters accumulate and the number of parents decreases; by measuring parents and daughters, we can calculate absolute ages.

Why is the geologic time scale important to geology? The geologic time scale enables geologists to reconstruct the chronology of events that have shaped the planet. The time scale has been instrumental in validating and studying plate tectonics and in estimating the rates of geological processes too slow to be monitored directly, such as the opening of an ocean over millions to hundreds of millions of years. In addition, the development of the time scale revealed that the planet is much older than early geologists and others had imagined and that since its beginning, it has undergone almost constant change as a result of gradual processes working throughout Earth's history. The creation of the geologic time scale paralleled the development of paleontology and the theory of evolution, one of the most revolutionary and powerful ideas in science.

KEY TERMS AND CONCEPTS

relative age (p. 218)

radiometric age (p. 218)

stratification (p. 219)

principle of original horizontality (p. 219)

principle of superposition (p. 219)

stratigraphic sequence (p. 220)

geologic time (p. 220)

paleontology (p. 222)

formation (p. 222)

unconformity (p. 223)

disconformity (p. 223)

nonconformity (p. 223)

angular unconformity (p. 223)

sequence stratigraphy (p. 224)

geologic time scale (p. 227)

eon (p. 227)

era (p. 227)

period (p. 227)

epoch (p. 228)

radiometric dating (p. 228)

radioactivity (p. 229)

half-life (p. 229)

EXERCISES

1. Name the geological periods, from youngest to oldest.

2. Give the absolute times of the beginnings of the Paleozoic, the Mesozoic, and the Cenozoic eras.

3. To what element does rubidium-87 decay radioactively?

4. Give the age of a sediment that might be dated by carbon-14.

5. What geologic events are implied by an angular unconformity?

6. What is the principle of superposition?

7. What is the principle of original horizontality?

8. How do angular unconformities differ from disconformities and nonconformities?

9. What property of fossils do geologists use to date the formations in which they are found?

10. How does the determination of the ages of igneous rocks help date fossils?

THOUGHT QUESTIONS

1. As you pass by an excavation in the street, you see a cross section showing paving at the top, soil below that, and bedrock at the base. You also notice that a vertical water pipe extends from a drain in the street into a sewer in the soil. What can you say about the relative ages of the various layers and the water pipe?

2. How would you be able to ascertain the relative ages of several volcanic ash falls exposed in an outcrop?

3. What evidence could you give to a friend to support the idea that a particular formation was many millions of years old?

4. Construct a diagram similar to the one in Figure 9.11 to show the following series of geological events: (a) sedimentation of a limestone formation; (b) uplift and folding of the limestone; (c) erosion of the folded terrain; (d) subsidence of the terrain and sedimentation of a sandstone formation.

5. Many fine-grained muds are deposited at a rate of about 1 cm per 1000 years. At this rate, how long would it take to accumulate a stratigraphic sequence half a kilometer thick?

6. What radioactive elements might you use to date a schist that is approximately one billion years old?

7. What geological event is dated by radioactive decay of a mineral in a schist?

8. What geological event is dated by radioactive decay of a mineral in a basalt?

9. Name a geological event that can be dated by both geological methods and radiometric dating.

10. Give an example of an angular unconformity, a disconformity, and a nonconformity from the Grand Canyon section shown in Feature 9.1.

11. Do you think it would be possible to radiometrically date a basalt on the Moon, which has a composition very like the basalts on Earth? What radioactive decay scheme might you use?

12. You would like to know the date when a dormant volcano in South America was last active. What methods could you use to determine this date?

SHORT-TERM TEAM PROJECTS: GEOLOGIC TIME

Comprehension of geologic time does not come easily. It is essential, however, for understanding rates of geologic change. John McPhee, a well-known geology author, popularized the calendar year as a way to express geologic time in human terms. For this project, you and a classmate will team up to create your own metaphor for geologic time.

The calendar-year metaphor involves a direct time-for-time conversion. You may compare geologic time to another time scale, such as a lifespan, or to other measures, such as distance, volume, or weight. A personal metaphor, such as comparing geologic time to the distance between your home states, is most enjoyable and easily remembered.

Your metaphor should include the following events: origin of Earth; oldest rocks; first life on the planet; transition to an oxygen-rich atmosphere; origin of multicellular life; first land plants; first land animals; appearance and extinction of dinosaurs; first hominids; first anatomically modern humans; three events of your choice from human history; your birthdays.

You will need to know the dates of the events, and you will need to calculate the percentage of geologic time between them. For example, if the Earth formed 4.5 billion years ago and the oldest rocks on Earth are about 3.8 billion years old, you can calculate that 0.7 billion years, or about 16 percent of geologic time, elapsed between the planet's formation and the solidification of its first known rocks: 4.5 - 3.8 = 0.7; 0.7/4.5 = 0.16. Calculate the percentage differences between the other geologic events and scale your metaphor accordingly.

Present your team project as a visual image. A distance-for-time metaphor, for example, might be represented with a map and markers along the way. Write a two-page description, including why you chose the metaphor and how to interpret the comparative time scales.

CARBONATE SEDIMENTS

See Chapter 7.

SUGGESTED READINGS

Berry, William B. N. 1987. *Growth of a Prehistoric Time Scale.* Palo Alto, Calif.: Blackwell Scientific Publications.

Brookins, Douglas G. 1990. *The Indoor Radon Problem.* New York: Columbia University Press.

Faure, Gunter. 1986. *Principles of Isotope Geology,* 2nd ed. New York: Wiley.

Isachsen, C. E., Samuel A. Bowring, Ed Landing, and Scott D. Samson. 1994. New constraint on the division of Cambrian time. *Geology* 22:496–498.

Palmer, Allison R. 1984. *Decade of North American Geologic Time Scale.* Geological Society of America, Map and Chart Series MC-50.

Simpson, George G. 1983. *Fossils.* New York: Scientific American Books.

Stanley, Stephen M. 1993. *Exploring Earth and Life Through Time.* New York: W. H. Freeman.

INTERNET SOURCES

Paleontology Without Walls

ⓘ http://www.ucmp.berkeley.edu/exhibit/ exhibits.html

Established by the University of California Museum of Paleontology, this site includes three major features: Phylogeny, Geologic Time, and Evolutionary Thought. "The family tree of life" can be searched for any taxon. Geologic Time includes some history of the time scale and a clickable time scale with links to stratigraphy, paleontology, and fossil locations.

The Burgess Shale Project

ⓘ http://scienceweb.dao.nrc.ca/burgess.html

The Burgess Shale contains a unique fossil record of early life on Earth. Information at this location explains the origin and significance of the Burgess fauna and contains related links for exploring geology (including the rock cycle and plate tectonics), paleontology, and evolution.

Geologic Time Scale

ⓘ http://www.lib.pdx.edu/~kohut/geotime/ geotime.html

Portland State University maintains this site, which includes a color-coded geologic time scale. Clicking on one of the geologic time periods expands the time scale to include the epochs and stages for that period.

Grand Canyon National Park

ⓘ http://www.kaibab.org/

This "unofficial" Grand Canyon National Park home page provides information about the geology of the canyon, a guided tour, maps, images, visitor information, and news articles related to the canyon and its environment.

Radon

ⓘ http://sedwww.cr.usgs.gov:8080/radon/ radonhome.html

Through this site, the U.S. Geological Survey provides information on radon, radon research, and geologic radon hazard assessments performed by USGS.

10

Intricately folded
sediments on the Tibetan
Plateau. Folding, faulting,
and uplift in the Himalayas
reflect the collision of
India with Asia.
(Ned Gillette/
The Stock Market.)

Folds, Faults, and Other Records of Rock Deformation

The eighteenth- and nineteenth-century scientists who laid the foundations of modern geology concluded that most sedimentary rocks were originally deposited as soft horizontal layers at the bottom of the sea and hardened over time. But they were puzzled that many hardened rocks were tilted, bent, or fractured (like those shown on the facing page). They wondered: What forces could have deformed these hard rocks in this way? Can we reconstruct the history of the rocks from the patterns of deformation found in the field? Today's geologists would add: How do rocks of all types become deformed and how does the deformation relate to plate tectonics? We will answer these questions in this and later chapters.

In this chapter we look at how geologists collect and interpret field observations to reconstruct the geologic history of a region.

They might ask: What kinds of rocks were originally deposited? What happened to them then? If they were deformed by tectonic forces, what was the nature of the deformation and what kinds of forces were involved? In their search for answers geologists found similar patterns of deformation around the globe. A few concepts can explain in general why and how rocks were deformed. But we must learn the features of deformed rocks before we can see order and then explain the patterns of deformation.

Folding and **faulting** are the most common forms of deformation in the sedimentary, metamorphic, and igneous rocks that make up Earth's crust.

Folds in rocks are like folds in clothing. Just as cloth pushed together from opposite sides bunches up in folds, layers of rock slowly compressed by forces in the crust are pushed into folds (Figure 10.1). Tectonic forces can also cause a rock formation to break and slip on both sides of a fracture, parallel to the fracture (Figure 10.2). This is a fault. Geologic folds and faults can range in size from centimeters to tens of kilometers. Many mountain ranges are actually a series of large folds or faults, or both, that have been weathered and eroded. Geologists now believe that the forces that move the large plates are ultimately responsible for the deformations found in most local areas.

FIGURE 10.1 An outcrop of originally horizontal rock layers bent into folds by compressional tectonic forces. *(Phil Dombrowski.)*

FIGURE 10.2 Small-scale faults. Once-continuous rock layers have been permanently displaced along fractures. Artificial cut near Mt. Carmel Junction, southern Utah. *(Tom Bean.)*

INTERPRETING FIELD DATA

Scientists are trained to be keen observers so that they can gather accurate information on which to base their theories. To figure out how rock formations are deformed, geologists need accurate information about the geometry of the beds they can see. A basic source of information is the outcrop, where the bedrock that underlies the surface everywhere is exposed, not obscured by soil or loose boulders. A sedimentary bed that has been bent into a fold is shown in Figure 10.1. Often, however, folded rocks are only partly exposed in an outcrop and show only as an inclined layer (Figure 10.3). The orientation of the layer is an important clue the geologist can use to piece together a picture of the overall deformed structure. It takes only two measurements to describe the orientation of a layer of rock exposed at a given location: the strike and the dip.

Measuring Strike and Dip

Figure 10.4 shows how the strike and the dip are observed and measured in the field. The **strike** is the direction of the intersection of a rock layer with a horizontal surface. The **dip,** which is measured at

FIGURE 10.3 Dipping limestone and shale beds on the coast of Somerset, United Kingdom. Children are walking along the strike of beds that dip to the left. Dip is the angle of steepest descent of the bed from the horizontal; strike is at right angles from the dip direction. *(Chris Pellant.)*

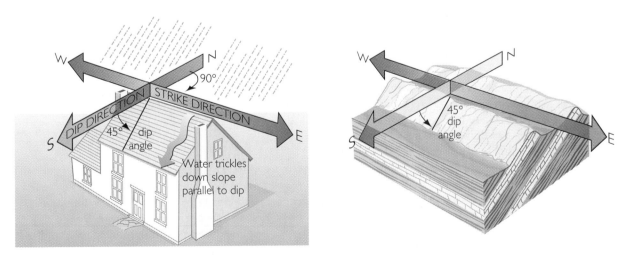

FIGURE 10.4 Geologists use the strike and dip of a formation to define its orientation at a particular place. (After A. Maltman, *Geological Maps: An Introduction* [New York: Van Nostrand Reinhold, 1990], p. 37.)

right angles to the strike, is simply the amount of tilting—that is, the angle at which the bed inclines from the horizontal. A geologist might describe a particular outcrop, for example, as "beds of coarse-grained sandstone striking north and dipping 30 degrees west."

Constructing a Geological Map and Cross Section

A convenient means of organizing information is a geological map, on which geologists record the locations of outcrops, the nature of their rocks, and the dips and strikes of inclined layers. (Appendix 4 gives more information on geological maps.) Also helpful in piecing together a geological story is a geological cross section, a diagram showing the features that would be visible if a vertical slice were made through part of the crust. A natural cross section can often be observed in the vertical face of a cliff, a quarry, or a road cut. A cross section can also be constructed from the information on a geological map. Figure 10.5 shows a simple geological map of an area where sedimentary rocks, originally horizontal, were bent into a fold and the cross section derived from it.

You probably noticed that the map and cross section in Figure 10.5 include portions of the fold that eroded away long ago. How does the geologist reconstruct the deformed shapes of the rock layers, even when erosion has removed parts of a formation? It's like putting together a three-dimensional jigsaw puzzle with missing pieces. Common sense and intuition play important roles in the process. The geologist may first notice such features as sedimentary layers and surmise that they were originally deposited as horizontal beds at the bottom of the sea. If the layers are now found to be inclined, as they are in Figure 10.5, the implication is that later events forced the rocks to be tilted or bent. The law of superposition (younger beds are laid down over older beds, as we saw in Chapter 9) tells the geologist which bed is the oldest (the bed numbered 1) and that the flanking, overlying beds are successively younger. Using the dip of the layers, the geologist constructs a cross section—a vertical cut as it would exist along the traverse marked A–B on the map. The geologist finds that the beds on both sides of formation 1 are identical. By linking them up with dashed lines that match the observed dips, the geologist reconstructs the boundaries of the portions of the beds removed by erosion. To complete the cross

section, the geologist projects the trend of the beds below the ground, even though the rocks cannot be seen.

Figure 10.5 tells the story of an ancient ocean, now gone, in which a succession of sedimentary rocks was deposited on the seafloor. These beds, once horizontal, were subjected to crustal forces of compression, bent into folds, and raised above sea level. Erosion could then remove a major portion of the section, leaving the present-day remnant portrayed by the map and cross section. The outcropping beds in Figure 10.3 probably have a similar history.

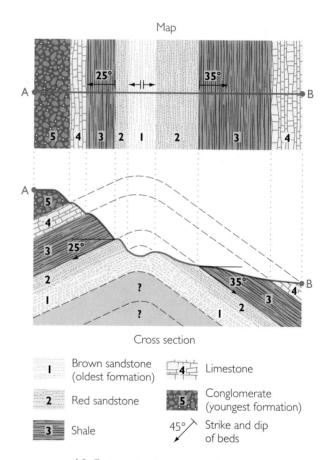

FIGURE 10.5 A geological map and a cross section derived from it. The arrangement of the layers indicates that formation 1 is at the bottom and therefore is the oldest; flanking formations are successively younger. The dashed lines reconstruct the eroded portion of the fold by connecting identical formations (shown by the same colors and numbers) and matching the observed dip.

HOW ROCKS BECOME DEFORMED

For years geologists were baffled by the problem of how rocks, which seem strong and rigid, could be distorted into folds by tectonic forces or broken along faults. The tectonic forces can be of three types: **compressive forces,** which squeeze and shorten a body; **tensional forces,** which stretch a body and tend to pull it apart; and **shearing forces,** which push two sides of a body in opposite directions. (To visualize shearing, think of what happens to a deck of cards held between your palms when your hands move parallel to each other but in opposite directions. The cards slide past one another, deforming the deck.) Figure 10.6 shows how rock bodies generally become deformed by these three types of forces. These same kinds of forces are active at the three types of boundaries between plates: compressive forces dominate at convergent boundaries, where plates collide; tensional forces dominate at divergent boundaries, where plates are pulled apart; and shearing forces dominate at transform faults at plate boundaries, where plates slide horizontally past each other.

What Determines Whether a Rock Bends or Breaks?

Although geology depends heavily on field observations, geologists also conduct laboratory experiments to discover why rock formations fold in one place and fracture in another. Experiments have been performed in which different types of rocks are squeezed at the low pressures and temperatures characteristic of conditions near the surface, as well as at temperatures and pressures high enough to simulate conditions at a depth of about 30 km in the Earth.

In one such experiment, the experimenters applied compressive force by pushing down with a piston on one end of a small cylinder of marble. At the same time, they applied pressure to all sides of the sample to simulate the confining pressure to which materials deep in the crust are subjected by the weight of the overlying rock. (Confining pressure is like the pressure you feel all over your body when you dive deep underwater.) Under low confining pressures like those at shallow depths in the crust, the marble sample deformed by fracturing. Under high confining pressures like those at greater crustal depths, the marble sample deformed

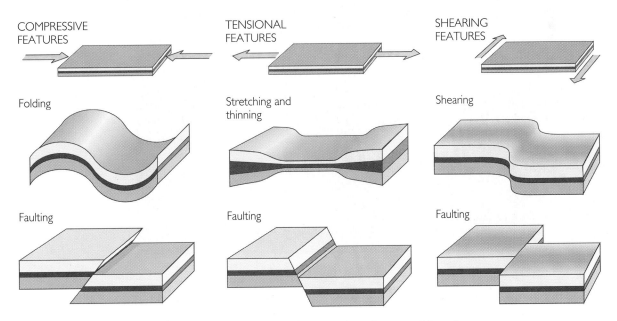

FIGURE 10.6 Rocks are deformed by folding or by faulting when they are subjected to different kinds of tectonic forces. Geologists see the pattern of deformation in the field and infer the nature of the forces that caused it.

slowly and steadily without fracturing. It behaved as a pliable or moldable material; that is, it deformed plastically into its shortened, bulging shape. Figure 10.7 shows the results of this experiment.

The investigators concluded that if a bed of this particular marble were subjected to tectonic forces near the surface, where confining pressure is relatively low, it would tend to deform by fracturing and faulting. If it were deeper than a few kilometers, however, the increased pressure would cause it to behave as a plastic material and change shape by gradually folding. Other experiments have shown that if rocks are hot when forces are applied, this kind of smooth, plastic deformation occurs more readily.

Brittleness and Ductility Under Natural Conditions

Rocks, like most solids, can be classified as **brittle** or **ductile** on the basis of the way they are deformed by forces. As forces are increased, a brittle material experiences little change until it breaks suddenly; ductile substances experience smooth and continuous plastic deformation. Glass that is near room temperature is a familiar brittle material, and modeling clay is a ductile one. Marble is brittle at shallow depths but ductile deeper in the crust.

Natural conditions are more complex, of course, than those under which laboratory experiments are performed. Tectonic forces are applied over millions of years, whereas a laboratory experiment may be performed in a few hours. Nevertheless, experiments shed some light on the way rocks respond to forces, and they give us more confidence in our interpretations of field evidence. When we see folds and fractures in the field, we can remember that some rocks are brittle, others are ductile, and the same rock can be brittle at shallow depths and ductile deep in the crust. The details vary from one type of rock to another. For example, laboratory experiments also teach us that we should expect most igneous rocks to be stronger (less deformable) than most sedimentary rocks, and **basement rocks** (old, underlying igneous or metamorphic rocks) to be more brittle than the ductile young sediments that may cover them.

(a)   (b)   (c)

FIGURE 10.7 Results of laboratory experiments conducted to discover how rocks, in this case marble, are deformed by compressive forces. (a) An undeformed sample. (b) A sample compressed under conditions representative of the shallow crust. (c) A sample compressed under conditions of the deeper crust. Samples (b) and (c) are shortened to 20 percent of their original length. Note that sample (b) has fractured. Thus we learn that marble is brittle at the laboratory equivalent of shallow depth. Sample (c), however, deformed smoothly. Marble is pliable or ductile at greater depth. *(M. S. Patterson, Australian National University.)*

HOW ROCKS FOLD

Folds and fractures are the signatures of deformation that geologists map in the field. They provide clues to the larger panorama of forces that result from plate tectonics. The term *fold* implies that a structure that originally was planar, such as a sedimentary bed, has been bent. The deformation may be produced by either horizontal or vertical forces in the crust, just as pushing in on opposite sides of a piece of paper or up from below may fold it. Folding is a common form of deformation observed in layered rocks, most typically in mountain belts. As we mentioned earlier, in many young mountain systems where erosion has not yet erased them, majestic, sweeping folds can be traced, some of them with dimensions of many kilometers (as in the photograph at the beginning of this chapter). On a much smaller scale, very thin beds can be crumpled into folds a few centimeters long (Figure 10.8). The bending can be gentle or severe, depending on the magnitude of the applied forces, the length of time they were applied, and the ability of the beds to resist deformation.

Types of Folds

Layered rocks can fold in several basic ways in response to compressional forces, depending on the properties of the rocks and the details of the forces. Geologists have developed a vocabulary to specify different types of folds and their parts. Upfolds, or arches, of layered rocks are called **anticlines;** downfolds, or troughs, are called **synclines** (Figure 10.9). The two sides of a fold are its **limbs.** The **axial plane** is an imaginary surface that divides a fold as symmetrically as possible, with one limb on either

FIGURE 10.8 Small-scale folds in a Precambrian iron formation, Beresford Lake area, Manitoba. Very thin layers have been crumpled into folds a few centimeters long. *(Geological Survey of Canada.)*

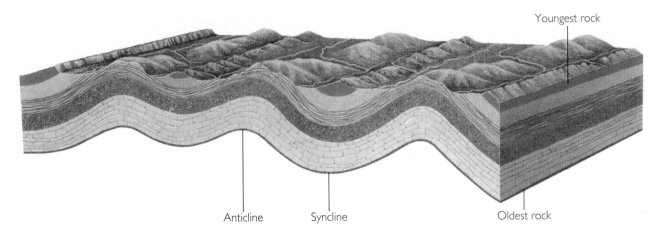

Youngest rock

Anticline   Syncline   Oldest rock

FIGURE 10.9 Anticlines fold upward; synclines fold downward.

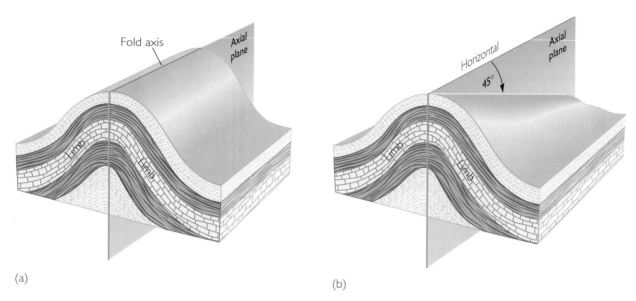

(a)

(b)

FIGURE 10.10 (a) Parts of a fold. (b) Parts of a plunging fold.

side of the plane. The line made by the length-wise interaction of the axial plane with the beds is the **fold axis.** A fold with a horizontal axis is shown in Figure 10.10a. If the axis is not horizontal, as in Figure 10.10b, the fold is called a **plunging fold.**

Not every fold has a vertical axial plane with limbs dipping symmetrically away from the axis, as in Figures 10.11 and 10.12a. With increasing horizontal force the folds can be thrown into **asymmetrical** shapes, with one limb dipping more steeply than the other (Figure 10.12b). This is a common situation. When the deformation is intense and one limb has been tilted beyond the vertical, the fold is said to be **overturned.** Both limbs of an overturned fold dip in the same direction, as shown in Figure 10.12c, but the order of the layers in the bottom limb is precisely the reverse of their original sequence; that is, older rocks are on top of younger rocks. The folds in Figure 10.13 have been over-turned to such an extent that the axial plane is ap-proaching the horizontal. One limb has been rotated into a completely upside-down sequence, with older beds on top of younger beds. (See also Figure 10.1.)

Follow the axis of any fold in the field and sooner or later the fold dies out, just as wrinkles do in a tablecloth. The fold appears to plunge into the ground as it disappears. Figure 10.14 diagrams the

geometry of **plunging anticlines** and **plunging synclines.** In eroded mountain belts, a zigzag pattern of outcrops may appear in the field after erosion has removed much of the surface rock. The eroded

FIGURE 10.11 A sharply folded symmetrical anticline in sandstone, Anza-Borrego Desert, California. *(Bill Evarts.)*

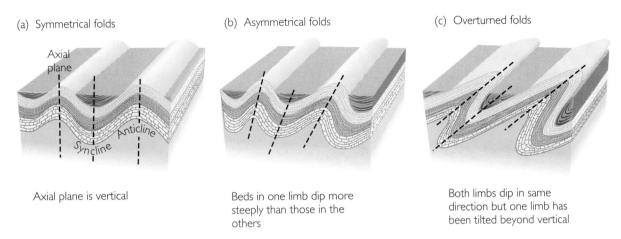

(a) Symmetrical folds

Axial plane

Syncline   Anticline

Axial plane is vertical

(b) Asymmetrical folds

Beds in one limb dip more
steeply than those in the
others

(c) Overturned folds

Both limbs dip in same
direction but one limb has
been tilted beyond vertical

FIGURE 10.12 (a) Symmetrical, (b) asymmetrical, and (c) overturned folds. Increasing horizontal force can deform symmetrical folds into asymmetrical and overturned folds.

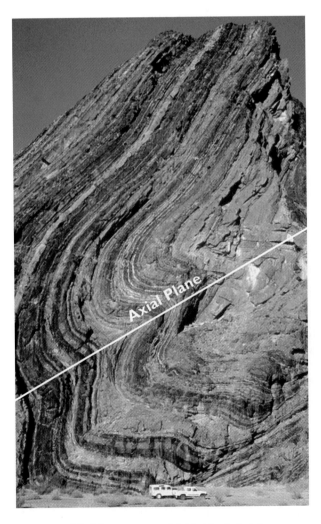

FIGURE 10.13 Overturned fold. *(Geological Survey of Israel.)*

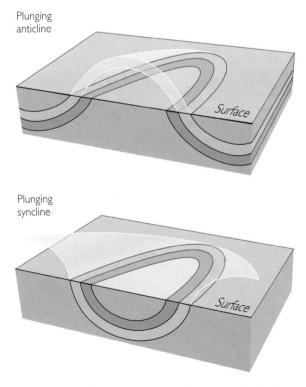

Plunging
anticline

Surface

Plunging
syncline

Surface

FIGURE 10.14 The geometry of plunging folds. Note the converging pattern of the layers where they intersect the surface. (After A. Maltman, *Geological Maps: An Introduction* [New York: Van Nostrand Reinhold, 1990], p. 84.)

Valley and Ridge belt of the Appalachians shows this characteristic pattern (Figure 10.15).

A **dome** is an anticlinal structure, a broad circular or oval upward bulge of rock layers. The flanking beds of a dome encircle a central point and dip radially away from it (Figure 10.16a). A **basin** is a synclinal structure, a bowl-shaped depression of rock layers in which the beds dip radially toward a central point (Figure 10.16b). Domes and basins are typically many kilometers in diameter. They are recognized in the field by outcrops with the characteristic circular or oval shapes seen in Figures 10.16b and 10.17.

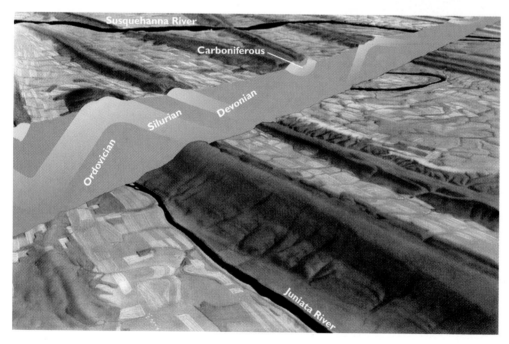

FIGURE 10.15 The erosional remnants of plunging folds show a characteristic zigzag pattern in this view of the Valley and Ridge belt of the Appalachian Mountains, 30 miles northwest of Harrisburg, Pennsylvania. In the drawing below, the imaginary trench reveals the subsurface structure. (From J. S. Shelton, *Geology Illustrated* [San Francisco: W. H. Freeman, 1966].)

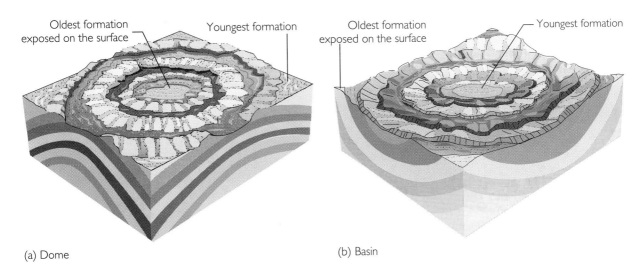

(a) Dome

(b) Basin

FIGURE 10.16 (a) The characteristic circular or elliptical outcrop pattern of a dome. The oldest bed is in the core; the flanking formations are successively younger and dip away from the core. (b) Outcrop pattern of an eroded basin. Youngest formations are in the core, flanked by successively older strata that dip toward the core.

FIGURE 10.17 Sinclair dome, an eroded dome in strata 6 miles east of Rawlings, Wyoming. The highway and railroad at the lower right suggest its dimensions. (*John S. Shelton.*)

Domes are very important in oil geology because oil is buoyant and tends to migrate upward through permeable rocks. If the rocks at the high point of a dome are not easily penetrated, the oil becomes trapped against them.

It is not entirely clear why domes and basins form. Some domes can be attributed to igneous rock that intrudes the crust, pushing the overlying sediments upward. Some basins are formed when a heated portion of the crust cools and contracts, causing the overlying sediments to subside. Others result when the crust is stretched by tectonic forces. The weight of sediments deposited in a shallow sea can depress the crust, forming a basin. There are many domes and basins in the central portion of the United States. The Black Hills of South Dakota are an eroded dome; much of the lower peninsula of Michigan is a sedimentary basin.

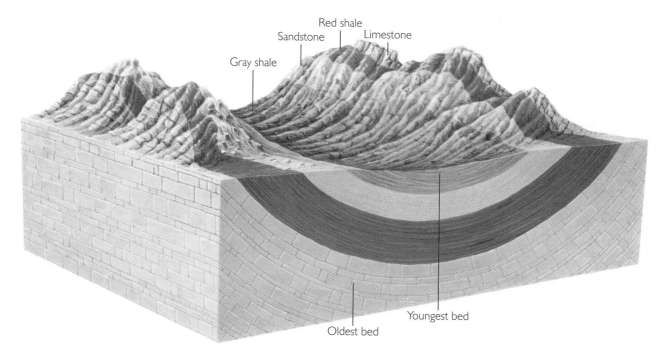

Gray shale
Sandstone
Red shale
Limestone
Youngest bed
Oldest bed

FIGURE 10.18 Geologists typically work from available surface outcrops of rock formations to reconstruct subsurface structures. This diagram shows the surface expression of eroded remnants of a syncline and the characteristic core of younger rocks flanked on both sides by older rocks dipping toward the core.

What Geologists Infer from Folds

Observations in the field seldom provide geologists with complete information. Either bedrock is obscured by overlying soils or erosion has removed much of the evidence of former structures. So geologists search for clues they can use to work out the relation of one bed to another. For example, in the field or on a map, an eroded anticline would be recognized by a strip of older rocks forming a core bordered on both sides by younger rocks dipping away, as we saw in Figure 10.5. An eroded syncline would show as a core of younger rocks bordered on both sides by older rocks dipping toward the core (Figure 10.18).

Folds typically occur in elongated groups. A strip of country in which the rock layers are folded—that is, a **fold belt**—suggests to a geologist that the region was compressed at one time by horizontal tectonic forces. The Valley and Ridge province of the Appalachians is a folded mountain belt (Figure 10.19). A satellite photograph taken from 200 miles above the Valley and Ridge province is shown in Figure 16.11. We will see in Chapter 21 that an ancient plate collision accounts for the wrinkling of the once-flat layers of sedimentary rocks in this region.

Appalachian plateau   **Valley and Ridge belt**   Great Valley | Reading Prong | Triassic lowland | Piedmont | Coastal Plain

FIGURE 10.19 The Valley and Ridge province of the Appalachian Mountains is the eroded remnant of a folded mountain belt. (After D. Johnson, *Stream Sculpture on the Atlantic Slope* [New York: Columbia University Press, 1931].)

HOW ROCKS FRACTURE: JOINTS AND FAULTS

We have seen that the way rocks deform depends on the kinds of forces to which they are subjected and the conditions that prevail. Some layers crumple into folds, and some fracture. There are two kinds of fractures, joints and faults. A **joint** is a crack along which no appreciable movement has occurred. A **fault** is a fracture with relative movement of the rocks on both sides of it, parallel to the fracture. Just like folds, joints and faults tell geologists something about the forces a region has experienced in the past.

Joints

Joints, which can be caused by tectonic forces, are found in almost every outcrop. Like any other brittle material, brittle rocks break more easily at flaws or weak spots when they are subjected to pressure. These flaws can be tiny cracks, fragments of other materials, or even fossils. Regional forces—compressional, tensional, or shearing—that have long since vanished may leave their imprint in the form of a set of joints.

Joints can also form as a result of nontectonic expansion and contraction of rocks when erosion has stripped away surface layers. The removal of those layers releases the confining pressure on underlying formations, allowing the rocks to expand and part at flaws. Joints can form in lava as a result of the contraction of the lava as it cools.

When a formation fractures at many places and develops joints, the joints are usually only the beginning of a series of changes that will significantly alter the formation. For example, joints provide channels through which water and air can reach deep into the formation and speed the weathering and weakening of the structure internally. If two or more sets of joints intersect, weathering may cause a formation to break into large columns or blocks (Figure 10.20).

Faults

Whereas folds usually signify that compression forces were at work, faults can be caused by all three types of forces: compressive, tensional, and shearing. These forces are particularly intense at plate boundaries. Faults are common features of mountain belts, which are associated with plate collisions, and of rift valleys, where plates are being pulled apart. Some

FIGURE **10.20** Eroded joints, Arches National Park, Utah. *(Willard Clay.)*

transform faults, such as the San Andreas fault of California (Figure 10.21), show such large horizontal displacements that the *offset,* or the relative movement of the two plates, may amount to hundreds of kilometers. Crustal forces can also be strong within plates and cause faulting in rocks far from plate boundaries.

FIGURE **10.21** View of the San Andreas Fault looking southeastward. The fault runs from top to bottom (dashed line) near the middle of the photo. Note the offset of the stream (Wallace Creek) as it crosses the fault, which is caused by the northward movement of the Pacific Plate with respect to the North American Plate. *(Gudmundar E. Sigvaldason/Nordic Volcanological Institute.)*

Geologists, like other scientists, develop a logical terminology to describe the features they observe. They define faults by the direction of relative movement, or slip, at the fracture (see Figure 10.22). The surface along which the formation fractures and slips is the *fault plane.* Two terms defined earlier, *dip* and *strike,* serve to describe the orientation of the fault plane. A *dip-slip fault* involves relative movement of the formations up or down the dip of the fault plane. A *strike-slip fault* is one in which the movement is horizontal, parallel to the strike of the fault plane. (A transform is a strike-slip fault that forms a plate boundary.) A movement along the strike and simultaneously up or down the dip is described as an *oblique-slip fault.* Dip-slip faults are associated with compression or tension, and strike-slip faults indicate

that shearing forces were at work. An oblique-slip fault suggests a combination of the two.

Faults need to be characterized further, because the movement can be up or down or right or left, as Figure 10.22 indicates. In a *normal fault,* the rocks above the fault plane move down in relation to the rocks below the fault plane, causing an extension of the section. (Figure 10.2 shows a small-scale normal fault in an outcrop. Figure 10.23 is an aerial view of a large-scale normal fault.) A *reverse fault,* then, is one in which the rocks above the fault plane move upward in relation to the rocks below, causing a shortening of the section. Reverse faulting results from compression. If, as we face a strike-slip fault, the block on the other side is displaced to the right, the fault is a *right-lateral fault;* if the block on the

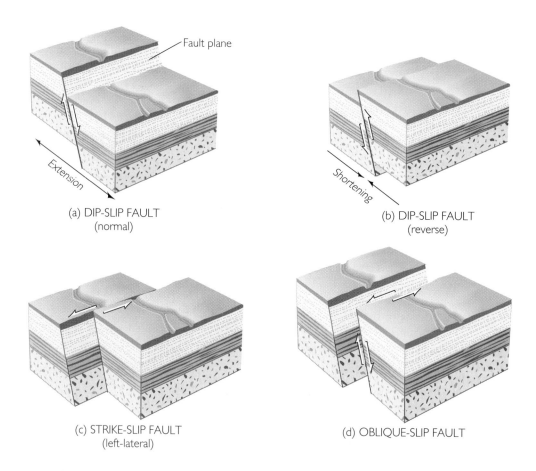

(a) DIP-SLIP FAULT
(normal)

(b) DIP-SLIP FAULT
(reverse)

(c) STRIKE-SLIP FAULT
(left-lateral)

(d) OBLIQUE-SLIP FAULT

FIGURE 10.22 Types of faults. (a) Normal faults, caused by tensional forces, result in extension. (b) Reverse faults, caused by compressional forces, result in shortening. (c) Strike-slip faults are associated with shearing forces. (d) Oblique slip suggests a combination of shear and compression or tension.

FIGURE 10.23 Normal fault in marble, Death Valley, California. *(Martin Miller.)*

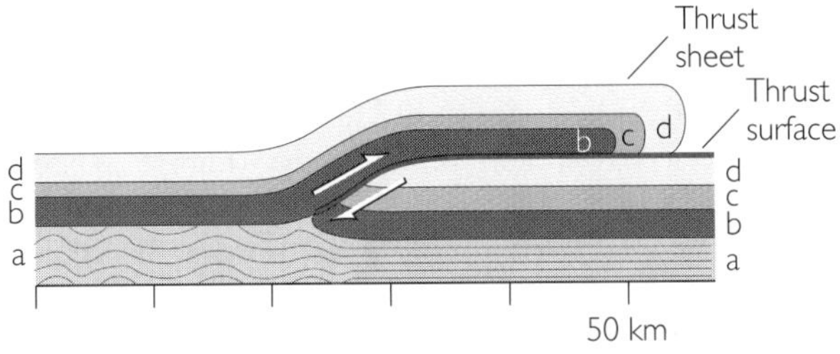

FIGURE 10.24 A large-scale overthrust sheet of a kind found in California and southern Nevada. Compressive forces have detached a sheet of rock layers—d, c, b—thrusting it a great distance horizontally over the section d, c, b, a.

other side of the fault is displaced to the left, it is a *left-lateral fault*. These movements result from shearing forces.

Finally, a reverse fault at which the dip of the fault plane is small, so that the overlying block is pushed mainly horizontally, is a *thrust fault* (Figure 10.24). Thrust faults at which one block has been pushed horizontally over the other are often found in intensely deformed mountain belts (Figure 10.25). These *overthrusts* are expressions of large-scale crustal compressive forces. In effect, the crust accommodates these forces by shortening, in this case by breaking, and one sheet overrides the other. Often the shortening may cover many tens of kilometers

FIGURE 10.25 The Keystone thrust fault of southern Nevada. Dark-colored Cambrian limestone has been thrust over light-colored Jurassic sandstone, younger by some 350 million years. *(John S. Shelton.)*

and involve multiple thrust faults. Think of a closed venetian blind. When the blind is raised so that the slats override one another, the blind is effectively shortened.

Tensional forces, which leave normal faults behind as evidence of their action, may split a plate apart. This splitting can result in the development of a **rift valley**—a depression where one block looks as though it had dropped between two flanking blocks that have been pulled apart (Figure 10.26). The tensional forces create a long, narrow trough bounded on each side by one or more parallel normal faults. The East African rift valleys, the rifts of mid-ocean ridges, the Rhine River valley, and the Red Sea rift (Figure 10.27) are famous rift valleys.

Geologists recognize faults in the field in several ways. If the relative movement is large, as in many transform faults, the formations currently facing each other across the fault probably differ in lithology and age. The offset section of what was once a single formation is often so far away that it cannot be found. When movements are smaller, offset features can be observed and measured. (For example, look again at the small-scale fault in Figure 10.2. See if you can match up the offset beds.) In establishing the time of faulting, geologists use a simple rule: a fault must be younger than the youngest rocks it cuts (the rocks had to be there before they could break) and older than the oldest undisrupted formation that covers it.

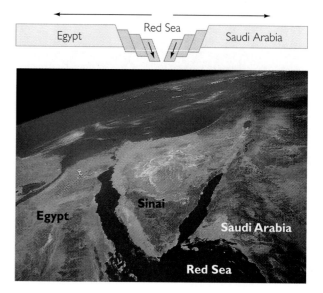

FIGURE 10.27 The African Plate, on which Egypt rides, and the Arabian Plate, bearing Saudi Arabia, are drifting apart. The tensional forces have created a rift valley, filled by the Red Sea. *(NASA/TSADO/Tom Stack.)* The diagram shows parallel normal faults bounding the rift valley in the crust beneath the sea.

UNRAVELING GEOLOGIC HISTORY

Usually the geologic history of a region is a succession of episodes of deformation and other geological processes. Let us take what appears to be a complicated example and see how some of the concepts introduced in this chapter lead to a simple interpretation. The cross sections in Figure 10.28 represent a few tens of kilometers in the Basin and Range provinces of Nevada and Utah that witnessed a succession of events: First came the deposition of horizontal layers of sediment, which then became tilted and folded by the horizontal forces of compression. They were then uplifted above sea level. There erosion gave them a new horizontal surface, which was covered unconformably by lava when forces deep in the Earth's interior caused a volcanic eruption. In the final stage, horizontal stretching resulted in normal faulting, which broke the crust into blocks. The geologist sees only the last stage (Figure 10.28e) but visualizes the entire sequence. Once the sedimentary beds are identified, the geologist starts with the knowledge that the beds must originally have been horizontal and undeformed at the bottom of an

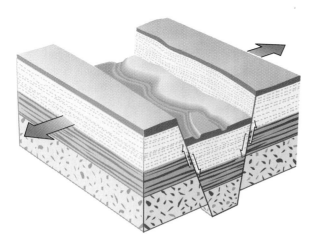

FIGURE 10.26 A downfaulted block, or rift valley, results from tensional forces.

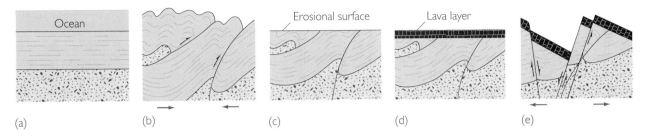

(a)   (b)   (c)   (d)   (e)

FIGURE 10.28 Stages in the development of the Basin and Range provinces of Nevada and Utah. (a) Horizontally stratified sediments (yellow) are deposited on the seafloor. (b) Horizontal compression (indicated by red arrows) causes folding and faulting. (c) Uplift above sea level is evidenced by the development of a new horizontal erosional surface. (d) Volcanic eruptions deposit sheets of lava over the eroded surface. (e) Stretching forces (indicated by red arrows) cause new, nearly vertical normal faults, breaking up the earlier features into blocks. A geologist sees only the last stage and attempts to reconstruct from the structural features all of the earlier stages in the history of a region. (After P. B. King, *The Evolution of North America* [Princeton, NJ: Princeton University Press, 1977].)

ancient ocean (Figure 10.28a). The succeeding events can then be reconstructed.

Present-day surface relief, such as we find in the Alps, the Rocky Mountains, the Pacific Coast Ranges, and the Himalayas, can be traced in large part to deformation that occurred over the past few tens of millions of years. These younger mountain systems still contain much of the information the geologist needs to piece together the history of deformation. Deformation that occurred hundreds of millions of years ago, however, no longer shows as prominent mountains. Erosion has left behind only the remnants of folds and faults in the old basement rocks of the continental interior (see Figure 10.15).

As we have seen, deformation—in the form of mountain belts with their structures of folds and faults, rift valleys, and strike-slip faults—leaves its unmistakable mark on the landscape. These topographic expressions are often guides to the deformation structures that shaped them. Even such relatively small-scale features as the shapes of hills and valleys and the courses of streams may be controlled by the complex interaction between underlying structures and erosion (Figure 10.29; see also Figure 10.21).

FIGURE 10.29 This scarp is a fresh surface feature that formed when a new reverse fault broke and caused a devastating earthquake in Armenia in 1988. In a few decades, the near-vertical scarp will erode into a gentler slope. *(Armando Cisternas, Université Louis Pasteur.)*

It is important to remember that topography is not determined by structure alone. Sometimes a valley forms in the trough of a syncline and a ridge forms at the crest of an anticline. However, we should not expect the crests of anticlines always to form ridges and the troughs of synclines always to become valleys. An important factor in the shaping of landforms upon stratified rocks is the amount of resistance the individual beds offer to weathering and erosion, as well as whether the layers are tilted, folded, or faulted.

We have seen that there is a pattern in the way rocks deform that relates to the forces ever present in Earth's crust. Plate movements play an important role in the generation of these forces. Geologists have learned to decipher this pattern, beginning with the formation of rocks and reconstructing their subsequent deformation and erosion.

SUMMARY

What do laboratory experiments tell us about the way rocks deform when they are subjected to crustal forces? Laboratory studies show that rocks vary in strength; that is, some resist deformation more than others in response to the forces to which they are subjected. Some deform as brittle materials, others as ductile. These qualities depend on the kind of rock, the temperature, the surrounding pressure, the magnitude of the force, and the speed with which it is applied.

What are some of the deformation structures that show up in rocks in the field? Among the geologic structures in rock formations that result from deformation are folds, domes, basins, joints, and faults.

What kinds of forces are involved in the formation of these structures? Folds are usually formed by compressive forces, as at boundaries where plates collide. Domes and basins are not fully understood. Some domes are formed by intrusion of magma deep in the crust. Basins can be formed when tensional forces stretch the crust or when a heated portion of the crust cools and contracts. The weight of sediments deposited in a basin can contribute to its deepening. Joints are fractures in rock caused by regional stresses or by the cooling and contraction of the rocks. Normal faults can be caused by tensional or stretching forces, such as those that occur at boundaries where plates diverge. Reverse faults and thrusts can be produced by compressive forces, such as those that occur at boundaries where plates converge. Shearing forces can produce strike-slip faults.

How do geologists reconstruct the history of a region? Geologists see the end results of a succession of events: deposition, deformation, erosion, volcanism, and so forth. They deduce the deformational history of a region by identifying and fixing the ages of the rock layers, recording the geometric orientation of the beds on maps, mapping folds and faults, and reconstructing cross sections of the subsurface consistent with the surface observations. They can ascertain the relative age of a deformation by finding a younger undeformed formation lying unconformably on an older deformed bed.

KEY TERMS AND CONCEPTS

folding (p. 244)
faulting (p. 244)
strike (p. 245)
dip (p. 245)
compressive forces (p. 247)
tensional forces (p. 247)
shearing forces (p. 247)
brittle rocks (p. 248)
ductile rocks (p. 248)

basement rocks (p. 248)
anticline (p. 249)
syncline (p. 249)
limb (p. 249)
axial plane (p. 249)
fold axis (p. 250)
plunging fold (p. 250)
asymmetrical fold (p. 250)
overturned fold (p. 250)

plunging anticline (p. 250)
plunging syncline (p. 250)
dome (p. 252)
basin (p. 252)
fold belt (p. 254)
joint (p. 255)
fault (p. 255)
rift valley (p. 258)

EXERCISES

1. Why do some rock layers fold and others break into faults when they are subjected to crustal forces?

2. What types of deformation structures would be expected at the three types of plate boundaries?

3. How would you identify a fault in the field? How would you tell whether it was a normal, reverse, or strike-slip fault?

4. If you found tilted beds in the field, how would you tell if they were part of an anticline or a syncline?

5. Draw a cross section of a rift valley and indi-cate by arrows the nature of the forces that produced it. Do the same for a thrust fault.

6. What was the direction of the crustal forces that deformed the Appalachian block depicted in Figure 10.19?

7. Draw a geological cross section that tells the following story: A series of marine sediments are deposited and subsequently deformed into folds and thrust faults. These events are fol-lowed by erosion. Volcanic activity ensues, and lava flows over the eroded surface. A final stage of high-angle faulting breaks the crust into sev-eral upheaved and downdropped rocks.

THOUGHT QUESTIONS

1. If you were asked to describe the geologic history of a region that had not yet been explored, how would you proceed?

2. Other than crustal forces, what might cause rocks to deform?

3. Although anticlines are upfolds and synclines are downfolds, we often find synclinal ridges and anticlinal valleys, as in Figures 10.9 and 10.19. Explain why. Try to devise a sequence of rock layers and series of events that could lead to each of these outcomes.

SUGGESTED READINGS

Hills, E. S. 1972. *Elements of Structural Geology,* 2nd ed. New York: Wiley.

Ramsay, J. F. 1967. *Folding and Fracturing of Rocks.* New York: McGraw-Hill.

Twiss, R. J., and E. M. Moores. 1992. *Structural Geology.* New York: W. H. Freeman.

INTERNET SOURCE

California Fault Parameters
🌐 **http://www.consrv.ca.gov/dmg/shezp/ fltindex.html**
The Division of Mines and Geology of the California Department of Conservation has compiled information on active faults in the state. Tabulated data include the fault length, average slip rate, maximum earthquake magnitude, return interval, and more. Maps of fault locations and a probabilistic seismic hazard map for California are linked to this site.

part

2

Surface
Processes

We live in natural landscapes shaped by rivers, glacial ice, the wind, and groundwater. We can and do alter our environment by building cities, cutting wide swaths for highways, and redirecting the flow of water. In the end, however, our existence depends on the basic geological processes that govern the dynamics of the land surface and on the vast bodies of water that cover much of the solid planet. The dynamics of Earth's surface are controlled by the Sun, whose radiant energy drives the atmosphere and oceans in a complex circulation pattern that ultimately produces our climate and transports water over the globe. Surface processes result from the interaction of the external solar heat engine with Earth's internal heat engine, which drives plate tectonics and raises mountains.

11

Landslide following the undercutting of soft sedimentary rocks by a flash flood. Rancho Mirage, California, July 1979. *(Joel Sternfeld, Pace/MacGill Gallery.)*

Mass Wasting

John McPhee, a well-known author of popular books on aspects of geology, was writing about the rain-soaked, stormy night of February 9, 1978, in the San Gabriel Mountains, which tower over the northern part of the Los Angeles area.

The Shields Canyon debris flow was a **mass movement**—one of many kinds of downhill movements of masses of soil, rock, mud, or other unconsolidated (loose and uncemented) materials under the force of gravity. The masses are not pulled down primarily by the action of an erosional agent, such as wind, running water, or glacial ice. Instead, mass movements occur when the force of gravity exceeds the strength (resistance to deformation) of the slope materials. Such movements are triggered by earthquakes, floods, or other provoking geological events. The materials then move down the slope, either at a slow or very slow rate or as a sudden, sometimes catastrophic, large movement. In various combinations of falling, sliding, and flowing, mass movements can displace small, almost imperceptible amounts of soil down a gentle hillside or can be huge landslides that dump tons of earth and rock on valley floors below steep mountain slopes.

Every year, mass movements take their toll of lives and property around the world. One of the most destructive occurred in the Andes Mountains of Colombia in 1985, when more than 20,000 people lost their lives in a giant mudflow of unconsolidated volcanic ash. Because mass movements are responsible for so much destruction, we want to be able to predict them, and we certainly want to avoid provoking them by unwise interference with natural processes. We cannot prevent most natural mass movements, but we can control construction and land development to minimize our losses.

Mass wasting includes all the processes by which masses of rock and soil move downhill under the influence of gravity, eventually to be carried away by other transporting agents. Mass wasting is one of the consequences of weathering and rock fragmentation. It is an important part of the general erosion of the land, especially in hilly and mountainous regions. Mass movements change the landscape by scarring mountainsides as great masses of material fall or slide away from the slopes. The material that moves ends up as tongues or wedges of debris on the valley floor, sometimes piling up and damming a stream running through the valley. The scars and debris deposits are clues to mass movements in the past. By reading those clues, geologists may be able to predict and issue timely warnings about new movements likely to occur in the future.

In mass movements, as in many other kinds of geological processes, human interference can have profound effects. Although human engineering works seem small compared with the natural world, they are significant. In the United States alone, just one activity, excavation for houses and buildings, breaks up and transports huge amounts of surface materials each year—up to 800 million tons, according to some calculations. This far exceeds the more

than 600 million tons moved annually in the United States by natural processes.

This chapter examines why masses move, what characteristics differentiate mass movements, and how weather, plate-tectonic settings, and human activity contribute to mass movements.

WHAT MAKES MASSES MOVE?

Field observations have led geologists to identify three primary factors that influence mass movements (Table 11.1):

- *The nature of the slope materials.* Slope materials may be solid masses of bedrock, regolith (the surface debris, including soil, formed by weathering), or sediment. These slope materials may be **unconsolidated**—loose and uncemented—or **consolidated**—compacted and bound together by mineral cements.

- *The amount of water in the materials.* The amount of water in the materials depends on how porous the materials are and on how much rain or other water they have been exposed to.

- *The steepness and instability of slopes.* The steepness and instability of slopes contribute to the tendency of materials to fall, slide, or flow under various conditions.

All three factors operate in nature, but slope stability and water content are most strongly influenced by human activity, such as excavation for building and highway construction. All three factors produce the same result: they lower resistance to movement, and then the force of gravity takes over and the slope materials begin to fall, slide, or flow.

The Nature of Slope Materials

Slope materials are highly variable in different kinds of terrain because they are so dependent on the details of the local geology. Thus, the metamorphic bedrock of one hillside may be badly fractured by foliation, while another slope only a few hundred meters away is composed of massive granite. Slopes of unconsolidated material, the least stable of all, may be found nearby.

UNCONSOLIDATED MATERIALS The way that the steepness and instability of slopes influence mass movements can be seen in the behavior of loose, dry sand. Children's sandboxes have made nearly every-

TABLE 11.1

FACTORS THAT INFLUENCE MASS MOVEMENTS			
NATURE OF SLOPE MATERIAL	**STEEPNESS OF SLOPE**	**AMOUNT OF WATER PRESENT**	**LIKELIHOOD OF MOVEMENT**
Loose sand or sandy silt	Angle of repose	Dry	Stable unless oversteepened by excavation
		Wet	May flow if sand is water-saturated
Unconsolidated mixture of sand, silt, and soil	Moderate	Dry	Stable unless oversteepened
		Wet	Prone to slump, slide, or flow
	Steep	Dry	Temporarily stable
		Wet	Very likely to slide or flow
Rock, jointed and deformed	Moderate to steep	Dry or wet	Rockfall or slide possible
Rock, massive	Moderate	Dry or wet	Stable
	Steep	Dry or wet	Rockfall or slide possible

one familiar with the characteristic slope of a pile of dry sand: the angle between the slope of any particular pile of sand and the horizontal is the same, whether the pile is a few centimeters or several meters high. For most sands, the angle is about 35°. If you scoop some sand from the base of the pile very slowly and carefully, you can increase the angle of slope a little, and it will hold temporarily. But if you then jump on the ground near the sandpile, the sand will cascade down the side of the pile, which again assumes its original slope of 35°. Although the cascading sand appears to move as a unit, much of the movement is by individual grains over and around one another. We can see this behavior of sand on the steepest slopes of sand dunes.

The original and resumed angle of the sandpile is its **angle of repose,** the maximum angle at which a slope of loose material will lie without cascading down. A slope that is steeper than the angle of repose is unstable and will tend to collapse to the stable angle.

The angle of repose varies significantly with a number of factors, one of which is the size and shape of the particles (Figure 11.1). Larger, flatter,

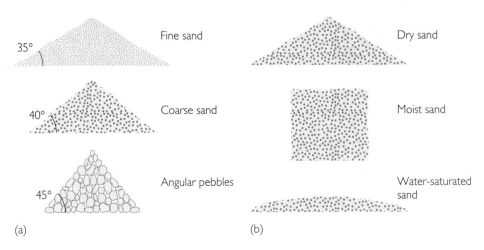

(a)   (b)

FIGURE 11.1 (a) The angle of repose of a mound of particles increases as the size of the particles increases and as their shapes become more angular. (b) The angle of repose depends on the amount of moisture between the particles. Moist sand sticks together, so it can have vertical sides, whereas water-saturated sand flows to a thin lens.

and more angular pieces of loose material remain stable on steeper slopes. The angle of repose also varies with the amount of moisture between particles. Damp sand's angle of repose is higher than that of dry sand because the small amount of moisture between the grains tends to bind them together so that they resist movement. The source of this binding tendency is **surface tension**—the attractive force between molecules at a surface (Figure 11.2a). Surface tension makes water drops spherical and allows a thin razor blade or paper clip to float on a smooth water surface (Figure 11.2b). Too much water keeps the particles apart and allows them to move freely over one another. Saturated sand, in which all the pore space is occupied by water, runs like a fluid and collapses to a flat pancake shape (Figure 11.2c). The surface tension that binds moistened sand allows beach sculptors to create elaborate sand castles (Figure 11.3). When the tide comes in and saturates the sand, the structures collapse.

CONSOLIDATED MATERIALS Consolidated dry materials, such as compacted and cemented sediments and vegetated soils, do not have the simple angles of repose characteristic of loose materials. The slopes of consolidated materials may be steeper and less regular, but they can become unstable when they are oversteepened or denuded of vegetation. See Figures 14.15 and 15.26 for examples of steep slopes in consolidated materials such as dense, tough clays that are compacted but not lithified. The attractive forces that bind together the particles of consolidated dry materials are of two types, cohesive and adhesive. The particles of consolidated sediments such as dense clays are bound together by cohesive forces associated with tightly packed particles. Cohesion in general is an attractive force between particles of a solid material that are close together (see Figure 11.2a for a diagram of these attractive forces in a liquid). Cohesion is the attractive force between particles of the same kind, whereas adhesion refers

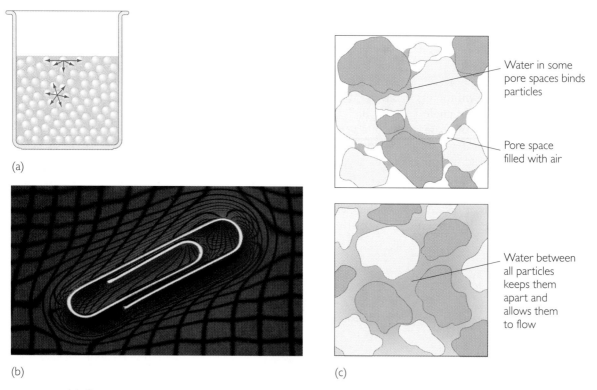

(a)

(b)

(c)

Water in some pore spaces binds particles

Pore space filled with air

Water between all particles keeps them apart and allows them to flow

FIGURE 11.2 (a) Molecules in the interior of a liquid are attracted in all directions, but molecules at the surface have a net inward attraction that results in surface tension. (b) Surface tension causes a paper clip to float on water; the water surface behaves as if it were an elastic membrane, stopping the clip from falling into the liquid. (c) In an unsaturated soil, the surface tension of a thin film of water adhering to particles binds them so that they resist movement. When the soil is saturated, water fills all the pore spaces, keeping particles apart and allowing them to flow.

FIGURE 11.3 Sand castles hold their shape because they are made of damp sand. The steepness of slopes is maintained by moisture between grains. *(Betty Crowell.)*

to the attractive force between materials of different kinds. Where the materials are both cohesive and weakly cemented by minerals precipitated in pore spaces, slopes may be very steep.

The resistance to movement resulting from adhesion, cohesion, cementation, and the binding action of plant roots is sometimes called "internal friction" because it is like the friction that resists the movement of any piece of matter against another. In a material with high internal friction, the particles are not so free to move as are loose particles such as sand. When these materials move, they tend to do so as a unit.

Water Content

Mass movement of consolidated materials usually can be traced to the effects of moisture, often in combination with other factors, such as loss of vegetation or oversteepening of the slope. When the ground becomes saturated with water, the material is lubricated, the internal friction is lowered, and the particles or larger aggregates can move past one another more easily. Water may seep into the bedding planes of muddy or sandy sediments, for example, and promote the slippage of beds past one another. When consolidated materials absorb large amounts of water, the pressure of the water in the pores of the material may be great enough to separate the grains and distend the mass. Then the material may start to flow like a fluid.

Soils become more susceptible to erosion and mass movement when they are stripped of vegetation, usually by burning or deforestation. When the soil is no longer bound by root systems, the slope

becomes more vulnerable to water invasion and thus potentially unstable, as it did in Yellowstone National Park when forest fires were followed by rains (Figure 11.4). Oversteepening of a slope of consolidated material, such as by a stream that undercuts part of a valley wall, makes the slope unstable in the same way as does steepening the angle of a sand pile. Sooner or later, the unstable valley wall will move down to assume a more stable angle.

Steepness and Instability of Slopes

Rock slopes range from relatively gentle slopes of easily weathered shales and volcanic ash beds to vertical cliffs of hard rocks such as granite. The stability of these slopes depends on the weathering and fragmentation of the rock. Shales, for example, tend to weather and fragment into small pieces that form a

FIGURE 11.4 A fire that stripped soil of vegetation and root systems in Yellowstone National Park caused a weakening of the soil that made it susceptible to erosion and mass movement. *(Grant Meyer.)*

FIGURE **11.5**
Weathered shale slope fragmented into a thin layer of rubble, Grand Canyon.
(Martin Miller.)

thin layer of loose rubble covering the bedrock (Figure 11.5). The resulting slope angle of the bedrock is similar to the angle of repose of loose, coarse sand. The weathered rubble gradually builds up to an unstable slope, and eventually some of the loose material will slide down. In many parts of the United States and Canada, these slides occur where highway builders have allowed unstable slopes to accumulate, leading to slides that close or restrict travel.

Limestones and hard, cemented sandstones in arid environments show contrasting behavior. In these rocks, which resist erosion and break into large blocks, there may be steep, bare bedrock slopes above and less steep slopes covered with broken rock below. The bedrock cliffs are fairly stable, except for the occasional mass of rock that falls and rolls down to the rock-covered slope below. That slope has a steep angle of repose but otherwise shows the same property of building to an unstable slope and then sliding down. Where such sandstones are interbedded with shale, slopes may be stepped. (The rocks in the Grand Canyon are an example; see the illustrations accompanying Feature 9.1.) As shale slides from under the sandstone beds, the harder beds are undercut, become less stable, and eventually fall as large blocks. If much water seeps into such an interbedded valley wall, the shale may be lubricated enough to enable movement of whole masses of shale and interbedded sandstone.

The structure of the beds influences their stability, especially when the dip of the beds (the angle the bedding makes with the horizontal; see Chapter 10) parallels the angle of the slope. Bedding planes may be zones of potential weakness because the adjacent beds differ in their mineral composition and texture or in their ability to absorb water. Such beds may become unstable, allowing masses of rock to slide along the weak bedding planes.

Triggers for Mass Movements

When the right combination of materials, moisture, and steepened slope angle makes a hillside unstable, a slide or flow is inevitable. All that is needed is a trigger. Sometimes a slide or debris flow, like the one in the San Gabriel Mountains described by John McPhee at the beginning of this chapter, is provoked by a heavy rainstorm. Many slides are set off by vibrations, such as those of earthquakes. Others may be precipitated by no single detectable event but just by gradual steepening that eventually results in the slope suddenly giving way. In some cases, we can take action to avert slides or debris flows (see Feature 11.1).

Geological reports can help to minimize the human costs of mass movements, but only if city planners and individual homeowners heed the reports and avoid building or buying in unstable areas. Most of the damage in the great Alaskan earthquake of March 27, 1964, was caused by the slides it triggered in the city of Anchorage. Mass movements of rock, earth, and snow wreaked havoc in residential areas, and there were major submarine slides along lakeshores and the seacoast. Huge landslides

Preventing Landslides

Many natural mass movements are so large or inevitable that we must learn to live with them. But we can prevent or minimize the effects of some movements—smaller ones and those provoked by human activity. Three main steps will help to prevent or decrease loss of life and damage to property.

1. We should avoid construction in areas that are prone to mass movements.

2. Where slopes are naturally stable, we should build in a way that does not make them unstable.

3. Because saturation is a critical factor, we must engineer water drainage so that slope materials will not become waterlogged and likely to slide or flow.

Avoidance of unwise building development requires good zoning regulations based on adequate geological assessment of the terrain. A geological survey will quickly reveal the conditions in an area that make construction unsafe, such as poorly vegetated slopes covered with unconsolidated material that absorbs water easily. In some areas where slopes and materials are unstable, the effort and expense required to engineer safe structures would be so great that the areas would be better off undeveloped. In other areas, no amount of engineering will counter nature's tendency for mass movements.

Architectural and landscape design can help ensure that relatively stable areas are not made unstable by stripping the vegetation that binds the soil or by artificially oversteepening the slopes. Such designs must adapt construction plans to fit the natural situation. If vegetation must be eliminated and slopes steepened during construction, the area should be properly graded with a low slope and replanted as soon as possible.

One essential measure is preventing a buildup of water in poorly drained soil or other unconsolidated material. Slopes that are otherwise stable may creep or slump after heavy rains if retaining walls or the construction itself prevents drainage of the water (see illustration). In areas where heavy rains tend to continue over long periods, storm drainage of potentially unstable slopes is vital. Good sewer systems are also part of slide prevention. A frequent culprit when soil becomes waterlogged is poorly designed drainage from septic tanks in a hillside home development.

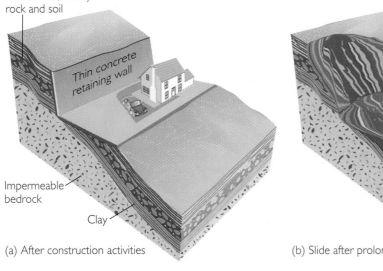

Weathered, rubbly rock and soil

Thin concrete retaining wall

Impermeable bedrock

Clay

(a) After construction activities

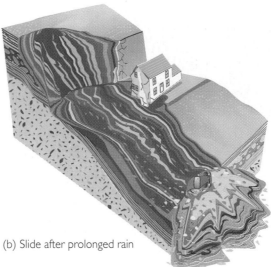

(b) Slide after prolonged rain

To build a house on some slopes is to court disaster. This slope is unstable, because it parallels the dip of the underlying beds and rests on a clay layer that would act as a lubricant if it became waterlogged. The slope in back of the house has been oversteepened, and the concrete retaining wall is too thin to hold it.

took place along the flat plains below the 30- to 35-m-high bluffs along the coast. The bluffs were composed of interbedded clays and silts. During the earthquake, the ground shook so hard that unstable, water-saturated sandy layers in the clay were transformed into fluid slurries, a process called *liquefaction*. Enormous blocks of clay and silt were shaken down from the bluffs and slid along the flat ground with the liquefied sediments, leaving a completely disrupted terrain of jumbled blocks and broken buildings (Figure 11.6). Houses and roads were carried along by the slides and destroyed. The whole process took only five minutes, beginning about two minutes after the first shock of the earthquake. At one locality, three people were killed and 75 homes destroyed.

Studies of the instability of those Alaskan slopes and the likelihood of earthquakes had indicated that the area was a prime candidate for landslides. A geological report issued more than a decade earlier had warned of the hazards of development, but the great scenic beauty of the area overwhelmed people's judgment. Many other equally hazardous slopes in earthquake-prone regions are being developed by people who willfully or mistakenly deny the high potential for dangerous mass movements.

CLASSIFICATION OF MASS MOVEMENTS

Although the popular press often refers to any mass movement as a landslide, geologists recognize many different kinds of mass movements, each with its own characteristics. We use the term *landslide* only in its popular sense, to refer to mass movements in general.

Geologists classify mass movements in accordance with several characteristics:

- The nature of the material (for example, whether it is rock or unconsolidated debris)

- The speed of the movement (from a few centimeters per year to many kilometers per hour)

- The nature of the movement: whether it is *sliding* (the bulk of the material moves more or less as a unit) or *flowing* (the material moves as if it were a fluid)

FIGURE **11.6** *(Left)* Landslide at Turnagain Heights, Alaska, triggered by the earthquake of 1964. *(Steve McCutcheon/Alaska Pictorial Service.)* *(Below)* Cross sections of the bluffs at Anchorage, Alaska, before and after the earthquake.

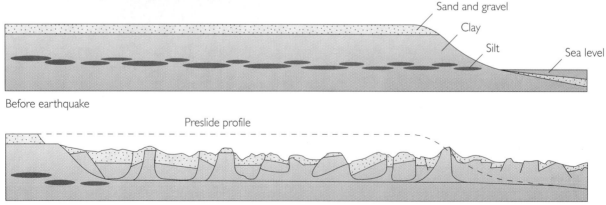

TABLE **11.2**

CLASSIFICATION OF MASS MOVEMENTS

DOMINANT MATERIAL	NATURE OF MOTION	VELOCITY		
		SLOW (1 CM/YEAR OR LESS)	MODERATE (1 KM/HR OR MORE)	FAST (5 KM/HR OR MORE)
Rock	Flow			Rock avalanche
	Slide or fall		Rockslide	Rockfall
Unconsolidated	Flow	Creep Solifluction	Earthflow Debris flow	Mudflow	Debris avalanche Debris flow
	Slide or fall		Slump	Debris slide

Table 11.2 summarizes the various kinds of mass movements geologists distinguish by means of these criteria. Shortly, we will describe the nature and speed of movements within two main categories, as determined by the dominant material in the movement.

Some movements have characteristics that are intermediate between sliding and flowing; most of the mass may move by sliding, for example, but parts of it along the base may move as a fluid. In this range of intermediate motions, mass movements are named for the dominant mechanism. Thus, a movement would be called a flow if that is the main type of motion, even though there is some sliding. Other movements overlap categories because of the greater range of transitional intermediate motions. It is not always easy to tell the exact mechanism of a move-

ment, because the nature of the movement must be reconstructed from the debris deposited after the event is over. Geologists are rarely present and prepared to observe and measure a mass movement as it takes place, and no one is about to experiment with large masses. Many slides and flows are difficult to deduce from the ancient geological record.

Rock Mass Movements

Rock movements encompass rockfalls, rockslides, and rock avalanches, several kinds of rapid downhill movements of small blocks or larger masses of bedrock (Figure 11.7). During a **rockfall,** newly detached individual blocks plummet suddenly in free fall from a cliff or steep mountainside. These angular blocks are broken from the outcrop by

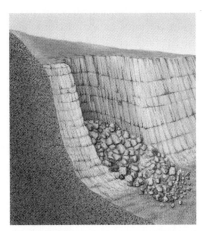
(a) **Rockfall:** Individual blocks plummet in a free fall from a cliff or steep mountainside.

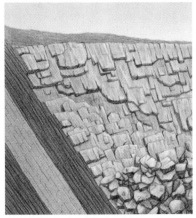
(b) **Rockslide:** Large masses of bedrocks move more or less as a unit in a fast downward slide.

(c) **Rock avalanche:** Large masses of broken rocky material flow, rather than slide, downward at high velocity.

FIGURE **11.7** Rock mass movements.

FIGURE 11.8
Rockslide, Elephant Rock, Yosemite National Park. *(Jeff Foott/DRK.)*

chemical and physical weathering and erosion. Weathering weakens the bedrock along joints until the slightest pressure, often exerted by the expansion of water as it freezes in a crack, is enough to trigger the rockfall.

People are rarely around to see rockfalls, although there were many eyewitnesses to a large rockfall in Yosemite National Park in the summer of 1996. Still, the evidence for their origin is clear from the mineral composition and texture of blocks in the rocky accumulation at the foot of the steep bedrock cliff. Referred to collectively as **talus,** these blocks of fallen rock can be matched with rock outcrops on the cliff. Individual rocks fall suddenly but very infrequently. Therefore, talus accumulates slowly, building up blocky slopes along the base of a cliff over long periods of time.

In many places, rocks do not fall freely but slide down a slope. These are **rockslides,** the fast movements of large masses of bedrock sliding more or less as a unit, commonly along downward-sloping bedding or joint planes (Figure 11.8).

Rock avalanches, like the more common snow avalanches, are flows rather than slides. They are composed of large masses of rocky materials that have broken up into smaller pieces by falling and sliding and then flow farther downhill at velocities of tens to hundreds of kilometers per hour. Rock avalanches are typically triggered by earthquakes and

are some of the most destructive mass movements. Their destructiveness comes from their large volumes (many are more than a half million cubic meters) and their fast transport of materials for thousands of meters at high velocities.

Most rock mass movements occur in higher mountainous regions; they are rare in lower hilly areas. Rock masses tend to move where weathering and fragmentation have attacked rocks already predisposed to breakage by structural deformation, relatively weak bedding, or metamorphic cleavage planes. In many of these regions, extensive talus accumulations have been built by infrequent but large-scale rockfalls and rockslides. The slopes most prone to an avalanche are those that are higher than 150 m and steeper than 25° and that are composed of rocks weakened by fractures and joints.

Unconsolidated Mass Movements

Most unconsolidated mass movements are slower than most rock movements. Although some unconsolidated materials move as coherent units, many flow like very viscous fluids, such as honey or syrup. (Viscosity, you will recall, is a measure of a fluid's resistance to flow.) The slower speed results largely from the lower slope angles at which these materials move. Figure 11.9 shows the range of unconsolidated mass movements. Unconsolidated material, often

(a) **Creep** is the downhill movement of soil or other debris at a rate of about 1–10 mm per year.

FIGURE **11.9** Unconsolidated mass movements.

(b) **Earthflow** is a movement of relatively fine-grained materials that travels up to a few kilometers per hour.

(c) **Debris flow** contains material that is coarser than sand and travels from a few kilometers per hour to many tens of kilometers per hour.

(d) **Mudflow** contains large quantities of water and may move at the rate of several kilometers per hour. Mudflows tend to move faster than earthflows or debris flows.

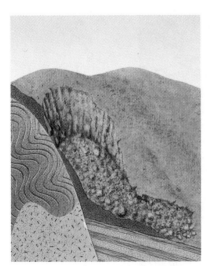

(e) **Debris avalanche** is the fastest unconsolidated flow, due to high water content and movement down steep slopes.

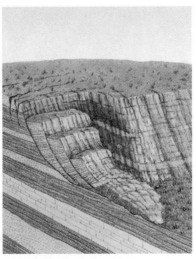

(f) **Slump** is a slow slide of unconsolidated material that travels as a unit.

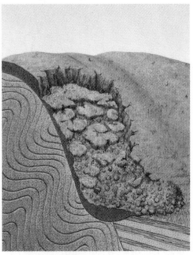

(g) **Debris slide** travels as one or more units and moves more quickly than a slump.

called debris, includes various mixtures of soil, broken-up bedrock, trees and shrubs, and materials of human construction, from fences to cars and houses.

The slowest unconsolidated mass movement is **creep,** the downhill movement of soil or other debris at a rate of about 1 to 10 mm per year, depending on the kind of soil, the climate, the steepness of the slope, and the density of the vegetation. Accumulations of debris from creep along gentle slopes build up so slowly that the rate of movement is difficult to measure over short time periods. The movement is a very slow deformation of the regolith, the upper layers of the regolith moving down the slope faster than the lower layers. Such slow movements may cause trees, telephone poles, and fences, which apparently are fixed firmly in the soil, to lean or move slightly downslope (Figure 11.10). The heavy weight of masses of soil creeping downhill can break poorly supported retaining walls and crack the walls and foundations of buildings.

Earthflows and debris flows are fluid mass movements that, for the most part, travel faster than creep, up to a few kilometers per hour, primarily because their resistance to flow is lower. **Earthflows** are fluid movements of relatively fine-grained materials, such as soils, weathered shales, and clay. **Debris flows** are fluid mass movements of rock fragments supported by a muddy matrix. They contain much material coarser than sand and tend to move more rapidly than earthflows, in some cases as much as 100 km per hour.

Mudflows are flowing masses of material mostly finer than sand, along with some rock debris, containing large amounts of water (Figure 11.11). Because of the large quantities of water, the mud offers less resistance to flow and thus tends to move faster than earth or debris. Many mudflows move at several kilometers per hour. Most common in hilly and semiarid regions, mudflows start after infrequent, sometimes prolonged, rains. The previously dry, cracked mud keeps absorbing water as the rain continues. In the process, its physical properties change; the internal friction decreases, and the mass becomes much less resistant to movement. The slopes, which are stable when dry, become unstable, and any disturbance triggers movement of waterlogged masses of mud. The mudflows travel down the upper valley slopes and merge on the valley floor. Where mudflows exit from confined valleys into broader, lower valley slopes and flats, they may splay out to cover large areas with wet debris. Mudflows can carry large boulders, trees, and even houses.

Debris avalanches are fast downhill movements of soil and rock that usually occur in humid mountainous regions. Their speed is due to the combination of high water content and steep slopes. Water-saturated debris may move as fast as 70 km per hour, a speed comparable to that of water flowing down a moderately steep slope. These flows carry with them everything in their paths. In 1962, a debris avalanche in the high Peruvian Andes traveled almost 15 km in about 7 minutes, engulfing most parts of eight towns and killing 3500 people. Eight years later, on May

FIGURE **11.10** A fence offset by creep in Marin County, California. *(Travis Amos.)*

FIGURE **11.11** An earthquake in Tadzhikistan in January 1989 produced 15-m-high mudflows on slopes weakened by rain. *(Vlastimir Shone/Gamma.)*

31, 1970, an earthquake in this same active subduction zone toppled a large mass of glacial ice at the top of one of the highest mountains, Nevado de Huascarán. As the ice broke up, it mixed with the debris of the high slopes and became an ice-debris avalanche. The avalanche picked up debris as it flowed down, increasing its speed to an almost unbelievable 200 to 435 km per hour. More than 50 million cubic meters of muddy debris roared down into the valleys, killing 17,000 people and wiping out scores of villages.

On May 30, 1990, an earthquake shook another mountainous area in northern Peru, again setting off mudflows and debris avalanches. It was just two days before a memorial ceremony scheduled to commemorate the disaster that had occurred 20 years earlier. In regions where unstable slopes build up and earthquakes are frequent because of the subduction of an oceanic plate beneath a continental one, there can be no doubt about the necessity of learning how to predict both earthquakes and the dangerous mass movements that follow. Feature 11.2 describes Japan's success in one such effort.

Mudflows and debris avalanches are easily triggered on slopes of volcanic cinder cones when the accumulations of unconsolidated ash and other erupted materials become saturated with rainwater. Some flows of volcanic debris are set off by events associated with the volcanic eruption itself (such as earthquakes), by downpours of rain and ash from the eruption cloud, and by sudden falls of volcanic debris. More than 55 such flows, many of them associated with eruptions, have occurred on the slopes of Mount Rainier, Washington, in the past 10,000 years, according to geologists of the U.S. Geological Survey. In 1980, a giant volcanic debris avalanche triggered by the eruption of Mount St. Helens, also in Washington, roared down the north slope of the mountain at a speed of about 200 km per hour and covered an area of more than 600 square kilometers below the mountain (Figure 11.12). (See Chapter 5 for more about the relationship of mass movements to volcanism.)

FIGURE **11.12** A volcanic debris avalanche filled in Spirit Lake after the 1980 eruption of Mount St. Helens. *(Russ Davies/The Image Bank.)*

11.2 LIVING ON EARTH

Reducing Loss from Landslides

In the winter of 1968–1969, the San Francisco Bay area sustained more than $25 million in landslide costs. The Bay region was hit hard again in the winter of 1992–1993, when prolonged and torrential rains saturated the ground, breaking a six-year drought.

In some countries, losses from landslides exceed those of all other natural hazards combined. In March 1987, for example, the vibrations of an earthquake in the mountains of Ecuador after a month of heavy rains triggered a landslide that killed at least 1000 people and ruptured the trans-Ecuadoran oil pipeline—the nation's most important source of income—at a cost of approximately $1.5 billion.

In the United States, landslides cause 25 to 50 deaths and $1 billion to $2 billion in property damage each year. Here, as in many other countries, losses in life and property will continue to increase as newly constructed homes,

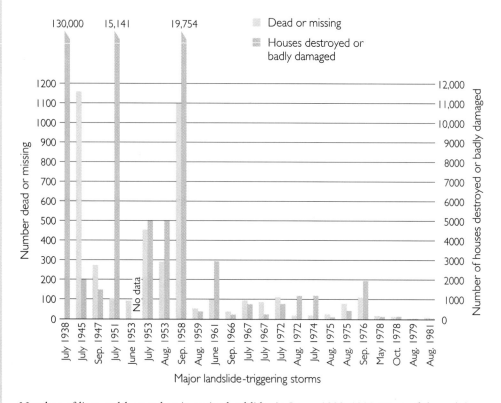

Number of lives and homes lost in major landslides in Japan, 1938–1981. Most of these slides were debris flows caused by a combination of heavy rainfall and steep slopes. Note the marked reduction in losses after the introduction of a national landslide control program in 1958. (From *Confronting Natural Disasters*, National Academy Press, 1987, based on data from the Ministry of Construction, Japan.)

roads, commercial buildings, and public works push into the hilly and unstable terrains that are most susceptible to landslides. The pressure for development stems from growing populations, the flight of city dwellers to the countryside, and ignorance or denial of the hazards and costs. According to the Federal Highway Administration, annual costs to repair major landslides on highways in the United States exceed $100 million.

Japan's landslide mitigation program is one model for other countries to follow. Much of the terrain of the Japanese Islands is a combination of hilly and mountainous topography, volcanic deposits, and deformed metamorphic rocks, a mixture that makes many slopes potential landslide hazards. After decades of severe losses of its citizens' lives and property, the Japanese government instituted land-use management controls in 1958. The program includes government control over how land can be used and what types of building construction are permitted. To minimize the possibility of potentially dangerous mass movements, the government also conducts research and puts into practice advanced engineering techniques, such as soil drainage networks. It forces adherence to strict building and grading regulations. In addition, the Japanese government sponsors a strong program of public education and broadcasts warnings of landslide danger to local populations. Experience and education have prepared people to evacuate hazardous areas at the first warning, and they know where to go.

The government's investment in these landslide control programs has been successful. As the accompanying graph indicates, major landslide disasters in 1938 resulted in 500 deaths and the destruction or damage of 130,000 homes. But in 1976, the worst year for landslides since the control program started in 1958, fewer than 125 lives were lost and only 2000 homes were destroyed or damaged.

A **slump** is a slow slide of unconsolidated material that travels as a unit. In most places, the slump slips along a basal surface that forms a concave upward shape, like a spoon. Faster than slumps are **debris slides,** in which the rock material and soil move largely as one or more units along planes of weakness, such as a waterlogged clay zone either within or at the base of the debris. During the slide, some of the debris may behave like a chaotic, jumbled flow. Such a slide may turn to mostly flow as it moves rapidly downhill and most of the material mixes as if it were a fluid.

Solifluction is a type of movement that occurs only in cold regions when water in the surface layers of the soil alternately freezes and thaws. When the surface zones thaw, the soil there becomes waterlogged (Figure 11.13). The water cannot seep downward, as it might in more temperate regions, because the deeper layers of soil are still frozen. Because the water cannot escape to the icy, impermeable deeper parts of the regolith or bedrock, water continues to accumulate and saturates the upper layers of soil so thoroughly that they ooze downhill, carrying broken rocks and other debris with them.

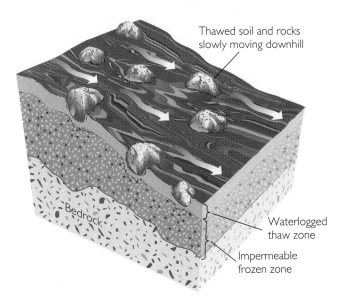

Thawed soil and rocks slowly moving downhill

Bedrock

Waterlogged thaw zone

Impermeable frozen zone

FIGURE 11.13 Solifluction operates only in cold regions. Thawed surface layers of soil become waterlogged and carry broken rocks and debris with them as they ooze over deeper layers that are still frozen.

Erosion of Mass Movement Materials

Materials found in talus slopes and the debris of slides, flows, and avalanches erode easily and extensively because they have already been broken into finer grain sizes and larger surface areas, which makes them weather more extensively. When these weathered materials reach lower slopes, they are transferred to small streams. As a result, few mass movements are preserved in the ancient rock record, although many geologists have seen the evidence of prehistoric mass movements of the last few thousand years. It is unusual to find mass movements millions of years old.

Marine Mass Movements

Although no one has been able to witness a submarine slide, we know that the seafloor is not exempt from mass movements because marine geologists can reconstruct such events from slide materials on the seafloor. For example, within the last 450,000 years, a mass measuring 4 by 5 km slid down one wall of the Mid-Atlantic Ridge rift valley. The slopes were weakened by faulting and hydrothermal metamorphism. The debris avalanche that formed by slope failure flowed more than 11 km into the rift valley, depositing a debris wedge of almost 20 billion cubic meters. Another submarine setting, the flanks of the Hawaiian volcano Mauna Loa, is covered by a large area of blocky debris that is the remains of a slide extending from the present shoreline seaward to depths of 5 km. The blocks and debris consist of mixed lavas and volcanic sediment. Marine volcanic slides are formed by sudden eruptions, either of submarine volcanoes or, as in this case, of a land volcano bordering the ocean.

UNDERSTANDING THE ORIGINS OF MASS MOVEMENTS

We can best understand how the steepness of a slope, the nature of the materials, and the materials' water content interact to create mass movements by looking at both natural mass movements and those provoked by human activities. To find the causes of modern slides, geologists dovetail eyewitness reports with geological investigations of the source of the slide and the distribution and nature of the debris dropped in the valley below. Causes of prehistoric slides can be inferred from geological evidence alone where the debris is still present and can be analyzed for size, shape, and composition.

Natural Causes of Landslides

In April 1983, about 4 million cubic meters of unconsolidated mud and debris slid down from a wall of Spanish Fork Canyon in Utah, burying a section of transcontinental railroad and blocking the major east-west highway through central Utah (Figure 11.14). The debris slide also dammed the river in the canyon and created a large lake that submerged the small town of Thistle, forcing its inhabitants to flee. Near-record depths of snow had accumulated at higher elevations in the surrounding Wasatch Mountains. A warm spring caused the snow to melt rapidly, and heavy rainstorms joined with meltwaters to set off debris flows and slides in unprecedented numbers in Spanish Fork and other canyons of the Wasatch Range. In this case, the nature of the materials—soil and rock that was structurally deformed, cracked, and weak—combined with the steep slopes of Spanish Fork to make the canyon susceptible to slides. The rain and melting snow caused a mass of regolith and weathered rock on the wall of the

FIGURE 11.14 A debris slide at Thistle, Utah, in 1983 blocked the railroad, highway, and river. *(Michael Collier.)*

canyon to become saturated, lubricating both the mass and the bedrock surface of the slide below it. Sooner or later, the mass had to give.

A classic bedrock landslide occurred early in this century in the Gros Ventre River valley of the mountainous Jackson Hole area of western Wyoming. This slide again illustrates how water, the nature of materials, and slope interact to produce slides. In the spring of 1925, melting snow and heavy rains made water abundant in the Gros Ventre River valley, both in the stream and in groundwater. One local rancher, out on horseback, was warned by the roar of the slide; he looked up to see the whole side of the valley racing toward his ranch. From the gate to his property, he watched the slide hurtle past him at about 80 km per hour and bury everything he owned.

About 37 million cubic meters of rock and soil slid down one side of the valley that day, surging more than 30 m up the opposite side and then falling back to the valley floor. Most of the slide was a confused mass of blocks of sandstone, shale, and soil, but one large section of the side of the valley, covered with soil and a forest of pine, slid down as a unit. The slide dammed the river, and a large lake grew over the next two years (Figure 11.15). Then the lake overflowed, breaking the dam and rapidly flooding the valley below.

The causes of the Gros Ventre slide were all natural; in fact, the stratigraphy and structure of the valley made a slide almost inevitable (Figure 11.16). On the side of the valley where the slide occurred,

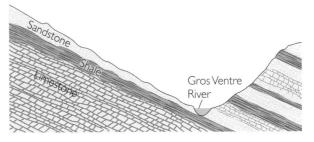

(a) Before slide

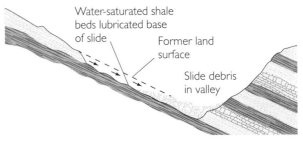

(b) After slide

FIGURE **11.16** Cross sections of the 1925 Gros Ventre slide. (After W. C. Alden, Landslide and flood at Gros Ventre, Wyoming, *Transactions of the American Institute of Mining, Metallurgical, and Petroleum Engineers*, 1928, pp. 345–361.)

a permeable, erosion-resistant sandstone formation dipped about 20° toward the river, paralleling the slope of the valley wall. Under the sandstone were beds of impermeable soft shale that became slippery when wet. The conditions became ideal for a slide when the river channel cut through most of the sandstone at the bottom of the valley wall and left it with virtually no support. Only friction along the bedding plane between the shale and sandstone kept the layer of sandstone from sliding. The river's removal of the sandstone's support was equivalent to scooping sand from the base of a sandpile—both cause oversteepening. The heavy rains and snow meltwater saturated the sandstone and the surface of the underlying shale, creating a slippery surface along the bedding planes at the top of the shale. No one knows what triggered the Gros Ventre slide, but at some point the force of gravity overcame friction and almost all of the sandstone slid down along the water-lubricated surface of the shale.

The formation of a dam on a river and the growth of a lake, as happened at both Thistle, Utah, and the Gros Ventre River valley, are common consequences of a landslide. Because most slide materials are permeable and weak, such a dam is soon breached when the lake water reaches a high level or

FIGURE **11.15** Landslide at Gros Ventre, Jackson Hole, Wyoming, 1925. This photo was taken after the slide had dammed the river and created a lake in which houses were submerged. The lake later broke through the dam, flooding the valley below. (*Jackson Hole Museum.*)

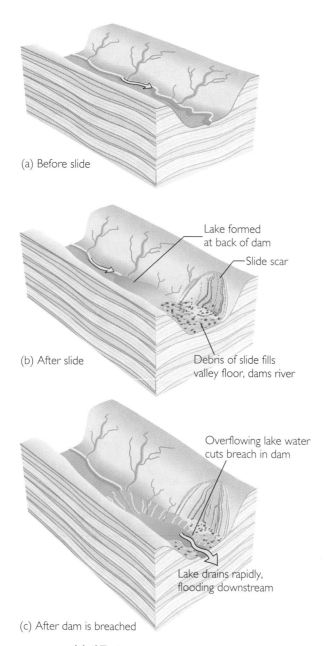

(a) Before slide

Lake formed at back of dam

Slide scar

(b) After slide

Debris of slide fills valley floor, dams river

Overflowing lake water cuts breach in dam

(c) After dam is breached

Lake drains rapidly, flooding downstream

FIGURE 11.17 Stages in the evolution of a river valley dammed by a landslide.

overflows. Then the lake drains suddenly, releasing a catastrophic torrent of water (Figure 11.17).

Plate-Tectonic Settings

There is hardly an area of geology that plate tectonics has not touched, and mass wasting is no exception. The connection between mass wasting and plate tectonics is made by considering topographic heights, steep slopes, the composition of surface materials, and rock-weakening effects. We can outline the effects by considering the basic plate-tectonic settings.

Perhaps the most obvious is the setting in which two plates converge, forming mountains. An ocean-continent plate convergence and its associated subduction zone, such as the Andes Mountains of South America, bring together high, steep mountain slopes and abundant volcanic eruptions, which produce ash that can easily become mudflows. Add to this the fracturing and deformation of rocks involved in mountain making and the many earthquakes along subduction zones, and all the conditions for frequent mass movements are met. A continent-continent plate collision, exemplified by the Himalayas, meets all of the same conditions except for volcanism.

Plate divergence, too, may be associated with mass movements, particularly at the steep slopes formed in continental rift valleys that initiate divergence. Transform fault plate boundaries, such as the San Andreas fault, are also the sites of frequent mass movements, given the steep slopes that may be found along the fault and the prevalence of earthquakes. We have already mentioned submarine landslides in the central rift valley of the Mid-Atlantic Ridge, a setting of plate divergence. Contrast these conditions with those of areas far from present or former plate boundaries, where topography may be relatively low, rocks may be relatively undisturbed, slopes are gentle, and earthquakes are rare.

Human Activities That Promote or Trigger Slides

Although the vast bulk of mass wasting is natural, human activities may promote or trigger landslides in vulnerable areas, as when human activities change natural slopes. Excavation at the base of a road-cut slope during work to widen a highway in Quebec in 1955 oversteepened the side of the road cut. The road was cut through soft formations of silt and clay that were susceptible to weakening by water. After a period of heavy rain, the steepened slope of the road cut became saturated with water and suddenly gave way in a debris flow that carried away buildings, roads, and people. Three deaths can be traced to this case of poor highway engineering.

Some geological settings are so susceptible to landslides that engineers may have to avoid construction projects in these areas entirely. A landslide in one such place ultimately killed 3000 people. The place was Vaiont, an Italian alpine valley, and the time was the night of October 9, 1963. A large reservoir in the valley was impounded by a concrete dam (the second highest in the world, at 265 m) and

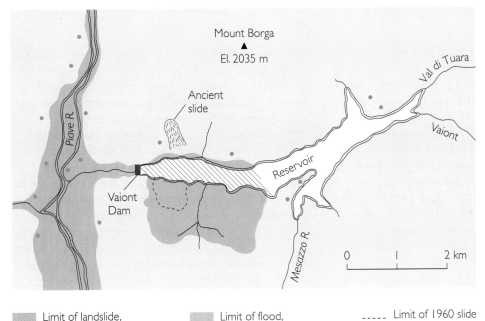

FIGURE **11.18** An ancient small slide on the north side of the walls of the Vaiont Dam reservoir, as well as a landslide in 1960 (shown by a dashed line in the orange area), warned of the danger of mass movement above the reservoir. In October 1963, a massive landslide that could have been predicted caused the water in the Vaiont reservoir to overflow the dam, flooding the downstream areas and killing 3000 people.

▨ Limit of landslide, October 9, 1963

▧ Area filled by slide, 1963

▨ Limit of flood, downstream from slide

- - - - - Limit of 1960 slide

• Cities and towns

bordered by steep walls of interbedded limestone and shale. A great debris slide of 240 million cubic meters (2 km long, 1.6 km wide, and more than 150 m thick) plunged into the deep water of the reservoir behind the dam. The debris filled the reservoir for a distance of 2 km upstream of the dam and created a giant spillover. In the violent torrent that hurtled downstream as a 70-m-high flood wave, 3000 people died.

Engineers had underestimated three warning signs at Vaiont (Figure 11.18):

▪ The weakness of the cracked and deformed layers of limestone and shale that made up the steep walls of the reservoir

▪ The scar of an ancient slide on the valley walls above the reservoir

▪ A forewarning of danger signaled by a small rockslide in 1960, just three years earlier

Although the landslide was natural and could not have been prevented, its consequences could have been much less severe. If the reservoir had been located in a geologically safer place, where the water was less likely to spill over its walls, damage might have been limited to a lesser loss of property and far fewer deaths. As noted earlier, we cannot prevent most natural mass movements, but we can minimize our losses through more careful control of construction and land development.

SUMMARY

What are mass movements, and what kinds of material are involved? Mass movements are slides, flows, or falls of large masses of material down slopes in response to the pull of gravity. Mass wasting is the erosion and sculpting of the land surface by mass movements. Mass movements may be imperceptibly slow or too fast to outrun. The masses consist of bedrock; consolidated material, including compacted sediment or regolith; or unconsolidated material, such as loose, easily disaggregated sediment or regolith. Rock movements include rockfalls, rockslides, and rock avalanches. Unconsolidated material moves by creep, slump, debris slide, debris avalanche, earthflow, mudflow, solifluction, and debris flow.

What factors are responsible for mass movements, and how are such movements triggered? The three factors that have the greatest bearing on the predisposition of material to move down a slope are (1) the steepness and instability of the slope, (2) the nature of the material, and (3) the water content of the material. Slopes become unstable when they become steeper than the angle of repose, the maximum slope angle that unconsolidated material will assume. Slopes in consolidated material may also become unstable when they are oversteepened or denuded of vegetation. Water absorbed by the material contributes to instability in two ways: by lowering internal friction (and thus resistance to

283

flow) and by lubricating planes of weakness in the material. Mass movements can be triggered by earthquakes or by sudden absorption of large quantities of water after a torrential rain. In many places, slopes build to a point of instability at which the slightest vibration will set off a slide, flow, or fall.

What factors are responsible for catastrophic mass movements, and how can such movements be prevented or minimized? Analysis of both natural mass movements and those induced by human activity shows that one of the main contributory factors is the oversteepening of slopes, either by natural erosional processes or by human construction or excavation. Because water content has such a strong effect on stability, absorption of water from prolonged or torrential rains is often an important factor. The structural attitude of the beds, especially when bedding dips parallel to the slope, can promote mass movements. Volcanic eruptions may produce a tremendous fallout of ash and other materials that build to unstable slopes. Slides or flows of the volcanic materials are triggered by earthquakes accompanying eruptions. Loss of life and damage to property from catastrophic mass movements can be prevented or minimized by avoidance of steepening or undercutting of slopes. Careful engineering can keep water from making material more unstable. In some areas that are extremely prone to mass movements, development may have to be restricted.

KEY TERMS AND CONCEPTS

mass movement (p. 265)

mass wasting (p. 266)

unconsolidated materials (p. 266)

consolidated materials (p. 266)

angle of repose (p. 267)

surface tension (p. 268)

rockfall (p. 273)

talus (p. 274)

rockslide (p. 274)

rock avalanche (p. 274)

creep (p. 276)

earthflow (p. 276)

debris flow (p. 276)

mudflow (p. 276)

debris avalanche (p. 276)

slump (p. 279)

debris slide (p. 279)

solifluction (p. 279)

EXERCISES

1. What role do earthquakes play in the occurrence of landslides?

2. What kinds of mass movements advance so rapidly that a person could not outrun them?

3. How does absorption of water weaken unconsolidated material?

4. What is the difference between a slide and a flow?

5. What is an angle of repose, and how does it vary with water content?

6. How does steepness of a slope affect mass wasting?

7. What is a debris flow, and how does it differ from a debris avalanche?

8. What is the typical history of a catastrophic mass movement?

9. What is a mudflow, and how is it formed?

10. What are the differences between a rockfall, a rockslide, and a rock avalanche?

THOUGHT QUESTIONS

1. You are planning a highway through hills made of unlithified sands and gravels and you want to minimize mass movements. What construction practices will you avoid?

2. Would a prolonged drought affect the potential for landslides?

3. What geological conditions might you want to investigate before you bought a house at the base of a steep hill of bedrock covered by a thick mantle of regolith?

4. You are excavating the base of a slope that is prone to landsliding, taking care not to oversteepen it, when heavy rain begins and continues to fall for three days. Why might you temporarily stop heavy construction trucks from driving near the slope?

5. Would you expect a talus slope of large blocks of granite to be prone to debris flows?

6. What evidence would you look for to indicate that a mountainous area had undergone a great many prehistoric landslides?

7. What factors would make the potential for mass movements in a mountainous terrain in the rainy tropics greater or lesser than the potential in a similar terrain in a desert?

8. What kind(s) of mass movements would you expect from a steep hillside with a thick layer of soil overlying unconsolidated sands and muds after a prolonged period of heavy rain?

9. What factors weaken a rock and enable gravity to start a mass movement?

10. Why would you expect that regions near a continental rift valley, where two plates are beginning to diverge, might have frequent mass movements?

LONG-TERM TEAM PROJECT: ENVIRONMENTAL CHANGES AND LAND USE

During the time in which you are enrolled in your geology class, a variety of natural and human-induced environmental changes may take place in your local area. There may be an earthquake, drought, flood, sinkhole collapse, or oil spill. Local officials may vote to open a new landfill, allow railroad transportation of hazardous materials through town, or develop the local waterfront.

In teams of four, prepare a series of news releases, approximately one per month, for your local media outlets in response to natural or human-induced environmental changes and to local land-use policy decisions. This project will require you to search for issues pertinent to geology in the local newspapers and broadcasts, and write responses from a geologist's perspective to the issues your team encounters.

Because you are writing for a nontechnical audience, you should make each news release brief (preferably only one page) and avoid the use of technical jargon. The title must grab the reader's attention. State the topic clearly in the first paragraph, describe the nature and significance of the event or issue, and emphasize its implications for the public.

Submit your news releases to your local newspapers and radio and television stations. Most local media outlets welcome newsworthy items about events in their community.

SUGGESTED READINGS

Bloom, Arthur L. 1991. *Geomorphology,* 2nd ed. Englewood Cliffs, N.J.: Prentice Hall.

Costa, John E., and Victor R. Baker. 1981. *Surficial Geology.* New York: Wiley.

Costa, John E., and G. F. Wieczorek. 1987. *Reviews in Engineering Geology,* vol. 7: *Debris Flows/Avalanches: Process, Recognition, and Mitigation.* Boulder, Colo.: Geological Society of America.

Eckel, E. B. (ed.). 1958. *Landslides and Engineering Practice.* Highway Research Board Special Report 29. National Academy of Sciences, National Research Council Publication 544.

Lundgren, Lawrence. 1986. *Environmental Geology.* Englewood Cliffs, N.J.: Prentice Hall.

Porter, Stephen C., and G. Orombelli. 1981. Alpine rockfall hazards. *American Scientist* 69:67–75.

Slosson, J. E., A. G. Keene, and J. A. Johnson (eds.). 1993. *Reviews in Engineering Geology,* vol. 9: *Landslides/Landslide Mitigation.* Boulder, Colo.: Geological Society of America.

Tyler, M. B. 1995. *Look Before You Build: Geologic Studies for Safer Land Development in the San Francisco Bay Area.* U.S. Geological Survey Circular 1130.

Voight, Barry (ed.). 1978. *Rockslides and Avalanches,* vol. 1: *Natural Phenomena.* New York: Elsevier.

INTERNET SOURCES

FEMA Fact Sheet: Landslides
🌐 **http://www.fema.gov/fema/landslif.html**
The Federal Emergency Management Agency (FEMA) provides fact sheets for landslides and most other natural hazards, including information on actions to take before, during, and after the event.

U.S. National Landslide Information Center
🌐 **http://geohazards.cr.usgs.gov/html_files/nlicsun.html**
Through this site, the U.S. Geological Survey provides a landslide inventory and fact sheets on landslide occurrence.

12

Lush vegetation flourishes in wetlands, such as those shown here at Big Springs in the Ozark Mountains. Wetlands store immense quantities of water, and they moderate large changes in water supply. Daily water flow at Big Springs is 277 million gallons. *(Garry D. McMichael / Photo Researchers.)*

The Hydrologic Cycle and Groundwater

Geologists who specialize in the science of **hydrology** study the movement and characteristics of water on and within land. Water moving in rivers and frozen in glacial ice is a major agent of erosion, which helps to shape the landscape of the continents. Water is essential to weathering, both as a solvent of minerals in rock and soil and as a transport agent that carries away the dissolved material. Water sinks into surface materials and forms large reservoirs, and it also provides the lubrication for many landslides and other forms of mass movement. Hydrothermal ore deposits are the products of hot water circulating over igneous bodies or through mid-ocean ridges.

This chapter is a survey of water in and on the Earth. Water is vital to all life on this planet. Humans cannot survive more than a few days without it, and even the hardiest desert plants and animals need some water. The amount of water we require for modern civilization is far greater than what we need for simple physical survival. Water is used in immense quantities for industry, agriculture, and such urban needs as sewage systems. The United States, one of the heaviest users of water in the world, has been steadily increasing its consumption since the nineteenth century. In the 35 years between 1950 and 1985 alone, we nearly tripled our water use, from 34 billion gallons a day to about 90 billion gallons a day. In the next five years, the figure almost quadrupled again, to 339 billion gallons by 1990. As we will see, this kind of growth cannot continue indefinitely.

Hydrology is becoming more important to all of us as the demand on limited water supplies increases. To protect those supplies while we satisfy our needs, we must understand not only where to find water but also how water supplies are renewed. With that knowledge, we can use and dispose of water in ways that do not endanger future supplies.

FLOWS AND RESERVOIRS

We can see water moving from one place to another in the rivers on Earth's surface, and we can see water stored on the surface in lakes and oceans. What is harder to see are the massive amounts of water stored in the atmosphere and underground and the flows into and out of these storage places. As water evaporates, it vanishes into the atmosphere as vapor. As rain sinks into the ground, it becomes **ground-water**—the mass of water stored beneath Earth's surface.

Each of the environments in which water is stored is a **reservoir,** a source or place of residence, in this case for water. Earth's main natural reservoirs for water are the oceans; glaciers and polar ice; underground waters; lakes and rivers; the atmosphere; and the biosphere. The distribution of water among these reservoirs is shown in Figure 12.1. The land reservoir includes lakes, rivers, and groundwater. The oceans are by far the largest reservoir. Although the total amount of water in rivers and lakes is relatively small, these reservoirs are important to human populations because they contain fresh water, ready for use. The amount of groundwater is more than 100 times the amount in rivers and lakes, but much of it is unusable because it contains large quantities of dissolved material.

Reservoirs have inflows by which they gain water, such as rain and river inflow, and outflows by which they lose water, such as evaporation and river

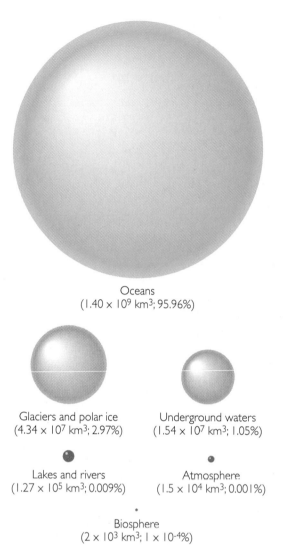

Oceans
(1.40×10^9 km³; 95.96%)

Glaciers and polar ice
(4.34×10^7 km³; 2.97%)

Underground waters
(1.54×10^7 km³; 1.05%)

Lakes and rivers
(1.27×10^5 km³; 0.009%)

Atmosphere
(1.5×10^4 km³; 0.001%)

Biosphere
(2×10^3 km³; 1×10^{-4}%)

FIGURE 12.1 Distribution of water in the Earth. The amounts of water present in various natural reservoirs are shown as spheres of comparative volumes. The content of each reservoir is given in cubic kilometers and as a percentage of the whole. (Revised from J. P. Peixoto and M. Ali Kettani, "The Control of the Water Cycle," *Scientific American*, April 1973, p. 46; E. K. Berner and R. A. Berner, *Global Environment,* Upper Saddle River, N.J.: Prentice Hall, 1996, pp. 2–4.)

outflow. If the inflow and outflow are equal, the size of the reservoir stays the same, even though water is constantly entering and leaving. Because of these movements, any given quantity of water spends a certain average time, called the residence time, in a reservoir. We will discuss reservoirs and residence time more fully in Chapter 24.

How Much Water Is There?

The world's total water supply is enormous, about 1.46 billion cubic kilometers distributed among the various reservoirs. If it covered the land area of the United States, it would submerge the 50 states under a layer about 145 km deep. This total is constant, even though the flows from one reservoir to another may vary from day to day, year to year, and century to century. Over these geologically short time intervals, there is no net gain or loss of water to or from Earth's interior, nor any significant loss of water from the atmosphere to outer space.

The Hydrologic Cycle

Water at or beneath the Earth's surface moves or "cycles" among the main reservoirs: the oceans, the atmosphere, and the land. The cyclical movement of water from the ocean to the atmosphere, through rain to the surface, through runoff and groundwater to streams, and back to the ocean is the **hydrologic**

cycle. This simplified description of the endless circulation of water and the amounts moved is illustrated in Figure 12.2. Within the range of temperatures found at the Earth's surface, water is able to shift among the three states of matter—liquid (water), gas (water vapor), and solid (ice). These transformations power some of the main flows from one reservoir to another in the hydrologic cycle. The external heat engine of the Earth, powered by the Sun, drives the hydrologic cycle, mainly by evaporating water from the oceans and transporting it as water vapor in the atmosphere. Under the right conditions of temperature and humidity, water vapor condenses to the tiny droplets of water that form clouds and eventually falls as rain or snow over the oceans and continents. Some of the water that falls on land soaks into the ground by **infiltration,** a process by which water enters rock or soil through joints or small pore spaces between particles. Some groundwater evaporates through the soil surface. Another part is absorbed by plant roots, carried up to the leaves, and returned to the atmosphere by

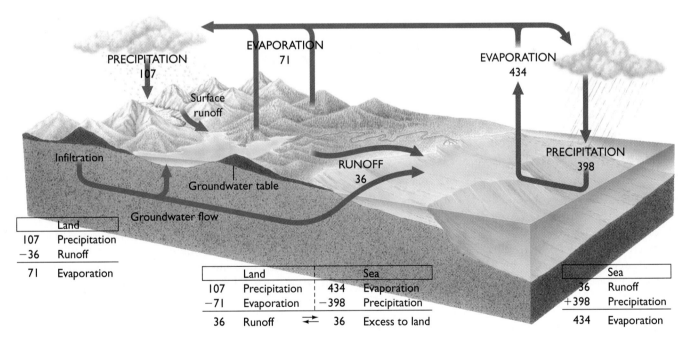

FIGURE 12.2 The hydrologic cycle. The movement of water into the atmosphere by evaporation from the oceans and continents is matched by precipitation as rain and snow. Evaporation from the oceans is balanced by surface runoff from the continents and rainfall over the oceans. The water-flow budgets of the oceans, the land, and the atmosphere are calculated at the bottom of the diagram. All figures are given in thousands of cubic kilometers per year. (Data from E. K. Berner and R. A. Berner, *Global Environment,* Upper Saddle River, N.J.: Prentice Hall, 1996, p. 3.)

transpiration, the release of water vapor from plants. Other groundwaters may return to the surface via springs that empty into rivers and lakes.

The rainwater that does not infiltrate the ground runs off the surface, gradually collecting into streams and rivers. The sum of all rainwater that flows over the surface is called **runoff.** Some runoff may later seep into the ground or evaporate from rivers and lakes, but most of it flows into the oceans.

Snowfall may be converted to ice in glaciers, which return the water to the oceans by melting and runoff or to the atmosphere by *sublimation,* the transformation from a solid (ice) directly to a gas (water vapor). The largest part of the water that evaporates from the oceans returns to them as rain and snow, commonly grouped together as precipitation. The remainder falls over the land and either evaporates or returns to the ocean as runoff.

Figure 12.2 shows how the total flows among reservoirs balance one another in the present system, which is affected by human activities. The land surface, for example, gains water by precipitation and loses the same amount of water by evaporation and runoff. The ocean gains water from runoff and precipitation and loses the same amount of water by evaporation. Thus, each reservoir stays constant. As you can see from Figure 12.2, more water evaporates from the oceans than falls on them as rain. This loss is balanced by the water returned as runoff from the continents. About one-third of the total precipitation on land (107,000 km^3) comes from the excess of evaporation over precipitation from the oceans (434,000 − 398,000 = 36,000 km^3), and that one-third returns to the ocean as runoff.

How Much Water Can We Use?

The global hydrologic cycle wields ultimate control over water supplies. Almost all of the water we use is fresh water—that is, water that is not salty, like the oceans. Small but steadily growing amounts of fresh water are produced by desalination (salt removal) of seawater in areas such as the arid Middle East, but in the natural world fresh water is supplied only by rain, rivers, lakes, some groundwaters, and water melted from snow or ice on land. All these waters are ultimately supplied by precipitation. Therefore, the practical limit to the amount of natural fresh water that we can ever envision using is the amount steadily supplied to the continents by precipitation. This steady supply also means that fresh water is a renewable resource. Although we may temporarily

deplete our supplies, precipitation will eventually rebuild them—although it may take hundreds or thousands of years to do so.

The division of precipitation into runoff, evaporation, and infiltration strongly affects our water supplies. We cannot use the fraction of precipitation that evaporates. We can use the fraction that infiltrates to become groundwater, but only if we can recover it by digging wells. Runoff is the most readily available fraction. The most desirable water supplies are those that are rapidly and continually replenished by runoff and infiltration.

Human activities constitute a massive interference in the operations of the natural hydrologic cycle. To mention only a few of the many significant effects:

- Evaporation is increased by the use of irrigation waters in dry areas.

- Runoff patterns are altered when water is diverted from one region to another, as in California.

- Paving that covers Earth's surface with highways, parking lots, and buildings decreases infiltration.

- Human contributions to global and local warming can lead to melting of glacial ice and changes in the balance of water in the other reservoirs.

As threats of water shortages loom, water usage enters the arena of public policy debate (see Feature 12.1). Later in this chapter, we look at some human activities that affect the quality of our water.

HYDROLOGY AND CLIMATE

For most practical purposes, local hydrology—that is, the amount of water there is in a region and the way it flows from one reservoir to another—is more important than global hydrology. The strongest influence on local hydrology is the climate, which includes both temperature and precipitation. Some of us live in warm areas where rain falls frequently throughout the year and water supplies—both at the surface and underground—are abundant. Others live in warm arid or semiarid regions where it rarely rains and water is a precious resource, or in icy climates that rely on meltwaters from snow and ice.

And some of us live in parts of the world where seasons of heavy rains, called monsoons, alternate with long dry seasons during which water supplies shrink, the ground dries out, and vegetation shrivels.

Wherever we come from, climate and the geology of the land strongly affect the amounts of water cycled from one place to another. As we deepen our understanding of global change, particularly of atmospheric warming, geologists are hard at work learning how climate affects the hydrologic cycle. They are especially interested in how climatic changes in precipitation and evaporation will affect water supplies by altering the amounts of infiltration and runoff that determine groundwater levels. Groundwaters in low-lying coastal regions may become salty as formerly fresh groundwaters are invaded by seawater, the result of a rise in sea level caused by global warming.

Humidity and Rainfall

Many differences in climate are related to the temperature of the air and the amount of water vapor it contains. The **relative humidity** is the amount of water vapor in the air, expressed as a percentage of the total amount of water that the air could hold at that temperature if saturated. When the relative humidity is 50 percent and the temperature is 15°C, for example, the amount of moisture in the air is one-half the maximum amount that the air could hold at 15°C.

Warm air can hold much more water vapor than cold air. When unsaturated warm air at a given relative humidity cools enough, it becomes supersaturated and some of the vapor condenses to form water droplets. When warm air in the atmosphere cools, the condensed water droplets form clouds. We can see clouds because they are formed of visible water droplets rather than invisible water vapor. When enough moisture has condensed in clouds and the droplets have grown so large that they are too heavy to stay suspended by air currents, they fall as rain.

Most of the world's rain falls in warm humid regions near the equator, where both the air and the surface waters of the oceans are warm. Under these conditions, a great deal of the water evaporates and humidity is high. When water-laden winds from these oceanic regions rise over nearby continents, the air cools and becomes supersaturated. The result is heavy rainfall over the land, even at great distances from the coast.

Landscape can alter patterns of precipitation, as when mountain ranges form "rain shadows," areas of low rainfall on the leeward (downwind) slopes. Moisture-filled air rising over high mountains cools and precipitates rain on the windward slopes, losing much of its moisture by the time it reaches the opposite, leeward slope (Figure 12.3). The air warms

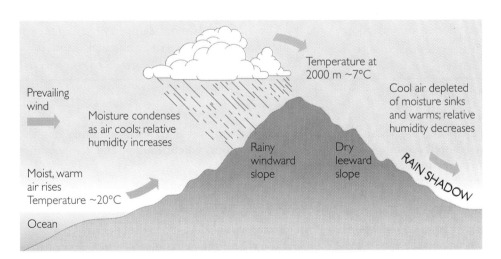

FIGURE 12.3 Rain shadows are areas of low rainfall on the leeward (downwind) slope of a mountain range. They form as prevailing winds carry moist, warm air to higher elevations, where temperatures are cooler. As the rising air cools, precipitation falls on the windward slope, leaving the leeward slope dry.

Water, a Precious Resource: Who Should Get It?

In the past, most people in the United States have taken their water supply for granted. The cost was generally low, and local, regional, and national governments assumed responsibility for quality and safety and managed supplies by constructing necessary wells, dams, and reservoirs. Scientific analyses of available supplies and user needs, however, indicate that many areas of the country will experience water shortages with increasing frequency. Accompanying these shortages will be growing conflict among the several sectors of consumers—residential, industrial, agricultural, and recreational—over who has the greatest rights to the water supply.

In recent years, widely publicized droughts and mandatory restrictions on water use, such as have occurred in California, have alerted the public that the nation faces major water problems. Public concern waxes and wanes, however, as periods of the hydrologic cycle of droughts and abundant rainfall come and go, and governments are not pursuing long-term solutions with the urgency they deserve.

Here are some facts to ponder:

- A human can survive with about 2 liters of water per day. In the United States, the per capita use for all purposes is about 6000 liters per day.

- Industry uses about 38 percent and agriculture about 43 percent of the water withdrawn from our reservoirs.

- Per capita domestic water use in the United States is two to four times greater than that of Western Europe, where users pay up to 350 percent more for their water.

- Although the states in the western United States receive one-fourth of the country's rainfall, per capita water use (mostly for irrigation) is 10 times greater than that of the eastern states, and at much lower prices. In California, for example, which imports most of its water, 85 percent of the water is for irrigation, 10 percent for municipalities, and 5 percent for industry. A 15 percent reduction in irrigation use would almost double the amounts of water available for use by cities and industries.

- The traditional ways of increasing water supply, such as building dams and reservoirs and drilling, have become extremely costly because most of the good (and therefore cheaper) sites have been used. The building of more dams to hold larger reservoirs carries environmental costs,

again as it drops to lower slopes. Because warm air can hold more total moisture before becoming saturated, the relative humidity declines, further decreasing the moisture available for rain. A rain shadow can be seen on the eastern side of the Cascade Mountains of Oregon. Moist winds blowing over the Pacific hit the mountain's western slopes, where rainfall is heavy. The eastern slopes, on the other side of the range, are dry and barren.

In contrast to tropical regions, polar climates tend to be very dry. The polar oceans are cold and so is the air, so evaporation from the sea surface is minimized and the air can hold little moisture.

Between the tropical and polar extremes are the temperate regions, where rainfall and temperatures are moderate.

Droughts

All climates can include droughts, periods of months to years when precipitation is much lower than normal. Drier regions are especially vulnerable to decreases in their water supplies during prolonged droughts. Lacking replenishment from precipitation, rivers may shrink and dry up, reservoirs may evaporate, and the soil may dry and crack while vegetation

such as flooding of inhabited areas and detrimental changes in river flows above and below dams. Factoring in these costs has led to delays or rejection of proposals for new dams.

■ Almost all the fresh water used in the United States eventually returns to the hydrologic cycle, but it may return to a reservoir that is not well located for human use, and the quality is often decreased. Recycled irrigation water often has increased salinity and is loaded with pesticides. Polluted urban waste water ends up in the oceans.

■ Global climatic change may lead to reduced rainfall in western states, exacerbating the problems there and making long-term solutions even more urgent.

Many analysts believe that the United States faces an allocation problem, not a supply problem. Some economists believe that shifting water usage from inefficient users to efficient users who contribute more to the economy might allow us to allocate water to places where it would do the most good economically. For example, 16 percent of California's water, enough to supply the needs of 30 million people, is used for subsidized irrigation

of alfalfa, a feed for horses and cattle. With different allocation policies based on market forces, more efficient users would have higher priority and would pay water suppliers more realistic market prices. Most scarce resources are sold at market prices, and according to some economists, charging market prices for water is a natural extension of this practice. In opposition to these beliefs are widely held views that emphasize the vital role of water in our personal as well as economic lives. It is important to allocate water in harmony with the needs of all of society, not just those who are able to pay steep prices.

But these conflicting views may exaggerate the difficulties. There is enough water available in the United States for the country to continue to grow economically and produce food cheaply. But institutional changes will be needed to alter our present haphazard allocation and pricing policies. New policies should promote conservation and efficiency. The decision-making process now dominated by economic considerations should include interested parties such as Native American tribes who may have treaty rights and wish to preserve traditional practices; recreational users; those interested in the environment; and, not least, scientists who understand hydrology.

dies. As populations grow, demands on reservoirs increase; a drought can deplete already inadequate water supplies.

The severest droughts in the last few decades have affected lands along the southern border of the Sahara Desert, where tens of thousands of lives have been lost to famine caused by breakdowns in agriculture and cattle grazing. As we will see in Chapter 14, this long drought has expanded the desert and effectively destroyed farming and grazing in the area.

Another prolonged but less severe drought affected most of California from 1987 until February 1993, when torrential rains broke the drought.

During the drought, groundwater and reservoirs dropped to their lowest levels in 15 years. Some control measures were instituted, but a move to reduce the extensive use of water supplies for irrigation encountered strong political resistance from farmers and the agricultural industry (see Feature 12.1).

Climates can be highly variable. The midwestern United States and parts of Canada experienced a severe but short-lived drought in 1988, when surface water supplies shrank and the Mississippi River was lowered by many feet and closed to traffic. By 1989, precipitation over the region had returned to normal.

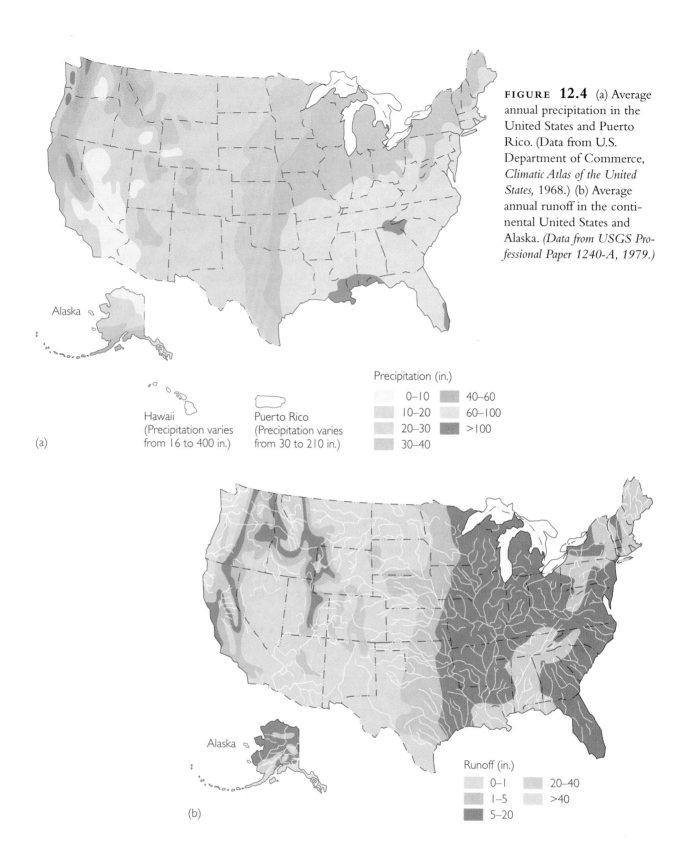

FIGURE 12.4 (a) Average annual precipitation in the United States and Puerto Rico. (Data from U.S. Department of Commerce, *Climatic Atlas of the United States,* 1968.) (b) Average annual runoff in the continental United States and Alaska. *(Data from USGS Professional Paper 1240-A, 1979.)*

Precipitation (in.)
0–10	40–60
10–20	60–100
20–30	>100
30–40	

Hawaii
(Precipitation varies from 16 to 400 in.)

Puerto Rico
(Precipitation varies from 30 to 210 in.)

(a)

Runoff (in.)
0–1	20–40
1–5	>40
5–20	

(b)

THE HYDROLOGY OF RUNOFF

A dramatic example of the connection between precipitation and local stream and river runoff can be seen when weather forecasters predict flash flooding after torrential rains. When levels of precipitation and runoff are measured over a large area, such as all of the states drained by a major river, and over a long period, such as a year, the relationship is less extreme but still strong, as the maps of precipitation and runoff shown in Figure 12.4 illustrate. When we compare them, we see that in areas of low precipitation—such as southern California, Arizona, and New Mexico—only a small fraction of precipitation ends up as runoff. In such dry regions, much of the precipitation is lost by evaporation and infiltration. In humid regions such as the southeastern United States, a much higher proportion of the precipitation runs off in rivers.

Even less obvious is the relationship between precipitation and the runoff from a large river flowing over a great distance. Such a river may carry large amounts of water from an area with high rainfall to an area with low rainfall. The Colorado River, for example, begins in an area of moderate rainfall in Colorado and then runs through an arid region in western Arizona and southern California.

Most surface runoff is carried by major rivers. The millions of small and intermediate-sized rivers carry about half of the world's entire runoff; the other half is carried by only about 70 major rivers. And almost half of that is carried by the Amazon of South America, which carries about 10 times more water than the Mississippi, the largest river of North America (Table 12.1).

Surface runoff collects and is stored in natural lakes and artificial reservoirs created by the damming of rivers. Wetland areas, swamps, and marshlands also act as storage depots for runoff (Figure 12.5). All these reservoirs, if their volumes are large enough, can absorb short-term inflows of major rainfalls,

TABLE 12.1

WATER FLOWS OF SOME GREAT RIVERS

	WATER FLOW (M³/S)
Amazon, South America	175,000
La Plata, South America	79,300
Congo, Africa	39,600
Yangtze, Asia	21,800
Brahmaputra, Asia	19,800
Ganges, Asia	18,700
Mississippi, North America	17,500

holding some of the water that would otherwise spill over the riverbanks. During dry seasons or droughts, the reservoirs continue to release water, either to streams or to water systems for human use. By smoothing out seasonal or yearly variations in runoff and releasing steady flows downstream, these reservoirs are important in flood control. For this reason, some geologists have worked to stop artificial draining of wetlands for real estate development, which could seriously damage a region's natural

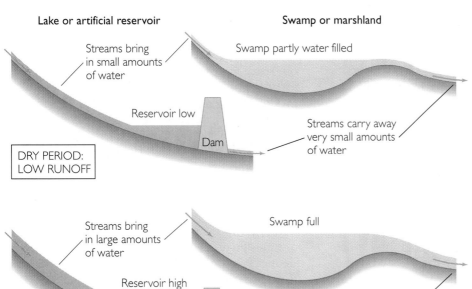

FIGURE 12.5
A swamp or marshland, like a natural lake or an artificial reservoir behind a dam, stores water during times of rapid runoff and slowly releases it during periods of little runoff.

flood regulation. Destruction of wetlands also threatens the biological diversity of many regions because wetlands are breeding grounds for a great many species of birds and invertebrates.

Wetlands are fast disappearing as the engineering of land developers continues. In the United States, more than half the original wetlands are gone. California and Ohio have kept only 10 percent of their original wetland areas. The movement to protect wetlands has spawned heated controversy. The legal definition of wetland has been argued for years and has become a political football. A 1995 scientific study of the question by the National Academy of Sciences has been attacked as "political" by opponents of regulation. Some politicians who object to regulations designed to protect wetlands have even asked for a 50 percent reduction in the extent of federally regulated wetlands.

GROUNDWATER

The enormous reservoir of groundwater stored beneath Earth's surface equals about 22 percent of all the fresh water stored in lakes and rivers, glaciers and polar ice, and the atmosphere. For thousands of years, people have drawn on this resource, either by digging shallow wells or by storing water that flows out on the surface at springs. Springs are direct evidence of water moving below the surface (Figure 12.6).

Groundwater forms as raindrops infiltrate soil and other unconsolidated surface materials, sinking even into cracks and crevices of bedrock. We tap groundwater by drilling wells and pumping the water to the surface. Well drillers in temperate climates know that they are most likely to find a good supply of water if they drill into porous sand or sandstone beds not far below the surface. Beds that store and transmit groundwater in sufficient quantity to supply wells are called **aquifers.**

How Water Flows Through Soil and Rock

If water moves into and through the ground, what determines where and how fast it flows? With the exception of caves, there are no large open spaces for pools or rivers of water underground. The only space available for water is the pore space between grains of sand and other particles that make up the

FIGURE 12.6 Groundwater exiting from a cliff; Vasey's Paradise, Marble Canyon, Grand Canyon National Park, Arizona. This is a dramatic example of a spring formed where hilly topography allows water in the ground to flow out onto the surface. *(Larry Ulrich.)*

soil and bedrock and the space in fractures. Some pores, however small and few, are found in every kind of rock and soil, but large amounts of pore space are most often found in sandstones and limestones.

You may recall from Chapter 7 that the amount of pore space in rock, soil, or sediment is its *porosity*—the percentage of its total volume that is taken up by pores. Porosity depends on the size and shape of the grains and how they are packed together. The more loosely packed the particles, the greater the pore space between the grains will be. In many sandstones, porosity is as high as 30 percent (Figure 12.7a). Minerals that cement grains reduce porosity (Figure 12.7b). The smaller the particles (Figure 12.7c) and the more they vary in shape (Figure 12.7d), the more tightly they will fit together. Porosity is higher in sediments and sedimentary rocks (10 to 40 percent) than in igneous or metamorphic rocks (as low as 1 to 2 percent). Pore space

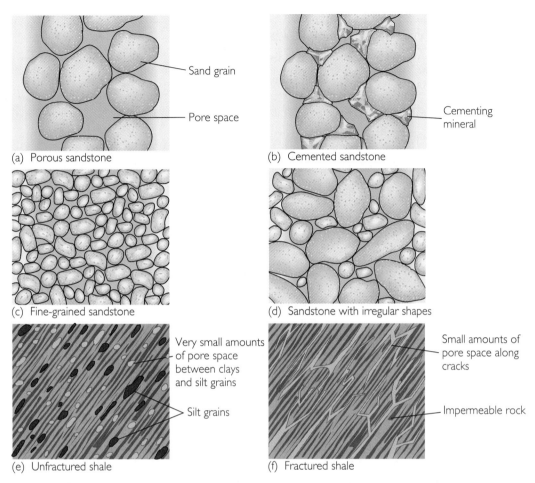

(a) Porous sandstone

Sand grain

Pore space

(b) Cemented sandstone

Cementing mineral

(c) Fine-grained sandstone

(d) Sandstone with irregular shapes

(e) Unfractured shale

Very small amounts of pore space between clays and silt grains

Silt grains

(f) Fractured shale

Small amounts of pore space along cracks

Impermeable rock

FIGURE 12.7 Pores in rocks are normally filled partly or entirely by water. (Pores in oil- or gas-bearing sandstones and limestones are filled with oil or gas.) (a) A highly porous, well-sorted sandstone has large amounts of pore space between grains. (b) A cemented sandstone has lower porosity because a cementing mineral has been precipitated in some of the pore space by fluids moving through the pores. (c) Porosity is reduced in fine-grained and poorly sorted sandstone. (d) Porosity is reduced in sandstone with irregularly shaped grains. (e) Unfractured shale has very low porosity (less than 10 percent). (f) Fractured shale has higher porosity, because some water can be stored in the fissures and cracks.

in limestones varies, depending on how many pores were created through dissolution by groundwater or during weathering. In most unfractured shales, porosity is well under 10 percent (Figure 12.7e) Fractured rocks may contain appreciable pore space—up to 10 percent of the volume—in their many cracks (Figure 12.7f). Porosity is highest of all—over 40 percent of volume—in soils and underlying loose sand and gravel layers.

Although porosity tells us how much water the rock can hold if all the pores are filled, it gives us no information about how rapidly water can flow through the pores. Water travels through a porous material by winding between grains and through cracks. The smaller the pore spaces and the more tortuous the path, the more slowly the water travels. The ability of a solid to allow fluids to pass through is its **permeability.** Generally, permeability increases as porosity increases, but not always. Permeability also depends on the sizes of the pores, how well they are connected, and how tortuous a path the water must travel to pass through the material.

TABLE **12.2**

POROSITY AND PERMEABILITY OF AQUIFER ROCK TYPES

ROCK TYPE	POROSITY (PORE SPACE THAT MAY HOLD FLUID)	PERMEABILITY (ABILITY TO ALLOW FLUIDS TO PASS THROUGH)
Gravel	Very high	Very high
Coarse- to medium-grained sand	High	High
Fine-grained sand and silt	Moderate	Moderate to low
Sandstone, moderately cemented	Moderate to low	Low
Fractured shale or metamorphic rocks	Low	Very low
Unfractured shale	Very low	Very low

Both porosity and permeability are important factors when one is searching for a groundwater supply. In general, a good groundwater reservoir will be a body of rock, sediment, or soil with both high porosity (so it can hold large amounts of water) and high permeability (so the water can be pumped from it easily). A rock with high porosity but low permeability may contain a great deal of water, but because the water flows so slowly, it is hard to pump it out of the rock. Table 12.2 summarizes the degree of porosity and permeability in various rock types.

The Groundwater Table

As well drillers bore deeper into soil and rock, the samples they bring up become increasingly wet. At shallower depths, the material is unsaturated; that is, the pores contain some air and are not completely filled with water. This level is called the **unsaturated zone**. Below it is the **saturated zone,** the level in which the pores of the soil or rock are completely filled with water. The saturated and unsaturated zones can be in either unconsolidated material or bedrock (Figure 12.8). The boundary between the two zones is the **groundwater table,** usually shortened to the "water table." When a hole is drilled below the water table, water from the saturated zone flows into the hole and fills it to the level of the water table.

Groundwater moves under the force of gravity, and so some of the water in the unsaturated zone may be on its way down to the water table. A fraction of the water, however, will remain in the unsat-

urated zone, held in small pore spaces by surface tension, the attraction between the water molecules and the surfaces of the particles. Surface tension, you may recall from Chapter 11, keeps the sand in a sandbox or on a beach moist, even though there are spaces below to which water could travel by gravity. Evaporation of water in pore spaces in the unsatu-

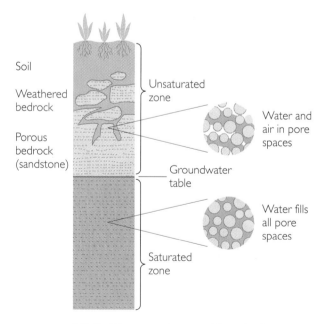

FIGURE **12.8** The groundwater table is the boundary between the unsaturated zone and the saturated zone. The saturated and unsaturated zones can be in either unconsolidated material or bedrock.

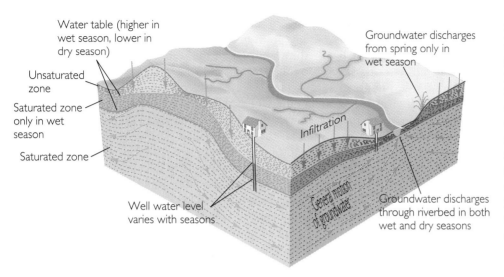

FIGURE 12.9 Dynamics of the groundwater table in permeable shallow formations in a temperate climate. Water enters the ground by infiltration of rain and melted snow and discharges at springs and rivers. The unsaturated zone varies in thickness as the water table rises in wet seasons and falls in dry seasons.

rated zone is slowed both by the effect of surface tension and by the relative humidity of the air in the pore spaces—possibly close to 100 percent.

If we were to drill wells at several sites and measure the elevations of the water levels in the wells, we could construct a map of the water table. A cross section of the landscape might look like the one shown in Figure 12.9. The water table follows the general shape of the surface topography, but the slopes are gentler. The water table is at the surface in river and lake beds and at springs. Under the influence of gravity, groundwater moves downhill from an area where the water-table elevation is high—under a hill, for example—to places where the water-table elevation is low, such as a spring where groundwater exits to the surface.

Water enters and leaves the saturated zone through recharge and discharge. **Recharge** is the infiltration of water into any subsurface formation, often by infiltration of rain or snow meltwater from the surface. Recharge also may take place through the bottom of a stream where the stream channel lies at an elevation above that of the water table (Figure 12.10). Streams that recharge groundwater in this way are called **influent streams,** and they are most characteristic of arid regions, where the water table is deep. **Discharge** is the opposite of recharge: the exit of groundwater to the surface. When a stream channel intersects the water table, water discharges from the groundwater to the stream. Such an **effluent stream** is typical of humid areas. Effluent streams continue to flow long after runoff has stopped because they are fed by groundwater. Thus, the reservoir of groundwater may be increased by influent streams and depleted by effluent streams.

During wet period

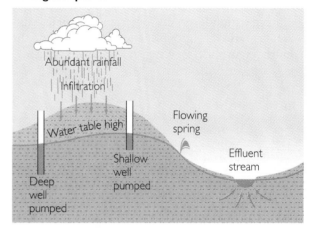

During dry period

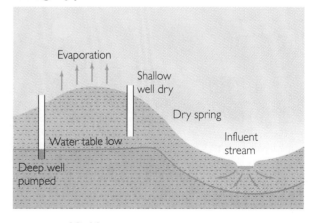

FIGURE 12.10 The depth of the water table fluctuates in response to the balance between water added from precipitation (recharge) and water lost by evaporation plus discharge from wells, springs, and streams. Streams become influent mainly in arid climates but also may do so after prolonged dry periods in temperate climates.

Artesian Flows

Groundwater may flow in unconfined or confined aquifers. In **unconfined aquifers,** the water travels through beds that extend with more or less uniform permeability to the surface, in both discharge and recharge areas. The level of the reservoir in an unconfined aquifer is the same as the height of the water table.

Many permeable aquifers, typically sandstones, are bounded above and below by shale beds of low permeability. These relatively impermeable beds are **aquicludes,** and groundwater either cannot flow through them or flows through them very slowly. When aquicludes lie both over and under an aquifer, they form a **confined aquifer,** in which water flow is contained under pressure.

The impermeable beds above a confined aquifer prevent rainwater from infiltrating downward into the aquifer. Instead, the confined aquifer is recharged by precipitation over the outcrop area; here rainwater enters the ground and travels down the aquifer (Figure 12.11). Water in a confined aquifer—known as an **artesian flow**—is under pressure. At any point in the aquifer, the pressure is equivalent to the weight of all the water in the aquifer above that point.

If a well is drilled into a confined aquifer at a point where the elevation of the ground surface is lower than that of the water table in the recharge area, the water will flow out of the well spontaneously. Such wells are called **artesian wells,** and they are extremely desirable because no energy is required to pump the water to the surface. The water is brought up by its own pressure.

In more complex geological environments, the water table may be more complicated. If there is a low-permeability clay layer—an aquiclude—in a high-permeability sand formation, the aquiclude may lie below the water table in a shallower aquifer and above the water table in a deeper aquifer (Figure 12.12). The water table in the shallower aquifer is called a **perched water table** because it is above the main water table in the lower aquifer. Many perched water tables are small lenses, but some extend for hundreds of square kilometers.

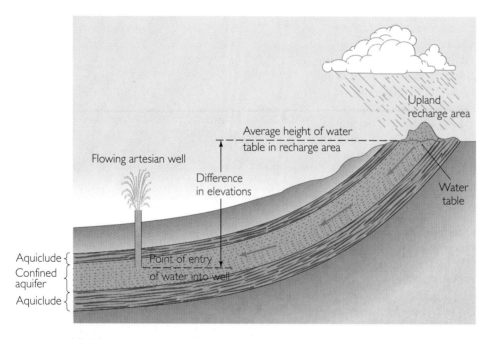

FIGURE 12.11 A confined aquifer is created when an aquifer is situated between two aquicludes (beds of low permeability). The artesian well flows in response to the difference in natural pressure (before the well was drilled) between the height of the water table in the recharge area and the bottom of the well. The actual pressure difference that governs the flow from the top of the well is the difference between the elevation of the water table and that of the top of the well. If the wellhead were as high as the water table in the recharge area, there would be no pressure difference and thus no flow.

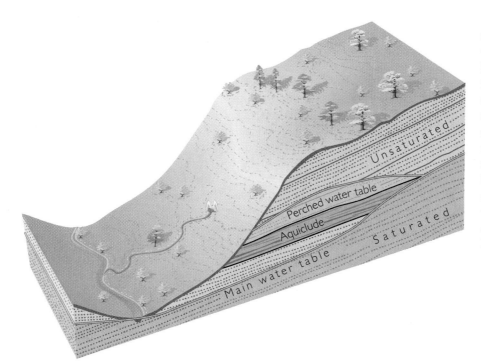

FIGURE **12.12** A perched water table is formed in geologically complex situations, in this case by a shale aquiclude located above the main water table in a sandstone aquifer. The dynamics of the perched water table's recharge and discharge may be different from those of the main water table. The main water table in this example can be recharged only from its lower outcrop slopes.

Balancing Recharge and Discharge

When recharge and discharge are balanced, the reservoir of groundwater and the water table remain constant in natural situations even though water is continually flowing through the aquifer. For recharge to balance discharge, rainfall must be frequent enough to equal the sum of the runoff from rivers and the outflow from springs and wells.

But recharge and discharge will not always be equal, because rainfall varies from season to season. Typically, the water table drops in drier seasons and rises during wet periods. A decrease in recharge, such as during a prolonged drought, will be followed by a longer term imbalance and lowering of the water table.

An increase in discharge, usually from increased well pumping, can produce the same imbalance. Shallow wells may end up in the unsaturated zone and dry up. When a well pumps water out of an aquifer faster than recharge can replenish it, the water level in the aquifer is lowered in a cone-shaped area around the well, called a cone of depression (Figure 12.13). The water level in the well is lowered to the depressed level of the water table. If the cone of depression extends below the bottom of the well, that well goes dry. If the bottom of the well is above the base of the aquifer, extending the well deeper into the aquifer may allow more water to be withdrawn, even at continued high pumping rates. If the well is deepened so much that the entire aquifer is tapped and the rate of pumping is maintained, however, the cone of depression can reach the bottom of the aquifer and deplete it. The aquifer will recover only if the pumping rate is reduced enough to give it time for recharge.

The extreme withdrawal of water not only can deplete the aquifer but may also cause another undesirable environmental effect: as the pressure of the water in pore space falls, materials formerly

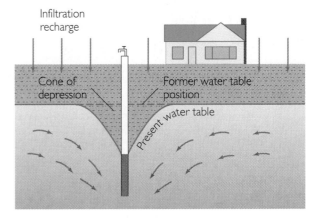

FIGURE **12.13** Excessive pumping in relation to recharge draws down the water table into a cone-shaped depression around a well. The water level in the well is lowered to the depressed level of the water table.

overlying the aquifer may be lowered, creating sink-like depressions (Figure 12.14). As water in some sediments is removed, the sediments undergo additional compaction; the loss of volume is reflected in the lowering of the surface. Subsidence attributable to this effect has occurred in Mexico City and Venice, Italy, as well as in numerous other regions of heavy pumping, such as the San Joaquin Valley in California. In these places, the rate of subsidence of the surface has reached almost 1 m every three years. Although a few experiments have attempted to reverse the subsidence by pumping water back into the groundwater system, they have not been very successful because most compacted materials do not expand easily to their former state. The best that has been done is to halt further subsidence by restricting pumping.

People who live near the ocean's edge may face a different problem when pumping rates are high in relation to recharge: the incursion of salt water into the well. Near shorelines or a little offshore, an underground boundary separates salt water under the sea from fresh water under the land. This boundary slopes down and inland from the shoreline in such a way that salt water underlies the fresh water of the aquifer (Figure 12.15a). A fresh groundwater lens under many ocean islands floats on a base of seawater. The fresh water floats because it is less dense than seawater (1.00 g/cm^3 versus 1.02 g/cm^3, a small but significant difference). Normally, the pressure of fresh water keeps the saltwater margin slightly offshore.

The balance between recharge and discharge in the freshwater aquifers maintains this freshwater-seawater boundary. As long as recharge by rainwater is at least equal to discharge by pumping, the well will provide fresh water. But if water is withdrawn faster than it is recharged, a cone of depression develops at the top of the aquifer, mirrored by an inverted cone rising from the freshwater-seawater boundary below. The cone of depression at the upper part of the aquifer makes it more difficult to pump fresh water, and the inverted cone below leads to an intake of salt water at the bottom of the well (Figure 12.15b). People living closest to the shore are the first affected. Some towns on Cape Cod in Massachusetts, on Long Island in New York, and in many other near-shore areas have had to post notices that town drinking water contains more salt than is considered healthful by environmental agencies. There is no ready solution to this difficulty other than to slow the pumping or, in some places, to recharge the aquifer artificially by funneling runoff into the ground.

You can see that a rise of sea level, which has been predicted as a result of global warming, would seriously alter the freshwater-seawater margin. As sea level rises, the margin also rises. Seawater can then invade coastal aquifers and turn fresh groundwater to salt water.

FIGURE **12.14** In Antelope Valley, California, overpumping of groundwater has led to fissures and sinklike depressions on Rogers Lakebed at Edwards Air Force Base, the landing site for the space shuttle. This fissure, formed in January 1991, is about 625 m long. *(Edwards Air Force Base.)*

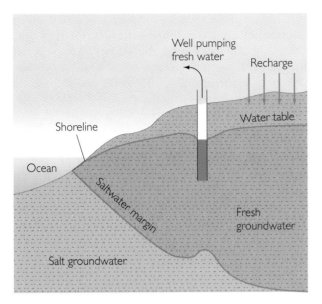

(a) Before extensive pumping

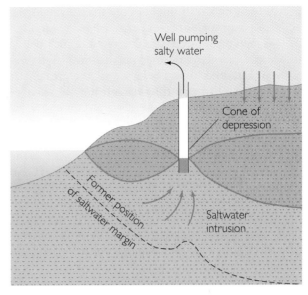

(b) After extensive pumping by many wells

FIGURE 12.15 The boundary between fresh and salty groundwater along shorelines is determined by the balance between recharge and discharge in the freshwater aquifers. (a) Normally, the pressure of fresh water keeps the saltwater margin slightly offshore. (b) Extensive pumping lowers the pressure of the fresh water, allowing the saltwater margin to move inland. This movement creates not only a cone of depression but an inverted cone of depression that brings salty water into the well. A well that formerly pumped fresh water now pumps salty water.

The Speed of Groundwater Flows

The balance between discharge and recharge is strongly affected by the *speeds* at which water moves in the ground. Most groundwaters move slowly, a fact of nature responsible for our groundwater supplies. If groundwater moved as rapidly as rivers do, aquifers would run dry after a period of time without rain, just as many small streams run dry. The slow-moving groundwater flow also makes rapid recharge impossible if groundwater levels are lowered by excessive pumping.

Although all groundwaters flow through aquifers slowly, some flow more slowly than others. In the middle of the nineteenth century, Henri Darcy, town engineer of Dijon, France, proposed an explanation for the difference in flows. While studying the town's water supply, Darcy measured the elevations of water in various wells and mapped the varying heights of the water table in the district. He calculated the distances the water traveled from well to well and measured the permeability of the aquifers. (Remember that permeability is the ease with which water passes through the pore spaces of the aquifer.) Here are his findings.

▪ For a given aquifer and distance of travel, the rate at which water flows from one point to another is directly proportional to the drop in elevation of the water table between the two points. As the difference in elevation increases, the rate of flow increases.

▪ The rate of flow for a given aquifer and given difference in elevation is inversely proportional to the flow distance the water travels. As the distance increases, the rate decreases. The ratio between the elevation difference and the flow distance is known as the **hydraulic gradient.** Just as a ball runs faster down a steeper slope than a gentler one, groundwater flows more quickly down a steeper hydraulic gradient. Groundwater in general does not run down the slope of the groundwater table but follows the hydraulic gradient of the flow, which may travel various paths below the water table.

▪ Darcy reasoned that the relationship between flow and hydraulic gradient should hold whether the water is moving through a porous sandstone aquifer or an open pipe. You might guess (correctly) that the water would move more quickly through a pipe than through the tortuous turns of pore spaces in an aquifer. Darcy recognized this factor and included a measure of permeability in his final equation, so that, other things being equal, the greater the permeability and thus the greater the ease of flow, the faster the flow.

Darcy's law, as this discovery is called, can be expressed in a simple relationship (Figure 12.16): the volume of water flowing in a certain time (Q) is proportional to the vertical drop (h) divided by the flow distance (l). The two remaining symbols are A, the cross-sectional area through which the water flows, and K, the hydraulic conductivity (a measure of permeability). (K also depends on the properties of the fluid, especially density and viscosity, which are important in dealing with fluids other than water.)

$$\frac{Q}{A} = K \times \frac{h}{l}$$

Velocities calculated by Darcy's law have been confirmed experimentally by measuring how long it takes a harmless dye introduced into one well to

reach another. In most aquifers, groundwater moves at a rate of a few centimeters per day. In very permeable gravel beds near the surface, groundwater may travel as much as 15 cm per day. This is still much slower than the speeds of 20 to 50 cm per second typical of river flows.

WATER RESOURCES FROM MAJOR AQUIFERS

Large parts of North America rely on groundwater for all their water needs. The demand on groundwater resources has grown as populations have increased and as uses such as irrigation have expanded (Figure 12.17). Many areas of the Great Plains and other parts of the Midwest rest on sandstone formations, most of which are confined aquifers like the one shown in Figure 12.11. Thousands of wells have been drilled into these formations, most of which transport waters over hundreds of kilometers and constitute a major resource. The aquifers are recharged from outcrops in the western high plains,

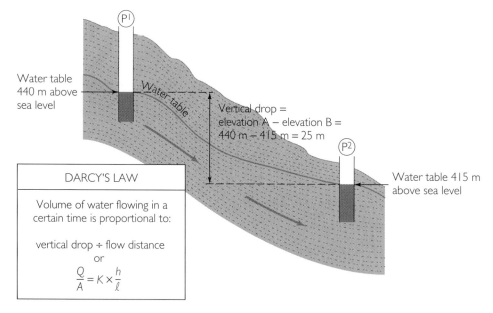

Water table 440 m above sea level

Water table

Vertical drop = elevation A − elevation B = 440 m − 415 m = 25 m

Water table 415 m above sea level

DARCY'S LAW

Volume of water flowing in a certain time is proportional to:

vertical drop ÷ flow distance

or

$$\frac{Q}{A} = K \times \frac{h}{\ell}$$

Q: Volume of water flowing in a given time

A: Cross-sectional area through which water flows

K: Hydraulic conductivity (a measure of permeability)

h: Vertical drop between two points

ℓ: Distance the flow travels

FIGURE **12.16** Darcy's law describes the rate of groundwater flow down a slope between two points, P^1 and P^2. The rate of flow is proportional to the difference in height between the high and low points of the slope (here shown as the drop in the elevation of the water table between the two points), divided by the flow distance between them (the hydraulic gradient) and K, a constant proportional to the permeability of the aquifer.

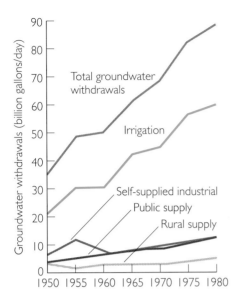

FIGURE 12.17 Billions of gallons of groundwater withdrawn in the United States, 1950–1980 (1 billion gallons per day = 3.785 × 10^6 m^3 per day). Public, industrial, and rural withdrawals rose moderately over this period; the major increase in total withdrawal came from increases in irrigation. These trends continue to the present. *(Data from USGS Water-Supply Paper 2250, 1984.)*

some very close to the foothills of the Rocky Mountains. From there, the water runs downhill in an easterly direction.

Darcy's law tells us that water flows, at rates proportional to the slopes of the aquifers, between their recharge areas and the areas of discharge from wells. In the western plains, the slopes are gentle and waters move slowly through the aquifers, recharging them at low rates. At first, many of these wells were artesian and the water flowed freely. As more wells were drilled, the water levels dropped, and the water had to be pumped to the surface. As extensive pumping withdraws water from some aquifers faster than the slow recharge from far away can fill them, the reservoirs are being depleted (see Feature 12.2).

Efforts to reduce excessive discharge have been supplemented by attempts to artificially increase recharge of aquifers in some areas. In Long Island, New York, for example, the water authority drilled a large system of recharge wells to put water into the aquifer from the surface. These wells pumped used water, which had been treated to purify it, back into the ground.

The water authority also constructed large shallow basins over natural recharge areas to catch and divert runoff to augment infiltration from surface waters, including storm and industrial waste drainage. Officials in charge of the program knew that urban development can decrease recharge by interfering with infiltration. As urbanization progresses, the impermeable materials used to pave large areas for streets, sidewalks, and parking lots prevent water from infiltrating the ground. Rainwater runoff increases, and the decrease in natural infiltration into the ground may deprive the aquifers of much of their recharge. One remedy is to catch and use the storm runoff, as the Long Island water authority did, in a systematic program of artificial recharge. The multiple efforts they expended helped rebuild the Long Island aquifer, though not to its original levels.

EROSION BY GROUNDWATER

Every year, thousands of people visit caves, either on tours of popular attractions such as Mammoth Cave, Kentucky, or in adventurous explorations of little-known caves. These underground open spaces are produced by the dissolution of limestone—or, rarely, other soluble rocks such as evaporites—by groundwater. The amounts of limestone that have dissolved to make caves may be huge. Mammoth Cave, for example, has tens of kilometers of large and small interconnected chambers, and the large room at Carlsbad Caverns, New Mexico, is more than 1200 m long, 200 m wide, and 100 m high. Limestone formations are widespread in the upper parts of the crust, but caves form only where these relatively soluble rocks are at or near the surface and enough carbon dioxide–rich water infiltrates the surface to dissolve extensive amounts of limestone.

As we saw in Chapter 6, the dissolution of limestone is enhanced by the atmospheric carbon dioxide contained in rainwater. Waters that infiltrate soils may pick up even more of the gas from the carbon dioxide given off by plant roots, bacteria, and other soil-dwelling organisms. As this carbon dioxide–rich water moves down to the water table, through the unsaturated zone to the saturated zone, it creates openings as it dissolves carbonate minerals. These openings enlarge as limestone dissolves along joints and fractures, forming a network of rooms and passages. Such networks form extensively in the saturated zone, where—because the caves are filled with water—dissolution takes place over all surfaces, including floors, walls, and ceilings.

12.2   LIVING ON EARTH

When Do Groundwaters Become Nonrenewable Resources?

For more than a hundred years, water from the Ogallala aquifer, a formation of sand and gravel, has supplied the cities, towns, ranches, and farms of western Texas and New Mexico (see map). The population of the region has climbed from a few thousand late in the nineteenth century to about a million today. The Ogallala continues to provide the irrigation water needed to support the agriculture that serves as the area's economic base, but the water pressure in the wells has declined steadily and the water table has dropped by 30 m or more.

The Ogallala aquifer of the southern plains is very slow to recharge naturally because rainfall is sparse, the degree of evaporation is high, and the recharge area is small. Pumping, primarily for irrigation, has been so extensive— about 6 billion cubic meters of water per year from 170,000 wells—that recharge cannot keep up. At the current rates of recharge, if all pumping were to stop, the water table would take several thousand years to recover its original position, with pressure restored. Some scientists have attempted artificial recharge by injecting water into the aquifer from shallow lakes that form in wet seasons on the high plains. These experiments have managed to increase recharge, but the aquifer is still in danger over the long term.

Estimates of the remaining supplies in the Ogallala indicate they will last only until the early years of the twenty-first century. As this valuable underground reservoir is drained, about 5.1 million acres of irrigated land in western Texas and eastern New Mexico will dry up—and so will 12 percent of the country's supply of cotton, corn, sorghum, and wheat and a significant fraction of the feedlots for the nation's cattle.

Other aquifers in the northern plains and elsewhere in North America are in a similar condition. In three major areas of the United States—Arizona, the high plains, and California—groundwater supplies are significantly depleted. The United States and other countries are essentially mining a nonrenewable resource in places where groundwater recharge is so slow. The water left in the aquifer steadily diminishes as we pump, just as a coal vein is depleted when it is mined.

It is likely that water-hungry regions all over the world will have to start serious water conservation measures, ranging from restrictions on water sprinklers and swimming pools to decreases in the huge amounts used for irrigation. As water use increases in relation to supply, we all must eventually adopt sensible conservation practices.

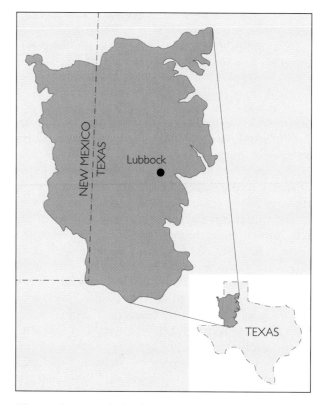

The southwestern high plains of Texas and New Mexico, underlain by the Ogallala aquifer. The blue region represents the aquifer. The general recharge area is located along the western margin of the aquifer. *(From USGS.)*

FIGURE 12.18 Chinese Theater, Carlsbad Caverns, New Mexico. Stalactites from the ceiling and stalagmites from the floor have joined to form a column. *(David Muench.)*

We can explore caves that were once dissolved below the water table but are now in the unsaturated zone as a result of a drop in the water table. In these caves, now air-filled, water saturated with calcium carbonate may drip from the ceiling. As each drop of water seeps through the cave's ceiling, some of the dissolved carbon dioxide it picked up as it passed through soils will evaporate, escaping to the cave atmosphere. As this happens, the calcium carbonate in the groundwater solution becomes less soluble, and each water droplet precipitates a small amount of calcium carbonate on the ceiling. These droplets accumulate, just as an icicle grows, in a long narrow spike of carbonate called a **stalactite** suspended from the ceiling. As the remainder of the drop hits the floor, more carbon dioxide escapes and another small amount of calcium carbonate is precipitated on the cave floor below the stalactite. These deposits also accumulate, forming a **stalagmite,** an inverted icicle-like deposit. Eventually, a stalactite and stalagmite may grow together to form a column (Figure 12.18).

Such caves may be populated by unusual, mostly eyeless, animals that are a curious by-product of the lack of light. Unusual species of bacteria, especially sulfate bacteria, have also been discovered. Some geologists think that the Carlsbad Caverns were formed partly by sulfate bacteria that produced sulfuric acid. The widespread distribution of enlarged cracks and openings in dissolved limestone terrains produces a large variation of permeability and hence high, irregularly distributed infiltration and groundwater flow rates.

In some places, dissolution may thin the roof of a limestone cave so much that it collapses suddenly, producing a **sinkhole**—a small, steep depression in the land surface above the cavernous limestone formation (Figure 12.19). Sinkholes contribute to a distinctive form of topography known as *karst,* named for a region in northern parts of the former Yugoslavia with an irregular terrain of hills and many sinkholes. **Karst topography** is characterized by sinkholes, caverns, and a lack of surface streams (Figure 12.20). Underground drainage channels replace the normal surface drainage system of small and large rivers. Short, scarce streams often end in sinkholes, detouring underground and sometimes reappearing miles away. Karst is most strongly advanced in regions with three characteristics:

- High-rainfall climate, with abundant vegetation (providing carbon dioxide–rich waters)
- Extensively jointed limestone formations
- Appreciable hydraulic gradients

Karst terrains are the scene of a number of environmental problems, including surface subsidence from collapse of underground space and potentially catastrophic cave-ins and sinkhole formations. In North America, karst topography is found in limestone terrains of Indiana and Kentucky and in the Yucatan Peninsula of Mexico. Karst is well developed

FIGURE 12.19 A large sinkhole formed by the collapse of a shallow underground cavern. Such collapses can occur so suddenly that moving cars are buried. Winter Park, Florida. *(Leif Skoogfors/Woodfin Camp.)*

on uplifted coral limestone of late Cenozoic terrains of tropical volcanic island arcs.

WATER QUALITY

Most residents of Canada and the United States take a supply of fresh, pure water for granted. But a growing number of people, fearful of contaminants in their water, are buying bottled spring water or installing home purifying systems. Almost all water supplies in North America are free of bacterial contamination, and the vast majority are chemically pure enough to drink safely. But earlier, we noted the problem some shoreline communities face, where heavy pumping has caused an incursion of seawater and unacceptable levels of sodium in the water. A more common problem exists in places where toxic wastes have polluted rivers and infiltrated aquifers from surface dumps.

Lead is a well-known pollutant derived from industrial processes. It is routinely eliminated from public water supplies by chemical treatment before the water is distributed through the water mains, yet the water supply systems inside private homes can add lead to water. In many older neighborhoods, where lead pipes are still common, some lead contamination occurs from old water mains or from lead pipes. In Boston, an astonishing 41 percent of 174 tap-water samples taken first thing in the morning had unsafe lead levels. Even in newer construction, the lead solder used to connect copper pipes and the metals used in faucets are small sources. Replacing old lead mains and substituting durable plastic pipe can reduce lead contamination. Even a simple practice helps: let the water run for a few minutes to clear pipes of waters in which they soaked overnight.

No easy solutions can rid the Earth of contaminants from radioactive waste. One of the major drawbacks to burying radioactive waste underground is that the radioactive material might be leached by groundwaters and find its way into water supply aquifers. Leakage into shallow groundwaters has already taken place from storage tanks and burial sites at the atomic weaponry plants in Oak Ridge, Tennessee, and Hanford, Washington.

Aside from toxic pollution, some groundwaters, although perfectly healthful to drink, simply taste bad. Some have a disagreeable taste of "iron" or are slightly sour. Other waters taste fine, but they are "hard"—it is difficult to make lather or soapsuds with them. Hard waters contain relatively large amounts of dissolved calcium carbonate and usually some magnesium carbonate, harmless to drink but not conducive to soaping and laundering. How do these differences in taste and quality arise in "safe" drinking waters?

Many of the highest quality, best tasting public water supplies come from lakes and artificial surface

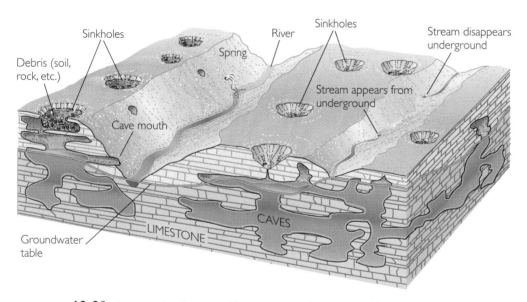

FIGURE 12.20 Some major features of karst topography: caves, sinkholes, and disappearing streams. Depending on the depth of the groundwater table, some caves may be wholly in the saturated zone and filled with water. Others may be partially filled with water, and shallow caves above the groundwater table may be empty.

reservoirs, many of which are simply collecting places for rainwater. But some groundwaters taste just as good. In general, these tend to be waters that pass through rocks that weather only slightly. Sandstones made up largely of quartz, for example, contribute little in dissolved substances, and thus waters passing through them taste fresh.

Other groundwaters pick up so-called impurities from natural sources, such as the ions dissolved from minerals weathering at the surface and small amounts of organic compounds dissolved from plants and animals (Figure 12.21). It is this load of dissolved substances that makes groundwaters "hard" or "soft," gives them their chemical composition, and contributes unpleasant taste. Waters that pass through waterlogged soils containing aromatic organic compounds and hydrogen sulfide, for example, may dissolve enough of these compounds to acquire a disagreeable taste. In greater quantities, some of these dissolved substances can be toxic.

Is the Water Drinkable?

Water that is agreeable to taste and is not dangerous to health is called **potable water.** The amounts of dissolved substances in potable waters are very small, usually measured by weight in parts per million (ppm). Potable groundwaters of good quality typically contain about 150 ppm total dissolved materials, since even the purest natural waters contain some dissolved substances derived from weathering. Only distilled water contains less than 1 ppm dissolved substances.

The many cases of groundwater contamination have led to the establishment of water quality standards based on medical studies. These studies have concentrated on the effects of ingesting average amounts of water containing various quantities of contaminant elements and compounds. For example, the Environmental Protection Agency has set the maximum allowable concentration of arsenic, whose poisonous nature is well known, at 0.05 ppm.

Groundwater is almost always free of solid particles when it seeps into a well from a sand or sandstone aquifer. The tortuous passageways of the rock or sand act as a fine filter for flows, removing small particles of clay and other solids and even straining out bacteria and large viruses. Limestone aquifers may have larger pores and so may filter less efficiently. Bacterial contamination at the bottom of a well has most often been introduced either from the surface by the pump materials or from nearby underground sewage disposal, often when septic tanks are located too close to the well.

We have learned in the last few decades that, contrary to expectations, bacteria can and do live at great depths (up to several thousand meters) in groundwaters, although there is no evidence that they are mobile. Most of these bacteria are supported by nutrients from organic matter buried with the original sediment, but geologists recently discovered bacteria that draw energy from hydrogen in rocks. The hydrogen is generated by geochemical reactions between groundwater and rocks such as basalt. These reactions, aside from serving as a source of energy for the bacteria, continue the weathering process underground.

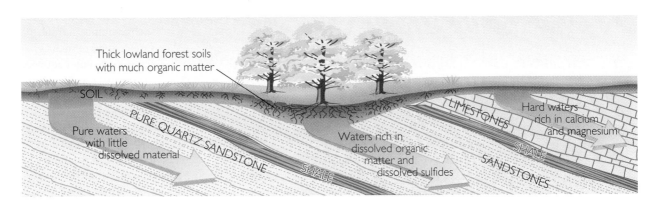

FIGURE 12.21 Groundwater picks up natural dissolved materials as it passes through rock and unconsolidated soils and sediments. Groundwaters passing through limestone dissolve carbonate minerals and carry away calcium, magnesium, and bicarbonate ions, making the water "hard." Pure quartz sandstones resist weathering and contribute almost no dissolved materials. Water coming through waterlogged forest or swampy soils may contain dissolved organic compounds and hydrogen sulfide.

Contamination of the Water Supply

Human activities can contaminate groundwaters (Figure 12.22). Chemical industry waste lagoons, leaking chemical storage barrels, sanitary landfill operations, and city garbage dumps can introduce hazardous or foul-tasting contaminants. Buried gasoline storage tanks can leak, and road salt inevitably drains into the soil and ultimately into aquifers. Rain can wash agricultural pesticides, herbicides, and fertilizers into the soil. From the soil, they percolate downward into aquifers. In some agricultural areas where nitrate fertilizers are heavily used, groundwaters may contain high quantities of nitrate. In one recent study, 21 percent of the shallow wells sampled for drinking water exceeded the maximum amounts (10 ppm) allowed in the United States. Such high nitrate levels pose a danger of "blue baby" syndrome (inability to maintain healthy oxygen levels) for infants 6 months and younger. As mentioned earlier, radioactive waste stored in steel tanks at the atomic weapons works at Hanford, Washington, has leaked into the ground and threatened the groundwater supply (see Feature 22.2). Keeping groundwaters free of the wide range of potential contaminants is becoming steadily more important as the demand for water increases and supplies shrink.

Septic tanks, widely used in some areas that lack full sewer networks, are settling tanks buried at shallow depths in which the solid wastes from house sewage are decomposed by bacteria. In a properly designed system, as the sewage water flows through and into the soil, any harmful bacteria and bits of solid waste and sludge are filtered out. Septic tank waters may also contain dissolved materials such as phosphate, nitrate, and toxic metals that cannot be filtered. Avoiding contamination of the water requires proper placement and drainage of septic tanks at sufficient distance from water wells in shallow aquifers.

Can we reverse the contamination of water supplies? The answer is a qualified yes, but the process is costly and very slow. The faster an aquifer recharges, the easier it will be to clean. If the recharge is fast, once we close off the sources of contamination, fresh water moves into the aquifer, and in a short time the water recovers its quality. Even fast recoveries, however, may take a few years. Contamination of slowly recharging reservoirs is more serious, because the rate of groundwater movement may be so slow that contamination from some distance may take a long time to show up. By the time it does, it is too late for rapid recovery. Even with cleaned-up recharge, some contaminated deep reservoirs hundreds of kilometers away from the recharge area may not respond until many decades later. When public water supplies are polluted by contaminants, we can pump the water and then treat it chemically to make it safe, but this is an expensive procedure. Alternatively, we can try to treat it while it remains under-

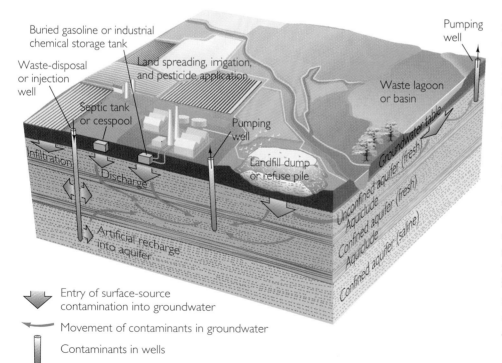

Buried gasoline or industrial chemical storage tank

Waste-disposal or injection well

Land spreading, irrigation, and pesticide application

Pumping well

Septic tank or cesspool

Pumping well

Waste lagoon or basin

Infiltration

Discharge

Landfill dump or refuse pile

Groundwater table

Unconfined aquifer (fresh)

Aquiclude

Confined aquifer (fresh)

Aquiclude

Confined aquifer (saline)

Artificial recharge into aquifer

⬇ Entry of surface-source contamination into groundwater

→ Movement of contaminants in groundwater

▮ Contaminants in wells

FIGURE 12.22 Human activities can contaminate groundwater. Contaminants from surface sources such as dumps and subsurface sources such as septic tanks enter aquifers via normal groundwater flow. Contaminants may be introduced to water supplies through pumping wells. Waste-disposal wells are designed to pump contaminants into deep saline aquifers, but they may accidentally leak into freshwater aquifers above. (Modified from U.S. Environmental Protection Agency.)

ground. One experimental and moderately successful procedure along these lines funneled contaminated water into a buried bunker full of iron filings that would react with contaminants and thus detoxify the waters. Contamination and recovery of aquifers is a problem for relatively shallow, fresh groundwaters. But are there deeper groundwaters that we could use?

WATER DEEP IN THE CRUST

All rocks below the groundwater table are saturated with water. Even in the deepest wells drilled for oil, some 8 or 9 km deep, geologists always find water in permeable formations. At these depths, waters move so slowly—probably less than a centimeter per year—that they have plenty of time to dissolve even very insoluble minerals from the rocks they pass through. Thus, dissolved materials become more concentrated in these waters than in near-surface waters, rendering them unpotable. For example, groundwaters that pass through salt beds, which are quick to dissolve, tend to become greatly enriched in sodium chloride.

At depths greater than 12 to 15 km, deep into the basement igneous and metamorphic rocks that everywhere underlie the sedimentary formations of the upper part of the crust, porosities and permeabilities are greatly reduced. Even these basement rocks are saturated, although the total amounts of water are extremely small because the porosity, distributed along small cracks and the boundaries between crystals, is so low (Figure 12.23). In some deeper regions of the crust that are undergoing active metamorphism, such as along subduction zones, hot concentrated waters play an important role in the chemical reactions by which metamorphic rocks are made, helping to dissolve some minerals and precipitating others (see Chapter 8). Even some mantle rocks are presumed to have very minute quantities of water.

Hydrothermal Waters

Natural hot springs occur in Yellowstone National Park in northwestern United States; Hot Springs, Arkansas; Banff Sulfur Springs, Alberta; Reykjavík, Iceland; and many other places in the world. In such locations, hydrothermal waters—hot waters deep in the crust—migrate rapidly upward without losing much heat and emerge at the surface, sometimes at boiling temperatures.

Hydrothermal waters are loaded with chemical substances dissolved from rocks at high tempera-

tures. As long as the water remains hot, the dissolved material can remain in solution. But as hydrothermal waters coming to the surface quickly cool, they may precipitate various minerals, such as opal (a form of silica) and calcite or aragonite (forms of calcium carbonate). Crusts of calcium carbonate that form at some hot springs build up to form the rock travertine, prized for its beauty as a polished stone used for buildings and tables. While still in the crust, hydrothermal waters are also responsible for depositing some of the world's richest metallic ores as they cool after migration, as we will see in Chapter 23.

Most of the hydrothermal waters of the continents derive from surface waters that percolated

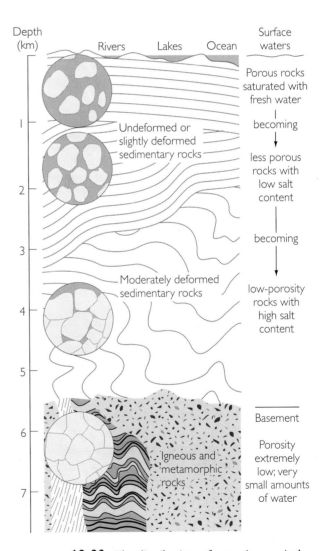

FIGURE 12.23 The distribution of water in a typical section of continental crust. Most water is at the surface or in sedimentary rocks buried at shallow depths. Porosity and water content generally decrease with increasing depth and greater structural deformation.

downward to deeper regions of the crust. These surface waters originate primarily as **meteoric waters**—rain, snow, or other forms of water from the atmosphere (from the Greek *meteōron,* "phenomenon in the sky," which also gives us the word *meteorology*). Meteoric waters may be very old; it has been determined that the water at Hot Springs, Arkansas, derives from rain and snow that fell more than 4000 years ago and slowly infiltrated the ground.

The other source of hydrothermal waters is water that escapes from a magma. In areas of igneous activity, sinking meteoric waters encounter hot masses of rocks, become heated, and then mix with water released from the nearby magma. The mixture of hydrothermal water then returns to the surface as hot springs or geysers (Figure 12.24). Whereas hot springs flow steadily, geysers erupt hot water and steam intermittently.

The theory explaining the intermittent eruptions of geysers is an example of geological deduction, for the dynamics of the underground hot water system is hidden from sight hundreds of meters below the surface. It is likely that geysers are connected to the surface by a system of very irregular and crooked fractures, recesses, and openings, in contrast to the more regular and direct plumbing of hot springs. The irregularity of the geyser plumbing, by sequestering some water in recesses, helps to prevent the bottom waters from mixing with shallower waters and therefore cooling. As the bottom waters heat up from contact with hot rock, they reach the boiling point, and steam starts to ascend and heat up the shallower waters. An eruption is triggered, followed by a return to quiet as the fractures slowly and irregularly refill with water.

Other hot springs come from meteoric waters that move downward into deep sedimentary rock formations, where they are heated by the normal increase in temperature with depth, and then return as hydrothermal waters to the surface. Many metallic ores and other mineral deposits in sedimentary rocks far from any igneous activity originated in this way.

In the search for new and clean sources of energy, geologists have turned to hydrothermal waters. Northern California, Iceland, Italy, and New Zealand have already harnessed the steam generated by hydrothermal activity in hot springs and geysers, using it to drive electricity-generating turbines. Hydrothermal waters may be put to wider use for power (see Chapter 22).

However important hydrothermal waters may be for power generation, ore deposits, and their supposed healing qualities, these waters do not contribute to surface water supplies, primarily because they contain so much dissolved material.

The Usable Waters of the Earth

This survey of water in and on the Earth leads us to an inescapable conclusion: the accessible and usable waters of the Earth are limited to the surface and the near-surface parts of the crust. Most of the water is in surface reservoirs: oceans, rivers, lakes, and glacial

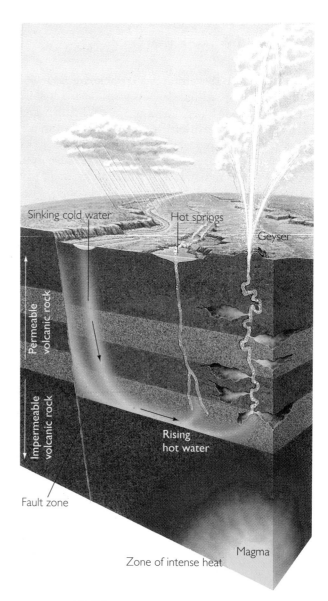

FIGURE 12.24 Circulation of water over a magma body produces geysers or hot springs. Cold rainwater soaks into the soil and filters down through permeable rocks. As it approaches the magma, it heats up and becomes less dense, thus setting up a circulation system that returns it to the surface. Hot springs rise more or less directly, while a geyser follows a much more irregular network of pores and cracks, which complicates the flow of water and leads to boiling and the production of steam and eruptions at intervals.

ice. The rest is stored in rocks, most of it at depths of not much more than 10 to 15 km. Although there are enormous quantities of fresh water stored in glacial ice, melting and transporting it has not proved practical so far. Thus, the overwhelming majority of the world's *usable* fresh waters are in the ground rather than on the surface. Over the long term, we must depend on the total precipitation that falls on the continents to replenish the water in all reservoirs. The rain is both the ceiling for our water supply and our standard for purity.

SUMMARY

How does water move around and in the Earth in the hydrologic cycle? Water moves in such a way that a constant balance is maintained among the major reservoirs of water at or near the Earth's surface: oceans, lakes, and rivers; glaciers and polar ice; and groundwater. Water is transferred to the atmosphere by evaporation from the oceans, by evaporation and transpiration from the continents, and by sublimation from glaciers. It is returned from the atmosphere to the oceans and continents by precipitation as rain and snow. Part of the precipitation that falls on land is returned to the ocean by runoff in rivers, and the remainder infiltrates the ground to become groundwater. Local variations in the evaporation-precipitation-runoff-infiltration balance arise because of differences in climate.

How does water move below the ground? Groundwater forms as rain infiltrates the surface of the ground and travels through pore spaces in the soil, sediment, or rock that serves as an aquifer. Water moves through the upper, unsaturated zone through the water table into the saturated zone. Groundwater moves downhill under the influence of gravity, eventually emerging at springs, where the water table intersects the ground surface. Over the long term, a groundwater aquifer is in dynamic balance between recharge and discharge. Groundwater may flow in unconfined aquifers, which are continuous to the surface, or in confined aquifers, which are bounded by aquicludes. Confined aquifers produce artesian flows and spontaneously flowing artesian wells. Darcy's law describes the groundwater flow rate in relation to the slope of the water table and the permeability of the aquifer.

What factors govern our use of groundwater resources? As population has grown in many areas, demand for groundwaters has greatly increased, particularly where irrigation is widespread. Many aquifers, such as those of the western plains of North America, have such slow recharge rates that continued pumping over the past century has reduced the pressure in artesian wells. As pumping discharge continues to be out of balance with recharge, such aquifers are being depleted, and there is no prospect of renewal for many years. Artificial recharge may help renew some aquifers, but conservation will be required to preserve others. Contamination of groundwater by sewage and industrial effluents reduces the potability of some waters and limits our resources.

What geological processes are affected by groundwater? Erosion by groundwater in humid limestone terrains produces karst topography, with caves, sinkholes, and disappearing streams. Heating of downward-percolating meteoric waters by magma bodies leads to a circulation that brings hydrothermal waters to the surface as geysers and hot springs. At great depths in the crust, more than 12 to 15 km, dense rocks have low porosities and hence contain extremely small quantities of water.

KEY TERMS AND CONCEPTS

hydrology (p. 287)
groundwater (p. 288)
reservoir (p. 288)
hydrologic cycle (p. 289)
infiltration (p. 289)
transpiration (p. 290)
runoff (p. 290)
relative humidity (p. 291)
aquifer (p. 296)
permeability (p. 297)
unsaturated zone (p. 298)

saturated zone (p. 298)
groundwater table (p. 298)
recharge (p. 299)
influent stream (p. 299)
discharge (p. 299)
effluent stream (p. 299)
unconfined aquifer (p. 300)
aquiclude (p. 300)
confined aquifer (p. 300)
artesian flow (p. 300)
artesian well (p. 300)

perched water table (p. 300)
hydraulic gradient (p. 303)
Darcy's law (p. 304)
stalactite (p. 307)
stalagmite (p. 307)
sinkhole (p. 307)
karst topography (p. 307)
potable water (p. 309)
meteoric waters (p. 312)

EXERCISES

1. What are the main reservoirs of water at or near the surface of the Earth?

2. How do mountains form rain shadows?

3. What is an aquifer?

4. What is the difference between the saturated and unsaturated zones of groundwater?

5. How do aquicludes make a confined aquifer?

6. How are recharge and discharge balanced to make a groundwater table stable?

7. How does Darcy's law relate groundwater movement to permeability?

8. How does dissolution of limestone relate to karst topography?

9. What are the sources of water in hot springs?

10. What are some common contaminants in groundwater?

11. Define the groundwater table.

THOUGHT QUESTIONS

1. If the Earth warmed so that evaporation from the oceans greatly increased, how would the hydrologic cycle of today be altered?

2. If you lived near the seashore and started to notice a slight salty taste to your well water, how would you explain the change in water quality?

3. Why would you recommend against extensive development and urbanization of the recharge area of an aquifer that serves your community?

4. If it were discovered that radioactive waste had seeped into groundwater from a nuclear processing plant, what kind of information would you need to predict the length of time before the radioactivity appeared in well water 10 km from the plant?

5. What geological processes would you infer are occurring at depth at Yellowstone National Park, which has many hot springs and geysers?

6. Why should communities ensure that septic tanks are maintained in good condition?

7. Why are more and more communities in cold climates restricting the use of salt to melt snow and ice on highways?

8. Your new house is built on soil-covered granite bedrock. Although you think that prospects for drilling a successful water well are poor because of the granite, the well driller familiar with the area says he has drilled many good water wells in this granite. What arguments might each of you offer to convince the other?

9. How might the hydrologic cycle have been quantitatively different during the maximum glaciation of 19,000 years ago, when a good deal of the continents was covered by ice?

10. You are exploring a cave and notice a small stream flowing on the cave floor. Where could the water be coming from?

LONG-TERM TEAM PROJECT

See Chapter 11.

SHORT-TERM TEAM PROJECT: FLOODING

In the U.S. Senate and House of Representatives, recent college graduates work as legislative assistants. Their main job is to brief a senator or a representative about pressing issues on which the official must cast an informed vote.

You and a partner are legislative assistants to a U.S. senator from South Dakota. The senator has been contacted by constituents in Pierre, S.D., which is located on the east bank of the south-flowing Missouri River and just south of the Oahe dam. The residents of Pierre have had repeated problems with winter flooding. The Army Corps of Engineers says that the flooding occurs because the Bad River, a tributary that flows across heavily farmed and cattle-grazed land into the Missouri River, picks up large quantities of sediment and dumps it where the two rivers join, making this area quite shallow. During the winter, thick ice forms on the slow-moving Missouri River, and liquid water must flow through a thin horizontal zone (parallel to the channel bottom) between the accumulated sediment and the thick ice. When water is released from above the dam in order to generate electricity, it backs up at the narrow passage and floods Pierre. The Corps' only suggestion to prevent the flooding is to build levees along the banks of the Missouri. The senator wants to know:

1. Is the sediment from the Bad River really the problem? Why or why not?

2. What might be done to curtail the sedimentation?

3. Would the construction of levees be a long-term solution to the problem? Why or why not?

4. Would simply dredging out the accumulated sediment in order to deepen the Missouri River channel provide a long-term solution?

5. What other solutions are there besides levee construction or dredging?

6. If the Corps has been able to control the flooding along the Mississippi River, doesn't that mean that it could deal with this problem up in South Dakota?

The senator wants these questions answered in a concise two- to three-page memo. In order to thoroughly prepare the senator to deal with the issue, your team should investigate the nature and quality of work undertaken by the Army Corps of Engineers and understand the geologic nature of the problem near Pierre.

SUGGESTED READINGS

Dolan, R., and H. G. Goodell. 1986. Sinking cities. *American Scientist* 74:38–47.

Dunne, Thomas, and Luna B. Leopold. 1978. *Water in Environmental Planning.* San Francisco: W. H. Freeman.

Frederick, K. D. 1986. *Scarce Water and Institutional Change.* Washington, D.C.: Resources for the Future.

Freeze, R. Allen, and John A. Cherry. 1979. *Groundwater.* Englewood Cliffs, N.J.: Prentice-Hall.

Heath, R. C. 1983. *Basic Groundwater Hydrology.* U.S. Geological Survey Water-Supply Paper 2220.

Jennings, J. N. 1983. Karst landforms. *American Scientist* 71:578–586.

Leopold, Luna B. 1974. *Water: A Primer.* San Francisco: W. H. Freeman.

National Research Council. 1993. *Solid-Earth Sciences and Society.* Washington, D.C.: National Academy Press.

U.S. Geological Survey. 1990. *Hydrologic Events and Water Supply and Use.* National Water Summary 1987, U.S. Geological Survey Water-Supply Paper 2350.

INTERNET SOURCES

EcoNet's Seas & Waters Directory
ⓘ http://www.igc.apc.org/igc/www.water.html
This gateway leads to EcoNet's Acid Rain Resources, Water Gopher, and Toxics, Hazards and Wastes Resources. Additional links to related sites for rivers, groundwater, aquatic conservation, water resources for cities, and marine and estuarine environmental issues are provided.

The Edwards Aquifer
ⓘ http://eardc.swt.edu/Edwards-info.html
Southwest Texas University supplies this excellent overview of the Edwards Aquifer, an important aquifer system in southern Texas. The site also provides good general information on hydrogeology, along with simulations and statistics.

National Climatic Data Center
ⓘ http://www.ncdc.noaa.gov/
The home page of the National Climatic Data Center is a source of climatic data for the entire United States and includes links to worldwide data. Access is provided to statistics on precipitation, temperature, atmospheric pressure, and climate models.

National Water Data Exchange
ⓘ http://h2o.usgs.gov/public/nawdex/nawdex.html
The National Water Data Exchange (NAWDEX) is a voluntary association of public and private organizations working to "improve the access to water data and water information." This site provides information on how to obtain data compiled by NAWDEX and links to on-line data.

Water Resources of the United States
ⓘ http://h2o.usgs.gov/
This site is a gateway for all U.S. Geological Survey water-related data. Groundwater data for many areas may be retrieved and real time hydrologic data from selected sites are linked to this Web page.

13

Stream flowing through a forest in Germany. Streams constantly change in depth, shape, and velocity in response to geological changes. *(Hans Reinhard/ Bruce Coleman.)*

Rivers: Transport to the Oceans

Rivers are the major geological agents operating on the surface of the land. As they erode bedrock and transport and deposit sand, gravel, and mud, streams of all sizes, from tiny rills to major rivers, are preeminent carvers of the landscape. In this chapter, we focus on how rivers accomplish their geological work: how water flows in currents; how currents carry sediment; how streams break up and erode solid rock; and, on a larger scale, how streams carve valleys and assume a variety of forms as they channel water downstream. Rivers are deeply embedded in the imagery of language. We picture rivers when we speak of flowing waters, babbling brooks, raging torrents. Our everyday language uses many different words to describe waterways, but geologists give more precise meanings to some of these terms. We reserve the word **stream** for any flowing body of water, large or small, and **river** for the major branches of a large stream system.

Almost every town or city in most parts of the world has some body of water flowing through it, and these streams have served as commercial waterways for barges and steamers and as water resources for resident populations and industries. The river Nile, for example, was vital to the agricultural economy of ancient Egypt and remains important to Egypt today. But living near a river also entails risk. When rivers flood, they destroy lives and property, sometimes on a huge scale.

Streams cover the greatest part of the Earth's surface and play the major role in shaping the face of the continental landscape. They erode mountains, carry the products of weathering down to the oceans, and deposit billions of tons of sediment along the way in bars and flood deposits. At their mouths, on the edges of the continents, they dump even greater quantities of sediment, building new land out into the oceans.

Worldwide, rivers carry about 16 billion tons of clastic sediment and an additional 2 to 4 billion tons of dissolved matter each year. Humans are responsible for much of the current river load. According to some estimates, prehuman sediment transport was about 9 billion tons per year, less than half the present value. Through agriculture and accelerated erosion, humans increase rivers' sediment load in some places, and through the construction of dams, which trap sediment behind the retaining walls, we lower the sediment load in other places.

Rivers carry back to the sea the bulk of the rainwater that falls on land, completing the hydrologic cycle. We begin by examining how running water moves in currents and how that movement enables streams to carry various kinds of sediment.

How Stream Waters Flow

All streams, large and small, share some basic characteristics of flowing fluids. We can illustrate two kinds of fluid flow using lines of motion called *streamlines*. In **laminar flow,** the simplest kind of movement, straight or gently curved streamlines run parallel to one another without mixing or crossing between layers (Figure 13.1a). The slow movement of thick syrup over a pancake, with strands of unmixed melted butter flowing in parallel but separate paths, is a laminar flow. **Turbulent flow** has a more complex pattern of movement, with streamlines mixing, crossing, and forming swirls and eddies (Figure 13.1b). Fast-moving river waters typically show this kind of motion. Turbulence—the degree to which there are irregularities and eddies in the flow—may be low or high. Whether a flow is laminar or turbulent depends on three factors:

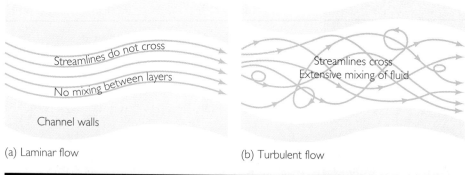

(a) Laminar flow

(b) Turbulent flow

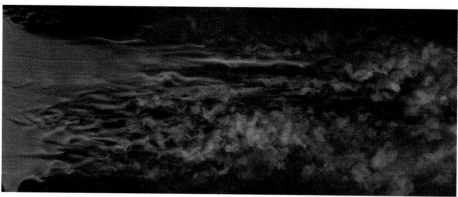

(c)

FIGURE **13.1** (a) In laminar flows, straight or gently curved streamlines run parallel without mixing or crossing. (b) In turbulent flows, streamlines mix, cross, and form swirls and eddies. (c) The transition from laminar to turbulent flow in water along a flat plate, revealed by injection of a dye. *(ONERA.)*

- Its velocity

- Its geometry (primarily its depth)

- Its **viscosity,** which is a measure of a fluid's resistance to flow. The more viscous (the "thicker") a fluid is, the more it resists flow. The higher the viscosity, the greater the tendency for laminar flow.

Viscosity arises from the attractive forces between the molecules of a fluid; these forces tend to impede the slipping and sliding of molecules past one another. The greater the attractive forces, the greater the resistance to mixing with neighboring molecules and the higher the viscosity. A cold syrup or a high-viscosity cooking oil, for example, is sluggish and laminar when it is poured. The viscosity of most fluids, including water, decreases as the temperature increases. Given enough heat, a fluid's viscosity may decrease sufficiently to change a laminar flow to a turbulent one.

Water has low viscosity in the common range of temperatures at the Earth's surface, and for this reason alone, most watercourses in nature tend to turbulent flow. In addition, the rapid movement of water in most streams makes them turbulent. In nature, we are likely to see laminar flows of water only in thin sheets of rain runoff flowing slowly down nearly level slopes and, in cities, in small flows in street gutters. Because most streams and rivers are broad and deep and their velocities high, their flows are almost always turbulent.

A stream may show turbulent flow over much of its width but be in laminar flow along its edge, where the flow is shallow and slow. The flow velocity is highest near the center of the stream, and we commonly refer to a rapid flow as a strong current.

STREAM LOADS AND SEDIMENT MOVEMENT

Erosion and Transport

Fluid flows vary in their ability to erode and carry sand grains and other sediment. Laminar flows of water can lift and carry only the smallest, lightest, clay-sized particles. Turbulent flows, depending on their speed, can move particles from clay size up to pebbles and cobbles. As turbulence lifts particles from the bed into the flow, it carries them downstream. It also rolls and slides larger particles along the bottom. A stream's **suspended load** includes all the material temporarily or permanently suspended

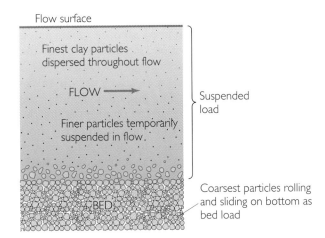

FIGURE 13.2 A current flowing over a bed of sand, silt, and clay transports particles in two ways: as bed load, the material sliding and rolling along the bottom; and as suspended load, the material temporarily or permanently suspended in the flow itself.

in the flow. Its **bed load** is the material the stream carries along the stream bottom by sliding and rolling (Figure 13.2).

The faster the current, the larger the particles carried as suspended and bed load. The ability of a flow to carry material of a given size is its **competence.** As the current increases in velocity and coarser particles are suspended, the suspended load grows. At the same time, more of the bed material is in motion, and the bed load also increases. As we would expect, the larger the volume of a flow, the more suspended and bed load it can carry. The total sediment load carried by a flow is its **capacity.**

Interactions between the velocity and volume of a flow affect both the competence and capacity of the stream. Along most of its length, the Mississippi River flows at moderate speeds and carries only fine to medium-sized particles, clay to sand, but it carries huge quantities of them. A small, steep, fast-flowing mountain stream, in contrast, may carry boulders, but only a few of them.

Settling from Suspension

A stream's ability to carry sediment depends on a balance between the uplifting forces that turbulence exerts on particles and the competing downward pull of gravity, which makes grains settle out of the current and become part of the bed. The speed with which suspended particles of various weights settle to the bottom is called the **settling velocity.** Small grains of silt and clay are easily lifted into the stream and settle slowly, so they tend to stay in suspension.

The settling velocity of larger particles, such as medium- and coarse-grained sand, is much faster. Most larger grains therefore stay suspended in the current only a short time before they settle.

The typical movement of sand grains is **saltation**—an intermittent jumping motion along the streambed. The grains are sucked up into the flow by turbulent eddies, move with the current for a short distance, and then fall back to the bottom (Figure 13.3). If you were to stand in a rapidly flowing sandy stream, you might see a cloud of saltating sand grains moving around your ankles. The bigger the grain, the longer it will tend to remain on the bed before it is picked up. Once it is in the current, it will settle quickly. The smaller the grain, the more frequently it will be picked up, the higher it will jump, and the longer it will take to settle.

Thus, turbulent flows transport sediment by suspension (clays), saltation (sands), and rolling and sliding along the bed (sand and gravel). To study how any one river carries sediment, geologists and hydraulic engineers measure the relationships between particle size and the force the flow exerts on the particles in the suspended and bed loads. Engineers use these data to calculate how much sediment a particular flow can move, and how rapidly. With this information, they can judge how to design dams and bridges or estimate how quickly artificial reservoirs behind dams will fill with sediment. Geologists also

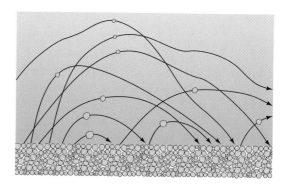

FIGURE 13.3 Saltation is an intermittent jumping motion of grains. In general, the smaller the particle, the higher it jumps and the farther it travels.

can infer the velocities of ancient currents from the sizes of grains in sedimentary rocks.

Figure 13.4 graphs the relationship between grain size (measured in millimeters, along the bottom of the graph) and velocity (measured in centimeters traveled per second, along the left side of the graph). The line between the yellow and gray areas is the velocity at which particles of various sizes settle to the bed. The line between the gray and blue areas is the velocity at which particles are eroded from the bed. The gray area is a transition zone between erosion and settling that depends on

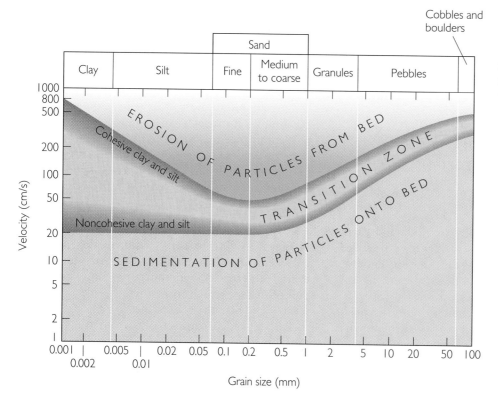

FIGURE 13.4 The relationship between particle size and velocity of the flow. (After F. Hjulstrom, as modified by A. Sundborg, "The River Klaralven," *Geografisk Annaler,* 1956.)

the depth of water in the flow and on factors other than grain size and current velocity. The line between gray and blue areas is for depths of 10 m and the line between yellow and gray areas is for depths of 0.1 m or less. The side of the graph to the right of the 0.2-mm grain size shows the steady increase of velocity required to erode and transport larger and larger grains. The side of the graph to the left of the 0.2-mm grain size shows an increase in velocity with decreasing grain size. This relationship exists because fine sediment particles that are cohesive—that is, they stick together, as many clay minerals do—are harder for the flow to lift from the streambed than noncohesive ones. The finer the cohesive particles, the greater the velocity required to erode them, as shown by the uppermost line on the left side of the graph. For these small grains, settling velocities are so slow that even a gentle current, about 20 cm per second, is able to keep the particles in suspension and transport sediment. This is reflected in the flat boundary between the yellow and gray zones for fine grain sizes.

Bedforms: Dunes and Ripples

When sand grains on a streambed are transported by saltation, they tend to form cross-bedded dunes and ripples (see Chapter 7). **Dunes** are elongated ridges of sand, ranging up to many meters high in large rivers. **Ripples** are very small dunes, from less than a centimeter to several centimeters high, whose long dimension is formed at right angles to the current. Although harder to observe, underwater ripples and dunes form in the same way, and just as commonly, as those formed by air currents on land. As sand grains move by saltation, they are eroded from the upstream side of ripples and dunes and deposited on the downstream side. The steady downstream transfer of grains across the ridges causes the ripple and dune forms to migrate downstream. The speed of this migration is much slower than the movement of individual grains and very much slower than the current. (We will look at ripple and dune migration in more detail in Chapter 14.)

The shape and migration speed of ripples and dunes change as the velocity of the current increases. At the lowest velocities, with few grains saltating, the sand bed of a stream is flat. At slightly higher velocities, the number of saltating grains increases. A rippled bed forms, and ripples migrate downstream (Figure 13.5). As the velocity increases further, the ripples grow larger and migrate faster until, at a certain point, ripples are replaced by dunes. Both ripples and dunes have a cross-bedded

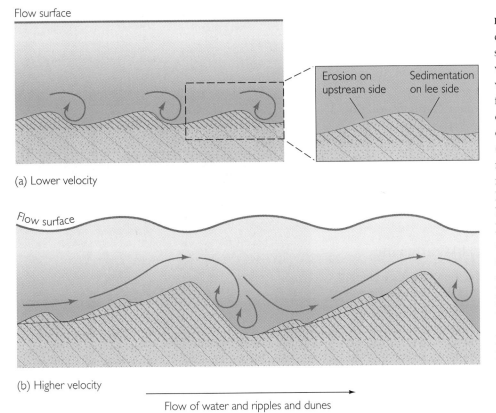

(a) Lower velocity

(b) Higher velocity

Flow of water and ripples and dunes

FIGURE 13.5 The change in the form of a sandy bed with increasing velocity. (a) At lower velocity, a rippled bed forms. Ripples migrate downstream and have a cross-bedded structure. (b) At higher velocity, rippled dunes form and migrate downstream. Dunes have the same cross-bedded structure as ripples. Because ripples migrate downstream faster than dunes, they tend to climb over the backs of dunes. Reverse eddies form in the lee of ripples and dunes. (After D. A. Simons and E. V. Richardson, "Forms of Bed Roughness in Alluvial Channels," *American Society of Civil Engineers Proceedings,* vol. 87, 1961, pp. 87–105.)

structure, and a reverse eddy forms in their lee. As the dunes grow larger, small ripples form, and—because they migrate more quickly than dunes—they tend to climb over the backs of the dunes. Very high velocities will wipe out the dunes and form a high-velocity flat bed below a dense cloud of rapidly saltating sand grains. Most of these grains hardly settle to the bottom before they are picked up again. Some are in permanent suspension.

HOW RUNNING WATER ERODES SOLID ROCK

We can easily see the rapid process of a current picking up loose sand from its bed and carrying it away, thus eroding the bed. At high water levels and during floods, streams can even scour and cut into unconsolidated banks, which then slump into the flow and are carried away. Gullies—valleys made by small streams eroding soft soils or weak rocks—cut their way headward into higher land. This headward erosion, which accompanies widening and deepening of the valleys, may be extremely rapid—up to several meters in a few years in easily erodible soils. We cannot so easily see the much slower erosion of solid rock. Running water erodes solid rock by abrasion, by chemical and physical weathering, and by the undercutting action of currents.

Abrasion

One of the major ways in which a river breaks apart and erodes rock is by slow abrasion. The sand and pebbles carried by the river create a sandblasting action that wears away even the hardest rock. On some river bottoms, pebbles and cobbles rotating inside swirling eddies grind deep **potholes** in the bedrock of the river bottom (Figure 13.6). At low water, we can see these pebbles and sand lying quietly at the bottom of exposed potholes.

Chemical and Physical Weathering

Chemical weathering, which alters a rock's minerals and weakens it along joints and cracks, helps destroy rocks in streambeds just as it does on the land surface. Physical weathering can be violent as the crashes of boulders and constant smaller impacts of pebbles and sand split the rock along cracks. Subjected to such impacts, rock is broken up much

faster in river channels than it is by slow weathering on a gently sloping hillside. Once large blocks of bedrock are loosened by impacts and weathering, strong upward eddies may pull them up and out by a sudden, violent plucking action.

Rock erosion is particularly strong at rapids and waterfalls. Rapids are places in a stream where the flow is extremely fast because the slope of the riverbed suddenly steepens, typically at rocky ledges. Because of the speed of the water and the great turbulence, blocks quickly break into smaller pieces and are carried away by the strong current.

Undercutting Action of Currents

The tremendous impact of huge volumes of plunging water and tumbling boulders quickly erodes rock beds at the bottom of waterfalls. Waterfalls also erode the underlying rock of the cliff that forms the falls. As erosion undercuts these cliffs, the upper beds

FIGURE 13.6 Potholes in river rock along McDonald Creek, Glacier National Park, Montana. The pebbles rotate inside the potholes, grinding deep holes in the bedrock. *(Carr Clifton/Minden Pictures.)*

FIGURE **13.7** This waterfall on the Iguaçú River, Brazil, is retreating upstream as falling water and sediment pound at the cliff's base and undercut it. From the center to the upper left, one can see the steep walls that are the remnant of the waterfall's retreat upstream. (*Donald Nausbaum.*)

collapse and the falls recede upstream (Figure 13.7). Erosion by falls is fastest where the rock layers are horizontal, with erosion-resistant rocks at the top and softer rocks, such as shales, making up the lower layers. Historical records show that the main section of Niagara Falls, perhaps the best known falls in North America, has been moving upstream at a rate of a meter per year.

STREAM VALLEYS, CHANNELS, AND FLOODPLAINS

As streams erode the Earth's surface—in some places bedrock, in others unconsolidated sediment—they create valleys. A stream **valley** encompasses the entire area between the tops of the slopes on both sides of the river. The cross-sectional profile of many river valleys is **V**-shaped, but many others show a low and broad profile like that in Figure 13.8. At the bottom of the valley is the **channel,** the trough through which the water runs. The channel carries all of the water during normal, nonflood times, and at low water levels, the stream may run only along the bottom of the channel. At high water levels, the stream occupies most of the channel. In broader valleys, a **floodplain**—a flat area about level with the top of the channel—lies on either side of the channel. This

is the part of the valley that is flooded when the river spills over its banks, carrying with it silt and sand from the main channel.

Stream Valleys

In high mountains, stream valleys are narrow and steep-walled, and the channel may occupy most or all of the valley bottom. A small floodplain may be

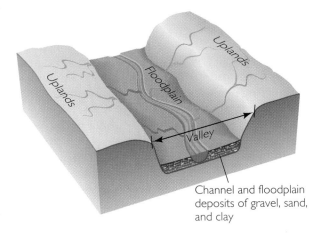

Channel and floodplain deposits of gravel, sand, and clay

FIGURE **13.8** A river flows in a channel that moves over a broad, flat floodplain in a wide valley eroded from uplands. Floodplains may be narrow or absent in steep valleys.

visible only at low water levels. In such valleys, the stream is actively cutting into the bedrock, a characteristic of tectonically active, newly uplifted highlands. In lowlands, where tectonic uplift has long since ceased, stream erosion of valley walls is helped by chemical weathering and mass wasting. With a long time to operate, these processes produce gentle slopes and floodplains many kilometers wide.

Channel Patterns

As stream channels make their way along the bottom of a valley, they may run straight in some stretches and snake their way along various irregular paths in others, sometimes splitting into multiple channels. The channel may flow along the center of the floodplain or hug one edge of the valley. In addition to straight stretches, the two other types of channel patterns are meandering and braided.

MEANDERS On a great many floodplains, channels follow curves and bends called **meanders,** so named for the Maiandros (now Menderes) River in Turkey, fabled in ancient times for its winding, twisting course. Meanders are normal in streams flowing on low slopes in plains or lowlands, where channels typically cut through unconsolidated sediments— fine sand, silt, or mud—or easily eroded bedrock. Meanders are less pronounced but still common where the channel flows on higher slopes and harder bedrock. In the latter terrain, meandering stretches of streams may alternate with long, relatively straight ones.

Some streams have incised meanders—deeply eroded, meandering V-shaped valleys that have practically no floodplains (Figure 13.9). Others may meander on somewhat wider floodplains bounded by steep, rocky valley walls. We are not sure why these two different patterns appear, but we do know that meandering is widespread not only in streams but also in a great many other kinds of flows. For example, the Gulf Stream, a powerful current in the western North Atlantic Ocean, meanders. Lava flows on Earth meander, and planetary geologists have found meanders in dry water channels and lava flows on Mars and, recently, in lava flows on Venus.

Meanders on a floodplain migrate over periods of many years, eroding the outside bank of bends, where the current is strongest (Figure 13.10a). Meanders shift position from side to side and also downstream, in a snaking motion something like that of a long rope being snapped (Figure 13.10b). As outside banks are eroded, curved sandbars, called

FIGURE 13.9 This section of the San Juan River, Utah, is a good example of an incised meander belt, a deeply eroded, meandering, V-shaped valley with practically no floodplain. *(Tom Bean.)*

point bars, are deposited along the inside banks, where the current is slower (Figure 13.10c). Migration may be rapid: some meanders on the Mississippi shift as much as 20 m per year. As meanders move, so do the point bars, building up an accumulation of sand and silt over the part of the floodplain across which the channel migrated (Figure 13.11).

As meanders migrate, sometimes unevenly, the bends may grow closer and closer together, until finally the river bypasses the next loop, often during a major flood. The river takes a new shorter course like that shown in Figure 13.10d, and in its abandoned path it leaves behind an **oxbow lake**—a crescent-shaped, water-filled loop.

Engineers sometimes artificially straighten and confine a meandering river, channeling it along a straight path with the aid of concrete abutments. The U.S. Army Corps of Engineers has been channeling the Mississippi River since 1878, and over a period of 13 years decreased the length of the lower Mississippi by 243 km. Part of the severity of the Mississippi River flood of 1993 has been ascribed to channelization and the height of the artificial river banks built by flood control engineers. Without channelization, floods are more frequent but less damaging. With it, flood damage may be catastrophic when a high flood does breach the artificially high banks, as it did in 1993. Channelization has also been criticized for destroying wetlands and

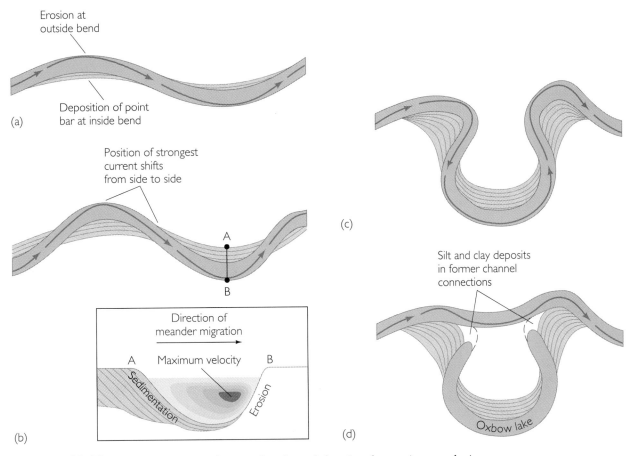

(a)

Erosion at outside bend

Deposition of point bar at inside bend

Position of strongest current shifts from side to side

A

B

(b)

Direction of meander migration

Maximum velocity

A   B

Sedimentation

Erosion

(c)

Silt and clay deposits in former channel connections

Oxbow lake

(d)

FIGURE 13.10 (a) A cross section of a meander channel showing the maximum velocity near the outside of a bend, where erosion is greatest, and the pattern of bedding on the point bar. (b) Meanders migrate over time, eroding the outsides of bends, where the current is strongest. (c) Continued migration accentuates the meanders, making them vulnerable to being cut off during a flood. (d) An oxbow lake evolves as a former meander is cut off.

much of the natural vegetation and animal life of the floodplain. Environmental concerns like these stimulated action to restore one channelized river, the Kissimmee in central Florida, to its original meandering course. If the U.S. Congress approves funds, work on the restoration may begin soon. If left to its own natural processes, the Kissimmee could take many decades or hundreds of years to restore itself.

BRAIDED STREAMS Some streams have many channels instead of a single one. The channels in these **braided streams** split apart and then rejoin, in a pattern resembling braids of hair. Braids are found in many settings, from broad valleys in lowlands to stream deposits in wide, downfaulted valleys adjacent to mountain ranges. They tend to form wherever rivers have large variations in volume of flow combined with a high sediment load and easily

FIGURE 13.11 A meandering river west of Anchorage, Alaska. The white areas are point bars formed at the insides of bends. *(Peter Kresan.)*

erodible banks. They are well developed, for example, in sediment-choked streams formed at the edges of melting glaciers (Figure 13.12).

CHANNEL STABILITY   The three channel patterns—straight, meandering, and braided—blend into one another without sharp divisions. Long, straight stretches of river show some slight curves and bends, and meandering rivers may have some straight stretches. Rivers may be braided in their upper courses and meandering lower down. Whatever the pattern, the channels are stable; some keep the same pattern for hundreds of years. The particular set of patterns that a stream shows throughout its length is a response to varying conditions of flow rates and volumes, sediment loads, and erodibility of banks.

The Stream Floodplain

As a channel migrates over the floor of a valley, it creates a stream floodplain. Point bars formed during migration build up the surface of the floodplain, as does sediment deposited by floodwaters when the stream overflows its banks. Erosional floodplains, covered with a thin layer of sediment, can form when a stream erodes bedrock or unconsolidated sediment as it migrates.

As floodwaters spread out over the floodplain, the velocity of the water slows and the current loses its ability to carry sediment. The speed of the flooding waters drops most quickly along the immediate borders of the channel; as a result, the current deposits much coarse sediment, typically sand and

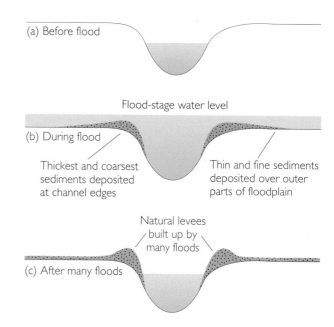

FIGURE 13.13   The formation of natural levees by river floods.

gravel, along a narrow strip at the edge of the channel. Successive floods build up **natural levees,** ridges of coarse material that confine the stream within its banks between floods, even when water levels are high (Figure 13.13). Where levees have built to a height of several meters and the channel is almost filled by the stream, the floodplain level is below the stream level. You can walk the streets of an old river town built on a floodplain, such as Vicksburg, Mississippi, and look up at the levee, knowing that the river waters are rushing by above your head.

During floods, finer sediments—silts and muds—are carried well beyond the channel banks, often over the entire floodplain, and are deposited there as floodwaters continue to lose velocity. Receding floodwaters leave behind standing ponds and pools of water. Here the finest clays are deposited as this standing water gradually disappears by evaporation and infiltration. Fine-grained floodplain deposits have been a major resource for agriculture since ancient times. The fertility of the floodplains of the Nile and other rivers of the Middle East, which contributed to the evolution of the early cultures that flourished there thousands of years ago, depended on frequent flooding. Today the great, broad floodplain of the Ganges River in northern India continues to play an important role in India's life and agriculture. Many other ancient and modern cities are also sited on floodplains (see Feature 13.1).

FIGURE 13.12   Chitina River, Alaska, a braided stream with numerous splittings and rejoinings of multiple channels. *(Tom Bean.)*

The Development of Cities on Floodplains

Floodplains are natural sites for urban settlements, for they combine easy transportation along a river with access to fertile agricultural lands. Such sites, however, are subject to the floods that formed the plains.

About 4000 years ago, cities began to dot river floodplains in Egypt along the Nile, in the ancient land of Mesopotamia along the Tigris and Euphrates rivers, and in Asia along the Indus River of India and the Yangtze and Huang Ho of China. Later, many of the capital cities of Europe were built on river floodplains: London on the Thames, Paris on the Seine, Rome on the Tiber. Floodplain cities in North America include St. Louis on the Mississippi, Cincinnati on the Ohio, and Montreal on the St. Lawrence.

Floods periodically destroyed portions of these ancient and modern cities on the lower parts of the floodplains, but each time they were rebuilt. Today, most large cities are protected by artificial levees that strengthen and heighten the river's natural levees. Extensive systems of dams can help control flooding that would affect these cities, but they cannot eliminate the risk entirely.

In 1973, the Mississippi went on a rampage with a flood that continued for 77 consecutive days at St. Louis. It reached a record 4.03 m above flood stage (the height at which the river first overflows the channel banks). In 1993, the Mississippi and its tributaries broke loose again, shattering the old record in a flood that has been officially designated the most devastating flood in U.S. history: 487 deaths and $15 billion to $20 billion in property damage. At St. Louis, the Mississippi stayed above flood stage for 144 of the 183 days between April and September. An unexpected result of this flood was widespread pollution as floodwaters leached agricultural chemicals from farmlands and then deposited them in flooded areas. Some geologists believe that the construction of levees and dikes has unnaturally confined the Mississippi. This confinement contributes to the record high floods because the river can no longer erode its banks and widen its channel to accommodate smaller portions of the additional water flowing during times of high discharge.

What are cities and towns in this position to do? Some have urged a halt to all construction and development on the lowest parts of the floodplains. Some have called for the elimination of federally subsidized disaster funds for rebuilding in such areas. Harrisburg, Pennsylvania, hit hard by a flood in 1972, turned some of its devastated riverfront area into a park. In a dramatic move after the 1993 Mississippi flood, the town of Valmeyer, Illinois, voted to move the entire town to high ground several miles away, to a new site chosen with the help of a team of geologists from the Illinois Geological Survey. Yet some people who have lived all their lives on floodplains want to stay and are prepared to live with the risk. The costs of protecting some river-bottom areas are prohibitive, and these places will continue to pose public policy problems.

Like many cities built on river floodplains, Liuzhou, China, is subject to flooding. This flood, in July 1996, was the largest recorded in the city's 500-year history. *(Xie Jiahua/China Features/Sygma.)*

STREAMS CHANGE WITH TIME AND DISTANCE

The flow of a stream appears steady when you look at it from a bridge for a few minutes or canoe along it for a few hours, but its volume and velocity at a single place may change appreciably from month to month and season to season. Streams are dynamic systems, moving from low to high waters and floods in a few years and reshaping their valleys over longer periods. Streams also change their flows and channel dimensions as they move downstream, from narrow valleys in their upland headwaters to broader floodplains in their middle and lower courses. Most of these longer term changes in streams are adjustments in the normal (nonflood) volume and velocity of flow as well as the depth and width of the channel.

Discharge

We measure the size of a stream's flow by its **discharge**—the volume of water that passes a given point in a given time as it flows through a channel of a certain width and depth. (Note that in Chapter 12, we also defined discharge as the volume of water leaving an aquifer in a given time. Both uses of the word are valid, because they are both volume of flow per unit time.) Discharge of a stream is commonly measured in cubic meters per second or cubic feet per second. Modest sized streams may vary

in discharge from about 0.25 to 300 m³/s. For example, a well studied medium sized river in Sweden (Kläraluen) has been measured to vary in discharge from 500 m³/sec at low water levels to 1320 m³/s at high water levels. The discharge of the Mississippi River at times is as low as 1400 m³/s and in times of flood may be more than 57,000 m³/s.

We find the discharge by multiplying the cross-sectional area (the width multiplied by the depth of the part of the channel occupied by water) by the velocity of the flow (distance traveled per second):

$$\text{discharge} = \begin{array}{c}\text{cross section}\\(\text{width} \times \text{depth})\end{array} \times \begin{array}{c}\text{velocity}\\(\text{distance traveled per second})\end{array}$$

This equation leads us to expect that if discharge is to increase, either velocity or cross-sectional area, or both, will have to increase. As you increase the discharge of a garden hose by turning up the water pressure, the cross-sectional area of the hose, measured by its diameter, cannot change, so the water comes out at higher speed. As discharge of a stream increases at a particular point, both the velocity and the cross-sectional area tend to increase. (The velocity is also affected by the slope of the channel and the roughness of the river bottom and sides, which we can neglect for the purposes of this illustration.) The cross-sectional area increases as the flow occupies more of the channel's width and depth (Figure 13.14).

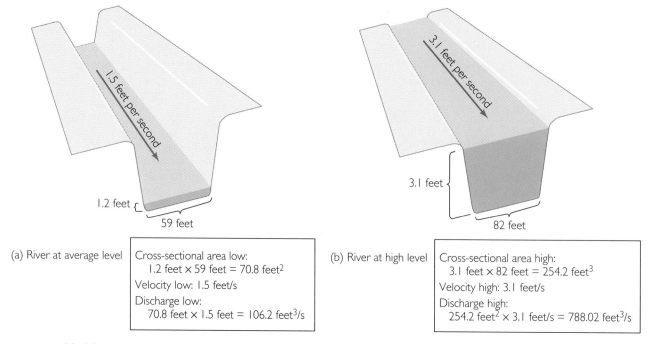

(a) River at average level

Cross-sectional area low:
  1.2 feet × 59 feet = 70.8 feet²
Velocity low: 1.5 feet/s
Discharge low:
  70.8 feet × 1.5 feet = 106.2 feet³/s

(b) River at high level

Cross-sectional area high:
  3.1 feet × 82 feet = 254.2 feet³
Velocity high: 3.1 feet/s
Discharge high:
  254.2 feet² × 3.1 feet/s = 788.02 feet³/s

FIGURE 13.14 Discharge depends on velocity and cross-sectional area. (a) A river at average discharge (b) at high discharge (Data from T. Dunne and L. B. Leopold, *Water in Environmental Planning*, San Francisco, W. H. Freeman, 1978)

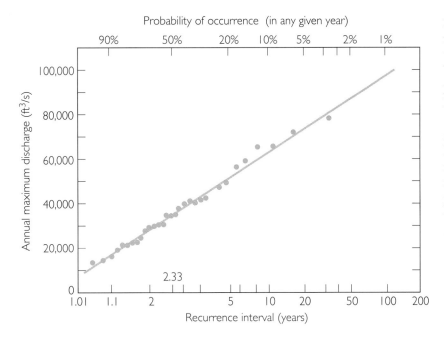

FIGURE 13.15 The flood frequency curve for annual floods on the Skykomish River at Gold Bar, Washington. This curve predicts the probability that a flood of a certain height will occur in any given year. For example, we can read from the graph that a high flood of 65,000 ft³/s has a probability of 10 percent and a recurrence interval of about 9.5 years, meaning that there is a 10 percent probability that a flood of this height will occur in any given year or, alternatively, that this high a flood is likely to recur every 9.5 years. (After T. Dunne and L. B. Leopold, *Water in Environmental Planning,* San Francisco, W. H. Freeman, 1978.)

Normal discharge in most rivers increases downstream as more and more water flows in **tributaries,** streams that discharge water into larger streams. As we have seen, increased discharge means that width, depth, or velocity must increase too. Velocity does not increase downstream as much as the increase in discharge leads us to expect, because of decreases in slope along the lower courses of a stream (decreasing slope reduces velocity). Where discharge does not increase significantly downstream and slope decreases greatly, a river will flow more slowly.

Floods

A flood is an extreme case of an increased discharge that results from a short-term imbalance between input and output. As the discharge increases, the flow velocity in the channel increases and the water gradually fills the channel. As the discharge continues to increase, the water floods over the banks. Rivers flood regularly, some at infrequent intervals, others almost every year. Some floods are large, with very high water levels lasting for days; at the other extreme are minor floods that barely break out from the channel before they recede. Small floods are more frequent, occurring on the average every 2 or 3 years. Large floods are generally less frequent, usually occurring only every 10, 20, or 30 years.

Because no one can know exactly how high—either in water height or in discharge—a flood will be in any given year, geologists state their predictions as *probabilities,* not certainties. For a particular stream,

for example, a geologist might state that there is a 20 percent probability that a flood of a given height—say, 3 m above bankfull stage—will occur in any one year. This chance corresponds to an average time interval—in this case, 5 years (20 percent = 1 in 5)—that we expect between two floods 3 m high. We speak of a flood of this height as a 5-year flood. The average time interval between the occurrence of two geological events of a given magnitude (in this case, a 3-m flood) is called a **recurrence interval.** A flood of greater magnitude—say, 6 m—on the same stream is likely to happen only once every 50 years, and it would therefore be called a 50-year flood. A graph of the annual probabilities and recurrence intervals for a range of flood heights in one river—the Skykomish, in Washington—is shown in Figure 13.15. Thus, on this river, we may expect an 18,000-ft³/s discharge, a very moderate level, to have a 90 percent probability of happening in any given year, corresponding to a recurrence interval of about 1.1 years. A high flood of 65,000 ft³/s has a probability of only 10 percent and a recurrence interval of about 9.5 years. When we speak of recurrence intervals, we must always remember that they express probabilities, not certainties.

The recurrence interval of floods of different heights varies from stream to stream. The interval depends on three factors:

- The climate of the region

- The width of the floodplain

- The size of the channel

In a dry climate, for example, the recurrence interval of a 3-m flood may be much longer than the recurrence interval of a 3-m flood on a similar stream in an area that gets intermittent rain. For this reason, individual graphs of recurrence intervals of major rivers are necessary if towns along these rivers are to be prepared to cope with floods of various heights (see Feature 13.2).

Longitudinal Profile and the Concept of Grade

We have seen that stream flow at any locality balances inputs and outputs, which temporarily get out of balance during floods. Studies of changes in discharge, velocity, channel dimensions, and topography, especially slope, along the entire length of a stream from headwaters to mouth, reveal a larger scale and longer term balance: a stream is in dynamic equilibrium between erosion of the streambed and sedimentation in the channel and floodplain over its entire length. This equilibrium is controlled by several factors:

- Topography (including slope)
- Climate
- Stream flow (including both discharge and velocity)
- Resistance of rock to weathering and erosion

A particular combination of factors—such as high topography, humid climate, high discharge and velocity, hard rocks, and low sediment load—would make the stream erode a steep valley into bedrock and carry downstream all sediment derived from that erosion. Conversely, downstream, where topog-

13.2 LIVING ON EARTH

Historic Floods as a Guide to Flood Control

A year seldom passes without reports of a major flood somewhere in the world. So frequent are disastrous floods that history records only the greatest of them. One of the largest of modern times was the great flood of the Huang Ho in China in 1931, which killed 4 million people. More recently, floods in Bangladesh near the mouths of the Ganges and Brahmaputra rivers led to the deaths of 300,000 people.

Advance warning of a flood can prevent many of these deaths. The success of such warnings was seen in Johnstown, Pennsylvania. One of the most famous floods in American history took place when a dam on the Conemaugh River failed in 1889, destroying homes and killing 2200 people. In 1977, Johnstown was hit again when another dam failed, but this time, although more property was lost than in the earlier flood, the death toll was reduced to 49.

Flooding is not restricted to a few dangerous localities. It is estimated that 20 million people in the United States alone live in areas subject to river and coastal flooding. In recent decades, flood-related property damage has exceeded $1.5 billion annually, in addition to the enormous suffering of families flooded out of their homes.

Flood-control dams are moderately effective for smaller floods but may offer little protection from very large ones. A system of 23 dams was built along the Susquehanna River in Pennsylvania after a flood in 1936, in an effort to contain large floods. In June 1972, Hurricane Agnes dropped 4 to 18 inches of rain on various localities, amounting to 3 trillion gallons of water over six days. Four to six inches of rain had fallen during the three previous weeks and the ground was already saturated. Most of the dams spilled over on the third day of the hurricane. The flood crested soon after at Harrisburg, the capital of Pennsylvania, at a meter higher than the previous record of 1936. Calculations have indicated that 522 new dams would have to be built to protect Harrisburg against another such flood. Clearly, the absurd expense of building so many dams and permanently flooding vast tracts of land behind them make this an unfeasible option.

Flood prediction depends on careful monitoring of weather conditions, such as prolonged storms or hurricanes. Heavy rain-

raphy is lower and the stream might be flowing over easily erodible sediments, the river would deposit bars and floodplain sediments, building up the elevation of the streambed by sedimentation.

We describe the slope of a river from headwaters to mouth by plotting the elevation of its streambed against distances from its headwaters. Figure 13.16 plots the slope of the Platte and South Platte rivers from the headwaters in central Colorado to the mouth in Nebraska. This smooth, concave-upward curve, which presents a cross-sectional view of the river, is its **longitudinal profile.** All streams, from small rills to large rivers, show this same general concave-upward profile, from notably steep near the stream's head to low, almost level, near its mouth.

Why do all streams, which can differ so markedly in detail, follow this profile? The answer lies in the combination of factors that control erosion and sedi-

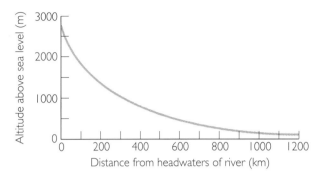

FIGURE 13.16 The longitudinal profile of the Platte and South Platte rivers from the headwaters of the South Platte in central Colorado to the mouth of the Platte at the Missouri River in Nebraska. All streams, whether small rills or large rivers, have this general shape in their longitudinal profile, although the curve may be steeper or shallower. (Data from H. Gannett, in "Profiles of Rivers in the United States," *USGS Water-Supply Paper 44,* 1901.)

fall, particularly when the ground may already be saturated with water from earlier rains or melting snows, predictably leads to increased discharge and eventual flooding. A detailed knowledge of how fast a particular river's discharge increases after rains can be used to determine the probability of flooding so that warnings can be given.

The combination of some dams for partial control and the careful and calculated raising of artificial levees can mitigate many floods, but it appears that we cannot count on such measures to eliminate the danger from large floods everywhere. Our efforts might be better spent on altering the pattern of urban development in the most hazardous areas.

These houses were part of the destruction left in the wake of the 1889 flood in Johnstown, Pennsylvania, which also claimed the lives of more than 2000 people. *(Johnstown Area Heritage Association.)*

mentation. Because streams run downhill, they all start at elevations higher than their lower courses and erode faster upstream than downstream. In their lower courses, with sediments inevitably derived from erosion of the upper courses, sedimentation becomes more significant. Differences in topography and the other factors may make the longitudinal profile steeper or shallower in the upper and lower courses of the stream, but the general shape remains concave upward.

The longitudinal profile is controlled at its lower end by a stream's **base level,** the elevation at which it disappears as a stream as it enters a large standing body of water, such as a lake or the ocean. Streams cannot cut below base level, for base level is the "bottom of the hill"—the lower limit of the longitudinal profile.

PROFILES CHANGE WHEN BASE LEVEL CHANGES Changes in natural base level affect the longitudinal profile in predictable ways. Figure 13.17 illustrates the longitudinal profile for the natural local and regional base levels of a river flowing into a lake and from the lake into the ocean. If the regional base level rises, as when sea level rises, the profile will show the effects of sedimentation as the river builds up channel and floodplain deposits to reach the new base-level elevation. Damming a river artificially can also create a new local base level, with similar effects on the longitudinal profile (Figure 13.18). The slope of the river upstream from the dam decreases, lowering the river's velocity and decreasing its ability to transport sediment. This causes the stream to deposit some of the sediment on the bed, which makes the concavity somewhat shallower than it was before the dam was built. Below the dam, the river, now carrying much less sediment, adjusts its profile to the new conditions and typically erodes its channel in the section just below the dam.

Falling sea levels also alter regional base levels and longitudinal profiles. The regional base levels of all streams flowing into the ocean are lowered, and their valleys are cut into former stream deposits. When the drop in sea level is large, as it was during the last glacial period, rivers erode steep valleys into coastal plains and continental shelves.

GRADED STREAMS Over a period of years, a stream's profile becomes stable as the stream gradually fills in low spots and erodes high spots, thereby producing the smooth curve that is the result of the balance between erosion and sedimentation. That balance is governed not only by the stream's base level but also by the elevation of its headwaters and by all the other factors controlling the equilibrium of the stream profile, as discussed earlier in the chapter. At equilibrium, the stream is a **graded stream,** one in which the slope, velocity, and discharge combine to transport its sediment load, with neither sedimentation nor erosion. If the conditions that give rise to a particular graded stream profile are changed, the stream's profile will change to reach a new equilibrium. This may involve changes in depositional and erosional patterns and alterations in the shape of the channel.

Over geologic times, where regional base level is constant, the longitudinal profile reflects the balance between tectonic uplift and erosion on the one hand and transport and deposition on the other. If uplift is dominant, typically in the upper courses of a stream, the profile is steep and expresses the dominance of erosion and transport. As uplift slows, the profile is lowered as the headwater region is eroded.

ALLUVIAL FANS One of the places in which a river must adjust suddenly to changed conditions is at a mountain front, where streams leave narrow mountain valleys for broad, relatively flat valleys. Along such fronts, typically at steep fault scarps, streams drop large amounts of sediment in cone- or fan-shaped accumulations called **alluvial fans** (Figure 13.19). This deposition results from the sudden

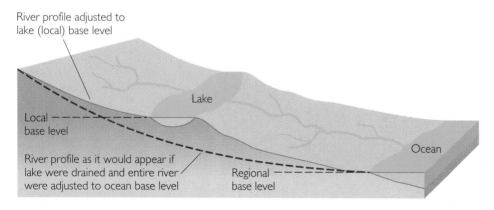

River profile adjusted to
lake (local) base level

Local
base level

Lake

River profile as it would appear if
lake were drained and entire river
were adjusted to ocean base level

Regional
base level

Ocean

FIGURE **13.17** A stream's base level controls the lower end of its longitudinal profile. The profiles illustrated here are for natural regional and local base levels of a river flowing into a lake and from the lake into the ocean. In each river segment, the profile adjusts to the lowest level it can reach.

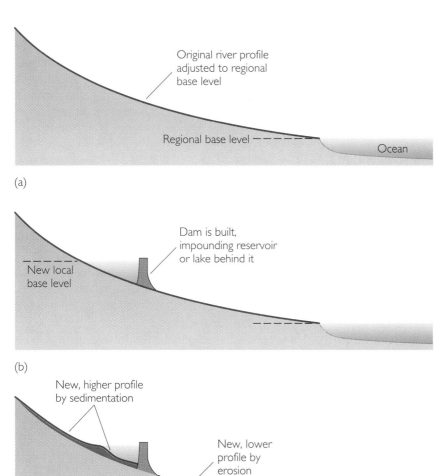

(a)

(b)

(c)

FIGURE 13.18 A change in the base level of a river caused by human intervention, such as the construction of a dam, alters a river's profile. (a) The original profile was adjusted to equilibrium with the regional base level. (b) The equilibrium is upset when a dam is built. The dam impounds a lake behind it and raises the local base level. (c) Upstream, the riverbed is adjusted to the new higher profile as sediment is deposited in the lake and in front of it; downstream, the bed is eroded to a new lower profile.

FIGURE 13.19 An alluvial fan (Tucki Wash) in Death Valley, California. Alluvial fans are large cone- or fan-shaped accumulations of sediment deposited when a stream must suddenly adjust to changed conditions, as at a mountain front. *(Michael Collier.)*

decrease in velocity as the channel widens greatly. To a minor extent, a lowering of slope below the front also slows the stream velocity. The surface of the alluvial fan normally shows a concave-upward profile connecting the steeper mountain part of the profile with the gentler valley or plains profile. Coarse materials, from boulders to sand, dominate on the steep upper slopes of the fan. Lower down, deposits are finer sands, silts, and muds. Fans from many adjacent streams along a mountain front may merge to form a long wedge of sediment whose appearance may mask the outlines of the individual fans that make it up.

TERRACES The role of uplift in changing the equilibrium of a stream valley is seen in the **terraces** that line many streams above the floodplain. These flat steplike surfaces mark former floodplains that existed at the higher level before regional uplift or an increase in discharge caused the stream to erode into the former floodplain. Terraces are made of floodplain deposits and are often paired, one on each side of the stream, at the same level (Figure 13.20). The sequence of events forming terraces starts when a stream creates a floodplain. Rapid uplift then changes the stream's equilibrium, causing it to cut down into the floodplain. In time, the stream reestablishes a new equilibrium at a lower level. It may then build another floodplain, which will also undergo uplift and be sculpted into another, lower pair of terraces.

EFFECT OF CLIMATE Climate also strongly affects the longitudinal profile, primarily through the way temperature and precipitation (and so vegetation) affect weathering and erosion (see Chapter 6). Warm temperatures and high rainfall promote weathering and erosion, and hence sediment transport by streams. There is evidence from the analysis of sediment transport over the entire United States that global change over the last 50 years is responsible for a general increase in stream flow. In the case of alluvial fans, although tectonics plays a dominant role in the formation of fans at fault scarps, short-term depositional upbuilding or trenching is the result of climatic change, primarily variations in temperature.

Lakes

Valued for their recreational and fishing uses, lakes are really accidents of the longitudinal profile, as one can see easily in cases where a lake has been formed behind a dam (see Figure 13.18). Lakes are large and small, ranging from ponds only a hundred meters across to the world's largest and deepest lake, the huge Lake Baikal in southwestern Siberia, which holds approximately 20 percent of the world's fresh waters in lakes and rivers. Lake Baikal is located in a continental rift zone, a typical plate-tectonic setting for lakes. The damming action in a rift valley is the result of faulting that blocks a normal exit of water.

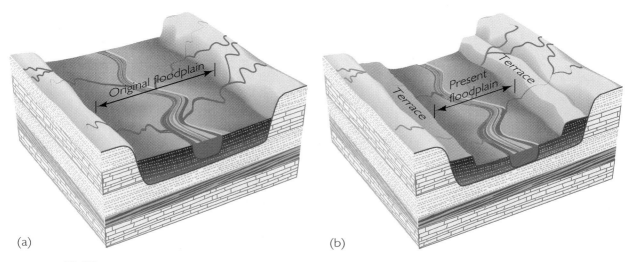

(a)   (b)

FIGURE 13.20 River terraces form when a river erodes into its floodplain and establishes a new floodplain at a lower level. The terraces are remnants of the former floodplain.

As a result, streams can flow in easily but cannot flow out until a sufficiently high water level allows them to exit. The abundance of lakes in the northern United States and Canada is a consequence of the disruption in drainage by glacial ice and glacial sedimentary debris. Sooner or later, if there is stability in tectonics and climate, such lakes will be drained as a new outlet is formed and the longitudinal profile becomes smooth. Lakes, even large ones, are small compared with oceans and thus are more likely to be affected by water pollution. Chemical and other industries have polluted Lake Baikal, and Lake Erie has had a high level of pollution for many years, although there has been some improvement recently.

Drainage Networks

Every topographic rise between two streams, whether it measures a few meters or a thousand meters, forms a **divide,** a ridge of high ground along which all rainfall is shed as runoff down one side of the rise or the other. All the divides that separate a stream and its tributaries from their neighbors define its **drainage basin,** an area of land surrounded by divides that funnels all its water into the network of streams draining the area (Figure 13.21). Drainage basins range from a small area, such as a ravine surrounding a small stream, to a great region drained by

FIGURE **13.22** The natural drainage basin of the Colorado River covers about 630,000 km², a large part of the southwestern United States. This basin is surrounded by divides that separate it from the neighboring drainage basins. *(After USGS.)*

a major river and its tributaries (Figure 13.22). A continent is divided into major drainage basins separated by major divides. In North America, the continental divide along the Rocky Mountains separates all waters flowing into the Pacific Ocean from all those going to the Atlantic.

Many divides tend to keep their places over long times as they are eroded to low ridges. But in some places, divides do change. If a stream on one side of a divide is able to erode and transport sediments much more rapidly than a stream on the opposite side, the divide may be eroded unevenly. At some places, the more active and competent stream may break through the divide and "capture" all or part of the drainage of its slower neighbor, a case of **stream piracy.** Stream piracy explains such odd landscapes as narrow valleys that have no active streams running in them.

Drainage Patterns

A map showing the courses of streams large and small, tributaries and main rivers, reveals a pattern of connections called **drainage networks.** If you followed a stream upstream, you would see that it steadily divides into smaller and smaller tributaries, in drainage networks that show characteristic

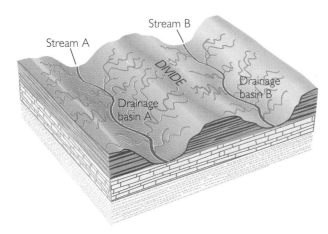

FIGURE **13.21** Stream valleys and drainage basins are separated by divides, which are ridges, gentle uplands, or mountain ranges.

branching patterns (Figure 13.23). Branching is a general property of many kinds of networks in which material is collected and distributed. The network of the human circulatory system, for example, distributes blood to the body through a branching system of arteries and collects it through a corresponding system of veins.

Perhaps the most familiar branching networks are those of trees and roots. Most rivers follow the same kind of irregular branching pattern, called **dendritic drainage,** from the Greek word for tree, *dendron*. This fairly random drainage pattern is typical of terrains where the bedrock is uniform, such as horizontally bedded sedimentary rocks or massive igneous or metamorphic rocks. A more orderly **rectangular drainage** pattern develops in a system of streams in which each straight segment of each stream takes on two perpendicular directions, usually

following fractures or joints in bedrock. A special rectangular pattern, **trellis drainage,** resembles the right-angled geometry of wooden trellises and forms when tributaries lie in the parallel valleys of steeply dipping beds in folded belts. This pattern evolves where bands of rock resistant to weathering alternate with bands that erode more rapidly in a series of anticlines and synclines. (Recall that anticlines fold upward and synclines fold downward.) This situation is found in terrains where sedimentary rocks have been deformed into parallel folds. The larger streams run along valleys eroded from the more easily eroded rocks and are joined at right angles by short tributaries running down from ridges of more resistant rocks. A final pattern, **radial drainage,** occurs when a system of streams runs in a radial pattern away from some central high point, such as a volcano or domal uplift.

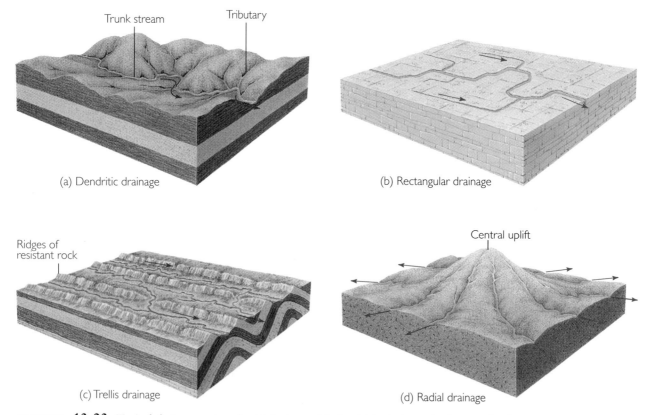

(a) Dendritic drainage

(b) Rectangular drainage

(c) Trellis drainage

(d) Radial drainage

FIGURE 13.23 Typical drainage networks. (a) A river with dendritic drainage is characterized by branches similar to the limbs of a tree. (b) In a typical rectangular drainage pattern developed on a strongly jointed rocky terrain, drainage tends to follow the joint pattern. (c) Trellis drainage develops in valley and ridge terrain, where rocks of varying resistance to erosion are folded into anticlines and synclines. (d) Radial drainage patterns develop on a single large peak, such as a large dormant volcano.

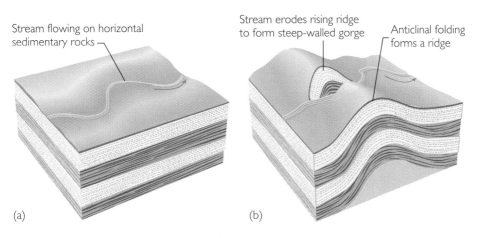

Stream flowing on horizontal sedimentary rocks

(a)

Stream erodes rising ridge to form steep-walled gorge

Anticlinal folding forms a ridge

(b)

FIGURE 13.24 (a, b) An antecedent stream forms as a stream cuts through a ridge being uplifted by tectonic forces. The stream maintains its course despite changes in the underlying rocks and in topography. (c) The Delaware Water Gap, located between Pennsylvania and New Jersey, is an antecedent stream. *(Michael P. Gadomski / Photo Researchers.)*

(c)

Drainage Patterns and Geologic History

We can observe directly or judge from historical records how most stream drainage patterns evolve, but others elude a simple explanation. Some streams, for example, cut through ridges to form steep-walled notches or gorges, in many places through bedrock that is resistant to erosion. What could cause a stream to cut a narrow valley directly through the ridge rather than running along the lowland on either side of it? The geologic history of the region provides the answers. If a ridge is formed by structural deformation while a preexisting stream is flowing over it, the stream may erode the rising ridge to form a steep-walled gorge, as in Figure 13.24. Such a stream is called an **antecedent stream** because it existed before the present topography was created, and it maintained its original course despite changes in the underlying rocks and in topography.

In another geological situation, a stream may be flowing in a dendritic drainage pattern over horizontal beds of sedimentary rocks and overlying folded and faulted rocks with varying resistance to erosion. Over time, the stream cuts down into the underlying rocks and erodes a gorge in the resistant bed. These **superposed streams** flow through resistant formations because their course was established at a higher level, on uniform rocks, before downcutting began. They tend to continue the

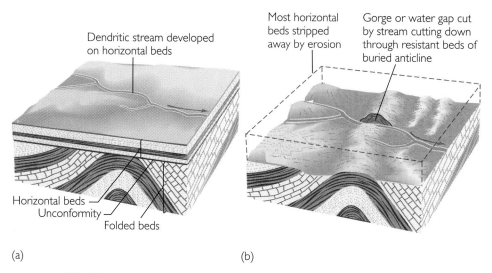

Dendritic stream developed on horizontal beds

Most horizontal beds stripped away by erosion

Gorge or water gap cut by stream cutting down through resistant beds of buried anticline

Horizontal beds ‒
Unconformity ‒
Folded beds

(a)   (b)

FIGURE 13.25 The development of a superposed stream by erosion of horizontal beds overlying folded beds of varying resistance to erosion. As the downcutting stream encounters a buried anticline, it erodes a narrow gorge, or water gap, in the resistant beds of the anticline.

pattern they developed earlier rather than adjusting to their new conditions. In our example, a dendritic pattern is forced onto a surface that would otherwise have evolved a rectangular network (Figure 13.25).

DELTAS: THE MOUTHS OF RIVERS

Sooner or later, all rivers end as they flow into a lake or an ocean, mix with the surrounding water, and—no longer able to travel downhill—gradually lose their forward momentum. The largest rivers, such as the Amazon and the Mississippi, can maintain some current many kilometers out to sea (Figure 13.26). Where smaller rivers enter a turbulent, wave-swept coast, the current disappears almost immediately beyond the river's mouth.

Delta Sedimentation

As its current gradually dies out, a river progressively loses its power to transport sediment. The coarsest material, normally sand, is dropped first, right at the mouth in most rivers. Finer-grained sands are dropped farther out, followed by silt, and still farther out by clay. As the floor of the lake or sea slopes to deeper water away from the shore, all the different sizes of dropped materials build up a depositional platform called a **delta.** (We owe the name *delta* to the Greek historian Herodotus, who traveled

through Egypt around 450 B.C. The roughly triangular shape of the sediments deposited at the mouth of the Nile prompted him to name it after the Greek letter Δ, delta.)

As rivers approach their deltas, where the slope profile is almost level with the sea, they reverse their normal upstream-branching drainage pattern. Instead of collecting more water from tributaries, they assume a pattern of **distributaries,** smaller rivers that receive water and sediment from the main channel, branch off *downstream,* and thus distribute the water and sediment into many channels. Materials dropped on top of the delta, normally sand, make up horizontal **topset beds.** Downcurrent, on the outer front of the delta, fine-grained sand and silt are deposited to form gently inclined **foreset beds,** which resemble large-scale cross-beds. Spread out on the seafloor seaward of the foreset beds are thin, horizontal **bottomset beds** of mud, which are eventually buried as the delta continues to grow. Figure 13.27 shows a typical large marine delta.

The Growth of Deltas

As the delta builds forward, the mouth of the river advances into the sea, leaving new land in its wake. Much of this land is a delta plain just a few meters above sea level. These plains include large areas of wetlands, which, as we noted in Chapter 12, are valuable for their water storage capacity and form the habitat of many diverse species of plants and animals. These wetlands have suffered a two-pronged attack in many areas. First, the extensive flood-

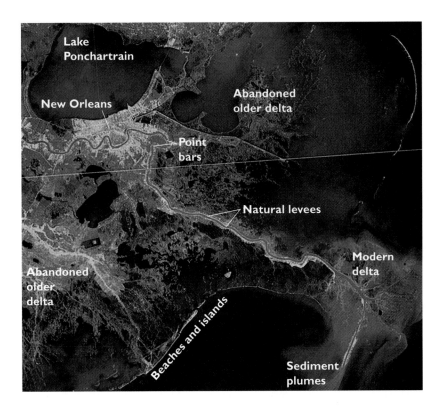

FIGURE 13.26 The infrared-sensitive film used to shoot this satellite image of the Mississippi delta causes the vegetation to appear red, relatively clear water to appear dark blue, and water with suspended sediment to appear light blue. At upper left are New Orleans and Lake Ponchartrain. Well-defined natural levees and point bars are at the center. At lower left are beaches and islands that were formed as sand from the river was transported from the delta by waves and currents. (From G. T. Moore, "Mississippi River Delta from Landsat 2," *Bulletin of the American Association of Petroleum Geologists,* 1979.)

control dams built since the 1930s have decreased sediment volumes brought to the edge of the delta, thereby reducing the sediment supply to the wetlands areas. At the same time, massive artificial levees have prevented the small but frequent floods that actually nourish the delta wetlands.

As deltas grow, they shift the flows from some distributaries to others having shorter routes to the sea. As a result of such shifts, the delta grows in one direction for some hundreds or thousands of years, then breaks out into a new distributary and begins to grow into the sea in another direction. A major

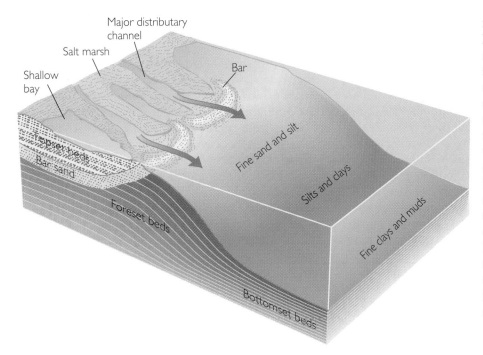

FIGURE 13.27 A typical large marine delta, many kilometers in extent, in which the foreset beds are fine-grained and deposited at a very low angle, normally only 4° to 5° or less. Sandbars form at the mouths of the distributaries, where the currents' velocity suddenly decreases. The delta builds forward by the advance of the bar and topset, foreset, and bottomset beds. Between distributary channels, shallow bays fill with fine-grained sediment and become salt marshes. This general structure is found on the Mississippi delta.

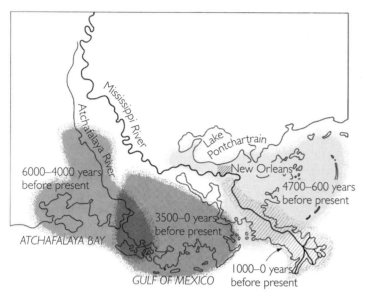

FIGURE 13.28 The Mississippi delta. Over the past 6000 years, the river has built its delta first in one direction and then in another as water flow shifted from one major distributary to another. The modern delta was preceded by deltas deposited to the east and west.

river such as the Mississippi or the Nile forms a large delta thousands of square kilometers in area. The delta of the Mississippi, like many other major river deltas, has been growing for millions of years. It started out, about 150 million years ago, around what is now the junction of the Ohio and the Mississippi rivers, at the southern tip of Illinois. It has advanced about 1600 km since then, creating almost the entire states of Louisiana and Mississippi as well as major parts of adjacent states. Figure 13.28 shows the growth of the Mississippi delta over the last 6000 years.

Effects of Waves, Tides, and Tectonics

Strong waves, shoreline currents, and tides affect the growth and shapes of deltas built into the sea. Waves and shoreline currents may move the sediment along the shore almost as rapidly as it is dropped by the river. The delta front then becomes a long beach shoreline with only a slight seaward bulge at the mouth. Where tidal currents move in and out, they redistribute deltaic sediment into elongate bars parallel to the direction of the currents, which in most places are at approximately right angles to the shore.

Where waves and tides are strong enough, deltas cannot form. The sediment brought down to the sea by the river is dispersed along shorelines as beaches and bars and is transported into deeper waters offshore. The east coast of North America lacks deltas for this reason. The Mississippi has been able to build out its delta because neither waves nor tides are very strong in the Gulf of Mexico.

Tectonics also exerts some control over where deltas form, for deltas require

- Uplift in the drainage basin that provides abundant sediment

- Crustal subsidence in the delta region that will accommodate huge tonnages of sediment

Two of the world's large deltas—the Mississippi and the Rhone (France)—derive their large sediment loads primarily from distant mountain ranges, the Rockies for the Mississippi and the Alps for the Rhone. Both are in the same plate-tectonic setting, a passive margin originally formed from a rifted continental margin.

There are few large deltas associated with active subduction zones. The reason may be that it is unusual for a large river, such as the Columbia of Washington and Oregon, to carry abundant sediment through a volcanic arc (the Cascade Range) to the sea. In addition, oceanic island arcs are too small in land area to provide much clastic sediment.

The continental convergence that elevated the Himalayas also formed the great deltas of the Indus and Ganges. Ultimately, plate-tectonic settings influence delta location and formation.

SUMMARY

How is flowing water in streams able to erode solid rock and to transport and deposit sediment? Any fluid can move in either laminar or turbulent flow, depending on its velocity, viscosity, and flow geometry. The turbulence that characterizes most streams is responsible for transporting sediment by suspension (clays), saltation (sands), and rolling and sliding along the bed (sand and gravel). The tendency for particles to be carried in suspension is countered by the gravitational force that leads them to settle to the bottom, measured by the settling velocity. When a stream flow slows, it loses its competence to carry sediment and deposits it, in many places as beds of rippled, cross-bedded sand. Running water erodes solid rock by abrasion; by chemical weathering that enlarges and opens cracks; by physical weathering as sand, pebbles, and boulders crash against rock; and by plucking and undercutting actions of currents.

How do stream valleys and their channels and floodplains evolve? As a stream flows, it carves a valley with steep to gently sloping walls and a more or less broad floodplain on either side of the channel. The channel may be straight, meandering, or braided. Although the channel carries all of the water and sediment during normal, nonflood times, as a stream increases its discharge to flood stage, it overflows its banks and floods. As floodwaters inundate the floodplain, the velocity slows, and the waters drop sediment that builds up natural levees and floodplain deposits. Recurrence intervals are a measure relating the probability that a flood of a given height will occur in any year to the interval of time between floods of that height.

How does a stream's longitudinal profile reflect the equilibrium between erosion and sedimentation? A stream is in dynamic equilibrium between erosion and sedimentation over its entire length. This equilibrium is affected by topography, discharge, velocity, and slope, such that a stream's longitudinal profile, always concave upward, reflects the elevation at its headwaters and the base level at its mouth in a lake or the ocean. Uplift at the upper end of a stream and the rise and fall of sea level at the lower end will change the profile. Alluvial fans form at mountain fronts, in response primarily to an abrupt widening of the valley and secondarily to a change in slope.

How do drainage networks work as collection systems and deltas as distribution systems for water and sediment? Each drainage basin is separated from its neighbors by a divide. Rivers and their tributaries constitute a branching-upstream drainage network that collects the water and sediment running off a specific drainage basin, which is separated from its neighbors by a divide. Drainage networks show various kinds of branching patterns: dendritic, rectangular, trellis, or radial, depending on topography, rock type, and structure. Near its mouth, as it forms its delta, a river tends to branch downstream into distributaries. Deltas are major sites of deposition of sediment, where rivers drop their sediment load as topset, foreset, and bottomset beds. Deltas are modified or even absent where waves, tides, and shoreline currents are strong. Tectonics controls delta formation by uplift in the drainage basin and subsidence in the delta region.

KEY TERMS AND CONCEPTS

stream (p. 317)
river (p. 317)
laminar flow (p. 318)
turbulent flow (p. 318)
viscosity (p. 319)
suspended load (p. 319)
bed load (p. 319)
competence (p. 319)
capacity (p. 319)
settling velocity (p. 319)
saltation (p. 320)
dune (p. 321)
ripple (p. 321)
pothole (p. 322)
valley (p. 323)

channel (p. 323)
floodplain (p. 323)
meander (p. 324)
point bar (p. 324)
oxbow lake (p. 324)
braided stream (p. 325)
natural levee (p. 326)
discharge (p. 328)
tributary (p. 329)
recurrence interval (p. 329)
longitudinal profile (p. 331)
base level (p. 332)
graded stream (p. 332)
alluvial fan (p. 332)
terrace (p. 334)

divide (p. 335)
drainage basin (p. 335)
stream piracy (p. 335)
drainage network (p. 335)
dendritic drainage (p. 336)
rectangular drainage (p. 336)
trellis drainage (p. 336)
radial drainage (p. 336)
antecedent stream (p. 337)
superposed stream (p. 337)
delta (p. 338)
distributary (p. 338)
topset bed (p. 338)
foreset bed (p. 338)
bottomset bed (p. 338)

EXERCISES

1. How does velocity determine whether a given flow is laminar or turbulent?

2. How does the size of a sediment grain affect the speed with which it settles to the bottom of a flow?

3. What kind of bedding characterizes a ripple or a dune?

4. How do braided and meandering river channels differ?

5. Why is a floodplain so named?

6. What is a natural levee, and how is it formed?

7. What is the discharge of a stream, and how does it vary with velocity?

8. How is a river's longitudinal profile defined?

9. What is the most common kind of drainage network developed over horizontally bedded sedimentary rocks?

10. What is a delta distributary?

LONG-TERM TEAM PROJECT

See Chapter 11.

SHORT-TERM TEAM PROJECT

See Chapter 12.

THOUGHT QUESTIONS

1. Why might the flow of a very small, shallow stream be laminar in winter and turbulent in summer?

2. Describe and compare the river floodplain and valley above and below the waterfall in Figure 13.7.

3. You live in a town on a meander bend of a major river. An engineer proposes that your town invest in new and higher artificial levees to prevent the meander from being cut off. Give the arguments, pro and con, for this investment.

4. In some places, engineers have artificially straightened a meandering stream. If such a straightened stream is then left free to adjust its course naturally, what changes would you expect?

5. If global warming produces a significant rise in sea level as polar ice melts, how would the longitudinal profiles of the world's rivers be affected?

6. In the first few years after a dam was built on it, a stream severely eroded its channel downstream of the dam. Could this erosion have been predicted?

7. Your hometown, built on a river floodplain, experienced a 50-year flood last year. What are the chances that another flood of that height will occur next year?

8. Define the drainage basin where you live in terms of the divides and drainage networks.

9. What kind of drainage network do you think is being established on Mount St. Helens since its violent eruption in 1980?

10. The Delaware Water Gap is a steep, narrow valley cut through a structurally deformed high ridge in the Appalachian Mountains. How could it have formed?

11. A major river, which carries a heavy sediment load, has no delta where it enters the ocean. What conditions might be responsible for the lack of a delta?

SUGGESTED READINGS

Chorley, Richard J., Stanley A. Schumm, and David E. Sugden. 1984. *Geomorphology.* New York: Methuen.

Dunne, Thomas, and Luna B. Leopold. 1978. *Water in Environmental Planning.* San Francisco: W. H. Freeman.

Leopold, Luna B. 1994. *A View of the River.* Cambridge, Mass.: Harvard University Press.

McPhee, John. 1989. *The Control of Nature.* New York: Farrar, Straus and Giroux.

Ritter, Dale F. 1986. *Process Geomorphology,* 2nd ed. Dubuque, Iowa: W. C. Brown.

Schumm, Stanley A. 1977. *The Fluvial System.* New York: Wiley-Interscience.

INTERNET SOURCES

Dartmouth Flood Observatory

ⓘ **http://www.dartmouth.edu/artsci/geog/floods/**
Dartmouth University maintains this source of information concerning the worldwide occurrence of large floods. Using satellite images and geographic information system software, observatory personnel map the extent of flooding immediately. Other features include a flood data base (for events since January 1, 1994), sample images, reports, current information, and links to related sites.

Daily Flood Summary

ⓘ **http://www.ca.uky.edu/agcollege/agweather/public/ WSHFLNNMC**
The University of Kentucky College of Agriculture provides data from the National Weather Service daily on flooding, flood warnings, and flood advisories throughout the United States.

Water Resources of the United States

ⓘ **http://h2o.usgs.gov/**
This site is a gateway for all U.S. Geological Survey water-related data. Surface water data for many areas may be retrieved and real time hydrologic data from selected sites are linked to this web page.

14

Giant elevated sand dunes bordering the desert plain of Namibia. Such dunes are formed by the wind in desert climates where sand is abundant. *(John Chard / Tony Stone Images.)*

Winds and Deserts

We all have been caught, at one time or another, in a wind so strong that it could have blown us over if we hadn't leaned into it or held onto something solid. London, England, which rarely gets strong winds, experienced a major windstorm on January 25, 1990. Winds blowing at more than 175 km per hour ripped roofs off buildings, blew trucks over, and made it virtually impossible to walk on the streets. The windstorm was essentially like a hurricane. Hurricanes generate torrential rains and high winds that erode the land while driving ocean waves that erode and transport huge tonnages of sediment on beaches and in shallow waters. Many winds are strong enough to blow sand grains into the air, as anyone who has ever been in a sandstorm can attest.

In this chapter, we focus on this potent force that shapes the surface of the land, particularly in deserts, where strong winds can howl for days on end. The wind is a major erosional and depositional agent, moving enormous quantities of sand, silt, and dust over large regions of the continents and oceans. The ability of the wind to erode, transport, and deposit sediment is much like that of water, for the same general laws of fluid motion that govern liquids govern gases as well. As we will see, however, there are differences that make the wind less powerful than water currents. In contrast to a stream, whose discharge depends on rainfall, wind works most effectively in the absence of rain. We will discuss Earth's deserts in particular detail because so many of the geological processes there are related to the work of the wind. The ancient Greeks called the god of winds Aeolus, and geologists today use the term **eolian** for the geological processes powered by the wind.

WIND AS A FLOW OF AIR

Wind is a horizontal flow of air—horizontal in relation to the surface of the rotating planet. As with water flows, we can describe air flows by streamlines. Although winds obey all the laws of fluid flow that apply to water in streams (see Chapter 13), there are some differences. In contrast to the flow of water in river channels, winds are generally unconfined by solid boundaries, except for the ground surface and narrow valleys. Air flows are free to spread out in all directions, including upward into the atmosphere.

Turbulence

Like the water flowing in rivers, air flows are nearly always turbulent. Recall from Chapter 13 that turbulence depends on three characteristics of a fluid: density, viscosity, and velocity. The extremely low density and viscosity of air—$\frac{1}{1000}$ the density and $\frac{1}{50}$ the viscosity of water—make it turbulent even at the velocity of a light breeze. Table 14.1 provides standard, internationally agreed upon descriptions of winds of various speeds and their effects on the surface of the sea.

As with water turbulence, the turbulence of air increases in proportion to the velocity of the flow. A light breeze will barely move tall grass, but the turbulence of a fresh wind can easily lift a hat. It takes a strong gale with turbulent gusts to rock a moving car. Turbulence also leads to sudden changes in the speed and direction of the wind; such changes can be strong enough to jolt and buck a large airplane.

Wind Belts

Winds vary in speed and direction from day to day, but over the long term, prevailing winds tend to come mainly from one direction. In temperate climates, the prevailing winds come from the west and are referred to as the westerlies (Figure 14.1). These belts of moderate temperature and rainfall are located at latitudes between polar and equatorial belts. (Latitude is the distance north or south of the equator,

---

TABLE 14.1

WIND SPEEDS		
WIND SPEED (KM/HR)	**DESCRIPTION**	**EFFECT OF WIND ON SEA SURFACE**
1	Calm	Mirror surface
1–19	Light to gentle breeze	Ripples and wavelets
20–49	Moderate to strong breeze	Moderate to large waves, whitecaps
50–88	Moderate to strong gale	High waves, foam, spray
89–117	Whole gale to storm	Very high waves, rolling sea
117	Hurricane	Sea white with spray and foam, low visibility

SOURCE: Modified from 1939 International Agreement; and N. Bowditch, *American Practical Navigator,* U.S. Navy Hydrographic Office Publication 9, 1958.

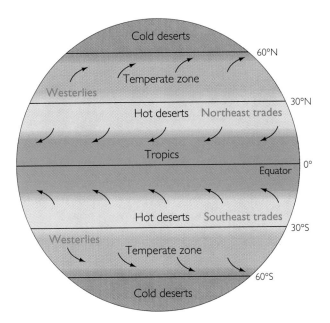

FIGURE 14.1 Circulation of Earth's atmosphere. In temperate zones, the prevailing wind belts come from the west. In the tropics, they blow from the east.

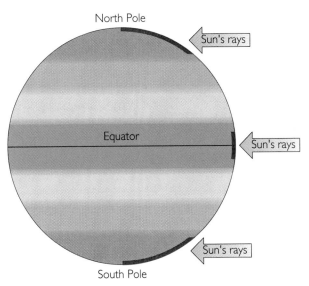

FIGURE 14.2 The angle of inclination of Earth's surface to the Sun's rays. The Sun's rays are almost perpendicular to Earth's surface at the equator. Therefore, heat is more concentrated in equatorial regions than near the poles, where the Sun's rays are spread out over greater areas.

measured by the angle that a radius of the globe at any point makes with the equatorial radius. The equator is at 0° latitude and the North Pole is at 90°.) In the tropics (low latitudes), the trade winds, or "trades" (named for an archaic use of the word *trade* to mean a track or course), blow from the east. This is where most of the world's deserts are.

Wind belts arise because the Sun warms a given amount of land surface most intensely at the equator, where the Sun's rays are almost perpendicular to the ground surface. The Sun heats the Earth less quickly at high latitudes and the poles because there the rays fall at an angle to the ground (Figure 14.2). Hot air, which is less dense than cold air, rises at the equator and flows toward the poles, gradually sinking as it cools. The cold, dense air at the poles then flows back along the surface of the globe toward the equator.

THE CORIOLIS EFFECT This simple circulatory pattern of air flow between the equator and the poles is complicated by Earth's rotation, which causes any current of air or water to be deflected eastward (to the right) in the northern hemisphere and westward (to the left) in the southern hemisphere. This effect on Earth's air flow is called the Coriolis effect, named after its discoverer.

As an example of the Coriolis effect, consider a small packet of air flowing north at the equator. This air also moves eastward, because of Earth's eastward rotation. At the equator, the Earth—and the packet of air—move eastward at about 1670 km per hour. As the air packet moves north, it keeps its eastward velocity because it is not fixed to the Earth's surface. At the same time, the Earth's surface north of the equator is moving more slowly than it is at the equator (as latitude lines decrease in length poleward). Thus, the air packet moves eastward more rapidly (1670 km per hour) than the Earth's surface moves below it (1450 km per hour at 30° latitude). To an observer in space, the air packet has a straight path while the Earth below rotates, but to an observer on Earth, the packet appears to have been deflected eastward (to the right). The deflection is proportional to current speed and latitude.

The Coriolis effect on atmospheric circulation leads to deflections of both northward and southward warm and cold air flows. For example, as surface winds blow southward into the hot equatorial belt, the wind is deflected to the right and hence blows from the northeast rather than the north. These are the northern trade winds. The northern hemisphere westerlies are the original northward flows deflected to the right and thus blow from the

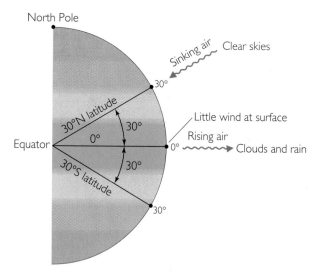

FIGURE 14.3 Latitude is the distance north or south of the equator, measured by the angle that a radius of the globe at any point makes with the equatorial radius. At the equator, there is little surface wind and the air rises, forming clouds and rain as it cools. At 30° latitude, north and south, the cooled air sinks, warms up, absorbs moisture, and yields clear skies. These two motions set up a horizontal circulation between the equator and the north and south latitudes.

southwest. Near the equator, the air is mainly rising, and so there is little wind at the surface. As the air rises it cools, causing the cloudiness and abundant rain of the tropics. At around 30°N and 30°S latitude, some of the high-altitude poleward flow starts to sink to the Earth's surface. As it sinks, the cold, dry air warms and absorbs moisture, producing clear skies and arid climates (Figure 14.3). This dry, sinking air overlies many of the deserts of the world, such as the Sahara.

WIND AS A TRANSPORT AGENT

The wind exerts the same kind of force on particles on the land surface that a river current exerts on its bed. Turbulence and forward motion combine to lift particles into the wind and carry them along, at least temporarily. Even the lightest breezes carry dust, the finest-grained material, usually consisting of particles less than 0.01 mm in diameter but often including somewhat larger silt particles. Wind can carry dust to heights of many kilometers, but only high-velocity winds can carry coarser particles, larger than 0.06 mm in diameter, such as sand grains. Moderate breezes can roll and slide these grains along a sandy

bed, but it takes a fresh wind to lift sand grains into the air flow. The wind usually cannot transport the largest particles, however, because of the low viscosity and density of air. As strong as winds can be, they can only rarely move large pebbles and cobbles the way rapidly flowing rivers do.

The Motion of Sand Grains in Winds

Sand moves in the wind by sliding and rolling along the surface and by saltation, the jumping motion that temporarily suspends grains in a current of water or air. Saltation in air flows works the same way it does in a river (see Figure 13.3), except that saltation is much more pronounced. Sand grains suspended in an air current often rise to heights of 50 cm over a sand bed and 2 m over a pebbly surface, much higher than grains of the same size can jump in water. The difference arises partly because air is less viscous than water and therefore does not inhibit the bounding action of the grains as much. In addition, the impact of falling grains as they hit the surface induces higher jumps. These collisions, cushioned hardly at all by the air, kick surface grains into the air in a sort of "splashing" effect. Impacts by saltating grains on the bed can push forward grains too large to be thrown up into the air, causing the sand bed to creep in the direction of the wind. A sand grain striking the surface at high speed can move forward another grain up to six times its own diameter. Surface creep moves grains less rapidly than saltation.

An almost inevitable consequence of the movement of sand along a bed by the wind is the formation of ripples and dunes much like those formed by water (Figure 14.4). Ripples in sand, like those under water, are transverse; that is, at 90° to the current. At low to moderate wind speeds, small ripples form. As speed increases, the ripples become larger. Ripples migrate in the direction of the wind over the backs of larger dunes. Because some wind is almost always blowing, a sand bed is almost always rippled to some extent.

How Much Can the Wind Carry, and How Far?

Most of us are familiar with rainstorms or snowstorms, high winds associated with heavy precipitation. We may be less aware of dry storms, during which high winds blowing for days on end carry enormous tonnages of sand, silt, and dust. The amount of windblown material the wind can carry depends on the sizes of the particles, the strength of the wind, and the surface material of the area over which the wind blows.

FIGURE 14.4 Wind ripples in sand at Stovepipe Wells, Death Valley, California. Although complex in form, these ripples are always transverse (at 90°) to the wind direction. *(Tom Bean.)*

WIND STRENGTH Figure 14.5 shows how much sand winds of various speeds can erode from a strip 1 m wide across a sand dune's surface. A strong wind of 48 km per hour can move half a ton of sand (about equivalent in volume to two large suitcases) in a single day from this small surface area. At higher wind speeds, the amounts that can be moved increase rapidly. No wonder entire houses can be buried by a sandstorm lasting several days.

SURFACE MATERIAL The wind can transport material only if sand and dust are available from surface materials, such as soil, sediment, or bedrock. Wet soils are too cohesive for wind to erode and transport. Whereas wind can carry sand grains weathered from a loosely cemented sandstone, it cannot erode grains from a granite or basalt. Most windblown sands are locally derived. Traveling mainly by saltation near the ground, most sand grains are buried in dunes after a relatively short travel distance, usually no more than a few hundred kilometers. The extensive sand dunes of major deserts such as the Sahara and the wastes of Saudi Arabia are exceptions. In those great sandy regions, sand grains may have traveled more than 1000 km.

WINDBLOWN DUST Air has a staggering capacity to hold dust. In large dust storms, 1 km³ of air may carry up to 1000 tons, equivalent to the volume of a small house. When such storms cover hundreds of square kilometers, they may carry more than 100 million tons of dust and deposit it in layers several meters thick (see Feature 14.1). In dusty regions, large quantities of dust remain suspended in the air even when winds are moderate, and they create a permanent haze. Even the thin air of Mars supports intense dust storms, some encircling the planet with red-orange dust particles.

Many dust particles are so small that they remain suspended in the atmosphere for a very long time, traveling long distances with the prevailing winds. Volcanic explosions such as that of Mount Pinatubo in the Philippine Islands in 1991 inject huge quantities of dust high into the atmosphere. The volcanic dust from Pinatubo circled the globe, and most of the finest-grained particles did not settle until 1994 or 1995. This volcanic dust, by partly reflecting the Sun's rays, temporarily cooled Earth's surface by a small amount for a few years.

Most nonvolcanic dust does not travel over the whole world, but much of it can travel for thousands of kilometers. The Sahara is one of the chief sources of dust in the world today, and its finer-grained particles have been traced as far away as England and across the Atlantic Ocean to Barbados in the Caribbean. It is estimated that wind annually transports

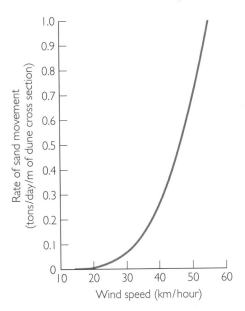

FIGURE 14.5 The amount of sand moved daily across each meter of width of a dune's surface in relation to wind speed. High-speed winds blowing for several days can move enormous quantities of sand. (After R. A. Bagnold, *The Physics of Blown Sand and Desert Dunes,* London, Methuen, 1941.)

14.1   LIVING ON EARTH

Droughts and Dust Bowls

On November 12 and 13, 1933, an enormous dust storm blew up in the southwestern plains of Texas, Oklahoma, and parts of adjacent states. Roads and houses were buried under thousands of tons of dust. The storm extended into the Midwest and as far east as New England, where tons of dust settled, turning snow-covered areas a dark brown.

The immediate cause of this great dust storm was a severe and prolonged drought in the early 1930s, coupled with farming practices that left the soil vulnerable to erosion. In the southern Great Plains, the rainfall is low, the soils are thin, and the winds are strong. But even after decades of intensive agriculture, the land had been fairly stable under these conditions until the long drought came. The killing dryness greatly reduced the vegetation, destroying the root systems that had held the soil together and leaving in its place parched, loosened dust. When the big windstorm of 1933 hit the plains, the strong winds picked up the dust and swirled it high in the atmosphere, choking the Plains states and giving the region a new name: the Dust Bowl.

During the later stages of the drought, even moderate winds blew clouds of dust. Refugees from the drought-stricken fields packed their belongings and migrated, many to California, in search of a livelihood. Their often tragic story was told by the Nobel Prize–winning novelist John Steinbeck in *The Grapes of Wrath* and sung by the folksinger Woody Guthrie in his "Dust Bowl Ballads."

The former Dust Bowl has been stable for more than 60 years, since the drought broke. Soil conservation methods have reduced erosion of the soil, and irrigation offsets the worst effects of short droughts. Nevertheless, the region remains vulnerable. Wind erosion of the soil in the plains was very high during dry periods in the 1950s and 1970s, though never so high as it was during the 1930s. Even in normal times, some plains are enveloped in a brownish dust haze during the plowing season, when the fields are dry and bare of vegetation. We cannot predict exactly how global climate changes will affect this area, only that another severe drought prolonged for several years will spell trouble.

From recent work, we know that in the past 10,000 years the climate of the western plains was arid for periods of time. Sand dunes covered large areas in a desert setting; now stabilized by vegetation, it will take only a few years of drought for them to become active again.

A dust storm in Elkhart, Kansas, May 21, 1937. *(Library of Congress.)*

FIGURE 14.6 Photomicrograph of a frosted, rounded grain of quartz from late Tertiary sand in southern Louisiana. The pits and etched pattern are characteristic of windblown sand in arid climates. *(David H. Krinsley.)*

260 million tons of material, mostly dust, from the Sahara to the Atlantic Ocean. As we noted earlier, however, most coarse-grained dust comes from areas not more than several hundred kilometers away.

MATERIALS CARRIED BY THE WIND Wind-blown sand may consist of almost any kind of mineral grain produced by weathering, but quartz grains are by far the most common because quartz is such an abundant constituent of many surface rocks, especially sandstones. In a few places, feldspar grains are abundant in eolian sands. Much more unusual are rock fragments of fine-grained shale or finely crystalline metamorphic and igneous rocks, which break down to a very small size under the continuous impacts of saltating grains.

Many windblown quartz grains have a frosted or matte (roughened and dull) surface like the inside of a frosted light bulb (Figure 14.6). Some of the grain frosting is produced by wind-driven impacts, but most is the result of slow, long-continued dissolution by dew. Even the tiny amounts of dew found in arid climates are enough to dissolve microscopic pits and hollows, creating the frosted appearance. Because frosting is not found in environments other than eolian, it is good evidence that a sand has been blown by the wind.

Windblown calcium carbonate grains accumulate where there are abundant fragments of shells and coral, as in Bermuda and on many coral islands in the Pacific Ocean. The White Sands National Monument in New Mexico is a prominent example of sand dunes made of gypsum sand grains eroded from evaporite bedrock.

Dust includes microscopic rock and mineral fragments of all kinds, especially silicates, as might be expected from their abundance as rock-forming minerals. Two of the most important sources of silicate minerals in dust are clays blown from soils in dry plains and volcanic dust from eruptions. Organic materials, such as pollen and bacteria, are also common. Charcoal is abundant downwind of forest fires; when it is found in buried sediments, it is an indication of forest fires in earlier geologic times. Since the beginning of the Industrial Revolution, we have been pumping new kinds of synthetic dust into the air, from ash produced by burning coal to the many solid chemical compounds produced by manufacturing processes, incineration of wastes, and auto exhausts.

WIND AS AN AGENT OF EROSION

By itself, the wind can do little to erode large masses of solid rock exposed at Earth's surface. It is only when the rock is fragmented by chemical and physical weathering that particles can be picked up by the wind. In addition, those particles must be dry, because wet soils and moist fragmented rock are held together by moisture. Thus, wind erodes most effectively in arid climates, where strong winds are dry and any moisture is quickly evaporated.

Deflation

As particles of dust, silt, and sand become loose and dry, blowing winds can lift them and carry them away, gradually lowering the surface of the ground in a process called **deflation** (Figure 14.7). Deflation,

FIGURE 14.7 A shallow deflation hollow in the San Luis Valley in Colorado. Wind has scoured the surface and eroded it to a slightly lower elevation. Deflation occurs in dry areas where the vegetation cover is broken. *(Breck P. Kent.)*

which can scoop out shallow depressions or hollows, occurs on dry plains and deserts and on temporarily dried-up river floodplains and lake beds. Firmly established vegetation, even the sparse vegetation of arid and semiarid regions, can retard it. In such places, deflation occurs slowly because the roots bind soil together and the stems and leaves break up the wind and shelter the ground surface. Deflation works fast, however, where the vegetation cover is broken, either naturally by killing drought or artificially by cultivation, construction, or tracks of motor vehicles.

When deflation removes the finer-grained particles from a mixture of gravel, sand, and silt in sediments and soils, it produces a remnant surface of gravel too large to be transported by wind. Over thousands of years, as deflation removes the finer-grained particles from successive deposits of streams, the gravel accumulates as a layer of **desert pavement,** a coarse, gravelly ground surface that protects the soil or sediments below from further erosion (Figure 14.8).

This theory of pavement formation is not completely accepted because a number of pavements seem not to have formed in this way. A new theory is that some desert pavements are formed by deposition of windblown sediments. The coarse rock pavement stays at the surface, while windblown dust infiltrates below the surface layer of pavement, is modified by soil-forming processes, and accumulates there.

FIGURE 14.8 (a) Desert pavement in the Kofa Mountains of Arizona, in the Sonoran Desert. *(David Muench.)* (b) According to the dominant view, the evolution of desert pavement begins when the wind blows away fine-grained materials from a heterogeneous soil or sediment, and deflation occurs. As the removal of the finer sand and silt particles lowers the surface, the coarse gravel that remains gradually becomes concentrated into a compact layer that prevents further erosion.

(a)

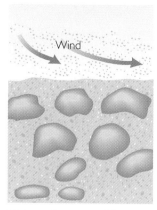

Mixture of coarse and fine particles at surface

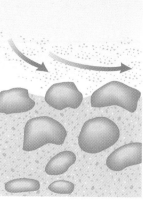

Wind gradually removes finer particles

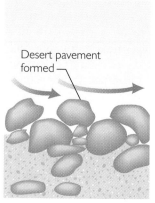

Desert pavement prevents further wind erosion

(b)

Sandblasting

A wind that contains blown sand is an effective natural **sandblasting** agent. The common method of cleaning buildings and monuments with compressed air and sand works on exactly the same principle: the impact of high-speed particles wears away the solid surfaces. Natural sandblasting mainly works close to the ground, where most sand grains are carried. Sandblasting rounds and erodes rock outcrops, boulders, and pebbles and frosts the occasional glass bottle.

Ventifacts are wind-faceted pebbles that show several curved or almost flat surfaces that meet at sharp ridges (Figure 14.9). Each surface or facet is made by sandblasting of the pebble's windward side. Occasional storms roll or rotate the pebbles, exposing a new windward side to be sandblasted to a plane. Many ventifacts are found in deserts and in glacial gravel deposits, where the necessary combination of gravel, sand, and strong winds is present.

On a much larger scale, eolian processes erode streamlined parallel ridges aligned with the direction of a strong prevailing wind. These ridges, called **yardangs,** are typically less than 10 m high and 100 m or more long, but very large ones may be 80 m high and more than 100 km long (Figure 14.10). Because they are found in the middle of extremely arid regions with little or no sand, they seem to be the products of abrasion by silt and dust. Exactly how the wind forms yardangs remains controversial.

FIGURE 14.9 These ventifacts, pebbles and cobbles, have been pitted and faceted by windblown sand. They were found in Sweetwater County, Wyoming. *(M. R. Campbell/USGS.)*

WIND AS A DEPOSITIONAL AGENT

When the wind dies down, it can no longer transport the sand, silt, and dust it has carried. The coarser material is deposited in variously shaped sand dunes

FIGURE 14.10 Yardangs outside the town of Minab, Iran. These parallel ridges, aligned with the direction of a strong prevailing wind, are found in extremely dry regions with little or no sand and may be the product of abrasion by silt and dust. *(Comstock.)*

ranging in size from low knolls to huge hills more than 100 m high. The finer silt and dust fall as a more or less uniform blanket of silt and clay. Geologists have observed these depositional processes working today and have linked them to the sediments' characteristics, primarily bedding and texture, to infer past climates and wind patterns from ancient sandstones and dust falls.

Where Sand Dunes Form

Sand dunes are found in relatively few environmental settings. Many of us have seen the dunes that form behind beaches along ocean coasts or large lakes. Some dunes are found on the sandy floodplains of large rivers in semiarid and arid regions. Most spectacular are fields of dunes in desert regions, which cover large expanses (Figure 14.11). Some reach heights of several hundred meters, truly mountains of sand.

All of the settings where dunes form have a ready supply of loose sand: beach sands along coasts, sandy river bars or floodplain deposits in river valleys, and sandy bedrock formations in deserts. Another common factor is wind power. On oceans and lakes, strong winds blow onshore off the water (Figure 14.12). Strong winds, sometimes of long duration, are common in deserts.

As we noted earlier, the wind cannot pick up wet materials easily, so most dunes are found in dry climates. The exception is dune belts along a coast, where sand is so abundant and dries so quickly in the wind that dunes can form even in humid climates. But in such climates, soil and vegetation begin to cover the dunes only a little way inland from the beaches, and the winds no longer pick up the sand.

Dunes may stabilize and become vegetated when the climate becomes more humid, then start moving again when the climate returns to more arid conditions. As noted earlier, there is geological evidence that during droughts two to three centuries ago and earlier, sand dunes in the western high plains of the United States and Canada were reactivated and migrated over the plains.

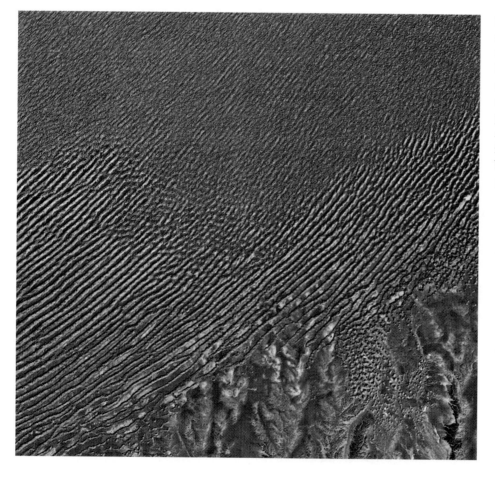

FIGURE **14.11** A vast expanse of sand dunes in the southern Arabian Peninsula. These linear dunes trend northeast-southwest, parallel with the prevailing northeast winds. At lower right is an eroding plateau of horizontally bedded sedimentary rock. *(ERIM.)*

FIGURE 14.12 Dunes formed in the back of a beach on the coast of central Peru. Winds, mainly onshore, blow sand back from the beach into a complex of parabolic dunes. *(Loren McIntyre.)*

How Sand Dunes Form and Move

How does a dune start? Given enough sand and wind, any obstacle, such as a large rock or a clump of vegetation, can start a dune. Streamlines of wind, like those of water, separate around obstacles and rejoin downwind, creating a wind-shadow zone downstream of the obstacle. Wind velocity is much lower in the wind-shadow zone than in the main flow around the obstacle, low enough to allow sand grains blown into the shadow to settle there. Because the velocity is low, these grains can no longer be picked up, and they accumulate as a **sand drift,** a small pile of sand in the lee of the obstacle (Figure 14.13). As the process continues, the sand drift itself becomes an obstacle. If there is enough sand and the wind continues to blow in the same direction long enough, the drift grows into a dune. Dunes may also grow by the enlargement of ripples, just as underwater dunes do.

As a dune grows, the whole mound starts to migrate downwind by the combined movements of a host of individual grains. Sand grains constantly saltate to the top of the low-angle windward slope, then fall over into the wind shadow on the lee slope, as shown in Figure 14.14. These grains gradually build up a steep unstable accumulation on the upper part of the lee slope. Periodically, the steepened buildup gives way and spontaneously slips or cascades down this **slip face,** as it is called, to a new

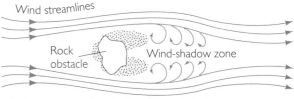

(a) Early stage: small drifts form in wind shadow

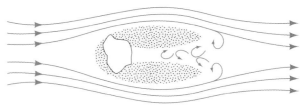

(b) Middle stage: large but separated drifts form in wind shadow

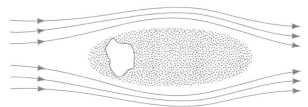

(c) Final stage: drifts coalesce into dune

FIGURE 14.13 Sand dunes may form in the lee of a rock or other obstacle. By separating the wind streamlines, the rock creates a wind shadow in which the eddies are weaker than the main flow. The windborne sand grains are thus allowed to settle and pile up in drifts that eventually coalesce into a dune. (After R. A. Bagnold, *The Physics of Blown Sand and Desert Dunes,* London, Methuen, 1941.)

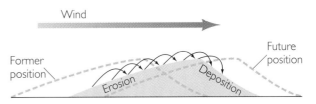

FIGURE 14.14 A ripple or dune advances by the movements of individual grains. The whole form moves forward slowly as sand erodes from the windward slope and is deposited on the leeward slope.

slope at a lower angle (Figure 14.15). If we overlook the short-term unstable steepenings in slope, the slip face maintains a stable, constant slope angle, its angle of repose. As we saw in Chapter 11, this angle increases with the size and angularity of the particles.

Successive slip faces deposited at the angle of repose create the cross-bedding that is the hallmark of windblown dunes. As dunes accumulate, interfere with one another, and become buried in a sedimentary sequence, the cross-bedding is preserved even though the original shapes of the dunes are lost. Sets of sandstone cross-bedding many meters thick are one of the lines of evidence that geologists use to infer high windblown dunes. From the directions of these eolian cross-beds, geologists can reconstruct wind directions of the past (see Figure 7.6).

As more sand accumulates on the windward slope of a dune than blows off onto the slip face, the dune grows in height. Heights of 30 m are common, and huge dunes in Saudi Arabia reach 250 m, which seems to be the limit. The explanation for limited dune height lies in the relationship of wind streamline behavior, velocity, and topography. Wind

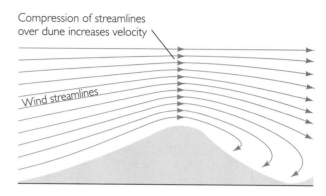

FIGURE 14.16 The height of a dune is limited by the behavior of wind streamlines. As the dune grows higher, the wind streamlines become compressed and their velocity increases. The increased velocity increases the wind's ability to transport sand grains. The rising dune eventually reaches a height at which the wind is so fast that sand grains blow off the dune as quickly as they are brought up the windward slope, and the dune stops growing vertically.

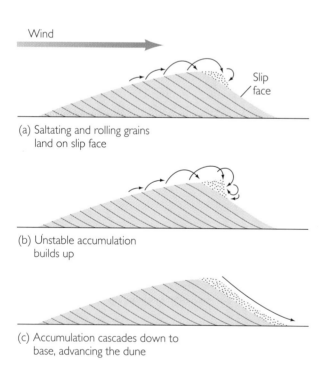

(a) Saltating and rolling grains land on slip face

(b) Unstable accumulation builds up

(c) Accumulation cascades down to base, advancing the dune

FIGURE 14.15 The formation of the slip face of a dune. In successive positions, (a) saltating, rolling, and sliding grains land on the slip face, which (b) accumulates an unstable slope as a result. Intermittently, this accumulation becomes so unstable that (c) it spontaneously slips to the base and a new, stable slope is formed at a lower angle.

streamlines advancing over the back of a dune become more compressed as the dune grows higher, as Figure 14.16 shows. As more air rushes through a smaller space, the wind velocity increases. Ultimately, the air speed at the top of the dune becomes so great that sand grains blow off the top of the dune as quickly as they are brought up the windward slope. When this balance is reached, the height of the dune remains constant.

Dune Types

A person standing in the middle of a large expanse of dunes might be bewildered by the seemingly orderless array of undulating slopes. It takes a practiced eye to see the dominant pattern, and it may even require observation from the air. Geologists know and agree that the general shapes and arrangements of sand dunes depend on the amount of sand available and the direction, duration, and strength of the wind. We cannot yet predict the specific shape a dune will take, however, or state with certainty the specific mechanisms by which a particular wind regime results in one dune form or another. We recognize four main types of dunes: barchans, transverse dunes, blowouts, and linear dunes.

BARCHANS   A **barchan** is a crescent-shaped dune, found often in groups or occasionally as a solitary dune, that moves over a flat surface of pebbles

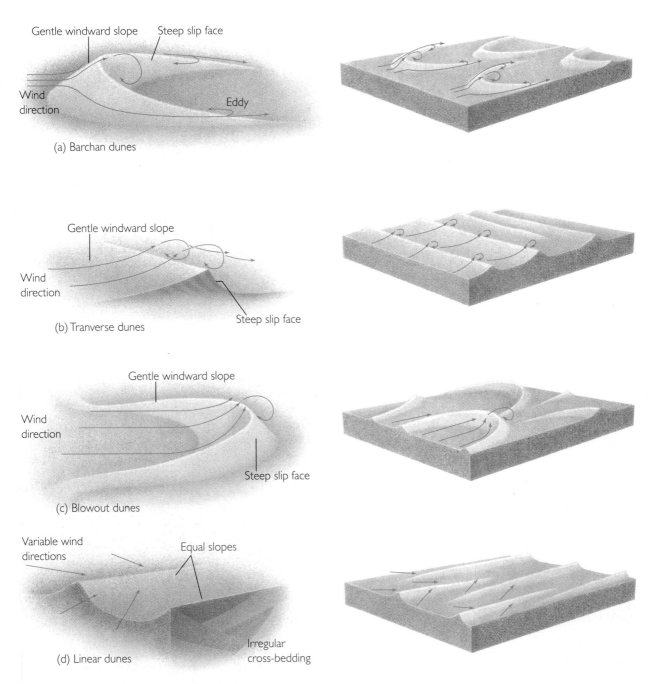

Gentle windward slope   Steep slip face

Wind direction

Eddy

(a) Barchan dunes

Gentle windward slope

Wind direction

Steep slip face

(b) Tranverse dunes

Gentle windward slope

Wind direction

Steep slip face

(c) Blowout dunes

Variable wind directions

Equal slopes

Irregular cross-bedding

(d) Linear dunes

FIGURE 14.17 Dune types in relation to prevailing winds. (a) Barchans are crescent-shaped dunes, usually but not always found in groups. The horns of the crescent point downwind. (b) Transverse dunes are long ridges oriented at right angles to the wind direction. (c) Blowout dunes are almost the reverse of barchans. The slip face of a blowout dune is convex downwind, while the barchan's is concave downwind. (d) Linear dunes are long ridges of sand oriented parallel to the wind direction.

or bedrock (Figure 14.17a). The points of the crescent are directed downwind, and the slip face is the concave curve advancing downwind. Barchans are the products of limited sand supply and unidirectional winds.

TRANSVERSE DUNES   Several barchans may coalesce to form irregular ridges whose long direction lies transverse (perpendicular) to the wind direction. These merged barchans may be transitional to **transverse dunes** (Figure 14.17b), long, wavy ridges oriented at right angles to the prevailing unidirectional wind. Transverse dunes form in arid regions where abundant sand dominates the landscape and vegetation is absent. Typically, sand dune belts behind beaches are transverse dunes formed by strong onshore winds. In temperate or humid areas, they are stabilized by vegetation at some distance from the beach.

BLOWOUT DUNES   If a section of such a dune belt is deflated and stabilizing vegetation is overwhelmed by sand, a parabola-shaped dune called a **blowout** (Figure 14.17c) will form at that point and migrate inland. In contrast to a barchan, which it superficially resembles, the arms of a blowout are directed upwind, and the slip face forms the convex curve advancing downwind.

LINEAR DUNES   **Linear dunes** are long, straight ridges more or less parallel to the general direction of the prevailing winds (Figure 14.17d). These dunes may reach heights of 100 m and extend many kilometers. The origin of linear dunes remains controversial, but most experts believe they are depositional forms created by winds of varying direction. Most areas covered by linear dunes have a moderate sand supply, a rough pavement, and winds that may shift but are always in the same general direction. During summer, for example, the winds blow from the southwest across linear dunes oriented east-west, depositing sand on a northeast-facing slip face. In winter, winds blow from the northwest and transport sand over the dune to a slip face facing southeast. The combination advances the dune in a general easterly direction while producing two different orientations of slip faces.

COMPOSITE FORMS   Extremely large, high, hilly forms are called **draas.** They are typically composites of superimposed dunes of several kinds, especially transverse and linear dunes, and in some places they reach heights of 400 m. Large dunes like these move much more slowly than small dunes, sometimes as slowly as 0.5 m per year. Where sand is abundant over wide areas and winds blow strongly, large fields of dunes are formed. The most extensive areas of this type are **ergs,** "seas of sand" found in major deserts such as those of Namibia (see the photograph at the opening of this chapter). An erg may cover as much as 500,000 km², twice the size of the state of Nevada.

Dust Falls and Loess

As the velocity of dust-laden wind decreases, the dust settles to form **loess,** a blanket of sediment composed of fine-grained particles. Beds of loess lack internal stratification; in compacted deposits more than a meter thick, loess tends to form vertical cracks and to break off along sheer walls during erosion (Figure 14.18). The vertical cracking may be caused by a combination of root penetration and uniform downward percolation of groundwater, but the exact mechanisms are still unknown.

Loess covers a huge amount of Earth's land surface, perhaps as much as 10 percent of it. Of this, more than a million square kilometers are in China (Figure 14.19). The great loess deposits of China spread over wide areas in the northwest; most are 30 to 100 m thick, although some exceed 300 m. The winds blowing over the Gobi Desert and the arid regions of central Asia were the source of the dust, which still blows over eastern Asia and the Chinese interior. Some of the loess deposits in China are 2 million years old, laid down at the same time as the elevation of the Himalayas and related uplifts of western China introduced rain-shadow and continental interior dry climates. These tectonic events

FIGURE 14.18 Pleistocene loess in Colorado, showing vertical cracking. In beds more than a meter thick, loess tends to form vertical cracks. Sheer walls then break away during erosion. *(H. E. Malde/USGS.)*

FIGURE 14.19 Comfortable dwelling caves hand-carved into steep cliffs of loess in central China. These deposits of windblown dust accumulated over the last 2.5 million years, reaching a thickness of up to 400 m. (*Stephen C. Porter.*)

were responsible for the cold, dry climates of the Pleistocene epoch in much of Asia. The cold, dry climates inhibited vegetation and dried out soils, causing extensive wind erosion and transportation.

The best known loess deposit in North America is in the upper Mississippi Valley. It is a legacy of recent glacial activity in the Pleistocene epoch, and the clues to its origin as windblown material deposited during melting of the large glaciers of that time were deciphered in the early part of this century. Geologists recognized that its pattern of distribution is that of a blanket of more or less uniform thickness on both hills and valleys, all in or near formerly glaciated areas. Changes in the regional thickness of the loess in relation to the prevailing westerly winds

confirm its eolian origin. The loess on the eastern sides of major river floodplains is 8 to 30 m thick, greater than on the western sides, and the thickness decreases downwind rapidly to 1 to 2 m farther east from the floodplains. The source of the dust was the abundant silt and clay deposited on extensive floodplains of rivers draining the edges of melting glaciers. Strong winds dried the floodplains, whose frigid climate and rapid rates of sedimentation inhibited vegetation, and blew up tremendous amounts of dust, which then settled to the east.

In China, North America, and elsewhere, soils formed on loess are fertile and highly productive. They also pose environmental problems, for they are easily eroded into gullies by small streams and are deflated by the wind when they are poorly cultivated.

Airborne dust has been measured far out to sea by oceanographic research vessels. Comparison of the composition of this dust with many of the components of deep-sea sediments in the same region indicates that windblown dust is an important contributor to oceanic sediment, as much as a billion tons each year. Much of this dust is of volcanic origin, and there are individual ash beds marking very large eruptions. Volcanic dust is abundant because much of it is very fine-grained and is injected high in the atmosphere, where it travels farther than much of the windblown dust from the continents.

THE DESERT ENVIRONMENT

The hot, dry deserts of the world are among the most hostile environments for humans, yet many of us are fascinated by the strange forms of animal and plant life and the bare rocks and sand dunes found there. Of all the environments of Earth's surface, the wind is best able to do its work of erosion and sedimentation in the desert.

Where Deserts Are Found

Rainfall is the major factor determining the location of the world's great deserts. The Sahara and Kalahari deserts of Africa and the Great Australian Desert get extremely low amounts of rainfall, normally less than 25 mm each year and in some places less than 5 mm. These deserts lie in the warmest regions of the globe, within 30° north and 30° south latitude of the equator. The deserts lie under virtually stationary areas of high atmospheric pressure; the Sun beats

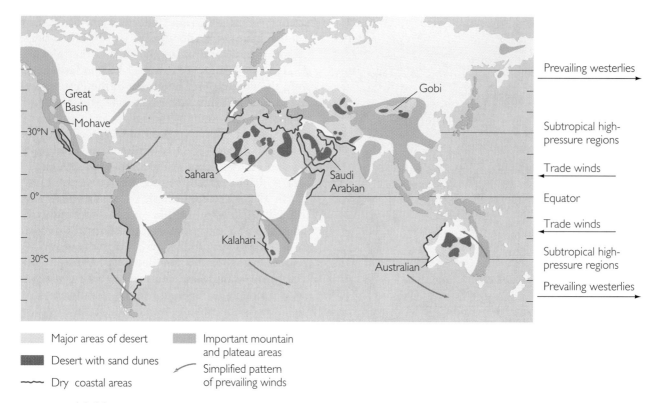

Major areas of desert

Desert with sand dunes

Dry coastal areas

Important mountain and plateau areas

Simplified pattern of prevailing winds

FIGURE **14.20** Major desert areas of the world (exclusive of polar deserts) in relation to prevailing wind directions and major mountain and plateau areas. Sand dunes are only a small proportion of the total desert area. (After K. W. Glennie, *Desert Sedimentary Environments,* New York, Elsevier, 1970.)

down through a cloudless sky week after week, and the air maintains extremely low humidity (Figure 14.20).

Deserts also form in mid-latitudes, those between 30° and 50° north and 30° and 50° south, in regions where rainfall is low because moisture-laden winds either are blocked by mountain ranges or must travel great distances from their source of moisture, the ocean. The Great Basin and Mohave deserts of the western United States, for example, lie in rain shadows created by the western coastal mountains. As we saw in Chapter 12, wind descending from the mountains warms and dries, leading to low precipitation. The Gobi and other deserts of central Asia are so far inland that the winds reaching them have precipitated all their ocean-derived moisture long before they arrive at the interior of the continent.

Another kind of desert forms in polar regions. Little precipitation is possible in these cold, dry areas because the frigid air can hold only extremely small amounts of moisture. The dry valley region of southern Victoria Land in Antarctica is so dry and cold that its environment resembles that of Mars.

Deserts, in a sense, are reflections of plate tectonics. The mountains that create rain shadows are made by collisions between converging continental and oceanic plates. The great distance separating central Asia from the oceans is a consequence of the size of the continent, a huge landmass assembled from smaller plates by continental drift. Large deserts are at low latitudes because continental drift powered by plate tectonics has moved them there from higher latitudes. If, in some future plate-tectonic scenario, the North American continent were to move south by more than 2000 km, the northern Great Plains of the United States and Canada would be converted to a hot, dry desert. Something like that has happened to Australia. Australia was once far to the south of its present position, and its interior had a moist, humid climate. Since then, Australia has moved northward into an arid subtropical zone, where its interior has become a desert. (We return to this topic in Chapter 20, where we describe the many changes that took place following the breakup of the supercontinent Pangaea.)

Changes in a region's climate, its population, or its population's behavior may transform semiarid lands into deserts by a process called **desertification.** Climatic changes that we do not fully understand may decrease precipitation for decades or even centuries. Following such a period of aridity, the region may return to milder, wetter conditions. It appears that for the past 10,000 years, climates of the Sahara have oscillated between more and less arid conditions. We have evidence from radar imaging of the Sahara from the space shuttle *Endeavor* that an extensive system of river channels existed a few thousand years ago. Now dry and buried by more recent sand deposits, these ancient drainage systems carried abundant running water across the northern Sahara during wetter climatic periods.

These oscillating climate changes were not the result of human activities, but such activities are responsible for some desertification today. The unrestrained growth of human populations and their agriculture and the expansion of animals' grazing activities may also result in the expansion of deserts. When population growth and periods of drought coincide, the results in semiarid regions can be disastrous (see Feature 14.2).

The opposite of desertification, "making the desert bloom," has been a slogan of some countries with desert lands. They are busy irrigating on a massive scale to convert marginal or desert areas to productive farmlands. The Great Valley of California, where much of North America's fruits and vegetables are grown, is a giant example. The growth of cities like Phoenix, Arizona, which depend on imported water, has produced islands of urbanization, complete with humid haze and air pollution, in an otherwise arid environment.

All told, arid regions amount to one-fifth of Earth's land area, about 27.5 million square kilometers. Semiarid plains account for an additional one-seventh. Given the reasons for the existence of large areas of deserts in the modern world—mountain building by plate tectonics, transport of continental regions to low latitudes by continental drift, and the climatic belts of the globe—we can be confident that, by the principle of uniformitarianism, extensive deserts have existed throughout geologic time.

Desert Weathering

As much as deserts differ from more humid regions, the same geological processes operate on both. Weathering and transportation work in the same way, but with a different balance in deserts than in temperate areas. In the desert, physical weathering predominates over chemical weathering. Feldspars and other silicates chemically weather to clay minerals, but slowly, for lack of the water required for the reaction to proceed. The little clay that does form is normally blown away by strong winds before it can accumulate. Slow chemical weathering and rapid wind transport combine to prevent the buildup of any significant thickness of soil, even where sparse vegetation binds some of the particles. Thus, soils are thin and patchy. Sand, gravel, rock rubble of many sizes, and bare bedrock are characteristic of much of the desert surface.

THE COLORS OF THE DESERT   The common rusty, orange-brown colors of weathered surfaces in the desert come from the ferric iron oxide minerals hematite and limonite. These minerals are produced by the slow weathering of iron silicate minerals such as pyroxene. The iron oxides, even when present only in small amounts, stain the surfaces of sands, gravels, and clays.

Desert varnish is a distinctive dark brown, sometimes shiny, coating found on many rock surfaces in the desert. It is a mixture of clay minerals with smaller amounts of manganese and iron oxides. Desert varnish is hypothesized to form very slowly from a combination of dew, chemical weathering that produces the clay minerals and iron and manganese oxides, and the sticking of windblown dust to exposed rock surfaces. The process is so slow that Native American inscriptions made in desert rock varnish hundreds of years ago still appear fresh, with a stark contrast between the dark varnish and the light unweathered rock beneath (Figure 14.21).

FIGURE 14.21 Petroglyphs scratched in desert varnish by early Native Americans; Newspaper Rock, Canyonlands, Utah. The scratches are several hundred years old and appear fresh, whereas the varnish accumulated over thousands of years. *(Peter Kresan.)*

14.2 INTERPRETING THE EARTH

Desertification on the Fringe of the Sahara

In 1984, a catastrophic famine struck Ethiopia and other countries bordering the southern Sahara, causing enormous human suffering. For decades, the region had been racked by a series of droughts, with a particularly devastating one in 1973. What were formerly vegetated semiarid lands used by small groups of migrating peoples had been degraded into barren lands incapable of supporting even a small population. The process of desertification had ruined the land, at least temporarily.

The reasons for desertification in sub-Saharan Africa are complex. This naturally semiarid region had become more desertlike when the decades of drought killed much natural vegetation. At the same time, as the population in the region grew and the people turned to ever more intensive grazing and agriculture, the easily erodible soil thinned and began to blow away in the fierce winds. Aridity and soil erosion have been so extreme that it is doubtful any conservation method could restore the original quality of the soil in the near future.

Some atmospheric scientists are now citing evidence of climate change oscillating between more and less arid conditions in the Sahara over the last 10,000 years. They believe the process of desertification may be largely natural, not attributable to the human population. They point to satellite and ground observation studies for the 10-year period from 1970 to 1980, which show little or no evidence of expansion of arid land radiating out of existing settlements.

If the climate does change slightly and the long dry period ends, additional rainfall will supplement human efforts to restore the land. But the countries now coping with this problem have limited economic resources. Other populations could face similar problems in the future. Some scientists estimate that about 35 percent of Earth's surface is potentially threatened by desertification. This area, on which more than 850 million people live, includes parts of the western United States and Mexico, central Asia, and Argentina.

Famine in Ethiopia is caused by the desertification of sub-Saharan Africa. People search for food and firewood, often in vain. *(Frilet/Sipa-Press.)*

Varnish accumulates over thousands of years. A geologically ancient (Miocene) varnish has been identified, but recognizing varnish as such in sandstones of the geological past is difficult.

STREAMS: THE PRIMARY EROSIVE AGENT

Wind is a more significant agent of erosion in the desert than elsewhere, but it cannot compete with the erosive power of streams. Even though it rains so seldom that most desert streams flow only intermittently, such streams do most of the erosional work in the desert when they do flow.

Even the driest desert gets occasional rain. In sandy, gravelly areas of deserts, infiltration of the occasional rainfall into soil and permeable bedrock temporarily replenishes groundwater in the unsaturated zone. There, some of it evaporates very slowly into open pore space between the grains. A smaller amount eventually reaches the groundwater table far below, in some places as much as hundreds of meters below the surface. Desert oases form when the groundwater table comes sufficiently close to the surface that roots of palms and other plants can reach it.

When rain falls as heavy cloudbursts, so much water falls in such a short time that even though a fraction soaks into the ground, infiltration cannot keep pace and the bulk of it runs off into streams. Unhindered by vegetation, the runoff is rapid and may cause flash floods along valley floors that have been dry for years. Thus, a large proportion of stream flows in the desert consist of floods. When such floods occur, their erosive power is great because most loose weathering debris is not held in place by vegetation. Streams may become so choked with sediment that they look more like fast-moving mudflows than like rivers. The abrasiveness of this sediment load moving rapidly at flood velocities makes such streams efficient eroders of bedrock valleys.

Desert Sediment and Sedimentation

ALLUVIAL SEDIMENTS

As sediment-laden flash floods dry up, they leave distinctive deposits on the floors of desert valleys. A flat fill of coarse debris often covers the entire valley floor, without the ordinary differentiation into channel, levees, and floodplains (Figure 14.22). The sediments of many other desert valleys clearly show the interfingering of stream-deposited channel and floodplain sediments and eolian sediments. The combination of stream and eolian processes in the past has formed extensive sheets of eolian sandstone separated by stream flood

surfaces and floodplain interdune sandstones. Large alluvial fans (see Chapter 13) are prominent features at mountain fronts in deserts because desert streams deposit much of their high sediment load on the fans. The rapid infiltration of stream water into the permeable fan material deprives the streams of the water required to carry the sediment load any farther downstream. Debris flows and mudflows make up large parts of the alluvial fans of arid, mountainous regions.

EOLIAN SEDIMENTS

By far the most dramatic sedimentary accumulations of the desert are sand dunes, dune fields, and ergs. Although films and TV lead one to think that deserts are mostly sand, actually only one-fifth of the desert areas of the world is covered by sand. The other four-fifths are rocky or

(a)

(b)

FIGURE **14.22** (a) Dry-wash flooding during a summer thunderstorm in Saguaro National Monument, Arizona. (b) The same dry wash a day after flooding. The coarse debris deposited by sudden desert floods may cover the entire valley floor. *(Peter Kresan.)*

FIGURE **14.23** A desert playa lake in Death Valley, California. Playa lakes are sources of evaporite minerals such as borax. Their waters may be deadly to drink because of high alkalinity or salinity. (*David Muench.*)

covered with desert pavement. Only a little more than one-tenth of the Sahara Desert is covered by sand, and sand dunes are far less common in the arid lands of the southwestern United States.

EVAPORITE SEDIMENTS   **Playa lakes** are permanent or temporary lakes that accumulate in arid mountain valleys or basins (Figure 14.23). As the lake waters evaporate, weathering products dissolved in them are concentrated and gradually precipitated. Playa lakes are sources of evaporite minerals: sodium carbonate, borax (sodium borate), and other unusual salts. The water of playa lakes may be deadly to drink because of high levels of alkalinity or salinity, which stem from evaporation and from salt contributed by desert streams. Large amounts of dissolved salts may also accumulate when the streams redissolve evaporite minerals deposited by evaporation from earlier runoff. If evaporation is complete, the lakes become **playas,** flat beds of clay that are sometimes encrusted with precipitated salts.

Desert Landscape

The landscape of the desert is one of the most varied on Earth. Large, low, flat areas are covered by playas, desert pavements, and dune fields. Uplands are rocky, cut in many places by steep river valleys and gorges. The lack of vegetation and soil makes everything seem sharper and harsher than landscapes in more humid climates (see Chapter 16). In contrast to the rounded, soil-covered, vegetated slopes found in most humid regions, the coarse, heterogeneously sized fragments produced by desert weathering form steep cliffs with masses of angular talus slopes at their bases (Figure 14.24).

Valleys in deserts have the same range of profiles as valleys elsewhere, but far more of them have steep walls caused by rapid erosion from mass movements and streams. Much of the landscape of deserts, like that of humid regions, is shaped by rivers, but the valleys—called **dry washes** in the western United States and **wadis** in the Near East—are usually dry. Even where water is scarce, it is the infrequent rainfall runoff in streams that does most of the basic work of erosion. The wind helps, but it rarely controls. Only in dune fields is the work of the wind the dominant process, and even there streams may flow in interdune areas and erode the edges of the dunes.

Desert streams are widely spaced because of the relatively infrequent rainfall, but drainage patterns are generally similar to those of other terrains, with one difference. Because of infrequent rainfall and water losses by evaporation and infiltration, many desert streams die out downstream or are dammed by dunes, never reaching across the desert to join larger rivers flowing to the oceans.

FIGURE **14.24** This desert landscape at Kofa Butte, Kofa National Wildlife Refuge, Arizona, shows the steep cliffs and masses of talus slopes produced by desert weathering. *(Peter Kresan.)*

A special type of eroded bedrock surface, called a **pediment,** is a characteristic landform of the desert. Pediments are broad, gently sloping platforms of bedrock left behind as a mountain front erodes and retreats from its valley (Figure 14.25a, b). The pediment spreads like an apron around the base of the mountains as thin deposits of alluvial sands and gravels accumulate (Figure 14.25c). Long-continued

FIGURE **14.25** Stages in the evolution of a typical pediment, an erosional form produced in arid mountainous settings. (a) First, a downfaulted lowland is formed. (b) Alluvial fan and floodplain sediments are deposited in the lowland. (c) The erosional surface of the pediment, produced by migrating streams, is covered by thin deposits of alluvial sands and gravels. (d) Long-continued erosion makes an extensive pediment with some mountain remnants. The mountain front retreats steadily but keeps the same steep slope angle, in contrast to the gradually rounded slopes that evolve in humid climates.

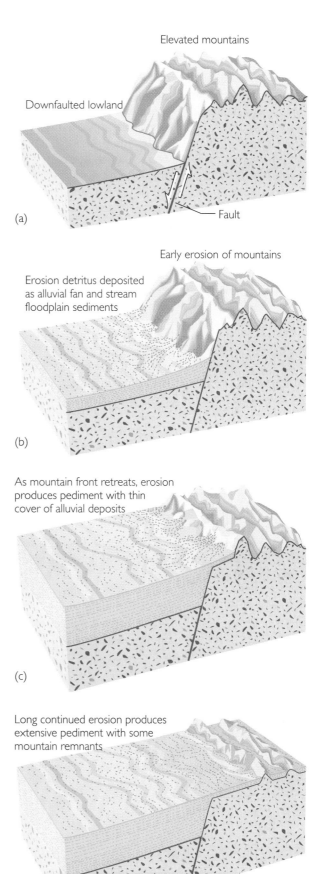

Elevated mountains

Downfaulted lowland

Fault

(a)

Early erosion of mountains

Erosion detritus deposited as alluvial fan and stream floodplain sediments

(b)

As mountain front retreats, erosion produces pediment with thin cover of alluvial deposits

(c)

Long continued erosion produces extensive pediment with some mountain remnants

(d)

erosion eventually forms an extensive pediment with some mountain remnants. A cross section of a typical pediment and its mountains would reveal a fairly steep mountain slope abruptly leveling into the gentle pediment slope (Figure 14.25d). Alluvial fans deposited at the lower edge of the pediment merge with the sedimentary fill of the valley below the pediment.

There is much evidence that the pediment is formed by running water that both cuts the erosional platform above and deposits an alluvial fan apron below. At the same time, the mountain slopes at the head of the pediment maintain their steepness as they retreat, instead of becoming the rounded, gentler slopes found in humid regions. We do not know how the specific rock types and erosional processes interact in an arid environment to keep slopes steep while the pediment enlarges.

Long-continued desert erosion also produces another kind of desert landscape made familiar by film and TV westerns made in southern California and nearby states. Exposed to desert erosion, horizontal sedimentary rocks may form **mesas**—table-like uplands capped by erosion-resistant beds bounded by steep-sided erosional cliffs. The capping bed of a mesa tends to maintain the upland-level elevation. Wherever it is cut through, however, erosion cuts down rapidly into the less resistant beds and creates cliffs (Figure 14.26a, b). As the cliffs retreat, the upland area shrinks, finally leaving only a few isolated mesas above the surrounding lowland (Figure 14.26c).

The desert landscapes of millions of years ago have now been obliterated by later erosion. Nevertheless, the preservation of desert sediments in the rock record has allowed geologists to reconstruct the time and extent of ancient deserts. One of the best studied is the Permian to Jurassic desert that once covered large areas of the western United States and Canada.

The deserts of the geologic past were like those of today in all essential respects. Thus, we know that arid climates and their consequences have dominated large sections of the continents for most of geologic time.

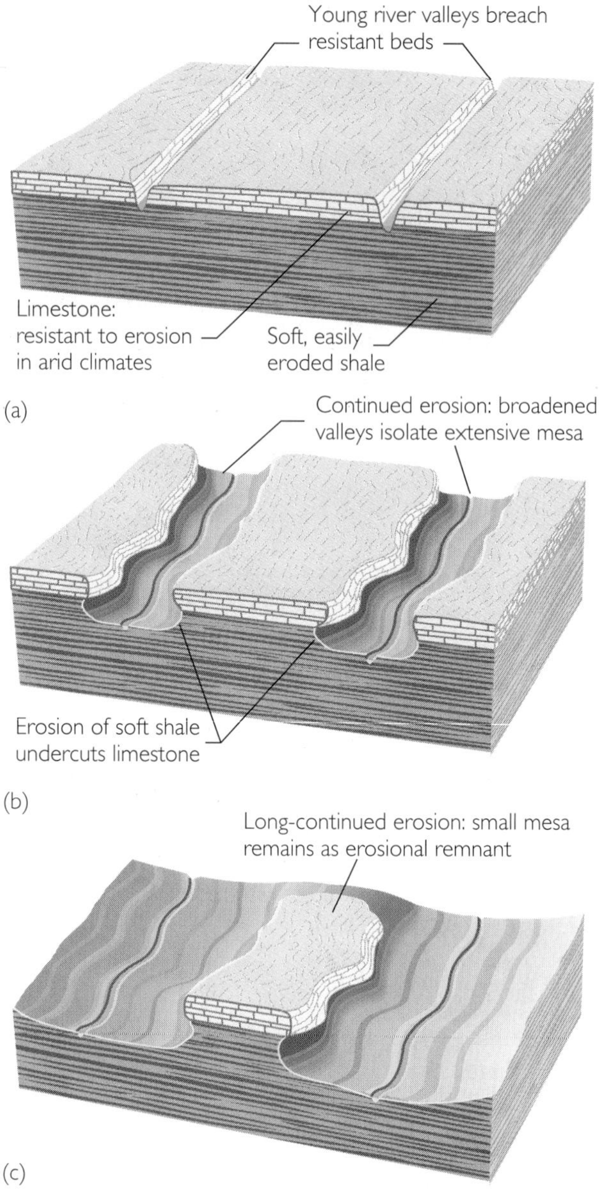

FIGURE **14.26** The evolution of a mesa. (a) The process is controlled by an erosion-resistant sedimentary bed overlying weak, easily erodible rocks. (b) The border of the mesa retreats as the lower beds are eroded, undercutting the resistant beds. (c) After long-continued erosion, a small mesa remains elevated above the lowland valley.

SUMMARY

How do winds erode and transport sand and finer-grained sediment? Wind, a horizontal flow of air, can pick up and transport dry sediment particles in the same way that running water transports sediment. Air flows are limited both in the size of particles they can carry (rarely larger than coarse-grained sand) and in their restricted ability to maintain particles in suspension for long times, because of air's low viscosity and density. Almost all air flows are turbulent, and high winds may reach velocities of 100 km per hour or more, enhancing the wind's ability to carry sediment in suspension. Windblown materials include volcanic ash, quartz and other mineral fragments such as clay minerals, and organic materials such as pollen and bacteria. The wind can carry great amounts of sand and dust. It moves sand grains primarily by saltation and carries the finer-grained silt and clay-sized particles (dust) in suspension. Winds blow sand into cross-bedded ripples and dunes. Deflation and sandblasting are the primary ways winds erode Earth's surface, thereby producing desert pavement and ventifacts.

How do winds deposit sand dunes and dust? In the course of transporting sand, the wind deposits it in dunes of various shapes and sizes in sandy desert regions, behind beaches, and along sandy floodplains, all of which are places with a ready supply of loose sand and moderate to strong winds. Dunes start as sand drifts in the lee of obstacles and grow to heights of many meters or even hundreds of meters. Dunes migrate downwind as sand grains saltate up the gentler windward slopes and fall over onto the steeper downwind slip faces. The various kinds of dunes—transverse, linear, barchan, and blowout—are responses to the speed of the wind, its constancy or variability of direction, and the abundance of sand. As the velocity of dust-laden winds decreases, the dust settles to form loess, a blanket of dust.

Loess layers in many recently glaciated areas were deposited by winds blowing over the floodplains of muddy streams formed by meltwater. Loess can accumulate to great thicknesses downwind of dusty desert regions.

How do wind and water combine to shape the desert environment and its landscape? Desert regions are formed in the rain shadows of mountain ranges, in subtropical regions of constant high pressure, and in the interiors of some continents. In all these places, originally moisture-laden winds become dry, and rainfall is rare. Weathering mechanisms are the same in deserts as in more humid regions, but in deserts physical breakdown of rocks is predominant, with chemical weathering at a minimum because of the lack of water. Most desert soils are thin, and bare rock surfaces are common. Streams may run only intermittently, but they are responsible for much of the erosion and sedimentation in the desert, carrying away heavy loads of coarse sediment and depositing them on alluvial fans and floodplains. Naturally dammed rivers in mountainous deserts can form playa lakes, which deposit evaporite minerals as they dry up. Desert landscapes are composed of dunes formed by eolian sedimentation; desert pavements; mesas; and pediments, which are broad, gently sloping platforms eroded from bedrock as mountains retreat while maintaining the steepness of their slopes.

KEY TERMS AND CONCEPTS

eolian (p. 346)

deflation (p. 351)

desert pavement (p. 352)

sandblasting (p. 353)

ventifact (p. 353)

yardang (p. 353)

sand drift (p. 355)

slip face (p. 355)

barchan (p. 356)

transverse dune (p. 357)

blowout (p. 358)

linear dune (p. 358)

draa (p. 358)

erg (p. 358)

loess (p. 358)

desertification (p. 361)

desert varnish (p. 361)

playa lake (p. 364)

playa (p. 364)

dry wash (p. 364)

wadi (p. 364)

pediment (p. 365)

mesa (p. 366)

EXERCISES

1. What is the range of materials and sizes of particles that the wind can move?

2. What is the difference between the way wind transports dust and the way it transports sand?

3. How is the ability of the wind to blow sedimentary particles linked to climate?

4. What are the main features of wind erosion?

5. Where do sand dunes form?

6. Name three types of sand dunes and show their relationship to wind direction.

7. What typical desert landforms are composed of sediment?

8. What are the geologic processes that form playa lakes?

9. What is desertification?

10. Where are loess deposits found?

THOUGHT QUESTIONS

1. You have just driven a truck through a sandstorm and discover that the paint has been stripped from the lower parts of the truck but the upper parts are barely scratched. What process is responsible, and why is it restricted to the lower parts of the truck?

2. What evidence might you find in an ancient sandstone that would point to its eolian origin?

3. Compare the heights to which sand and dust are carried in the atmosphere and explain the differences or similarities.

4. Trucks continually have to haul away sand covering a coastal highway. What do you think might be the source of the sand? Could its encroachment be stopped?

5. What features of a desert landscape would lead you to believe it was formed mainly by streams, with secondary contributions from eolian processes?

6. Which of the following options would be a more reliable indication of the direction of the wind that formed a barchan dune: crossbedding, or the orientation of the dune's shape on a map? Why?

7. What factors determine whether sand dunes will form on a stream floodplain?

8. There are large areas of sand dunes on Mars. From this fact alone, what can you infer about conditions on the Martian surface?

9. What aspects of an ancient sandstone would you study to show that it was originally a desert sand dune?

10. What kinds of landscape features would you ascribe to the work of the wind and/or to the work of streams?

11. How does desert weathering differ from or resemble weathering in more humid climates?

12. What evidence would cause you to infer that dust storms and strong winds were common in glacial times?

LONG-TERM TEAM PROJECT

See Chapter 11.

SUGGESTED READINGS

Bagnold, Ralph A. 1941. *The Physics of Blown Sand and Desert Dunes.* London: Methuen.

Brookfield, M. E., and S. Thomas (eds.). 1983. *Eolian Sediments and Processes.* New York: Elsevier.

Cooke, R. U., A. Warren, and A. Goudie. 1993. *Desert Geomorphology.* England: UCL Press.

Haff, P. K. 1986. Booming dunes. *American Scientist,* 74:376–381.

Idso, Sherwood B. 1976. Dust storms. *Scientific American* (October):108.

Mabbutt, J. A. 1977. *Desert Landforms.* Cambridge, Mass.: MIT Press.

Pye, K. 1989. *Aeolian Dust and Dust Deposits.* New York: Academic Press.

Sheridan, D. 1981. *Desertification of the United States.* Washington, D.C.: Council on Environmental Quality.

INTERNET SOURCES

The Day of the Black Blizzard

🛈 **http://www.discovery.com/DCO/doc/1012/world/history/dustbowl/dustbowlopener.html**

Discovery Channel Online presents this Web site on the American Dust Bowl of the 1930s. Following the events of one day as viewed by residents of a Kansas town, images and audio clips document one of the great dust storms.

Bright Edges of the World

🛈 **http://www.nasm.edu:1995/**

An electronic exhibit produced by the National Air and Space Museum of the Smithsonian Institution, this site focuses on the arid, semiarid, and subhumid regions of the world. Major features include "What Are Drylands?", "Drylands: A Global Asset," and "Drylands: A Global Concern."

Tracking Environmental Change in West Africa

🛈 **http://edcwww.cr.usgs.gov/doc/edchome/usaid/tap.html**

"The U.S. Geological Survey and the USAID Mission to Senegal have defined a cost-shared project ($1.7 million over 3 years) to monitor rapid changes in West Africa's natural and agricultural resources over the last 50 years. The collaborative project also aims to better understand how humans and climate influence these changes." This site summarizes the findings of the project.

The Winds of Mars

🛈 **http://cass.jsc.nasa.gov/pub/publications/slidesets/winds.html**

This NASA site is an introduction to eolian activity and landforms on Mars. Images from the *Mariner 9* and *Viking* missions to Mars include detailed descriptions.

Wind Erosion Research

🛈 **http://www.weru.ksu.edu/**

The U.S. Department of Agriculture—Agricultural Research Service Wind Erosion Laboratory at Kansas State University "serves as a focal point for wind erosion research in the United States." The site includes general information about wind erosion, wind erosion simulation models, and a data archive.

15

Grand Plateau Glacier,
Glacier Bay National Park,
Alaska. These rivers of ice
flow downhill in valleys,
like streams of water. In
the middle background,
tributary glaciers join the
main glacier.
(Fred Hirschmann.)

Glaciers: The Work of Ice

The view of Earth's vast blue oceans and white clouds from space dramatically emphasizes the water covering our planet. But satellite pictures of polar regions and white-peaked mountain ranges also show us that much of the water is frozen. In this chapter, we take a close look at Earth's icy regions, the ways they change, how fast they change, and how human activities affect such changes. We also examine the effect glaciers have as they carry and deposit loads of sediment, leaving their mark on Earth's surface as they advance and retreat. Ice is an important geological agent shaping the land surface in Earth's frigid regions.

About 10 percent of Earth's land surface is covered by glacial ice, much of which moves slowly and steadily outward from the centers of the polar ice caps and downward from the mountain peaks. Over the short term, the ice melts at the edges of glaciers at about the same rate as it advances, and so the total ice area remains the same.

Over a longer span of time, a cooling climate can cause the ice sheets to expand as melting fails to keep pace with the accumulation and movement of glacial ice. As recently (geologically speaking) as 20,000 years ago, snow and ice covered almost three times as much of the land surface as they do now. Yet only a few decades from now, ice on Earth may shrink significantly as global warming melts the ice faster than it can accumulate and move. The global impact could be immense, as melting ice raises sea level and submerges low-lying cities and as climatic zones migrate, changing temperate zones to semi-arid ones and vice versa. There is no doubt that a knowledge of Earth's icy regions is an intensely practical subject.

Glaciers erode steep-walled valleys, scrape bedrock surfaces, and pluck huge blocks from their rocky floors. In the relatively short geological time span of the recent ice ages, glaciers carved far more topography than did rivers and the wind. The debris of glacial erosion is enormous—ice transports huge tonnages of sediment, carrying it to the edge of the glacier, where it is deposited or carried away by meltwater streams. The widespread effects of glacial erosion and sedimentation over the Earth affect

- Water discharge and sediment loads of major river systems
- Quantity of sediment delivered to the oceans
- Erosion and sedimentation in coastal areas and on shallow continental shelves by creating changes in sea level

Many of the landscapes of mountain belts were sculpted by glaciers that have since melted away. The landscapes of large portions of continental lowlands were shaped relatively recently by Pleistocene glaciers that extended far into what are now temperate regions. As these glaciers melted, they left sediments and erosional features as evidence of their former extent.

Because snow and prolonged cold are necessary for the formation of glaciers, glacial sediments and erosional forms are indicators of past and present climates. The former large extent of glaciers is evidence of a recent past in which the climate over large regions was much colder than it is today. At some times much farther back in the geologic past, glaciers covered regions that are now tropical jungles. At other times, parts of the globe now covered with ice were warm and humid, and there were no polar ice caps. The former extent of glaciers can be used not only to trace climates of the past but also to predict future changes in climate.

ICE AS A ROCK

What is ice? To a geologist, a block of ice is a rock, a mass of crystalline grains of the mineral ice. Like most rocks, ice is hard, but it is much less dense than most rocks. It shares with igneous rocks an origin as a frozen fluid. Like sediments, it is deposited in layers at the surface of the Earth and can accumulate to great thicknesses. Like metamorphic rocks, it is transformed by recrystallization under pressure. Glacial ice forms by the burial and metamorphism of the "sediment" snow. The rock is formed as the loosely packed snowflakes—each a single crystal of the mineral ice—age and recrystallize into a solid mass (Figure 15.1). The unusual characteristic of ice as a mineral is its extremely low melting temperature, hundreds of degrees lower than the temperatures at which most minerals melt.

What Is a Glacier?

A mass of ice, like any other mass of rock or soil at Earth's surface, can move downhill. **Glaciers** are

FIGURE **15.1** A typical mosaic of crystals of glacier ice. Each small area of uniform color is a single crystal as seen in polarized light. The tiny circular and tubular spots are bubbles of air. (*Z. Xie/Lanzhou Institute of Glaciology, Academia Sinica, People's Republic of China.*)

large masses of ice on land that show evidence of being in motion or of once having moved.

We divide glaciers, on the basis of size and shape, into two basic types: valley glaciers and continental glaciers.

VALLEY GLACIERS Many skiers and mountain climbers are familiar with **valley glaciers,** sometimes called *alpine glaciers.* These rivers of ice form in the cold heights of mountain ranges where snow accumulates, usually in preexisting valleys, and they flow down the bedrock valleys. The photograph at the beginning of this chapter shows a valley glacier. Most of these glaciers occupy the complete width of the valley and may bury its bedrock base under hundreds of meters of ice. In warmer, low-latitude climates, valley glaciers may be found only at the heads of valleys on the highest mountain peaks. In colder, high-latitude climates, valley glaciers may descend many kilometers, down the entire length of a valley; in some places, they may extend as broad lobes into lower lands bordering mountain fronts. Where valley glaciers flow down coastal mountain ranges, they may terminate at the ocean's edge, where masses of ice break off and form icebergs.

CONTINENTAL GLACIERS A **continental glacier** is much larger than a valley glacier and is an extremely slow moving, thick sheet of ice (hence sometimes called an *ice sheet*). The world's largest ice sheets today are those that cover much of Greenland and Antarctica (Figure 15.2). The glacial ice of Greenland and Antarctica is not confined to mountain valleys but covers practically their entire land surface. In Greenland, 2.8 million cubic kilometers of ice cover 80 percent of the island's total area of 4.5 million square kilometers. The upper surface of the ice sheet resembles an extremely wide convex lens. At its highest point, in the middle of the island, the ice is more than 3200 m thick (Figure 15.3). From this central area, the ice surface slopes to the sea on all sides. At the mountain-rimmed coast, the

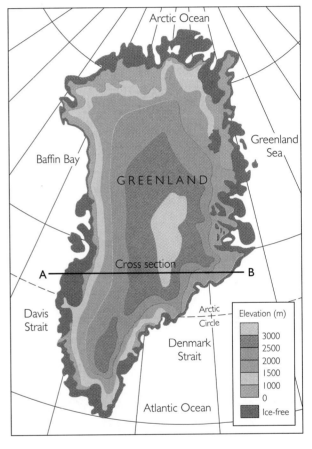

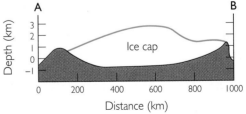

FIGURE 15.2 Sentinel Range, Antarctica. These mountains stick up through the thick ice of the Antarctic continental glacier, a huge expanse. *(Betty Crowell.)*

FIGURE 15.3 The extent of the glacial ice cap and the elevation of the ice surface on Greenland. The generalized cross section of south-central Greenland, A–B, shows the lenslike shape of the ice cap. The ice moves down and out from the thickest section. (From R. F. Flint, *Glacial and Quaternary Geology,* New York: Wiley, 1971.)

ice sheet breaks up into narrow tongues resembling valley glaciers that wind through the mountains to reach the sea. At the sea, the ice breaks off to form icebergs.

As large as the Greenland glacier is, it is dwarfed by the Antarctic ice sheet (Figure 15.4). Ice blankets 90 percent of the Antarctic continent, covering an area of about 12.5 million square kilometers and reaching thicknesses of about 3000 meters. In Antarctica, as in Greenland, the ice domes in the center and slopes down to the margins. In places, thinner sheets of ice floating on the ocean are attached to the main glacier on land. Of these, the best known is the Ross Ice Shelf, a thick layer of ice about the size of Texas that floats on the Ross Sea.

Ice caps are the masses of ice formed at Earth's North and South Poles. Most of the Arctic ice cap, formed at the lowest north latitudes, lies over water

and is generally not referred to as a glacier. Almost all of the Antarctic ice cap lies over land, the continent of Antarctica, and is considered to be a continental glacier.

How Glaciers Form

A glacier starts with abundant winter snowfall that does not melt away in the summer. The snow is gradually converted to ice, and when the ice is thick enough, it begins to flow.

FIRST REQUIREMENT: LOW TEMPERATURES
For glaciers to form, temperatures must be low enough to keep snow on the ground year-round. These conditions are found at high latitudes (polar and subpolar regions) and high altitudes (moun-

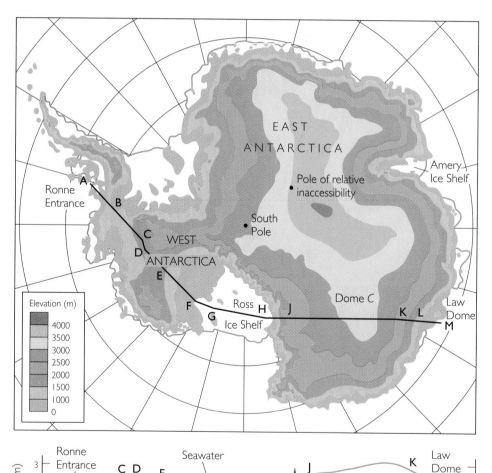

FIGURE 15.4 This contour map and the cross section of Antarctica show the topography of the continental ice sheet that covers the entire continent and the land beneath it. Ice shelves are shown in white. (After Uwe Radok, "The Antarctic Ice," *Scientific American,* August 1985, p. 100; based on data from the International Antarctic Glaciological Project.)

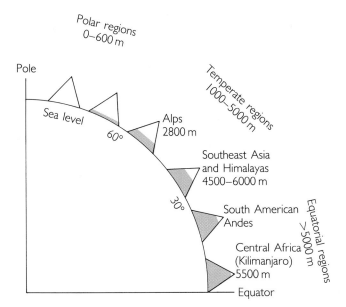

FIGURE 15.5 The height of the snow line, the altitude above which snow does not melt completely in summer, varies with latitude, from at or near sea level in polar regions to heights of more than 5500 m at the equator.

Cold climates are not necessarily the snowiest. For example, Nome, Alaska, has a polar Arctic climate with a yearly average maximum temperature of 9°C, but it gets only 4.4 cm of precipitation yearly, virtually all of it as snow. Compare these figures with those for Caribou, Maine. Caribou has a cool temperate climate with a yearly average maximum temperature of 25°C and an annual precipitation of 90 cm. The average annual snowfall is a whopping 310 cm. Nevertheless, the conditions around Nome, where little of the snow melts, are better for the formation of glaciers than are those in Caribou, where all that snow melts in the spring. And in arid climates, glaciers are unlikely to form at all, unless the temperature is so frigid all year that practically no snow melts and all is preserved, as it is in Antarctica.

Glacial Growth: Accumulation

A fresh snowfall is a fluffy mass of loosely packed snowflakes. As the small, delicate, loosely packed crystals age on the ground, they shrink and become equant grains, as in Figure 15.6. During this

tains). As we noted in Chapter 14, high latitudes are cold because the angle between the Sun's rays and Earth's surface increases toward the poles; the higher the angle, the smaller the amount of solar energy the area receives. High altitudes are cold, as the first high-flying airplane pilots discovered, because the lowest 10 km of the atmosphere grows steadily cooler as height above the ground surface increases. As a result, the height of the snow line—the altitude above which snow does not completely melt in summer—varies. Even in warm climates, glaciers will form if the mountains are high enough. Near the equator, glaciers form only on mountains that are higher than about 5500m. This minimum altitude steadily decreases toward the poles, where snow and ice stay year-round even at sea level (Figure 15.5).

SECOND REQUIREMENT: ADEQUATE AMOUNTS OF SNOW
Snow and glacier formation requires moisture as well as cold. Because moisture-laden winds tend to drop most of their snow on the windward side of a high mountain range, the leeward side is likely to be dry and unglaciated. Most of the high Andes Mountains of South America, for example, lie in a belt of prevailing easterly winds. Glaciers form on the moist eastern slopes, but the dry western side has little snow and ice.

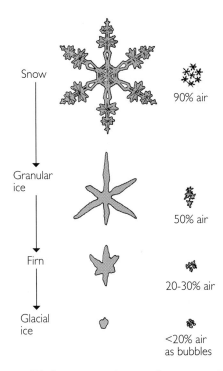

FIGURE 15.6 Stages in the transformation of snow crystals, first to granular ice, then to firn, and finally to glacial ice. A corresponding increase in density accompanies this transformation as air is eliminated from the crystals. (After H. Bader et al., "Der Schnee und seine Metamorphose," *Beitrage zur Geologie der Schweiz,* 1939.)

transformation, the mass of snowflakes is compacted to form a denser, granular snow. As new snow falls and buries the older snow, the granular snow further compacts to an even denser form, called **firn** (from a Germanic word applied to anything related to the previous year). Further burial and aging produce solid glacial ice as the smallest grains recrystallize, cementing all the grains together. Think of snow as a sediment that is transformed by burial into a metamorphic rock called ice. The whole process may take only a few years but is more likely to take 10 to 20 years.

A typical glacier grows slightly during the winter, as snow falls on the glacial surface and is converted to ice. The amount of snow added to the glacier annually is its **accumulation.**

As glacial ice accumulates, it entraps and preserves valuable relics of Earth's past. In 1992, Italian and Austrian scientists announced the discovery of the body of a prehistoric human preserved in alpine ice on the border between their two countries. In northern Siberia, extinct animals such as the woolly mammoth, a great elephantlike prehistoric creature that once roamed icy terrains, have been found frozen and preserved by ancient ice. Much smaller evidence in the form of dust particles and bubbles of atmospheric gases is also preserved in glacial ice (see Figure 15.1). Chemical analysis of air bubbles in very old, deeply buried Antarctic and Greenland ice, for example, tells us that levels of atmospheric carbon dioxide were lower during the last glaciation than they have been since the glaciers retreated.

Glacial Shrinkage: Ablation

When ice accumulates to a thickness sufficient for movement to begin, the formation of the glacier is complete. Ice, like water, flows downhill under the pull of gravity. The ice may move down a mountain valley or down from the dome of ice at the center of a continental ice sheet. In either case, it brings the glacier into lower altitudes where temperatures are warmer.

The total amount of ice that a glacier loses each year is called **ablation.** Four mechanisms are responsible for the loss of ice:

- Melting—As the ice starts to melt, the glacier loses material.

- **Iceberg calving**—Pieces of ice break off and form icebergs when a glacier moves to a shoreline (Figure 15.7).

FIGURE 15.7 Iceberg calving at Glacier Bay National Park, Alaska. Calving occurs when huge blocks of ice break off at the edge of a glacier that has moved to a shoreline, as this one has. Here, the glacier is projecting over the water. *(Tom Bean.)*

- **Sublimation**—In cold climates, ice can be transformed directly from the solid to the gaseous state.

- Wind erosion—Strong winds can erode the ice primarily by melting and sublimation.

Glacial shrinkage results from warming and melting at the glacier's leading edge. Thus, even though a glacier is advancing downward or outward from its center, the ice edge may be retreating. The main mechanisms by which glaciers lose ice are melting and calving.

Glacial Budgets: Accumulation Minus Ablation

The difference between accumulation and ablation, the net budget, gives the growth or shrinkage of a glacier (Figure 15.8). When accumulation minus ablation is zero over a long period, the glacier remains of constant size, even as it continues to flow downslope from the area where it formed. Such a glacier is accumulating snow and ice in its upper reaches and, in equal measure, ablating in its lower parts. If accumulation exceeds ablation, the glacier grows; if ablation exceeds accumulation, the glacier shrinks.

Glacial budgets vary from year to year. Some show trends of growth or shrinkage in response to climatic variation over periods of many decades. Yet over the past several thousand years, many glaciers have remained constant, on the average. Now that many scientists are becoming concerned about the effects of global warming on Earth's climate, geologists have proposed that glacial budgets be carefully monitored. Glacial shrinkage in certain areas may be a good early warning of local or regional climate change. In 1995, for example, satellites gave us images of the ice shelves of the Antarctic Peninsula extending from West Antarctica (see Figure 15.4 for locations). These images showed extensive retreats of shelf ice and the calving of an 80-km-long iceberg. This shrinkage corresponds to the warming of the west side of Antarctica by 2.5°C over the past 50 years. We can see a similar pattern of retreat, though on a smaller scale, in a small Antarctic Peninsula ice shelf between 1936 and 1992 (Figure 15.9). Some scientists worry that global warming could possibly lead to extensive calving of icebergs, which in turn would lead to major rises of sea level.

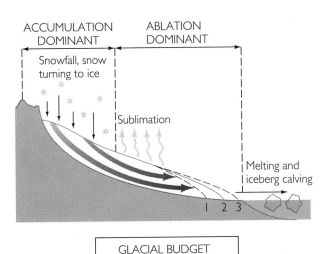

FIGURE **15.8** Accumulation of a glacier takes place mainly by snowfall near the colder upper regions. Ablation takes place mainly in the warmer lower regions by sublimation or melting and iceberg calving. The difference between accumulation and ablation is the glacial budget.

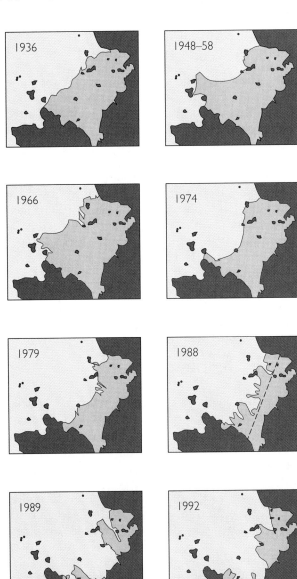

FIGURE **15.9** Successive stages in the retreat of a small ice shelf off the Antarctic Peninsula between 1936 and 1992. The area shown in the rectangle is about 90 km by 70 km. As the first three cells indicate, the shelf grew slightly between 1936 and 1966, but then it retreated dramatically over the next 26 years, shrinking to its much smaller state by 1992. (Modified from D. G. Vaughan and C. S. M. Doake, 1996, "Recent Atmospheric Warming and Retreat of Ice Shelves on the Antarctic Peninsula," *Nature*, 379:328–330.)

Glaciers: Movable Feasts for Water-Poor Regions?

As a glacier melts, great quantities of meltwater flow out from under the ice and from its edges. These meltwaters are the primary sources of cold-water streams that flow in mountain valleys below glaciers. If a glacial valley becomes dammed by glacial debris, a lake may form at the terminus of the glacier.

The abundance of meltwaters points up the large amounts of fresh water tied up in glacial ice. Glaciers are possible sources of fresh water for water-poor regions, if the transportation from source to users can be made economical. Some geologists have even suggested that icebergs might be towed through the ocean like barges. Although this idea may not be practicable very soon, it might conceivably be put to use in the future.

HOW GLACIERS MOVE

When ice builds to a thickness sufficient to make gravity overcome the ice's resistance to movement, normally at least several tens of meters, it starts to move and thus becomes a glacier. As glaciers move, the ice deforms and flows slowly downhill in the same kind of laminar flow as thin, slow water streams (see Chapter 13). It is important to understand glacial flow, for the motion of glaciers is responsible for the immense amount of ice's geologic work. In fact, it was seeing the results of glacial movement—erosion, transportation, and sedimentation—that first clued scientists to the fact that ice does move. Unlike the readily observed rapid flow of a river, glacial motion is so slow that from day to day the ice seems not to move at all, giving rise to the expression "moving at a glacial pace."

The rate of glacial movement increases as the slope steepens or the ice thickens. Even on a flat surface, such as a continental lowland, ice will flow outward if it builds up to a great thickness. Just as a viscous fluid flows on a flat surface—honey on a slice of bread, for example—a continental glacier will flow and spread out as its thickness increases. But how does ice, a solid, flow as if it were a slow-moving, viscous liquid?

Mechanisms of Glacial Flow

Glaciers flow primarily by two mechanisms, plastic flow and basal slip. In plastic flow, ice deforms and slides internally on a microscopic scale. In basal slip,

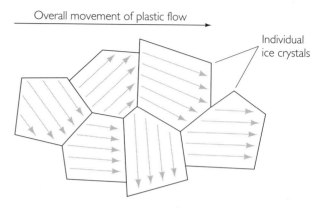

FIGURE 15.10 The plastic flow of a glacier is accomplished by small slips, over short intervals of time, along the microscopic planes of a great number of ice crystals. Taken together, these movements add up to a large movement of the whole mass of ice. Under stress in the interior of a thick glacier, the ice recrystallizes so that the planes become parallel to the flow direction.

ice slides downslope along the base of a glacier, like a brick sliding down an inclined board.

MOVEMENT IN DRY GLACIERS: PLASTIC FLOW Under the great pressure within a glacier, individual crystals of ice slip tiny distances, on the order of a ten-millionth of a millimeter, over short intervals of time (Figure 15.10). This movement is known as **plastic flow,** because it is like the plastic deformation of deeply buried rock (discussed in Chapter 10). The total of all the small movements of the enormous number of ice crystals that make up the glacier amounts to a large movement of the whole mass of ice. To visualize this process, think of a random pile of decks of playing cards, each deck held together by a rubber band. The whole pile can be made to shift by many small slips between cards in the individual decks. As ice crystals grow under stress deeper in the glacier, their microscopic slip planes become parallel, making flow rates faster.

Plastic flow dominates in bitterly cold regions, where the ice throughout the glacier, including its base, is well below the freezing point. The basal ice is frozen to the ground, and most of the movement of these cold, dry glaciers takes place above the base by plastic deformation. What little movement there is rips up any detachable pieces of bedrock or soil (Figure 15.11).

MOVEMENT IN WET GLACIERS: BASAL SLIP The other mechanism of glacial movement is **basal slip,** the sliding of a glacier along its base. The

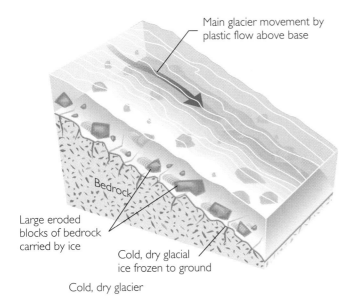

Main glacier movement by
plastic flow above base

Bedrock

Large eroded
blocks of bedrock
carried by ice

Cold, dry glacial
ice frozen to ground

Cold, dry glacier

FIGURE 15.11 Plastic slip occurs in exceptionally cold regions, where the glacier's basal ice is frozen to the ground.

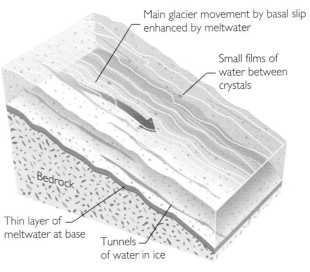

Main glacier movement by basal slip
enhanced by meltwater

Small films of
water between
crystals

Bedrock

Thin layer of
meltwater at base

Tunnels
of water in ice

Wet glacier

FIGURE 15.12 Basal slip occurs when meltwater forms at the glacier's base, creating a lubricant.

amount and kind of basal slip vary, depending on the temperature at the boundary between ice and ground in relation to the melting point of ice. At the base of a glacier, the ice is under tremendous pressure from the weight of the overlying ice. Because the melting point of ice decreases as pressure increases, the ice at the base of a glacier may melt, creating a lubricant in the form of meltwater. Meltwater can form even at temperatures lower than the freezing point at the surface. This is the same effect that makes ice skating possible. The weight of the body on the narrow skate blade provides enough pressure to melt just a little ice under the blade, which lubricates the blade so that it can slide easily along the surface. Similarly, the melting at the ice-ground boundary creates a lubricating layer of water on which the overlying ice can slide.

The temperature of the ice at the base of a glacier depends partly on the surface temperature of the ice and partly on the flow of heat from the ground below. In moderately cold regions, the air temperature at the surface is not too low and the flow of heat from the ground is higher than normal. Given these conditions, the temperature at the base of the glacier may be high enough for some melting to occur. Moderately cold areas are typical of glacier-filled valleys in temperate regions. In such areas, the ice may be at the melting point not only at the bottom of the glacier but also at higher places, particularly near the glacier's surface if the air is warmer than freezing. Plastic flow contributes a small amount of internal heating from the friction generated by microscopic slips of crystals. In these "wet" glaciers, water is found in the ice as

small drops between crystals and as pools in tunnels in the ice (Figure 15.12). The water throughout the glacier eases internal slip between layers of ice. In addition, ice may melt a little at the bottom and then refreeze, each time moving downhill a little bit.

CREVASSES The upper parts of glaciers (shallower than about 50 m) have little pressure on them. At these low pressures, the ice behaves as a rigid, brittle solid, cracking as it is dragged along by the plastic flow of the ice below. These cracks, called **crevasses,** break up the surface ice into many small and large blocks in places where glacier deformation is strong (Figure 15.13). Crevasses are more likely to occur

FIGURE 15.13 Crevasses, Franz Josef Glacier, South Island, Westland National Park, New Zealand. Crevasses are localized where the glacier valley curves or at sudden changes of slope. *(John Turner/ Tony Stone Images.)*

where the ice drags against bedrock walls, at curves in the valley, and where the slope steepens sharply. The movement of brittle surface ice at these places is a "flow" resulting from the slipping movements between these irregular blocks, similar in some ways to the microscopic slips of ice crystals but on a much larger scale.

Flow Patterns and Speeds

Most valley glaciers move at a speed that varies with the depth of the ice and with the glacier's position in relation to the valley walls. These glaciers flow partly by basal slip and partly by plastic flow within the body of ice. Strong frictional forces at the glacier's base and sides where it is in solid contact with the rock of the valley inhibit the movement of the ice, as shown in Figure 15.14.

The different speeds within alpine valley glaciers were first measured over a century ago by Louis Agassiz, a Swiss zoologist and geologist. As a young professor, not yet 30, he and his students established

FIGURE 15.15 Beardmore Glacier, Antarctica, showing fast flow lines from the middle background to the foreground. This glacier is about 25 km wide and flows from the Antarctic Plateau into the Ross Ice Shelf. *(Wolfgang Bayer/Bruce Coleman.)*

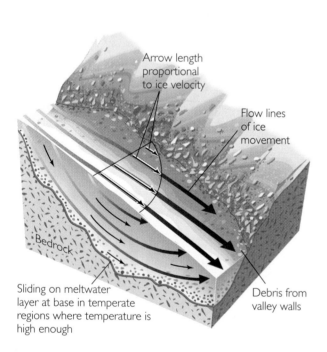

FIGURE 15.14 Ice flow in a typical valley glacier of a temperate region. As a result of frictional forces, the rate of movement decreases toward the base. The glacier moves most rapidly at its center. Friction along the valley walls reduces velocity at the edge of the flow.

a camp on a glacier to monitor its movement. They pounded stakes into the ice and measured their changes in position over a few years. The most rapid movement, about 75 m in one year, proved to be along the center line of the glacier. The deformation of long vertical tubes pounded deep into the ice later demonstrated that ice at the base of the glacier moved more slowly than ice in the center.

In modern times, geologists have used satellites and airborne radar to map the shapes and overall movements of glaciers. They have found, for example, that the Antarctic continental glacier at the South Pole is moving northward at a rate of 8 to 9 m per year, only a fraction of the highest speed Agassiz measured. The ice of some valley glaciers may move at a uniform speed. Such movement occurs, however, only where the climate permits motion as a single unit, and the glacier slides entirely by basal slip along a lubricating layer of meltwater next to the ground.

ICE STREAMS   Parts of the West Antarctic glacier flow rapidly in ice streams 25 to 80 km wide and 300 to 500 km long (Figure 15.15). These streams reach velocities of 0.3 to 2.3 m per day, compared with the flow of 0.02 m per day in the adjacent ice sheet. Boreholes in the ice have revealed that the base of the ice stream is at the melting point and that meltwater is mixed with soft sediment. One current theory is that rapid movement of the ice streams is related to deformation of the water-saturated basal sediment. Similar ice streams may form during a warming of the climate that leads to ice breakup and rapid deglaciation. The ice streams would contribute to retreat of glaciers and instability of the West Antarctic ice sheet.

SURGES   A **surge,** a sudden period of fast movement of a valley glacier, sometimes occurs after a long period of little movement. Surges may last more than two or three years, and during that time the ice may speed along at more than 6 km per year, a thousand times the normal velocity of a glacier. Although the mechanism of surges is not fully understood, it appears that a surge follows a buildup of water pressure in meltwater tunnels at or near the base of the glacier. This pressurized water greatly enhances basal slip.

FLOW IN CONTINENTAL GLACIERS   Continental glaciers in polar climates, where basal slip may be minor or absent, show the highest rates of movement in the center of the ice. Pressure there is very high, and the only frictional retarding forces are between layers of ice moving at different speeds in laminar flow. Near the surface, where pressure is least, the ice moves more slowly (Figure 15.16).

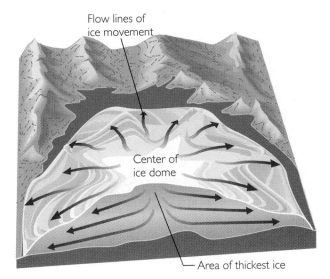

Flow lines of ice movement

Center of ice dome

Area of thickest ice

FIGURE 15.16 The flow of a continental glacier. The ice moves down and out from the thickest section, as shown by arrows.

GLACIAL LANDSCAPES

Just as you cannot see your own footprint in the sand while your foot is covering it, you cannot see the effects of an active glacier at its base or sides. Only when the ice melts is its geological work of erosion and sedimentation revealed. We can infer the physical processes driven by moving ice from the topography of formerly glaciated areas and the distinctive landforms left behind.

Glacial Erosion and Erosional Landforms

The capacity of glaciers to erode solid rock is amazing. A valley glacier only a few hundred meters wide can tear up and crush millions of tons of bedrock in a single year, enough to keep a fleet of 300 large dump trucks busy 365 days. The glacial ice erodes this heavy load of sediment from the rock floor and sides of a glacier and carries it in the ice to the ice front, where it is deposited as the ice melts. Estimates of the size of these deposits demonstrate that ice is a far more efficient agent of erosion than is water or wind. We can also compare rates and amounts of sediment deposited in glacial and interglacial periods. From such studies, we know that the total amount of sediment deposited in the world's oceans was several times larger during recent glacial times than during nonglacial periods. This difference is attributable to the great efficiency of erosion by glaciers in comparison to water and wind erosion.

At its base and sides, a glacier engulfs jointed, cracked blocks and breaks and grinds them against the rock pavement below. This grinding action fragments rocks into a great range of sizes, from boulders as big as houses to fine silt- and clay-sized material called **rock flour.** Freed from the ice at the melting edge of a glacier, this finely pulverized material dries and becomes dust. The wind can blow this dust for great distances, ultimately depositing it as the loess that is so common during glacial periods (see Chapter 14).

As a glacier drags rocks along its base, they scratch or groove the pavement as they grind against it. **Striations,** as such abrasions are termed, are strong evidence of glacial movement. Their orientation shows us the direction of ice movement, an especially important factor in the study of continental glaciers, which lack obvious valleys. By mapping striations over wide areas formerly covered by

FIGURE 15.17 Glacial polish, striations and grooves on bedrock, Glacier Bay National Park, Alaska. Striations are evidence of the direction of ice movement and are especially important clues for reconstructing the movement of continental glaciers. *(Carr Clifton.)*

continental glaciers, we can reconstruct their flow patterns (Figure 15.17).

Small hills of bedrock, called by the French term *roches moutonées* (literally, "sheep rocks") for their resemblance to a sheep's back, are smoothed by the ice on their upcurrent side and plucked—like the rough surface of a loaf of bread pulled apart—by the advancing ice to a rough, steep slope on the downcurrent side (Figure 15.18). These contrasting slopes also indicate the direction of ice movement.

A valley glacier carves a series of erosional forms as it flows from its origin to its lower edge. At the head of the glacial valley, the plucking and tearing

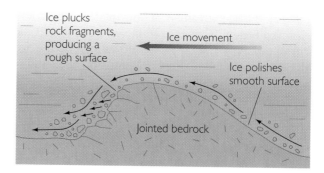

FIGURE 15.18 A roche moutonée is a small bedrock hill, smoothed by the ice on the upcurrent side and plucked to a rough face on the downcurrent side as the moving ice pulls fragments from joints and cracks.

action of the ice tends to carve out an amphitheater-like hollow, called a **cirque,** usually shaped like one half of an inverted cone (Figure 15.19). With continued erosion, cirques at the heads of adjacent valleys gradually meet at the mountaintops, producing sharp, jagged crests called **aretes** along the divide.

As a valley glacier moves down from its cirque, it excavates a valley or deepens a preexisting river

FIGURE 15.19 Valley glaciers with high snowfields and cirques in the background. Tributary glaciers are flowing down steep valleys and joining with the master glacier in the foreground. Harvard Glacier, Alaska. *(B. Washburn.)*

valley, creating a characteristic **U-shaped valley** (Figure 15.20). Glacial valley floors are flat and their walls steep, in contrast to the V-shaped valleys of many mountain rivers.

Glaciers and rivers differ not only in the shape of the valleys they create but also in the way their tributaries form junctions. Although the ice surface is level at the point where a tributary glacier joins the main valley glacier, the floor of the tributary valley may be much shallower than that of the larger main valley. When the ice melts, the tributary valley is left as a **hanging valley,** one whose floor lies

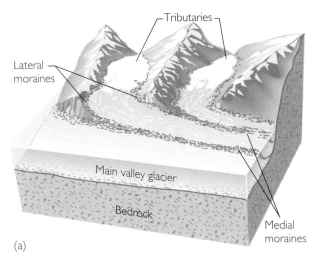

(a)

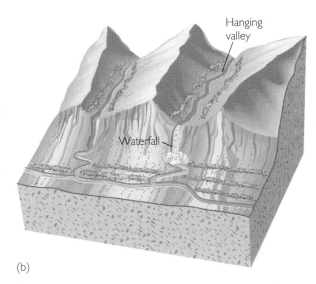

(b)

FIGURE 15.21 The creation of hanging valleys and their waterfalls by tributary valley glaciers. (a) During glaciation, tributary glaciers enter a major glacier at different floor levels. (b) After glaciation ends, the ice melts and the region is left with hanging valleys. A stream then forms a waterfall as it hurtles over the side of the hanging valley.

FIGURE 15.20 U-shaped glacial valley, Glacier National Park, Montana. The U-shaped valley is a characteristic form created by valley glaciers. *(Steve Kaufman/DRK.)*

high above the main valley floor (Figure 15.21). After the ice has gone and rivers occupy the valleys, the junction is marked by a waterfall as the stream in the hanging valley plunges over the steep cliff separating it from the main valley below.

Unlike rivers, valley glaciers at coastlines may erode their valley floors far deeper than sea level. When the ice retreats, these steep-walled valleys, which still maintain a U-shaped profile, are flooded with seawater (Figure 15.22). These glaciated arms of the sea, called **fjords,** create the spectacular rugged scenery for which the coasts of Alaska, British Columbia, and Norway are famous.

Glacial Sedimentation and Sedimentary Landforms

Glaciers transport eroded rock materials of all kinds and sizes downstream, eventually depositing them where the ice melts. Ice is a most effective transporter of debris because the material it picks up does not settle out, as does the load carried by a river. Like water and wind currents, ice has both a competence (an ability to carry particles of a certain size) and a capacity (a total amount of sediment it can transport). (Chapter 13 covers competence and capacity in more detail.) The competence of ice is extremely high: it can carry huge blocks many meters across that no other transporting agent could budge. Ice also has a tremendous carrying capacity. Some ice is so full of rock material that it is dark and looks like sediment cemented with ice.

When glacial ice melts, it deposits a poorly sorted, heterogeneous load of boulders, pebbles, sand, and clay. A wide range of particle sizes is the hallmark that differentiates glacial sediment from the much better sorted material deposited by streams and winds. To early geologists who were not aware of its glacial origins, the heterogeneous material was puzzling. They called it *drift,* because it seemed to have drifted in somehow from other areas. The term **drift** is now used for *all* material of glacial origin found anywhere on land or at sea.

Some drift is deposited directly by melting ice. This unstratified and poorly sorted sediment is known as **till,** and it may contain all sizes of fragments—clay, sand, or boulders (Figure 15.23). The large boulders often contained in till are called **erratics** because of their seemingly random composition, so very different from that of local rocks.

Other deposits of drift are laid down as the ice melts and releases sediment. Meltwater streams flowing in tunnels within and beneath the ice and in streams at the ice front may pick up, transport, and deposit some of the material. Just like any other waterborne sediment, this material is stratified and well sorted and may be cross-bedded. Drift that has been caught up and modified, sorted, and distributed by meltwater streams is called **outwash.** As mentioned earlier, fine-grained material from outwash floodplains may be blown by strong winds and deposited as loess.

FIGURE 15.22 A fjord, a glacial valley flooded with seawater. This view looks inland at McCarthy Fjord, Kenai Fjords National Park, Alaska. *(Peter Kresan.)*

FIGURE 15.23 Till deposited during the Pleistocene epoch on the eastern side of the Sierra Nevada in California. Note the differences in particle sizes and the lack of stratification. *(Martin Miller.)*

ICE-LAID DEPOSITS An accumulation of rocky, sandy, and clayey material carried by the ice or deposited as till is a **moraine.** There are many kinds of moraines, each named for its position with respect to the glacier that formed it. One of the most prominent in size and appearance is an *end moraine,* formed at the ice front. As the ice flows steadily downhill, it brings more and more sediment to its melting edge, where the unsorted material accumulates as a hilly ridge of till. *Terminal moraines* are end moraines that mark a glacier's farthest advance, and they are a geologist's best guide to the former extent of a valley or continental glacier.

As noted earlier, a glacier erodes rock and unconsolidated material from the sides of its valley. It gains additional material from mass movements as the glacier undercuts the valley walls above the ice. The eroded dirt and rock, incorporated into the ice as a dark, sediment-laden strip where the glacier scrapes and moves along the sides of the valley, are *lateral moraines* (Figure 15.24). The lateral moraines of joining glaciers merge to form a *medial moraine* in the middle of the larger flow below the junction. Lateral and medial moraines, like end moraines, are left behind as ridges of till after the glacier melts away.

A *ground moraine* is a layer of glacial drift deposited beneath the ice. Ground moraines range from thin and patchy, with exposed bedrock knobs and pavements, to thicknesses great enough to bury bedrock completely. Regardless of the shape or location, moraines of all kinds consist of till.

FIGURE 15.24 Lateral moraine (dark brown running from center to lower right of photograph), Kennicott Glacier, Wrangell–St. Elias National Park, Alaska. This morainal material is scraped and eroded from the side walls of the glacial valley. *(Tom & Pat Leeson/DRK.)*

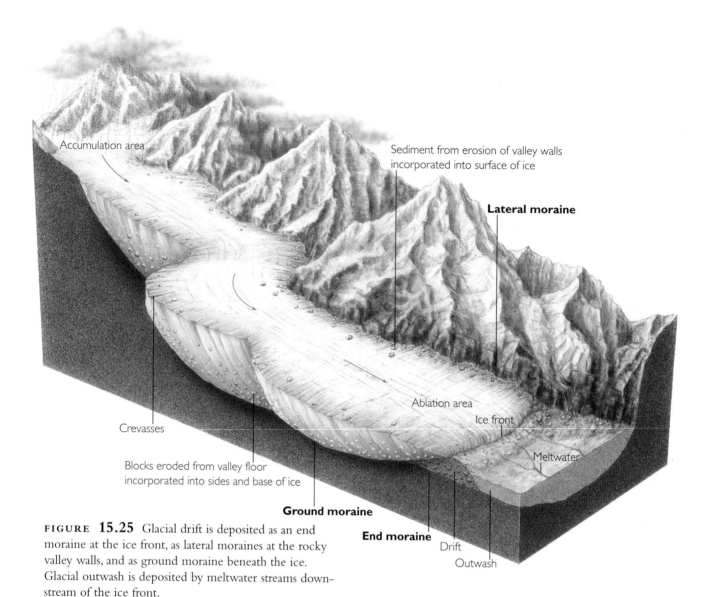

FIGURE 15.25 Glacial drift is deposited as an end moraine at the ice front, as lateral moraines at the rocky valley walls, and as ground moraine beneath the ice. Glacial outwash is deposited by meltwater streams downstream of the ice front.

Figure 15.25 illustrates the processes by which various kinds of moraines form as a glacier works its way through a valley.

Some continental glacier terrains display prominent landforms called **drumlins,** large, streamlined hills of till and bedrock that parallel the direction of ice movement (Figure 15.26). Usually found in clusters, drumlins are shaped like long, inverted spoons with the gentle slopes on the downstream sides. They may be 25 to 50 m high and a kilometer long. We do not yet fully understand the precise origin of drumlins. One hypothesis is that drumlins form under glaciers found in temperate regions that have water at their base. These glaciers move over a plastic mixture of subglacial sediment and water. Complex flow patterns of such mixtures are revealed by curving and looping striations on bedrock. The hypothesis proposes that this plastic mass would be subjected to increased pressure when it encounters a bedrock

FIGURE 15.26 Drumlins (foreground) in Dodge County, Wisconsin. Drumlins are composed of till. Some have a bedrock base with a streamlined shape formed by ice movement. The steep side commonly faces the direction from which the ice came, but the reverse is also found. *(Tom Bean/DRK.)*

FIGURE **15.27** Retreat of the ice margin has revealed a kame and an esker formed by a stream that ran under the glacier. The esker is the winding ridge of sand and gravel running from the upper right to the lower left. At the former ice front *(lower left),* the esker ends in a kame as a delta built out into a lake. The grooves at right are in soft glacial sediment and are not bedrock striations. *(B. Washburn.)*

knob or other obstacle and would then lose water and solidify, forming a streamlined mass. An alternative hypothesis proposes that drumlins are formed by ice erosion of an earlier accumulation of till.

WATER-LAID DEPOSITS Deposits of outwash from glacial meltwaters take a variety of forms. **Kames** are small hills of sand and gravel dumped near or at the edge of the ice (Figure 15.27). Some kames are deltas built into lakes at the ice edge; when the lake drains, these deltas are preserved as flat-topped hills. Kames are often exploited as commercial sand and gravel pits.

Silts and clays are deposited on the bottom of a lake at the edge of the ice in a series of alternating coarse and fine layers called varves (Figure 15.28). A **varve** is a pair of layers formed during one year by seasonal freezing of the lake surface. In the summer, when the lake is free of ice, coarse silt is laid down as abundant meltwater streams flow from the glacier into the lake. In the winter, when the surface of the lake is frozen, the water below is stilled and the finest clays settle, forming a thin layer on top of the coarse summer layer.

Some lakes formed by continental glaciers were huge, many thousands of square kilometers in extent. The till dams that created these lakes were sometimes breached and carried away at a later time,

FIGURE **15.28** Pleistocene varved clay, from an excavation in Stockholm, Sweden. The light layers are the coarse sediments deposited in a lake during warm seasons. The dark layers are the fine clays deposited when the lake was frozen in winter. *(John S. Shelton.)*

causing the lakes to drain rapidly and creating huge floods. In eastern Washington, an area called the scablands is covered by broad, dry stream channels, relics of torrential floodwaters draining from glacial Lake Missoula. From giant current ripples, sandbars, and the coarseness of the gravels found there, geologists have estimated flow velocities as high as 30 m per second and a discharge of 21 million cubic meters per second, as compared with ordinary river flows whose velocities are measured in tens or hundreds of centimeters per second and whose discharges range from a few cubic meters per second to the Mississippi in flood at 57,000 cubic meters per second.

Eskers are another form of meltwater-laid deposits. These long, narrow, winding ridges of sand and gravel are found in the middle of ground moraines (see Figure 15.27). They run on for kilometers in a direction roughly parallel to the direction of ice movement. The well-sorted, water-laid character of esker materials and the sinuous, channellike course of the ridge suggest their origin. Eskers were deposited by meltwater streams flowing in tunnels along the bottom of a melting glacier. The tunnels themselves were opened by water seeping through crevasses and cracks in the ice.

Glaciated terrains are dotted with **kettles,** hollows or undrained depressions that often are steep-sided and may be occupied by ponds or lakes. Modern glaciers, which may leave behind huge isolated blocks of ice in outwash plains as they melt, offer the clue to the origin of kettles. A block of ice a kilometer in diameter could take 30 years or more to melt. During that time, the melting block might have been partly buried by outwash sand and gravel carried by meltwater streams, usually braided, coursing around it. By the time the block had melted completely, the margin of the glacier would have retreated so far from the region that little outwash would reach the area. The sand and gravel that formerly surrounded the block of ice would now surround a depression. If the bottom of this kettle lies below the groundwater table, a lake will form. Figure 15.29 shows the evolution of such a kettle.

Permafrost

The ground is always frozen in very cold regions, where the summer temperature never gets high enough to melt more than a thin surface layer. Perennially frozen soil, or **permafrost,** today covers as much as 25 percent of Earth's total land area. In addition to the soil itself, permafrost includes aggregates of ice crystals in layers, wedges, and irregular masses. The proportion of ice to soil, as well as the permafrost's thickness, varies from region to region.

In Alaska and northern Canada, permafrost may be as much as 300 to 500 m thick. The ground be-

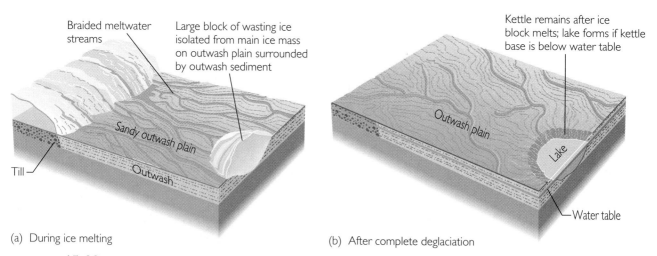

(a) During ice melting

(b) After complete deglaciation

FIGURE 15.29 Evolution of an outwash kettle. (a) As a glacier retreats, it may leave behind large blocks of wasting ice that are gradually buried by outwash from the receding ice front. (b) After the front has retreated far enough from the region, outwash sedimentation stops, the ice block melts, and a depression remains. If the depression is deep enough to intersect the groundwater table, groundwater fills it and forms a lake.

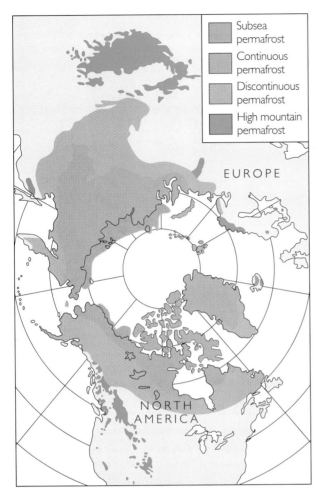

FIGURE 15.30 The North Pole is at the center of this map of the Northern Hemisphere showing the distribution of permafrost. The large area of high mountain permafrost at the top of the map is on the Tibetan Plateau. *(After a map by T. L. Pewe, Arizona State University.)*

Permafrost covers about 82 percent of Alaska and 50 percent of Canada, as well as great parts of Siberia (Figure 15.30). Beyond the polar regions, it is present in high, mountainous areas, especially on the Tibetan Plateau. Permafrost extends downward several hundreds of meters in shallow marine areas off Arctic coasts, presenting difficult engineering problems for offshore oil drillers.

ICE AGES: THE PLEISTOCENE GLACIATION

In the middle of the nineteenth century, the same Louis Agassiz who measured the speed of a Swiss alpine glacier immigrated to the United States and became a professor at Harvard University, continuing his studies in geology and other sciences. For his research, Agassiz visited many places in the northern parts of Europe and North America, from the mountains of Scandinavia and New England to the rolling hills of the American Midwest. In all these diverse regions, Agassiz recognized the signs of glacial erosion and sedimentation (Figure 15.31). In flat plains country, he saw moraines that reminded him of the end moraines of valley glaciers. The heterogeneous material of the drift, including erratic boulders, convinced him of its glacial origin.

low the permafrost layer, insulated from the bitterly cold temperatures at the surface, remains unfrozen, heated from below by the internal heat of the Earth. Permafrost is a difficult material to handle in engineering projects—such as roads, building foundations, and the Alaska oil pipeline—because permafrost melts as it is excavated. The meltwater cannot infiltrate the still-frozen soil below the excavation, so it stays at the surface, waterlogging the soil and leading it to creep, slide, and slump. Engineers decided to build part of the Alaska pipeline aboveground when an analysis showed that in places the pipeline would thaw the permafrost around it and lead to unstable soil conditions.

FIGURE 15.31 Irregular hills alternate with lakes in a terrain of glacial till, Coteau des Praires, South Dakota. *(John S. Shelton.)*

15.1   INTERPRETING THE EARTH

Future Changes in Sea Level and the Next Glaciation

What would happen if the 25 million cubic kilometers of the planet's water now tied up as ice were to melt? The change in sea level would be catastrophic, for the melting of all existing ice would raise the oceans by about 65 m. Major cities of the world—London, New York, Los Angeles, Tokyo, and many others—would be flooded. The bulk of the low-lying areas of the continents, where most of the world's population lives, would be submerged beneath shallow seas.

It is not likely that all of Earth's ice will melt for many millions of years to come. But it is very possible that a warming of climate in the next century will have two effects that could cause the sea level to rise.

- The melting of glacial ice would by itself produce a small rise.

- The extremely small expansion of water that accompanies an increase in temperature would be multiplied by the huge amount of ocean water warmed, causing an additional expansion of the oceans and a rise in sea level.

A global warming of just a few degrees might cause sea level to rise by a few meters. Though this amount may seem negligible, it is not, for at times of storms and tidal surges, this higher sea level would be enough to cause major flooding of low-lying coastal regions. Millions of people in Bangladesh, for example, live on lands only a few meters above sea level on the deltas of the Ganges and Brahmaputra rivers. A rise of sea level of only 1 m would be devastating for these people, who have already suffered catastrophic floods in which thousands of lives have been lost.

We can expect the Earth to go back into a period of glaciation over the next 10,000 years, since the current interglacial age has already lasted 10,000 years and the average length of interglacial periods is about 20,000 years. The general scenario is predictable. As polar and temperate regions grow colder, valley glaciers will enlarge. In areas of northern North America, Europe, and Asia, snow will stop melting in the summers. Ice will gradually accumulate in large continental glaciers, and it will eventually start to flow into the colder parts of formerly temperate regions, engulfing everything in its path. At the same time, sea level will drop, leaving the seaports of the world high and dry. A mass migration of people and animals into warmer regions would follow. Forest and plant populations would quickly alter in response to their new climates. It is hard to envision how the human population of the future will cope with such enormous changes.

The Bay of Bengal, Bangladesh. Low-lying Bangladesh is subject to disastrous flooding in the event of a rise in sea level, which could result from global warming. (*James P. Blair/National Geographic.*)

The areas covered by these glacial deposits were so vast that the ice that produced them must have been a continental glacier larger than Greenland or Antarctica (Figure 15.32). Eventually, Agassiz and others convinced geologists and the general public that a great continental glaciation had effectively extended the polar ice caps far into regions that now enjoy temperate climates. For the first time, people began to talk about ice ages. It was also apparent that the glaciation occurred in the relatively recent past, for the drift was soft, like freshly deposited sediment. We now know the age accurately from the carbon-14 (radioactive carbon) dating of logs buried in the drift. The drift of the last glaciation was deposited during one of the most recent epochs of geologic time, the Pleistocene, which lasted from 1.8 million to 10,000 years ago.

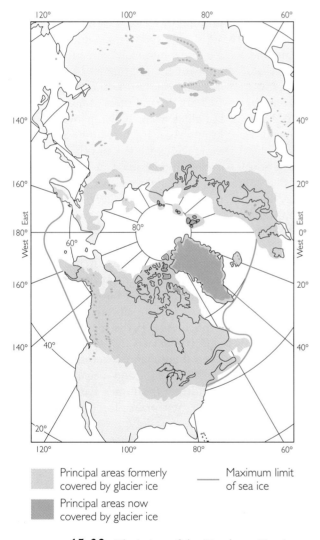

FIGURE 15.32 Glaciation of the Northern Hemisphere. The range of glaciation was established from glacial drift, which is widespread in areas that now enjoy a temperate climate. (After R. F. Flint, *Glacial and Quaternary Geology,* New York: Wiley, 1971.)

Legend:
Principal areas formerly covered by glacier ice
Principal areas now covered by glacier ice
Maximum limit of sea ice

Multiple Glacial Ages: Continental and Oceanic Records

It was not long before it became clear that there were multiple glacial ages during the Pleistocene, with warmer interglacial intervals between them. As geologists mapped glacial deposits about a century ago, they became aware that there were several layers of drift, the lower ones corresponding to earlier glacial ages. Between the older layers of glacial material were well-developed soils containing fossils of warm-climate plants. These soils were evidence that the glaciers retreated as the climate changed from frigid to warm. By the early part of this century, scientists believed that four distinct glaciations had affected North America and Europe during the Pleistocene epoch.

This idea was overthrown in the later twentieth century, when geologists and oceanographers examining oceanic sediment found fossil evidence of warming and cooling of the oceans. The sediments presented a much more complete geological record of the Pleistocene than that found in the continental glacial deposits. Glaciations earlier than the last were sparse and poorly preserved in the glacial deposits, and the information they yielded was fairly general—that warmer intervals corresponded to interglacial times and colder intervals to glacial ages. The ocean sediments contained more specific clues. The fossils buried in Pleistocene and earlier ocean sediments were of foraminifera, small, single-celled marine organisms that secrete shells of calcium carbonate (calcite, $CaCO_3$) (see Figure 7.19). These shells differ in their proportion of ordinary oxygen, oxygen-16, and the heavy oxygen isotope, oxygen-18 (see Chapter 2). The ratio of oxygen-16 to oxygen-18 found in the calcite of a foraminifer's shell depends on the temperature of the water it lived in. Different ratios in the shells preserved in various layers of sediment revealed the temperature changes in the oceans through time in the Pleistocene epoch

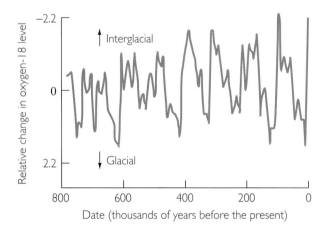

FIGURE 15.33 Relative changes in the ratio of oxygen-18 to oxygen-16 in the calcite of foraminifera in deep-sea sediment cores. This ratio changes in response to increases and decreases in the temperature of the water in which the foraminifera make their home and the extent to which evaporation and growth of glaciers enrich oxygen-18 in the ocean. Thus, this graph also indicates interglacial conditions (the peaks) and glacial conditions (the valleys). (After T. E. Graedel and P. F. Crutzen, *Atmospheric Change*, New York: W. H. Freeman, 1993.)

(Figure 15.33). During glaciations, marine sediments are enriched in oxygen-18 because oxygen-16 is preferentially evaporated from the oceans and trapped in glacial ice.

Isotopic analysis of shells also allowed measurement of another glacial effect: the oxygen isotope ratio of the ocean changes as a great deal of water is withdrawn from it by evaporation and is precipitated as snow to form glacial ice. The lighter oxygen-16 isotope has a greater tendency to evaporate from the ocean surface than does the heavy oxygen-18 isotope, which means that more of the heavy isotope is left behind. Thus, scientists could trace the growth and shrinkage of continental glaciers, even in parts of the ocean where there may have been no great change in temperature—around the equator, for example. From this analysis of marine sediments, we have learned that there were many shorter, more regular cycles of glaciation and deglaciation than geologists had recognized from the glacial tills of the continents alone.

Changes in Sea Level

You may recall from our coverage of the hydrologic cycle in Chapter 12 that the sea is the source of much of the water that falls as snow on the continents. Thus, during the Pleistocene epoch, the more water tied up as ice in continental glaciers, the less water there was in the ocean. As continental glaciers grew, sea level dropped everywhere on the globe. Accumulation of a large amount of ice—say, 400,000 km^3—corresponds to only a small drop in sea level—about 1 m—depending on the density of the ice and the configuration of the shallow ocean and low-lying land areas. But it has been estimated that during the maximum extent of the most recent glaciation, 18,000 years ago, sea level dropped about 130 m, corresponding to an enormous volume of ice, about 70 million cubic kilometers, or almost three times the amount of ice on Earth today.

Figure 15.34 shows the temperatures of the oceans and the ice distribution and climatic zones of nonglaciated regions of the continents 18,000 years ago, estimated from isotope ratios and the detailed paleontologic study of foraminifera. Masses of ice from 2 to 3.5 km thick were built up over North America, Europe, and Asia. In the Southern Hemisphere, the Antarctic ice expanded and the southern tips of South America and Africa were covered with ice. The continents were slightly larger than they are today because the continental shelves surrounding them, some more than 100 km wide, were exposed by the large drop in sea level. The map shows that the climatic zones at low latitudes may not have been very different from today's, but many temperate regions were heavily forested. According to the latest information, this map may have to be revised to show some cooling at low latitudes, but we are not sure by how much. As it is, the map remains the best synthesis of glacial climates that we have today.

Rivers extended their channels across the newly emergent continental shelves and began to erode channels in the former seafloor. Early humans roamed these low coastal plains, for early cultures, such as those of prehistoric Egypt, were evolving in the lands beyond the ice sheets.

Climate Change and Causes of Glacial Epochs

Ever since Agassiz's time, geologists have pondered the causes of ice ages. Today the discussion continues, heightened by concern over possible short-term warming trends that will melt part of the world's glaciers and long-term cooling trends that might expand them (Feature 15.1). Theories of ice ages

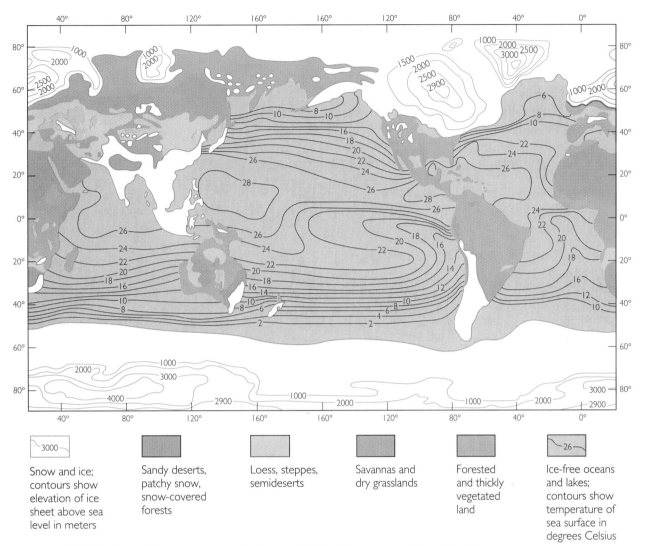

FIGURE 15.34 Sea surface temperatures (in degrees Celsius), ice extent, and ice elevation on a day in August 18,000 years ago. Continental outlines reflect the lowering of sea level 85 m below present levels. This map shows tropical temperatures similar to those of today, but some recent information based on studies of tropical corals suggests that the tropics may have been 5°C cooler. Nevertheless, this map remains overall the best estimate of ocean temperatures in a glacial age. (After CLIMAP, "The Surface of the Ice-Age Earth," *Science,* vol. 191, 1976.)

center on climate change and possible factors in the atmosphere and oceans that could provoke glaciation at one time and then reverse course to cause deglaciation. Such theories must also take into account the fact that the Pleistocene glaciation was not unique in Earth's history. Since the early part of this century, we have known from glacial striations and lithified ancient tills, called **tillites,** that glaciations covered parts of the continents numerous times in the geologic past, long before the Pleistocene.

Theories of ice ages attempt to explain two phenomena:

- The general cooling of polar regions, which led to the polar ice caps. Evidence from glacial sediments in Antarctica and the ocean indicate that the polar ice caps of today began forming about 10 million years ago, in the Miocene epoch, and grew slowly until they culminated in the full-blown glaciers of the Pleistocene.

■ The causes of alternating glacial and interglacial ages, which follow a more or less regular pattern throughout the Pleistocene.

A favored explanation for the general cooling of polar regions lies in the positions of the continents in relation to one another and to the poles. These positions are determined by continental drift driven by plate tectonics. For most of Earth's history, there were no extensive land areas in the polar regions, and there were no ice caps. The oceans circulated through the open polar regions, transporting heat from the equatorial regions and helping the atmosphere to distribute temperatures fairly evenly over the globe. When large land areas drifted to positions that obstructed efficient transport of heat by the oceans, the differences in temperature between the poles and equator increased. As the poles cooled, the ice caps formed. We can predict that many millions of years from now, if the large polar and subpolar lands of the northern continents and the Antarctic continent drift away from the poles, the ice caps will melt away.

So far, the alternation between glacial and interglacial ages has been explained best in terms of astronomical cycles. The shape of Earth's orbit around the Sun changes cyclically, sometimes being more circular and at other times more elliptical. The degree of ellipticity of Earth's orbit around the Sun is known as *eccentricity*. Thus, a circular orbit has low eccentricity, and a more elliptical orbit has high eccentricity (Figure 15.35a). The time interval of one cycle of this variation from low to high eccentricity is about 100,000 years. The angle of tilt of Earth's rotation axis cycles between 21.5° (and 24.5°, with a time period of about 41,000 years, again changing slightly the heat received from the Sun (Figure 15.35b). Also, Earth wobbles slightly on its axis of rotation, giving rise to a *precession,* or wobble, with a time period of about 23,000 years (Figure 15.35c). This, too, changes the amount of heat that Earth receives from the Sun.

Careful calculations of these orbital motions were first worked out in the 1920s and 1930s by Milutin Milankovitch, a Yugoslav geophysicist. His calculations demonstrated that glacial and interglacial stages correspond to variations in the heat Earth receives from the Sun due to the sum of all of Earth's orbital changes. Recent work on this theory predicts regular changes in climate, with a long-term periodic glaciation every 100,000 years and shorter-term ones about every 40,000 and 20,000 years (see Figure 15.36). The changes in climate are expected to be greater when the long-term periods are multiples of short-term ones.

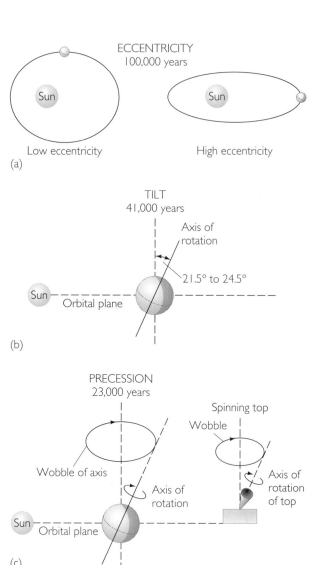

FIGURE 15.35 Three components of orbital variations in Earth's path around the Sun and their time intervals for a single cycle. (a) *Eccentricity* is the degree of ellipticity of Earth's orbit around the sun. (b) *Tilt* is the angle between the axis of rotation and the vertical to the orbital plane. (c) *Precession* is the wobble of the axis of rotation. One can easily imagine this motion by thinking of the wobble of a spinning top.

Evidence from Boreholes in the Ice

As the climatic interpretation of the marine sediment record of multiple glaciations was confirmed and grew more detailed, geologists searched for better continental records of glaciation. This search led to the development of new technologies for drilling deep boreholes and recovering ice cores in the present glacial ice sheets in Antarctica and Greenland. In the Antarctic, Russian geologists at Vostok Station have been drilling ever deeper cores since the 1970s; in 1993, they reached a depth of 2755 m. Several groups of U.S. and European geologists have drilled a series of boreholes in the Greenland ice.

The ice cores from these boreholes have given us a detailed stratigraphic record of yearly ice layers produced from the conversion of snow to ice. At the same time, the oxygen isotope and carbon dioxide content of the ice layers has yielded a record of temperature variations, so that the glacial and interglacial intervals could be determined from the ice (Feature 15.2). An additional kind of information came from bubbles of air trapped from the time the ice was formed. This air is a sample of the atmosphere of

thousands of years ago. Analyzing these tiny bubbles for various gases, Swiss and French scientists have discovered that carbon dioxide in the atmosphere was lower during glacial periods and higher during warm interglacial periods. This is just what one would expect from the greenhouse effect, a striking confirmation of the theory of global climate change discussed more fully in Chapter 24. The changes in carbon dioxide and the shifts from glacial to interglacial periods do not seem to have taken place at precisely the same times, but some of the time differences may be attributable to errors in measurement.

Ice cores from a mountain glacier 6 km high in the Peruvian Andes have recently caused us to revise our view of tropical climates during the last ice age. Although past work on deep-sea sediments suggested that the tropical oceans were warm and hardly affected by glaciation at higher latitudes, the Peruvian core indicates that tropical climates had sharp oscillations in temperature at the end of the most recent ice age, just as did the regions at higher latitudes.

Climate Since the Last Glaciation

We have learned from ice cores in Greenland that there was a complicated series of events before, during, and following the relatively abrupt deglaciation that terminated the most recent glaciation. A fast warming seems to have taken place about 14,500 years ago. About 1000 years later, the climate started to cool again and reached cold glacial values about 12,500 years ago. This temporary return to glacial conditions, called the "Younger Dryas" period, lasted about 1000 years. Then, about 11,700 years ago, the temperature increased again, by about 6°C, and deglaciation proceeded to the present state of polar ice caps and glaciers.

Climatic events of the past thousand years have been deduced from ice cores and confirmed by human history. From 1400 to 1650 A.D., the Earth went into a "Little Ice Age," during which the Baltic Sea froze over, and ice on the Thames River in England—which has not frozen over in modern times—reached thicknesses of several inches. There is no agreement among geologists and atmospheric scientists on the causes of these events, but it seems certain that we will be finding out much more about climate change in relation to glaciation as the nations of the world become increasingly concerned about the impact of future changes in climate.

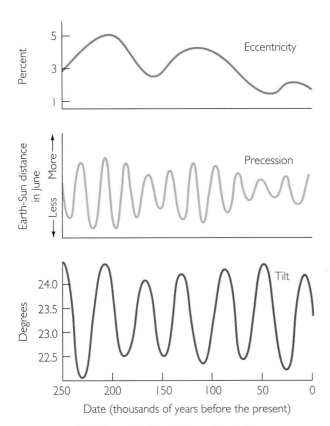

FIGURE **15.36** Orbital variations that influence climate: eccentricity, precession, and tilt. According to one theory, glacial and interglacial stages correspond to variations in the heat that the Earth receives from the Sun as a result of these orbital variations. (After T. E. Graedel and P. F. Crutzen, *Atmospheric Change,* New York: W. H. Freeman, 1993.)

Vostok and GRIP: Ice-Core Drilling in Antarctica and Greenland

At the Vostok science station in the frozen Antarctic, Russian scientists have been working year-round since 1960 to discover the climatological history of Earth that is hidden in the ice and gain insight into global climate change. In the 1970s, scientists at Vostok drilled boreholes 500 to 952 m deep in the eastern Antarctic ice and brought up a set of ice cores showing layers produced by annual cycles of ice formation from snow. Careful counting of the layers, working from the top down, revealed the age of the ice, much as tree rings reveal how old a tree is. This time-stratigraphic record of the ice proved to correlate with the temperatures thought to prevail when the layers formed. A high ratio of oxygen-18 to oxygen-16 in an ice layer indicates that the ice formed when the atmospheric temperature was relatively high. In addition, high concentrations of carbon dioxide, methane, and other "greenhouse gases," so called for their warming effect on the atmosphere, also suggest that the ice layer formed during a period of atmospheric warming.

By the 1990s, ice-borers at Vostok had extended the depth of drilling to 2755 m, penetrating ice not only from the last glacial period but also from the interglacial period before it and thereby accumulating a stratigraphic record for the last 160,000 years. These cores showed that eastern Antarctica, where the ice cores were drilled, was colder and drier during the last glaciation than during the 11,000 years of the current post-glacial period. Furthermore, variations in the oxygen isotope ratios of ice layers add to other evidence that variations in Earth's orbit control the cyclical alternation of glacial and interglacial periods. The carbon dioxide in ice lay-

ers formed during glacial intervals showed a notable decrease associated with climate cooling, but this decrease comes a little later than the time of first cooling indicated by oxygen isotope ratios. We do not fully understand this lag as yet. On the other hand, an increase in carbon dioxide does coincide with the warming of the atmosphere.

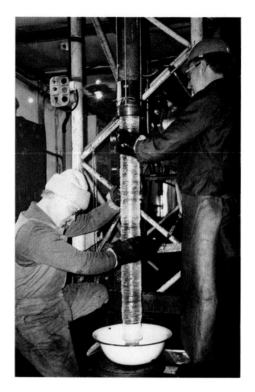

Russian scientists at Vostok science station in Antarctica carefully removing an ice core from a drill. The layers produced by annual cycles of ice formation are clearly visible. (*R. J. Delmas, Laboratorie de glaciologie et géophysique de l'environnement, Centre National de la Recherche Scientifique.*)

SUMMARY

How do glaciers form, and how do they move? Glaciers form where climates are cold enough that snow, instead of melting away in summer, becomes transformed by recrystallization into ice. As snow accumulates, the ice thickens, either at the tops of mountain valley glaciers or at the domed centers of continental ice sheets, until it becomes so heavy that gravity starts to pull it downhill. During periods of stable climate, the size of a glacier remains constant as it loses ice by melting, sublimation, and iceberg calving in the ablation zone, where the glacier loses ice, and is replenished by snow in the

Meanwhile, in the Arctic, a group of scientists in 1992 completed two years of drilling of the Greenland ice cap at the top of 3 km of ice. The Greenland Ice Core Project, better known as GRIP, has been an outstanding example of international scientific cooperation in Europe. GRIP is sponsored by the European Science Foundation, which includes the governments of Belgium, Denmark, France, Germany, Iceland, Italy, Switzerland, and the United Kingdom. The GRIP core penetrated ice layers of the last glacial period (11,000 to 115,000 years before the present) and the last interglacial period (the Eemian, 115,000 to 135,000 years before the present), and stopped in layers going back 235,000 years, near the end of the preceding interglacial period, as shown in the diagram below.

The GRIP core was complemented by a parallel core drilled 30 km to the west of the GRIP site by the United States Greenland Ice-Sheet Project, known as GISP2. The GRIP and GISP2 ice cores agreed very closely back 110,000 years, confirmed the glacial chronology at Vostok, and gave a good picture of the changes in climate and atmospheric composition of the past quarter-million years. Analysis of the cores also corrected earlier suggestions that the climates of the Eemian interglacial period were unstable and changed appreciably on a short time scale. The GRIP and GISP2 findings also confirmed the remarkably rapid climate oscillations during the last glacial period. Both the GRIP and the Vostok cores yielded precise oxygen-18 profiles that show close correlation with those of deep-sea sediments, giving further detail and further supporting the chronology of glacial-interglacial climate change. It has become clear also that temperature changes between

glacial and interglacial periods were much larger than scientists had thought.

These triumphs have not been won without difficulties. Extreme care had to be taken to avoid melting and contaminating the ice cores during coring operations, transportation to the laboratories, and storage until scientists could make the complex analyses of the ice and the air bubbles trapped in it. The scientists also had to be watchful against misleading results caused, for example, by the reaction of carbon dioxide with impurities in the ice. It is a tribute to the patience, ingenuity, and care exerted by these hardy bands of scientists that the scientific community worldwide has recognized the great importance of glacial ice-core research to understanding the history and predicting the future of global climate change.

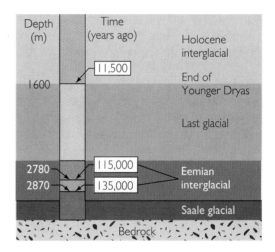

Comparison of ice depths and time before the present for recent glacial and interglacial intervals. (After J. W. C. White, "Don't Touch That Dial," 1993, *Nature,* 364:186.)

accumulation zone, where the glacier gains ice. During a warming period, a glacier shrinks as ablation exceeds accumulation; conversely, a glacier expands as accumulation outpaces ablation in a cooling climate. Glaciers move by a combination of plastic flow and basal slip. Plastic flow dominates in very cold regions, where the glacier's base is frozen to the ground. Basal slip is more important in warmer climates, where melting at the glacier's base lubricates the ice's movement over rock. The speed of a glacier's movement varies with the level and position of the ice.

How do glaciers erode bedrock, transport and deposit sediment, and shape the landscape? Glacial erosion is efficient at scraping, plucking, and grinding bedrock into sizes ranging from boulders to finely ground rock flour. Valley glaciers erode cirques and aretes at their heads, U-shaped and hanging valleys in their main courses, and fjords where they end at the ocean, eroding their valleys below sea level. Glacial ice has a high competence and capacity to carry abundant sediments of all sizes, transporting huge quantities to the ice front, where melting releases them. The sediments may be directly deposited from the melting ice as till or picked up by meltwater streams and laid down as outwash. Characteristic depositional forms laid down by ice are moraines and drumlins; kames, eskers, and kettles are laid down by meltwater. Permafrost is perennially frozen soil in very cold regions.

What are ice ages and what causes them? Glacial drift of Pleistocene age is widespread over high-latitude regions that now enjoy temperate climates, evidence that continental glaciers once expanded far beyond the polar regions. Studies of the geologic ages of glacial deposits on land and sediments of the seafloor show that the Pleistocene glacial epoch consisted of multiple advances (glacial intervals) and retreats (interglacial intervals) of the continental ice sheets. Each advance corresponded to a global lowering of sea level that exposed large areas of continental shelf; during interglacial intervals, sea level rose and submerged the shelves. Although the causes of glaciation remain uncertain, it appears that the general cooling of the globe leading to glaciation was the result of continental drift that gradually moved continents to positions where they obstructed the general transport of heat from the equator to the polar regions. A favored explanation for the alternation of glacial and interglacial intervals is the effect of astronomical cycles, by which very small periodic changes in Earth's orbit and axis of rotation alter the amount of sunlight received at the Earth's surface. There is also evidence that decreased levels of carbon dioxide in the atmosphere diminish the greenhouse effect and lead to glaciation. In contrast, the current increase in atmospheric carbon dioxide resulting from the burning of fossil fuels may cause global warming.

KEY TERMS AND CONCEPTS

glacier (p. 372)
valley glacier (p. 373)
continental glacier (p. 373)
firn (p. 376)
accumulation (p. 376)
ablation (p. 376)
iceberg calving (p. 376)
sublimation (p. 376)
plastic flow (p. 378)
basal slip (p. 378)
crevasse (p. 379)

surge (p. 381)
rock flour (p. 381)
striation (p. 381)
cirque (p. 382)
arete (p. 382)
U-shaped valley (p. 383)
hanging valley (p. 383)
fjord (p. 384)
drift (p. 384)
till (p. 384)
erratic (p. 384)

outwash (p. 384)
moraine (p. 385)
drumlin (p. 386)
kame (p. 387)
varve (p. 387)
esker (p. 388)
kettle (p. 388)
permafrost (p. 388)
tillite (p. 393)

EXERCISES

1. How are valley glaciers distinguished from continental glaciers?

2. How is snow transformed into glacial ice?

3. How does glacial growth or shrinkage result from the balance between ablation and accumulation?

4. What are the mechanisms of glacier flow?

5. How do glaciers erode bedrock?

6. Why are striations evidence of a former glaciation?

7. Name three kinds of glacial sediment.

8. Name three landforms made by glaciers.

9. How does glaciation affect sea level?

10. What variations in Earth's orbit influence climate?

11. How does carbon dioxide in the atmosphere affect climate?

THOUGHT QUESTIONS

1. Why is glacial ice stratified in many places?

2. Some parts of a glacier contain much sediment, and others very little. What accounts for the difference?

3. Contrast the kinds of till you would expect to find in two glaciated areas, one a terrain of granitic and metamorphic rocks, the other a terrain of soft shales and loosely cemented sands.

4. What geological evidence would you search for if you wanted to know the direction of glacial movement in a continentally glaciated region?

5. You are walking over a winding ridge of glacial drift. What evidence would you look for to discover whether you were on an esker or an end moraine?

6. One of the dangers of exploring glaciers is the possibility of falling into a crevasse. What topographic features of a valley glacier or its surroundings would you use to infer that you were on a part of the glacier that was badly crevassed?

7. You are scouting out possible commercial sand and gravel deposits in an area that was once near the edge of a continental glacier. What landforms will you look for?

8. You live in New Orleans, not far from the mouth of the Mississippi River. What might be your first indication that the world was entering a new glacial age?

9. In the century just before Columbus's voyage in 1492, the Native Americans living in the northern United States and Canada must have noticed a significant change in climate. What kinds of changes in their surroundings might they have noticed, and how long did such changes last?

10. Some geologists think that one result of continued global warming could be the shrinkage and collapse of the West Antarctic ice sheet. How might this affect the populations of North America and Europe?

11. There is evidence from drilling that there is liquid water at the base of some glaciers. What kinds of glaciers and what factors might be responsible for the melting of ice at the bottom of these glaciers?

SUGGESTED READINGS

Broecker, W. S., and G. H. Denton. 1990. What drives glacial cycles? *Scientific American* (January):48–56.

Covey, C. 1984. The Earth's orbit and the ice ages. *Scientific American* (February):58.

Denton, George H., and T. J. Hughes. 1981. *The Last Great Ice Sheets.* New York: Wiley.

Flint, Richard F. 1971. *Glacial and Quaternary Geology.* New York: Wiley.

Hambrey, M. J., and J. Alean. 1992. *Glaciers.* Cambridge: Cambridge University Press.

Imbrie, John, and K. P. Imbrie. 1979. *Ice Ages: Solving the Mystery.* Short Hills, N.J.: Enslow Press.

LaChapelle, E. R., and A. S. Post. 1971. *Glacier Ice.* Seattle: University of Washington Press.

Price, R. J. 1973. *Glacial and Fluvioglacial Landforms.* New York: Macmillan.

Sharp, R. P. 1988. *Living Ice. Understanding Glaciers and Glaciation.* Cambridge: Cambridge University Press.

INTERNET SOURCES

Blackcomb Glaciers
ⓘ **http://www.whistler.net/glacier/**
Sponsored by the Whistler Resort near Vancouver, B.C., this site includes a general introduction to glaciers, and specific information about the Blackcomb and Horstman glaciers.

Monitoring Glacier Changes from Space
ⓘ **http://sdcd.gsfc.nasa.gov/GLACIER.BAY/hall.science.txt.html**
The Laboratory for Hydrospheric Processes at NASA's Goddard Space Flight Center is the home for this study of glacial retreat and advance. Climatic and remotely sensed data from satellites are used to monitor the ice position at Vatnajokull, Iceland; Pasterze Glacier, Austria; Glacier Bay, Alaska; and College Fjord, Alaska. This site is part of a larger resource, "A Multi-Media Tour of Glacier Bay, Alaska."

16

The Grand Teton Range, Wyoming. This landscape is dominated by high mountains elevated by tectonics. The meandering Snake River in the foreground has eroded the valley below the mountains. (*Tony Waltham.*)

Landscape Evolution

From high, snowcapped peaks to broad, rolling plains, Earth's landscape comprises a great array of landforms, large and small. The challenge to geologists is to explain how such varied landscapes are formed in different locales. Although the study of landscape is one of the oldest branches of geology, in many ways it is still one of the most difficult to reduce to a set of simple principles that can be applied to any terrain. Nevertheless, we have learned a great deal. Through slow changes, imperceptible on a human time scale, landscapes evolve as weathering, erosion, transportation, and deposition combine to sculpt the land surface. Erosion dominates in high mountains, sedimentation in low-lying plains.

Erosion and sedimentation are themselves powered by tectonics. Uplifts of Earth's crust raise mountains and expose them to intense erosion, while subsidence of the crust creates low areas that become the sites of deposition.

On a smaller scale, the deformation of the crust into folds and faults controls the particular forms of many hills and valleys.

Landscape also depends on geologic history. That history determines what kinds of rocks, with varying resistance to erosion, are exposed at the surface and how the rocks have been affected by structural deformation, igneous activity, and metamorphism. The geologic history may encompass millions of years of tectonic stability, allowing weathering and erosion to wear away the land to lower, more level landscapes.

In this chapter, we look closely at the way major geologic forces—geologic history, tectonics, climate, weathering, erosion, and sedimentation—interact in the dynamic process that carves the landscape.

TOPOGRAPHY, ELEVATION, AND RELIEF

We begin our study of landscape evolution with the facts of any terrain that are obvious as one looks at Earth's surface: topography, elevation, and relief. But today, looking is not enough: for such a study, we need to take photographs from the ground, an airplane, or a satellite and use them to make maps.

Accurate maps of Earth's surface are vital for studies of hydrology (where and how big are surface reservoirs), ecology (how are forests and farmlands distributed), glaciology (where and how big are glaciers), atmospheric circulation (how the shape of the ground surface affects winds), and most other aspects of Earth sciences. Given that importance, how can we objectively describe, or map, the varied surface of the Earth, from the heights of its mountain peaks to the depths of its valleys?

We begin with **topography,** the varying heights that give shape to Earth's surface. We compare heights to sea level, the average height of the oceans around the globe. We can then express the altitude, or vertical distance above sea level, as **elevation.** A topographic map shows the distribution of elevations in an area and most commonly represents this distribution by **contours,** lines that connect points of equal elevation (Figure 16.1). The more closely spaced the contour lines, the steeper the slope.

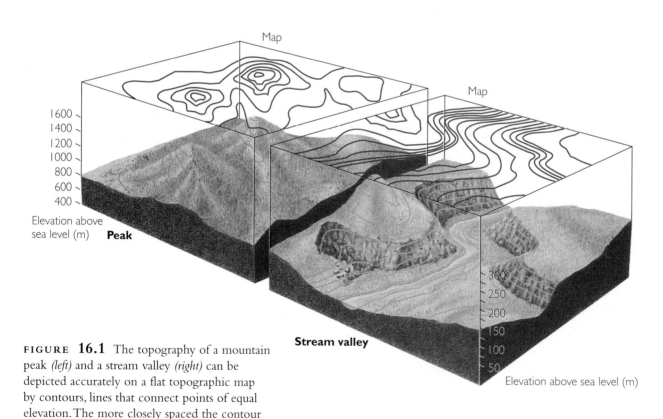

FIGURE 16.1 The topography of a mountain peak *(left)* and a stream valley *(right)* can be depicted accurately on a flat topographic map by contours, lines that connect points of equal elevation. The more closely spaced the contour lines, the steeper the slope. (After A. Maltman, *Geological Maps: An Introduction*, New York, Van Nostrand Reinhold, 1990, p. 17.)

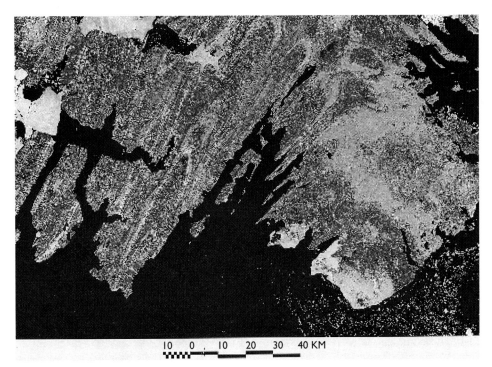

FIGURE **16.2** Satellite view of Bathurst Island, north of Hudson Bay in far northern Canada. The zigzag topography of ridges and valley is controlled by structural deformation into anticlines and synclines. The surface is bare, eroded, and glaciated. *(Earth Satellite Corp.)*

How do we construct a map, which is a geometric representation in two dimensions, of these three-dimensional topographies (see Appendix 4)? We associate geometry with Euclid, a Greek who systematized and taught geometry as a logical structure in Egypt in the third century B.C. But the rudiments of geometry had evolved long before then, arising in part from the needs of surveyors to determine land boundaries. From the beginnings of geology, centuries ago, geologists learned to survey topography and construct maps to plot and record geologic information. Although land surveying based on age-old methods is still used for some purposes, modern mapmakers rely on aerial photographs, radar imaging, and other technologies that enable them to discern not only elevation but also other topographic properties (Figure 16.2).

One of these additional properties of topography is **relief**, the difference between the highest and the lowest elevations in a particular area (Figure 16.3). To estimate the relief of an area from contours on a topographic map, we subtract the elevation of the lowest contour, usually at the bottom of a river valley, from that of the highest, at the top of the highest hill or mountain. Relief is a measure of the roughness of a terrain. The higher the relief, the more rugged the topography. Most regions of high elevation also have high relief, and most areas of low elevation have low relief, although there are excep-

tions. Mountains that rise steeply from the seashore, like some of those on the Pacific coast of North America, may be at relatively moderate elevations yet have high relief—a great distance between the base at sea level and the top of the mountain. Other regions, like the Tibetan Plateau of the Himalayas, may lie at high elevations but have relatively low relief.

Relief is specific to an area. At one extreme is relief on a global scale, from Mount Everest, the highest mountain in the world at an elevation of 8854 m, to the shores of the Dead Sea in Israel and

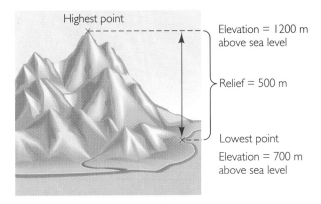

FIGURE **16.3** Relief is the difference between the highest and lowest elevations in a region.

Jordan, the lowest land area in the world at 392 m below sea level. That global relief of 9246 m dwarfs the modest relief, about 800 m, of areas in the Appalachian Mountains and relief values of a few thousand meters in areas of the Rocky Mountains.

If we fly over the United States, we can see many sorts of topography. Figure 16.4 is a computer-processed digital map of the 48 contiguous states that shows the details of large- and small-scale landforms. The advantages of computerized imaging become clear when one compares this map with the hand-drawn relief map of North America shown in Figure 16.15. The new digital map is superior to a hand-drawn map in its greater accuracy and more faithful representation of large and small features. It also avoids the obscuring of topography by clouds, roads and buildings, and vegetation. Perhaps most important is that it provides an overall view of the entire contiguous United States while at the same time showing features as small as 2.5 km across. The moderate elevations and relief of the elongate ridges and valleys of the Appalachian Mountains contrast with the low elevations and relief of the midwestern plains. Even more striking is the contrast between

the plains and the Rocky Mountains. As we examine these different types of topography more closely, we can characterize them not only by elevation and relief but also by their landforms: the steepness of their slopes, the shapes of the mountains or hills, and the forms of the valleys.

LANDFORMS: FEATURES SCULPTED BY EROSION AND SEDIMENTATION

The **landforms** of a region—the characteristic landscape features that attained their shapes through the processes of erosion and sedimentation—are guides to the region's geologic structure and history. Rivers, glaciers, and the wind leave their marks in a variety of forms on the surface of the Earth: rugged mountain slopes, broad valleys, floodplains, dunes, and many others. Geologists have learned to recognize the shapes of these marks of erosion and sedimentation, but even the most ordinary landforms are not easy to define precisely.

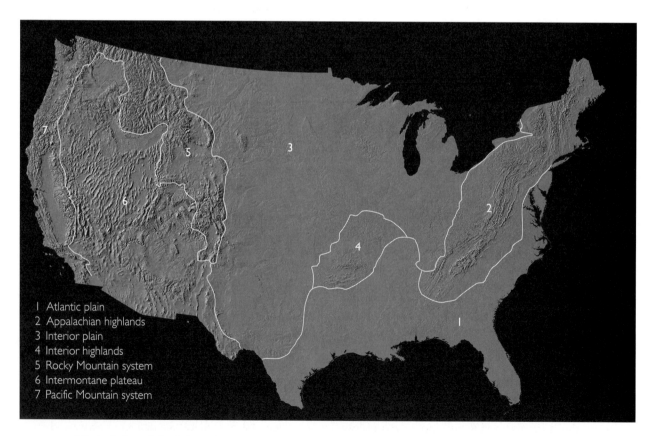

1 Atlantic plain
2 Appalachian highlands
3 Interior plain
4 Interior highlands
5 Rocky Mountain system
6 Intermontane plateau
7 Pacific Mountain system

FIGURE **16.4** A digital shaded-relief map of landforms in the contiguous United States. *(Gail P. Thelin and Richard J. Pike, USGS, 1991.)*

FIGURE 16.5 Most mountains are found in mountain ranges, not as individual peaks. In this glacially sculpted terrain in southern Argentina, all the peaks are sharp arêtes. *(Galen Rowell/Peter Arnold.)*

Mountains and Hills

We have used the word *mountain* many times in this book, yet in the varied usage of the word, both in everyday life and in geology, we can define it no more precisely than to say that a mountain is a large mass of rock that projects well above its surroundings. Most mountains are found with others in mountain ranges, where peaks of various heights are easier to distinguish than distinct separate mountains (Figure 16.5). Where mountains rise as single peaks above the surrounding lowlands, they are usually isolated volcanoes or erosional remnants of former mountain ranges.

We distinguish between mountains and hills only by size and custom. Elevations that would be called mountains in lower terrains are called hills in regions of high elevation. In general, however, landforms more than several hundred meters above their surroundings are called mountains.

Mountains and hills are reflections of tectonic activity caused directly or indirectly by plate movements. The more recent the activity, the more likely the mountains are to be high. The Himalayas, the highest mountains in the world, are also the youngest. The steepness of the slopes in mountainous and hilly areas generally correlates with elevation and relief, the steepest slopes generally being found in high mountains with great relief. The slopes of mountains lower in elevation and relief are less steep and rugged. The slopes are gentler in hills and may be barely noticeable in plains.

Plateaus

A **plateau** is a large, broad, flat area of appreciable elevation above the neighboring terrain. Smaller plateaus may be called tablelands. In the western United States, a small, flat elevation with steep slopes on all sides is called a **mesa** (Figure 16.6). Most plateaus have elevations of less than 3000 m, but the Altiplano of Bolivia is at an elevation of 3600 m and the extraordinarily high Tibetan Plateau has an average elevation of almost 5000 m. Many plateaus owe their flatness to their floors of undeformed sedimentary rocks or layers of lava flows. Plateaus form where tectonic activity produces a general uplift, rather than through deformation by lateral compression.

FIGURE 16.6
A mesa in Monument Valley, Arizona. The flat tops are held up by erosion-resistant beds. *(Raymond Siever.)*

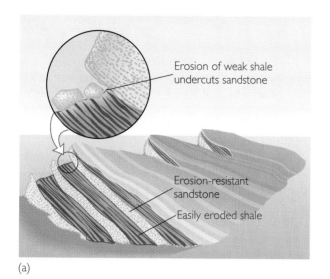

Erosion of weak shale undercuts sandstone

Erosion-resistant sandstone

Easily eroded shale

(a)

(b)

FIGURE 16.7 (a) Cuestas are formed where gently dipping beds of erosion-resistant rock, such as sandstone, are undercut by erosion of an easily eroded underlying rock, such as shale. (b) Cuestas formed on structurally tilted sedimentary rocks in Dinosaur National Monument, Colorado. *(Martin Miller.)*

Structurally Controlled Cliffs

The folds and faults produced by rock deformation during mountain building leave their marks on Earth's surface. **Cuestas** are asymmetrical ridges in a tilted and eroded series of beds of alternating weak and strong resistance to erosion. One side of a cuesta has a long, gentle slope determined by the dip of the erosion-resistant bed. The other side is a steep cliff formed at the edge of the resistant bed where it is undercut by erosion of a weaker bed beneath (Figure 16.7). Much more steeply dipping or vertical beds of hard strata erode more slowly to form **hogbacks,** which are steep, narrow, more or less symmetrical ridges (Figure 16.8). Steep cliffs are also produced by nearly vertical faults in which one side rises higher than the other.

Structurally Controlled Ridges and Valleys

In young mountains, during the early stages of folding and uplift, the upfolds (anticlines) form ridges and the downfolds (synclines) form valleys (Figure 16.9). But as tectonic activity moderates and erosion bites deeper into the structures, the anticlines may form valleys and the synclines ridges. This happens where the rocks—typically sedimentary rocks such as limestones, sandstones, and shales—exert strong control on topography by their variable resistance to erosion. If the rocks beneath an anticline are mechanically weak, as shales are, the core of the anticline may be eroded to form an anticlinal valley (Figure 16.10). In a region that has been eroded for many

millions of years, a pattern of linear anticlines and synclines produces a series of ridges and valleys like those of the Appalachian Mountains in Pennsylvania and adjacent states (Figure 16.11).

FIGURE 16.8 Hogbacks are narrow ridges formed by layers of erosion-resistant sedimentary rocks that are tectonically turned up so that the beds are vertical or nearly so. These hogback ridges are in the Rocky Mountains near Denver, Colorado. *(Tom Till/ DRK.)*

FIGURE 16.9 Valley and ridge topography formed on a folded terrain of sedimentary rock. The deformation is so recent (Pliocene) that erosion has not yet significantly modified the original structural forms of anticlines (ridges) and synclines (valleys). Zagros Mountains, Iran. *(NASA.)*

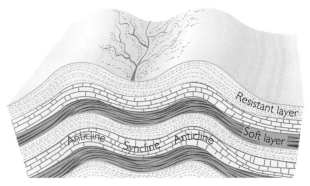

(a) Early stage of folding: ridges over anticlines, streams in synclines

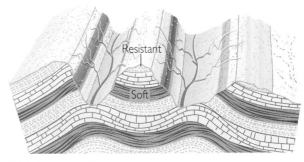

(b) Later stages of erosion: ridges may overlie synclinal axes if capped by resistant beds

FIGURE 16.10 Two stages in the development of ridges and valleys in folded mountains. (a) In early stages, ridges are formed by anticlines. (b) In later stages, the anticlines may be breached and ridges may be held up by caps of resistant rocks as erosion forms valleys in less resistant rocks.

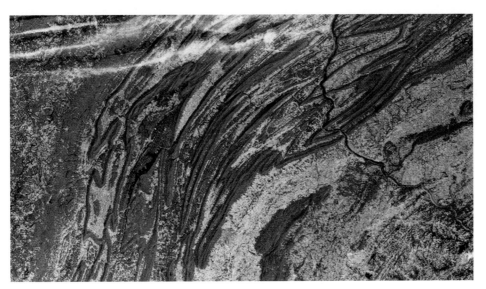

FIGURE 16.11 The Appalachian Valley and Ridge province shows the tectonically controlled topography of linear anticlines and synclines produced by millions of years of erosion. The prominent ridges, shown in red-orange, are held up by erosion-resistant sedimentary rocks that have been folded into an intricate series of anticlines and synclines. *(Earth Satellite Corp.)*

River Valleys

Observations of river valleys in various regions led to one of the important early theories of geology: the idea that river valleys were created through erosion by the rivers that flowed in them. Geologists could see that sedimentary rock formations on one side of a valley matched the same kinds of formations on the opposite side, leading them to conclude that the formations had once been deposited as a continuous sheet of sediment. The river had removed enormous quantities of the original formation by breaking up the rock and carrying it away.

River valleys have many names—canyons, gulches, arroyos, and gullies—but all have the same general geometry. A vertical cut through a young mountain river valley with little or no floodplain shows a simple V-shaped profile (Figure 16.12). A broad, low river valley with a wide floodplain shows a cross section that is more open but still distinct from the U-shaped profile of a glacial valley. The widths of river valleys vary in regions of different general topography and type of bedrock, ranging from the narrow gorges of mountainous belts and erosion-resistant rock types to the wide, shallow valleys of plains and easily eroded rock types. Between these extremes, the width of a valley generally corresponds to the erosion state of the region, being somewhat broader in mountains that have begun to be lowered and rounded from erosion and much broader in low-lying hilly topography.

A **badland** is a deeply gullied topography resulting from the fast erosion of easily erodible shales and clays, such as those of the South Dakota bad-lands (Figure 16.13). Practically the entire area is a proliferation of gullies and valleys with little flat land between them.

Tectonic Valleys

Many valleys formed primarily by tectonic processes are long, narrow, relatively flat-floored, and bounded on one or both sides by faults. This general shape results from downward movement of the valley caused by crustal subsidence along these faults. Rivers, and often lakes, occupy these tectonic valleys, as the San Joaquin and Sacramento rivers do in the Great Valley of California. The rivers of a tectonic valley may deposit a great deal of sediment eroded from nearby mountains, filling in the bedrock valley with a broad alluvial plain. This flat floor merges with higher coalesced alluvial fans sloping along the faulted boundaries of the valley. Alternatively, the rivers that run in these valleys may modify the structurally controlled valley form by eroding the valley walls to gentler angles or by eroding the valley deeper than would happen by tectonics alone. A rift valley is a special type of tectonic valley formed by the incipient or active spreading apart of lithospheric plates. The great African rift valleys are occupied by large lakes. The river Jordan and the Dead Sea are in another rift valley.

Another type of lowland that reflects tectonic activity is a **basin.** These depressions, of various shapes from circular to elongate, form where Earth's surface has subsided as a result of tectonic move-

Narrow mountain canyon   Open valley in hills   Broad, flat valley in lowlands

FIGURE 16.12 River valleys vary from narrow, V-shaped profiles in mountains to broader profiles in lowlands.

FIGURE 16.13 Gully erosion in the badlands of South Dakota, formed in easily eroded sedimentary rocks. Sage Creek Wilderness, Badlands National Park. *(Willard Clay.)*

ment. Basins are often found between mountain ranges. One of the large basins of the United States is the Great Basin, occupying much of western Nevada just east of the Sierra Nevada.

The Origins of Landforms

As we have seen in previous chapters, there are many other kinds of landforms, both erosional and depositional, that result from the activity of groundwater, winds, and ice. Chapters 12, 14, and 15, for example, discuss landforms such as karst topography, dunes, and moraines.

Landforms are primarily the products of erosion, transportation, or sedimentation, alone or in some combination. But tectonics also helps sculpt Earth's landforms: the profile of a river valley in tectonically elevated high mountains differs from that of a river valley in tectonically stable low plains. And bedrock lithology plays a role, too: river valleys cut through easily erodible sediments and rocks are broader and have gentler slopes than valleys that pass through erosion-resistant rocks.

A third factor that has a strong influence on landscape is climate. A hot, dry desert, subject to eolian and fluvial processes, has a landscape very different from that of a polar region characterized by glacial ice and a frigid climate. The topography of a region, itself the result of tectonics and erosion, strongly influences the rate of weathering and erosion. How all these factors interact tells us how major geologic forces sculpt the surface of the Earth.

FACTORS THAT CONTROL LANDSCAPE

Broadly speaking, landscape is controlled by the interaction of Earth's internal and external heat engines. The internal engine drives tectonics, elevates mountains and volcanoes, and lowers tectonic valleys and basins. The external heat engine, powered by the Sun, wears away the mountains and fills in the basins with sediment. Sunlight causes the motions of the atmosphere that produce climate, the different temperature regimes of the globe, and the rainwater that runs off the continents as rivers.

Tectonics Proposes, Erosion Disposes

In describing the formation and erosion of mountains, we might say that tectonics proposes and erosion disposes. Tectonics, driven by plate motions, elevates mountains and lowers tectonic valleys and basins. Erosion lowers the land surface and carves bedrock into valleys and slopes. The interaction between the two is a **negative-feedback process.** In this kind of process, one action produces an effect (the feedback) that tends to slow the original action and stabilize the process at a lower rate. For example, if you are thirsty, you will drink a glass of water quickly at first. But as the drinking reduces your thirst (the feedback), you will drink more slowly, until finally your thirst is completely satisfied and you stop drinking. The process of drinking has stabilized at a rate of zero.

A similar negative-feedback process begins when strong tectonic action elevates mountains. Tectonic uplift provokes an increase in erosion rate (Figure 16.14). The higher the mountains grow, the faster erosion wears them down. As long as mountain building continues, elevations stay high or increase. When mountain building slows, perhaps because of a change in the rate of plate-tectonic movements, the mountains rise more slowly or stop rising entirely. As growth slows or stops,

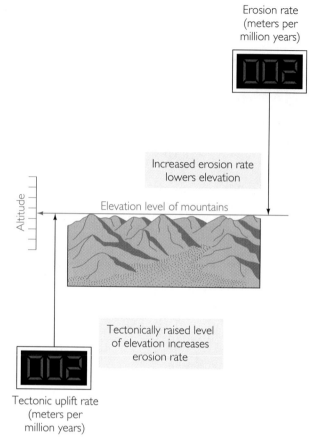

FIGURE 16.14 A negative-feedback loop relates uplift and erosion to surface elevation. Tectonic uplift causes an increase in erosion rate, which in turn lowers the surface elevation. The elevation is thus a balance between tectonic uplift and erosion rate.

erosion starts to dominate and elevations begin to decrease. This explains why old mountains, like the Appalachians, are relatively low compared to the much younger Rockies. As the lowering of mountains proceeds, the erosion also slows, and the whole process eventually tapers off. Elevation is thus a balance between tectonic uplift and erosion rate.

Topography Influences Weathering and Erosion

Topography exerts a strong control over weathering and erosion. High elevation and relief enhance the fragmentation and mechanical breakup of rocks, partly by promoting freezing and thawing. Also, fragmented debris on mountains moves quickly downhill in slides and other mass movements, exposing fresh rock to attack by the weather. Rivers run faster in mountains than in lower topography, and where the climate is cool, mountain glaciers scour bedrock and erode deep valleys. Chemical weathering plays an important part in the erosion of high mountains, but the mechanical breakup of rocks is so rapid that most of the debris appears to be almost unweathered. The products of chemical decay, dissolved material and clay minerals, are carried down from the steep slopes of mountains as soon as they are formed. The intense erosion that occurs in high elevations produces a topography of steep slopes; narrow, deep river valleys; and narrow floodplains and drainage divides.

A contrasting scene exists in lowlands, where weathering and erosion are slow and the clay mineral products of chemical weathering accumulate as thick soils. Mechanical breakup occurs in lowlands, too, but its effects are small compared with those of chemical weathering. Most rivers run on broad floodplains in lowlands and do little mechanical cutting into bedrock. Glaciers, which grind and erode bedrock at high elevations, are absent in lowlands except in cold polar regions. Even on lowland deserts, strong winds merely facet and round rock fragments and outcrops rather than breaking them up. A lowland thus tends to have a gentle topography with rounded slopes, rolling hills, and flat plains.

Climate and Topography Interact

You may recall from Chapter 6 that the effects of climate on weathering include freezing and thawing, expansion and contraction due to heating and cooling, and the chemical dissolving action of water. Rainfall and temperature, the components of climate, affect weathering and erosion through the rain that falls on bedrock and soil, the infiltration of water into the soil, mass wasting, and rivers, all of which help to break up the rock and mineral particles eroded from slopes and carry them downhill. Mountain climbers are very aware of the relationship of climate to topography. As one climbs higher, it gets colder, even in the tropics, and vegetation becomes less abundant. First one passes the timberline, above which no trees grow, and at higher elevations there is hardly any vegetation other than lichens, leafless primitive plants that can survive under these harsh conditions. Lichens aid weathering somewhat, but the general effects of vegetation on weathering are minimal at high elevations. High mountain landscapes are steep and jagged, consisting of bare rock cliffs and talus slopes. Erosion by glaciers contributes to the rugged landscape.

Topography also has other effects on climate. For example, mountains cause rain shadows, which are dry areas on the leeward slopes of mountain ranges (see Chapter 12). Also, the interiors of large continents such as Asia, far from any oceans, tend to have relatively low rainfall.

Latitude Mimics Elevation

The climate becomes cooler as latitude increases, and the corresponding effects on weathering and erosion are much the same as those caused by increasing elevation. At low latitudes near the equator, where tropical rain forests are found, the abundant water, lush vegetation, and warm temperatures promote rapid weathering and soil development. Because these forests are mostly in relatively low-lying areas, mechanical erosion is minimal. The result is a topography of low, rounded hills or, where river sedimentation is rapid, extensive river-bottom plains, such as the bottomlands of the Amazon River in the Brazilian jungle.

Where water is scarce at low latitudes, as in deserts such as the Sahara, chemical weathering is slow, and thin soils and the absence of vegetation promote rapid mechanical erosion. Desert landscapes tend to have rough topography, with steep slopes at higher elevations and gentler slopes at lower elevations where the intermittent rivers deposit layers of gravel and sand.

Chemical and mechanical erosion are more balanced in temperate climates. At lower elevations, moderate rainfall and temperatures promote abundant vegetation, particularly forests, and slopes are moderate. Rugged topography is the rule only at the higher elevations, where coldness and dryness prevail.

At high latitudes near the poles, where temperatures are low and moisture may be present mostly as snow, chemical weathering of rocks is restricted, but fragmentation and mechanical erosion are extensive. Bare rock and talus slopes with rugged topography are found at high and moderate elevations. Valley bottoms and lower slopes are covered with permanently frozen soil, and vegetation is less abundant than in temperate regions. Large glaciers may erode deep valleys and deposit extensive sheets of erosional debris at their melting edges.

In summary, climate, topography, and latitude—and their interactions—have strong effects on landscape through their enhancement or suppression of chemical and physical weathering and erosion. If one factor can be singled out as dominant in landscape evolution, it is topography, and that leads us back to tectonics. In all latitudes and climates, the slopes of high mountains are steep and rugged and lowlands tend to have gentler topographies. The effects of folding and faulting on the structure of the terrain are recognizable in any climate or latitude. All these effects on landscape can be seen in North America.

THE FACE OF NORTH AMERICA

In Chapter 3, we described the kinds of outcrops we might see on a trip across North America. The nature of those outcrops, from sea cliffs to creek bottoms, and the major relief patterns of the continent illustrate clearly that landscape and landforms are the products of tectonics and climate (Figure 16.15; see also Figure 16.4). The major mountain chains are the Appalachians in the east and the Rockies in the west. Less familiar to many of us are the high mountain chains of the far north, in Alaska and the Canadian Northwest Territories.

These mountain chains are the records of present and former plate boundaries. Along the Pacific coast, the San Andreas fault—the boundary between the North American and Pacific plates—is active today, as is the subduction of the Gorda and Juan de Fuca plates under North America. The other mountain chains offer evidence of movements along plate boundaries that have long since been sutured and immobilized. The high elevations of the western and northern mountains make them steep and rugged even in temperate climates. The Appalachians are much lower, with a gentler topography.

Between the mountain belts are the prairies and plains. In the United States, the western plains slope gently eastward from the foothills of the Rockies; the midwestern and southern lowlands slope southward. These slopes are clearly indicated by the paths of the major rivers, like the Missouri flowing eastward from the Rockies and the Ohio flowing westward from the Appalachians. Both join the Mississippi River, which flows south to the Gulf of Mexico. A low coastal plain lies along much of the central and southern Atlantic coast and the Gulf coast, where crustal subsidence allows continental shelf sedimentation and sea level intermittently rises and falls.

North of the Great Lakes are the low-relief plains and lakes of Ontario and Quebec. To the west are the prairies of Manitoba, Saskatchewan, and Alberta. Most of the plains of the United States and southern Canada are floored by horizontal sedimentary rocks of Paleozoic age. To the north, the lowlands of central and northern Canada are floored mostly by deeply eroded and weathered Precambrian metamorphic and igneous rocks. In Canada, the rivers of the interior lowlands drain eastward to the Atlantic or northward to the Arctic. All of these lowlands are provinces of the interior of the North American Plate, now far

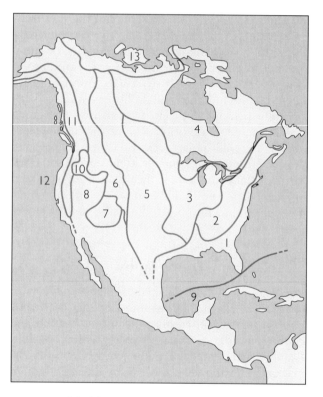

FIGURE 16.15 The map above shows a generalized division of North America into landform provinces: 1. Atlantic Coastal Plain; 2. Appalachian Mountains and Plateaus; 3. Central Lowland; 4. Canadian Shield; 5. Great Plains; 6. Rocky Mountains; 7. Colorado Plateau; 8. Basin and Range; 9. Central America; 10. Columbia Plateau; 11. Interior Mountains and Plateaus; 12. Pacific Border; 13. Arctic Lowlands. Compare with Figure 16.4. The hand-drawn relief map of North America at the right clearly shows the regional terrain textures on which physiographic divisions of the continent are based. (Landform province map after W. L. Graf, "Geomorphic Systems of North America," *Geological Society of America Centennial Special Volume 2*, 1987, p. 3. Landform relief map-diagram by Erwin Raisz; copyright © by Erwin Raisz, reprinted with permission by Raisz Landform Maps, Melrose, Mass.)

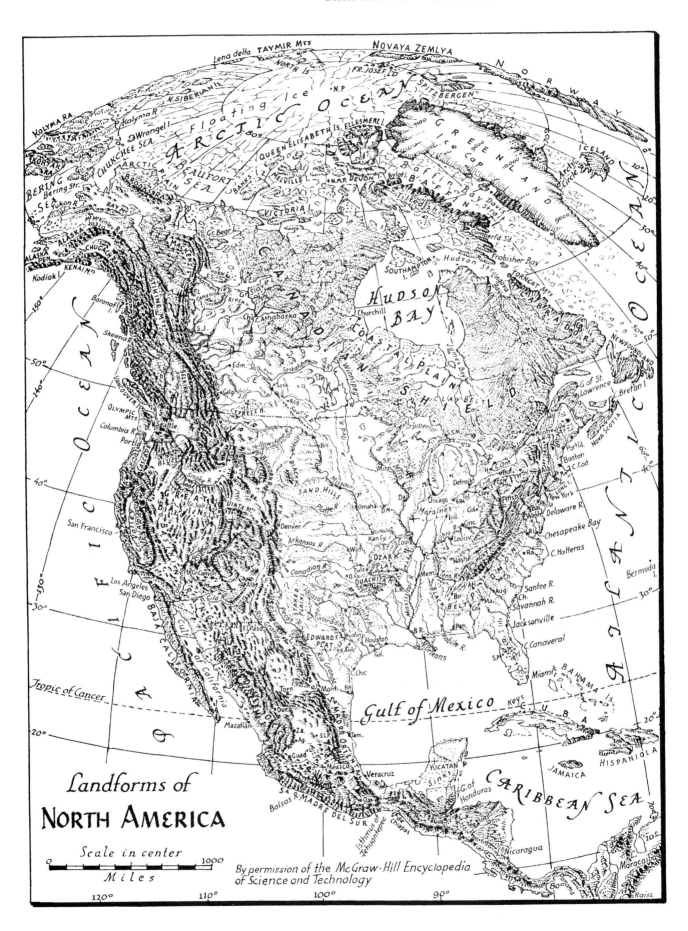

Landforms of
NORTH AMERICA

Scale in center
0   1000
Miles

By permission of the McGraw-Hill Encyclopedia
of Science and Technology

from any plate boundary. But in the Precambrian, 600 million or more years ago, this region was the site of active plate tectonics (Figure 16.16).

The Basin and Range province, a region of many smaller chains of mountains alternating with elongate basins, lies between the Rocky Mountains and the Sierra Nevada (see Figure 16.15). The topography and landscape of the Basin and Range are the products of recent tectonic activity interacting with a semiarid to arid climate. West of the Basin and Range are the igneous rock terrains of the Sierra Nevada and the Cascades, which are young, high, rugged ranges elevated in the Mesozoic and Cenozoic eras by the convergence of the North American Plate and present and former plates to the west. The western edge of the continent is marked to the south by the Coast Ranges east of the Great Valley of California and to the north by the volcanic Coast Ranges of Oregon, Washington, and British Columbia.

These maps of North America emphasize the dominance of tectonics but also suggest the importance of age. The highest and ruggedest mountains are the young western and northern ranges. The

FIGURE **16.16** A low-lying erosional plain developed on a structurally heterogeneous terrain of the Canadian Shield after hundreds of millions of years of tectonic stability. The grid of low, rounded ridges is formed by erosion-resistant igneous rock dikes. *(R. S. Hildebrand/ Geological Survey of Canada.)*

lower and older eastern mountains are rounded and gentle. The oldest of all, the expanses of the Precambrian terrain of northern and central Canada, are worn down to a low-lying plain, even though the rocks are complexly deformed. To take age into account in the formation of landscape, we need to know how landscapes evolve over geologic time.

THE EVOLUTIONARY PATH OF LANDSCAPES

The relationship between geologic age and mountain landscape was emphasized in one of the early theories of landscape development, which held sway in the first part of the twentieth century.

Early Views: Davis's Cycle of Erosion

William Morris Davis, a Harvard geologist of the time, studied mountains and plains all over the world. Davis proposed a *cycle of erosion* that progresses from the high, rugged, tectonically formed mountains of youth to the rounded forms of maturity and the worn-down plains of old age and tectonic stability. Erosion eventually wears down the landscape to a relatively flat surface, leveling all structures and differences in bedrock. Davis saw the flat surfaces of extensive unconformities as evidence of such plains in past geologic times. Here and there, an isolated hill might stand as an uneroded remnant of former heights.

For decades, Davis's cycle was accepted by many geologists, partly because they thought they could find many examples of what seemed to be the different stages of youth, maturity, and old age. There were competing proposals, some of which emphasized somewhat different courses of evolution of slopes and landscape in various climates, but orderly evolution was at the heart of these models, too. Most geologists at that time accepted Davis's assumption that mountains were elevated suddenly over short geologic times and then stayed tectonically fixed as erosion slowly wore them down. We now envision mountain building as an uneven, long-drawn-out, intermittent process driven by continuing plate motions.

The Current View: A Balance Between Uplift and Erosion

If tectonics continues over tens or even hundreds of millions of years, and Davis's scheme is no longer acceptable, how can we decipher the ways in which landscape evolves? Current views of landscape evolution emphasize the balance between erosion and tectonic uplift (Figure 16.17). If uplift is faster, the mountains will rise; if erosion is faster, the mountains will be lowered. When tectonics dominates, mountains are high and steep, and they remain so as long as the balance is in favor of tectonics. When erosion exceeds uplift, slopes become lower and more rounded. Because few areas of the world remain tectonically quiescent as long as 100 million years, the perfectly flat erosion plain that Davis proposed could form only rarely in Earth's history.

Convergence Leads to Uplift

What controls the rate and duration of tectonic activity? Uplift begins when two plates converge. The convergence of two oceanic plates leads to arcs of volcanic islands, whose topography is dominated by active volcanism, with erosion lagging. The convergence of an oceanic and a continental plate leads to a topography like that of the Pacific coast of South America, where the volcanic arc is on land. Extensive volcanism and lateral compression led to the construction of a high mountain chain, the Andes, while sedimentation and subsidence in the forearc region between the Andes and the Peru-Chile Trench offshore formed a narrow coastal plain. The convergence of two continental plates leads to the highest mountains, such as the Himalayas, which are bordered by thick alluvial plains below the mountains and by great deltas where major rivers enter the sea.

The rate of uplift varies with the speed of plate convergence. For reasons geologists do not fully understand, convergence may slow to a halt for periods of time and then resume. At other times, plate movements accelerate and then slow. High rates of convergence lead to rapid uplifts that outstrip erosion. Erosion dominates when convergence slows. Erosion may win out permanently if, as happens, the geometry of plate motions changes and new plate boundaries are created far from the mountains.

Erosion May Promote Uplift

In the past few years, geologists have been considering the possible role of erosion in promoting tectonic uplift. Changes in climate may significantly increase weathering and erosion rates and thus strip great weights of rock from the surface. This may lead to a response of the interior that uplifts the surface and raises mountains. (This topic is explored in more detail in the discussion of isostasy in Chapter 19.)

Climate Plays a Modifying Role

Geologists now recognize that climate plays an important role in modifying landscape evolution. A

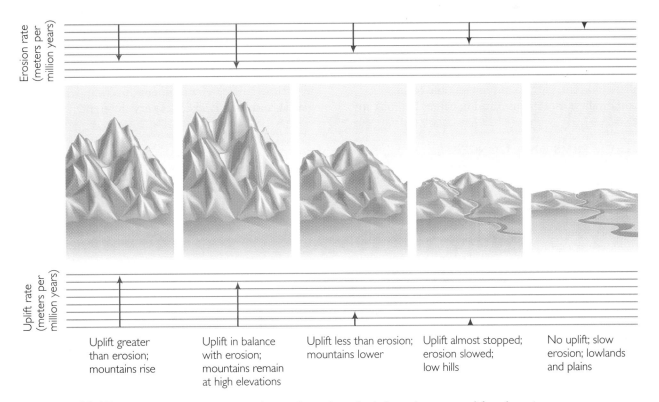

FIGURE 16.17 The stages of landscape evolution depend on the balance between uplift and erosion.

desert landscape evolves differently from a temperate landscape in a similar tectonic situation: the desert's slopes remain steeper even as the topography matures. Glacial landscapes are the unique products of ice, a special erosion and transportation agent that operates only in frigid climates.

Geologic History Is Reflected in Landscape

Although there are broad categories of landscape, each region has its own geologic history. That history determined the rock types at the surface and their structures, both of which strongly affect the course of landscape evolution. The latitude of a region is also important. A region may move from low to higher latitudes by continental drift; the different climate in the new location may modify or transform the landscape. There are other, poorly understood reasons for changes in climate. We have recently discovered, for example, that in the relatively recent geologic past, some areas of the Sahara Desert were much more humid than they are now. Thus landscape is the mirror of geologic history. To see clearly in that mirror takes a great deal of geological work and understanding.

Summary

What are the principal components of landscape?
Landscape is described in terms of topography: elevation, the altitude of the surface of the Earth above sea level; relief, the difference between the highest and lowest spots in a region; and the varied landforms produced by erosion and sedimentation by rivers, glaciers, mass wasting, and the wind. The most common landforms are mountains and hills, plateaus, and structurally controlled cliffs and ridges, all produced by tectonic activity modified by erosion. Landforms may be erosional or sedimentational: river and glacier valleys and badlands are primarily erosional; river floodplains, dune belts, and glacial moraines are primarily sedimentational.

What are the major factors that control landscape?
Landscape is determined by tectonics, erosion, climate, and the type of bedrock. Tectonics, driven by plate motions, elevates mountains and lowers tectonic valleys and basins. Erosion carves bedrock into valleys and slopes. Climate affects weathering and erosion and makes glacial and desert landscapes possible. The varying resistance of bedrock types to erosion accounts for differences in slope and valley profiles, steep slopes being found in rocks with greater resistance to erosion.

How do landscapes evolve? Landscapes begin their evolution with tectonic uplift, which in turn stimulates erosion. While tectonic activity remains dominant, mountains are high and steep. As tectonic activity slows, erosion becomes more important; the land surface is lowered and slopes are rounded. As erosion becomes dominant, the former mountains are worn down to gentle hills and broad plains. In the unusual event of long-continued tectonic quiescence, the land surface may end as a level plain. Climate and bedrock type strongly modify the evolutionary path in various surface environments, making desert, glacial, and karst landscapes very different.

Key Terms and Concepts

topography (p. 402)	landform (p. 404)	hogback (p. 406)
elevation (p. 402)	plateau (p. 405)	badland (p. 408)
contour (p. 402)	mesa (p. 405)	basin (p. 408)
relief (p. 403)	cuesta (p. 406)	negative-feedback process (p. 410)

Exercises

1. What is topographic relief, and how is it related to altitude?

2. Give three examples of landforms.

3. How does geologic structure control topography?

4. Compare weathering and erosion in topographically high and low areas.

5. How does climate affect topography?

6. In what regions of North America do active plate-tectonic movements currently affect landscape?

7. What kind of landscape evolves from the convergence of a continental plate and an oceanic plate?

8. How does the balance between tectonics and erosion affect mountain heights?

9. What is the difference between a cuesta and a hogback? Do you suppose there are or are not gradual transitions from one to the other?

10. What kind of landscape would you be likely to find along a continental rift valley?

THOUGHT QUESTIONS

1. The heights of two mountain ranges lie at different elevations, A at about 8 km and B at about 2 km. Without knowing anything else about them, could you make an intelligent guess about the relative ages of the mountain-building processes that formed them?

2. If you were to climb 1 km, from a river valley to a mountaintop, in two different regions, one where the mountaintops were about 2 km high and the other where they were about 6 km high, which would probably be the most rugged climb?

3. A young mountain range of uniform age, rock type, and structure extends from a far northern frigid climate through a temperate zone to a southern tropical rainy climate. How would the landscape of the mountain range differ in each of the three climates?

4. What differences in landscape would you expect between a coastal plain made of young, unlithified sediments and slightly lithified sedimentary rocks and an interior plain whose bedrock is old, very lithified sedimentary rocks?

5. Describe the main landforms in a low-lying humid region where the bedrock is limestone.

6. In what landscapes would you expect to find lakes?

7. What landforms are characteristic of glaciated lowlands?

8. Find your home region on the map in Figure 16.15 and, using anything you know about its geologic history, speculate on the origin of its landscape.

9. What changes could you predict for the landscape of the Himalaya Mountains in the next 100 million years?

10. What changes in the landscape of the Rocky Mountains of Colorado might result from a change in the present temperate but somewhat dry climate to a warmer climate with a large increase in rainfall?

LONG-TERM TEAM PROJECT

See Chapter 11.

SUGGESTED READINGS

Bloom, Arthur L. 1991. *Geomorphology*, 2nd ed. Englewood Cliffs, N.J.: Prentice Hall.

Chorley, Richard J., Stanley A. Schumm, and David E. Sugden. 1984. *Geomorphology*. London: Methuen.

Davis, William M. 1909. *Geographical Essays*. Boston: Ginn.

Hunt, Charles B. 1974. *Natural Regions of the United States and Canada*. San Francisco: W. H. Freeman.

Strain, P., and F. Engle. 1992. *Looking at Earth*. Atlanta: Turner Publishing.

Sullivan, W. 1984. *Landprints: On the Magnificent American Landscape*. New York: New York Times Book Co.

INTERNET SOURCE

Color Landform Atlas of the United States

🌐 **http://fermi.jhuapl.edu/states/states.html**

Links for every state provide access to two maps and additional sources of information about the state. The topographic map is optimized to show the landforms, with the same color shading used for every state. The county map suppresses the topography to emphasize the county boundaries. This site is located at the Johns Hopkins University Applied Physics Laboratory.

The Pacific Ocean, with the volcanic island of Honolulu, Hawaii, in the background. The wind filling the boat sails is also creating waves as it blows over the surface of the water. *(Bob Abraham / The Stock Market.)*

17

The Oceans

For most of human history, the 71 percent of Earth's surface covered by the oceans was a mystery. The large populations who lived at the edge of the sea knew well the force of the waves and the rise and fall of tides, but they could only guess at the nature of the seafloor deeper than the shallowest coastal waters. By the middle of the nineteenth century, occasional oceangoing ships made scientific observations of water depths, the plants and animals of the sea, and the chemistry of seawater. Then, in 1872, H.M.S. *Challenger,* a small wooden British warship converted and fitted out for the first specifically scientific study of the seas, left England for a four-year voyage over the world's oceans. The 50 thick volumes of reports from that expedition gave the public its first knowledge of the Mid-Atlantic Ridge, one of the longest, highest mountain ranges in the world—all of it underwater. The *Challenger* expedition discovered great areas of submerged hills and flat plains, extraordinarily deep trenches, and submarine volcanoes.

Today, more than a century after that pioneering voyage, hundreds of oceanographic research vessels from many countries ply the seas in search of answers to the questions first raised by the early discoveries. What tectonic forces raised the submarine mountain ranges and depressed the trenches? Why are some areas flat and others hilly? Although oceanographers made many important discoveries in the first half of this century, the answers to most of these questions had to await the plate-tectonics revolution of the late 1960s. In fact, it was geological and geophysical observations of the ocean floors, not the continents, that led to the theory of plate tectonics.

In this chapter, we examine what these early and more modern researchers have discovered about Earth's oceans, with their waves and tides; submerged mountains and valleys; underwater volcanoes; and many kinds of rocks, sediments, and chemical components. First, though, we define some terms.

We refer to the oceans both as the five major oceans (Atlantic, Pacific, Indian, Arctic, and Antarctic) and as the single connected body of water called the **world ocean.** The term *sea* includes both the oceans and smaller bodies of water set off somewhat from the world ocean. Thus, the Mediterranean Sea is narrowly connected with the Atlantic Ocean by the Straits of Gibraltar and with the Indian Ocean by the Suez Canal. Other seas, such as the North Sea and the Atlantic Ocean, are broadly connected. In the world ocean, seawater—the salty water of the oceans and seas—is remarkably constant in its general chemical composition from year to year and from place to place. This is so because the oceans maintain an equilibrium determined by the general composition of river waters entering the sea, the composition of the sediment brought into the oceans, and the formation of new sediment in the ocean (see Chapter 24).

We begin our exploration of the oceans with shorelines, where we can observe the constant motion of ocean waters and their effects on the shore.

THE EDGE OF THE SEA: WAVES AND TIDES

Coasts, the broad regions where land meets sea, present striking contrasts of landscape. On the coast of North Carolina, for example, long, straight, sandy beaches stretch for miles along low coastal plains (Figure 17.1). In New England, by contrast, rocky cliffs bound elevated shores, and the few beaches that occur are made of gravel (Figure 17.2). Many of the seaward edges of islands in the tropics, such as those in the Caribbean Sea, are coral reefs, the delight of divers. As we will see, tectonics, erosion, and sedimentation work together to create this great variety of shapes and materials.

FIGURE 17.1 A long, straight, sandy beach on South Pea Island, North Carolina. *(Peter Kresan.)*

FIGURE 17.2 Small pocket beach *(right foreground)* of pebbles and cobbles. Acadia National Park, Maine. *(Ric Ergenbright Photography.)*

The major geological forces operating at the **shoreline,** the line where the water surface intersects the shore, are waves and tides. Together, they erode even the most resistant rocky shores. Waves and tides create currents, which transport sediment produced by erosion of the land and deposit it on beaches and in shallow waters along the shore.

Wave Motion: The Key to Shoreline Dynamics

Centuries of observation have taught us that waves are changeable. During quiet weather, waves roll regularly into shore with calm troughs between them. In the high winds of a storm, however, waves are everywhere, moving in a confusion of shapes and sizes; they may be low and gentle far from the shore, yet become high and steep as they approach land. High waves can break on the shore with fearful violence, shattering concrete seawalls and tearing apart houses built along the beach. To understand the dynamics of shorelines and to make sensible decisions about shore development, we need to understand how waves work. The behavior of waves has long been observed by seafarers on sailing ships, who wanted to know which wave conditions would speed their way and which could endanger the ship.

Waves are created by the wind blowing over the surface of the water, transferring the energy of motion from air to water. As a gentle breeze of 5 to 20 km per hour starts to blow over a calm sea surface, ripples—little waves less than a centimeter high—take shape (see Table 14.1). As the speed of the wind increases to about 30 km per hour, the ripples grow to full-sized waves. Stronger winds create larger waves and blow off their tops to make whitecaps. The height of the waves increases as

- The wind speed increases.

- The wind blows for longer times.

- The distance over which the wind blows the water increases.

Storms blow up large, irregular waves that radiate outward from the storm area, like the ripples moving outward from a pebble dropped into a still pond. As the waves travel out from the storm center in ever-widening circles, they become more regular, changing to low, broad, rounded waves called **swell,** which can travel hundreds of kilometers. Several storms at different distances from a shoreline, each producing its own pattern of swell, account for the often irregular intervals between waves approaching the shore.

Waves travel as a form, but the water stays in the same place. If you have seen waves in an ocean or a large lake, you have probably noticed how a piece of wood or other light material floating on the water moves a little forward as the top of a wave passes and then a little backward as the trough between waves passes. While moving back and forth, the wood stays in roughly the same place, and so does the water around it.

Small water particles at the surface or beneath the waves move in circular vertical orbits. At any given point along the path of a wave, all the water particles are at the same relative positions in their orbits regardless of their depth. The radii of the orbits are large near the water surface, but they gradually decrease to

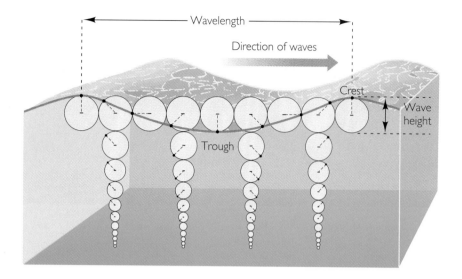

zero at some depth below, as shown in Figure 17.3. This is so because the greater the distance the water particle is from the surface, the farther it is from the force of the wind. And the farther the particle is from the force of the wind, the more important the viscous forces resisting water movement become. A wave form is made as many water particles move to the top of the orbit, and the wave advances as the particles continue around the orbit. The trough is created as the particles reach the bottom of the orbit.

We describe a wave form in terms of three characteristics (see Figure 17.3):

- **Wavelength,** the distance between crests

- **Wave height,** the vertical distance between the crest and the trough

- **Period,** the time it takes for successive waves to pass

We measure the velocity at which a wave moves forward by using a basic equation:

$$V = \frac{L}{T}$$

where V is the velocity, L is the wavelength, and T is the period. Thus, a typical wave with a length of 24 m and a period of 8 seconds would have a velocity of 3 m per second. The periods of most waves range from just a few seconds to as long as 15 or 20 seconds, with wavelengths varying from about 6 m to as much as 600 m. Consequently, wave velocities vary from 3 to 30 m per second. At a depth of about one-half the wavelength, orbital

motion stops and wave motion ceases. That is why deep divers and submarines are unaffected by the waves at the surface.

The Surf Zone

Swell becomes higher as it approaches the shoreline. There it assumes the familiar sharp-crested wave shape. These waves are called *breakers* because, as the waves come closer to shore, they break and form **surf,** a foamy, bubbly surface. The offshore belt, along which breaking waves collapse as they approach the shore, is the **surf zone.** Breaking waves pound the shore, eroding and carrying away sand, weathering and breaking up solid rock, and destroying structures built close to the shoreline.

The transformation from swell to breakers starts where the bottom shallows to less than one-half the wavelength of the swell. At that point, the small orbital motions of the water particles just above the bottom become restricted because the water can no longer move vertically. Right next to the bottom, the water can only move back and forth horizontally. Above that, the water can move vertically just a little, combining with the horizontal motion to give a flat elliptical orbit rather than a circular one (Figure 17.4). The orbits become more circular the farther they are from the bottom.

The change from circular to elliptical orbits slows the whole wave, because the water particles take longer to travel around ellipses than around circles. While the wave slows, its period remains the same because the swell keeps coming in from deeper water at the same rate. From the wave equation, we

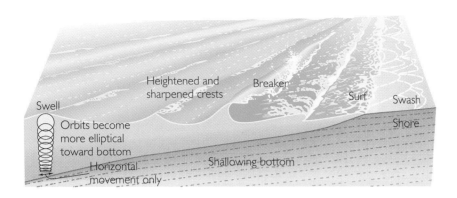

FIGURE **17.4** Formation of a breaking wave as swell meets a shallowing bottom. Note that the orbits of water particles become more elliptical as they approach a shallow bottom. As water particles take longer to travel around their ellipses, the whole wave slows. Waves become more closely spaced until, heightened and sharpened, they break with a crash in the surf zone.

know that if the velocity decreases and the period remains constant, the wavelength must also decrease. The typical wave we used as our example earlier might, while keeping the same period of 8 seconds, change to a length of 16 m and thus a velocity of 2 m per second. Thus, the waves become more closely spaced, higher, and steeper, and their wave crests become sharper.

As a wave rolls toward the shore, it becomes so steep that the water can no longer support itself, and the wave breaks with a crash in the surf zone (see Figure 17.4). Gently sloping bottoms cause the waves to break farther out, and steeply sloping bottoms make waves break closer to shore. Where rocky shores are bordered by deep water, the waves break directly on the rocks with a force amounting to hundreds of tons per square meter, throwing water high into the air. It is not surprising that concrete

seawalls built to protect buildings along the shore quickly start to crack and must be repaired constantly.

After breaking at the surf zone, the waves, now reduced in height, continue to move in, breaking again right at the shoreline (Figure 17.5). They run up onto the sloping front of the beach, forming an uprush of water called **swash.** The water then runs back down again as **backwash.** Swash can carry sand and, if the waves are high enough, large pebbles and cobbles. The backwash carries the particles back down again.

The motion of the water back and forth near the shore is strong enough to carry sand grains and even gravel. Fine sand can be moved by wave action in water up to about 20 m deep. Large waves caused by intense storms can scour the bottom at much greater depths, down to 50 m or more.

FIGURE **17.5** Waves obliquely approaching the shoreline and being refracted to be almost parallel to the shoreline. The white color shows the surf zone. Oceanside, California. *(John S. Shelton.)*

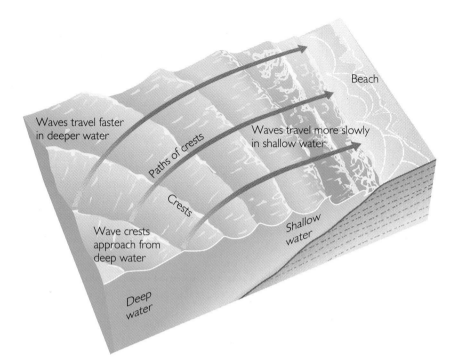

Wave Refraction

Far from shore, the lines of swell are parallel to one another but are usually at some angle to the shoreline. As the waves approach the beach over a shallowing bottom, the rows of waves gradually bend to a direction more parallel to the shore (Figure 17.6). This bending of lines of wave crests as they approach the shore from an angle is called **wave refraction.** It is similar to the bending of light rays in optical refraction, which makes a pencil half in and half out of water appear to bend at the water surface. Water refraction begins as a wave approaches the shore at an angle. The part of the wave closest to the shore encounters the shallowing bottom first, and the orbits of the water particles in that part of the wave become more elliptical. As this happens, the front of the wave slows. Then the next part of the wave meets the bottom and also slows. Meanwhile, the parts closest to shore have moved into even shallower water and slowed even more. Thus, in a continuous transition along the wave crest, the line of waves bends toward the shore as it slows.

Wave refraction results in more intense wave action on projecting headlands and less intense action in indented bays, as Figure 17.7 illustrates. The water becomes shallow more quickly around headlands than in the surrounding deeper water on either side. Waves are refracted around headlands—that is, they are bent toward the projecting part of the shore

from both sides. The waves converge around the point of land and expend proportionately more of their energy breaking there than at other places along the shore. Thus, erosion by waves is concentrated at headlands and tends to wear them away more quickly than it does straight sections of shoreline.

The opposite happens as a result of wave refraction in a bay. The waters in the center of the bay are deeper, so the waves are refracted on either side into shallower water. The energy of wave motion is diminished at the center of the bay, making bays good harbors for ships.

Although refraction makes waves more parallel to the shore, many waves still approach at some small angle. As the waves break on the shore, the swash moves up the beach slope perpendicular to this small angle. The backwash runs down the slope at a similar small angle but in the opposite direction, like a parabola. The combination of the two motions results in a trajectory that moves the water a short way down the beach (Figure 17.8). Sand grains carried by swash and backwash are thus moved along the beach in a zigzag motion known as **longshore drift.**

Waves approaching the shoreline at an angle can also cause a **longshore current,** a shallow-water current that is parallel to the shore. The water that moves with swash and backwash in and out from the shore at an angle creates a zigzag path of water particles that adds up to a net transport along the shore

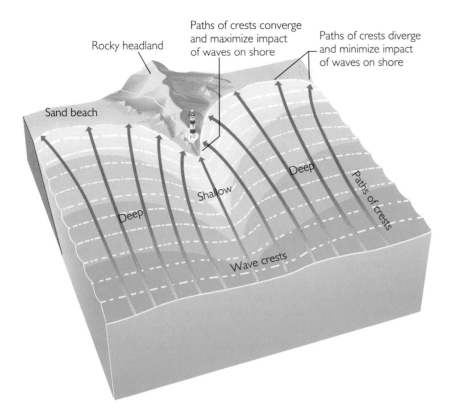

Rocky headland

Paths of crests converge and maximize impact of waves on shore

Paths of crests diverge and minimize impact of waves on shore

Sand beach

Deep

Shallow

Deep

Paths of crests

Wave crests

FIGURE 17.7 Wave refraction around a headland and bay. Wave energies are concentrated at headlands and dispersed at bays.

in the same direction as the longshore drift. Much of the net transport of sand along many beaches comes from this kind of current. Longshore currents are prime determiners of the shapes and extent of sand bars and other depositional shoreline features. At the same time, because of their ability to erode loose sand, longshore currents may remove much sand

from a beach. Longshore drift and longshore currents working together are potent processes in the transport of large amounts of sand on beaches and in very shallow waters. In deeper but still shallow waters (less than 50 m), longshore currents—especially those running during large storms—strongly affect the bottom.

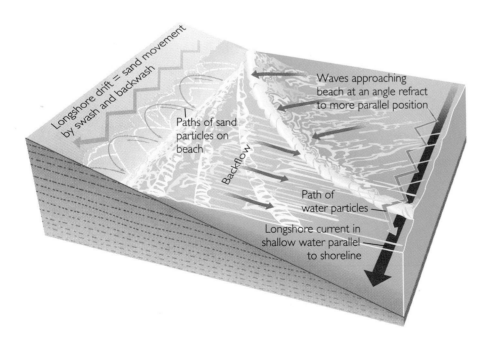

Longshore drift = sand movement by swash and backwash

Paths of sand particles on beach

Waves approaching beach at an angle refract to more parallel position

Backflow

Path of water particles

Longshore current in shallow water parallel to shoreline

FIGURE 17.8 Longshore drift is the zigzag movement of sand grains thrown up on the shore by waves approaching at an angle. Waves approaching the shore obliquely also cause a zigzag movement of water particles in shallow water, giving rise to a longshore current.

Some types of currents related to longshore currents can pose a threat to unwary swimmers. A *rip current,* for example, is a strong flow of water moving perpendicularly outward from the shore. It occurs when a longshore current builds up along the shore and the water piles up imperceptibly until a critical point is reached. At that point, the water breaks out to sea, flowing through oncoming waves in a fast current. Swimmers can avoid being carried out to sea by swimming parallel to the shore to get out of the rip.

The Tides

The twice-daily rise and fall of the sea that we call **tides** have been known to mariners and shoreline dwellers for thousands of years. During that time, many observers also noticed a relationship among the position and phases of the Moon, the heights of the tides, and the times of day at which the water reaches high tide. Not until the seventeenth century, however, when Isaac Newton formulated the laws of gravitation, did we begin to understand that the tides result from the gravitational pull of the Moon and the Sun on the water of the oceans.

THE MOON, THE SUN, GRAVITY, AND THE TIDES

The Earth and the Moon attract each other strongly with a gravitational force that is slightly greater on the sides of the bodies that face each other. The gravitational attraction between any two bodies decreases as they get farther apart. Thus, the tide-producing force varies on different parts of the Earth, depending on whether they are closer to or farther from the Moon.

Because the average distance between Earth and the Moon is constant over time, the gravitational attraction of the Moon to Earth must be exactly balanced by the centrifugal force due to the rotation of Earth. (Centrifugal force is a force that repels a body away from an axis around which it rotates.) This centrifugal force is the same everywhere on Earth. The gravitational attraction of the Earth, however, varies from place to place depending on its distance from the Moon. Because centrifugal force must equal the average gravitational attraction, at the point closest to the Moon, the gravitational attraction will be greater than the centrifugal force, resulting in a net force toward the Moon. At the point farthest from the Moon, the gravitational attraction will be less than the centrifugal force, and the net force will be away from the Moon. For other points on Earth's surface, the tide-producing force is the vector sum (taking direction as well as magnitude into account) of the centrifugal and gravitational forces.

The net tidal force causes two bulges of water on the Earth's oceans, one from the side nearest the Moon, where the net force is toward the Moon, and the other on the side farthest from the Moon, where the net force is away from the Moon (Figure 17.9). The net gravitational attraction between the oceans and the Moon is at a maximum on the side of Earth facing the Moon and at a minimum on the side facing away from the Moon. As Earth rotates, the bulges of water stay approximately aligned. One

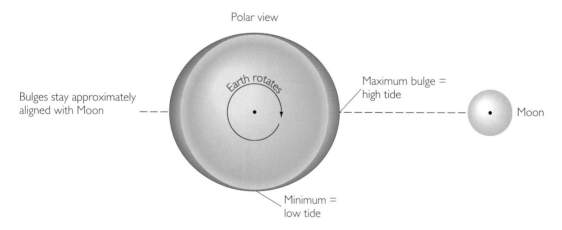

FIGURE 17.9 The Moon's gravitational attraction causes two bulges of water on the Earth's oceans, one on the side nearest the Moon and the other on the side farthest from the Moon. As the Earth rotates, these bulges remain aligned and pass over Earth's surface, forming the high tides.

always faces the Moon, the other is always directly opposite. These bulges passing over the rotating Earth are the high tides.

The Sun, although much farther away, has so much mass (and thus so much gravity) that it, too, causes tides. Sun tides are a little less than half the height of Moon tides. Sun tides are not synchronous with Moon tides (Figure 17.10). Sun tides come as the Earth rotates once every 24 hours, the length of a solar day. The rotation of the Earth with respect to the Moon is a little longer because the Moon is moving around the Earth, giving a lunar day of 24 hours and 50 minutes. In that lunar day, there are two high tides, with two low tides between them.

When the Moon, Earth, and Sun line up (see Figure 17.10a), the gravitational pulls of the Sun and the Moon reinforce each other. This produces the **spring tides,** which are the highest tides; they derive their name from their height, not from the season. They appear every two weeks at full and new Moon. The lowest tides, the **neap tides,** come in between, at first- and third-quarter Moon, when the Sun and Moon are at right angles to each other with respect to the Earth (see Figure 17.10b).

Although the tides occur regularly everywhere, the difference between high and low tides varies in different parts of the ocean. As the tidal bulges of water rise and fall, they also move along the surface of the ocean, encountering obstacles, such as continents and islands, that hinder the flow of water. In the middle of the Pacific Ocean—in Hawaii, for example, where there is minimal constriction and obstruction of the flow of the tides—the difference between low and high tides is only 0.5 m. On the Pacific coast near Seattle, where the coast along Puget Sound is very irregular, the tides are constricted through narrow passageways, and the difference between the two tides is about 3 m. Extraordinary tides occur in a few places, such as the Bay of Fundy in eastern Canada, where the tidal range can be more than 12 m. Because many people living along the shore need to know when tides will occur, governments publish tide tables showing predicted tide heights and times; these tables combine local experience with knowledge of the astronomical motions of Earth and the Moon with respect to the Sun.

Tides may combine with waves to cause extensive erosion of the shore and destruction of shoreline property. Intense storms passing near the shore during a spring tide may produce **tidal surges,** waves at high tide that can overrun the entire beach and batter sea cliffs. Tidal surges are not to be confused with the commonly but incorrectly termed "tidal waves." Although there are no such waves associated

with the tides, there are unusually large ocean waves called *tsunamis* (a Japanese word), which are caused by undersea events such as earthquakes, landslides, and the explosion of oceanic volcanoes (see Figure 18.20 and Feature 18.1).

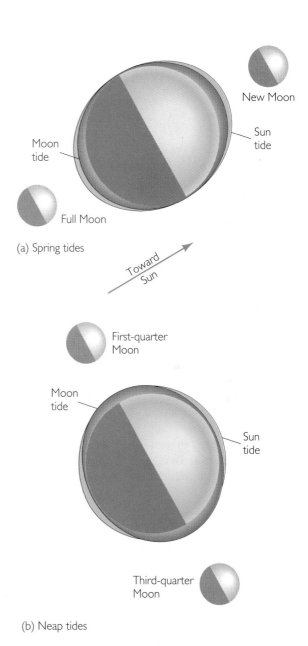

(a) Spring tides

(b) Neap tides

FIGURE 17.10 The relative positions of the Earth, Moon, and Sun determine the heights of high tide during the lunar month. (a) At new and full Moon, Sun and Moon tides reinforce each other and make the highest (spring) high tides. (b) At first- and third-quarter Moon, Sun and Moon tides are in opposition, causing low (neap) tides.

FIGURE **17.11** Tidal flats, such as this one at Mont-Saint-Michel, France, may be extensive areas covering many square kilometers but most often are narrow strips seaward of the beach. When a very high tide advances on a broad tidal flat like Mont-Saint-Michel's, it may move so rapidly that areas are flooded faster than a person can run. The beachcomber is well advised to learn the local tides before wandering. *(Thierry Prat/Sygma.)*

TIDAL CURRENTS   Tides moving near shorelines cause currents that can reach speeds of a few kilometers per hour. As the tide rises, the water flows in toward the shore as a **flood tide,** moving into shallow coastal marshes and up small streams. As the tide passes the high stage and starts to fall, the **ebb tide** moves out, and low-lying coastal areas are exposed again. Such tidal currents meander across and cut channels into **tidal flats,** the muddy or sandy areas that lie above low tide but are flooded at high tide (Figure 17.11).

SHORELINES

Waves, longshore currents, and tidal currents interact with the rocks and tectonics of the coast to shape shorelines into a multitude of forms. We can see these factors at work in the most popular of shorelines, beaches.

Beaches

A beach is a shoreline made up of sand and pebbles. Beaches may change shape from day to day, week to week, season to season, and year to year. Waves and tides sometimes broaden and extend a beach by depositing sand and sometimes narrow it by carrying sand away.

Many beaches are straight stretches of sand that range from a kilometer to more than a hundred kilometers long; others are smaller crescents of sand between rocky headlands. Belts of dunes border the landward edge of many beaches; bluffs or cliffs of sediment or rock border others. Beaches may have tide terraces—flat, shallow areas between the upper beach and an outer bar of sand—on their seaward sides (Figure 17.12).

The Structure of Beaches

Figure 17.13 shows the major parts of a beach, all of which may not be present at all times on any particular beach. Farthest out is the **offshore,** bounded by the surf zone, where the bottom begins to become shallow enough for waves to break. The **foreshore** includes the surf zone; the tidal flat; and, right at the shore, the swash zone, a slope dominated by the

FIGURE **17.12** Tide terrace. At low tide, the outer ridge (a sandbar at high tide) is exposed. Also exposed is the shallow depression between the ridge and the upper beach, which is rippled by the tidal flow in many places. *(James Valentine.)*

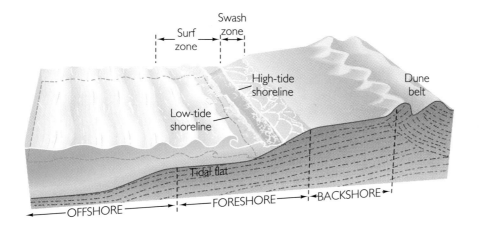

FIGURE 17.13
A profile of a beach, showing its major parts.

swash and backwash of the waves. The **backshore** extends from the swash zone up to the highest level of the beach.

THE SAND BUDGET OF A BEACH   A beach is a scene of incessant movement. Each wave moves sand back and forth with swash and backwash. Both longshore drift and longshore currents move sand down the beach. At the end of a beach and to some extent along it, sand is removed and deposited in deep water. In the backshore or along sea cliffs, sand and pebbles are freed by erosion and replenish the beach. The wind that blows over the beach trans-

ports sand, sometimes offshore into the water and sometimes onshore onto the land.

All these processes together maintain a balance between adding and removing sand, resulting in a beach that may appear to be stable but is actually exchanging its material on all sides. The sand budget of a beach—the inputs and outputs by erosion and sedimentation—is illustrated in Figure 17.14. At any point along the stretch, the beach gains sand by the inputs: from erosion of material along the backshore; from longshore drift and longshore current; and from rivers that enter the sea along the shore, bringing in sediment. The beach loses sand from the out-

INPUTS	OUTPUTS
Sediments eroded from backshore cliffs by waves	Sediments transported to backshore dunes by offshore winds
Sediments eroded from upcurrent beach by longshore drift and current	Sediments transported downcurrent by longshore drift and current
Sediments brought in by rivers	Sediments transported to deep water by tidal currents and waves

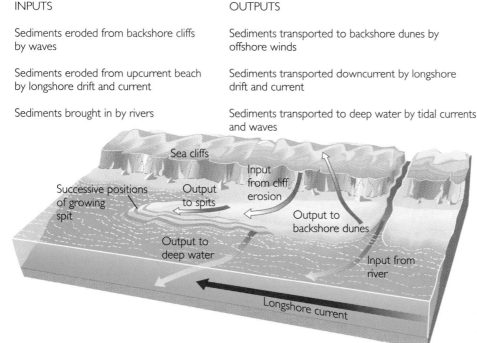

FIGURE 17.14 The beach budget is a balance between inputs and outputs of sand by erosion and sedimentation.

17.1 LIVING ON EARTH

Preserving Our Beaches

Orrin Pilkey of Duke University, a geologist and oceanographer, is in the forefront of scientists concerned about saving our beaches and halting massive development on fragile shorelines. Battered by the waves eroding the shoreline, many houses could be saved by building concrete buttresses, seawalls, and other structures designed to save shoreline property, but these structures would destroy the beach. Pilkey, a well-known researcher on coastal processes, is an advocate for the beaches of the Carolinas, which have come under heavy pressure from commercial developers. Knowing how the beach system works, he believes it is foolish to try to interfere with the natural process by which beaches remain in dynamic equilibrium with the waves and currents.

Humans are altering this equilibrium on more and more beaches by building cottages on the shore; paving beach parking lots; erecting seawalls; and constructing groins, piers, and breakwaters. A *groin* is a barrier built out from shore to prevent sand movement and beach erosion. The consequence of poorly thought-out construction is the shrinkage of the beach in one place and its growth in another, usually where no one wants it. The classic example is a narrow groin built out from shore at right angles to it. Over the following months and

A house tilts as sand is eroded from the beach and cliffs *(left background)* by storm waves. Nantucket, Massachusetts. *(Steve Rose/ Rainbow.)*

puts: winds carry sediments to backshore dunes, longshore drift and current carry it downcurrent, and deep water transports it by currents and waves during storms.

If the total input balances the total output, the beach is in equilibrium and keeps the same general form. If input and output are not balanced, the beach either grows or shrinks. Temporary imbalances are natural over weeks, months, or years. A series of large storms, for example, might move large amounts of sand from the beach to somewhat deeper waters on the far side of the surf zone, narrowing the beach. Then, in a slow return to equi-

librium over weeks of mild weather and low waves, the sand might move onto shore and rebuild a wide beach. Without this constant shifting of the sands, beaches might be unable to recover from trash, litter, and some kinds of pollution. Within a year or two, even oil from spills will be transported or buried out of sight, although the tarry residue may later be uncovered in spots. Beaches would clean up rapidly if the littering were to stop.

SOME COMMON FORMS OF BEACHES   We can now account for some common beaches. Long, wide, sandy beaches grow where sand inputs are

years, the sand disappears from the beach on one side and greatly enlarges the beach on the other side—much to the surprise of the builders. As landowners and developers bring suit against one another and against state governments, trial lawyers take the issue of "sand rights"—the beach's right to the sand it naturally contains—into the courts.

The disappearance and enlargement of beaches are the predictable results of a longshore current. The waves, current, and drift bring sand toward the groin from the upcurrent direction (usually the dominant wind direction). Stopped at the groin, they dump the sand there. On the downcurrent side of the groin, the current and drift pick up again and erode the beach. On this side, however, replenishment of sand is sparse because the groin blocks the current. As a result, the beach budget is out of balance, and the beach shrinks. If the groin is removed, the beach relaxes to its former state.

The only way to save a beach is to leave it alone. Even if concrete walls and piers can be kept in repair with large expenditures of money, many times at public expense, the beach itself will suffer. Along some beaches, resort hotels truck in sand to replace that lost, but that expensive solution is temporary, too. Sooner or later, we must learn to let the beaches remain in their natural state.

Construction of groins along a shore to control erosion of a beach may produce erosion downcurrent of the groin and loss of parts of the beach *(right of groin)* while sand piles up on the other side *(left of groin)*. Longshore current flows from left to right. *(Philip Plisson / Explorer.)*

abundant, often where soft sediments make up the coast. Where the backshore is low and the winds blow from onshore, wide dune belts border the beach. If the shoreline is tectonically elevated and the rocks are hard, cliffs line the shore, and any small beaches that evolve are composed of material eroded from the cliffs. Where the shore is low-lying, sand is abundant, and tidal currents are strong, extensive tidal flats are laid down and are exposed at low tide.

What happens if one of the inputs is cut off—for example, by a concrete wall built at the top of the beach to prevent erosion? Because erosion sup-

plies sand to the beach as one of the inputs, preventing it cuts the sand supply and so shrinks the beach. Attempts to save the beach may actually destroy it (see Feature 17.1).

Erosion and Deposition at Shorelines

The topography of the shoreline, like that of the interior, is a product of tectonic forces elevating or depressing the Earth's crust, erosion wearing it down, and sedimentation filling in the low spots. Thus, the factors directly at work are

FIGURE **17.15** The Twelve Apostles, a group of stacks at Port Campbell, Australia, developed in horizontal beds of sedimentary rock. These remnants of shore erosion are left as the shoreline retreats. (*Kevin Schafer.*)

- Uplift of the coastal region, which leads to erosional coastal forms

- Subsidence of the coastal region, which produces depositional coastal forms

- The nature of the rocks or sediments at the shoreline

- Changes in sea level, which affect the drowning or emergence of a shoreline

- The average and storm wave heights

- The heights of the tides, which affect both erosion and sedimentation

EROSIONAL COASTAL FORMS   Erosion is active at tectonically uplifted rocky coasts. Along these coasts, prominent cliffs and headlands jut into the sea, alternating with narrow inlets and irregular bays with small beaches. Waves crash against rocky shorelines, undercutting cliffs and causing huge blocks to fall into the water, where they are gradually worn away. As the sea cliffs retreat by erosion, isolated remnants called **stacks** are left standing in the sea, far from the shore (Figure 17.15). Erosion by waves also planes the rocky surface beneath the surf zone and creates a **wave-cut terrace,** sometimes visible at low tide (Figure 17.16). Wave erosion continuing over long periods may straighten shorelines, as headlands retreat faster than recesses and bays.

Where relatively soft sediments or sedimentary rocks make up the coastal region, the slopes are gentler and the heights of shoreline bluffs are lower. Waves efficiently erode these softer materials; erosion of bluffs on such shores may be extraordinarily rapid. The high sea cliffs of soft glacial materials along the Cape Cod National Seashore in Massachusetts, for instance, are retreating about a meter each year. Since Henry David Thoreau walked the

FIGURE **17.16** Wave-cut terrace at Bolinas Point, California. The cliffs have retreated from left to right as erosion planed a terrace at low-tide level. (*John S. Shelton.*)

FIGURE 17.17 Aerial view of the southern tip of Cape Cod, Monomoy Point, Massachusetts. This spit has advanced into deep water to the south *(foreground)* from the main body of the Cape to the north *(background)*. *(Steve Dunwell/The Image Bank.)*

DEPOSITIONAL COASTAL FORMS Sediment builds up in areas where tectonic subsidence depresses the crust along a coast. Such coasts are characterized by long, wide beaches and wide, low-lying coastal plains of sedimentary strata. Shoreline forms include sandbars, low-lying sandy islands, and extensive tidal flats. Long beaches grow longer as longshore currents carry sand to the downcurrent end of the beach. There it builds up, first as a submerged bar, then rising above the surface and extending the beach by a narrow addition called a **spit** (Figure 17.17).

Long sandbars offshore may build up and become **barrier islands,** forming a barricade between open ocean waves and the main shoreline (Figure 17.18). Barrier islands are common, especially along low-lying coasts of easily erodible and transportable sediments or poorly cemented sedimentary rocks where longshore currents are strong. Some of the most prominent barrier islands are found along the coast of New Jersey, at Cape Hatteras, and along the Texas coast of the Gulf of Mexico, where one, Padre Island, is 130 km long. As the bars build up above the waves, vegetation takes hold, stabilizing the islands and helping them resist wave erosion during storms. Barrier islands are separated from the coast by tidal flats or shallow lagoons. Like beaches on the main shore, barrier islands are in dynamic equilibrium with the forces shaping them. If their equilibrium is disturbed by natural changes in climate or wave and current regimes, or by real estate development, they may be disrupted or devegetated, leading to increased erosion and even disappearance. Other barrier islands may grow larger and more stable.

entire length of the beach below those cliffs in the mid-nineteenth century and wrote of his travels in *Cape Cod,* about 6 km² of coastal land has been eaten by the ocean, equivalent to about 150 m of beach retreat.

It has been estimated that more than 70 percent of the total length of the world's sand beaches has retreated in recent decades, at a rate of at least 10 cm per year, and 20 percent of the total length has retreated at a rate of more than 1 m per year. Much of this movement may be traced to the damming of rivers, which decreases the sediment supply to the shoreline.

FIGURE 17.18 Partially developed barrier islands separating the Gulf of Mexico *(right)* from the shallow waters and tidal flats of the main shoreline of western Florida. Barrier islands are easily disturbed and made subject to erosion by natural changes and real-estate development. *(Richard A. Davis, Jr.)*

Over the course of hundreds of years, shorelines may undergo significant changes. Hurricanes and other intense storms, such as the "storm of the century" that hit the East Coast in March 1993, may form new inlets or elongate spits or may breach them. Such changes have been documented by remapping at various time intervals from aerial photographs. The shoreline at Chatham, Massachusetts, at the elbow of Cape Cod, has changed enough over the past 160 years that a lighthouse has had to be moved. Figure 17.19 illustrates the numerous changes that have taken place in the configuration of the bars to the north and the long spit of Monomoy Island, as well as several breaches of the bars. Many homes are now at risk in Chatham, but there is little the residents or the state can do to prevent these beach processes from taking their natural course.

Changes in Sea Level

Shorelines are sensitive to changes in sea level, which can change the approach of waves, alter tidal heights, and affect the path of longshore currents.

Rise and fall of sea level can be local, a result of tectonic subsidence or uplift, or global, the result, for example, of continental glacial melting or growth (see Chapter 15).

Sea-level changes are detected by geologic studies of wave-cut terraces. These terraces, which were cut at sea level, are now elevated, showing that, at least locally, sea level has dropped in the intervening period (Figure 17.20). A new technique uses satellite altimeter measurements to determine the altitude of the sea surface relative to the carefully determined orbit of the satellite (see Feature 17.2). A study shows that over a two-year period, these measurements can determine differences in the altitude of the sea surface with the remarkable degree of precision of 4 mm. This precision has allowed us to estimate a global sea-level change of about 4 mm per year. Much of this rise may be due to short-term variations, but researchers hope that this method will give us reliable information on rises in sea level caused by global warming. Recent studies indicate that human activities over the globe may be affecting sea level, producing the small rise recorded over the twentieth century. These activities include impoundment of

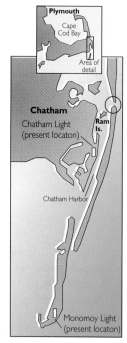

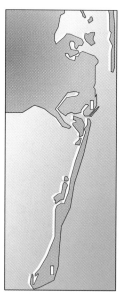
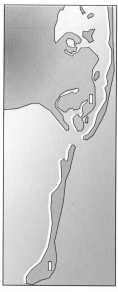
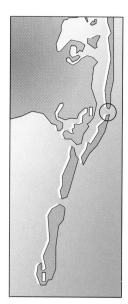

1830–1850 Circle shows approximate location of 1846 breach in barrier spit. Ram Island later disappears

1870–1890 The beach south of the inlet breaks up and migrates southwest toward the mainland and Monomoy

1910–1930 The southern beach has disappeared, and its remnants soon will connect Monomoy to the mainland

1950–1970 The northern beach steadily grows with cliff sediment; Monomoy breaks from the mainland

1987 140-year cycle begins again with Jan. 2 breach in the barrier spit across from the Chatham Light

FIGURE 17.19 Changes in the shoreline at Chatham, Massachusetts, at the elbow of Cape Cod, over the past 160 years. (After Cindy Daniels, *Boston Globe*, February 23, 1987.)

FIGURE 17.20 Raised shoreline terraces caused by coastal emergence indicate a fall in sea level since the terrace cuts were formed at the higher sea level. (*John S. Shelton.*)

surface waters behind dams, draining of wetlands, deforestation, groundwater withdrawal, and surface water diversion. According to one estimate, such activities may have contributed a part of the observed sea-level rise, which may register 0.3 to 1.1 m by the end of the twenty-first century. Others dispute the magnitude of that estimate, but it is clear that sea level may be at least slightly affected by human works, a conclusion that few could have foreseen.

During periods of lowered sea level, areas that were offshore are exposed to agents of erosion. Rivers extend their courses over formerly submerged regions and cut valleys into newly exposed coastal plains. When sea level rises, flooding the lands of the backshore, river valleys are drowned, marine sediments build up along former land areas, and erosion is replaced by sedimentation. Today, long fingers of the sea indent many of the shorelines of the northern and central Atlantic coast. These long indentations are former river valleys that were flooded as the last glacial age ended about 10,000 years ago and the sea level rose.

Drowned river valleys are one kind of **estuary,** which is a coastal body of water connected to the ocean and also supplied with fresh water from a river. The fresh water comes down the river and mixes with seawater in the estuary long before it reaches the main shoreline. At its upper reaches—in some places, many kilometers upstream from the mouth—an estuary is fresh. Downstream, it becomes saltier as it gradually mixes with seawater. By the time it nears the main shoreline, the estuary is entirely seawater.

The shorelines of the world serve as our barometers of impending change as they respond to altered global and local conditions. For example, if global warming causes sea level to rise, we will first see the effects on our beaches. The pollution of our inland waterways sooner or later arrives at our beaches, as sewage from city dumping and oil from ocean tankers wash up on the shore. And as real estate development and construction along shorelines expands, we will see the continuing contraction and even disappearance of some of our finest beaches.

Profound as the geological changes in shorelines may be, they are dwarfed by the geologic processes at work in the vast bulk of the oceans where the water is deep and hides the active seafloor below. In the rest of this chapter, we explore some of the important geologic processes of the deep ocean. We begin with the unusual tools available for measuring and mapping the seafloor.

SENSING THE FLOOR OF THE OCEAN

The best way to see the seafloor is directly from a deep-diving submersible. Pioneered by the French oceanographer Jacques-Yves Cousteau, these small ships can observe and photograph at great depths. With their mechanical arms, they can break off pieces of rock, sample soft sediment, and catch specimens of exotic deep-sea animals. Newer robotic submersibles are guided by scientists on the mother ship above (Figure 17.21). But submersibles are expensive to build and operate, and they cover small areas at best.

FIGURE 17.21 The *Benthic Explorer.* This small robotic vehicle can explore the seafloor while being directed from shipboard. (*T. Kleindinst/Woods Hole Oceanographic Institution.*)

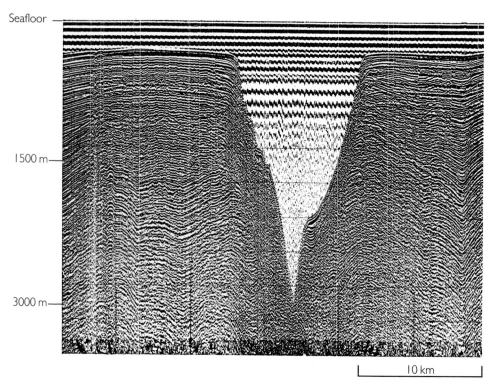

Seafloor —

1500 m —

3000 m —

|← 10 km →|

FIGURE **17.22** An echo-sounding profile of the Congo submarine canyon off the Republic of Zaire on the west coast of Africa. The bottom of the canyon at this point is about 3000 m below the seafloor of the continental shelf; at the top, the canyon is more than 10 km wide. The wavy lines below the seafloor surface are sound reflections from bedding planes in continental shelf sediments, somewhat deformed because of mild tectonic disturbance. *(K. O. Emery, Woods Hole Oceanographic Institution.)*

For most work, today's oceanographers use instrumentation to sense the seafloor topography indirectly from a ship at the surface. One shipboard instrument is an echo sounder, which sends out pulses of sound waves. When the sound waves are reflected back from the ocean bottom, they are picked up by sensitive microphones in the water. By measuring the interval between the time the pulse leaves the ship and the time it returns as a reflection, and by figuring in the speed of sound in water, oceanographers can compute the depth. The result is an automatically traced profile of the bottom topography. (Figure 17.22 shows an example.) Echo sounding is

also used to probe the stratigraphy of sedimentary layers beneath the ocean floor (see Feature 19.1).

Many other instruments are lowered to the bottom to detect such properties as the magnetism of the seafloor, the shapes of undersea cliffs and mountains, and heat coming from the interior. One of the most spectacular of these instruments are the underwater radar beams used to map the details of topography of the bottom (Figure 17.23). Several different systems use one or more beams that cut varying swaths both ahead and to the side, thus illuminating topography with a detail and accuracy not possible with echo sounders. On a larger scale, underwater cameras can

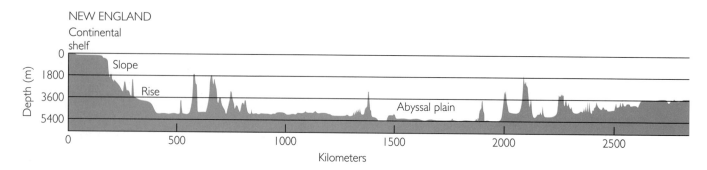

FIGURE **17.24** Topographic profile of the floor of the Atlantic Ocean from New England *(left)* to Gibraltar *(right)*. (After B. C. Heezen, "The Origin of Submarine Canyons," *Scientific American,* August 1956, p. 36.)

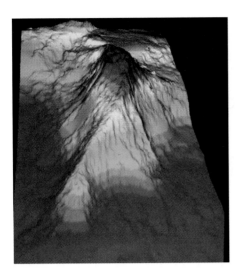

FIGURE **17.23** Loiki Seamount just south of the Big Island of Hawaii, as imaged by computer-enhanced side-scan radar. Loiki is the newest in the string of hot-spot volcanoes that form the Hawaiian island chain. *(Ocean Mapping Development Center, University of Rhode Island.)*

photograph the details of the seafloor surface and the organisms that inhabit the deep. Combined instrument packages that include radar, photography, gravity, and magnetic measurements are towed near the bottom to give even greater accuracy. To be a marine geologist today places one at the center of a beehive of high technology.

Since 1968, the United States–sponsored Deep Sea Drilling Program and its successor, the international Ocean Drilling Program, have sunk hundreds of drill holes to depths of many hundreds of meters below the seafloor. Cores obtained from these drill holes have given us an unprecedented three-dimensional picture of the seafloor and provided

samples for detailed physical and chemical studies. (In Chapter 20, we discuss the role of these drilling programs in the study of plate tectonics.)

PROFILES OF TWO OCEANS

An Atlantic Profile

Just as we profiled North America in Chapter 3 by taking a hypothetical trip across the continent, we can imagine driving a deep-diving submarine along the floor of the Atlantic Ocean from North America to Gibraltar. A topographic profile of the floor of the Atlantic is shown in Figure 17.24.

Starting from the coast of New England, we would descend from the shoreline to depths of 50 to 200 m and travel along the **continental shelf,** a broad, flat, sand- and mud-covered platform that is part of the continent but slightly submerged. After traveling about 50 to 100 km across the shelf, down a very gently inclined surface, we would find ourselves at the edge of the shelf, where we would start down a steeper incline, the **continental slope.** This slope, covered mostly with mud, descends at an angle of about 4°, a drop of 70 m over a horizontal distance of 1 km, which would feel like a noticeable grade if we were driving on land.

The continental slope is irregular and marked by gullies and **submarine canyons,** deep valleys eroded into the slope and the shelf behind it (see Figure 17.22). On the lower parts of the slope, at depths of around 2000 to 3000 m, the incline becomes gentler. Here it merges into a more gradual incline called the **continental rise,** an apron of muddy and sandy sediment extending into the main ocean basin. The rise is broken by an occasional shallow channel or canyon.

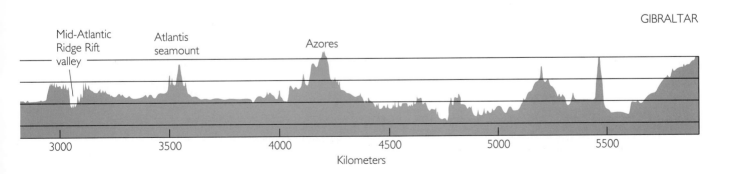

GIBRALTAR

Mid-Atlantic
Ridge Rift
valley

Atlantis
seamount

Azores

3000   3500   4000   4500   5000   5500

Kilometers

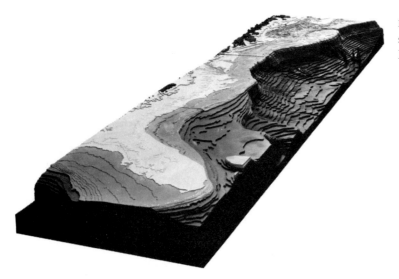

FIGURE 17.25 The Atlantic continental shelf, slope, and rise off the eastern coast of North America.

Figure 17.25 depicts a continental shelf, slope, and rise.

The continental rise is hundreds of kilometers wide, and it grades imperceptibly into a wide, flat **abyssal plain** that covers large areas of the ocean floor at depths of about 4000 m to almost 6000 m (Figure 17.26). These plains are broken by occasional submerged volcanoes, mostly extinct, called **seamounts** (see Figure 17.23). A few seamounts extend to the surface as islands. The Azores are such volcanic islands. In the Caribbean Sea, many of the volcanic islands are capped with coral reefs and other kinds of limestone.

As we travel along the abyssal plain, we gradually climb into a province of low abyssal hills whose slopes are covered with fine sediment. Continuing

up the hills, the sediment layer becomes thinner and outcrops of basalt appear beneath it. As we rise along this steep, hilly topography to depths of about 3000 m, we are climbing the flanks and then the mountains of the Mid-Atlantic Ridge.

Abruptly, we come to the edge of a deep, narrow valley (about 1 km wide) at the top of the ridge, a narrow cleft marked by active volcanism. This is a rift valley where two plates separate. As we cross the valley and climb the east side, we are moving from the North American Plate to the Eurasian Plate (see the plate map in the front endpapers).

Continuing east, we find the same topography as on the west side of the ridge, only in reverse order, for the ocean floor is symmetrical on either side of the ridge. Again we pass over abyssal hills; gradually

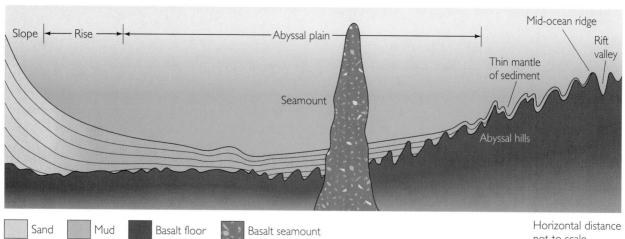

FIGURE 17.26 Profile from continental rise to Mid-Atlantic Ridge.

FIGURE 17.27

Artist's representation of the North Atlantic Ocean floor from echo-sounding data. Depths shown are in feet below sea level. (Detail from *World Ocean Floor* by Heinrich C. Berann, based on bathymetric studies by Bruce C. Heezen and Marie Tharp. Copyright © Marie Tharp.)

descend to an abyssal plain; then ascend to the continental rise, slope, and shelf off the coast of Europe.

Oceanographic research vessels made many such trips, crisscrossing the oceans with echo sounders. By the late 1970s, researchers had mapped the floor of the oceans. On the map shown in Figure 17.27, we can see some of the details of individual ocean-floor provinces of the North Atlantic and get an impression of the strange submarine landscape. This map is still somewhat of an approximation, for many of the areas of the seafloor have never been visited by an echo-sounding research ship, and the mappers had to interpolate some of the details between widely separated ship tracks. The next generation of maps would be based on satellite data (Feature 17.2).

Some of the characteristics of oceans are related to plate-tectonic movements. The Atlantic, bisected by the Mid-Atlantic Ridge, is primarily a spreading ocean and has only a small subduction zone in the Caribbean Sea. In contrast, the Pacific, crossed by the East Pacific Rise, shows a large number of subduction zones that are currently narrowing this ocean.

The entire network of ridges, trenches, and transform faults that bound the world's ocean plates can be seen in the map in Feature 17.2. This map was constructed not from echo sounding but from satellite data kept secret by the Navy during the Cold War and only recently made available to all oceanographers. A Navy satellite called Geosat, which repeatedly orbited the Earth, used a radar altimeter to measure the altitude of the sea surface. The bumps and depressions in the sea surface on this map mimic the highs and lows of the seafloor. For example, a seamount that rises 1500 m from the seafloor produces a bump in the sea surface of about 1.5 m. From such data, we can visualize ridges, trenches, plateaus, and abyssal plains that exist where no ship has ever gone. Marine geologists are now busy studying this map and using the data from the satellite to make detailed studies of seafloor topography that we never knew existed.

A Pacific Profile

Just as continents' profiles may differ, a profile of the Pacific Ocean shows features not seen in the Atlantic profile. If we were to travel westward from South America beginning on the west coast of Peru or Chile, we would, as before, cross a continental shelf. This shelf, however, is only a few tens of kilometers wide and more than 100 m deep. At the edge of the shelf, the continental slope is much steeper and extends down to 8000 m as we enter the Peru-Chile

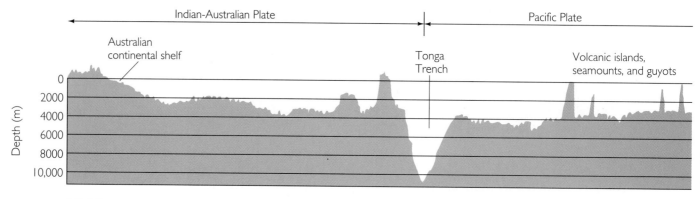

FIGURE 17.28 Topographic profile of the floor of the Pacific Ocean traveling westward from South America *(right)* to Australia *(left)*. (Data synthesized from various sources.)

Trench (Figure 17.28). This long, deep, narrow depression in the seafloor is the surface expression of the subduction of the Nazca Plate, a small plate in the eastern Pacific, under the South American Plate.

Continuing across the trench and up onto the higher hilly region of the Nazca Plate, we soon come to a mid-ocean ridge, the East Pacific Rise. The East Pacific Rise is lower than the Mid-Atlantic Ridge, but it has the characteristic central rift valley and outcrops of basalt. On the west side of the East Pacific Rise, we cross over to the Pacific Plate and drive on westward over its broad central regions. Eventually we come to another trench, the Tonga. This is one of the deepest places in all the oceans, almost 11,000 m deep. Here, in the middle of the ocean, the Pacific Plate subducts beneath the Indian-Australian Plate. On the west side of the trench, an arc of volcanic islands rises from the deep seafloor and erupts basalt and andesite. Leaving the island arc, we return to the deep seafloor, now on the Indian-Australian Plate, and soon come to the continental rise, slope, and shelf of Australia, similar to the east coast of North America.

CONTINENTAL MARGINS

The shorelines, shelves, and slopes of the continents are together called **continental margins.** The profiles of the oceans show that there are two types of continental margins: passive and active. A **passive margin** is a continental borderland far from a plate boundary. A good example is the broad region, associated with a spreading ocean, off the east coast of North America (Figure 17.29). Such margins are called passive (implying quiescence) because volcanoes are absent and earthquakes are few and far be-

tween. In contrast, **active margins** are associated with subduction zones and transform faults. The volcanic activity and frequent earthquakes give these narrow and tectonically deformed continental margins their name. One active margin at a subduction zone is off the west coast of South America (see Figure 17.28). This margin includes the trench offshore, the narrow shelf, and the active volcanic belt of the Andes Mountains. As the subducting Nazca Plate bends below the continent, it heats up, and magmas form and work their way to the surface.

The continental shelves of passive margins consist of essentially flat-lying, shallow-water sediments, both terrigenous and carbonate, several kilometers thick (see Figure 17.29). Although the same kinds of sediment are found on active margin shelves, they

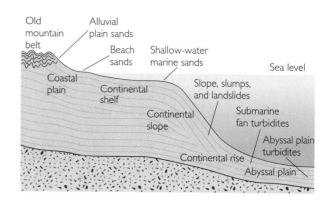

FIGURE 17.29 A profile of the Atlantic passive continental margin off southern New England. (After K. O. Emery and E. Uchupi, *Atlantic Continental Margin of North America,* American Association of Petroleum Geologists, 1972.)

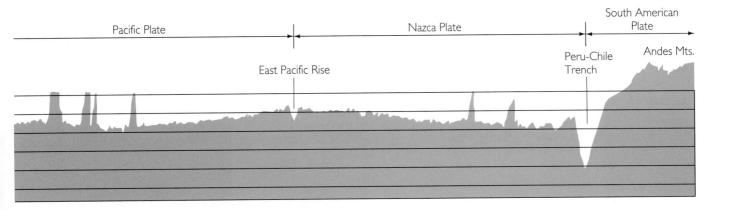

are more likely to be structurally deformed and to include ash and other volcanic materials.

Continental Shelf

The continental shelf is one of the most economically valuable parts of the ocean. Georges Bank off New England and the Grand Banks of Newfoundland, for example, have been among the world's most productive fishing grounds for all of this century. In recent years, the continental shelf, especially off the Gulf coast of Louisiana and Texas, has housed huge oil-drilling platforms. These are some of the reasons why in 1982 most of the world's nations (although not the United States) signed the international Law of the Sea treaty governing territorial and economic rights of nations.

Continental shelves are broad and relatively flat at passive continental margins and are narrow and uneven at active margins. As we noted earlier, because continental shelves lie at shallow depths, they are subject to exposure and submergence as a result of changes in sea level. During the Pleistocene glaciation, all of the shelves now at depths of less than 100 m were above sea level, and many of their features were formed then. At that time, shelves at high latitudes were glaciated, producing an irregular topography of shallow valleys, basins, and ridges. The shelves in lower latitudes were left more regular, broken by occasional stream valleys.

Continental Slope and Rise: Turbidity Currents

The waters of the continental slope and rise are too deep for the seafloor to be affected by waves and tidal currents. As a consequence, muds, silts, and sands that have been carried across the shallow continental shelf come to rest as they are draped over the slope. The slope shows signs of the slumping of sediment and the erosional scars of gullies and submarine canyons. Deposits of sands, silts, and muds on both slope and rise indicate active sediment transport in these deep waters. For some time, geologists were puzzled over what kind of current might cause both erosion and sedimentation on the slope and rise at such great depths.

The answer proved to be a **turbidity current**—a flow of turbid, muddy water down a slope. Because of its suspended load of mud, the turbid water is denser than the overlying clear water and flows beneath it. Turbidity currents were first noticed more than a century ago where the Rhone River enters Lake Geneva in Switzerland. The muddy river water enters the clear water of the lake and flows as a distinct current along the sloping bottom to the level floor of the lake, where it fans out over the bottom.

Turbidity currents can both erode and transport sediment, and their role in ocean processes was first understood by Philip Kuenen, a Dutch geologist and oceanographer. In 1936, Kuenen produced and filmed such currents in his laboratory by pouring muddy water into the end of a long, narrow tank with a sloping bottom. He showed that these currents could move at many kilometers per hour and that the speed was proportional to the steepness of the slope and the density of the current. Kuenen reasoned that because of its speed and turbulence, a turbidity current could erode and transport large quantities of sand down the continental slope. He then proposed the idea, revolutionary for that time, that turbidity currents operate widely in the ocean, especially on continental slopes, at depths well below any possible wave or tidal action.

17.2 TECHNOLOGY AND EARTH

Charting the Seafloor by Satellite

The rich geology of the seafloor—its ridges, trenches, seamounts, and transform boundaries—became apparently only after decades of ship soundings. Our knowledge of regions where few ships travel remains fragmentary. Recently, however, scientists have developed a tool that enables a satellite to "see through" the ocean and chart the topography of the seafloor, gathering data in mere months.

The new method makes use of an altimeter mounted on a satellite. The altimeter sends pulses of radar beams that are reflected back from the ocean below, giving measurements of the distance between the satellite and the sea surface with a precision of a few centimeters. The height of the sea surface depends not only on waves and ocean currents but also on changes in gravity caused by the topography and composition of the underlying seafloor. The gravitational attraction of a seamount, for example, can cause water to "pile up" above it, producing a bulge in the sea surface as much as 5 m above average sea level. Similarly, the diminished gravity over a deep-sea trench would show as a depression of the sea surface of as much as 60 m. In this way, features of the ocean floor can be inferred from satellite data and displayed as if the seas were drained away.

In the satellite photograph shown here, shallow regions, deep regions, and intermediate depths can be distinguished. Also clearly visible are the raised stripe between Europe and North America that marks the Mid-Atlantic Ridge and its associated transform boundaries, the trail of a hot spot in the Pacific marked by the Emperor-Hawaiian seamount chain, and the major deep-sea trenches at subduction boundaries. New features not revealed by ship surveys have already been found, and future surveys may reveal even deeper structures, such as convection currents in the mantle.

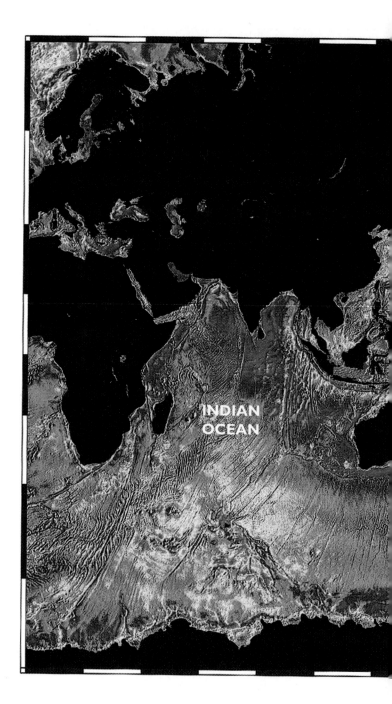

INDIAN OCEAN

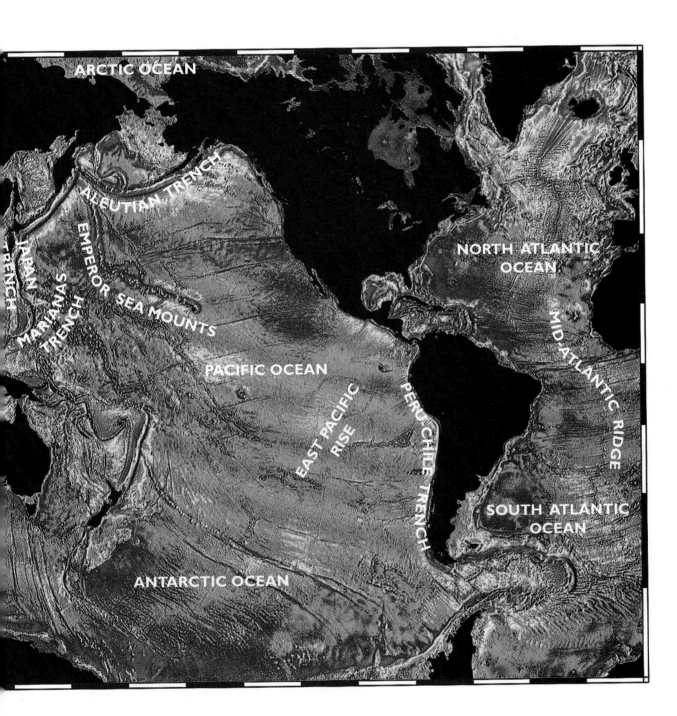

Marine gravity anomaly from satellite altimetry. *(D. T. Sandwell and W. H. F. Smith/Geological Data Center, Scripps Institution of Oceanography, University of California, San Diego.)*

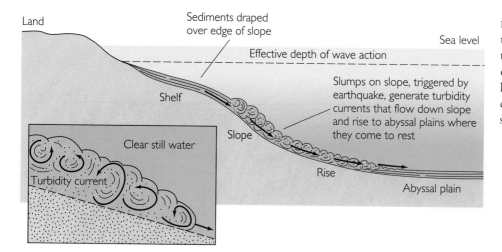

Land

Sediments draped over edge of slope

Sea level

Effective depth of wave action

Shelf

Slumps on slope, triggered by earthquake, generate turbidity currents that flow down slope and rise to abyssal plains where they come to rest

Slope

Rise

Abyssal plain

Clear still water

Turbidity current

FIGURE 17.30 How a turbidity current forms in the ocean. These currents can erode and transport large quantities of sand down the continental slope.

Turbidity currents start when occasional earthquakes trigger slumps of the sediment draped over the edge of the continental shelf and onto the continental slope (Figure 17.30). The sudden slump, or submarine landslide, throws mud into suspension, creating a dense, turbid layer of water near the bottom. This turbid layer starts to flow, accelerating down the slope.

As the turbidity current reaches the foot of the slope and the gentler incline of the continental rise, the current slows and some of the coarser sandy sediment starts to settle, often forming a **submarine fan,** a deposit something like an alluvial fan on land. Many currents continue across the rise, cutting channels in the submarine fans (Figure 17.31). Eventually, the currents reach the level bottom of the ocean basin, the abyssal plain, where they spread out and come to rest in graded beds of sand, silt, and mud called **turbidites.**

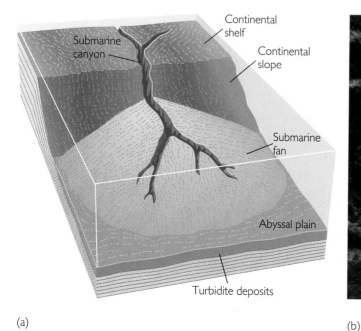

Continental shelf

Submarine canyon

Continental slope

Submarine fan

Abyssal plain

Turbidite deposits

(a)

(b)

FIGURE 17.31 (a) Submarine canyons and fans are formed by turbidity currents that start on the continental shelf or slope and erode canyons in them, leading to channels on the fan. (b) Sandfall at the head of a submarine canyon at the edge of the continental shelf. These falls generate sandy flows, like turbidity currents, that lay down fans of sandy sediment at the foot of the continental slope. *(U.S. Navy.)*

17.3 INTERPRETING THE EARTH

Hot Springs on the Seafloor

Much of the heat coming from Earth's interior is thought to be dissipated when cold seawater percolates into the many fissures associated with seafloor spreading along mid-ocean ridges. The cold water sinks several kilometers, encounters hot basalt, surges upward, and emerges on the seafloor as hot springs, rich in dissolved minerals and gases leached from the magma. This hypothesis was first verified in 1977; subsequently, spectacular discoveries of such hydrothermal vents at many places along different mid-ocean ridges have been reported. The Galápagos Islands rift zone spreading center was the first to be discovered, followed by reports of vents at several other places on the East Pacific Rise, on the Gorda and Juan de Fuca ridges in the Pacific off the shores of Oregon and Washington, and at several places along the Mid-Atlantic Ridge. Vents have also been explored in the western Pacific, at the Mariana Trough, and in several other locations. It is now certain that vents are widely distributed over the oceans, especially along mid-ocean ridges.

The hot springs seem to take two main forms. In the Galápagos Islands rift zone, the springs flow gently from cracks at maximum temperatures of about 16°C. On the East Pacific Rise, near Baja, California, superheated water (380°C) spouts forcefully from mineralized chimneys. The chimneys are built up from the dissolved minerals that precipitate around the hot jet as it mixes with the near-freezing waters on the ocean bottom. The deep-diving submarine *Alvin,* not built for such high temperatures, was nearly destroyed when it first approached a superheated vent.

Hydrothermal vents represent a major ore-forming process, possibly an important new source of such minerals as sulfide ores rich in zinc, copper, and iron. Once every 8 million years, the entire ocean cycles through such hydrothermal systems, and this process profoundly affects ocean chemistry by transporting elements from the interior to the ocean. The ecology of these vents is completely different from that of the dark, near-freezing, barren ocean bottom at great depths. Dense colonies of exotic life forms populate the warm water surrounding the vents. Among them are new species of giant worms, clams, and crabs. The vent animals live on unusual primitive bacteria called the *Archaea* that draw energy from the hydrogen sulfide, carbon dioxide, and oxygen in the vent water rather than from the Sun, as the species at the ocean surface do. The Archaea have been discovered to be a major group of organisms, on a parallel with bacteria and multicellular life forms. A new chapter of ocean science was opened by the deep-sea explorers who discovered the hot springs of the seafloor and verified the dominant role of circulating seawater in cooling the new crust formed at ocean ridges.

A plume of hot, mineral-laden water spouts from a hydrothermal vent on the East Pacific Rise. *(D. B. Foster/Woods Hole Oceanographic Institution.)*

For more than a decade after Kuenen's findings, some scientists challenged the existence of turbidity currents, but dramatic confirmation of the hypothesis came from an unexpected source. Transatlantic telegraph cables that had been laid on the ocean floor from North America to Europe were known to break periodically. In 1929, following an earthquake, a particularly large series of cable breaks was reported on the Atlantic continental slope and rise off the Grand Banks of Newfoundland.

In 1952, oceanographers Bruce Heezen and Maurice Ewing of Columbia University, impressed with Kuenen's work, plotted the exact times and positions of these breaks. Could slope slumps and turbidity currents explain the breaks? A rapid breaking of the cables high on the slope was followed by a sequence of breaks going down the slope, farther and farther from the center of the earthquake. If the breaks downslope had been caused by earthquake waves, they should have occurred much earlier. The only reasonable explanation was that the earthquake triggered a slump, which activated a turbidity current fast and powerful enough to snap the cables as it raced down the slope and rise. Later, graded beds were recovered from the seafloor in the path of the presumed flow. As researchers plotted similar patterns of cable breaks in other places, they realized that the turbidity current theory had been confirmed.

Today, we recognize turbidity currents as important agents of erosion and sedimentation on continental margins. They are found in trenches along active margins as well as on the continental slopes and rises of passive margins. They form the abyssal plains that cover large areas of the ocean floor.

Submarine Canyons

Submarine canyons are deep valleys eroded into the continental shelf and slope. They were discovered near the beginning of the twentieth century and were first mapped in detail in 1937. Almost immediately after ocean turbidity currents were hypothesized, they were proposed to be the erosive agents that incised submarine canyons into many continental shelves (see Figure 17.31). This idea, like the currents themselves, remained controversial for many years. Even though submarine canyons have been mapped in detail and their walls and floors amply photographed and sampled, they have been among the most perplexing topographic features of the seafloor. When the canyons were first discovered, some geologists thought they might have been formed by rivers. But this hypothesis soon proved impossible as the complete explanation. Most of the canyon floors are thousands of meters deep, far below the approximately 100-m depth to which rivers could erode during the maximum lowering of sea level in the ice ages. Even so, there is no question that the shallower parts of some canyons were river channels during periods of low sea level.

Although other types of currents have been proposed, turbidity currents are now the favored explanation for the deeper parts of submarine canyons. Evidence supporting this conclusion comes in part

from a comparison of modern canyons and their deposits with well-preserved similar deposits of the past, particularly the pattern of turbidites deposited on submarine fans.

THE FLOOR OF THE DEEP OCEAN

The deep seafloor is constructed primarily by volcanism related to plate-tectonic motions and secondarily by sedimentation in the open sea. When plates grow by spreading from a mid-ocean ridge, huge quantities of basalt well up from the mantle, forming the oceanic lithosphere. As the plates spread away from the ridge, the basaltic lithosphere cools and contracts, lowering the seafloor. While this is happening, the basalt surface receives a steady rain of sediment from surface waters and gradually becomes mantled with deep-sea muds and other deposits.

Mid-Ocean Ridges

Mid-ocean ridges are the sites of the most intense volcanic and tectonic activity on the deep seafloor. The main rift valley is the center of the action. The valley walls are faulted and intruded with basalt sills and dikes (Figure 17.32), and the floor of the valley is covered with flows of basalt and talus blocks from the valley walls, mixed with a little sediment settling from surface waters. Mid-ocean ridges are offset at many places by transform faults that laterally displace the rift valleys (see Figure 17.27).

Hydrothermal springs form on the rift valley floor as seawater percolates into cracks and fractures in the basalt on the flanks of the ridge, is heated as it moves down to hotter basalt, and finally exits at the valley floor, where it boils up at temperatures as high as 380°C (see Feature 17.3). Some springs are "black smokers," full of dissolved hydrogen sulfide and metals leached from the basalt by the hot waters. Others are "white smokers" with a different composition and lower temperatures. Hydrothermal springs on the seafloor produce mounds of iron-rich clay minerals, iron and manganese oxides, and large deposits of iron-zinc-copper sulfides.

Hills and Plateaus

The floor of the deep oceans away from mid-ocean ridges is a landscape of hills, plateaus, sediment-floored basins, and seamounts. Most of the thousands of volcanoes are submerged, but some rise to the sea surface. Seamounts and volcanic islands may be isolated, in clusters, or in chains. They may be formed along a mid-ocean ridge or where a plate overrides a mantle hot spot. Many seamounts have flat tops, the result of erosion of an island volcano when it was above sea level. These **guyots,** as they are called, are submerged because the plate they were riding on cooled, contracted, and subsided as it passed away from the hot spot that produced the upwelling basalt from the mantle.

Abyssal hills, plateaus, and low ridges are all accumulations of volcanic rock. Many of these features form when the seafloor first opens at a rift valley.

Others form as volcanic chains are created over hot spots. The traces of transform-fault offsets of the mid-ocean ridges interrupt the abyssal seafloor with long ridges coupled with parallel valleys, both roughly at right angles to the ridge (see Figure 17.27 and Feature 17.2). One of the largest plateaus, and probably the oldest (Jurassic-Cretaceous boundary), is the Shatsky Rise in the northwest Pacific Ocean, about 1600 km southeast of Japan. The Shatsky Rise is thought to have accumulated from the outpouring of basalt at a former hot spot, the product of a very large mantle plume or set of plumes.

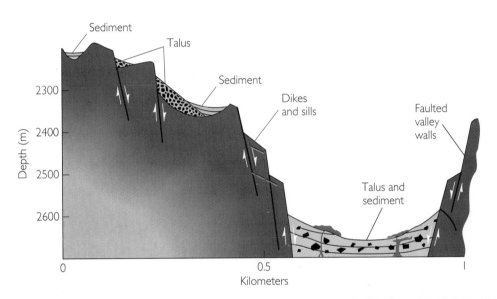

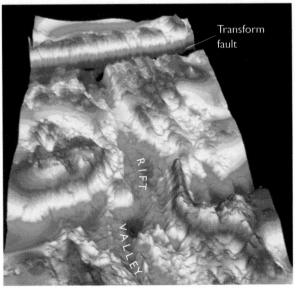

FIGURE **17.32** *Above:* A profile of the central rift valley of the Mid-Atlantic Ridge in the FAMOUS (French-American Mid-Ocean Undersea Study) area southwest of the Azores Islands. The deep valley, where most of the basalt is extruded, is faulted. (After ARCYANA, "Transform Fault and Rift Valley from Bathyscaph and Diving Saucer," *Science,* vol. 190, 1975, p. 108.) *Right:* Computer-generated image (based on data from multibeam sidescan sonar) of the Mid-Atlantic Ridge at about 31° south latitude reveals a 19-km-wide central valley studded with volcanic cones. The valley is bounded by mountains about 2000 m high. A transform fault intersects the valley at the top of the image. (K. C. Macdonald and P. J. Fox, "The Mid-Ocean Ridge," *Scientific American,* vol. 262, 1990, pp. 72–79.)

FIGURE 17.33 Some of the Maldive Islands, coral atolls of the Indian Ocean southwest of the tip of southern India, each surrounded by a circular coral reef that protects the lagoons. In the background is a reef without a central island. *(Guido Alberto Rossi/The Image Bank.)*

Although the deep seafloor is tectonically active at mid-ocean ridges, subduction zones, and hot spots, it is quiescent over much of the vast areas of abyssal hills and abyssal plains. In these areas, the only geological activity is the slow rain of sediment from surface waters, a feature that has led to serious consideration of some parts of the deep seafloor as a waste repository (see Feature 17.4).

Coral Reefs and Atolls

For more than 200 years, coral reefs have attracted explorers and travel writers. Ever since Charles Darwin sailed the oceans on the *Beagle* from 1831 to 1836, these reefs have been a matter of scientific discussion, too. Darwin was one of the first to analyze the geology of coral reefs, and his theory of their origin is still accepted today.

The coral reefs Darwin studied were **atolls,** islands in the open ocean with circular lagoons enclosed within a more or less circular chain of islands (Figure 17.33). Coral reefs also form at continental margins, such as those of the Florida Keys. The outermost part of a reef is a slightly submerged, wave-resistant reef front, a steep slope facing the ocean. The reef front is composed of the interlaced skeletons of actively growing coral and calcareous algae, forming a tough, hard limestone. Behind the reef front is a flat platform extending into a shallow lagoon. An island may lie at the center of the lagoon. Parts of the reef, as well as a central island, are above water and may become forested. A great many plant and animal species inhabit the reef and the lagoon.

Coral reefs are generally limited to waters less than about 20 m deep, for below that depth seawater does not transmit enough light to enable reef-building corals to grow. (Exceptions are some kinds of individual—noncolonial—corals that grow in much deeper waters.) Darwin explained how coral reefs could be built up from the bottom of the dark, deep ocean. The process starts with a volcano building up to the surface from the seafloor (Figure 17.34). As the volcano temporarily or permanently becomes dormant, coral and algae colonize the shore and build **fringing reefs,** coral reefs similar to atolls that grow around the edges of a central volcanic island. Erosion may then lower the volcanic island almost to sea level.

Darwin reasoned that if such a volcanic island were to slowly subside beneath the waves, actively growing coral and algae might keep pace with the subsidence, continuously building up the reef so that the island remained. In this way, the volcanic island would disappear and we would be left with an atoll. More than 100 years after Darwin proposed his theory, deep drilling on several atolls penetrated volcanic rock below the coralline limestone and confirmed the theory. And some decades later, the theory of plate tectonics explained both volcanism and the subsidence that resulted from plate cooling and contraction.

PHYSICAL AND CHEMICAL SEDIMENTATION IN THE OCEAN

Almost everywhere that oceanographers search the seafloor, they find a blanket of sediment. The muds and sands mantle and cover the topography of basalt originally formed at mid-ocean ridges. The ceaseless sedimentation in the world's oceans modifies the structures formed by plate tectonics and creates its own topography at sites of rapid deposition. The sediment is mainly of two kinds: terrigenous muds and sands eroded from the continents, and biochemically precipitated shells of organisms that live in the sea. In parts of the ocean near subduction zones, sediments derived from volcanic ash and lava flows are abundant. In tropical arms of the sea where evaporation is intense, evaporite sediments are deposited.

Sedimentation on Continental Margins

Terrigenous sedimentation on the continental shelf is produced by the same forces that form beaches: waves and tides. The waves of large storms and hurricanes move sediment over the shallow and moderate depths of the shelf, and tidal currents also flow over the shelf. The waves and currents distribute the sediment brought in by rivers into long ribbons of sand and layers of silt and mud.

Biochemical sedimentation on the shelf results from the buildup of layers of the calcium carbonate shells of clams, oysters, and many other organisms living in shallow waters. Most of these organisms cannot tolerate muddy waters and are found only where terrigenous materials are minor or absent, such as along the extreme southern coast of Florida or off the coast of Yucatan in Mexico. Here, coral reefs thrive and organisms build up large thicknesses of carbonate sediment.

Deep-Sea Sedimentation

Far from the continental margins, fine-grained terrigenous and biochemically precipitated particles suspended in seawater slowly settle from the surface to the bottom. These open-ocean sediments, called **pelagic sediments,** are characterized by great distance from continental margins, fine particle size, and a slow settling mode of deposition. The terrigenous materials are brownish and grayish clays, which accumulate on the seafloor at a very slow rate, a few millimeters every 1000 years. A small fraction, about 10 percent, may be blown by the wind to the open ocean. Winds from the Sahara Desert, for example,

FIGURE 17.34 Evolution of a coral reef from a subsiding volcanic island, first proposed by Charles Darwin in the nineteenth century.

(a) Volcano rises from ocean floor

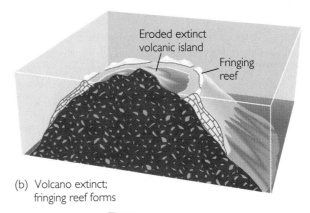

(b) Volcano extinct; fringing reef forms

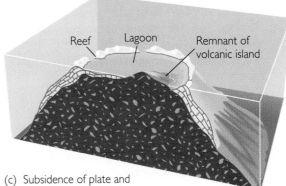

(c) Subsidence of plate and volcanic island as reef builds up

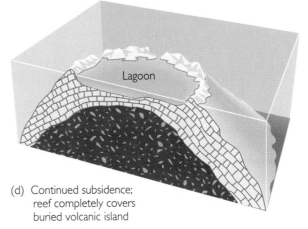

(d) Continued subsidence; reef completely covers buried volcanic island

17.4 LIVING ON EARTH

The Oceans as a Deep Waste Repository

People have been using the oceans as a garbage dump for millennia, since the first time someone in a dugout canoe threw refuse over the side. Today, enormous quantities of garbage, sewage, and industrial waste find their way into the sea from ships and coastal communities. Only now, as hypodermic syringes and other medical wastes as well as all sorts of other refuse wash ashore, are some nations beginning to restrict such dumping.

In 1995, environmental groups registered a strong protest against Shell Oil Company's bid to dispose of a whole marine storage platform by sinking it to the bottom of the North Atlantic. The bid became the center of a dispute when it was held up as an example of improper disposal on the seafloor of material containing toxic metals. The waste platform would have included "hundreds of kilograms" of cadmium, mercury, zinc, lead, and nickel. Shell eventually dropped the plan, but some marine geochemists have pointed out that these same metals enter the ocean at hydrothermal vents on mid-ocean ridges at a rate of 0.5 million to 5 million tons each year. These vast tonnages dwarf the small amounts that would have been leached from the sunken platform.

Some oceanographers are evaluating how we might safely dump a variety of hazardous materials, from highly toxic sewage sludges to high-level radioactive wastes, which are considered candidates for burial in the deep ocean. The goal is to deposit the waste in some location where it could remain undisturbed for the hundreds of thousands of years required for radioactive materials to decay enough to be harmless to humans and animals.

The most likely sites are in the middle of an oceanic plate, far from plate boundaries with their tectonic and volcanic activity. One suggestion is to design strong drums of corrosion-resistant material that would be embedded many meters below the seafloor in the accumulated sediment. Proponents of this idea believe the sediment would form a seal around the drums and retard or prevent any leakage that might occur after long burial. For this idea to work economically and safely, we must be able to convert the waste material to a form, such as a dense glass or ceramic, that could be loaded into the drums and would resist chemical attack by seawater or sediment pore waters.

Critics of deep-ocean burial point out that the dangers to ocean life, even at depths where few organisms interact with the bulk of the ocean's populations, are incalculable should the drums prove susceptible to leakage. Their arguments have gained support from the recent revelation that Russian ships have been dumping low-level radioactive waste in the Sea of Japan since the mid-1950s and still continue to do so. Such a violation of a prudent dumping policy has made the public chary of allowing any seabed disposal at all. No agreement has been reached, and the argument continues because waste disposal on land presents different but equally valid hazards.

Plastic debris dumped at sea has washed up onto this beach on the Gulf of Mexico. (*Robert Visser, Greenpeace.*)

FIGURE 17.35 Scanning electron micrograph of oceanic ooze. Shown here are shells of both carbonate- and silica-secreting unicellular organisms. *(Scripps Institution of Oceanography, University of California, San Diego.)*

blow much dust, silt, and fine sand into the eastern Atlantic off the coast of Africa. Windblown volcanic ash may be deposited downwind from subduction-zone volcanoes.

Within the pelagic sediments, the most abundant biochemically precipitated particles are the shells of foraminifera, tiny single-celled animals that float in the surface waters of the sea. These calcium carbonate shells fall to the bottom after their occu-

pants die. There they accumulate as **foraminiferal oozes,** sandy and silty sediments composed of foraminiferal shells (Figure 17.35). Other carbonate oozes are made up of shells of different microorganisms, called *coccoliths.*

Foraminiferal and other carbonate oozes are abundant at depths of less than about 4 km, but they are rare on the deeper parts of the ocean floor. This cannot be because of a lack of shells, for the surface waters are full of them everywhere and the living foraminifera are unaffected by the bottom far below. The explanation for the absence of carbonate oozes below a certain depth, called the **carbonate compensation depth (CCD),** is that the shells dissolve in deep seawater (Figure 17.36). Because of the nature of the physical circulation of the oceans, the deeper waters of the ocean differ from shallower waters in three ways:

- They are colder—colder, denser polar waters sink beneath warmer tropical waters and travel toward the equator along the bottom.

- They contain more carbon dioxide—not only do colder waters absorb more carbon dioxide than warmer waters, but any organic matter they are carrying tends to be oxidized to carbon dioxide during their long circulation travels.

- They are under higher pressure—this pressure results from the greater weight of overlying water.

These three factors make calcium carbonate more soluble in deep waters than in shallow ones. As the shells of dead foraminifera fall to the bottom below

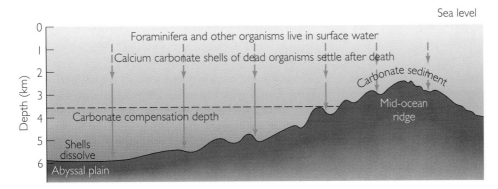

FIGURE 17.36 The carbonate compensation depth is the level in an ocean below which the calcium carbonate of foraminifera and other carbonate-shelled organisms that have settled from surface waters dissolves. As the shells of dead organisms settle into the deep waters, they enter an environment undersaturated in calcium carbonate and so dissolve.

the CCD, they enter an environment undersaturated in calcium carbonate, and they dissolve.

Another kind of biochemically precipitated sediment, **silica ooze,** is produced by sedimentation of the silica shells of diatoms and radiolaria. Diatoms are green, unicellular algae found in abundance in the surface waters of the oceans. Radiolaria are unicellular organisms that secrete shells of silica instead of calcium carbonate. After burial on the seafloor, silica oozes are cemented into the siliceous rock, chert.

Some components of pelagic sediments are formed by chemical reactions of seawater with sediment on the seafloor. The most prominent examples are manganese nodules, which are black, lumpy accumulations ranging from a few millimeters to many centimeters across. These nodules cover large areas of the deep ocean floor, as much as 20 to 50 percent of the Pacific. Rich in nickel and other metals, they are a potential commercial resource if we find some economical way to mine them from the seafloor.

DIFFERENCES IN THE GEOLOGY OF OCEANS AND CONTINENTS

Our studies of the ocean floor have given us an appreciation of the differences between continental geology and submarine geology. Whereas tectonics and erosion shape the continents, volcanism and sedimentation predominate in the ocean. Volcanism creates island groups, such as the Hawaiian Islands, in the middle of the oceans; arcs of volcanic islands near deep oceanic trenches; and mid-ocean ridges. Sedimentation shapes much of the rest of the ocean floor. Soft sediments of mud and calcium carbonate blanket the low hills and plains of the sea bottom and accumulate on oceanic plates as they spread from mid-ocean ridges. As the plates move farther and farther from a ridge, they accumulate more and more sediment. Eventually the plates are swallowed by subduction zones, which destroy the oceanic sediment record by metamorphism and melting.

The oceans have no folded and faulted mountains like those on the continents. Instead, plate-tectonic deformation is restricted to the faulting and volcanism found at mid-ocean ridges and at subduction zones. Weathering and erosion are much less important in the oceans than on the land because there are no efficient fragmentation processes, such as freezing and thawing, or major erosive agents, such as streams. Deep-sea currents can erode and transport sediment but cannot effectively attack basaltic plateaus or hills.

Because tectonic deformation, weathering, and erosion are minimal over much of the seafloor, many more details of the geologic record are preserved in layers of oceanic sediment than in continental sediments. But the oldest parts of the rock record are continuously erased by subduction. The oldest sediments preserved on today's ocean floor are Jurassic, about 150 million years old; they lie at the western edge of the Pacific Plate. In the next million years, they too will disappear down a subduction zone. It takes about 150 million years, on average, for the crust created at mid-ocean ridges to spread across an ocean and come to a subduction zone.

This brief survey of the oceans shows that they are geologically complex, with distinctive structures, topographies, and sediments. Our knowledge of the oceans is still in its infancy; much of what we know was discovered only in the past few decades. Much more remains to explore as we invent new ways to sound, map, and sample the ocean floor.

SUMMARY

What processes shape shorelines? At the edge of the sea, waves and tides, interacting with tectonics, control the formation and dynamics of shorelines, from beaches and tidal flats to uplifted rocky coasts. Waves are generated by winds blowing over the sea; as the waves approach the shore, they are transformed into breakers in the surf zone. Wave refraction results in longshore currents and longshore drift, which transport sand along beaches. Tides, generated by the gravitational attraction of the Moon and Sun on the water of the oceans, are agents of sedimentation on tidal flats.

What are the major components of continental margins? Continental margins are made up of shallow continental shelves; continental slopes that descend more or less steeply into the depths of the ocean; and continental rises, gently sloping aprons of sediment deposited at the lower edges of the continental slopes and extending to abyssal plains farther out in the ocean. Waves and tides affect the continental shelves, but continental slopes are shaped by turbidity currents, deep-water currents formed as slumps and slides on the continental slope create turbid suspensions of muddy sediment in bottom waters.

Turbidity currents also produce submarine fans, submarine canyons, and abyssal plains. Active continental margins form where oceanic lithosphere is subducted beneath a continent, and passive continental margins form where rifting and seafloor spreading carry continental margins away from plate boundaries.

How is the deep seafloor formed? The deep seafloor is constructed by volcanism at mid-ocean ridges and at oceanic hot spots such as Hawaii and by deposition of fine-grained clastic and biochemically precipitated sediments. Mid-ocean ridges are the sites of seafloor spreading and the extrusion of basalt, which produces new oceanic lithosphere.

Deep-sea trenches form as oceanic lithosphere is pulled downward into a subduction zone. Isolated, submerged seamounts and guyots, volcanic islands, plateaus, and abyssal hills are all accumulations of volcanic rock, most of which are mantled by sediment. Coral reefs form as volcanic islands are fringed with reefs constructed by organisms, which then continue to build the island and keep it at the surface when the volcano becomes extinct and subsides below sea level. Pelagic sediments consist of reddish-brown clays and foraminiferal and silica oozes composed of the biochemically precipitated calcium carbonate and silica shells of microscopic organisms living in surface ocean waters.

KEY TERMS AND CONCEPTS

world ocean (p. 420)
coast (p. 420)
shoreline (p. 421)
swell (p. 421)
wavelength (p. 422)
wave height (p. 422)
period (p. 422)
surf (p. 422)
surf zone (p. 422)
swash (p. 423)
backwash (p. 423)
wave refraction (p. 424)
longshore drift (p. 424)
longshore current (p. 424)
tide (p. 426)
spring tide (p. 427)
neap tide (p. 427)

tidal surge (p. 427)
flood tide (p. 428)
ebb tide (p. 428)
tidal flat (p. 428)
offshore (p. 428)
foreshore (p. 428)
backshore (p. 429)
stack (p. 432)
wave-cut terrace (p. 432)
spit (p. 433)
barrier island (p. 433)
estuary (p. 435)
continental shelf (p. 437)
continental slope (p. 437)
submarine canyon (p. 437)
continental rise (p. 437)
abyssal plain (p. 438)

seamount (p. 438)
continental margin (p. 440)
passive margin (p. 440)
active margin (p. 440)
turbidity current (p. 441)
submarine fan (p. 444)
turbidite (p. 444)
guyot (p. 447)
atoll (p. 448)
fringing reef (p. 448)
pelagic sediments (p. 449)
foraminiferal ooze (p. 451)
carbonate compensation depth
  (CCD) (p. 451)
silica ooze (p. 452)

EXERCISES

1. How are ocean waves formed?

2. How does wave refraction concentrate erosion at headlands?

3. Where do turbidity currents form?

4. How has human interference affected some beaches?

5. Where and how is the deep seafloor created by volcanism?

6. What plate-tectonic process is responsible for deep-sea trenches?

7. Describe the formation of a coral reef in the open ocean.

8. What are pelagic sediments?

9. Along what kinds of continental margins do we find broad continental shelves?

10. What features of the seafloor are associated with plate divergences and with plate convergences?

11. What processes in the ocean are responsible for foraminiferal and silica oozes?

THOUGHT QUESTIONS

1. Over a 100-year period, the southern tip of a long, narrow, north-south beach has become extended about 200 m to the south by natural processes. What shoreline processes would have caused this extension?

2. After a period of calm along a section of the eastern shore of North America, a severe storm with high winds passes over the shore and out to sea. Describe the state of the surf zone before, during, and after the storm passes.

3. Why would you want to know the timing of high tide if you wanted to observe a wave-cut terrace?

4. How might plate tectonics account for the contrast between the broad continental shelf off the eastern coast of North America and the narrow, almost nonexistent shelf off the western coast?

5. A major corporation hires you to determine whether New York City's garbage can be dumped at sea within 100 km offshore. What kinds of places would you explore, and what are your concerns?

6. There is very little sediment to be found on the floor of the central rift valley of the Mid-Atlantic Ridge. Why might this be so?

7. A plateau rising from the deep ocean floor to within 2000 m of the surface is mantled with foraminiferal ooze, whereas the deep seafloor below the plateau, about 5000 m deep, is covered with reddish-brown clay. How can you account for this difference?

8. Bermuda is a coral reef island in the western Atlantic Ocean. Describe the rocks you might encounter if you were to drill a very deep hole into this island.

9. What kinds of sediment might you expect to find in the Peru-Chile Trench off the coast of South America?

10. You are describing a deep-sea sediment core from the base up. In the bottom portion, most of the sediment is a foraminiferal ooze; midway to the top, however, the ooze disappears and the core is brownish gray. What could be responsible for this change?

11. You are studying a sequence of sedimentary rocks and discover that the beds near the base are marine shallow-water sandstones and mudstones. Above this is an unconformity, overlying which are nonmarine sandstones. Above this group of beds there is another unconformity, and overlying it are marine beds similar to those at the base. What could account for this sequence?

12. What are the chief differences between the Atlantic and Pacific oceans in terms of topography, tectonics, volcanism, and other ocean processes?

LONG-TERM TEAM PROJECT: COASTAL PROTECTION

In 1996, the Institute of Marine and Coastal Sciences at Rutgers University published a report that calls for a major shift in policy with regard to coastal protection. Rather than defend and rebuild coastal beaches and shore properties in high hazard areas, the report urges the New Jersey Department of Environmental Protection to advocate the removal of storm-damaged buildings and a policy of letting nature reclaim the beach. The report also suggests that shore towns be encouraged not to zone vulnerable areas for development. Some geologists consider this policy very wise.

Working in teams of three or four, pick a nearby or favorite vacation beach. If feasible, each group should visit their beach at the beginning of the semester and after each storm event during the semester to measure and record the width of the beach and other observations of how the beaches changes. Other sources of information might include the state's department of environmental protection, the town's chamber of commerce, and USGS Web sites. Using records available from the U.S. Army Corps of Engineers under the Freedom of Information Act, investigate the approach that has been used to maintain and preserve the local shore that you have chosen to study. Has the Corps chosen to construct seawalls or groins? Alternatively, has the Corps recommended pumping sand from offshore back onto the beach? What measures has the state taken to protect its shoreline, and what effect has this had on the physical appearance of the beach? Has the town chosen to let shore processes proceed naturally? Using your knowledge of coastal processes, your observations, if possible, and informa-

tion about changes in your beach from other sources, assess the wisdom of the approach taken to preserve the beach of your study.

SUGGESTED READINGS

Anderson, Roger. 1986. *Marine Geology*. New York: Wiley.

Cone, J. 1991. *Fire Under the Sea*. New York: William Morrow.

Davis, R. A. 1994. *The Evolving Coast*. New York: W. H. Freeman.

Dolan, R., and Lins, H. 1987. Beaches and barrier islands. *Scientific American* (July):146.

Hardisty, J. 1990. *Beaches: Form and Process*. New York: Harper Collins Academic.

Hollister, C. D., A. R. M. Nowell, and P. A. Jumars. 1984. The dynamic abyss. *Scientific American* (March):42.

Macdonald, K. C., and B. P. Luyendyk. 1981. The crest of the East Pacific Rise. *Scientific American* (May):100.

Nunn, P. D. 1994. *Oceanic Islands*. Oxford: Blackwell.

Pilkey, Orrin H., and W. J. Neal. 1988. Coastal geologic hazards. In R. E. Sheridan and J. A. Grow (eds.), *The Geology of North America*, vol. I-2, *The Atlantic Continental Margin: U.S.* Boulder, Colo.: Geological Society of America, pp. 549–556.

Schlee, J. S., H. A. Karl, and M. E. Torresan. 1995. *Imaging the Sea Floor*. U.S. Geological Survey Bulletin 2079.

INTERNET SOURCES

FEMA Fact Sheet: Hurricanes
ⓘ **http://www.fema.gov/fema/hurricaf.html**
The Web site of the Federal Emergency Management Agency (FEMA) provides fact sheets for hurricanes and most other natural hazards, including information on actions to take before, during, and after the event.

Ocean Planet
ⓘ **http://seawifs.gsfc.nasa.gov/ocean-planet.html**
A companion Web site to an exhibit at the Museum of Natural History of the Smithsonian Institution, Ocean Planet includes all the text and most of the images of the traveling exhibit. Features include sections on ocean resources, oceans in peril, sea people, and ocean science.

USGS Center for Coastal Geology
ⓘ **http://stimpy.er.usgs.gov/**
Located in St. Petersburg, Florida, this U.S. Geological Survey field office provides studies of the south Atlantic and Gulf coasts of the United States with sections on hurricanes and "Coasts in Crisis!"

USGS Marine and Coastal Geology Program
ⓘ **http://marine.usgs.gov/**
Its introduction describes the purpose of this site as "investigating geologic issues of marine and coastal areas under the themes of Environmental Quality and Preservation, Natural Hazards and Public Safety, Natural Resources, and Earth-Sciences Information and Technology." The program Web contains links to a number of topics, including sea stack erosion, Lake Erie shoreline changes, studies of San Francisco Bay, and more.

USGS Woods Hole Field Center
ⓘ **http://woodshole.er.usgs.gov/**
The Woods Hole, Massachusetts, field office of the U.S. Geological Survey Marine and Coastal Geology program focuses on environmental quality and preservation, natural hazards and public safety, natural resources, and earth sciences information and technology. The site includes links to active research projects, research results, and on-line data resources.

Coastal Zone Hazards Maps
ⓘ **http://www.geo.duke.edu/psds_hazmaps.htm**
Duke University provides this site, which focuses on coastal hazards in North Carolina. Hazard maps include risk zones and hazard type.

USGS Western Region Marine and Coastal Surveys
ⓘ **http://www-marine.wr.usgs.gov/docs/wrmcst.html**
Located in Menlo Park, California, this site provides "What's New" summaries of current research activities on the U.S. western coast. A notable feature is the "brief description of marine geology research methods."

NOAA VENTS Program
ⓘ **http://www.pmel.noaa.gov/vents/home.html**
The VENTS Program of the National Oceanographic and Atmospheric Administration (NOAA) was established to study "the oceanic impacts and consequences of submarine hydrothermal venting." Features include a "What's New" section, information organized by discipline (geophysics, geology, physical and chemical oceanography) and by site (all in the Pacific), and "Special Interests" topics.

Internal Processes, External Effects

Earth's atmosphere, oceans, and crust originate in the deep interior. So do the forces that deform the rocky crust of our planet. These tectonic forces, driven by the energy supplied by Earth's internal heat, create plates and keep them in motion. We see the effects of tectonic forces in volcanic eruptions and earthquakes, as well as in the grandeur of mountain belts. By studying plate motions, earthquakes, volcanoes, and the deformation of the crust, geologists can infer the properties of the deep interior and the nature of the forces at work there.

18

A highway in Kobe, Japan, that collapsed in the 1995 earthquake. The Kobe earthquake, which origi- nated under a densely populated area, killed more than 5000 people and destroyed more than 100,000 homes. Japan is one of the most active seismic areas in the world. *(Asaki Shimbun/Sipa-Press.)*

Earthquakes

Hardly have they set foot in the city . . . when they feel the earth tremble under their feet; the sea rises boiling in the port and shatters the vessels that are at anchor. Whirlwinds of flame and ashes cover the streets and public squares; the houses crumble, the roofs are tumbled down upon the foundations, and the foundations disintegrate; thirty thousand inhabitants of every age and of either sex are crushed beneath the ruins. . . .

After the earthquake, which had destroyed three-quarters of Lisbon, the country's wise men . . . decided that the spectacle of a few persons burned by a slow fire in a great ceremony is an infallible secret for keeping the earth from quaking.

—Voltaire, *Candide* (1759)

When Voltaire set out to describe the destruction of the great earthquake of 1755 in Lisbon, Portugal, he was able to draw on the detailed reports of contemporary observers (Figure 18.1). The Lisbon earthquake is remembered not only for its tragic proportions but also for its role in helping to usher in the Enlightenment, the period in the eighteenth century during which observation and rational explanation of natural events began to gain favor over irrationality and superstition.

No description, however vivid, can replace the personal experience of the violent ground movement of a major earthquake. Outdoors, you would be knocked down and the loud noise of buildings being destroyed would be deafening. Bridges would sway and collapse. Indoors, you would be shaken out of bed or thrown wall to wall in a hallway. Furniture would slide all over a room, chandeliers would sway and fall, window glass would break and spray you with shards, dishes and groceries would crash to the floor from shelves. A poorly constructed building would collapse, the floors above falling on you to crush or entrap you.

The collapse of buildings is the major cause of earthquake casualties. Fire, landslides, and dam failures contribute to the havoc. The video pictures of the death and destruction that recorded earthquakes over the past two decades in Mexico, Armenia, Iran, Japan, California, and the Philippines brought the dimensions of such cataclysms to hundreds of millions of viewers worldwide for the first time.

This chapter will examine how seismologists (scientists who study earthquakes) locate, measure, and try to predict earthquakes. We will see how the pattern of earthquakes helped lead to the development of plate-tectonics theory. Most earthquakes occur at plate boundaries, the intensely strained zones where plates collide, split apart, or slide past each other. We will find that the destructiveness of earthquakes can be reduced when governments act on seismologists' estimates of where and when earthquakes are likely to occur. Governments can mandate that buildings, dams, bridges, and other structures be located and designed to withstand earthquakes and can evacuate people in dangerous areas. A long-term goal of seismologists is to increase the accuracy of their earthquake predictions; timely warnings could save millions of lives.

WHAT IS AN EARTHQUAKE?

An **earthquake** is a shaking or vibration of the ground. An earthquake occurs when rocks being deformed suddenly break along a fault. The two blocks of rock on both sides of the fault slip suddenly, setting off the ground vibrations. This slippage occurs most commonly at plate boundaries—regions of the Earth's crust or upper mantle where most of the ongoing deformation takes place.

The earthquake on the San Andreas fault that devastated San Francisco in 1906 received the most detailed study of any earthquake up to that time. Based on the observations, the investigators of that catastrophe advanced the **elastic rebound theory** to explain why earthquakes occur.

To visualize what happens in an earthquake, according to the theory, imagine the following experiment carried out across a fault between two hypothetical crustal blocks (Figure 18.2). Suppose that surveyors had painted straight lines on the ground,

FIGURE 18.1 A contemporary fanciful print portraying the destruction of the Portuguese coastal city of Lisbon by the earthquake and violent sea waves of November 1, 1755. T. C. Lotters geographischem Atlas, nach 1755 (Jan Kozak, Prague). *(Suggested by L. C. Pakiser, USGS.)*

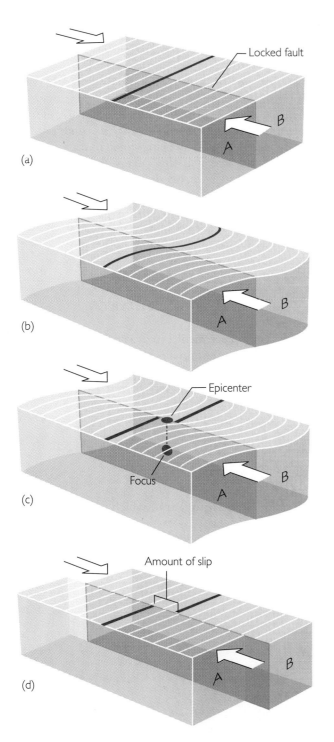

(a)

(b)

(c)

(d)

running perpendicular across a fault from one block to the other, as in Figure 18.2a. The two blocks are being pushed in opposite directions by plate motions. Because they are pressed together by the weight of the overlying rock, however, friction locks them together along the fault. They do not move, just as a car does not move when the emergency brake is engaged. Therefore, instead of slipping along the fault, the blocks are deformed or bent elastically near the fault, as shown by the bent lines in Figure 18.2b. Elastic deformation means that the blocks would spring back and return to their undeformed shape if the fault were suddenly to unlock. As the plate movements continue to push the blocks in opposite directions, the slow movement of deformation continues, perhaps for decades. The strain in the rocks, evidenced by the bending of the survey lines, builds up until the frictional bond that locks the fault can no longer hold at some point on the fault, and it breaks. The rupture extends over a section of the fault, and the blocks slip suddenly along the fault (Figure 18.2c). Figure 18.2d shows that after the earthquake the two blocks have rebounded or sprung back to their undeformed state, the imaginary bent lines have become straight, and the two blocks have been displaced. The distance of the displacement is called the **slip.**

The point at which the slip initiates is the **focus** of the earthquake (Figure 18.2c). The **epicenter** is the geographic point on the Earth's surface directly above the focus. For example, you might hear in a news report: "Seismologists at the California Institute of Technology report that the epicenter of last night's destructive earthquake in California was located 35 miles southeast of Los Angeles. The depth of the focus was 10 kilometers." Although the focus of the 1988 Armenian earthquake was 10 km below the surface, the fault broke through to the surface and produced the fault scarp shown in Figure 10.29. Faulting in a major earthquake can extend for

FIGURE 18.2 The elastic rebound theory of an earthquake. (a) Two crustal blocks, A and B, are slowly forced to slide past each other. (b) Friction along the fault prevents slip, and the crust is deformed. (c) Strain builds up until the "frictional lock" is broken at the focus; the blocks rebound or spring back to their predeformational state as a rupture occurs. The focus of an earthquake is the site of an initial slip on the fault. The epicenter is the point on the surface directly above the focus. (d) The rupture spreads and an earthquake slip occurs over a section of the fault. The imaginary white lines painted on the surface portray the deformation before the earthquake and the rebound triggered by the earthquake.

FIGURE 18.3 The great San Francisco earthquake of 1906 was caused by slip along the San Andreas fault. In this photograph, taken soon after the earthquake, near Bolinas, California, a fence that crossed the fault was offset, showing a slip of nearly 3 m. *(G. K. Gilbert.)*

as much as 1000 km, and the slip of the two blocks can be as large as 15 m (Figure 18.3).

When the blocks slip suddenly at the time of the earthquake, intense vibrations called **seismic waves** (from the Greek *seismos,* meaning "shock" or "earthquake") travel outward from the focus much as waves ripple outward from the spot where a stone is dropped in a still pond (Figure 18.4). Near the epicenter, the waves can cause the ground to shake violently.

The elastic strain that slowly builds up over decades when two blocks are pushed in opposite directions is analogous to the strain exerted on a rubber band when it is slowly stretched. The sudden release of strain in an earthquake, signaled by slip along a fault and the release of seismic waves, is analogous to the violent rebound or springback movement that occurs when the rubber band breaks. Elastic energy is actually being stored in the stretched rubber band, and it is this energy that is suddenly released in the backlash. In the same way, elastic energy accumulates and is stored over many decades in rocks under strain. The energy is released at the moment the fault ruptures and is radiated as seismic waves in the few minutes of an earthquake.

STUDYING EARTHQUAKES

As in any experimental science, instruments and field observations provide the basic data used to study earthquakes. These data enable the investigator to analyze the seismic waves that originate in earthquakes, locate earthquakes, determine their sizes and numbers, and understand their relationships to faults.

Seismographs

The modern **seismograph,** which records the seismic waves generated by earthquakes, is the most important tool for studying earthquakes and probing

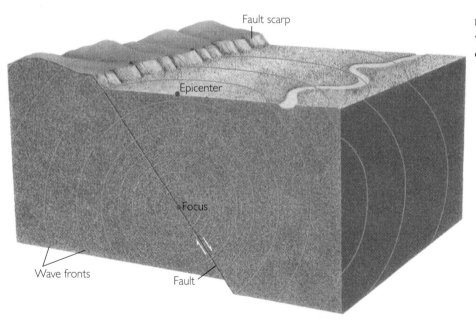

FIGURE 18.4 Seismic waves radiate from the focus of an earthquake.

the deep interior of Earth. The seismograph is to the Earth scientist what the telescope is to the astronomer—a tool for peering into inaccessible regions. The ideal seismograph would be a device fixed to a stationary frame not attached to the Earth, so that when the ground shook, its movements could be measured by the changing distance between the seismograph, which did not move, and the vibrating ground, which did.

As yet, we have no way to establish a seismograph that is not attached to the Earth, although modern space technology may remove this limitation (Figure 18.5). So we compromise. A mass is attached to the Earth so loosely that the ground can vibrate up and down or side to side without causing much motion of the mass.

One way to achieve this loose attachment is to suspend the mass from a spring (Figure 18.6a). When seismic waves move the ground up and down, the mass tends to remain stationary because of its inertia (an object at rest tends to stay at rest), but the mass and the ground move relative to each other because the spring can compress or stretch. In this way, the

vertical displacement of the Earth caused by seismic waves can be recorded by a pen on chart paper.

Another way to achieve loose attachment of the mass is with a hinge. A seismograph that has its mass suspended on hinges like a swinging gate (Figure 18.6b) can record the horizontal motions of the ground. In modern seismographs, the most advanced electronic technology is used to amplify ground motion before it is recorded. These instruments

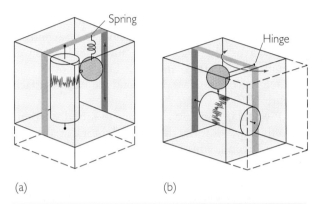

(a)   (b)

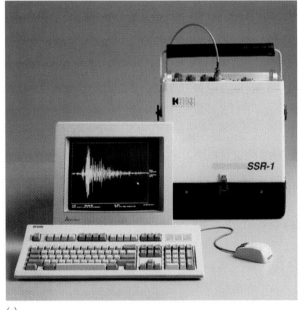

(c)

FIGURE 18.5 One of the 250 Global Positioning System (GPS) stations in a network that collects satellite observations along faults in southern California. These instruments use signals from GPS satellites orbiting Earth to detect small displacements of the surface from which strain changes can be computed. Observations of strain changes may help scientists to evaluate future probabilities of earthquakes. *(Jet Propulsion Laboratory/NASA.)*

FIGURE 18.6 Seismographs record (a) vertical or (b) horizontal motion. Because of its loose coupling to the Earth through the spring (a) or hinge (b) and its inertia, the mass does not keep up with the motion of the ground. The pen traces the differences in motion between the mass and the ground, in this way recording vibrations of seismic waves. A typical observatory has instruments to measure three components of ground motion: up-down, horizontal east-west, and horizontal north-south. Modern seismographs (c) amplify the vibrations electronically. *(Kinemetrics.)*

(Figure 18.6c) can detect ground displacements as small as 10^{-8} cm—an astounding feat, considering that such small displacements are of atomic size.

Seismic Waves

Install a seismograph anywhere, and within a few hours it will record the passage of seismic waves generated by earthquakes somewhere on Earth. The waves will have traveled from the earthquake focus through the Earth and arrived at the seismograph in three distinct groups. The first waves to arrive are called primary waves or **P waves.** The secondary or **S waves** follow. Both P and S waves travel through the Earth's interior. Finally come the **surface waves,** which travel around the Earth's surface (Figure 18.7).

P waves in rock are analogous to sound waves in air, only P waves travel through solid rock at about 5 km per second, which is about 14 times faster than sound waves travel through air. Like sound waves, P waves are *compressional waves,* so called because they travel through solid, liquid, or gaseous materials as a succession of compressions and expansions; that is, *contractions and relaxations* (Figure 18.8a). P waves can be thought of as push-pull waves: they push or pull particles of matter in the direction of their path of travel.

S waves travel through solid rock at about half the speed of P waves. They are called *shear waves* because they push material at right angles to their path of travel (Figure 18.8b). Shear waves do not exist in liquids or gases.

Surface waves are confined to the Earth's surface and outer layers because, like waves on the ocean, they need a free surface to ripple in order to exist (Figure 18.9). Their speed is slightly less than that of S waves. One type of surface wave sets up a rolling motion in the ground, while another type shakes the ground sideways.

Seismic waves have been felt and their destructiveness witnessed throughout human history, but not until the close of the nineteenth century were seismologists able to devise instruments to record such waves so that they could be analyzed. Seismic waves enable us to locate earthquakes and determine the nature of faulting, and they provide our most important tool for probing the Earth's deep interior.

Locating the Epicenter

If we can locate the epicenters of the thousands of earthquakes that occur each year, we will have taken the first step toward understanding them. The principle involved in locating a quake's epicenter is quite similar to the principle used to deduce the distance to a lightning bolt on the basis of the time interval between the flash of light and the sound of thunder—the greater the distance to the bolt, the larger the time interval. Because light travels faster than sound, the lightning flash may be likened to the P waves of earthquakes and the thunder to the slower S waves.

The time interval between the arrival of P and S waves depends on the distance the waves have traveled from the focus. This relationship is established by recording seismic waves from earthquakes or underground nuclear explosions that are at a known distance from the seismograph. To determine the approximate distance to an epicenter, seismologists read from a seismogram the amount of time that elapsed between the arrival of the first P waves and

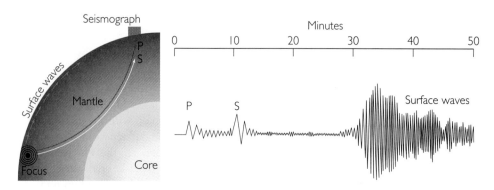

FIGURE 18.7 Seismographic recording of P, S, and surface waves from a distant earthquake. The cross section shows the paths followed by the three types of waves.

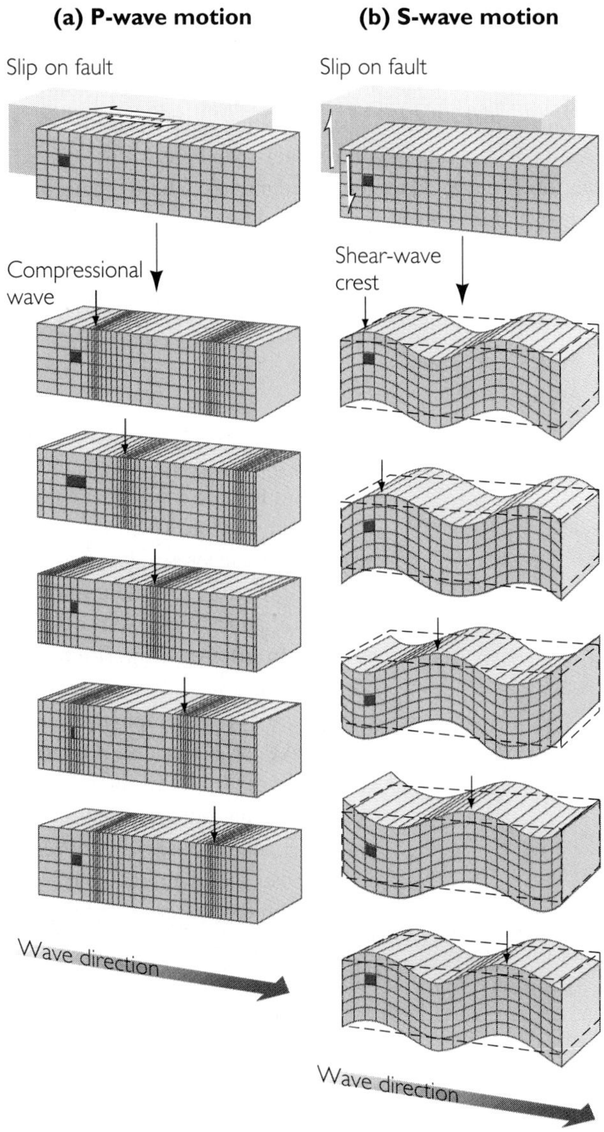

(a) P-wave motion

Slip on fault

Compressional wave

Wave direction

(b) S-wave motion

Slip on fault

Shear-wave crest

Wave direction

FIGURE **18.8** Comparison of P-wave and S-wave motions.

(a) Stages in the deformation of a block of material with the passage of compressional, or P, waves through it. The undeformed block is shown at the top. In the sequence from top to bottom, a wave of compression, marked by an arrow, moves through the block with the P-wave velocity. The compression is followed by an expansion, shown by the changing shape of the marked square. Any small piece of matter shakes back and forth in the direction that the wave travels in response to alternating compressions and expansions. A sudden push (or pull) in the direction of wave propagation, indicated by slip on the fault (white arrows), would set up P waves.

(b) Stages in the deformation of a block of material with the passage of shear, or S, waves through it. A wave crest, marked by an arrow, moves through the block with the S-wave velocity. The marked square shows the shearing deformation (from a square to a parallelogram in the figure) that occurs as the shear wave passes through. A small piece of matter shakes at right angles to the path of the wave. Slip on a fault (white arrows) at right angles to the direction of wave propagation would set up S waves.

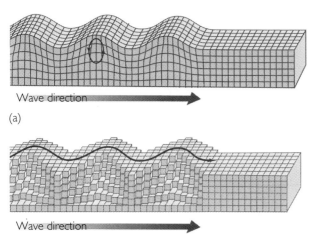

Wave direction

(a)

Wave direction

(b)

FIGURE **18.9** In one type of surface wave (a), the ground vibrates in a rolling, elliptical motion that dies down with depth beneath the surface. In another type of surface wave (b), the ground shakes sideways, with no vertical motion. (After Bruce A. Bolt, *Nuclear Explosions and Earthquakes: The Parted Veil,* San Francisco, W. H. Freeman, 1976, p. 49.)

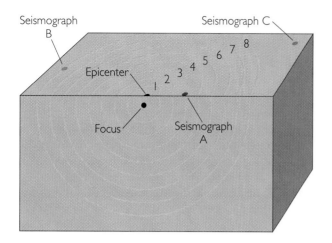

(a) The contour numbers in this diagram give the time interval (in minutes) between the arrival of first P and first S waves successive distances from the epicenter of an earthquake. Because P waves travel about twice as fast as S waves, the further the waves travel, the wider becomes the interval between the arrival of the different waves. Seismographic station A, for example, which is closer to the epicenter, recorded a 3-minute interval; whereas more distant station B recorded an 8-minute interval. The intervals between P and S waves are a critical factor in interpreting seismographic readings.

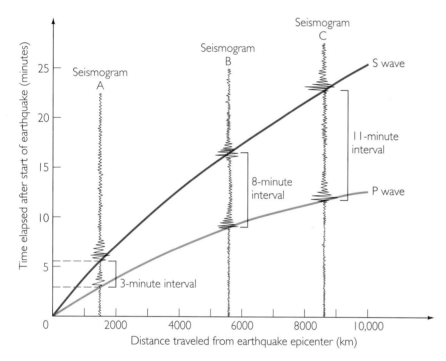

(b) Seismic time-travel curves, like the ones shown in this graph, are basic tools for determining the distance of an earthquake from the seismograph that records it. Geologists at station A, where a 3-minute interval between P and S waves was recorded, can match this interval to the corresponding space between the P and S curves on the graph and determine that their distance from the epicenter was 1500 km. In the same way, seismic recordings at station B, with an 8-minute interval between P and S waves, yields a distance of 5600 km, and at station C, with an 11-minute interval, a distance of 8600 km.

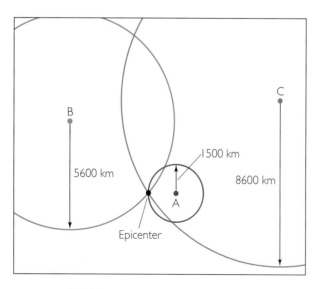

(c) Knowing the distance from the epicenter of three different stations, geologists can pinpoint the location of the epicenter using a map and some simple geometry. They draw three circles, each one centered on one of the three stations and each having a radius equal to the station's distance from the epicenter. The epicenter lies at the single point where the three circles intersect. (The epicenter and depth of focus are now determined by a computer that simulates this graphical method.)

FIGURE **18.10** From readings at different seismographic stations, geologists can locate the epicenter of an earthquake.

the later arrival of the S waves. Then they use a table or a graph like that shown in Figure 18.10 to determine the distance from the seismograph to the epicenter. If they know the distances from three or more stations, they can pinpoint the epicenter. They can also deduce the time of the shock at the epicenter because the arrival time of the P waves at each station is known, and from a graph or table it is possible to determine how long the waves took to reach the station. Actually, the process just described is now carried out with repeated iterations in a computer until the data from a large number of seismographic stations agree on where the epicenter is, the time the earthquake began, and the depth of the focus below the surface.

Measuring the Size of an Earthquake

The ability to locate earthquakes is only one step on the way to understanding them. The seismologist must also determine their size, or magnitude. Other variables being equal (distance to the epicenter, the area's geology, density of population, strength of buildings), an earthquake's magnitude is the main factor in its destructiveness. The sizes and numbers of earthquakes in a given area indicate the degree of tectonic activity.

RICHTER MAGNITUDE   In 1935 Charles Richter, a California seismologist, devised a simple procedure that is now used all over the world to measure the size of an earthquake. Richter studied astronomy as a young man and learned that astronomers assign each star a magnitude—a measure of its brightness. Adapting this measure to earthquakes, Richter assigned each earthquake a number (now called the **Richter magnitude**) as a measure of its size. The Richter magnitude depends on the amplitude (size) of the ground movement caused by seismic waves. To simplify matters, Richter chose to base his measurement on the largest ground motion recorded by a seismograph (Figure 18.11).

Just as the brightnesses of stars vary over a huge range, so do the sizes of earthquakes. For this reason, Richter needed to compress his scale. He did so by taking the logarithm of the largest movement of the ground. With this procedure, two earthquakes that differ in size of ground motion by a factor of 10 differ in magnitude by 1 Richter unit (Table 18.1). The ground motion of an earthquake of magnitude 3, therefore, is 10 times that of an earthquake of magnitude 2. Similarly, a magnitude 6 earthquake produces ground motions that are 100 times greater than those of a magnitude 4 earthquake. The energy released as seismic waves by an earthquake increases even faster, by a factor of 33 for each unit. As Table 18.1 suggests, there are limits to the sizes of detected earthquakes, but there are no well-established limits to the possible sizes of earthquakes. Large earthquakes are less frequent than small ones. Worldwide, about 800,000 earthquakes take place each year in the magnitude range of 2.0 to 3.4, while a shock of magnitude 8 or larger occurs only once every 5 or 10 years.

A number of adjustments had to be worked into the calculation of Richter magnitude. The most important one takes into account that just as sound weakens with increasing distance from its source, the seismograph reading must be adjusted by a factor that takes into account the weakening of seismic waves as they spread away from the focus. Richter provided tables to make this adjustment. Thus, seismologists all over the world can study their records and in a few minutes come up with nearly the same value for the magnitude of an earthquake no matter how close or far away their instruments are from the focus.

MOMENT MAGNITUDE   Although the Richter magnitude has become a household word and is an important concept historically, seismologists prefer a

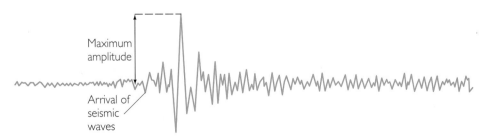

FIGURE 18.11 The maximum amplitude of the ground shaking indicated on the seismographic record is used to assign a Richter magnitude to an earthquake.

TABLE **18.1**

EARTHQUAKE MAGNITUDES AND ENERGY

MAGNITUDE	ENERGY RELEASED (MILLIONS OF ERGS)	ENERGY EQUIVALENCE AND EFFECT
−2	600	100-watt light bulb left on for a week
−1	20,000	Smallest earthquakes detected
0	600,000	Seismic waves from 1 pound of explosives
1	20,000,000	A two-ton truck traveling 75 miles per hour
2	600,000,000	Not felt but recorded
3	20,000,000,000	Smallest earthquake commonly felt
4	600,000,000,000	Seismic waves from 100 tons of explosives
5	20,000,000,000,000	
6	600,000,000,000,000	Damage varies from slight to great, depending on quality of construction
7	20,000,000,000,000,000	
8	600,000,000,000,000,000	1906 San Francisco (magnitude = 8.3)
9	20,000,000,000,000,000,000	Largest recorded earthquake (magnitude = 8.9); destruction nearly total
10	600,000,000,000,000,000,000	Approximately all the energy used in the United States in one year

SOURCE: Modified from U.S. Geological Survey.

measure of the size of an earthquake that reflects what happened at the earthquake source rather than how much the ground shakes at a distant point. Therefore, they developed a new magnitude scale that depends on the product of the slip of the fault when it broke, the area of the fault break, and the rigidity or stiffness of the rock. This new measure is called the **moment magnitude.**

Moment magnitude is closely related to the physical reality of the amount of energy radiated by the earthquake. Although both the Richter method and the moment method produce roughly the same numerical values, the moment magnitude is more closely related to the cause of an earthquake than to its effect and is therefore more useful to scientists.

OTHER FACTORS IN EARTHQUAKE INTENSITY

Earthquake magnitude does not necessarily describe the destructiveness of a particular earthquake because a magnitude 8 earthquake 2000 km from the nearest city might cause no human or economic damage, while a magnitude 6 quake immediately beneath a city could cause serious damage.

The **Modified Mercalli Intensity Scale** is an old earthquake intensity scale devised to assign a measure of the destructiveness of an earthquake rather than its magnitude or physical size. It is a qualitative scale based on the observed effect on people and damage to structures. In this scale, earthquake intensities range from I (very weak, not felt by people) to XII (total damage). Destructiveness depends on the magnitude, the distance to the epicenter, the strength of buildings and structures, the nature of the soil and bedrock on which foundations rest, and other local conditions. For example, some unconsolidated soils actually liquefy when shaken by intense seismic waves, which causes the foundations of buildings to fail, whereas foundations resting on bedrock are more secure. The Mercalli intensity assigned to the San Francisco earthquake of 1906 was XI.

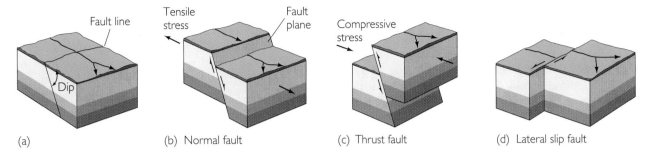

(a)   (b) Normal fault   (c) Thrust fault   (d) Lateral slip fault

FIGURE 18.12 The three main types of fault movements that initiate earthquakes, and the stresses that cause them: (a) situation before movement takes place; (b) normal fault due to tensile stress; (c) thrust (or reverse) fault due to compressive stress; (d) lateral slip (or strike-slip) fault due to shearing stress.

Determining Fault Mechanisms from Earthquake Data

When an earthquake occurs, seismologists analyze seismograms from many stations to determine the quake's epicenter and magnitude. Members of a team then go to the epicenter to examine the faulting. Mainly, they want to see how the orientation of the fault plane and the direction of slip fit into the regional pattern of crustal forces. Was the earthquake the result of normal, thrust, or strike-slip faulting (Figure 18.12)? What can be seen in the field?

If nothing can be seen in the field, the focus was so far below the surface that the fault does not show. Still, seismologists can deduce the kind of faulting that occurred below the surface from information in the seismograms. This ability is especially convenient, for very few earthquake faults break through to the surface so that the slip direction and fault orientation can be observed directly.

By now, so many seismographs have been established around the world that they literally surround the focus of any earthquake. Seismologists have found that in some directions from an earthquake, the very first seismic movement of the ground recorded by a seismograph, the P wave, is a push (an upward motion of the trace on the seismogram) away from the focus. On seismographs located in other directions, the initial ground movement may be a pull (downward motion of the trace) toward the focus. These differences reflect the fact that the slip on a fault looks like a push if you view it from one direction but like a pull if you view it from another (Figure 18.13). The "pushes" and "pulls" can be divided into four sections based on the positions of the stations, as shown in the illustration. One of the two boundaries is the fault orientation. The slip is inferred from the arrangement of pushes and pulls. In this manner, without surface evidence, geologists can deduce from their instruments whether the crustal forces that triggered the earthquake were compressional, tensional, or shear.

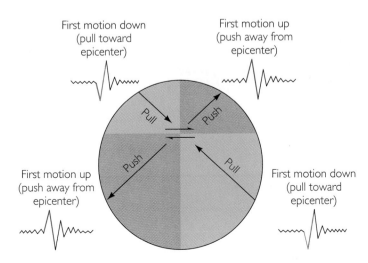

First motion down (pull toward epicenter)

First motion up (push away from epicenter)

First motion up (push away from epicenter)

First motion down (pull toward epicenter)

FIGURE 18.13 The first motion of P waves arriving at seismographic stations is used to determine the orientation of the fault plane and the directions of slip. If the fault and slip were ⇑ or ⇌, the first motions would be the same. Seismologists can choose between the two possibilities because they know the direction of the fault from additional information. For example, other epicenters of earthquakes on the same fault align along the fault and provide a trace of its direction.

THE BIG PICTURE: EARTHQUAKES AND PLATE TECTONICS

With the tools for locating earthquakes, measuring their magnitudes, and deducing their fault mechanisms, seismologists have discerned patterns in the distribution of earthquakes of various types. They have explained these patterns within the framework of plate-tectonics theory—and thereby provided tremendous support for the theory.

Earthquake Belts at Plate Boundaries

The **seismicity chart** in Figure 18.14 shows the locations of the epicenters of almost 30,000 earth-

quakes that occurred over a six-year period. Earthquakes that originated at depths greater than about 100 km are shown in Figure 18.15. Seismologists have known for decades that earthquakes tend to occur in "belts." One of the best known belts corresponds to the "Ring of Fire" surrounding the Pacific Ocean. In recent years, it has become possible to detect the more numerous small earthquakes and to improve methods of locating epicenters. Seismic belts can now be defined so accurately that they can be correlated with geologic features.

SHALLOW-FOCUS EARTHQUAKES AT DIVERGENT BOUNDARIES Almost all divergent plate boundaries occur on the seafloor. The narrow belts of mid-ocean earthquakes coincide with mid-ocean

FIGURE **18.14** Epicenters of some 30,000 earthquakes with focal depths between 0 and 700 km, recorded over a six-year period. (*Base map courtesy of Peter W. Sloss, NOAA-NESDIS-NGDC.*)

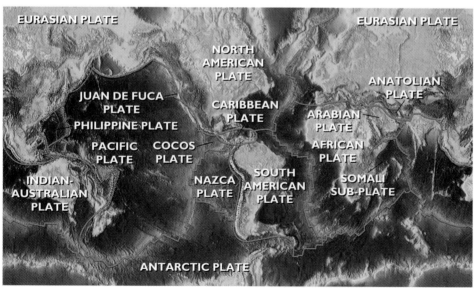

FIGURE **18.15** Subset of earthquakes from Figure 18.14 with focal depths greater than 100 km. These deep-focus earthquakes typically originate near margins where plates collide and thus serve to identify such plates. (*Base map courtesy of Peter W. Sloss, NOAA-NESDIS-NGDC.*)

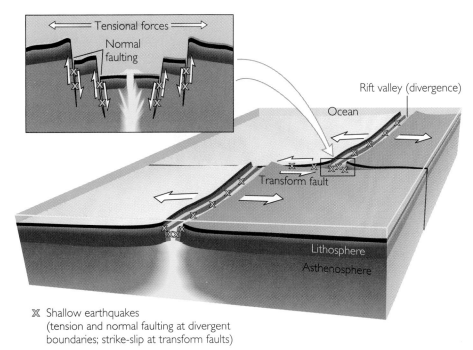

Tensional forces

Normal faulting

Rift valley (divergence)

Ocean

Transform fault

Lithosphere

Asthenosphere

FIGURE 18.16 Earthquakes associated with two types of plate boundaries: divergent boundaries at ocean ridges and transform faults.

✕ Shallow earthquakes
(tension and normal faulting at divergent boundaries; strike-slip at transform faults)

ridge crests, as shown in Figure 18.16. When the topography of mid-ocean ridges is examined in detail, the ridges are often found to be segmented, the segments being offset by transform faults. Earthquake epicenters also line the transform faults between the offset ridge segments. Moreover, the fault mechanisms of the ridge-crest earthquakes, revealed by analysis of the first P-wave motion, are normal, with the faults striking parallel to the trend of the ridges. Normal faulting indicates that tensional (or pull-apart) forces are at work. This is why rift valleys develop in the ridge crests. Seismologists, using the independent evidence of their seismic records, found that mid-ocean ridges define plate boundaries where plates are being pulled apart. They also found that the earthquakes that coincide with transform faults show strike-slip mechanisms—just as one would expect where plates slide past each other in opposite directions. Seismology gives elegant support to the concept that plates spread apart at mid-ocean ridge crests.

DEEP-FOCUS EARTHQUAKES AT CONVERGENT BOUNDARIES
The earthquakes that originate at depths greater than about 100 km (see Figure 18.15) are found to coincide with continental margins or island chains that are adjacent to ocean trenches and young volcanic mountains. The west coast of South America and the chains of islands that make up Japan and the Philippines are such areas. These features define a boundary where plates collide (see

Figure 18.17). Almost all deep-focus earthquakes occur along the inclined plane of a subducted plate, where it plunges back into the mantle beneath an overriding plate. An analysis of the initial P waves of these earthquakes reveals that the earthquakes in the area are produced by thrust faulting, signifying that compressive forces are at work, as would be the case at a collision boundary.

SHALLOW-FOCUS EARTHQUAKES WITHIN PLATES
Although most earthquakes occur at plate boundaries, the seismicity map (see Figure 18.14) shows that a small percentage originate within plates. Their foci are relatively shallow, and the majority occur on continents. Among these earthquakes are some of the most destructive in American history: New Madrid, Missouri (1812); Charleston, South Carolina (1886); Boston, Massachusetts (1755). Apparently, strong crustal forces can still occur and cause faulting within the lithospheric plates, far from modern plate boundaries.

This global correlation among topography, geology, and seismicity provided essential data for understanding the many complex phenomena that occur at plate boundaries. It may seem straightforward now to draw a line through the seismic belts and so define plate boundaries, but this important advance could not have been made without the knowledge explosion in seismic data and the synthesizing minds of a few scientists who sifted through those data in the 1960s.

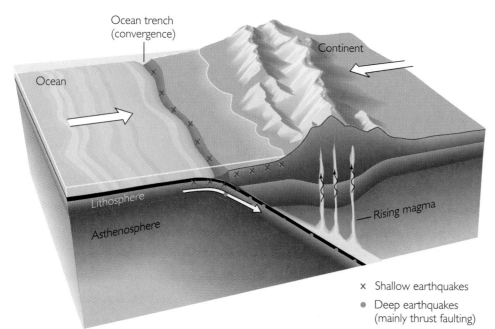

FIGURE 18.17 Shallow- and deep-focus earthquakes at convergent plate boundaries. Earthquakes occur because of compression forces. Alignment of deep-focus earthquakes along an inclined plane led to the discovery of subducted plates.

Interestingly, the increase in the number of seismic observatories and the use of computers to store and analyze seismic data were stimulated by research done during negotiations for a nuclear test-ban treaty in the 1960s. The purpose was to determine whether small underground nuclear explosions could be detected and distinguished from earthquakes so that a violator of the treaty could not claim that a nuclear test was an earthquake. The seismologists were successful in distinguishing between the two.

Figure 18.17 shows the distribution of deep- and shallow-focus earthquakes at plate boundaries.

EARTHQUAKE DESTRUCTIVENESS

Each year 800,000 little tremors that are not felt by humans are recorded by instruments around the world. About 100 earthquakes occur each year with Richter magnitudes between 6 and 7. A great earthquake, one with a Richter magnitude exceeding 8, occurs somewhere in the world about once every 5 to 10 years.

It is a fortunate fact that most earthquakes are small. Damage to buildings and other structures near the epicenter begins at magnitude 5 and increases to nearly total destruction when the magnitude is greater than 8. Two California earthquakes—the 1989 Loma Prieta earthquake (magnitude 7.1), which occurred on the San Andreas fault some 80 km south of San Francisco, and the 1994 North-ridge earthquake (magnitude 6.9)—were among the costliest disasters in the history of the United States. Damage amounted to more than $6 billion in the Loma Prieta earthquake and $20 billion in the Northridge quake because of their proximity to areas of dense population. Sixty people were killed in each event, but the death toll would have been hundreds of times higher if stringent earthquake-resistant building codes had not been in place (Figure 18.18).

FIGURE 18.18 Many more buildings like this one would have collapsed in the Northridge, California, earthquake of 1994 if the newer buildings in the area had not been constructed according to stringent codes for earthquake resistance. (© 1994 Chronmo Sohn/Sohn/Photo Researchers, Inc.)

As damaging as these earthquakes were, they released less than a hundredth the energy of some truly great earthquakes, such as those in San Francisco (1906, 8.3), Tokyo (1923, 8.2), Chile (1960, 8.6), Alaska (1964, 8.6), and China (1976, 7.8). One California politician announced that the Loma Prieta earthquake had the positive benefit of warning Californians to prepare for the truly great earthquake that is bound to occur there.

Destructive earthquakes are even more frequent in Japan than in California. The recorded history of destructive earthquakes in Japan, going back 2000 years, has left an indelible impression on the Japanese people. Perhaps that is why Japan is the best prepared of any nation in the world to deal with earthquakes in terms of public education, building codes, warning systems, and other mitigation measures. Despite this preparation, more than 5000 people were killed in a devastating (magnitude 7.2) earthquake that struck Kobe on January 16, 1995 (Figure 18.19). The casualties and the enormous failure of structures (50,000 buildings destroyed) were due in part to the less stringent building codes that were in effect before 1980, when much of the city was built, and the location of the earthquake epicenter so close to the city.

In too many parts of the world (not Japan), authorities take a fatalistic attitude, as though nothing could be done in advance to mitigate the losses of an earthquake. They are wrong.

How Earthquakes Cause Their Damage

Earthquakes cause destruction in several ways. Ground vibrations can shake structures so hard that they collapse. The ground accelerations near the epicenter of a very great earthquake can approach and even exceed the acceleration of gravity, so an object lying on the surface can literally be thrown into the air. Very few structures built by human hands can withstand such severe shaking, and those that do are severely damaged. People can find the strong motion of an earthquake traumatic in itself, especially when plaster and brick are falling. Building collapse caused the enormous casualties in Morocco in 1960, China in 1976, Armenia in 1988, and Kobe, Japan, in 1995.

Fires ignited by ruptured gas lines or downed electrical power lines are especially dangerous. Damage to water mains in an earthquake can make fire fighting all but impossible. Of the 99,000 fatalities in the Tokyo earthquake of 1923, 38,000 were due to fire.

Powerful seismic waves also take their toll on landforms and underlying soils (see Chapter 11). Avalanches were responsible for the deaths in a 1970 earthquake in Peru and a 1949 earthquake in the former Soviet Union. Unstable, wet soils behave like a liquid when they are subjected to intense seismic shocks, and the ground simply flows away, taking buildings, bridges, and everything else with it.

FIGURE 18.19 These buildings in Kobe, Japan, toppled during an earthquake in 1995. *(Reuters/Corbis-Bettmann.)*

Soil liquefaction destroyed the residential area of Turnagain Heights, near Anchorage, Alaska, in the 1964 earthquake (see Figure 11.6); the Nimitz Freeway near San Francisco in the 1989 Loma Prieta earthquake; and areas of Kobe in the 1995 earthquake.

Earthquakes that occur near coasts occasionally generate the towering waves commonly called tidal waves but more accurately named **tsunamis** (the Japanese term), which travel across the ocean at speeds of up to 800 km per hour and form walls of water that can be higher than 20 m when they reach the coast (Figure 18.20). Tsunamis cause tremendous damage when they sweep over low-lying coastal areas (Figure 18.21). Undersea avalanches triggered by earthquakes and volcanic explosions can also cause tsunamis (see Feature 18.1).

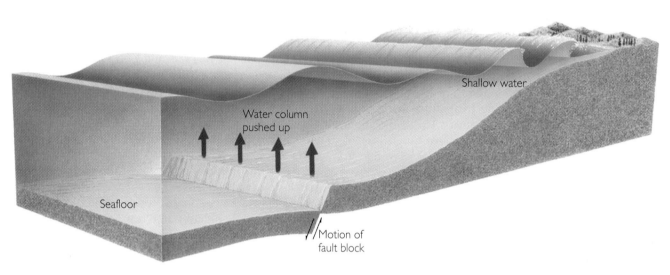

FIGURE 18.20 Generation of a tsunami by fault movements caused by an earthquake on the seafloor. Movement of the seafloor due to an earthquake produces a surge of water, which oscillates and flows out as a long sea wave, or tsunami. Such a wave is only a few meters high on the deep sea but can increase in height manyfold when it enters shallow coastal waters.

FIGURE 18.21 Destruction at Maumere, Flores Island, Indonesia, caused by a tsunami generated during an earthquake on December 12, 1992. Although a warning system exists to alert people on distant coasts to the danger of tsunamis, it cannot yet function rapidly enough to help residents in the epicenter region. *(Reuters/Bettmann.)*

18.1 LIVING ON EARTH

Tsunamis

A tsunami is a sea wave that is triggered by an undersea event such as an earthquake or landslide or by the eruption of an oceanic volcano. (A popular name for a tsunami is a "tidal wave"—an unfortunate usage, because tsunamis have nothing to do with tides.) These events push or displace a large piece of the adjacent ocean, and this disturbance is transformed into a wave that travels across the ocean at speeds of up to 800 km per hour. In mid-ocean, where the water is deep, tsunamis are hardly noticeable, but when they approach shallow coastal waters, the waves steepen and pile up until they become destructively high, perhaps more than 20 m. As we mentioned in Chapter 5, the tsunami generated by the explosion of Krakatoa in Indonesia in 1883 reached 40 m in height and drowned 36,000 people on nearby coasts. In 1960, an earthquake off the coast of Chile excited a tsunami that crossed the Pacific and caused property damage and loss of life in Japan. An earthquake off the coast of Alaska in 1964 triggered a tsunami that caused tens of millions of dollars of damage in nearby settlements; it was still destructive when it reached Crescent City, California, where it killed several people.

Disastrous tsunamis are a serious threat to 22 countries surrounding the Pacific Ocean. They have also been known to damage Atlantic coastal countries, though they occur less frequently in the Atlantic than in the Pacific. Because tsunamis can be so destructive, programs have been initiated (and should be expanded) to reduce their damage. The Pacific Tsunami Warning Center rapidly locates subsea earthquakes, estimates their potential for initiating a tsunami, and quickly warns countries that may be in danger. A warning may be broadcast as much as a few hours before the arrival of the tsunami, allowing time for the evacuation of coastal populations. A more difficult situation occurs when a tsunami from a nearby earthquake arrives so quickly that there is no time to warn nearby communities. Some measure of protection can be achieved in such cases by building barrier walls to block the inundation of the sea, but such construction is expensive and is being tried to a significant degree only in Japan.

Some do's and don'ts if you live in an area subject to tsunamis:

- If you feel a strong earthquake, move quickly away from the coastal lowlands to higher ground.

- If you hear that an earthquake has occurred under the ocean or a coastal region, be prepared to move to higher ground.

- Tsunamis sometimes signal their arrival by a precursory rise or fall of coastal water, and this natural warning should be heeded.

- People have lost their lives by going down to the beach to watch for a tsunami. Don't make the same mistake.

- Follow the advice of your local emergency organization.

Tsunami barrier in the town of Taro, Japan. *(Courtesy of Taro, Japan.)*

Table 18.2 lists the human losses of some historical earthquakes. This record should serve to alert authorities to take steps to reduce the destructiveness of earthquakes.

Mitigating the Destructiveness of Earthquakes

Can anything be done to reduce earthquake hazards? A good first step is to prepare a map showing the likelihood of an earthquake based on the number that have occurred in the past. A seismic-risk map of the United States is shown in Figure 18.22. You may be surprised to find that you live in a zone where there is risk of earthquake damage.

EARTHQUAKE PROTECTION PROGRAMS A seismic-risk map provides a basis for organizing local earthquake protection programs consonant with the degree of danger. In high-risk areas, building codes

TABLE 18.2

DEATHS CAUSED BY SEVERE EARTHQUAKES

YEAR	PLACE	ESTIMATED DEATHS	YEAR	PLACE	ESTIMATED DEATHS
856	Corinth, Greece	45,000	1939	Erzincan, Turkey	40,000
1038	Shansi, China	23,000	1948	Fukui, Japan	5,000
1057	Chihli, China	25,000	1949	Ecuador	6,000
1170	Sicily	15,000	1949	Khait, USSR	12,000
1268	Silicia, Asia Minor	60,000	1950	Assam, India	1,500
1290	Chihli, China	100,000	1954	Northern Algeria	1,500
1293	Kamakura, Japan	30,000	1956	Kabul, Afghanistan	2,000
1456	Naples, Italy	60,000	1957	Northern Iran	2,500
1531	Lisbon, Portugal	30,000	1960	Southern Chile	5,700
1556	Shen-shu, China	830,000	1960	Agadir, Morocco	12,000
1667	Shemaka, Caucasia	80,000	1962	Northwestern Iran	12,000
1693	Catania, Italy	60,000	1963	Skopje, Yugoslavia	1,000
1693	Naples, Italy	93,000	1968	Dasht-e Bayaz, Iran	11,600
1731	Peking, China	100,000	1970	Peru	20,000
1737	Calcutta, India	300,000	1972	Managua, Nicaragua	10,000
1755	Northern Persia	40,000	1976	Guatemala	23,000
1755	Lisbon, Portugal	30,000–60,000	1976	Tangshan, China	250,000
1783	Calabria, Italy	50,000	1976	Philippines	3,100
1797	Quito, Ecuador	41,000	1976	New Guinea	9,000
1822	Aleppo, Asia Minor	22,000	1976	Iran	5,000
1828	Echigo (Honshu), Japan	30,000	1977	Romania	1,500
1847	Zenkoji, Japan	34,000	1978	Iran	15,000
1868	Peru and Ecuador	25,000	1980	Algeria	3,500
1875	Venezuela and Colombia	16,000	1980	Italy	4,000
1896	Sanriku, Japan	27,000	1981	Iran	3,000
1897	Assam, India	1,500	1982	West Arabian Peninsula	2,800
1898	Japan	22,000	1983	Turkey	1,400
1906	Valparaiso, Chile	1,500	1985	Mexico	30,000
1906	San Francisco	500	1986	El Salvador	1,200
1907	Kingston, Jamaica	1,400	1987	Ecuador	1,000
1908	Messina, Italy	160,000	1988	Nepal	~1,000
1915	Avezzano, Italy	30,000	1988	China	~1,000
1920	Kansu, China	180,000	1988	Armenia	25,000
1923	Tokyo, Japan	99,000	1990	Iran	40,000
1930	Apennine Mountains, Italy	1,500	1990	Philippines	1,700
1932	Kansu, China	70,000	1993	Latur, India	>11,000
1935	Quetta, Baluchistan	60,000	1993	Iran	55,000
1939	Chile	30,000	1995	Kobe, Japan	5,094

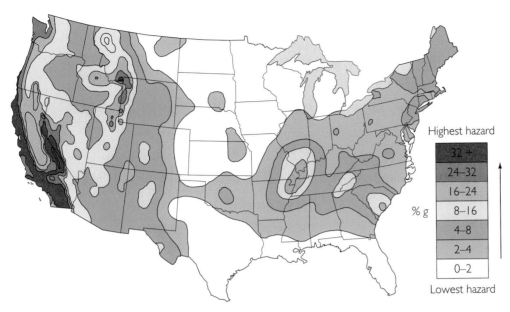

FIGURE **18.22**
Expected level of earth-quake-shaking hazards. The levels of ground shaking for different regions are shown by numbers and contour lines that express the maximum amount of shaking likely to occur at least once in a 50-year period as a percentage of the acceleration of gravity (*g*). Damage be-gins to occur at about 10 percent *g* (yellow). Major damage occurs when the ground shaking ex-ceeds 20 percent *g* (light orange). (*Modified from USGS chart.*)

Highest hazard

32 +	
24–32	
16–24	
% g	8–16
4–8	
2–4	
0–2	

Lowest hazard

should require engineers to design structures that can resist earthquake damage. Figure 18.23 shows something that is unwise—a residential area built on an active fault zone, the San Andreas, the most dangerous in the United States. Construction on unstable soils or in avalanche-prone areas should also be regulated. As a homeowner or tenant, you should inspect your home for hazards. The house should be bolted firmly to the foundation. Appliances connected to gas lines should be bolted down, and the lines should be flexible to avoid breaking and becoming a source of fire. Shelves should be attached to walls and heavy objects placed at low levels. Beds should not be near windows. If your home was constructed before modern building codes were introduced, added bracing and structural support may be required. See Feature 18.2 for steps you can take to protect yourself and your family in an earthquake.

EARLY-WARNING INDICATORS   A few decades ago, only astrologers, mystics, and religious zealots were concerned with earthquake prediction. Today seismologists in many countries are actively working on this problem and can claim a few successes. In February 1975, an earthquake was predicted five hours before it occurred near Haicheng, in northeast China. The Chinese seismologists used what they considered to be premonitory (precursor) events to make their predictions, such as the occurrence of swarms of tiny earthquakes and a rapid deformation of the ground several hours before the main shock. They also took seriously an ancient piece of peasant wisdom: that animals, as if sensing coming danger,

FIGURE 18.23 Housing tracts constructed in recent years within the San Andreas fault zone, San Francisco Peninsula. The white line indicates the approximate fault trace, along which the ground ruptured and slipped about 2 m during the earthquake of 1906. (*R. E. Wallace, USGS.*)

18.2 LIVING ON EARTH

Protection in an Earthquake

BEFORE AN EARTHQUAKE

At home

Have a battery-powered radio, flashlight, and first-aid kit in your home. Make sure everyone knows where they are stored. Keep batteries on hand.

Learn first aid.

Know the location of your electric fuse box and the gas and water shut-off valves (keep a wrench nearby). Make sure all responsible members of your family learn how to turn them off.

Don't keep heavy objects on high shelves.

Securely fasten heavy appliances to the floor, and anchor heavy furniture, such as cupboards and bookcases, to the wall.

Devise a plan for reuniting your family after an earthquake in the event that anyone is separated.

At school

Urge your school board and teachers to discuss earthquake safety in the classroom and secure heavy objects from falling. Have class drills.

At work

Find out if your office or plant has an emergency plan. Do you have emergency responsibilities? Are there special actions for

you to take to make sure that your workplace is safe?

DURING AN EARTHQUAKE

Stay calm. If you are indoors, stay indoors; if outdoors, stay outdoors. Many injuries occur as people enter or leave buildings.

If you are indoors, stand against a wall near the center of the building, or get under a sturdy table. Stay away from windows and outside doors.

If you are outdoors, stay in the open. Keep away from overhead electric wires or anything that might fall (such as chimneys, parapets, and cornices on buildings).

Don't use candles, matches, or other open flames.

If you are in a moving car, stop away from overpasses and bridges and remain inside until the shaking is over.

At work

Get under a desk or sturdy furniture. Stay away from windows.

In a high-rise building, protect yourself under sturdy furniture or stand against a support column.

Evacuate if told to do so. Use stairs rather than elevators.

At school

Get under desks, facing away from windows.

If on the playground, stay away from the building.

From Bruce A. Bolt, *Earthquakes,* newly revised and expanded ed. (New York: W. H. Freeman, 1993), pp. 220–222.

behave strangely just before an earthquake. For example, snakes come to the surface from their underground lairs and fish in ponds become agitated. Several million people, prepared in advance by a public education campaign, evacuated their homes and factories in the hours before the shock. Although many towns and villages were destroyed and a few hundred lives were lost, Western scientists who have

since visited the region estimate that tens of thousands of lives were saved.

Chinese seismologists claim to have successfully predicted other earthquakes, but unfortunately they were able to provide only a long-term warning (within five years) of the great Tangshan earthquake of August 1976—not accurate enough to save the 250,000 people estimated to have lost their lives. A

Public demonstration of earthquake safety in Japan. (*Yves Gellie/Matrix.*)

If on a moving school bus, stay in your seat until the driver stops.

AFTER AN EARTHQUAKE

Check yourself and people nearby for injuries. Provide first aid if needed.

Check water, gas, and electric lines. If damaged, shut off valves.

Check for leaking gas by gas odor (*never use a match*). If it is detected, open all windows and doors, shut off gas meter, leave immediately, and report to authorities.

Turn on the radio for emergency instructions. Do not use the telephone—it will be needed for high-priority messages.

Do not flush toilets until sewer lines are checked.

Stay out of damaged buildings.

Wear boots and gloves to protect against shattered glass and debris.

Approach chimneys with caution.

At school or work

Follow the emergency plan or instructions given by someone in charge.

Stay away from beaches and waterfront areas where tsunamis could strike, even long after the shaking has stopped.

Do not go into damaged areas unless authorized. Martial law against looters has been declared after a number of earthquakes.

Expect aftershocks: they may cause additional damage.

few smaller earthquakes are claimed to have been predicted in the United States and the former Soviet Union, but most go unpredicted.

Scientists in the United States, Japan, China, Russia, and other countries are engaged in an intensive search for premonitory indicators to help them predict the time and place of a destructive earthquake. Possible indicators being examined include

▪ A rapid tilting of the ground or other form of surface deformation.

▪ An unusual aseismic slip (smooth, slow sliding along a fault rather than the sudden slip of an earthquake) on a fault.

▪ An episode of stretching of the crust across a fault. Such tensional strain might serve to pull apart two blocks in contact along a fault. This stretching would tend to reduce the friction between the two facing blocks, partially "unlocking" the fault.

■ Changes in the physical properties of rock in the vicinity of a fault, such as its ability to conduct an electric current or its P-wave velocity. Detectable changes might show up hours or days before an earthquake.

■ Changes in the level of water in wells. Such changes might precede an earthquake because the porosity of rock may increase or decrease as a result of premonitory changes in strain.

■ An unusual increase in the frequency of smaller earthquakes in a region before a main shock. Figure 18.24 shows how such an indicator could be used to predict when a great earthquake might occur.

Actually, all these phenomena have been observed in various combinations, but not with the consistency and reliability that a useful prediction method requires.

SEISMIC GAP Ask a seismologist to predict the time of the next great earthquake and the response is likely to be "The longer the time since the last big one, the sooner the next great shock." This simple statement is the basis of the **seismic gap method.** The idea is that earthquakes result from the accumulation of strain caused by the steady motion of the plates along faults. On reaching some critical level of strain, the brittle lithosphere breaks. The cycle of slow accumulation of strain and its sudden release in an earthquake recurs again and again. The average interval varies from place to place. According to the seismic gap method, the most likely place for an earthquake to occur is at a locked portion of a

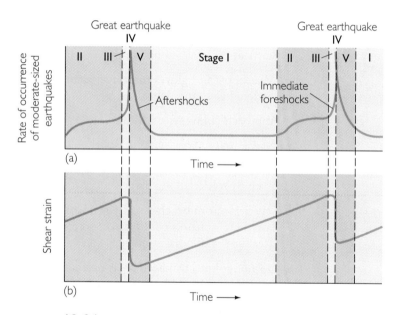

FIGURE 18.24 Model of cycle of repeating great earthquakes along a plate boundary such as the San Andreas fault. (a) Great earthquakes (Stage IV) are followed by several years of declining aftershocks (Stage V) and a relatively long quiescent interval (Stage I) during which tectonic strain gradually increases. Most of the 50- to 500-year intervals between great earthquakes are characterized by low levels of seismic activity (Stage I). Stage II, a gradually increasing level of earthquakes, occurs some decades before the main shock. In Stage III, seismic activity increases sharply and can be felt as foreshocks signaling the approach of a great quake. Stage III may be measured in years, days, or hours before the main shock (Stage IV). Some seismologists believe that northern California is entering stage II and that southern California may be entering stage III. (b) Movement of plates results in buildup of shear strain along the fault, which is released suddenly at the moment of a great earthquake (Stage IV). Notice the steady buildup of shear strain during the seismically quiescent period between quakes. (After C. B. Raleigh, K. Sieh, L. R. Sykes, and D. L. Anderson, "Forecasting Southern California Earthquakes," *Science,* vol. 217, 1982, pp. 1097–1104.)

fault where an earthquake is "due"—that is, where the time since the last earthquake has reached or exceeds the average interval between earthquakes in this location.

The average earthquake recurrence interval can be estimated in several ways. We can estimate the timing of great earthquakes several thousand years in the past by finding soil layers that are offset by fault displacements and then dating these layers. In another method, the interval is given by the number of years of steady plate motion it takes to accumulate, or "store," the fault displacements that occurred in earlier earthquakes. For example, it would take 150 years to accumulate a fault displacement of 6 m if the plate motion were 4 cm per year. Although these methods give similar results, the uncertainty of the prediction is large—as much as 50 percent of the average recurrence time. The time of prediction uncertainty can amount to many decades, 75 years in the example just cited. Although the method has predicted six earthquakes within a few years of their occurrence, it has failed to predict others. The concurrent use of the seismic gap method and one or more of the premonitory indicators described earlier, however, could sharpen the prediction.

The part of the San Andreas fault that runs through southern California provides a good example of how these methods might be applied. The recurrence time between great earthquakes in this region, measured by the two methods, is 100 to 150 years. The last great earthquake occurred in 1857; therefore an earthquake can be expected at any time—tomorrow or some decades from now.

WORRISOME DEVELOPMENTS A puzzling new pattern of earthquakes has arisen recently. Since 1971, southern California has had a barrage of moderate-sized earthquakes (magnitudes less than 7), not on the San Andreas fault but on a system of thrust faults beneath Los Angeles. These faults cause concern because they are in the middle of a population center—much closer than the San Andreas fault, which is 50 km away from Los Angeles. The closeness of these thrust faults to the city can result in ground shaking that is more intense than some earthquake-resistant buildings can withstand. For this reason, such earthquakes can cause great damage. Many of these faults are aptly termed "blind thrusts" because they are not visible on the surface and are not otherwise known until an earthquake occurs. The Northridge earthquake was on a blind thrust. How these recently activated thrust faults relate to plate movements and to the nearby San

Andreas strike-slip fault is a question geologists are investigating.

Another source of concern is the evidence that geologists have gathered in northern California, Oregon, Washington, and British Columbia that many great earthquakes have occurred in this region over the past few thousand years, the most recent some 300 years ago. They found such features of destructive earthquakes as ground subsidence and tsunamis. The subsidence left a record of flooded, dead, coastal forests. The ancient tsunamis left behind sheets of sand. Although the severe earthquake potential of this region was recognized only recently, it should not come as a surprise. The Pacific Northwest is a convergent plate boundary (see front endpapers). The Juan de Fuca Plate is being subducted under the North American Plate at a rate of about 4.5 m per century, not as a continuous motion but as an accumulation of repeated slips accompanying great earthquakes. Geologists now recognize that this densely populated region is being set up for another great earthquake.

A reliable method of predicting earthquakes within a few days and with very few false alarms has yet to be found. All that seismologists can do today is provide estimates of probability. One estimate, issued in 1994, calculates an 86 percent chance that a destructive earthquake, with magnitude exceeding 7, will strike southern California in the next 30 years. This prospect is particularly worrisome because of the high density of population in the region. A 1980 government report estimates that if an earthquake on the order of the 1857 earthquake (whose magnitude exceeded 8) were to occur, 10,000 to 15,000 people would die, 50,000 would be hospitalized, and $17 billion in property would be destroyed. These estimates might be too low or too high by a factor of 2 or 3. If earthquakes could be predicted so that a warning could be issued hours or days before the shock, the casualties could be reduced significantly.

Since radio waves can outrace seismic waves, it is technically feasible to radio a warning that an earthquake has occurred tens of seconds before the destructive seismic vibrations arrive. Some scientists have recently proposed that installation of an automated system could save lives and property. Gas lines could be shut off, nuclear reactors deactivated, people alerted to seek safety under tables, and so forth.

In an informal poll of 25 experts working on earthquake prediction, some thought it would never be achieved, but most expected that sometime within the next 10 to 100 years scientists would find a way to predict earthquakes—one of nature's most feared hazards. Let us hope so.

SUMMARY

What is an earthquake? An earthquake is a shaking of the ground caused by seismic waves that emanate from a fault that breaks suddenly. When the fault breaks, the strain built up over years of slow deformation by tectonic forces is released in a few minutes as seismic waves.

Where do most earthquakes occur? In keeping with plate-tectonic theory, most earthquakes originate in the vicinity of plate boundaries. The small number that occur far from plate boundaries demonstrate the power of tectonic forces to cause faulting within existing plates.

What governs the type of faulting that occurs in an earthquake? In most earthquakes, the fault mechanism is determined by the kind of plate boundary: normal faulting under tensile stress occurs at boundaries of divergence, thrust faulting under compressive stress at boundaries of convergence, and strike slip along transform faults.

What is earthquake magnitude and how is it measured? Earthquake magnitude is a measure of the size of an earthquake. The Richter magnitude is determined from the amplitude of the ground motions measured when seismic waves are recorded on seismographs. The moment magnitude scale is based on the product of the slip of the fault, the area of the fault break, and the rigidity of the rock.

What are the three types of seismic waves? Two types of waves travel through Earth's interior: P (primary) waves, which are conducted by all forms of matter and move fastest, and S (secondary) waves,

which are conducted only by solids and move at about half the speed of P waves. Surface waves need a free surface to ripple. They move more slowly than the interior waves.

What causes the destructiveness of earthquakes? Ground vibrations can damage or destroy buildings and other structures and can also trigger avalanches. Fires are a serious threat following an earthquake. Earthquakes on the seafloor can excite tsunamis, which sometimes cause widespread destruction when they reach shallow coastal waters.

What can be done to mitigate the damage of earthquakes? Construction in earthquake zones can be regulated so that buildings and other structures will not be located on soils that are unstable and will be strong enough to withstand destructive vibrations. People living in earthquake-prone areas should be informed about what to do when an earthquake occurs, and public authorities should plan ahead and be prepared with emergency supplies, rescue teams, evacuation procedures, fire-fighting plans, and other steps to minimize the consequences of a severe earthquake.

Can scientists predict earthquakes? Scientists can characterize the degree of risk in a region, but they cannot consistently predict earthquakes with the degree of accuracy that would be needed to alert a population hours or days in advance. They are searching for premonitory phenomena that may be used to pinpoint the time and place of an earthquake more accurately.

KEY TERMS AND CONCEPTS

earthquake (p. 460)
elastic rebound theory (p. 460)
slip (p. 461)
focus (p. 461)
epicenter (p. 461)
seismic wave (p. 462)

seismograph (p. 462)
P wave (p. 464)
S wave (p. 464)
surface wave (p. 464)
Richter magnitude (p. 467)
moment magnitude (p. 468)

Modified Mercalli Intensity Scale (p. 468)
seismicity chart (p. 470)
tsunami (p. 474)
seismic gap method (p. 480)

EXERCISES

1. What is an earthquake? Describe three scales for measuring its size. Which is preferable? How many earthquakes cause serious damage each year?

2. How does the distribution of earthquake foci correlate with the three types of plate boundaries?

3. What kinds of earthquake faults occur at the three types of plate boundaries?

4. Destructive earthquakes occasionally occur within plates, far from plate boundaries. Why?

5. Seismograph stations report the following S − P time differences for an earthquake: Dallas,

S – P = 3 minutes; Los Angeles, S – P = 2 minutes; San Francisco, S – P = 2 minutes. Use a map of the United States and travel-time curves (see Figure 18.10) to obtain a rough epicenter.

6. At a place along a boundary fault between the Nazca Plate and the South American Plate, the relative plate motions are 11.1 cm per year. The last great earthquake, in 1880, showed a fault slip of 12 m. When should local residents begin to worry about another great earthquake?

THOUGHT QUESTIONS

1. Taking into account the possibility of false alarms, reduction of casualties, mass hysteria, economic depression, and other possible consequences of earthquake prediction, do you think the objective of predicting earthquakes is a worthwhile goal?

2. Would you rather live on a planet without earthquakes (implying no plate tectonics)?

LONG-TERM TEAM PROJECT

See Chapter 11.

SUGGESTED READINGS

Bolt, Bruce A. 1993. *Earthquakes*, newly revised and expanded ed. New York: W. H. Freeman.

Earthquakes and Volcanoes. A bimonthly periodical. Washington, D.C.: U.S. Government Printing Office.

Jackson, D. D., and others. 1995. Seismic hazards in southern California: Probably earthquakes, 1994–2024. *Bulletin of the Seismological Society of America* 85:379–439.

Normile, D. 1995. Quake builds case for strong codes. *Science* 267:444–446.

Pakiser, L. C. 1991. *Earthquakes*. Washington, D.C.: U.S. Government Printing Office.

Raleigh, C. B., K. Sieh, L. R. Sykes, and D. L. Anderson. 1982. Forecasting southern California earthquakes. *Science* 217:1097–1104.

U.S. Geological Survey. 1995. *Reducing Earthquake Losses throughout the United States*. Fact Sheets 096-95, 097-95, 224-95, 225-95. Available from Earthquake Information Hot Line, USGS, MS 977, Menlo Park, CA 94025. Also from http://quake.wr.usgs.gov.

Wuethrich, Bernice. 1994. It's official: Quake danger in northwest rivals California's. *Science* 265:1802–1803.

INTERNET SOURCES

USGS Pasadena Office
🛈 **http://www-socal.wr.usgs.gov/USGS/**
The USGS field office in Pasadena, California, is the source of much of the earthquake information for southern California. Data on the "Last Significant Earthquake" in southern California, "Past", "Present", and "Future" seismic activity are the major features of this site.

National Earthquake Information Center
🛈 **http://wwwneic.cr.usgs.gov/**
The National Earthquake Information Center, located in Golden, Colorado, is the gateway for worldwide seismic information.

Caltech Seismological Laboratory
🛈 **http://www.gps.caltech.edu/seismo/seismo. page.html**
The "Seismo Lab" at Caltech features the earthquake "Record of the Day" (and previous "Records of the Day"), data on southern California earthquake mechanisms, current research projects at the laboratory, and links to other seismology-related Web servers.

Tsunami
🛈 **http://www.geophys.washington.edu/tsunami/ intro.html**
The University of Washington Geophysics Program provides "Tsunami!: The WWW Tsunami Information Resource," a gateway for general tsunami information and research.

Seismology and Earthquake Information
🛈 **http://www.geophys.washington.edu/SEIS/ welcome.html**
Information on earthquakes in the Pacific Northwest and catalogs and maps of recent earthquakes in the Pacific Northwest, United States, and worldwide are the major features of this site at the University of Washington.

19

Artist's rendering of
seismic waves bouncing
off Earth's molten core,
revealing an irregular
boundary between the
core and the bottom of
the mantle.

(*Tomo Narashima.*)

Exploring Earth's Interior

Volcanism and deformation bring rocks to the surface of the Earth from depths as great as 50 to 100 km. By studying these rocks, geologists can make inferences about some of the properties of the Earth at those depths. Geologists studying Earth to its full depth of 6400 km, however, have no tangible material to analyze. Instead, they look for phenomena at Earth's surface that reflect the properties and behavior of matter deep inside the Earth. One of the major ways that geologists study the interior is by measuring the speed of seismic waves, which varies with the kinds of materials they pass through. Geologists also learn about the interior by measuring the flow of heat from great depths to the surface and by measuring the properties of Earth's magnetic field.

In Chapter 18, we saw one result of seismic waves: the destruction they cause when they shake the ground. In this chapter, we will see another facet of seismic activity: how geologists use the behavior of seismic waves to infer the kinds of materials that make up Earth's interior. We will also see how heat flow and magnetism provide clues to the structure and composition of the interior.

EXPLORING THE INTERIOR WITH SEISMIC WAVES

Geologists know that waves of different types have a common feature: the velocity at which they travel depends on the material they pass through. Experiments show that light waves travel more quickly through air than through glass, and seismic P and S waves travel more rapidly through basalt than through granite. Remember that P (primary) waves are compressional waves and S (secondary) waves are shear waves, and that compressional waves move faster than shear waves when they pass through the Earth. Geologists calculate the velocity of a P or S wave by measuring the time of travel and dividing

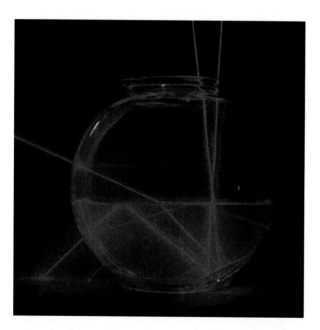

FIGURE 19.1 In this experiment, two beams of light enter the bowl of water from the top. Both beams are reflected from the water-air interface at the bottom (the effect of the thin glass container is too small to see). One of the reflected waves is bent downward (refracted) as it passes from the water to the air. The other is reflected again at the water-air interface. *(Susan Schwartzenberg/ The Exploratorium.)*

the distance traveled by this elapsed time. Any differences detected for different paths can be used to infer the properties of materials the waves have encountered along these paths. For example, the first experiments measuring the travel time of P waves from earthquakes or explosions found two P waves, one traveling faster than the other. This was how Earth's crust was discovered. The slow wave could be attributed to a path through the crust, the fast wave to a path through the mantle below.

The concepts of travel times and paths sound simple enough, but complications arise when waves encounter more than one material. At the boundary between two materials, some of the waves bounce off—that is, they are reflected—and others are transmitted into the second material, just as light is partly reflected and partly transmitted when it strikes a windowpane. The waves that cross the boundary between two materials bend, or refract, because their velocity is not the same in the second material as in the first. Figure 19.1 shows a laser light beam whose path bends as it goes from air into water, much as a P or S wave bends as it travels from one material to another. Seismologists study how seismic waves are refracted and reflected, and this behavior reveals certain features of Earth's interior.

Paths of Seismic Waves in the Earth

If Earth were made of a single material with constant properties from the surface to the center, P and S waves would travel along a straight line through the interior from the focus of an earthquake to a distant seismograph. But in reality, the waves do not follow a straight path, from which geologists infer that the Earth is layered and is made up of many materials that conduct P and S waves at different velocities. The waves bend as they go from layer to layer, and as a result their paths through the interior are curved.

WAVES TRAVELING THROUGH THE EARTH Figure 19.2a shows the curved paths of P waves as they travel from an earthquake focus to distant points where they may be recorded on a seismograph. Geologists have determined these wave paths and their corresponding travel times by studying the seismographic records of earthquakes all over the world.

Follow the path of a wave that just misses the core and emerges at the surface at an angular distance of 105°. Next, follow the waves that strike the core. Note that they are bent downward when they enter the core and are bent again when they leave.

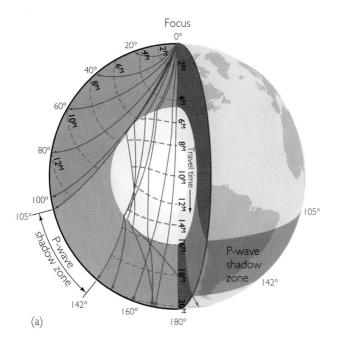

(a)

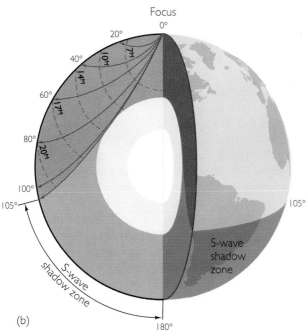

(b)

Because of this bending at the mantle-core boundary, none of these waves emerge at the surface before 142° angular distance from the focus. Thus, no P wave reaches the surface between 105° and 142°—as though the core had cast a shadow across this zone. This is in fact called the **shadow zone.**

Discovery of the shadow zone led geologists to surmise that the Earth has a core made of a material different from that of the overlying mantle. They could even conclude that the core is liquid, because the waves are bent downward rather than upward when they enter the core, much as a light beam bends downward after it enters water (see Figure 19.1). This means that the waves travel more slowly in the core than in the mantle. P waves are known to move much more slowly in liquids than in solids, so it was reasonable to guess that the existence of the shadow zone implies a molten core. This guess was verified by one other piece of evidence: the behavior of S waves (Figure 19.2b). Whenever S waves strike the core, they fail to emerge from the other side and are not seen at all beyond the 105° angular distance from the focus. Only fluids are known not to transmit shear waves; therefore, geologists concluded that a molten core must be blocking S waves from reaching these distances.

WAVES REFLECTED IN THE EARTH   Because a reflected wave implies a boundary between two materials, a good way to find boundaries in Earth's interior is to find reflections. Let us see what happens to P and S waves when they bounce off the boundary between two layers. In Figure 19.3, for example,

FIGURE 19.2 (a) The pattern of P-wave paths through Earth's interior. Dashed lines show the progress of wave fronts through the interior at 2-minute intervals. The P-wave shadow zone extends from 105° angular distance from the focus to 142°. P waves cannot reach the surface within this zone because of the way they are bent when they enter and leave the core. (b) The larger S-wave shadow zone extends from 105° to 180°. Although S waves strike the core, they cannot travel through a fluid and therefore never emerge beyond 105° from the focus.

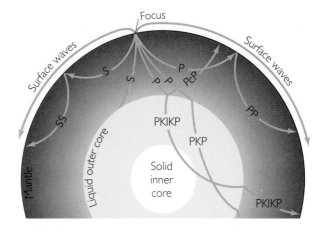

FIGURE 19.3 P and S waves radiate from an earthquake focus in many directions. A PcP wave bounces off the core. PP or SS waves are reflected from Earth's surface. A PKP wave is transmitted through the liquid outer core. A PKIKP wave traverses the solid inner core.

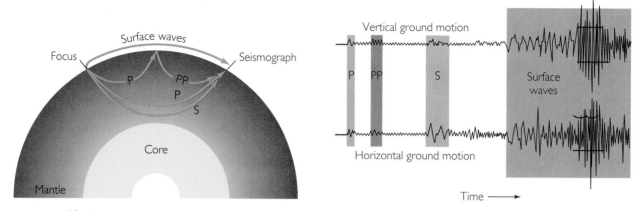

FIGURE 19.4 Seismographic recording of P, PP, S, and surface waves from a distant earthquake.

follow the wave PcP as it bounces like a radar beam from the Earth's core back to the surface. Because the velocity of a P wave is known, the round-trip travel time for PcP is used to determine the depth to the core, just as the time it takes an echo to return can be used to measure the distance to the wall on the other side of a valley. P waves can also bounce off the Earth's surface back into the Earth, as shown by the PP wave in Figure 19.3. Similarly, S waves that reflect back into the Earth are called SS. The P waves that penetrate the outer core (called PKP) or the inner core (called PKIKP) are useful for exploring those regions. Figure 19.4 shows seismograms in which the P, PP, S, and surface waves are indicated. They are differentiated from one another by their amplitude and frequency. In general, an earthquake generates all the wave types shown in Figures 19.2, 19.3, and 19.4. A seismograph at a given distance from the focus may not show them all, however, because their amplitudes vary with distance. One practical application of reflected seismic waves is their use in exploring for oil (see Feature 19.1). Another is in measuring the thickness of a glacier or, in the case of P waves from a mechanical source, the depth of the ocean.

Composition and Structure of the Interior

Thousands of sensitive seismographs and highly accurate clocks enable seismologists around the world to measure the travel times of P, S, and surface waves precisely. Nuclear explosions set off underground at nuclear test sites also excite seismic waves and add valuable data to those derived from earthquakes.

From these measurements, seismologists can plot travel-time curves of the kind shown in Figure 18.10 for the various kinds of seismic waves. Travel times are the basic data of seismology.

The travel times of compressional and shear waves depend on their velocities in the materials they pass through in Earth's interior. The key to making travel times a useful geological tool is to learn how to convert them to a graph or table that shows how the velocity of seismic waves changes with depth in the Earth. Solving this problem is something like guessing which of several possible routes a driver took on a trip from Los Angeles to San Francisco and how fast he traveled along the route if you know that the trip took 6.1 hours and know the speed limits along the way.

The solution that seismologists have come up with is shown in Figure 19.5. The illustration shows both the changes in the velocities of P and S waves with depth and what this information suggests about the structure of Earth's interior. The major layers that make up Earth's interior are a very thin outer crust, a mantle of rock extending to a depth of 2900 km, a liquid outer core extending another 2200 km, and a solid inner core extending to the center of the Earth at a depth of 6370 km.

THE CRUST The crust, the outermost layer, has been explored extensively by means of seismic waves. It varies in thickness (Figure 19.6)—thin under oceans (about 5 km), thicker under continents (about 40 km), and thickest under high mountains (ranging up to 65 km). P waves move through crustal rocks at about 6 to 7 km per second. By sampling various types of materials from the crust and mantle and by measuring the velocities of waves through these materials in the laboratory, we can

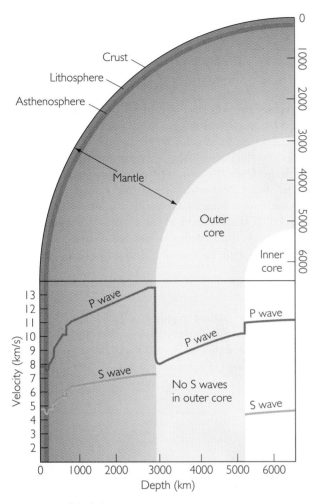

FIGURE 19.5 Changes in P- and S-wave velocities with depth in the Earth reveal the sequence of layers that make up Earth's interior.

compile a library of seismic speeds through all sorts of materials that compose the Earth. The velocities of P waves in igneous rocks, for example, are as follows:

▪ Felsic (granitic): 6 km per second

▪ Mafic (gabbro): 7 km per second

▪ Ultramafic (peridotite): 8 km per second

These velocities differ because velocity depends on the elasticity, rigidity, and density of a material, all of which change with chemical composition and crystal structure.

We know from the correlations of wave velocity and rock type that the continental crust consists mostly of granitic rocks, with gabbro appearing near the bottom, and that no granite occurs on the floor of the deep ocean. The crust there consists entirely of basalt and gabbro. Below the crust, the velocity of P waves increases abruptly to 8 km per second. This is an indication of a sharp boundary between crustal rocks and underlying mantle, as shown in Figure 19.6. The velocity of 8 km per second indicates that the deeper rocks are probably the denser ultramafic rock peridotite. The boundary between the crust and the mantle is called the **Mohorovičić discontinuity** (**Moho** for short), after the Yugoslav seismologist who discovered it in 1909. The indications that the crust is less dense than the underlying mantle are consistent with the theory that the crust is made up of lighter materials that floated up from the mantle.

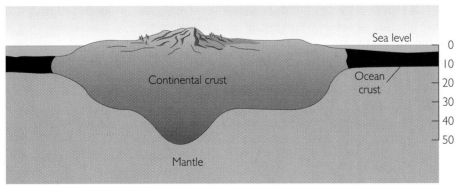

Horizontal distance not to scale

FIGURE 19.6 Seismic waves reveal the boundary between the crust and underlying mantle and variations in the thickness of the crust. Relatively light continental crust projecting into the denser mantle serves as a buoyant root providing "flotation" for the continent. The root is deeper under mountains, where flotation is required to support the heavier load, in accordance with the principle of isostasy.

Finding Oil with Seismic Waves

Exploration for oil is an important application of seismology. In offshore prospecting, a ship tows a sound source and underwater phones. P waves (called sound waves in a fluid or gas) are generated by a pneumatic device that works like a balloon burst. The sound waves bounce off rock layers below the seafloor and are picked up by the phones. In this way, subsurface sedimentary structures that trap oil—such as faults, folds, and domes—are "mapped" by the reflected waves. This technique is used extensively to explore the submerged continental shelves and shallow seas for oil and gas deposits.

Oceanographers use this method to study the sedimentary layers on the continental slope and on the floor of the deep sea. Geologists working on land also use P waves generated by an explosion or a mechanical device in their search for oil. The techniques developed by the oil industry are now also used in basic geological research to map deep geologic structures beneath mountain ranges, basins, and volcanoes. Changes in crustal thickness; the location of subducted plates; and the extent of thrust sheets, magma chambers, and other structures can be studied in this way.

Right: Seismic method of prospecting offshore can reveal structures in subsurface sedimentary rocks that could be traps for oil or gas. *(Photo by Halliburton Geophysical Services.)*

The idea that continents are less dense than the mantle and float on it, as a life jacket or an iceberg floats on the ocean, is the **principle of isostasy.** Life jackets float because the lightweight flotation material inside them is less dense than seawater. Icebergs float because ice is less dense than seawater. The flotation, or buoyancy, occurs because the volume of the life jacket or of the ice below the sea surface weighs less than the volume of water it displaces. Bigger icebergs stand higher above the sea surface but also have a deeper root below the surface to provide greater buoyancy. Continents float because the large volume of less dense continental crust that projects into the denser mantle provides the buoyancy, as shown in Figure 19.6. Note that the crust is thicker under a mountain because a deeper root is needed to float the additional weight of the mountain.

The idea that continents float suggests that there is a liquid in which they float. Yet from seismological data, we know that the mantle just beneath the crust is solid rock. How can continents float on solid rock? As we saw earlier, rocks, which we know to be solid and strong over the short term (seconds or years), are weak over the long term (thousands to millions of years) and flow very slowly like a viscous fluid when forces are applied. The principle of isostasy implies that over long periods, the mantle has little strength and behaves like a viscous liquid when it is forced to support the weight of continents and mountains.

Isostasy also implies that as a large mountain range forms, it slowly sinks under gravity and the crust bends downward. When enough of a root bulges into the mantle, the mountain floats. If a mountain range is reduced by erosion, the weight on the crust is lightened and less root is needed for flotation. As the mountain erodes down, the root floats up until both disappear. This is occurring in Scandinavia at the present time, but not because a mountain is eroding. When the load of a continental glacier was removed at the end of the last ice age, the root provided too much flotation (see Feature 19.2). As a result, Scandinavia is slowly rising as its root gradually floats up. This process, called *isostatic rebound,* will continue until Scandinavia reaches the level it held before the last ice age.

Earth's topography shows a bimodal distribution; that is, on the average, most of the elevation of the solid surface tends to be either about 4 to 5 km below sea level or up to 1 km above it. The lower

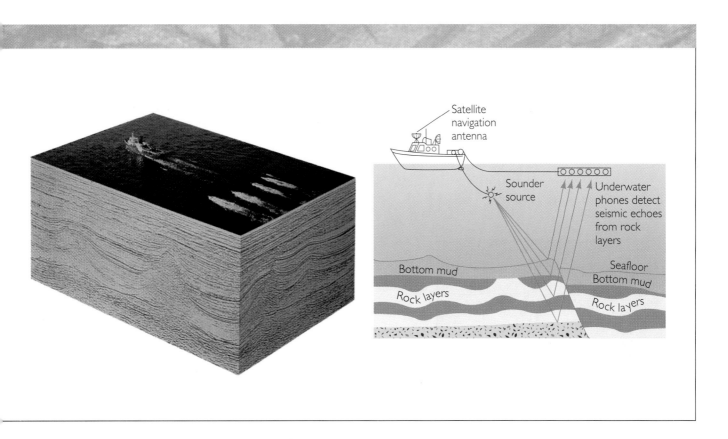

elevation characterizes the seafloor. The higher elevation is typical of continents worn down to near sea level by erosion. Isostasy is at work here in that the continental crust is lighter than oceanic crust and "floats" higher.

THE MANTLE The upper mantle rock peridotite, identified by the velocity of P waves that travel through it, is mainly made up of olivine and pyroxene, the two silicates of magnesium and iron mentioned in Chapters 2 and 3. Laboratory experiments show that olivine and pyroxene change their properties and forms with changes in pressure and temperature. The conditions of the upper mantle can cause them to begin to melt. At greater depths, pressure can force the atoms of these minerals closer together into more compact structures. Both of these phenomena occur in the mantle, as we will now see.

In the 1970s, geologists the world over concentrated research efforts on an international project to study the upper 1000 km of the Earth. They made many discoveries, especially about zones at various depths in the mantle, as revealed by changes in the velocity of S waves (Figure 19.7).

The outermost zone, the lithosphere (zone a in Figure 19.7), is a slab up to 100 km thick in which the continents are embedded. S waves pass easily through the lithosphere without being absorbed—a sign of solidity. Earth's tectonic plates are large fragments of the lithosphere.

In the zone below the lithosphere, the velocity of S waves decreases and the waves are partially absorbed. Laboratory experiments show that both of these phenomena are characteristic of S waves passing through a solid that contains a small amount of liquid. The asthenosphere (zone b in Figure 19.7), or zone of weakness, is just such a partially fluid solid. It rises close to the surface at mid-ocean ridges where plates separate and is found at depths up to about 100 km elsewhere. Geologists believe that the asthenosphere contains a small quantity of melt, perhaps a few percentage points. This idea fits nicely with the evidence (cited in Chapter 5) that the asthenosphere is the probable source of much basaltic magma. It is also consistent with other speculations about plate tectonics, especially the picture of solid lithospheric plates moving more easily because the underlying asthenosphere is partially molten and weak. The asthenosphere ends at a depth of about

200 km, where the velocity of S waves increases to a value that fits that of solid peridotite.

From about 200 to 400 km (zone c in Figure 19.7), the velocity of S waves increases gradually with depth. The increase is as much as we would expect from the increasing pressure but not enough to make us think that any material other than peridotite is present. About 400 km below the surface (zone d in Figure 19.7), however, the velocity of S

waves increases rapidly. This increase is too great to be explained by a simple change in chemical composition. What could explain it is a change in crystal structure—that is, a repacking of atoms more closely—which could be caused by the high pressure at this depth. Geologists turned to the laboratory to test this theory. When scientists squeezed olivine, the major constituent of peridotite, they found that when the temperatures and pressures reached values corresponding to depths of about 400 km, the atoms took up a more compact arrangement and assumed a crystal structure known to have a higher P and S wave velocity. Thus, the laboratory results were consistent with the seismological observations (see the discussion of olivine-spinel-perovskite in Chapter 2).

In the region from 450 to 650 km (zone e in Figure 19.7), properties change little as depth increases. Near 670 km (zone f in Figure 19.7), however, the velocity of S waves increases again, this time so greatly that atoms must be packed even more closely.

Two entirely different approaches, the analysis of seismic waves and the study of materials at high pressures, led to the same conclusion about the collapse of olivine to more compact materials at these depths.

The lower mantle, extending from 700 km to the core at a depth of 2900 km, changes little in composition and crystal structure with depth, as revealed by the fact that the velocity of S waves increases gradually in this region.

THE CORE Although Earth's core is 2900 km from the surface, it is still within reach of seismic waves. Earlier in this chapter, we saw that geologists infer from the slower movement of P waves through the core and the complete blockage of S waves that the core is fluid. But it is not fluid to the very center of the Earth. P waves that penetrate to depths of 5100 km suddenly speed up, a discovery made by a Danish seismologist, Inge Lehman. Her interpretation of this increase in velocity was that Earth's innermost core is solid (see Figure 19.5).

Geologists still know nothing about the composition of the core from direct observation. But information derived from astronomical data, laboratory experiments, and seismological data have led them to form some definite ideas. First, to be consistent with the theory that the core is made up of material that sank during the initial formation and differentiation of the Earth, geologists searched for substances that were dense. Second, because the core contains one-third of Earth's mass, geologists considered substances that were abundant in the universe. Astronomers study abundances of elements in the cosmos, and

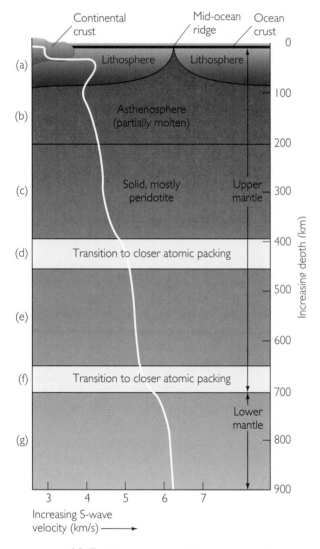

FIGURE **19.7** The structure of the upper mantle, the outermost 900 km of the Earth, is illustrated by a plot of S-wave velocity against depth. Changes in velocity mark the strong lithosphere, the weak asthenosphere, and two zones in which changes occur because increasing pressure forces a rearrangement of the atoms into denser or more compact crystalline structures. (After D. P. McKenzie, "The Earth's Mantle," *Scientific American,* September 1983, p. 66.)

19.2   INTERPRETING THE EARTH

The Uplift of Scandinavia: Nature's Experiment with Isostasy

If you depress a cork floating in water with your finger and then release it, the cork pops up almost instantly. A cork floating in molasses would rise more slowly; the drag of the viscous fluid would slow the process. If we could perform a similar experiment on the Earth, we could learn much about how isostasy works, in particular about the viscosity of the mantle and how it affects rates of uplift and subsidence. How convenient it would be if we could push the crust down somewhere, remove the depressive force, and then sit back and watch the depressed area rise!

Nature has been good enough to perform this experiment for us in less time than it takes to build and erode a mountain. The force is the weight of a continental glacier—an ice sheet 2 to 3 km thick. Such ice sheets can appear with the onset of an ice age in the geologically short period of a few thousand years. The crust is depressed by the ice load, and a downward bulge develops on its underside to the extent needed to provide buoyant support. At the onset of a warming trend, the glacier melts rapidly. With the removal of the weight, the depressed crust begins to rebound. We can discover the rate of uplift by dating ancient beaches that are now well above sea level. Such raised beaches can tell us how long ago a particular stretch of land was at sea level.

Such depression and uplift have occurred in Norway, Sweden, Finland, Canada, and elsewhere in glaciated regions. The ice cap retreated from these regions some 10,000 years ago, and the land has been rising ever since.

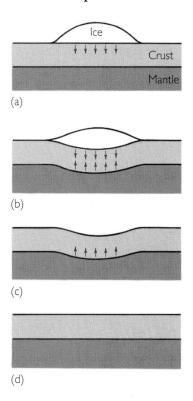

Isostasy and postglacial uplift. (a) A continental glacier grows, forming a weight on the crust. (b) The crust sags, and its projection into the mantle below becomes a root that supports the ice load isostatically. (c) The glacier melts, but the root remains because of the viscosity of the mantle. (d) Buoyancy of the root leads to slow uplift. As the root floats up, the surface assumes its original level. Arrows depict the direction of forces exerted by the ice load and the root. Scandinavia is now between stages (c) and (d). (The illustration is not to scale: the crust is about 40 km thick; a 3-km-thick glacier would produce a root about 1 km thick.)

their data point to iron as a plentiful heavy element. Laboratory measurements of the speed of P waves in liquid and solid iron at pressures and temperatures corresponding to those of the inner core approximately fit the observed speed of P waves in the inner core.

From these observations, geologists concluded that Earth's core is composed mostly of iron, molten in the outer core and solid in the inner core, as determined by the temperatures and pressures at which iron changes from a liquid to a solid. This conclusion is buttressed by discoveries of meteorites that are made almost entirely of iron and that presumably came from the breakup of a planetary body that also had an iron core. Studies of seismic waves that graze the core indicate that the boundary between the mantle and the core—previously thought to be smooth—is actually rough, with a topography of about 5 km. This is a puzzle waiting for an explanation.

In recent years, seismologists found that when the P wave PKIKP (see Figure 19.3) travels through the inner core in a north-south direction, it moves about 4 seconds faster than it does when traveling in the east-west direction. Geologists conducting laboratory experiments with new equipment that enables them to study the properties of iron at pressures and temperatures reached in the inner core can now explain this result. They find that iron is anisotropic under the conditions of the inner core; that is, it has a crystal structure in which the speed of P waves depends on the direction the waves travel through the crystal. These experimenters propose that the inner core is made up of iron crystals aligned in the same direction and therefore showing

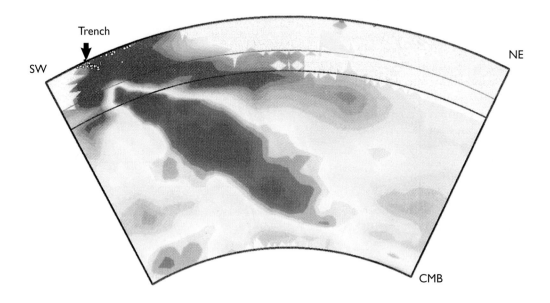

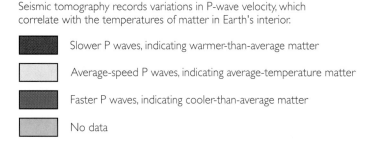

Seismic tomography records variations in P-wave velocity, which correlate with the temperatures of matter in Earth's interior.

Slower P waves, indicating warmer-than-average matter

Average-speed P waves, indicating average-temperature matter

Faster P waves, indicating cooler-than-average matter

No data

FIGURE 19.8 Seismic tomography scan of a cross section of the mantle. This scan provides firm evidence of the subduction of a large, ancient oceanic plate known as the Farallon Plate (the blue slab). The modern Juan de Fuca and Cocos plates are the only remnants of the Farallon Plate, which has been subducted since the Jurassic period. The cross section runs vertically from the crust to the core-mantle boundary (labeled CMB) and horizontally from the convergent boundary in Central America, on the southwest, to the Caribbean, on the northeast. The blue slab, interpreted as subducted, cool, oceanic lithosphere, has penetrated the mantle almost to the CMB. The two black lines at constant depth represent the locations of the two zones of transition to closer atomic packing shown in Figure 19.7. *(Courtesy of Rob Van der Hilst, Massachusetts Institute of Technology.)*

the properties of a single giant crystal. The direction could have been forced by Earth's magnetic field or by the stresses generated by Earth's rotation. When the scientists calculate the directional dependence of P waves passing through such a giant crystal, they can account for the 4-second difference. This discovery of an inner core composed of a giant crystal of iron will undoubtedly help explain some of the properties of Earth's magnetic field.

Seismological observations and laboratory measurements of the properties of materials combine to paint a picture of Earth's interior: a zoned planet with a metallic iron core and a rocky mantle. The mantle includes two transition zones in which atoms are forced into closer packing, a partially molten asthenosphere, and most of the lithosphere. A thin, lightweight crust—the end product of the differentiation process—caps the mantle.

The future holds even greater promise for exploration of Earth's interior as new tools are developed. The newest tool is an adaptation of the computerized axial tomography (CAT) scanners used in medicine to reconstruct images of organs as a computer calculates small differences in x rays that sweep the organ in many directions. Geologists are using seismic waves that sweep the mantle to construct images of pieces of subducted slabs, the rising plumes of hot spots, and other discrete structures down to Earth's center. In this work, hot matter is correlated with slower P- and S-wave velocities, and cool matter with faster velocities (Figure 19.8).

EARTH'S INTERNAL HEAT

The evidence of Earth's internal heat is everywhere: volcanoes, hot springs, and the elevated temperatures in mines and boreholes. Even global plate motions, earthquakes, and the uplift of mountains are driven by this internal heat.

Earth's interior is hot for several reasons, as we mentioned in Chapter 1. Briefly, the violent origin of Earth by the infalling of chunks of matter made the interior hot, and disintegration of the radioactive elements uranium, thorium, and potassium also produced a significant amount of heat.

Heat Flow from Earth's Interior

As soon as the Earth heated up, it began to cool, and it is cooling to this day as heat flows from the hot interior to the cool surface. The Earth cools in two main ways: slowly by conduction and more rapidly by convection.

CONDUCTION   Think of heat as energy in transit from a hot place to a cool place. Heat energy exists in a material as the vibration of atoms; the higher the temperature, the more intense the vibrations. The **conduction** of heat occurs when thermally agitated atoms and molecules jostle one another, mechanically transferring the vibrational motion from a hot region to a cool one (Figure 19.9).

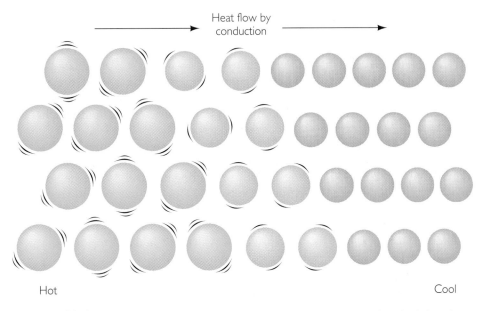

Heat flow by conduction

Hot

Cool

FIGURE 19.9 Heat flow by conduction through a solid. Heat applied at the left induces thermal agitation of the atoms. Heat is conducted as the vibrations gradually spread to the right.

Materials vary in their ability to conduct heat. Metal is a better conductor than plastic (think of how rapidly the metal handle of a frying pan heats up in comparison with one made of plastic). Rock and soil are very poor heat conductors; that is why underground pipes are less susceptible to freezing than those above ground and why underground vaults have a nearly constant temperature despite large seasonal temperature changes at the surface.

Because of the poor conductivity of rock, a lava flow 100 m thick would take about 300 years to cool from 1000°C to surface temperatures. Similarly, heat that entered one side of a plate of rock 400 km thick would take about 5 billion years to reach the other side, longer than the Earth has existed. In other words, if the 4.5-billion-year-old Earth cooled by conduction only, heat from depths greater than about 400 km would not yet have reached the surface. The mantle, which was molten in Earth's early history (as we saw in Chapter 1), would still be liquid. We know from seismic waves that this is not the case. Therefore, we must look for some means of removing heat from the interior that is more efficient than conduction to account for cooling of the Earth and solidification of the mantle over the past 4 billion years. Convection is that mechanism.

CONVECTION Convection occurs when a heated fluid, either liquid or gas, expands and rises because it becomes less dense than the surrounding material. Convection moves heat more efficiently than conduction because the heated material itself moves, carrying its heat with it. Colder material flows in to take the place of the hot rising fluid, is itself heated, and then rises to continue the cycle. This is the process by which water is heated in a kettle (see Figure 1.13). Because liquids conduct heat poorly, a kettle of water would take a long time to heat to the boiling point if convection did not distribute the heat rapidly. Convection moves heat when a chimney draws, when warm tobacco smoke rises, or when clouds form on a hot day.

Although solids generally cool only by conduction, convection can occur in solids that "flow" over longer periods. The silicone compound known as Silly Putty demonstrates how a solid can flow. Silly Putty can be bounced like a ball or broken by a sudden blow, but overnight a ball of it flows into a pancake shape under its own weight. In the case of the Earth, the mantle behaves as a rigid solid over the short term, from seconds to years. But when forces are applied over millions of years, at conditions of high pressure and temperature, the mantle behaves as an extremely viscous fluid and "creeps" or flows. Thus, convection in the mantle is indeed possible and prompts geologists to debate some key questions: Is convection an important process by which heat is transferred within the Earth? Is convection occurring now? Has it occurred at any time in the past?

EFFECTS OF CONVECTION It turns out that seafloor spreading and plate tectonics are direct evidence of convection at work. The rising hot matter under mid-ocean ridges builds new lithosphere, which cools as it spreads away; eventually, it sinks back into the mantle, where it is resorbed (Figure 19.10). This is convection; heat is carried from the interior to the surface by the motion of matter.

Some geologists believe that only the upper few hundred kilometers of the mantle are subject to the convection that drives plates, as in Figure 19.10. This would imply that the upper and lower mantles do not mix. Others think that the whole mantle is involved. Still others believe that the rising narrow plumes beneath hot spots (see Chapter 5) provide the driving force for convection. Regardless of the specifics, geologists now believe that the movement of heat from the interior to the surface as the seafloor spreads is an important mechanism by which Earth has cooled over geologic time. They also recognize that the subduction of ocean crust into the mantle is a recycling mechanism, whereby materials that resided at the surface for a few hundred million years are reinjected into the mantle. They puzzle over the questions: How far down does the subducted slab go before it is assimilated, that is, heats up and mixes with the rest of the mantle? Does it sink down to the core or is it stopped by the transition zones? Perhaps seismic tomography will answer these questions.

The British geologist Arthur Holmes was among the first to propose convection as the driving mechanism of continental drift. When he advanced this theory in the 1930s, Holmes was 30 years ahead of his time; corroboration had to wait for the extensive exploration of the seafloor that began after World War II, which led in 1963 to the concept of seafloor spreading.

Although the heat energy transferred to the surface from the interior is enough to create and move plates and to raise mountains and make earthquakes,

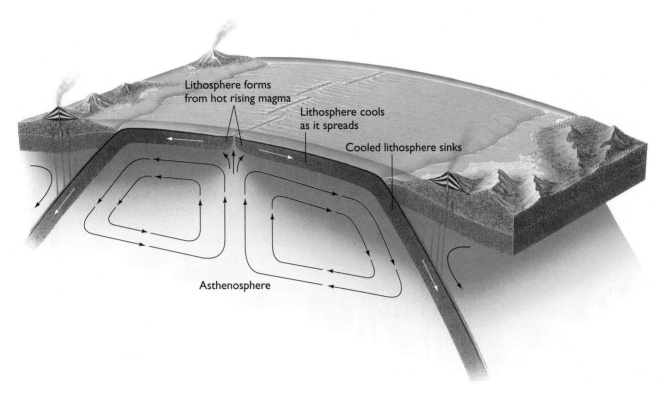

FIGURE 19.10 Some geologists believe that plate-tectonic movements can be explained by convection in the upper mantle. Hot matter rises and spreads laterally under the ocean ridges; it cools as it spreads and solidifies to form the cold, rigid lithosphere. The descending matter in the subduction zone is the cooled lithosphere. Other geologists believe that convection occurs in the entire mantle.

it is puny in comparison with the energy received from the Sun. The Sun delivers 5000 times the energy received from the interior. Moreover, solar heat has geological consequences, for it is the dominant controller of climate; it drives the atmosphere and hydrosphere, causing rain and winds—the chief agents of erosion. In a real sense, Earth's internal heat engine builds mountains, and its external heat engine, the Sun, destroys them.

Temperatures in the Earth

Just how hot Earth gets in its interior is a matter of considerable importance to geologists. Temperature and pressure determine whether matter is solid or molten, the degree to which solid matter can creep, and how atoms are packed together in crystals. The higher the temperature at depth, the more rapidly convecting matter will move.

At present, all geologists can do is draw certain conclusions from the limited information they have about temperature. They know that the average increase of temperature with depth, as measured in boreholes and mines, is about 2 or 3°C per 100 m. How can they estimate temperatures in the Earth at depths greater than those they can reach with a thermometer—that is, below about 8 km? They can't simply assume that the rising temperatures they observe as depth increases near the surface continue at the same rate all the way to the center of the Earth. At that rate, temperatures near the center would be so high (tens of thousands of degrees) that most of the interior would be molten, which seismology tells us is not the case.

One possible **geotherm,** or temperature-depth curve, arrived at by some geologists is illustrated in Figure 19.11. They combined the temperature of lava that originates in the mantle and emerges from volcanoes, laboratory data on the temperatures at which rocks and iron begin to melt, and information from seismology to infer the geotherm from the surface to the very center of the Earth, where they believe the temperature rises to between 4000 and 5000°C. Other geologists believe their laboratory experiments suggest that the temperature at the center may be as high as 6000–8000°C and the melting point even higher. More experiments will be required to reconcile these differences.

THE INTERIOR REVEALED BY EARTH'S MAGNETIC FIELD

Earth has a magnetic field, and geologists have learned how to use its characteristics as tools to examine Earth's interior, particularly the core and the lithosphere.

The Earth as One Big Magnet

In 1600, William Gilbert, physician to Queen Elizabeth I, first explained how a magnetic compass works. He offered the proposition that "the whole Earth is a big magnet" whose field acts on the small magnet of the compass needle to align it in the north–south direction.

Earth's magnetic field behaves as if a small but powerful permanent bar magnet were located near the center of the Earth and inclined about 11° from the Earth's axis of rotation, as shown in Figure 19.12. Magnetism can be visualized as lines of force of a magnetic field that indicate the presence of a magnetic force at each point in space. A compass needle that is free to swing under the influence of this magnetic force rotates into a position parallel to the local line of force, approximately in the north-south direction.

Unfortunately, although a good description of the magnetic field can be given if we assume a permanent magnet at the center of the Earth, this model has a fatal defect. Laboratory experiments show that heat destroys magnetism, and materials lose their permanent magnetism when temperatures exceed about 500°C. Material below depths of about 20 or 30 km in the Earth, therefore, cannot be magnetized because the temperatures are too high.

Another way to create a magnetic field is with electric currents. Dynamos in power plants make electricity by means of an electrical conductor in the form of a coil of copper wire rotated through a magnetic field. The rotation is driven by steam or falling water. Where inside the Earth is there a dynamo with the capacity to generate enough current to explain the magnetic field we observe at the surface?

Scientists theorize that the place to look might be the Earth's liquid iron outer core. Because liquid iron can move readily and iron is a good conductor, the core might be the moving conductor required of a dynamo. Scientists speculate that the liquid iron is

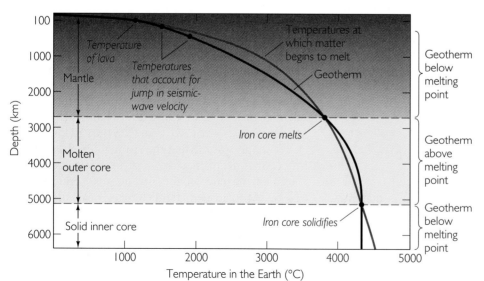

FIGURE **19.11** Increase of temperature with depth in the Earth is inferred from studies of volcanoes and seismic waves and from laboratory experiments. Note that the geotherm is below the melting point in the solid mantle and inner core and is above the melting point in the fluid outer core.

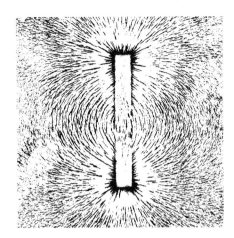

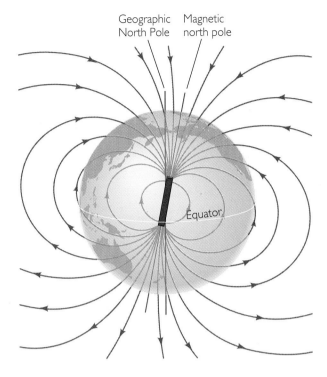

Geographic
North Pole   Magnetic
north pole

Equator

FIGURE **19.12** *Left:* The magnetic field of a bar magnet is revealed by the alignment of iron filings on paper. (From *PSSC Physics,* 3rd ed. Lexington, Mass.: D. C. Heath, 1971.) *Right:* Earth's magnetic field is much like the field that would be produced if a giant bar magnet were placed at the Earth's center and slightly inclined (11°) from the axis of rotation. Lines of magnetic force produced by such a bar magnet are shown. A compass needle points to the north magnetic pole because it orients in the direction of the local line of force.

stirred into convective motion by heat generated from radioactivity in the core. By a process not completely understood, this motion is thought to produce both the electric currents and the magnetic field needed to sustain a dynamo in the core. A magnetic field emanates to the surface from the dynamo in the core. This idea offers the best explanation so far of Earth's magnetic field. The existence of the magnetic field itself is in turn further evidence that Earth's outer core is liquid iron.

Paleomagnetism

In the early 1960s, an Australian graduate student found a fireplace in an ancient campsite where the aborigines had cooked their meals. The stones were magnetized. He carefully removed several stones that had been baked by the fires, first noting their physical orientation. Then he measured the direction of the stones' magnetization and found that it was exactly the reverse of Earth's present magnetic field.

He proposed to his disbelieving professor that as recently as 30,000 years ago, when the campsite was occupied, the magnetic field was the reverse of the present one—that is, a compass needle would have pointed south rather than north!

Scientists have discovered how to determine the direction of Earth's magnetic field in the past, not just thousands but millions of years before there were instruments to record it. Recall that high temperatures destroy magnetism. An important property of many very hot, magnetizable materials is that as they cool below about 500°C they become magnetized in the direction of the surrounding magnetic field. The reason is that groups of atoms of the material align themselves in the direction of the magnetic field when the material is hot. Once the material cools, these atoms are locked in place and therefore are always magnetized in the same direction. This is called **thermoremanent magnetization,** because the magnetization is "remembered" by the rock long after the magnetizing field has disappeared. Thus, the Australian student was able to determine the

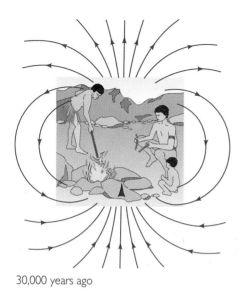 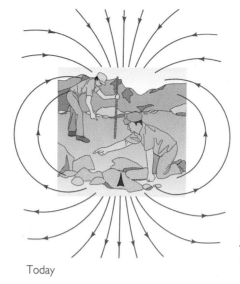

30,000 years ago   Today

FIGURE **19.13** Earth's magnetic field 30,000 years ago was the reverse of today's. One way we know this is from the discovery of reversely magnetized rocks found in the fireplace of an ancient campsite. The rocks, cooling after the last fire, became magnetized in the direction of the ancient magnetic field, leaving a permanent record of it, just as a fossil leaves a record of ancient life.

direction of the field when the stones cooled after the last fire and took on the magnetization of Earth's magnetic field of that time (Figure 19.13).

Imagine an ancient volcano erupting, say 100 million years ago. When the lava solidified and cooled, it became magnetized, leaving us with a permanent record of the geomagnetic field in mid-Cretaceous time, just as a fossil leaves a record of ancient life.

Some sedimentary rocks can also take on a type of remanent magnetization. Recall that marine sedimentary rocks are formed when particles of sediment that have settled through the ocean to the seafloor become lithified. Magnetic grains among the particles—chips of the mineral magnetite, for example—would become aligned in the direction of Earth's magnetic field as they fell through the water, and this orientation would be incorporated into the rock when the particles became lithified. The **depositional remanent magnetization** of a sedimentary rock would be a result of the parallel alignment of all these tiny magnets, as if they were compasses pointing in the direction of the field prevailing at the time of deposition (Figure 19.14).

Ancient magnetism, called **paleomagnetism** or **fossil magnetism,** has become an important tool for understanding the history of the Earth. Scientists collect old rocks on every continent and determine their magnetism and ages to reconstruct the history of the magnetic field. The oldest magnetized rocks found so far indicate that 3.5 billion years ago, Earth had a magnetic field not unlike the present one. The presence of magnetism in rocks that old implies that a fluid core probably existed for at least three-fourths of Earth's 4.5-billion-year history.

Magnetic Stratigraphy

The Australian student's discovery of a reversely magnetized rock is consistent with worldwide observations of rocks demonstrating that Earth's magnetic field reverses periodically. The compass needle on

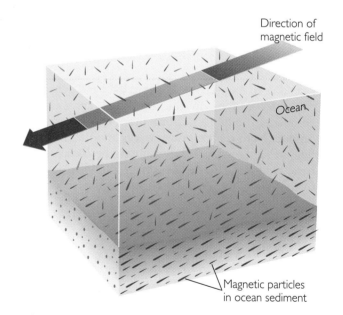

FIGURE **19.14** Newly formed sedimentary deposits can become magnetized in the same direction as the contemporaneous magnetic field of the Earth. Magnetic mineral grains transported to the ocean with other erosion products become aligned with the Earth's magnetic field while settling through the water. This orientation is preserved in the lithified rock, which thus "remembers" the field that existed at the time of deposition.

which navigators and hikers depend is not so stable as it seems. Erratically, but roughly every half-million years, Earth's magnetic field changes polarity, taking perhaps a few thousand years to die down and then build up in the opposite direction. Reversals are clearly indicated in the fossil magnetic record of layered lava flows, as shown in Figure 19.15. Each layer of rocks from the top down represents a progressively earlier period of geologic time, whose age can be determined by radiometric dating methods. The direction of remanent magnetism can be obtained for each layer, and in this way the time sequence of flip-flops of the field—that is, the **magnetic stratigraphy**—can be deduced. The detailed history of reversals over more than 5 million years has been worked out in this way (Figure 19.16). This information is of use to archeologists and anthropologists as well as geologists. For example, the magnetic stratigraphy of continental sediments has been used to date sediments containing the remains of predecessors of our own species.

About half of all rocks studied are found to be magnetized in a direction opposite to that of Earth's present magnetic field. Apparently, then, the field has flipped frequently over geologic time, and normal (same as now) and reversed fields are equally likely. Normal and reversed magnetic epochs (each of which is named after an outstanding specialist in magnetism) seem to last on the order of a half-million years. Superimposed on the major epochs are transient, short-lived reversals of the field, known as magnetic events, which may last anywhere from several thousand to 200,000 years. The Australian graduate student apparently found a new reversal event within the present normal magnetic epoch.

The cause of reversals remains for future scientists to explain. We will see in Chapter 20, however, that their occurrence has made possible an important discovery that enables geologists to determine rates of seafloor spreading and extend the history of reversals back to more than 100 million years.

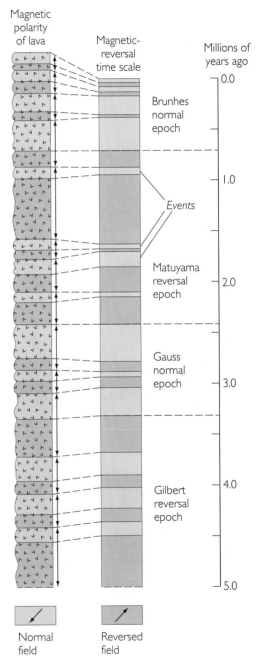

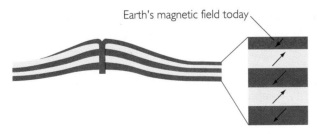

FIGURE 19.15 Lava beds become magnetized in the direction of the Earth's magnetic field existing at the time the beds solidified and cooled. In this way, they preserve the record of reversals of Earth's magnetic field. The modern flow at the top shows the direction of the field today. Underlying beds record the directions of ancient fields.

FIGURE 19.16 Magnetic polarities of lava flows are used to construct the time scale of magnetic reversals over the past 5 million years. Within epochs of magnetic polarity, there are short-term flip-flops of the field called *events*. In no one place is the entire sequence found; the sequence is worked out by patching together the ages and polarities from lava beds all over the world.

SUMMARY

What do seismic waves reveal about the layering in the interior of the Earth? Seismic waves reveal that beneath Earth's felsic crust lies a denser ultramafic mantle. The crust and outer mantle to a depth of about 70 km make up the slablike lithosphere, which is broken into large, mobile plates. Beneath the lithosphere lie a partially molten asthenosphere, a primary source of basaltic magma; two transition zones where atoms are forced into closer packing by high pressures; a thick lower mantle; a fluid outer core, made mostly of iron; and an inner core made mostly of solid iron.

What is the principle of isostasy and what evidence supports it? The principle of isostasy proposes that continents float on the denser mantle, supported by a buoyant root that projects into the mantle. Mountains require even deeper roots to support their weight. Seismic waves reveal the existence of these roots.

Where does the energy that drives geological processes come from? Earth heated up in the process of becoming a planet, and its temperature was in-

creased further by the heat released by the disintegration of radioactive elements. The cooling process takes place primarily by convection in the mantle. The convective movements are responsible for the motions of plates. Most geological activities, such as mountain making, volcanism, and earthquakes, occur at the boundaries where plates collide, separate, or slide by each other. The Sun's energy is responsible for climate, wind, and rain—all controlling factors in the erosion of rocks.

What is paleomagnetism and what is its importance? Geologists have discovered that rocks can become magnetized in the direction of Earth's magnetic field at the time they are formed. This remanent magnetization of rocks can be preserved for millions of years. Paleomagnetism tells us that Earth's magnetic field has reversed (flipped back and forth) over geologic time. The chronology of reversals has been worked out so that the direction of remanent magnetization of a rock formation is often an indicator of stratigraphic age.

KEY TERMS AND CONCEPTS

shadow zone (p. 487)
Mohorovičić discontinuity (Moho)
  (p. 489)
principle of isostasy (p. 490)
conduction (p. 495)

convection (p. 496)
geotherm (p. 498)
thermoremanent magnetization
  (p. 499)
depositional remanent
  magnetization (p. 500)

paleomagnetism (fossil magnetism)
  (p. 500)
magnetic stratigraphy (p. 501)

EXERCISES

1. How does the speed of P waves differ in granite, gabbro, and peridotite?

2. What evidence suggests that the asthenosphere is probably partially molten?

3. What evidence indicates that Earth's outer core is molten and composed mostly of iron?

4. What is the depth to the core and how do we know it?

5. What is the difference between heat conduction and convection?

6. How is convection in the mantle related to plate tectonics?

7. How can a solid rock creep?

8. How can a mountain float on the mantle when both are composed of rock?

9. How do rocks become magnetized when they form?

10. Was the direction of Earth's magnetic field what we think of as normal or reversed 4.7 million years ago?

THOUGHT QUESTIONS

1. Mars and the Moon show no evidence of tectonic plates or of their motions. What does that observation imply about the state and temperature of the interiors of these planetary bodies?

2. How does the existence of Earth's magnetic field, iron meteorites, and the abundance of iron in the cosmos support the idea that Earth's core is mostly iron and the outer core is liquid?

3. How would you use seismic waves to find a chamber of molten magma in the crust?

4. How would seismic tomography answer the question: How deep do the subducted slabs go before they are assimilated?

SUGGESTED READINGS

Bolt, B. A. 1982. *Inside the Earth.* San Francisco: W. H. Freeman.

Bolt, B. A. 1993. *Earthquakes and Geological Discovery.* New York: Scientific American Library.

Bott, M. H. P. 1982. *The Interior of the Earth: Its Structure, Constitution, and Evolution,* 2nd ed. London: Edward Arnold.

Ernst, W. G. 1990. *The Dynamic Planet.* New York: Columbia University Press.

Fowler, C. M. R. 1990. *The Solid Earth.* Cambridge: Cambridge University Press.

McKenzie, D. P. 1983. The Earth's mantle. *Scientific American* (September):66.

Olson, P., P. G. Silver, and R. W. Carlson. 1990. The large-scale structure of convection of the Earth's mantle. *Nature* 344:209–215.

Wysession, M. 1995. The inner workings of the Earth. *American Scientist* 83:134–146.

INTERNET SOURCES

Earth's Interior
🅘 **http://www.seismo.unr.edu/ftp/pub/louie/class/100/interior. html**
Starting with the basics, this resource provides detailed information on a range of topics pertinent to Earth's interior.

Earth's Interior and Plate Tectonics
🅘 **http://bang.lanl.gov/solarsys/earthint.htm**
This is an excellent starting point for exploring the title topics. The site includes numerous illustrations, links to more detailed information on selected topics, and a glossary.

20

A three-dimensional
perspective of the remote
Karakax Valley in the
northern Tibetan Plateau
of western China, created
by combining two satellite-
based radar images. A
collision between India
and Eurasia contributed to
the uplift of this area, with
elevations from 4000 m in
the valley to over 6000 m
at the peaks of the Kunlun
Mountains. The diagonal
crease below the top of
the image is the Altyn
Tagh fault. *(Eric Fielding/
Jet Propulsion Laboratory,
NASA.)*

Plate Tectonics: The Unifying Theory

Geologists believe that Earth's lithosphere is broken into about a dozen plates, which slide by, collide with, or separate from each other as they move over Earth's interior. Plates are created where they separate and are recycled where they collide, in a continuous process of creation and destruction. Continents, embedded in the lithosphere, drift along with the moving plates. The theory of plate tectonics describes the movement of plates and the forces acting between them. It also explains the distribution of many large-scale geologic features—mountain chains, structures on the seafloor, volcanoes, and earthquakes—that result from movements at plate boundaries. Plate tectonics provides the conceptual framework for this book, and indeed for much of geology. This chapter examines the development and implications of plate-tectonic theory.

FROM CONTROVERSIAL HYPOTHESIS TO RESPECTABLE THEORY

Just a few years after plate-tectonic theory was proposed, a young geologist who had achieved prominence because of his work on plate tectonics and a well-known older geologist met during an international scientific meeting in Moscow. The setting was a party in the apartment of a Soviet geologist, and the conversation was well lubricated by vodka. The din of cocktail-party chatter stopped suddenly when the younger man called out to his older colleague, "Dr. ———, everyone tells me how brilliant you were in your younger days. If that's the case, why didn't you discover seafloor spreading and plate tectonics twenty years ago?"

The explosive response of the older man needn't be recorded, but the question, properly generalized, is indeed thought-provoking. Why did plate tectonics, which explains a broad range of phenomena and unifies much of geological thought, arrive so late?

Continental Drift

Some of the basic ideas of plate tectonics have a long history of support by a few visionary scientists and rejection by most of the established scientific leaders. The initial concept of **continental drift**—large-scale movements of continents over the globe—had been around for a long time (see Feature 1.1). In 1596, the Flemish geographer Abraham Ortelius pointed out the jigsaw-puzzle fit of the coasts on both sides of the Atlantic, as if the Americas, Europe, and Africa were at one time assembled together and had subsequently drifted apart. By the close of the nineteenth century, the Austrian geologist Eduard Suess put some of the pieces of the puzzle together and postulated that the combined present-day southern continents had once formed a single giant continent, Gondwanaland. In 1915, Alfred Wegener, a German meteorologist, cited as further evidence of the breakup and drift of continents the remarkable similarity of rocks, geologic structures, and fossils on opposite sides of the Atlantic. In the years that followed, Wegener postulated a supercontinent called **Pangaea** (Figure 20.1), Greek for "all lands," that began to break up in the Mesozoic era, some 200 million years ago, into the continents as we know them today, with ocean filling the widening gaps between them.

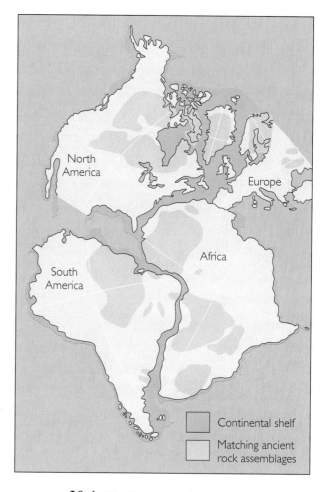

FIGURE 20.1 The jigsaw-puzzle fit of continents bordering the Atlantic Ocean is a feature noted by scientists since the late sixteenth century. From this observation, Alfred Wegener in 1915 postulated a former supercontinent, which he called Pangaea. Wegener cited as additional evidence the similarity of geologic features on opposite sides of the Atlantic. The matchup of ancient crystalline rocks is shown by orange in adjacent regions of South America and Africa and of North America and Europe. (Geographic fit from data of E. C. Bullard; geological data from P. M. Hurley.)

After about a decade of spirited debate, the controversy remained unresolved, and the theory of continental drift was ignored by all but a few geologists in Europe, South Africa, and Australia. They pointed not only to geographic matching but also to geological similarities in rock ages and trends in geologic structures on opposite sides of the Atlantic. They also offered significant arguments, accepted now as good evidence of drift, based on fossil and climatological data. Fossils of a late Paleozoic reptile, for example, are found only in Africa and South America, suggesting that the two continents were joined at the time (Figure 20.2). In addition, vertebrates and land plants on different continents showed similarities in evolution up to the postulated breakup time but followed divergent evolutionary paths thereafter, presumably because of the isolation and changing environments of the separating continental fragments. The distribution of Permian glacial deposits in South America, Africa, India, and Australia is difficult to explain in terms of separate glaciers, some close to the equator, but if the southern continents are reassembled into Gondwanaland in the South Polar region, a single continental glacier could account for all the glacial deposits.

Thermal Convection and Seafloor Spreading

Although much of the evidence for continental drift was good, it was not yet sufficient. As yet, no one had come up with a plausible driving force that would have split Pangaea and moved the continents apart. In 1928, Arthur Holmes invoked the mechanism of thermal convection in the mantle as the driving force of continental drift. Holmes proposed that subcrustal convection currents "dragged the two halves of the original continent apart, with consequent mountain building in the front where the currents are descending, and the ocean floor development on the site of the gap, where the currents are ascending."

Holmes came close to expressing the modern notions of plates, divergence, and subduction when he speculated that a subcrustal basaltic layer serves as a conveyor belt that carries a continent along to the place where the belt turns downward into the mantle, leaving the continent resting on top. Nevertheless, he recognized that "purely speculative ideas of this kind, specially invented to match the requirements, can have no scientific value until they acquire support from independent evidence."

Convincing evidence began to emerge as a result of extensive exploration of the seafloor after World War II. The mapping of the Mid-Atlantic Ridge and the discovery of the deep, cracklike valley, or rift, running down its center sparked much speculation. In the early 1960s, Harry Hess of Princeton University and Robert Dietz of the University of California suggested that the seafloor separates along the rifts in mid-ocean ridges and that new seafloor forms by upwelling of hot mantle materials in these cracks, followed by lateral spreading. Thus was born the theory of **seafloor spreading.**

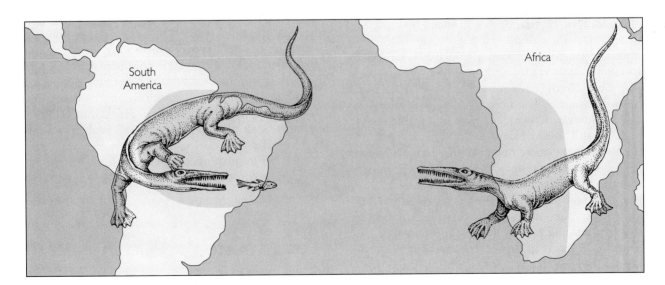

FIGURE **20.2** Fossils of the late Paleozoic reptile *Mesosaurus* are found in South America and Africa and nowhere else in the world. If *Mesosaurus* could swim across the South Atlantic Ocean, it could have crossed other oceans and should have spread more widely. That it did not suggests that South America and Africa must have been joined at that time. (After A. Hallam, "Continental Drift and the Fossil Record," *Scientific American,* November 1972, pp. 57–66.)

Acceptance of Plate Tectonics

It remained for the next generation of geologists to broaden the concept of continental drift and seafloor spreading into the more general theory of plate tectonics. Beginning about 1967, they extended the ideas of Hess and of the Canadian geologist J. T. Wilson about the mobility of the lithosphere by identifying the separate lithospheric plates and discussing their relative motions and the phenomena that occur at their boundaries. By the end of the 1960s, the evidence for plate tectonics had become so persuasive in its abundance that most Earth scientists embraced the theory. Textbooks were revised, and specialists began to consider the implications of the new concept for their own fields. From time to time in the history of science, a fundamental concept appears that unifies a field of study by pulling together diverse theories and explaining a large body of observations. Such a concept in physics is the theory of relativity; in chemistry, the nature of the chemical bond; in biology, DNA; in astronomy, the Big Bang; and in geology, plate tectonics.

Why did it take the scientific establishment almost four centuries to move from skepticism about continental drift to acceptance of plate tectonics? Scientists work in different styles. Scientists with particularly inquiring, uninhibited, and synthesizing minds often are the first to perceive great truths. Although their perceptions may turn out to be false, these individuals are often the first to see the great generalizations of science. Most scientists, however, proceed more cautiously and wait out the slow process of gathering supporting evidence. Continental drift and seafloor spreading were slow to be accepted largely because the audacious ideas came far ahead of the firm evidence. The oceans had to be explored, a new worldwide network of seismographs had to be installed and used, magnetic stratigraphy had to be painstakingly worked out, and the deep sea had to be drilled before the majority could be convinced. Scientists in a well-known European laboratory compiled a list (in good humor) of the names of Earth scientists in the order of the date they accepted seafloor spreading as a confirmed phenomenon. The names of scientists of distinction appear at both the top and the bottom of the list.

As graduate students before the acceptance of plate tectonics, your authors were taught that the continents and ocean basins were in fixed positions with respect to one another and that continental drift was a wild hypothesis not to be taken seriously. We now know that on the geologic time scale, the continents move horizontally and the seafloor is far from permanent.

SEAFLOOR DYNAMICS   The present ocean basins are being created by seafloor spreading and consumed by subduction on a time scale of about 200 million years. This is only about 4 percent of the age of the Earth, a measure of how geologically young the seafloor is. Geologists have drilled into the floor of the deep oceans of the world in efforts to find remnants of seafloor older than 200 million years—without success. In 1990, the oldest oceanic rocks were found after a 20-year search in the western Pacific. They were of mid-Jurassic age, only some 175 million years old. Older oceanic rocks don't exist; they have been subducted. Small fragments of older seafloor are found embedded in continents—relics of old oceans that have disappeared.

CONTINENTAL DYNAMICS   Continents are more enduring than the seafloor. They are too buoyant to be subducted. They may be fragmented, moved, aggregated, and deformed by the movement of plates, but the pieces survive. The old core of North America, for example, was assembled by plate collisions about 2 billion years ago from pieces of even older continents, some as old as 4 billion years. Continents can be eroded and fragmented, but they can also grow over time by the gradual accumulation of materials along their margins. New continental strips can be added on here and there from time to time as plates separate and collide, fragment, move about, and reassemble.

Whereas seafloor spreading is a relatively simple mechanism and is reasonably well understood, the evolution of continents is enormously complex. Just how plate tectonics explains continental geology is now receiving much attention. Continental rock assemblages, volcanism, metamorphism, the evolution of mountain chains—all are being reexamined in the framework of plate tectonics. Most geologists now believe that the geology of continents has been dominated by plate tectonics for at least 3.5 billion years of geologic history.

THE MOSAIC OF PLATES

According to the theory of plate tectonics, the lithosphere is broken into a dozen or so rigid plates (Figure 20.3; see also the front endpapers). The plates slide over a partially molten, weak asthenosphere, and the continents, embedded in some of the moving plates, are carried along passively. Continental drift is basically a consequence of plate movements. Plate tectonics works because Earth's rigid lithosphere en-

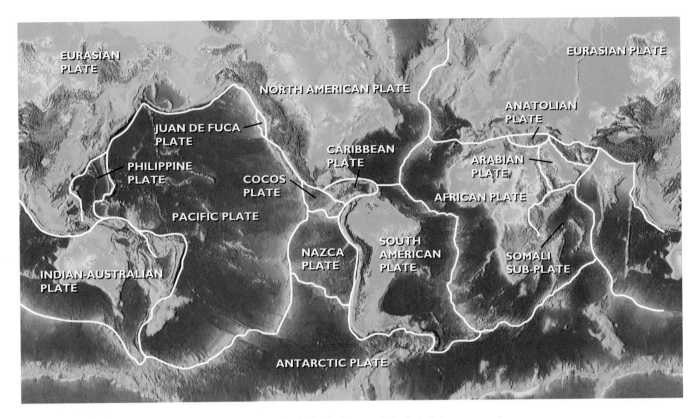

FIGURE 20.3 Computer-generated, color-shaded relief map of Earth in Mercator projection, showing plate boundaries. The image is generated from a digital data base of land and seafloor elevations. The shading simulates illumination from the west. (*Peter W. Sloss, NOAA-NESDIS-NGDC.*)

ables plates with horizontal dimensions of thousands of kilometers to move as distinct, rigid units with very little buckling or breaking, except at boundaries.

Earlier chapters described three types of plate boundaries according to the relative motions of adjacent plates:

1. Boundaries of divergence or spreading, typically ocean ridges and rifted continents

2. Boundaries of convergence or collision, typically deep-sea trenches, mountain ranges, and magmatic belts

3. Transform boundaries

Divergent Boundaries

At divergent boundaries, where plates move apart, partially molten mantle material rises and fills the gap between them. This material becomes new lithosphere added to the trailing edges of the diverging plates.

SEAFLOOR PLATE SEPARATION On the seafloor, the boundary between separating plates is marked by a mid-ocean ridge that exhibits active basaltic volcanism, shallow-focus earthquakes, and normal faulting caused by tensional or stretching forces created by the pulling apart of two plates. Figure 20.4a shows what happens there. (A detailed portrait of the Mid-Atlantic Ridge may be seen in Figure 17.27.) The process by which plates separate and ocean crust is created is called seafloor spreading. The Mid-Atlantic Ridge and the East Pacific Rise are spreading centers that have created millions of square kilometers of seafloor. As we mentioned in Chapter 1, Iceland is an exposed segment of the Mid-Atlantic Ridge. It provides an opportunity to view the process of plate separation and seafloor spreading directly (see Figure 1.16).

CONTINENTAL PLATE SEPARATION Early stages of plate separation can be found on continents. Such sites are characterized by long, downfaulted rift valleys, basaltic volcanic activity, and shallow-focus

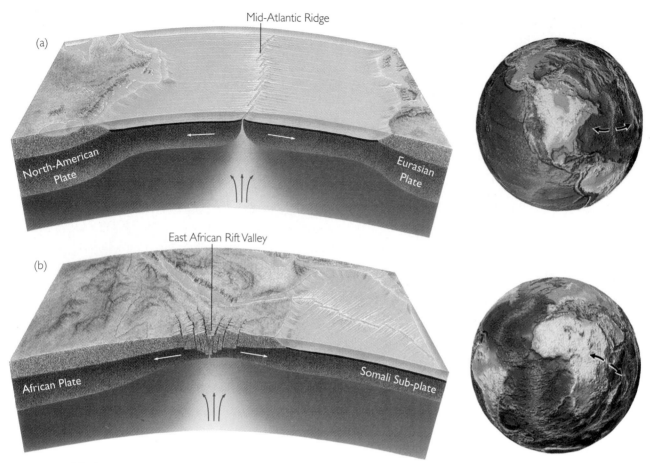

FIGURE 20.4 (a) Rifting and seafloor spreading along the Mid-Atlantic Ridge create a mid-ocean volcanic mountain chain and a coincident earthquake belt. (b) Initiation of rifting and plate separation within a continent. Characteristic features are rift valleys, with multiple normal faults; volcanism; and earthquakes. *(Color-shaded relief globes by Peter W. Sloss, NOAA-NESDIS-NGDC.)*

earthquakes (Figure 20.4b). The Great Rift Valley of East Africa, between the African Plate and the Somali Subplate, is thought to represent an early stage of plate separation within a continent. Sometimes, continental rifting may slow down or stop before the continent splits apart and a new ocean basin opens up. The East African Rift Valley and the Rhine Valley in Europe, both of which are still mildly active, may be such places. The Red Sea and the Gulf of California are rifts that are farther along in spreading; because of their greater width and depth, they have been flooded by the ocean. The Arabian Peninsula is splitting away from Africa at the Red Sea (Figure 20.5a), and Baja California is separating from the Mexican mainland at the Gulf of California (Figure 20.5b).

Convergent Boundaries

Plates collide along convergent boundaries. The overridden plate is subducted, or thrust downward into the mantle below, where eventually it is recycled. Collision and subduction produce deep-sea trenches, adjacent mountain ranges of folded and faulted rocks, and magmatic belts. The magmatic belt can be a mountain range on land or a chain of volcanic islands, called an **island arc,** on the seafloor. Once subducted lithosphere is heated, water and other volatiles "boil off." These join hot materials in the wedge of mantle above and induce melting. This is the source of magma that feeds and builds the overlying chain of volcanoes. The collision of two plates generates very large forces in the

(a)

FIGURE 20.5 (a) The Red Sea *(lower right)* divides to form the Gulf of Suez on the left and the Gulf of 'Aqaba on the right. The Arabian Peninsula on the right, splitting away from Africa on the left, has opened these great rifts, which are now flooded by the sea. The Nile River *(left center)* flows north into the Mediterranean Sea *(top). (Earth Satellite Corporation.)* (b) The Gulf of California, an opening ocean due to plate motions, marks a widening rift between Baja California and the Mexican mainland. *(Worldsat International/Photo Researchers.)*

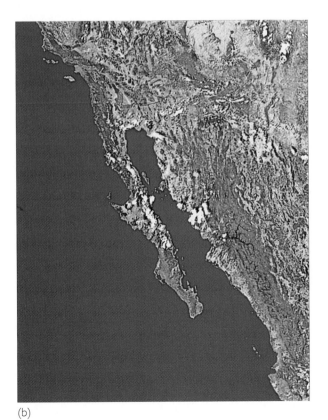

(b)

region, and in a general way these forces must result in the faulting that triggers the shallow- and deep-focus earthquakes that occur in subduction zones, as explained in Chapter 18. By the time the lithospheric plate has migrated from a divergent to a convergent boundary, it has cooled to become denser than the mantle below. Some geologists believe that the weight of the sinking part of the plate helps to pull the entire plate down and thus serves as an important part of the driving mechanism of plate tectonics.

A convergent boundary that exhibits all of these features is the one between the Nazca Plate and the South American Plate (see Figures 20.3 and 20.6a and the front endpapers). The Peru-Chile deep-sea trench offshore of these countries, the Andes Mountains with their many volcanoes, and some of the world's largest shallow- and deep-focus earthquakes occur here.

When a plate collision involves two continents, both tend to remain afloat. The collision of India and Asia is a good example. In this case, the Eurasian Plate is overriding the Indian Plate, creating a double thickness of crust and forming the highest mountain range in the world, the Himalayas. The

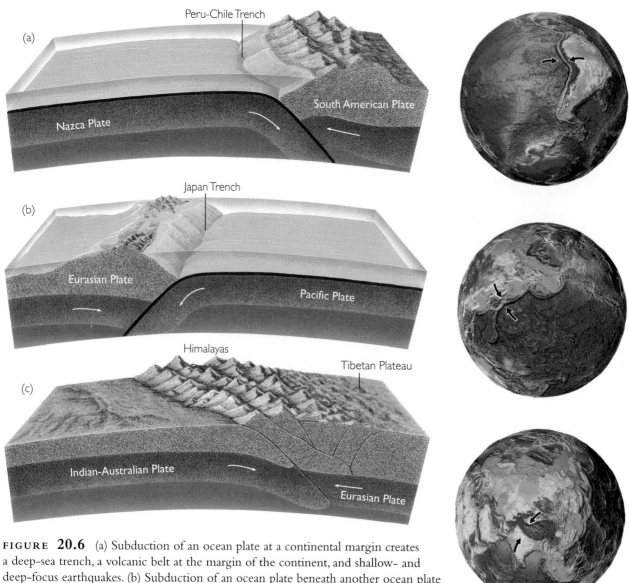

FIGURE 20.6 (a) Subduction of an ocean plate at a continental margin creates a deep-sea trench, a volcanic belt at the margin of the continent, and shallow- and deep-focus earthquakes. (b) Subduction of an ocean plate beneath another ocean plate forms a volcanic island arc. (c) A continent-continent plate collision creates multiple thrusting and folding, double thickening of the continental crust, and high mountains. *(Color-shaded relief globes by Peter W. Sloss, NOAA-NESDIS-NGDC.)*

Himalayas are supported isostatically by a crustal root projecting into the denser mantle below.

Figure 20.6 shows types of plate collisions.

Transform Boundaries

At transform boundaries, plates slide past each other, neither creating nor destroying lithosphere. Transform faults occur where the continuity of a divergent boundary is broken and offset, as Figure 20.7 shows. The San Andreas fault in California, where the Pacific Plate slides by the North American Plate, is a prime example of a transform boundary on land. Shallow-focus earthquakes with horizontal slips occur on transform boundaries.

Combinations of Plate Boundaries

Each plate is bounded by some combination of these three kinds of boundaries. As the plate maps show, the Nazca Plate in the Pacific is bounded on three sides by zones of divergence, along which new lithosphere forms, and on one side by the Peru-Chile subduction zone, where lithosphere is consumed. Most of the North American Plate is bounded by the divergent Mid-Atlantic Ridge on the east, the San Andreas fault and other transform boundaries on the west, and zones of subduction and transform that run from Oregon to the Aleutians on the northwest. Figure 20.8 depicts some relationships among modern plates, oceans, and continents.

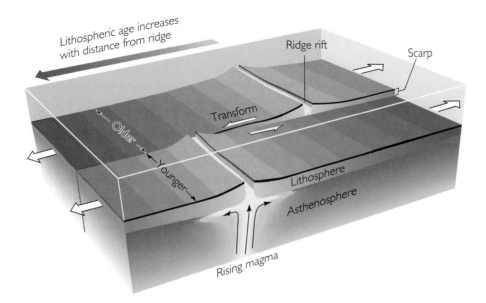

FIGURE 20.7 On the seafloor, plates move in opposite directions across a transform boundary between offset sections and in the same direction across other segments of the transform. A scarp can occur because the transform fault juxtaposes older seafloor with younger seafloor. Older seafloor is colder and contracts, sinking to a lower topographic level than younger seafloor.

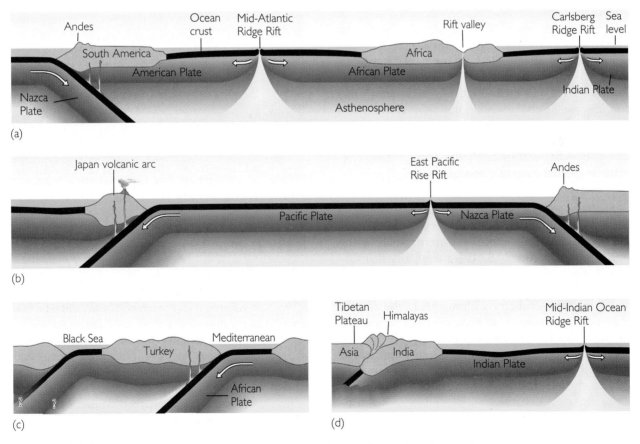

FIGURE 20.8 Several plates and the boundaries between them as they exist today. Seafloor spreading in the Atlantic (a), Pacific (b), and Indian (d) oceans is depicted, along with plate collisions of the types ocean–ocean (Japan, b), ocean–continent (South America, b, and Turkey, c), and continent–continent (India, d).

RATES OF PLATE MOTION

How fast do plates move? Do some plates move faster than others, and if so, why? Is the velocity of plate movements today the same as it was in the geologic past? Geologists have tried to answer these questions in recent years to gain a better understanding of plate tectonics. In doing so, they have developed some ingenious methods to study the motions of plates.

The Seafloor as a Magnetic Tape Recorder

During World War II, extremely sensitive instruments were developed to detect submarines by the magnetic fields emanating from their steel hulls. Geologists made some slight modifications to these instruments and towed them behind research ships so that they could measure the local magnetic field created by magnetized rocks beneath the sea. In Chapter 19, we saw that rocks can become magnetized by the Earth's magnetic field in whatever direction the field is oriented at the time the rocks formed. The present direction of the magnetic field is referred to as "normal." The opposite orientation is referred to as "reverse." In the geologic past, Earth's magnetic field switched back and forth erratically between normal and reverse. If the research ship was above rocks magnetized in the normal direction, geologists found a positive local magnetic field, or a *positive anomaly;* reversely magnetized rocks below the seafloor created a *negative anomaly.*

MAGNETIC ANOMALY PATTERNS Steaming back and forth across the ocean, seagoing scientists discovered magnetic anomaly patterns, such as the one shown in Figure 20.9, with a regularity that surprised them. In many areas, long, narrow bands of positive and negative magnetic anomalies showed an almost perfect symmetry with respect to the crest of the mid-ocean ridge. This peculiar magnetic pattern puzzled scientists for several years until two Englishmen, F. J. Vine and D. H. Mathews—and, independently, two Canadians, L. Morley and A. Larochelle—made a startling proposal in 1963. They reasoned that the positive and negative magnetic bands correspond to bands of rock on the seafloor that were magnetized during ancient episodes of normal and reversed magnetism of Earth's field. If this were the case, the magnetic bands provided evidence in support of the theory of

seafloor spreading, which had already been proposed (Figure 20.10). The scientists argued that the ocean progressively widens as new seafloor is created along a crack on the crest of a mid-ocean ridge. Magma flowing up from the interior solidifies in the crack and becomes magnetized in the direction of Earth's field at the time. As the seafloor splits and moves away from the ridge, approximately half of the newly magnetized material moves to one side and half to the other, forming two symmetrical magnetized bands. Newer material fills the crack, continuing the process. In this way, the seafloor acts like a tape recorder that encodes, by magnetic imprinting, the history of the opening of the oceans in terms of the history of reversals of Earth's magnetic field.

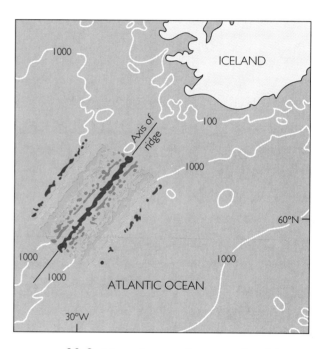

FIGURE 20.9 Magnetic anomaly pattern found in an oceanographic survey over the Reykjanes Ridge, a part of the Mid-Atlantic Ridge southwest of Iceland. The spaces between the colored bands show where the survey ship found negative magnetic anomalies corresponding to rock formations on the seafloor below that are reversely magnetized. The bands shown in color indicate where the ship found positive anomalies. The rocks below the colored bands are magnetized in the normal direction, that is, similar to the present-day direction. The almost perfect symmetry of this pattern with respect to the ridge axis puzzled geologists. When the pattern was explained (see Figure 20.10), it provided strong support for the concept of seafloor spreading.

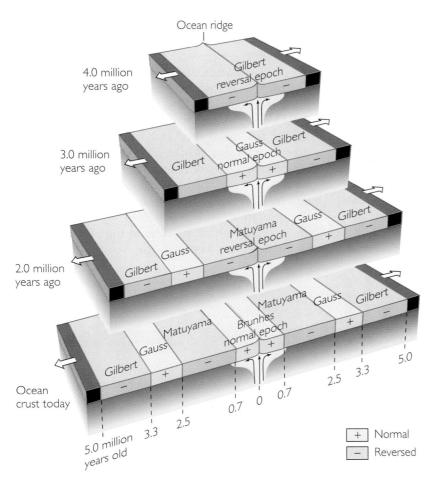

FIGURE 20.10 Bands of magnetized crust form successively as each new piece of seafloor is intruded, cools, and becomes magnetized in the normal or reversed direction of the magnetic field extant at that time. As plate separation continues, the newly magnetized crust is pushed out on both sides and gradually moves outward with the separating plates. The pattern of normal (+) and reversed (−) magnetic bands on the seafloor follows the succession of magnetic reversals over time worked out on land (compare with Figure 19.16).

INFERRING SEAFLOOR AGES AND SPREADING VELOCITY Remember that the ages of reversals have been worked out from magnetized lavas on land (see Figure 19.16). Using this known sequence of reversals over time, geologists could assign ages to the bands of magnetized rocks on the seafloor. Because they now knew the age of a band of magnetized rocks on the seafloor and knew the distance from a mid-ocean ridge crest where the magnetized rocks were created, they could calculate how fast the ocean opened up—that is, the velocity of plate movements. Rocks on the crest of the ridge would be modern, hence normally magnetized, because they were extruded during the current normal magnetic epoch. Conversely, magnetized rocks corresponding to a magnetic epoch of about 1 million years ago would have been displaced some distance from the ridge—say, about 20 km on each side of the ridge crest if the plates are spreading apart at a rate of 2 cm per year, or 40 km if the spreading rate is 4 cm per year. Using this method, geologists found that the highest rate of seafloor spreading is 10 to 12 cm per year at the East Pacific Rise, and the lowest is 2 cm per year at the Mid-Atlantic Ridge.

The normal-reversal time scale can be followed through many oscillations of Earth's magnetic field. The corresponding magnetic bands on the seafloor, which can be thought of as age bands, extend from the ridge crests across the ocean basins over a time span exceeding 100 million years.

The power and convenience of using seafloor magnetism to work out the history of ocean basins cannot be overemphasized. Simply by steaming back and forth over the ocean, measuring the magnetic field of the magnetized rocks of the seafloor, and correlating the pattern of reversals with the time sequence worked out by the methods just described, geologists determined the ages of various regions of the seafloor without even examining rock samples. In effect, they learned how to "replay the tape."

The simplicity and elegance of seafloor magnetism made it an attractive tool. But it was an indirect method in that rocks were not recovered from the seafloor and their ages were not directly determined in the laboratory. Corroborative evidence of seafloor spreading and plate movement would still be needed to convince the remaining skeptics. Deep-sea drilling supplied it.

Deep-Sea Drilling

In 1968, a program of drilling into the seafloor was launched (see Feature 20.1). This joint project of major oceanographic institutions and the National Science Foundation aimed to drill through, retrieve, and study seafloor sediments from many places in the world's oceans. Using hollow drills, scientists

brought up cores containing sections of seafloor sediments and underlying volcanic crust. In some cases, the drilling penetrated thousands of meters below the seafloor surface. Geologists thus had an opportunity to work out the history of the ocean basin from direct evidence.

One of the most important facts geologists sought was the age of each sample. Because sedimentation begins as soon as ocean crust forms, the age of the oldest sediments in the core, those immediately on top of the basaltic crust, tells the geologist how old the ocean floor is at that spot. The age is obtained from the fossil skeletons of foraminifera, single-celled animals that live in the ocean and sink to the bottom when they die (see Chapter 17). It

20.1   TECHNOLOGY AND EARTH

Drilling in the Deep Sea

The deep-sea drilling vessel *Joides Resolution* is 143 m long, and amidships it carries a drilling derrick 61 m high. In size and capacity to drill in the deepest ocean, it is the only ship of its kind in the world. It can lower drill pipe thousands of meters to the seafloor and drill thousands of meters into the sediments and underlying basaltic crust.

Before the ship could accomplish such a feat, a technological breakthrough was required. A way had to be found to hold the ship stationary during the drilling process, regardless of currents and winds; otherwise, the drill pipe would break off. The problem was solved by development of a positioning device that uses sound waves transmitted by acoustic beacons planted on the seafloor. Any change in the ship's position is sensed by a computer that monitors changes in the time of arrival of the sound pulses from each beacon. The same computer controls the speed and steering to keep the vessel stationary.

Deep-sea drilling was the answer to those who said, when lunar exploration started, "Better to explore the ocean's bottom than the back side of the Moon." We ended up

doing both. The deep-sea drilling program, now known as the Ocean Drilling Project, is more than 25 years old and has become international in scope.

Scientists on board a drilling ship take samples from cores of sediment recovered from the seafloor. These samples can be analyzed to reveal the history of ocean basins and ancient climatic conditions. *(Texas A&M University.)*

was found that the oldest sediments in the cores become older with increasing distance from mid-ocean ridges and that the age of the seafloor at any one place agrees almost perfectly with the age determined from magnetic reversal data. This agreement validated magnetic dating of the seafloor and clinched the concept of seafloor spreading. What a triumph for the scientists who discovered this tool!

Isochrons

Figure 20.11 shows the ages of the seafloor of the world's oceans as determined from the fossil and magnetic reversal data. Each colored band represents a span of time that gives the age of the crust within the band. The boundaries between bands are contours that connect rocks of equal age and are called **isochrons.** Isochrons show the time that has elapsed, and therefore the amount of spreading that

has occurred, since the crustal rocks were injected as magma into a mid-ocean rift. Notice how the seafloor becomes progressively older on both sides of the ridge rifts, where seafloor spreading originates. For example, the distance from a ridge axis to a 139.6-million-year isochron (boundary between green and blue bands) indicates the extent of new ocean floor created over that time span.

The more widely spaced isochrons (the wider color bands) of the eastern Pacific signify faster spreading rates than those in the Atlantic. No sediments older than the Jurassic period, about 200 million years ago, have been found. This observation attests to the youth of the seafloor in comparison with the continents. Over a period of 100–200 million years in some plates and tens of millions of years in others, the ocean lithosphere thickens and cools, becomes denser than the underlying mantle, and sinks into it in the process of subduction.

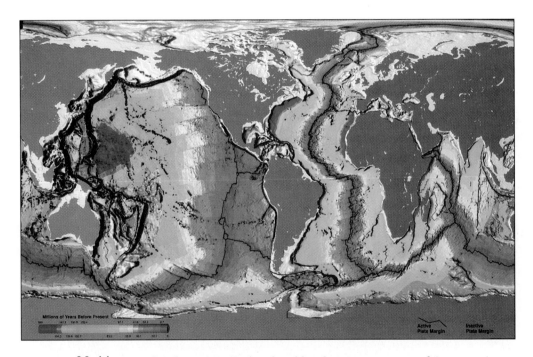

FIGURE 20.11 Age of seafloor crust. Each colored band represents a span of time covering the age of the crust within the band. The boundaries between bands are contours of equal age called isochrons. Isochrons give the age of the seafloor in millions of years since its creation at mid-ocean ridges. The dark gray color indicates land. The light gray indicates shallow water over continental shelves. Mid-ocean ridges, along which new seafloor is extruded, coincide with the youngest seafloor (dark red). The Atlantic Ocean is symmetrical about the Mid-Atlantic Ridge. Asymmetry of the pattern in the Pacific is caused partly by subduction in the Aleutian Trench south of Alaska, in the Peru-Chile Trench along the west coast of South America, and in many trenches in the western Pacific. *(Peter W. Sloss, NOAA-NESDIS-NGDC.)*

Figure 20.12 shows the directions of plate motions and summarizes the plates' relative velocities—that is, their velocities with respect to one another. (If you stretched a rubber band between two plates and anchored the ends on each plate, the change in the rubber band's length over time would indicate the relative velocity between the two plates. If you had an instrument anchored to a single plate that could measure the change in its position over time, you would be recording the absolute velocity of the plate. The trail of extinct volcanoes away from a hot spot fixed in the mantle, described in Chapter 5, is one way to measure the absolute velocity of a plate.)

Geologists have noted that the fast-moving plates, as measured by the relative velocities at their divergent boundaries (the Pacific, Nazca, Cocos, and Indian) are being subducted along a large fraction of their boundaries. In contrast, the slow-moving plates (the North and South American, African, Eurasian, and Antarctic) have large continents embedded in

them and do not have significant attachments of downgoing slabs. An attractive hypothesis to explain these observations is that rapid plate motions are caused by the pull exerted by large-scale downgoing slabs, and slow plate motions are caused by the drag associated with embedded continents.

The Geometry of Plate Motion

The direction of the movement of one plate in relation to another depends on geometric principles that govern the behavior of rigid bodies. Earth's plates behave as rigid bodies in that the distances between three points on the same plate—say, New York, Miami, and Bermuda on the North American Plate—do not change, no matter how the plate moves. But the distance between, say, New York and Lisbon increases because the two cities are on different plates that are being separated along a narrow

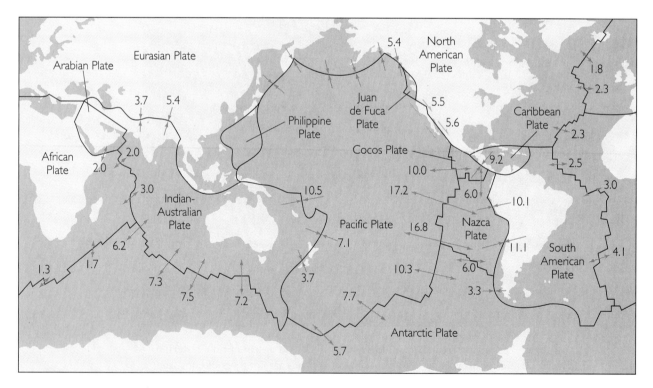

FIGURE 20.12 Relative velocities (in centimeters per year) and directions of plate separation and convergence. Opposed arrowheads indicate convergence. Diverging arrowheads indicate plate separation at ocean ridges. Parallel arrowheads, as along the San Andreas fault in California, indicate transform faults, where plates slide past each other. Spreading is fastest between the Pacific and Nazca plates and slowest between the North American and Eurasian plates. (Data from C. Demets, R. G. Gordon, D. F. Argus, and S. Stein, Model Nuvel-1, 1990.)

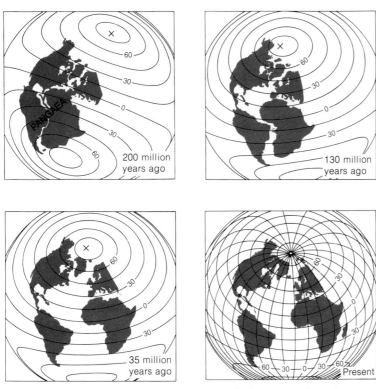

X = Ancient geographic pole

FIGURE 20.13 Plate movements have led to the northward drift of the continents and the opening of the Atlantic Ocean over the past 200 million years. The central Atlantic, the Caribbean, and the Gulf of Mexico began to form about 200 million years ago, in Triassic time, when Pangaea began to break up and Africa and South America drifted away from North America. The South Atlantic opened about 150 million years ago with the separation of South America from Africa. As the continents drifted apart, they also migrated in a northerly direction to their present positions. Note that the equator passed through the southern parts of the United States and Europe in Triassic time. (After J. D. Phillips and D. Forsyth, "Plate Tectonics, Paleomagnetism, and the Opening of the Atlantic," *Bulletin of the Geological Society of America,* vol. 83, no. 6, 1972, pp. 1579–1600.)

zone of spreading on the Mid-Atlantic Ridge. Two primary geometric principles govern the sliding of plates on our planet:

1. **Transform boundaries reveal the direction of plate movement.** No overlap, buckling, or separation occurs along typical transform boundaries. The two plates merely slide past each other without creating or destroying plate material. Look for a transform boundary if you want to deduce the direction of plate motions, because the orientation of the fault is the direction in which one plate slides with respect to the other, as Figure 20.7 shows.

2. **Isochrons reveal the positions of plates in earlier times.** Isochrons on the seafloor are roughly parallel and are symmetrical with respect to the ridge axis along which they were created. Figure 20.11 illustrates this observation. Since each isochron was at the plate boundary of separation at an earlier time, isochrons that are of the same age but on opposite sides of an ocean ridge can be brought together to show the positions of the plates and the configuration of the continents embedded in them as they

were in that earlier time. For example, using isochrons, geologists can reconstruct the opening of the Atlantic Ocean and, with other information, the northward drift of continents, as shown in Figure 20.13.[1]

By applying such geometric principles, geologists are able to deduce spreading rates from spreading directions and magnetic anomalies and can work out the history of the motions of all the lithospheric plates. Some results have already been pictured in Figures 20.11 and 20.12. Geologists are also searching for other ways to measure the motions of plates. If the hot spots discussed in Chapter 5 turn out to be fixed in the mantle below plates, then the string of extinct volcanoes trailing from the hot spot records the movement of individual plates as they glide over the mantle (see Figure 5.30).

---

[1]Archeologists believe that the builders of the Great Pyramid of Egypt designed it to aim due north. Today, it is aimed slightly east of north. Did the ancient Egyptian astronomers make a mistake in orienting the pyramid 40 centuries ago? Probably not. Over this period, Africa drifted enough to rotate the pyramid out of alignment with true north.

Satellite Measurement of Plate Motions

We used to depend solely on ships to explore the topography and structure of the seafloor. But ships would require decades to explore every corner of every ocean. Fortunately, another new geological tool made possible by advances in technology is now available to speed up the process. It is a method of imaging the seafloor from an orbiting satellite that bounces radar beams off the surface of the ocean. The height of the sea surface reflects the topography of the seafloor below (see Feature 17.2). All the major structures associated with seafloor spreading are revealed as if the water were drained away. Satellite altimetry mapping, as the system is known, has located mid-ocean ridges, deep-sea trenches, and transform boundaries, many not previously found by ship surveys.

An even newer method of measuring plate motions is the Global Positioning System (GPS), which uses radio signals from 24 satellites, each with a precise atomic clock on board. A ground-based radio receiver no bigger than this book receives the signals from five or more satellites at the same time and detects the differences in the travel time of the signals from each satellite. A tiny microprocessor in the receiver can use these time differences to locate the receiver to within one centimeter or so.[2] A GPS instrument used to monitor earthquakes in southern California is shown in Figure 18.5.

Teams of scientists using data from GPS receivers placed on different plates find that their measurements of plate motions over a period of a few years agree in magnitude and direction with those found by magnetic anomaly and isochronic methods. In a sense, the satellite serves as an outside observer, independently validating the theories and methods of Earth-bound geologists as they reconstruct plate motions from the geologic record. These experiments indicate that plate motions are remarkably steady over periods of time ranging from a few years to millions of years. Plate motions are now being measured on a yearly basis using GPS in many places over the globe. For example, the fastest plate motion in the world has been found in a small region of the boundary between the Pacific and Australian plates, near the Samoan Islands. The convergence is 24 cm per year.

ROCK ASSEMBLAGES AND PLATE TECTONICS

The only record we have of past geologic events is the incomplete one found in the rocks that have survived erosion or subduction. Because only seafloor younger than 200 million years has survived subduction, we must focus on the continents to find the old rocks that provide the evidence for most of Earth's history. Some of the methods of reading the rock record have been described in earlier chapters. They include interpreting unconformities, faults, and other structures; finding evidence for uplift and erosion; deducing the environment in which sediments were deposited; and reconstructing the original condition of rocks that have been deformed or metamorphosed. Chapter 4 investigated the relationship between plate boundaries and various types of igneous rocks. Here we explore in more detail the types and combinations of rocks that are characteristic of the various kinds of plate boundaries. Geologists who study continents use these characteristic **rock assemblages** to identify ancient episodes of plate separations and collisions.

Rocks at Divergent Boundaries

Before the advent of plate tectonics, geologists were puzzled by unusual assemblages of rocks that were characteristic of the seafloor but were found on land. Known as **ophiolite suites,** these assemblages consist of deep-sea sediments, submarine basaltic lavas, and mafic igneous intrusions (Figure 20.14). Using data gathered from deep-diving submarines, dredging, deep-sea drilling, and seismic exploration, geologists can now explain these rocks as fragments of oceanic crust that were transported by seafloor spreading and then raised above sea level and thrust onto a continent in an episode of plate collision.

ROCK ASSEMBLAGES ON THE SEAFLOOR The creation of oceanic crust along the axis of a mid-ocean ridge is intimately related to the separation of plates. The process of seafloor creation is poorly understood, but it is known to involve magmatism, seawater circulation, and tectonic activity. Figure 20.15 is a highly schematic and simplified picture of what

---

[2]GPS receivers will soon be in every automobile, as part of a navigating system that will lead the driver to a specific street address. It is interesting that the scientists who developed the atomic clocks used in GPS did so for research in fundamental physics and had no idea that they would be creating a multibillion-dollar industry. Along with the transistor, laser, and other examples, GPS technology demonstrates the serendipitous manner in which basic research repays the society that supports it.

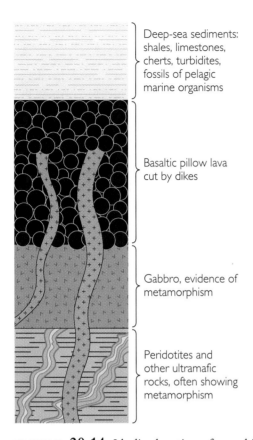

Deep-sea sediments: shales, limestones, cherts, turbidites, fossils of pelagic marine organisms

Basaltic pillow lava cut by dikes

Gabbro, evidence of metamorphism

Peridotites and other ultramafic rocks, often showing metamorphism

FIGURE 20.14 Idealized section of an ophiolite suite. The combination of deep-sea sediments, submarine pillow lavas, sheeted basaltic dikes, and mafic igneous intrusions indicates a deep-sea origin. Ophiolites are fragments of ocean lithosphere emplaced on a continent as a result of plate collisions.

may be happening, based in part on studies of ophiolites found on land and information gleaned from ocean drilling and sound-wave profiling of the kind shown in Feature 19.1. On some of the more complete ophiolite sequences preserved on land, we can literally walk across rocks that used to lie along the Moho of the ocean crust. Ocean drilling has penetrated to the gabbro layer of the seafloor, but not to the Moho below. And sound-wave profiles have found several small magma chambers or lenses similar to the one shown in Figure 20.15.

As the plates separate, hot mantle rises and begins to melt. By the time the mantle material reaches shallow depths, it is a mush of perhaps 85 percent crystals and 15 percent basaltic melt. The mush fills a shallow, thin, lens-shaped magma chamber, from which it erupts in vertical sheets of dikes. Dikes intrude dikes to form a structure that has been likened to a pack of cards standing on end. Basalt spilling out onto the seafloor freezes as pillow lavas (see Figure 5.4)—the characteristic type of undersea volcanism—and forms a cover over the sheeted dikes. As the mush cools and solidifies at its boundaries, it forms the coarse-grained basaltic rock gabbro as the layer below the sheeted dikes. A thin blanket of deep-sea sediments begins to cover the newly formed ocean crust.

As the seafloor spreads, the layers of sediments, lavas, dikes, and gabbros are transported away from the mid-ocean ridge, where this characteristic sequence of rocks that make up the oceanic crust is assembled—almost like a production line. According

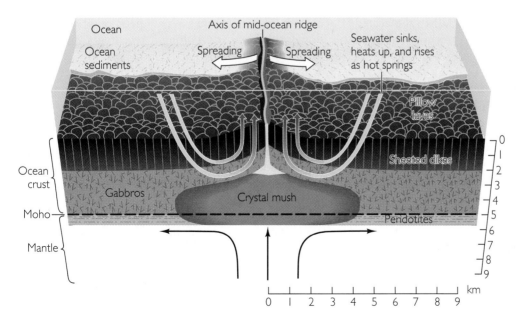

FIGURE 20.15
A highly schematic and simplified sketch of how ocean crust might form at mid-ocean ridges. A hot mush of crystals and melt (red) rises from the mantle at the axis of a mid-ocean ridge. The melt is concentrated in a small shallow lens (yellow) from which dikes of magma erupt repeatedly to form seafloor pillow lavas and underlying sheeted dikes. The mush cools to form the gabbro layer of the crust as the plates spread apart.

FIGURE **20.16**
Precambrian ophiolite
suite, northern Quebec,
Canada. The basalt pillows
and other rocks typical of
plate separation in this
2-billion-year-old slice of
seafloor are compelling
evidence that plate
tectonics was occurring
in Precambrian time.
*(M. St. Onge/Geological
Survey of Canada.)*

to this hypothesis, the mush-and-magma lenses along the length of the ridge are periodically replenished by fresh injections from the mantle below to keep the process going. The scale of this "factory" where ocean crust is created is a few kilometers wide and deep, and it extends intermittently along the thousands of kilometers of mid-ocean ridge.

Massive ore bodies rich in iron, copper, and other minerals are found along mid-ocean ridges. The ores are formed when seawater sinks through porous volcanic rocks; becomes heated; and leaches these elements from the lavas, dikes, gabbros, and perhaps even the underlying mush. When the heated seawater enriched with dissolved minerals rises and reenters the cold ocean, the ore-forming minerals precipitate (see Figure 23.14).

Ophiolite suites such as the one in Figure 20.16 preserve on land fragments of ancient seafloor.

ROCK ASSEMBLAGES AT INTRACONTINENTAL RIFTS When plate separation begins within a continent, the ancient site of rifting can often be found by another characteristic assemblage of rocks and structures. These include rifts and downdropped blocks of continental crust, volcanic intrusions, and thick sedimentary basins along continental margins. The process of rifting begins when the continental crust and underlying lithosphere are stretched and thinned by the forces of plate separation (Figure 20.17a). A long, narrow rift develops, with great downdropped crustal blocks. Hot ductile mantle

rises and fills the space created by the thinned lithosphere and crust, initiating the volcanic eruption of basaltic rocks in the rift zone.

If the divergence continues until the two segments of continent separate, the widening rift is flooded by the sea and a new ocean basin forms and grows (Figure 20.17b). The receding continental margins subside gradually as the underlying lithosphere cools and contracts, forming offshore basins that can receive sediments eroded from the adjacent land. These are **continental shelf deposits**—sedimentary rock assemblages that are laid down in an orderly sequence under the quiet conditions of a slowly subsiding basin along a receding continental margin. The sedimentary basins off the Atlantic coasts of North and South America, Europe, and Africa are products of this process. These basins began to form when the supercontinent Pangaea split about 200 million years ago and the American plates separated from the European and African plates. Figure 20.17c shows the wedge-shaped deposit of sediments underlying the Atlantic continental shelf and margin of the United States, which were formed in this way. Because the trailing edge of the continent subsides slowly, the offshore basins continue to receive sediments for a long time. The load of the growing mass of sediment further depresses the crust, so that the basins can receive still more material from land. The result of these two effects is that the deposits can accumulate in an orderly fashion to thicknesses of 10 km or more (Figure 20.17d).

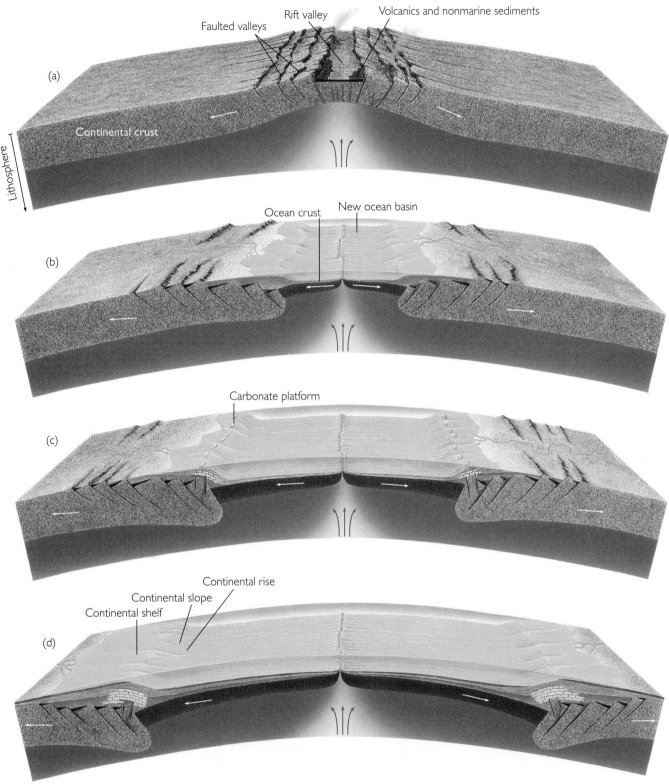

Vertical distances not to scale

FIGURE 20.17 The development of sedimentary basins on a rifted continental margin. A rift develops in Pangaea as hot mantle materials upwell and the ancient continent stretches and thins. Volcanics and Triassic nonmarine sediments are deposited in the faulted valleys (a). Seafloor spreading begins (b). The lithosphere cools and contracts under the receding continental margins, which subside below sea level. Evaporites, deltaic deposits, and carbonates (c) are deposited and then covered by Jurassic and Cretaceous sediments derived from continental erosion (d). The Atlantic margins of Europe, Africa, and North and South America have histories similar to this.

Rocks at Ocean–Ocean Convergences

Orogeny means "mountain making," particularly by the folding and thrusting of rock layers, often with accompanying magmatic activity. Plate tectonics has provided new insights into this fundamental process of geology and the rock assemblages associated with it.

When an ocean plate collides with and overrides another ocean plate, several complex processes are set in motion (Figure 20.18). The ocean sediments of the downgoing plate are mostly scraped onto the margin of the overriding plate while the cool, subducted lithosphere descends into the hot mantle below. At depths of 50 to 100 km, the subducted slab encounters temperatures in the range of 1200 to 1500°C. Some of the basalt and sediment in the slab melts, releasing water and other materials. The released water induces peridotite, the major constituent in the wedge of mantle above the subducted plate, to melt. (Laboratory experiments show that the addition of small amounts of water can lower the temperature at which mantle rocks melt by several hundred degrees.)

The heated and now buoyant mantle material rises, and further melting occurs as the pressure eases. The resultant mixture of melt from the slab and melt from the peridotitic mantle becomes an ultramafic magma. The contribution of slab melt to the magma is inferred because trace elements known to be present in ocean crust and sediments are found in the magma. Most of the ultramafic magma accumulates at the base of the crust of the overriding plate, and some of it intrudes into the crust. Fractional crystallization occurs, and some crustal rocks are assimilated into the magma (Chapter 4). In this way, the ultramafic magma evolves into mafic and more silicic magmas and lavas such as basalt, andesite, and dacite. The intrusions and volcanic eruptions build an arc-shaped chain of volcanic islands on the seafloor. The West Indies, the Aleutians, the Philippines, and the Marianas are islands of this sort.

Between the island arc and the deep-sea trench some 200 km seaward is an area called the forearc (see Figure 20.18). This area includes a forearc basin—a depressed zone that fills with sediments derived from the arc—and an accretionary wedge formed from the pile of sediments and ocean crust scraped off the descending plate.

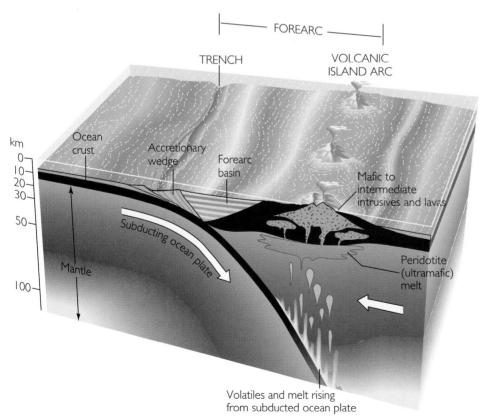

FIGURE 20.18 Rock assemblages associated with the collision of two ocean plates and subduction. Water, other volatiles, and melt rise from the heated subducted plate and cause melting of peridotite in the overlying wedge of mantle. Ocean crust is intruded by magma to form a volcanic arc, which erupts mafic and intermediate lavas. A forearc basin fills with arc-derived sediments. An accretionary wedge builds up from ocean sediments and crust scraped off the descending plate. (After National Research Council, "Margins," *Proceedings of Workshop of National Research Council,* Washington, D.C., National Academy of Sciences, 1989.)

Rocks at Ocean–Continent Convergences

The rock assemblages that form when a plate carrying a continent on its leading edge collides with and overrides a subducting oceanic plate are shown in Figure 20.19.

MOUNTAIN BELTS AND MAGMATISM   A mountain belt develops on the continental margin, built up by multiple intrusions of igneous rock and eruptions of lava. Multiple thrusts may occur, with sections of rock thrust on top of one another, contributing to the mountain-building episode.

As when one ocean plate collides with another, magma develops primarily from the melting of peridotite in the wedge of mantle above the subducted plate. The magma produces basaltic and andesitic lavas, along with a few dacites and rhyolites, and granitic batholiths that intrude into the continental crust. The assemblage of igneous rocks is more silicic than those found on island arcs, perhaps because the magma derived from the mantle assimilates the melting continental crust.

Metamorphic rocks are found in these magmatic belts, typically the result of recrystallization under high temperatures and low pressures. These conditions occur because the hot fluids rise close to the surface, delivering much heat to this low-pressure environment.

MÉLANGE OF METAMORPHOSED SEDIMENTS
Thick sediments, many of them turbidites, eroded from the continent rapidly fill the adjacent depressions in the seafloor around the subduction zone. As it descends, the cold oceanic slab stuffs the region below the inner wall of the trench (the wall closer to land) with these sediments and with deep-sea sediments and ophiolite shreds scraped off the descending plate. Regions of this sort between the magmatic arc on the continent and the offshore trench are enormously complex and variable. The deposits are all highly folded, intricately sliced, and metamorphosed. They are difficult to map in detail but are recognizable by their distinctive mix of materials and structural features. Such a chaotic mixture has been called a **mélange.** The metamorphism is the kind characteristic of high pressure and low temperature, because the material may be carried relatively rapidly to depths as great as 30 km, where recrystallization occurs in the environment

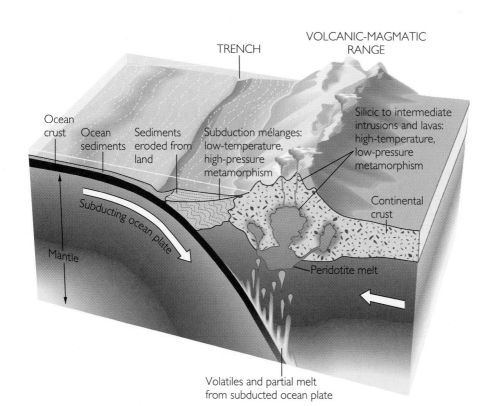

FIGURE 20.19 Rock assemblages associated with ocean–continent plate collisions and subduction: ocean trenches, mélange deposits, magmatic belts, metamorphism, and volcanism. The drawing is not to scale; the thickness of the lithosphere is about 70 km, the depth of the ocean trench is about 10 km, and the distance from trench to magmatic belt is 300 to 400 km.

of the still-cold subducting slab. Somehow, as part of the subduction process, the material rises back to the surface.

EVIDENCE OF ANCIENT OCEAN-CONTINENT COLLISIONS Find a paired belt of mélange and magmatism and you have a relic left behind by an ancient episode of subduction. The mélange formed offshore near where the subducted plate turned downward. The magmatic belt formed landward, from the melt that rose from greater depths along the downthrust plate.

The essential elements of these rock assemblages of collision have been found in many places in the geologic record. One can see mélange in the Franciscan formation of the California Coast Ranges and magmatism in the parallel belt of the Sierra Nevada to the east. This paired belt marks the Mesozoic collision between the North American Plate and the Farallon Plate, which has disappeared by subduction (see Figure 19.8). The location of mélange on the west and magmatism on the east shows that the now-absent Farallon Plate was the subducted one, overridden by the North American Plate on the east.

Other paired belts can be found along the continental margins framing the Pacific basin—in Japan, for instance. The central Alps were produced by the convergence of a Mediterranean plate with the European continent. The Andes Mountains (from which the name of the volcanic rock andesite is derived), near the west coast of South America, are products of a collision between ocean and continental plates. Here the Nazca Plate collides with and is subducted under the South American Plate.

Rocks at Continent-Continent Convergences

Because plates may have continents embedded in them, a continent can collide with another continent, as shown in Figure 20.20. Because continental crust is buoyant, both continents may resist subduction and stay afloat. A wide zone of intense deformation develops at the boundary where the continents grind together. This boundary is marked by a mountain range in which sections of deep- and shallow-water sediments are found highly folded and broken by multiple thrust faults. The buildup of thrust sheets, one atop another, results in a much-thickened continental crust in the collision zone. Often, there is a magmatic belt within the mountain range. The remnant of such a zone left behind in the geologic record is called a **suture.** Ophiolites are often found near the suture; they are relics of an ancient ocean that disappeared in the convergence of two plates.

A prime example of a collision of continents is the Himalayas, which began to form some 50 million years ago when a plate carrying India collided with the Eurasian Plate. The collision continues: India is moving into Asia at a rate of a few centimeters per year, and the uplift is still going on, together with faulting and many great earthquakes.

Geologists now believe that much of the geology of continents can be explained by episodes of continental rifting and plate separation and by continental plate collisions.

Rocks at Transform Boundaries

At a transform boundary between two oceanic plates (see Figure 20.7), the rocks on either side are of the same type (oceanic crust and sediment) but of different ages and water depths because of differing distances from the mid-ocean ridges where they originated. When a transform fault occurs on land, as in the case of the San Andreas fault, the rocks on either side of the transform boundary are more likely to be of different types and ages. Unlike seafloor rocks, rocks found on land are highly variable from place to place, and a large fault offset is likely to juxtapose rocks that formed under different circumstances of intrusion, metamorphism, deposition, deformation, and erosion.

MICROPLATE TERRANES AND PLATE TECTONICS

Geologists have argued for decades about blocks as large as hundreds of kilometers across within orogenic belts of continents, with assemblages of rocks that are alien to their surroundings. These **microplate terranes**[3]—sometimes called displaced, exotic, or suspect terranes—contrast sharply with adjacent provinces in the assemblages of rocks, the nature of folding and faulting, and the history of magmatism and metamorphism. Fossils indicate that

---

[3]The familiar term *terrain* is synonymous with region, area, or territory. *Terrane* is used to signify a region in which rocks that were formed elsewhere have been attached after being carried great distances by plate movements.

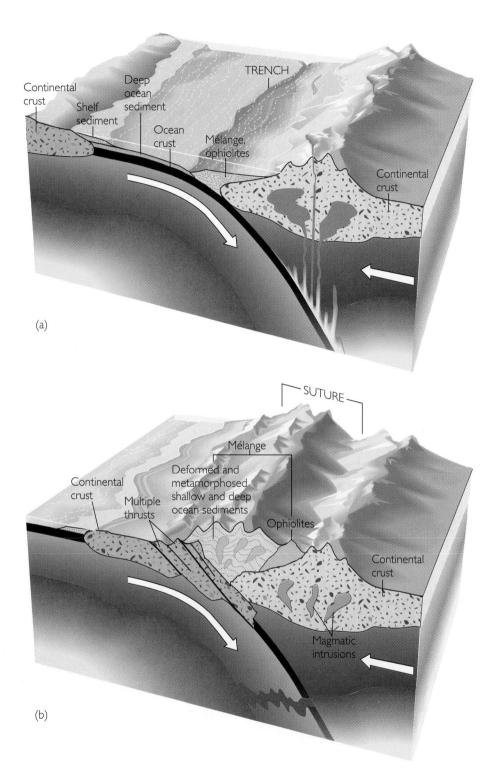

(a)

(b)

FIGURE 20.20 (a) A plate carrying a continent is subducted under a plate with a continent at its leading edge. (b) The two continents collide. The continent at left breaks into several thrust sheets, thickening the continental crust and raising a high mountain range. Other rocks associated with plate convergence are caught up in the collision zone: magmatic intrusions, deformed and metamorphosed shelf and deep-water sediments, and ophiolite fragments. The Tertiary-period collision of India and Asia is an example.

these blocks originated in environments and at times different from those of the surrounding area. Microplate terranes are now believed to be fragments of other continents, seamounts, island volcanic arcs, or slices of ocean crust that were swept up and plastered onto a continent when plates collided or remnants left behind when a continent split apart and separated.

The Appalachian orogenic belt, ranging from Newfoundland to the southeastern United States, contains microplate terranes—slices of ancient Europe, Africa, oceanic islands, and crust welded onto

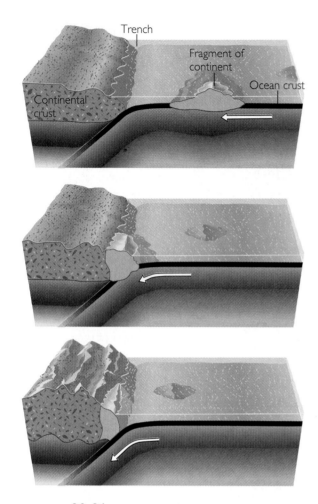

FIGURE 20.21 Accretion of a microplate terrane. An oceanic island arc or a fragment of continent is carried into a plate collision zone. Instead of being subducted, it is welded onto the overriding plate. Because the fragment may have originated thousands of kilometers away, its rock assemblages differ from those of the surrounding geological terrain.

North America during ancient collisions (Figure 20.21). Most of Florida is probably a piece of Africa left behind when North America and Africa parted about 200 million years ago. Florida's oldest rocks and fossils are more like those in Africa than like those found in the rest of the United States. As many as 100 areas of the Cordilleran orogenic belt of western North America have been identified as microplate terranes added on to the continent (Figure 20.22). Microplate terranes have also been found in Japan, Southeast Asia, China, and Siberia, but their original locations have yet to be worked out. Over the past 4 billion years (the age of the oldest known continental rocks), the continents have grown at an average rate of about 2 km^3 per year, most commonly by terrane accretion and magmatism.

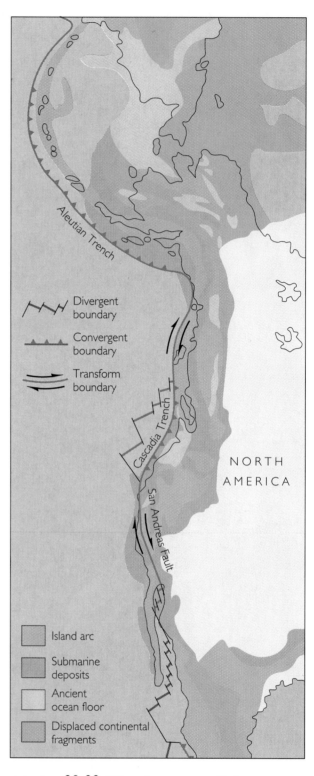

FIGURE 20.22 Microplate terranes added to western North America in the past 200 million years. They are made up of island arcs, ancient seafloor crust (ophiolites, ultramafics), continental fragments, and submarine deposits (mélange, subduction-zone deposits), as indicated by different colors in the map. (After D. R. Hutchison, "Continental Margins," *Oceanus,* vol. 35, Winter 1992–1993, pp. 34–44; modified from work of D. G. Howell, G. W. Moore, and T. J. Wiley.)

THE GRAND RECONSTRUCTION

One of the great triumphs of modern geology is the reconstruction of events that led to the assembly of the supercontinent Pangaea and its later fragmentation into the continents we know today. Pangaea was the only continent existing at the close of the Paleozoic era, some 250 million years ago. It stretched from pole to pole (see Figure 20.13) and was made up of smaller continents that collided during the Paleozoic era—not the same continents we know today, but continents that existed earlier in the Paleozoic.

The ocean-floor record of these events has been destroyed by subduction, so we must rely on the older evidence preserved on continents to identify and chart the movements of these paleocontinents. Old mountain belts such as the Appalachians of North America and the Urals, which separate Europe from Asia, help us locate ancient collisions of the paleocontinents. In many places, alien rock assemblages reveal ancient episodes of rifting and subduction. Rock types and fossils also indicate the distribution of ancient seas, glaciers, lowlands, mountains, and climatic conditions. A knowledge of ancient climates enables geologists to locate the latitudes at which the continental fragments formed, which in turn helps us to assemble the jigsaw fragments of ancient continents. Like a compass frozen in time, the fossil magnetism of a continental fragment records its ancient orientation and position.

The Assembly of Pangaea

One of the latest efforts to depict the pre-Pangaean configuration of continents is shown in Figure 20.23. It is truly impressive that modern science can recover

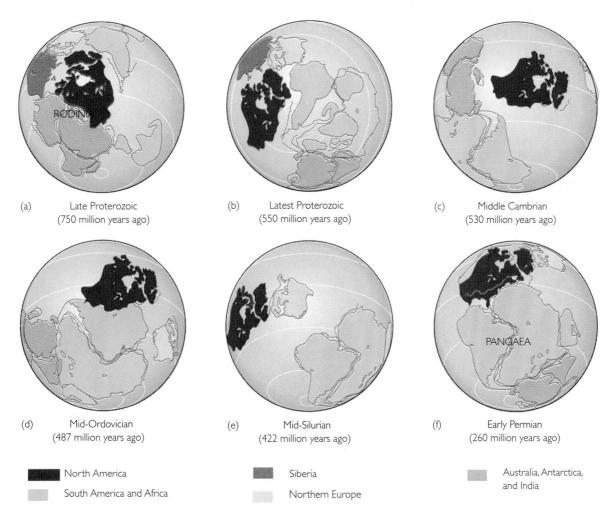

(a) Late Proterozoic (750 million years ago)

(b) Latest Proterozoic (550 million years ago)

(c) Middle Cambrian (530 million years ago)

(d) Mid-Ordovician (487 million years ago)

(e) Mid-Silurian (422 million years ago)

(f) Early Permian (260 million years ago)

North America
South America and Africa
Siberia
Northern Europe
Australia, Antarctica, and India

FIGURE 20.23 Assembly of Pangaea. One reconstruction of the changing positions of the continents from the late Proterozoic to the early Permian. (I. W. D. Dalziel, "Earth Before Pangaea," *Scientific American,* January 1995, pp. 58–63.)

the geography of this strange world of hundreds of millions of years ago. The evidence of rock assemblages, fossils, climate, and magnetism led scientists to portray the breakup of a supercontinent, Rodinia, 750 million years ago, and to follow its fragments over the next 500 million years as they drifted and reassembled into the supercontinent known as Pangaea. Geologists are continuing to sort out more details of this complex jigsaw puzzle, whose individual pieces change shape over geologic time.

The Breakup of Pangaea

More is known about the **breakup of Pangaea** than about the breakup of its predecessor supercontinent, Rodinia, because much of the evidence is still available on the seafloor. Isochrons, for example, provide the time and the directions of the seafloor spreading that rifted the continent and moved its fragments. Figure 20.24 reconstructs this record of the breakup of Pangaea as we now understand it.

Figure 20.24a shows the world as it looked at the close of the Paleozoic era. Pangaea was an irregularly shaped landmass surrounded by a universal ocean called Panthalassa ("all seas"), the ancestral Pacific. The Tethys Sea, between Africa and Eurasia, was the ancestor of part of the Mediterranean. Permian glacial deposits have been found in widely separated areas, such as South America, Africa, India, and Australia. This distribution is explained by postulating a single continental glacier flowing over the South Polar regions of Gondwanaland in Permian time, before the breakup of the continents.

FIGURE 20.24 The breakup of Pangaea. (After R. S. Dietz and J. C. Holden, "The Breakup of Pangaea," *Scientific American,* October 1970, p. 30.)

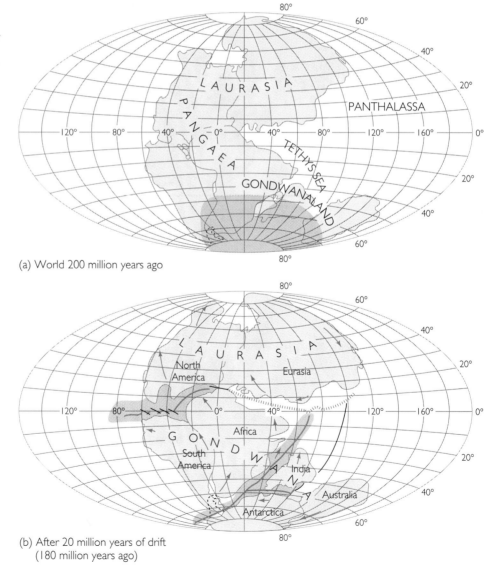

(a) World 200 million years ago

(b) After 20 million years of drift
(180 million years ago)

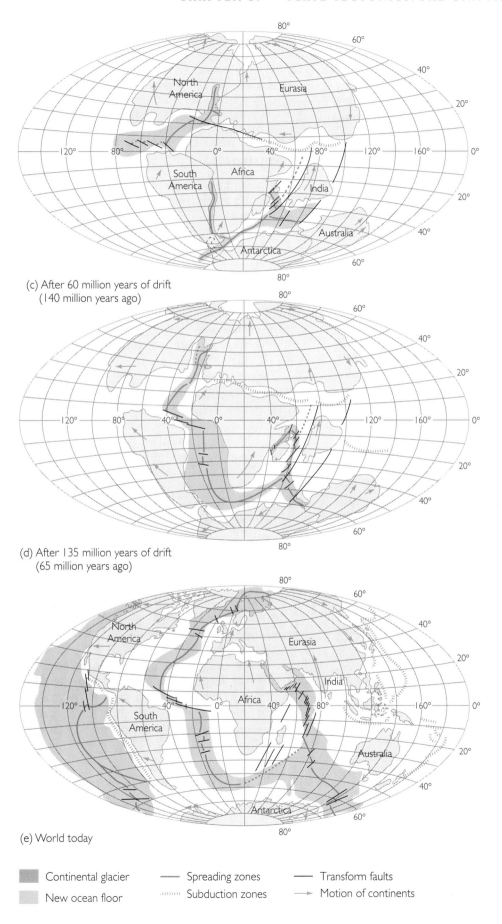

(c) After 60 million years of drift
(140 million years ago)

(d) After 135 million years of drift
(65 million years ago)

(e) World today

Continental glacier —— Spreading zones —— Transform faults

New ocean floor ⠒⠒⠒ Subduction zones → Motion of continents

The breakup of Pangaea was signaled by the opening of rifts from which basalt poured. Rock assemblages that are relics of this great event can be found today in Triassic dikes and sills from Nova Scotia to North Carolina and in the great Palisades sill along the Hudson River. The radioactivity in these rocks tells us that the breakup and the beginning of drift occurred about 200 million years ago.

The geography of the world in the early Jurassic period, after 20 million years of drift, is shown in Figure 20.24b. The Atlantic had partially opened, the Tethys had contracted, and the northern continents (Laurasia) had all but split away from the southern continents (Gondwana). New ocean floor also separated Antarctica-Australia from Africa–South America. India was off on a trip to the north.

By the end of the Jurassic period, 140 million years ago, drift had been under way for 60 million years. The big event at that time was the splitting of South America from Africa, which signaled the birth of the South Atlantic (Figure 20.24c). The North Atlantic and Indian oceans were enlarged, but the Tethys Sea continued to close. India continued its northward journey.

The close of the Cretaceous period, 65 million years ago, saw a widened South Atlantic, the splitting of Madagascar from Africa, and the close of the Tethys to form an inland sea, the Mediterranean (Figure 20.24d). After 135 million years of drift, the modern configuration of continents becomes discernible.

The modern world, produced over the past 65 million years, is shown in Figure 20.24e. India has collided with Asia, ending its trip across the ocean, although it is still pushing northward. Australia has separated from Antarctica. Nearly half of the present-day ocean floor was created in this period.

Figure 20.8 summarizes the relationships of the modern American, African, Eurasian, and Indian plates. Most of the modern Pacific Ocean basin consists of the Pacific Plate on the west side of the East Pacific Rise spreading zone, as can be seen in Figures 20.8b and 20.11. But where is the corresponding plate that should be on the east side of the rise? The implication is that an area equal to most of the present Pacific Ocean has disappeared by subduction under the Americas in the past 130 million years. Some 10,000 km or more of Pacific seafloor may have been thrust under North America. It has been estimated that in the past 180 million years alone, lithosphere equivalent to the surface area of the Earth has been subducted; in a sense, buried in "lithospheric graveyards" in the mantle. These graveyards can be recognized by seismic tomography, which reveals them as regions with relatively higher seismic velocities that reflect their cooler temperatures (see Figure 19.8). They are located in the deep mantle mainly under North and South America, eastern Asia, and other sites adjacent to plate collision boundaries.

Implications of the Grand Reconstruction

Hardly any branch of geology remains untouched by this grand reconstruction of the continents. Economic geologists are using the fit of the continents to find mineral and oil deposits by correlating the formations in which they occur on one continent with their predrift continuations on another continent. Paleontologists are rethinking some aspects of evolution in the light of continental drift, as we mentioned earlier. Geologists are broadening their focus from regional mapping to the world picture, for the concept of plate tectonics provides a way to interpret such geologic processes as sedimentation and orogeny in global terms. For example, one of the longest mountain belts in Earth's history was formed when the ancient continents collided to form Pangaea. The belt was torn apart with the opening of the modern Atlantic Ocean. The modern remnants are now on different continents: the Caledonian mountain belt that runs along the northwest margin of Europe, and the Appalachian belt of North America.

Oceanographers are reconstructing currents as they might have existed in the ancestral oceans to understand the modern circulation better and to account for the variations in deep-sea sediments that are affected by such currents. Scientists are "forecasting" backward in time to describe temperature, winds, the extent of continental glaciers, and the level of the sea as they were in predrift times. They hope to learn from the past so they can predict the future better—a matter of great urgency because of the possibility of greenhouse warming triggered by human activity. What better testimony to the triumph of this once outrageous hypothesis than its ability to revitalize and shed light on so many diverse topics?

THE DRIVING MECHANISM OF PLATE TECTONICS

Up to this point, everything we have discussed might be called descriptive plate tectonics. But a description is not an explanation. We will not fully understand plate tectonics until we can explain why plates move. The International Geodynamics Project, a global program of coordinated experiments, enlisted the efforts of thousands of scientists in many countries to discover the underlying cause of plate motions.

In Chapter 19, we described the mantle as a hot solid capable of flowing like a liquid at a speed of a few centimeters per year, about the rate at which your fingernails grow. The plates of the lithosphere somehow move in response to the flow in the underlying mantle. As is generally the case when we have an abundance of data in search of a theory, many hypotheses have been advanced. Figure 20.25 shows some of these ideas. Some scientists believe that plates are pushed by the weight of the ridges at the zones of spreading or pulled by the weight of the sinking slab at subduction zones (Figure 20.25a). Others hold that the plates are dragged along by currents in the underlying asthenosphere (Figure 20.25b). Hot spots, the jets of hot matter rising from

the deep interior, may spread laterally when they reach the lithosphere, dragging the plates. Downward return flow consists of subducted lithosphere sinking to great depth, possibly as far as the core-mantle boundary (Figure 20.25d).

We agree with those who view the process not piecemeal, but as a highly complex convective flow involving rising hot, partially molten materials; a cool surface boundary layer; and sinking cool, solid materials under a variety of conditions ranging from melting to solidification and remelting. A significant part of the mantle must be involved, because seismic tomography (see Chapter 19) shows slabs to penetrate to great depths (Figure 20.25d). Among the tasks left to the next generation of Earth scientists is the incorporation of such important details as the shapes of plates, the history of their movements, and the formation and growth of continents into an explanation involving convective currents in the interior.

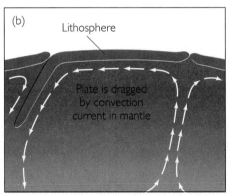
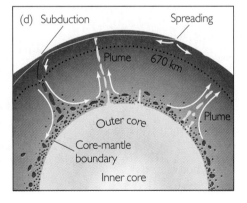
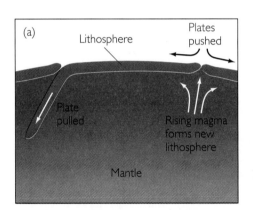
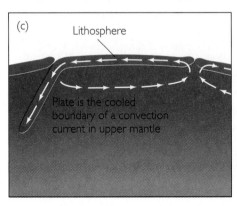

FIGURE 20.25 Four competing driving mechanisms of plate tectonics. (a) The plates are pushed by the weight of the ridges at the centers of spreading or pulled by a cool, heavy, downgoing slab, or both. (b) Plates are dragged by convection currents in the mantle. (c) A plate is the cooled upper boundary region of a circulating convection current in the hot, plastic upper mantle. The circulation includes melting, solidification, and remelting. (d) Jetlike thermal plumes rise from great depth, possibly from the bottom of the mantle, cause hot spots, and spread laterally, dragging the plates. Downward return flow may involve cool slabs of subducted ocean lithosphere sinking to these depths under gravity, fragmenting and partially melting along the way.

SUMMARY

What is the theory of plate tectonics? According to the theory of plate tectonics, the lithosphere is broken into about a dozen rigid moving plates. Three types of plate boundaries are defined by the relative motion between plates: boundaries of divergence, boundaries of convergence, and transform boundaries.

What are some of the geologic characteristics of plate boundaries? In addition to earthquake belts, many large-scale geologic features are associated with plate boundaries, such as narrow mountain belts and chains of volcanoes. Boundaries of convergence are recognized by deep-sea trenches, inclined earthquake belts, mountains and volcanoes, and paired belts of mélange and magmatism. The Andes and the trenches of the west coast of South America are modern examples. Ancient convergences may show as old mountain belts, such as the Appalachians or the Urals. Divergent boundaries typically show as seismic, volcanic mid-ocean ridges (such as the Mid-Atlantic Ridge). Transform boundaries, along which plates slide past each other, can be recog-

nized by their topography, seismicity, and offsets in magnetic anomaly bands.

How can the age of the seafloor be determined? The age of the seafloor can be measured by means of magnetic anomaly bands and the stratigraphy of magnetic reversals worked out on land. The procedure has been verified and extended by deep-sea drilling. Isochrons can now be drawn for most of the Atlantic and for large sections of the Pacific, enabling geologists to reconstruct the history of the opening and closing of these oceans. On the basis of this method and of geological and paleomagnetic data, geologists can sketch the fragmentation of Pangaea over the last 200 million years into the present-day configuration of continents.

What is the mechanism that causes plates to form and move? Although plate motions can now be described in some detail, the driving mechanism is still a puzzle. Almost all hypotheses involve some form of convection in the mantle, with hot matter rising and cool matter sinking, driven by Earth's internal heat.

KEY TERMS AND CONCEPTS

continental drift (p. 506)

Pangaea (p. 506)

seafloor spreading (p. 507)

island arc (p. 510)

isochron (p. 517)

rock assemblage (p. 520)

ophiolite suite (p. 520)

continental shelf deposits (p. 522)

mélange (p. 525)

suture (p. 526)

microplate terrane (p. 526)

breakup of Pangaea (p. 530)

EXERCISES

1. Give a geographic example of each of the three types of plate boundaries.

2. What evidence suggests that Pangaea ever existed?

3. How can the rate of motion between plates be calculated?

4. What is the rate of separation between the North American and Eurasian plates? Between the Nazca and Pacific plates?

5. What kinds of rocks would you expect to find near a boundary of divergence? Near a boundary of convergence?

6. What happens when two continents collide in a boundary of convergence?

7. What are the circumstances that led to the formation of the Himalayas, the Andes, and the Urals?

8. What are the driving forces of plate tectonics?

THOUGHT QUESTIONS

1. What would Earth be like if plate tectonics did not exist?

2. If plate tectonics explains so much of geology, why was it not until the 1960s that most geologists accepted the concept?

3. Would you characterize plate tectonics as a hypothesis, a theory, or a fact? Why?

4. Can you conceive of an experiment that would demonstrate what mechanism provides the driving force of plate tectonics?

SUGGESTED READINGS

Anderson, R. N. 1986. *Marine Geology*. New York: Wiley.

Beloussov, V. V. 1979. Why I do not accept plate tectonics. *EOS* 60:207–210. (See also comments on this paper by A. M. S. Senger and K. Burke on the same pages.)

Courtillot, V., and G. E. Vink. 1983. How continents break up. *Scientific American* (July):42.

Cox, A. (ed.). 1973. *Plate Tectonics and Geomagnetic Reversals*. San Francisco: W. H. Freeman.

Dalziel, I. W. D. 1995. Earth before Pangaea. *Scientific American* (January):58–63.

Hallam, A. 1973. *A Revolution in the Earth Sciences: From Continental Drift to Plate Tectonics*. New York: Oxford University Press (Clarendon Press).

Kearey, P., and F. J. Vine. 1990. *Global Tectonics*. Oxford, England: Blackwell Scientific Publications.

Macdonald, K. C., and others. 1991–1992. Mid-ocean ridges. *Oceanus* 34 (no. 4).

Menard, H. W. 1986. *Islands*. New York: Scientific American Library.

Phillips, J. D., and D. Forsyth. 1972. Plate tectonics, paleomagnetism, and the opening of the Atlantic. *Bulletin of the Geological Society of America* 83 (no. 6):1579–1600.

Sclater, J. G., and C. Tapscott. 1979. The history of the Atlantic. *Scientific American* (June):156–174.

Smith, Deborah K., and J. R. Cann. 1993. Building the crust at the Mid-Atlantic Ridge. *Nature* 365:707–714.

Solomon, S. C., and D. R. Toomey. 1992. The structure of mid-ocean ridges. *Ann. Rev. Earth and Planet. Sci.* 20:329–364.

Uyeda, S. 1978. *The New View of the Earth*. San Francisco: W. H. Freeman.

Wessel, G. R. 1986. *The Geology of Plate Margins*. Geological Society of America, Map and Chart Series MC-59.

INTERNET SOURCES

JOIDES Resolution Page
🌐 **http://ocean.saddleback.cc.ca.us/oceanpub/hotstuff/resolut.html**
A "Cyber-Tour of the World's Most Advanced Deep Sea Drilling Ship: The JOIDES Resolution" is provided by Saddleback Community College in Mission Viejo, California. The operations of the ship and living and working conditions are described and illustrated with text and images.

NGDC Marine Geology and Geophysics Division
🌐 **http://www.ngdc.noaa.gov/mgg/mggd.html**
This division of the National Geophysical Data Center "compiles and maintains extensive databases in both coastal and open ocean areas. Key data types include bathymetry, geophysics (gravity, magnetics, seismic reflection), sediment thickness, data derived from ocean drilling and seafloor sediment and rock samples, and data from the Great Lakes." Links to all these topics are available at this gateway.

Ocean Drilling Program
🌐 **http://www-odp.tamu.edu/**
Texas A&M University is the home for this gateway to information on the Ocean Drilling Program (ODP). Major topic areas are "What Is ODP?" (including a history of the ODP, a drill site map, and links to related sites), "Cruise Information" (including ship schedule, participating in a cruise, and scientific cruise summaries), and "How To" do a variety of tasks, such as access ODP publications and obtain core data.

Plate Motion Calculator
🌐 **http://manbow.ori.u-tokyo.ac.jp/tamaki-html/plate_motion.html**
The Plate Motion Calculator at the Ocean Research Institute of the University of Tokyo models present-day relative motion or absolute motion across any selected plate boundary and provides the returning velocity and direction for any location specified by latitude and longitude.

Plate Tectonics
🌐 **http://volcano.und.nodak.edu/vwdocs/msh/ov/ovpt.html**
The University of North Dakota's Volcano World is the home for this general overview of plate tectonics.

21

Folds exposed in 2-billion-year-old limestone and shale beds. Wopmay orogenic belt, northern Canada. *(Robert S. Hildebrand, Geological Survey of Canada.)*

Deformation of the Continental Crust

I f a movie camera snapped one frame of the Earth's surface every thousand years, almost 50 million years of geologic time could be condensed to a 35-minute film. Geologic movements that stretch over thousands of years, far too slow to be observed in a human lifetime, could be studied and understood. The film would show continents drifting apart to open one ocean and colliding to close another. We could see regions of the Earth being uplifted, tilted, faulted, and folded to create plateaus, mountains, rift valleys, and other geologic structures. In other places, subsidence would lower portions of a continent beneath the sea, forming basins in which sediments would accumulate. Actually, such movies have been made with the use of animation, and some simulations of plate movements are also available for your personal computer. Each frame of these films is a sketch based on an interpretation of the geologic record at a particular time.

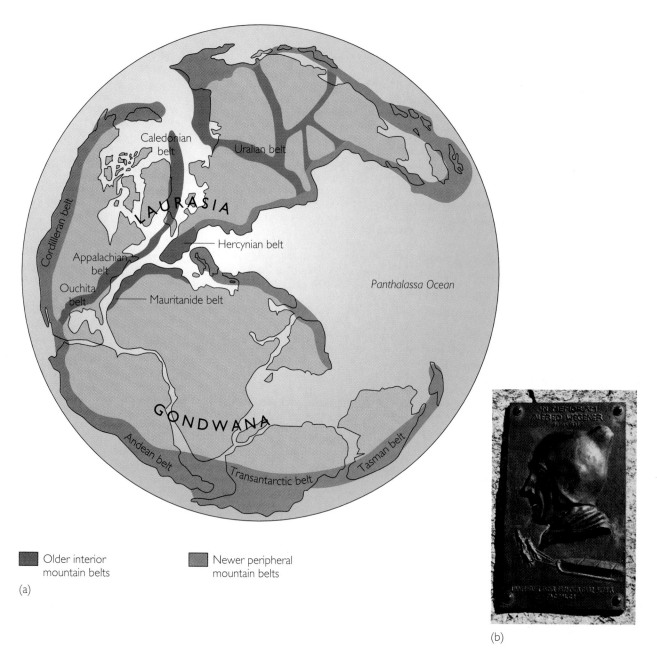

Older interior mountain belts

Newer peripheral mountain belts

(a)

(b)

FIGURE 21.1 (a) Pangaea, the supercontinent, as it was assembled 290 million years ago. Continental collisions produced the interior mountains—the Appalachian-Caledonian, Hercynian, and Uralian belts (purple). These ancient sutures, sites of vanished oceans, are evidenced today by relic mountains with multiple thrusts and folds containing slivers of seafloor preserved in ophiolite remnants. Subduction of ocean floor, still going on today, marks the peripheral Cordilleran, Andean, Tasmanian, and other orogenic belts (orange). (J. B. Murphy and R. D. Nance, "Mountain Belts and the Supercontinent Cycle," *Scientific American,* April 1992, p. 84.) (b) Memorial to Alfred Wegener, considered the grandfather of plate tectonics, near the spot on Greenland where he died during an expedition to investigate continental drift. In 1912, Wegener postulated that all present-day continents had once been assembled into the supercontinent that he called Pangaea. *(Courtesy of Siegfried J. Bauer, Institute for Meteorology and Geophysics, University of Graz, Austria.)*

Most of Earth's surface consists of oceans whose underlying crust is younger than 200 million years. Plate-tectonics theory explains that ocean basins are being continuously created at mid-ocean ridges, where new lithosphere is formed from melt that rises from the mantle below. The lithosphere spreads, cools, and is destroyed where it sinks at subduction zones. The existing seafloor, therefore, provides a record of only about 4 percent of Earth's 4.5-billion-year history. Only the continents have rocks as old as 4 billion years, and it is there that we must turn to look back over most of geologic time—but not all of it. It is generally believed that any relic of crust that might have formed in the first 500 million years of Earth's history would have been destroyed by the intense meteoritic bombardment that was going on throughout the solar system at that time.

The 4-billion-year span of geologic evolution recorded in the continental crust is complex, but we are beginning to understand it better. Geologists use concepts from plate tectonics to interpret old geologic features, such as eroded mountain belts and ancient rock assemblages, in terms of closing ocean basins and colliding continents (Figure 21.1). They now appreciate that deformation takes place in a rigid skin—the lithosphere, only about 100 km thick, paper thin compared to the 6300 km of malleable mantle and core beneath it. In recent years geologists have also learned that uplift of mountain ranges can affect world climate, change the chemistry of the oceans, and influence the location of oil deposits and other resources.

This chapter will look at some 4 billion years of deformation of the continental crust. The geologic fabric of continents exhibits a pattern: eroded remnants of very old deformed rocks in the interior, more recent deformation in the mountain systems closer to margins. Mountain building occurs where plates collide, as sediments of the continental margin are thrown into a series of folds and faults. A subducted plate melts and magma rises into the deformed belt. Plate movements bring in foreign fragments and weld them to the deformed belt. Upward and downward movements within the continent create interior basins and domes and lift old, worndown mountains high again. Offshore, downward movements create basins on the continental shelves.[1]

---

[1]We draw on the work of B. C. Burchfiel of MIT, one of the geologists who is reinterpreting the geology of the continental crust in the framework of plate tectonics.

SOME REGIONAL TECTONIC STRUCTURES

About one-third of the Earth's surface is covered by continents. The rocks that make up the continental crust can be grouped in two distinct categories:

1. *Undeformed sedimentary rocks:* The veneer of sedimentary rocks deposited in an orderly process and not yet significantly deformed; and

2. *Deformed rocks:* The deformed regions of sedimentary, igneous, and metamorphic rocks that have been subjected to intense crustal forces during various geologic periods.

Most of the continental crust, either exposed rocks or basement buried beneath the layered sedimentary (or sometimes volcanic) cover, falls into the second category; that is, rocks that have been deformed and altered by crustal forces, such as the rocks shown in the photograph at the beginning of this chapter. Thus, **orogeny**—the mountain-building processes of folding, faulting, magmatism, and metamorphism—and the evolution of continents are intimately related.

The geologic fabric of the continents is not random. Rocks that reveal the most ancient episodes of deformation tend to be found in the interior of continents, which are now generally stable and eroded flat (see Figure 16.4, showing the flattened interior of the continental U.S.). External to these old terrains are the more recently active mountain belts, where most of the present-day mountain systems are found. They occur as long, narrow topographic features at the margins of continents, such as the Cordillera, which runs down the western edges of North and South America, and the Appalachian belt, which trends southwest to northeast on the eastern margin of North America. Similarly, the Alpine-Himalayan chain runs across the margins of southern Europe and Asia.

The continents have grown both by the aggregation of small and large fragments in collisions (terrane accretion; Chapter 20) and by the addition of new materials during subduction of ocean plates. The key to deciphering the process by which the ancient crust was deformed lies in the younger peripheral belts, because there much of the record of deformation is still preserved uneroded in the rocks.

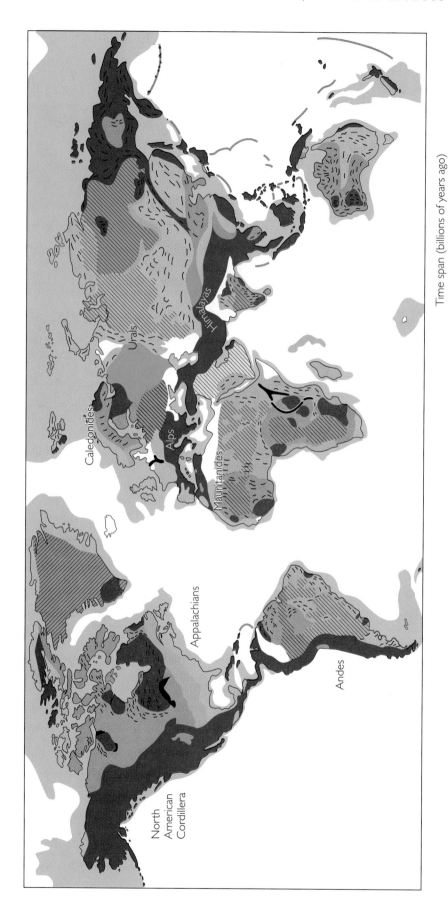

FIGURE 21.2 Most of the rocks that make up the continental crust underwent episodes of deformation during different geologic periods. In places the rocks are present beneath sedimentary or volcanic rock or ice (diagonal lines). Colors show the time spans of major deformational episodes. The oldest deformed rocks tend to be in the interiors of continents; younger and more recently deformed rocks are found near margins. Also shown are continental rocks beneath the oceans in continental margins and plateaus (blue), volcanic arcs (red), and continental rifts (black). Short curved lines mark structural trends. (After B. C. Burchfiel, "The Continental Crust," *Scientific American*, September 1983, p. 30.)

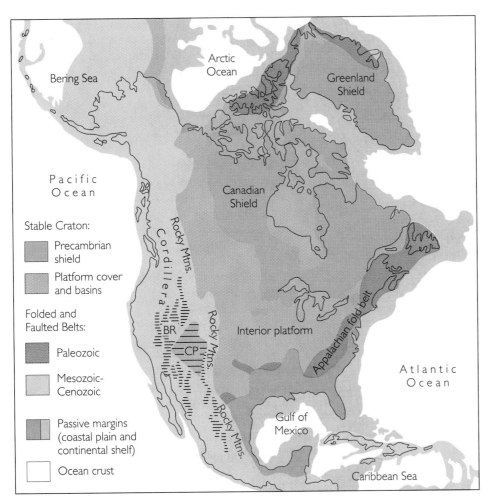

FIGURE 21.3 Major tectonic features of North America: Canadian Shield; interior platform; Cordilleran orogenic belt, including Basin and Range (BR and short dashes); Colorado Plateau (CP); Appalachian fold belt; coastal plain and continental shelves of passive margins. (After A. W. Bally, C. R. Scotese, and M. I. Ross, *Geology of North America,* vol. A, Boulder, Colo., Geological Society of America, 1989.)

The map in Figure 21.2 summarizes this world pattern of deformed continental rocks, colored according to the geologic period in which the deformation occurred.

In the following sections, it will be useful to refer to Figure 16.4 in order to see the location and physiography (surface landforms) of the regional tectonic structures of the United States, when these are discussed.

THE STABLE INTERIOR

Cratons are the extensive, flat, tectonically stable interiors of the continents, composed of ancient rocks that underwent intense deformational episodes in Precambrian time and have been relatively quiescent since then. Typically, the cratons include large areas, called **shields,** that consist of very old, exposed crystalline basement rocks. The Canadian Shield (blue in Figure 21.3) is typical. It is dominated by granitic and metamorphic rocks, such as gneisses, together with highly deformed metamorphosed sedimentary and volcanic rocks. These rock assemblages indicate intense mountain-building episodes in Precambrian time followed by a long period of stability, indicated by a lack of evidence of more recent deformation. This primitive region contains one of the oldest records of geologic history, much of it still known only in outline. It is famous for

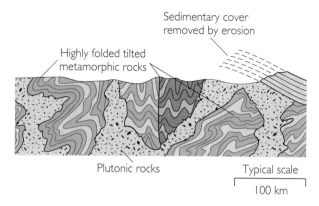

FIGURE 21.4 The highly deformed and metamorphosed rocks in the Canadian Shield indicate that intense orogenic (mountain-building) episodes took place in Precambrian times, before conditions stabilized.

major deposits of iron, gold, copper, and nickel. Figure 21.4 is an idealized cross section of the Canadian Shield. Other shields are found in Scandinavia and Finland, Siberia, central Africa, Brazil, and Australia.

South of the Canadian Shield is a sediment-covered, almost level region, called a **platform** (the "interior platform" in Figure 21.3). It forms the central stable region and the Great Plains of the United States. This platform is in a sense a sub-surface continuation of the Canadian Shield, for it contains similar Precambrian basement rocks, but here they are covered by a veneer of Paleozoic sedimentary rocks typically less than about 2 km thick.

The North American platform sediments were laid down on the deformed and eroded Precambrian basement under a variety of conditions. The rock assemblages indicate sedimentation in extensive shallow inland seas (marine sandstones, limestones, shales, deltaic deposits, evaporites) and deposition on alluvial plains or in lakes or swamps (nonmarine sediments, coal deposits). Many of the continent's deposits of uranium, coal, oil, and gas are contained in the platform's sedimentary cover.

Within the platform broad **sedimentary basins** are defined by roughly oval depressions where the sediments are thicker than on the surrounding platform. The Michigan Basin, a circular area of about 500,000 km² that covers much of the lower peninsula of Michigan, subsided throughout much of Paleozoic time and received sediments more than 3 km thick in its central, deepest part. The sandstones and other sedimentary rocks, laid down under quiet conditions, have remained unmetamorphosed and only slightly deformed to this day. The sedimentary

formations in this basin have been likened to a pile of saucers. Some geologists believe that basins subsided in an episode of stretching and thinning of the continental lithosphere.

OROGENIC BELTS

The details of orogeny preserved in the younger mountain belts helps us to understand the incomplete record found in the eroded remnants of Precambrian mountains that now make up the basement rocks of shields and platforms. Fringing the great stable interior of the North American continent, for example (see Figure 21.3), are younger orogenic belts, regions that were deformed by folding and faulting and were subjected to plutonism and metamorphism at various times in the Paleozoic, Mesozoic, and Cenozoic eras.

Most geologists now believe that orogenies—that is, periods of mountain building—involve plate collisions. Most of the motion of collision is absorbed by the subduction of one of the plates. However, when two continental plates collide, a basic tenet of plate tectonics—the rigidity of plates—must be modified. Continents are buoyant fragments of crust embedded atop the plates on which they ride. Both colliding fragments of continental crust tend to float and resist being subducted with the plates that carry them. Instead, the collisional forces are so great that the continental crust loses its rigidity, deforms, and breaks in a variety of ways. By a combination of intense folding and faulting, the crust absorbs much of the motion of the collision within a zone of intense deformation extending hundreds of kilometers into the continent. The faulting breaks the crust into multiple thrust sheets up to 20 km thick, stacked one above the other by subhorizontal (low-angle) thrusting (Figure 21.5). The individual thrust sheets are often themselves deformed and metamorphosed. Wedges of continental shelf sediments can be detached from the basement on which they were deposited and thrust inland. Figure 21.6,

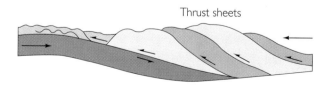

FIGURE 21.5 When plates bearing continents collide, the continental crust can break into multiple thrust fault sheets stacked one above the other.

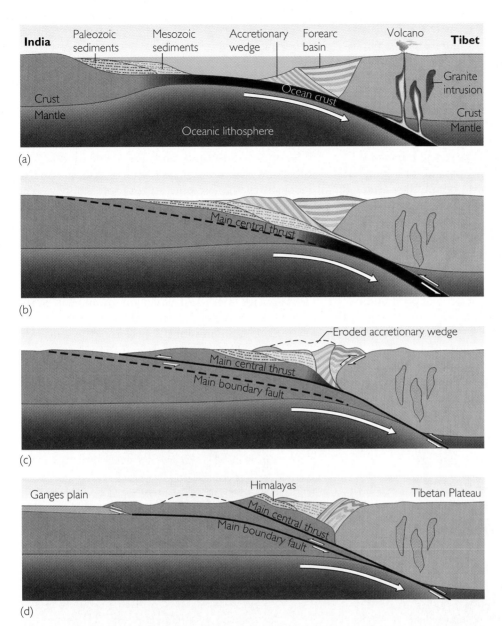

FIGURE 21.6 The Himalayan orogeny, simplified and vertically exaggerated. (a) Subduction of the Indian Plate under the Eurasian Plate began some 60 million years ago. The Indian Plate moved northward, carrying the Indian subcontinent with its leading edge of Paleozoic and Mesozoic continental shelf sediments. Magma rising from the subducted plate formed granitic intrusions and volcanoes, thickening the crust. An accretionary wedge was formed from the pile of sediments and oceanic crust scraped off the descending plate. A forearc basin formed behind the wedge, trapping sediments eroded from Tibet. (b) The Indian subcontinent collided with Tibet sometime between 40 and 60 million years ago. India was too buoyant to be subducted into the mantle, and the Indian crust broke in a thrust fault—the main central thrust. (c) As the collision continued, with the motion taken up along the thrust fault, a slice of Indian crust and shelf sediments was stacked onto the oncoming subcontinent. The accretionary wedge and the forearc basin sediments were thrust northward onto Tibet. (d) About 10 to 20 million years ago a new thrust fault developed—the main boundary fault. A second slice of crust was stacked onto India, lifting up the first slice. The two overthrust slices make up the bulk of the Himalayas, including the Paleozoic sediments that are found in the peaks. (After P. Molnar, "The Structure of Mountain Ranges," *Scientific American,* July 1986, p. 70.)

depicting the orogeny that formed the Himalayas, shows most of these features. Figure 21.7 is a photograph of thrust faulting in the Tibetan Himalayas. The Himalayan orogeny is still going on, with India moving into Asia at a rate of about 3 to 5 cm per year (see Feature 21.1).

We saw in Chapter 20 that a significant fraction of an orogenic belt can be made up of a succession of displaced or microplate terranes, fragments of crust that may have traveled thousands of kilometers from distant parts of the world before they accreted to the orogenic belt. These microplate terranes can be fragments of oceanic crust (ophiolites), island arcs, or pieces of continental crust carried by the subducting plate but driven up over the edge of the continent rather than down into the mantle (see Figure 20.22).

Orogeny thickens the continental crust and extends it in several ways, in addition to stacking of multiple thrust sheets. One is by microplate accretion. Another is by the addition of batholiths and volcanic rocks derived from melting of the mantle. Continents can grow by a succession of orogenies over geologic time. In Figures 21.1, 21.2, and 21.3 we can see this growth in the pattern of younger orogenies fringing older ones.

Case History: The Appalachians

The old, eroded Appalachian Mountains are a classical fold-and-thrust belt that extends along eastern North America from Newfoundland to Alabama. The rock assemblages and structures found there today are the primary data geologists have used to reconstruct the tectonic history of the belt. The Appalachians consist of a sequence of physiographic subregions. Moving eastward from the mildly deformed plateaus in the west, we encounter regions of increasing deformation (Figure 21.8).

1. *Valley and Ridge province.* Thick Paleozoic sedimentary rocks laid down on an ancient continental shelf were folded and thrust to the northwest by compressional forces from the southeast. The rocks show that deformation occurred in three orogenic episodes, one at the end of the Ordovician, one at the end of the Devonian, and one in Permian-Carboniferous time.

FIGURE 21.7 The Himalayan Mountains in east-central Tibet. The dark-colored rocks forming the crest of the ridge are early Cenozoic sedimentary rocks thrust over younger (light-colored) Cenozoic sedimentary rocks. Overthrusting of several kilometers reflects a shortening of the crust in Tibet, after India collided with Eurasia. *(Peter Molnar.)*

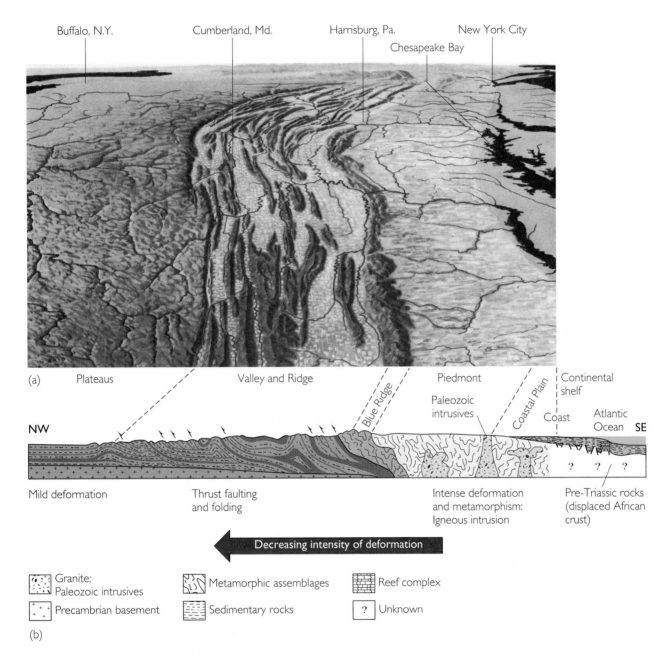

FIGURE 21.8 The Appalachian Mountain region of the south and central eastern United States, with an aerial view to the northeast and an idealized cross section. The major physiographic subregions from west to east in order of increasing intensity of deformation are the following: the Appalachian plateaus, slightly uplifted, with mildly deformed sediments; the Valley and Ridge province, consisting of long mountain ridges and valleys and erosional features in a fold and thrust belt of sedimentary rocks; the Blue Ridge, consisting of high mountains of Precambrian crystalline rock; the Piedmont, rough to gentle hilly terrain of metamorphosed sedimentary and volcanic rock; and the coastal plain and continental shelf, a terrain of low hills grading to flat plains on sediments extending offshore. Multiple plate collisions in the Paleozoic folded the sediments of the Valley and Ridge province and transported them towards the interior of the continent along great thrust sheets. The Blue Ridge and Piedmont sheets were thrust westward over younger sediments. The modern coastal plain and shelf developed after the Triassic-Jurassic splitting of North America from Africa and the opening of the modern North Atlantic Ocean. (After S. M. Stanley, *Earth and Life Through Time,* New York, W. H. Freeman, 1985.)

2. *Blue Ridge province.* These eroded mountains are composed largely of Precambrian and Cambrian crystalline rock, showing much metamorphism. The Blue Ridge rocks were not intruded and metamorphosed in place, but were thrust as sheets over the sedimentary rocks of the Valley and Ridge province (see Figure 21.8b).

3. *Piedmont.* This region contains Precambrian and Paleozoic metamorphosed sedimentary and volcanic rocks intruded by granite, all now eroded to low relief. Volcanism began in the late Precambrian and continued into the Cambrian. The Piedmont was thrust over Blue Ridge rocks along a major thrust fault, overriding them to the northwest. At least two episodes of deformation are evident, one at the end of the Devonian, the other in the early Carboniferous.

4. *Coastal Plain.* Relatively undisturbed sediments of Jurassic age and younger are underlain by rocks similar to those of the Piedmont. The continental shelf is the offshore extension of the Coastal Plain.

Figure 21.9 shows a possible plate-tectonic reconstruction of southern Appalachian history consistent with present-day geologic features. In late Precambrian time (a), a supercontinent split apart, leaving what was to be North America on one side of the rift, Africa on the other, and the ancestral Atlantic Ocean in between, which grew as the plates separated. A continental fragment that would become the Blue Ridge and part of the Piedmont was left between the diverging continents at this time. A marginal sea separated the fragment from North America. Continental shelf sediments were deposited along the margins of the receding plates.

In early Cambrian time (b), plate convergence began to close the ancestral Atlantic. (Africa is out of view to the right of Figure 21.9b, on the other side of the Atlantic.) The ancient Atlantic seafloor was subducted under North America and the continental fragment, and an island arc developed adjacent to the subduction zone. The fragment was somewhere between the island arc and North America.

Convergence continued from the middle Cambrian (21.9c) through the Ordovician to early Silurian (21.9d), when the continental fragment collided with North America. The fragment was thrust as sheets of crystalline rock over the younger North American shelf sediments to become what are now the Blue Ridge and part of the Piedmont.

A second orogenic episode occurred near the close of the Devonian (e), when the island arc collided with and was sutured to North America.

Evidence of this event is seen today in the presence in the Piedmont of metamorphosed sedimentary and volcanic rocks of the type characteristic of an island arc. This collision pushed the Blue Ridge and Piedmont sheets farther west. The Atlantic continued to close. (Africa reappears in Figure 21.9e, on its way to a collision with North America.)

The Atlantic closed completely in Carboniferous and Permian time, when Africa slammed into North America to form the supercontinent Pangaea. The extensive folding and thrusting in the Valley and Ridge province occurred during this culminating orogeny. The accumulated thrusting of all three deformational episodes may have pushed the Blue Ridge and Piedmont sheets more than 250 km westward over the continental shelf sediments of ancestral North America.

The most recent chapter in this story—the breakup of Pangaea and the opening of the modern North Atlantic—began some 200 million years ago in Triassic-Jurassic times, when rifting occurred east of the Piedmont. Africa split away from North America, leaving behind a piece of itself that today is the pre-Triassic rocks (see Figure 21.8b) under the Coastal Plain and continental shelf (21.9f). The present-day shelf deposits began to develop on the margins of North America and Africa as the plates once more began to drift apart. Although the details of Figure 21.9 will change as more knowledge is obtained, most experts agree that as a result of the multiple collisions and rifting over a span of some 500 million years, North America grew by the accretion of a continental fragment, an island arc, a piece of Africa, other foreign fragments, and a modern continental shelf formed from sediments eroded from the landmass to the west. Compare the reconstruction in Figure 21.9f and the modern geologic cross section of Figure 21.8b to see how the plate-tectonic events relate to the rock assemblages and folding and thrusting found today in the southern Appalachians.

Similar sequences of successive collisions—in which the rock layers of the continental crust were compressed and deformed by folding and breaking into a stack of thrust sheets—were also important in the orogenies that formed the North American Cordillera; the Alps; the Urals of Russia; the Caledonides of Scotland; and, as we have seen, the Himalayas. The Urals, separating Europe and Asia, represent an ancient suture, or boundary line, between colliding continental plates in the middle of a continent that has not rifted since it formed. The Mauritanide mountain belt of western Africa appears to be the mirror image of the Appalachians —not a surprising result for the other side of a collision boundary (see Figure 21.1).

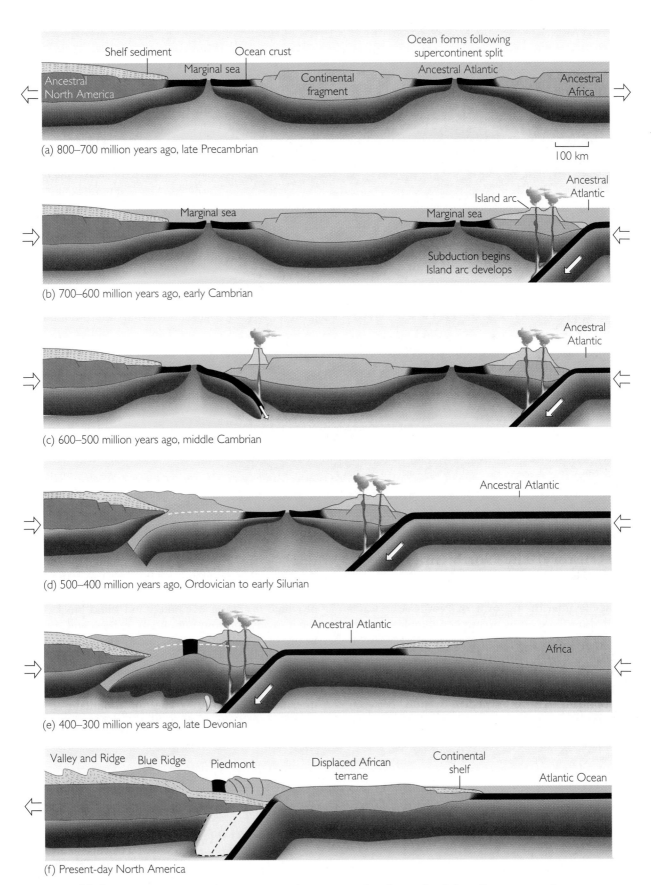

(a) 800–700 million years ago, late Precambrian

Shelf sediment Ocean crust Ocean forms following supercontinent split

Ancestral North America Marginal sea Continental fragment Ancestral Atlantic Ancestral Africa

100 km

(b) 700–600 million years ago, early Cambrian

Marginal sea Marginal sea Island arc Ancestral Atlantic

Subduction begins Island arc develops

(c) 600–500 million years ago, middle Cambrian

Ancestral Atlantic

(d) 500–400 million years ago, Ordovician to early Silurian

Ancestral Atlantic

(e) 400–300 million years ago, late Devonian

Ancestral Atlantic Africa

(f) Present-day North America

Valley and Ridge Blue Ridge Piedmont Displaced African terrane Continental shelf Atlantic Ocean

FIGURE **21.9** The Appalachian orogenic belt can be interpreted as the result of plate divergences and convergences from late Precambrian to the present. In the southern Appalachians, successive collisions resulted in subduction and overthrusting of the continental margin of North America. (After F. A. Cook et al., "Thin-Skinned Tectonics in the Crystalline Southern Appalachians," *Geology*, 7, 1979, pp. 563–567.)

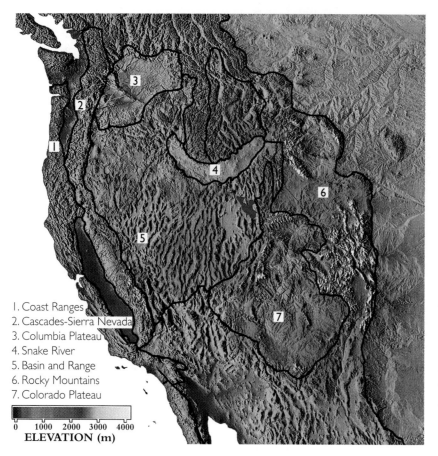

1. Coast Ranges
2. Cascades-Sierra Nevada
3. Columbia Plateau
4. Snake River
5. Basin and Range
6. Rocky Mountains
7. Colorado Plateau

0   1000   2000   3000   4000
ELEVATION (m)

FIGURE 21.10 Color-shaded relief map of the western United States. Computer manipulation of digitized elevation data produced an image in which the major structural provinces and tectonic history of the area are clearly visible, as if illuminated by a light source low in the west. The long, linear ridges of the Basin and Range contrast with the somewhat smoother Colorado Plateau and the rugged and complex relief of the Rocky Mountains and the Cascade and Coast ranges. The low, flat floor of the Snake River is punctuated by individual volcanic cones. The Columbia Plateau stands out as a relatively flat, low-lying basin between the Cascades and the Rocky Mountains. The smooth, sea-level floor of the Great Valley of California is bounded on the east by the Sierra Nevada and on the west by the conspicuous, linear trend of the San Andreas fault system. *(Courtesy of David Simpson, IRIS Consortium.)*

Case History: The North American Cordillera

The stable interior platform of North America is bounded on the west by a younger complex of several types of orogenic zones (Figure 21.10). This is the region of the North American Cordillera, a mountain belt extending from Alaska to Guatemala, which contains some of the highest peaks on the continent. Across its middle section, between San Francisco and Denver (Figure 21.11), the Cordilleran system is about 1600 km wide and includes several contrasting physiographic provinces: the Coast Ranges along the Pacific Ocean; the lofty Sierra Nevada and Cascades; the Basin and Range province (a region of faulted and tilted blocks that form many narrow mountain ranges and valleys from the California-Nevada border to western Utah); the high tableland of the Colorado Plateau; and the rugged Rocky Mountains, which end abruptly at the edge of the Great Plains on the stable interior. Some of the types of mountains found in orogenic belts such as the Cordillera and the Appalachians are depicted in Figure 21.12.

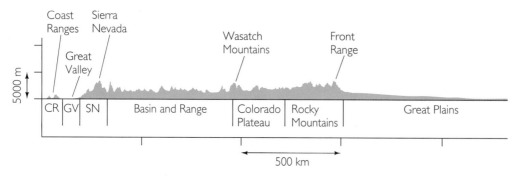

FIGURE 21.11 Cross section of the Cordillera from San Francisco to Denver. *(Modified from David Simpson.)*

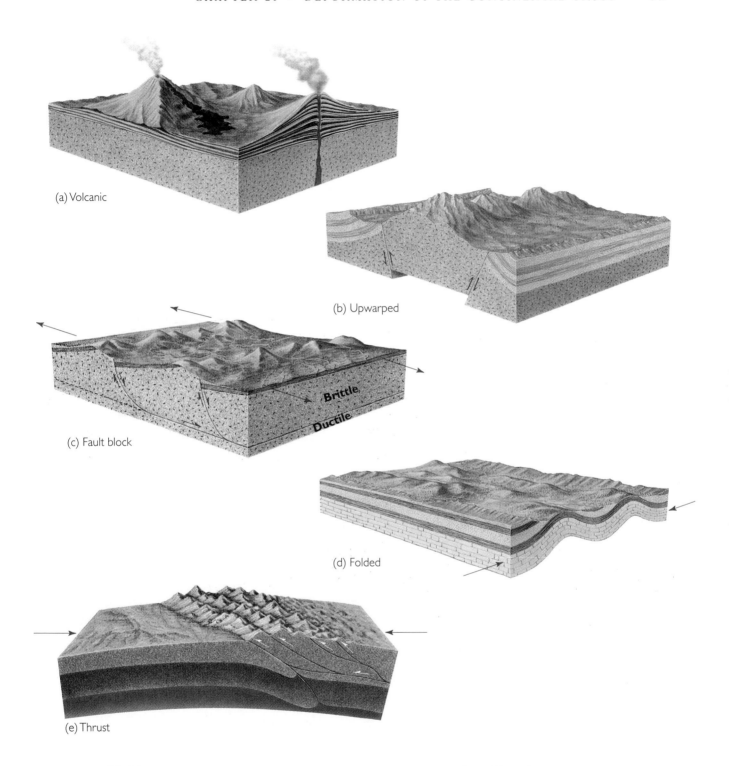

(a) Volcanic

(b) Upwarped

(c) Fault block

Brittle

Ductile

(d) Folded

(e) Thrust

FIGURE 21.12 Mountain structures vary in form and origin. (a) Mountains formed by volcanic action, such as the Cascades, extending from northern California through western Oregon and Washington into British Columbia. (b) Upwarped mountains, with reverse faults, such as the Front Range of the Rocky Mountains in Colorado. (c) Mountain ranges formed from tilted fault blocks, bounded by normal faults, with valleys between ranges, such as occur in the mountains of the Basin and Range province. (d) Mountains resulting from folded layers of rock, such as the Appalachian fold belt. (e) Mountains formed from stacked or overlapping thrust faults, such as occur at boundaries of plate convergence. Examples are the Himalayas and the southern Appalachians.

21.1   INTERPRETING THE EARTH

The Collision Between India and Eurasia

Some 50 million years ago the Indian subcontinent, riding on the subducting Indian plate, collided with Eurasia. The collision slowed India's advance, but the plate continued to drive northward. So far, India has penetrated about 2500 km into Eurasia.

Since continental crust does not subduct readily, geologists must explain what happened to a piece of crust that was as wide as India and 2500 km long. Using earthquake data and satellite photographs, Peter Molnar and Paul Tapponier prepared this tectonic map and offered a startling proposal to account for the vast area of Eurasia displaced by the collision.

The Himalayas, the world's highest mountains, were formed from overthrust slices of the old north portion of India, stacked one atop the other (Figures 21.6b and 21.6c). This process took up some of the compression.

Horizontal compression in Tibet found some relief in vertical expansion; this process contributed to the uplift of the high Tibetan Plateau. Compression by thrust faulting is the pattern of deformation in the Tien Shan Mountains. But these and other zones of compression could account for perhaps only half of India's penetration into Eurasia. Molnar and Tapponier account for the other half by suggesting that China and Mongolia were pushed eastward, out of India's way, like toothpaste squeezed from a tube. The movement took place along the enormously long Altyn Tagh fault and other strike-slip faults shown on the map. The mountains, plateaus, faults, and great earthquakes of Asia, thousands of kilometers from the Indian-Eurasian suture, are thus influenced by the continuing collision of the two continents.

Key to map

┈┬┬┬┬┬ Visible normal fault	←●→ Earthquake indicating strike-slip faulting	Uplift resulting from compression
←●→ Earthquake indicating normal faulting	⇄ Strike-slip fault	Extension zones
┬┬┬┬ Visible thrust fault	⟋ Visible faults with undetermined slip direction	· Small earthquakes
●→ Earthquake indicating thrust faulting	⫽⫽⫽ Folds	• Large earthquakes
●→ Earthquake indicating thrust faulting, showing direction of overthrust sheet		

The history of the Cordillera is a complicated one, with details that vary along its length. It is a story of the interaction of the Pacific Plate, a completely subducted plate called the Farallon Plate (see Figure 19.8), and the North American Plate over the past billion years. Several thousand kilometers of the Pacific Plate were subducted eastward under North America over this time span. All of the phenomena of plate collisions occurred: the accretion of continental and oceanic fragments (some geologists now suspect that as much as 70 percent of the Cordilleran orogenic belt may consist of microplate terranes); thrusting and folding of deep-water and shelf deposits; volcanism and the intrusion of granitic plutons; metamorphism; and the reworking or overprinting of features of one episode of

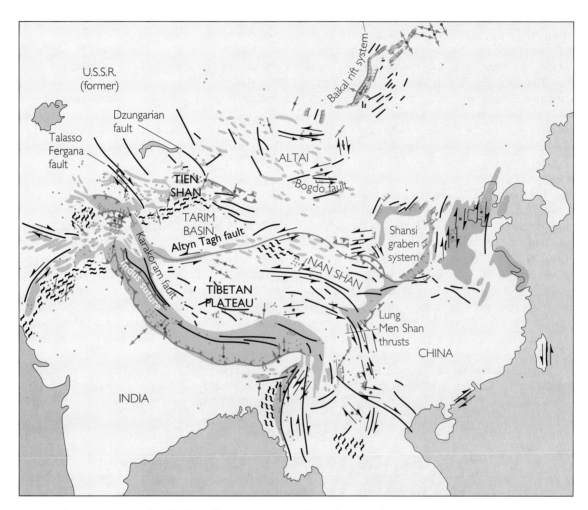

Tectonic features associated with the collision between India and Eurasia: large-scale faulting, uplift, and earthquakes. (After P. Molnar and P. Tapponier, "The Collision Between India and Eurasia," *Scientific American,* April 1977, p. 30.) See key to map at left.

orogeny by later orogenic events. Together these processes have led to the juxtaposition of deformed rock assemblages that vary in age and origin and the overprinting that characterize the Cordillera.

The Cordilleran system is topographically higher and more extensive than the Appalachians because its main orogeny was more recent, having occurred in the last half of the Mesozoic era and early Tertiary period. The younger an orogeny, the less time for erosion to wear it down. The form and height of the Cordillera we see today are also manifestations of even more recent events in the Tertiary and Quaternary periods, over the last 15 or 20 million years, which resulted in **rejuvenation** of the mountains; that is, the mountains were raised again and brought back to a more youthful stage. At that time, the central and southern Rockies attained much of their present height as a result of a broad regional upwarp, or uplift (see Figure 21.12b). The Rockies were raised 1500 to 2000 m as Precambrian basement rocks and their veneer of later-deformed sediments were pushed above the level of their surroundings. Stream erosion was accelerated, the mountain topography sharpened, and the canyons deepened. Other examples of rejuvenation by upwarping include the Adirondacks of New York, the Labrador Highlands, and the mountains of Scandinavia and Finland. Rejuvenated deformed belts are also seen in the modern topography of the Alps, the Urals, and the Appalachians. The cause of rejuvenation is a matter of current debate among geologists.

The late structural imprinting responsible for the present-day features of the Basin and Range province and the tilted uplift of the Sierra Nevada of California (Figure 21.13) also occurred in the Tertiary and Quaternary periods, during a spectacular episode of faulting in a region extending southeast from southern Oregon to Mexico and encompassing Nevada, western Utah, and parts of eastern California, Arizona, New Mexico, and western Texas. Thousands of nearly vertical faults sliced the crust into innumerable upheaved and downdropped blocks, forming hundreds of nearly parallel **fault-block mountain ranges** (see Figure 21.12c) separated by alluvium-filled rift valleys bounded by normal faults. Some geologists think that the explanation is to be found in a crustal stretching or extension caused by flow in the mantle below. Other fault-block mountains are the Wasatch Range of Utah, the Teton Range of Wyoming, and the tilted edges of the rift valleys of east Africa and the Dead Sea–Jordan Valley rift of Israel.

The Colorado Plateau seems to be an island of the central stable region, cut off from the interior by the Rocky Mountain orogenic belt. Since the late Precambrian it has been a stable area, experiencing no thick basin deposits and no major orogeny. Its rock formations, exposed in the Grand Canyon (see Feature 9.1), reflect mainly up-and-down movements.

COASTAL PLAIN AND CONTINENTAL SHELF

The Atlantic coastal plain and the continental shelf, its offshore extension (see Figure 21.3), began to develop in the Triassic period with the rifting that preceded the opening of the modern Atlantic Ocean. The rift valleys formed basins that trapped a thick series of nonmarine sediments. As these deposits were accumulating, they were intruded by basaltic sills and dikes. The Connecticut River valley and the Bay of Fundy are such sediment-filled rift valleys (Figure 21.14).

FIGURE 21.13 The Sierra Nevada (skyline) viewed from the Panamint Mountains, California. The Sierra Nevada fault scarp (steep cliffs and slopes formed by faulting) is in the distance. In the middle ground are the Argus Mountains, formed by a late Quaternary upfaulting of early Quaternary basalts. A small fault-block valley, far side up, cuts the alluvium of Panamint Valley (below the center of the picture). *(W. B. Hamilton, U.S. Geological Survey.)*

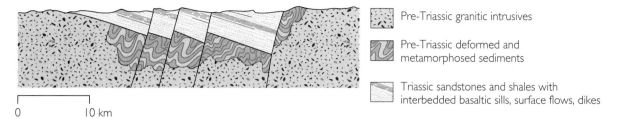

Pre-Triassic granitic intrusives

Pre-Triassic deformed and metamorphosed sediments

Triassic sandstones and shales with interbedded basaltic sills, surface flows, dikes

0   10 km

FIGURE 21.14 Triassic rift valleys of Connecticut. Tilted fault blocks formed rift valleys. Nonmarine sediments were trapped in basins formed by rift valleys. Basaltic flows intruded and covered these deposits.

In the early Cretaceous period, the deeply eroded sloping surface of the Atlantic coastal plain and continental shelf began to subside and to receive sediments from the continent. Cretaceous and Tertiary sediments up to 5 km thick filled the slowly subsiding trough, and even more material was dumped into the deeper water of the continental rise (see Figure 20.17). This living offshore sedimentary basin (it's still receiving sediments) may be the site of some future orogeny. If the present stage of opening of the Atlantic is reversed some millions of years from now, the sediments in this basin will be folded and faulted in the same process that produced the Appalachians.

The coastal plain and shelf of the Gulf of Mexico are continuous extensions of the Atlantic ones, interrupted only briefly by the Florida peninsula. The Mississippi, Rio Grande, and other rivers that drain the interior of the North American continent have delivered sediments to fill a trough some 10 to 15 km deep running parallel to the coast (Figure 21.15). The Gulf coastal plain and shelf are a rich reservoir of petroleum and natural gas. The Atlantic shelf is now being actively explored for these and other resources.

REGIONAL VERTICAL MOVEMENT

So far our discussion of crustal movements has emphasized orogeny, which originates in plate collisions and involves compressive deformation (folding, thrust faulting), intrusion of magma, volcanism, and metamorphism. All over the world, however, sedimentary rock sequences record another kind of history: gradual downward and upward movements of the crust without significant deformation, called **epeirogeny.**

Although many of the vertical movements are connected with orogeny, epeirogenic movements are slow and intermittent and commonly affect large regions without producing extensive folding or faulting. Fossil trees and other plants embedded in coal deposits now mined deep in the Earth tell of earlier times when they grew on the surface. Great thicknesses of sediments deposited on the seafloor and then buried hundreds or thousands of meters beneath it have been raised to hundreds and thousands of meters above sea level, where they are now found. These deposits give evidence of continued slow subsidence during sedimentation. In many cases, later elevation to their present position above

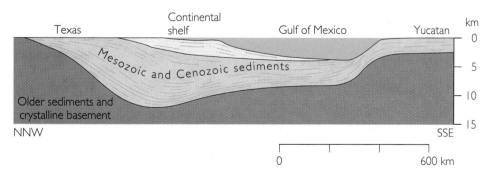

FIGURE 21.15 Inferred thicknesses of Mesozoic and Cenozoic sediments in the Gulf of Mexico from Texas to Yucatan. Maximum thickness near the Texas coast may exceed 10 km. (After P. B. King, *The Evolution of North America,* Princeton, N.J., Princeton University Press, 1977.)

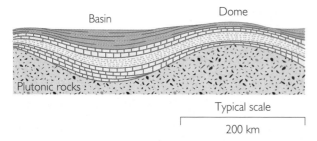

FIGURE 21.16 Idealized section of basin and dome structures. Basins and domes provide evidence of vertical movements in the relatively undeformed, stable interior platform of the United States.

the sea can be attributed to simple uplift, with no significant disturbance of the deposits.

A typical product of epeirogenic downward movement is a basin in the stable interior, such as the Michigan Basin. Upward movement with little or only moderate faulting or folding can produce broad uplands and plateaus. The Colorado Plateau and the Black Hills of South Dakota were both produced by general upward movements. The Black Hills structure is an oval dome—an area of uplift that slopes more or less uniformly in all directions from the highest point—rising more than 2 km above the surrounding Great Plains. (Figure 21.16 compares a basin and a dome.) The region in which the Black Hills are located was once much lower; it was a terrain of Precambrian igneous and metamorphic rocks that subsided beneath the sea and was covered by a blanket of Paleozoic and Mesozoic sediment more than 2 km thick. Sometime between the late Cretaceous and the Oligocene, the whole area of about 50,000 km² was uplifted, as if it were pushed up from below, without being extensively crumpled or broken. Subsequent erosion has stripped away the sediments that overlay the central part of the uplift, exposing the Precambrian igneous and metamorphic rocks below. Sculpted into the granitic rock of Mount Rushmore in the Black Hills are the world's largest presidential statues. Here is also where the famous Homestake gold mine was discovered in the Precambrian rocks near Lead, South Dakota (Figure 21.17).

Geologists have no ready explanation for most of these slow and broad epeirogenic movements, but they have some hypotheses that may account for some of them (Figure 21.18). The uplift of Finland and Scandinavia (see Feature 19.2) and the raised beaches of northern Canada (Figure 21.19) represent the slow upward recovery of the crust after the removal of the glacial load that had depressed it (Figure 21.18a). The formation of deep basins on

both sides of mid-ocean ridges is believed to be caused by the cooling and contraction of the newly formed ocean plate (Figure 21.18b). Heating of the lithosphere from below can result in thinning and upwarping (Figure 21.18c, top panel). Movements in the mantle can stretch and thin the lithosphere above, without breaking the plate. This may explain the subsidence of basins within continents (Figure 21.18c, middle panel). If stretching continues and a rift occurs, two continents result with an ocean growing between them. The sediment-filled basins on continental margins (such as those on the east coasts of the Americas and the west coasts of Europe and Africa) reflect the subsidence of the edge of the continent after rifting. This subsidence is caused by contraction as the margin cools and erodes during its withdrawal from the rift (Figure 21.18c, bottom panel), as we explained in Chapter 20. Intrusion of magma can thicken the continental crust and result in uplift (Figure 21.18d). In general, however, warping within continents or anywhere far from plate margins is not completely understood.

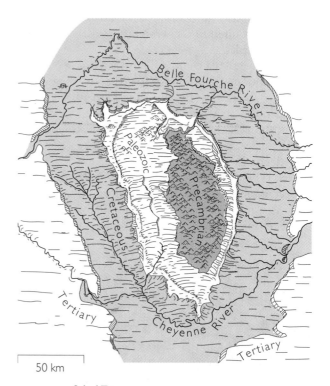

FIGURE 21.17 The Black Hills of South Dakota, a dome structure in the Great Plains. Erosion of the central part exposes the core of Precambrian igneous and metamorphic rocks and the succession of Paleozoic and Mesozoic sediments that slope away from the center. (After C. B. Hunt, *Geology of Soils,* San Francisco, W. H. Freeman, 1972.)

(a) Recovery

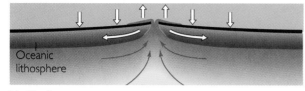

(b) Cooling

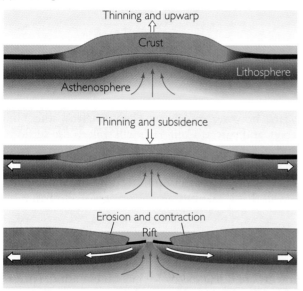

(c) Heating

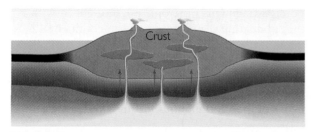

(d) Intrusion

FIGURE 21.18 Some proposed mechanisms for vertical movements (not to scale). (a) A glacial ice load buckles the crust; slow uplift follows removal of the ice. (b) Lithosphere forms at mid-ocean ridges, the ridge crest is uplifted, and the seafloor subsides as the plate cools and contracts. (c) Thinning and upwarping of continental lithosphere as a result of heating. Stretching can thin the continental crust as it subsides to form a basin within the continent. If the plate rifts, two continents result, with an ocean growing between them. Receding edges of the continents erode at the top and cool and contract within, forming subsiding continental margins. (d) The crust thickens with the addition of magma.

FIGURE 21.19 Raised beaches, evidence of upward recovery of the crust after removal of glacial load. Baffin Island, Northwest Territories, Canada. (*L. M. Cumming, Geological Survey of Canada.*)

Up-and-down movement of large regions is not restricted to the geologic past. The city of Venice, for example, is slowly sinking into the Adriatic Sea at a rate of 4 mm per year. Although tectonic downwarping is the primary cause of the subsidence, withdrawal of water and natural gas from the underlying sediments was hastening the death of this beautiful city (Figure 21.20). Venice has stopped withdrawing water and gas, but the tectonic subsidence continues.

FIGURE 21.20 Venice is slowly subsiding into the Adriatic Sea. The water now reaches the base of the columns of this old building. The raised sidewalk, of recent construction, will be awash in several decades if subsidence continues. (*R. Frassetto, National Research Council of Italy.*)

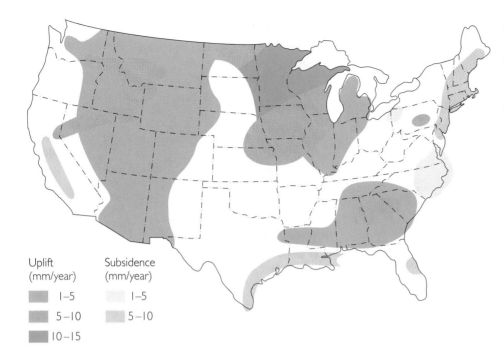

FIGURE **21.21** Areas of present-day uplift and subsidence in the United States. *(After S. P. Hand, National Oceanic and Atmospheric Administration.)*

Uplift
(mm/year)

Subsidence
(mm/year)

1–5	1–5
5–10	5–10
10–15	

Topographic surveys across the United States have revealed the pattern of vertical movements shown in Figure 21.21. Large regions are sinking or rising at rates of 1 to 15 mm per year. Not all of these movements will continue. If the pattern persists for the geologically short time of only a million years, however, much of New England, the Gulf coast, and part of California will sink to the seafloor, and plateaus a few kilometers high will grow from the lowlands in the Midwest. These seemingly slow rates of displacement, unnoticed by the inhabitants, are all that it takes to produce plateaus and basins.

SUMMARY

What are the major belts of deformation of North America? Most of the continental crust can be divided into belts that have been deformed during various geologic periods. North America contains a large central stable region that has been relatively undisturbed (except for gentle vertical movements and erosion) since episodes of intense deformation in the Precambrian. Surrounding the stable interior are younger orogenic belts—the mountainous Cordillera and Appalachians, which were deformed in plate collisions at various times in the Paleozoic, Mesozoic, and Cenozoic eras. Continents can grow by a succession of orogenies over geologic time in several ways: by sedimentation and buildup of continental shelves, by the addition of batholiths and volcanic rocks derived by melting in the mantle in subduction zones, and by microplate accretion.

What events typify an orogeny caused by plate convergence? A typical orogenic episode is preceded by a stage in which plates separate and an ocean widens between them. Continental margins subside and accumulate sediments. Plate convergence initiates deformation in a belt that extends hundreds of kilometers from the collision site. The marginal sediments are disrupted by folding and faulting. Thrust sheets 10 to 20 km thick slide over one another over distances of tens to hundreds of kilometers. Foreign fragments (microplates), brought in with the subducting plate, accrete to the continent. Intrusion of batholiths and metamorphism typically accompany orogeny. The mountains erode after the orogeny ends. A renewed stage of uplift or block faulting that again raises the regions accounts for many of the mountainous features we see today.

What are epeirogeny and orogeny? Forces within the crust can deform large regions of the continents. Some regional movements are simple up-and-down displacements without severe deformation of the rock formations (epeirogeny); examples are the Colorado Plateau and the postglacial uplift of Scandinavia and central Canada. In other cases, horizontal forces connected mainly with plate collisions can produce extensive and complex folding and faulting (orogeny), as in the Cordillera and Appalachians of North America, the Alps of Europe, and the Himalayas of Asia. The Rockies and the Alps are mountains that were eroded to low relief only to be rejuvenated by recent broad regional uplift.

KEY TERMS AND CONCEPTS

orogeny (p. 539)

craton (p. 541)

shield (p. 541)

platform (p. 542)

sedimentary basin (p. 542)

rejuvenation (p. 551)

fault-block mountain range (p. 552)

epeirogeny (p. 553)

EXERCISES

1. Evidence of vertical crustal movements is often found in the geologic record. Give some examples of such evidence.

2. Describe the geologic evolution of a region that leads to the formation of mesas, buttes, and tablelands bounded by high cliffs.

3. From 1925 to 1977 the San Joaquin Valley of California subsided 9 m. Guess why.

4. Summarize the stages of a typical orogenic episode in a series of sketches with legends.

5. Draw a rough topographic profile of the contiguous United States from San Francisco to Washington, D.C., and label the major geologic features.

6. Summarize the main features (orogenic belts, shields, platforms, coastal plains, shelves) of the major structural regions of a continent other than North America.

7. Why is the North American Cordillera topographically higher than the Appalachians?

8. Are the interiors of continents usually younger or older than the margins? Why?

THOUGHT QUESTIONS

1. How would you recognize a microplate? How could you tell if it originated far away or nearby?

2. How would you identify a region where active orogeny is taking place today? Give an example.

3. Would you prefer to live on a planet with orogenies or without them? Why?

4. How could the uplift of a mountain range affect climate, the chemistry of the oceans, and the location of oil and mineral deposits?

SUGGESTED READINGS

Bally, A. W., and A. R. Palmer (eds.). 1989. *The Geology of North America: An Overview.* Boulder, Colo.: Geological Society of America.

Burchfiel, B. C. 1983. The continental crust. *Scientific American* (September):130.

Cook, F. A., D. S. Albaugh, L. D. Brown, S. Kaufmann, J. E. Oliver, and R. D. Hatcher, Jr. 1979. Thin-skinned tectonics in the crystalline southern Appalachians. *Geology* 7:563–567.

Jones, D. L., A. Cox, P. Coney, and M. Beck. 1982. The growth of western North America. *Scientific American* (November):70.

Kearey, P., and F. J. Vine. 1990. *Global Tectonics.* London: Blackwell Scientific Publications.

Murphy, J. B., and R. D. Nance. 1992. Mountain belts and the supercontinent cycle. *Scientific American* (April):84.

Taylor, S. R. 1995. The geochemical evolution of the continental crust. *Reviews of Geophysics* 33:241–265.

Twiss, R. J., and E. M. Moores. 1992. *Structural Geology.* New York: W. H. Freeman.

INTERNET SOURCES

Black Hills Stratigraphy

🔘 **http://hills.net/~misteree/Strat.html**

This is a good description of the geology of the Black Hills, including geologic and relief maps, and stratigraphic sections.

250 Million Years Through Geologic History

🔘 **http://www.geog.psu.edu/McCracken/McCracken HTML/McCrackenTop.html**

This "geological history tour" from Bellefonte to Snow Shoe, Pennsylvania, crosses the transition from the Valley and Ridge province to the Allegheny Mountains. Geologic maps, cross sections, and a road log describe the geology along the route. The Web site is located at Pennsylvania State University.

Conserving Earth's Bounty

Earth's internal processes have given us our life-sustaining environment: a breathable atmosphere, oceans, rich soils, a moderate climate. Humans living during the Stone Age survived in this environment at subsistence levels. Humankind progressed to a better quality of life when we learned how to recover and use Earth's minerals. These resources create wealth and comfort by providing the materials and energy needed to grow and process food, build structures, transport things, and manufacture goods of all kinds. Profligate use of Earth's finite resources without regard for the fragility of the Earth system can lead to depletion of those resources and the hazardous accumulation of wastes; it can also trigger climatic change, with serious consequences. The challenge to humankind is to use Earth's resources wisely and equably to ensure a sustainable future.

This wind farm in California is using wind-mills to generate electricity. Wind energy is one form of renewable solar energy. *(Comstock.)*

Energy Resources from the Earth

Humans have steadily increased the amounts of coal, oil, natural gas, and uranium we take from the Earth to power our complex societies. How do these fuels form? Where are they found? Who does and will control them? How long will the supplies of these vital nonrenewable resources last, and what will we do when they are exhausted? Such questions have aroused concern about the environmental impact of the residues of the fuels we use so recklessly, about the need to conserve our resources, and about the development of substitutes for them. These concerns reflect a new and deeper understanding that we cannot continue to draw wealth from the Earth indefinitely without concern for our habitat and the generations to come.

We rely on oil, coal, natural gas, and uranium, resources derived from the Earth, for the energy to run our factories, construct our buildings, heat our homes, generate our electricity, produce our food, and fuel our transportation. In our increasingly systematic search of the globe for new sources of the fuels we depend on, we use our geological knowledge of how known natural deposits form to determine where we may find more of them. At the same time, we are becoming more sensitive to the finiteness of Earth's resources and the delicacy of its environment. We are paying more attention to extracting and using natural resources more efficiently, without damaging the environment. We are beginning to think about how we can change our use of resources to achieve **sustainable development**—development that will preserve the prospects of future generations.

RESOURCES AND RESERVES

Two major questions arise in all discussions of materials we draw from the Earth: How much is left? How long will it last? The amount left consists of more (we hope) than reserves. **Reserves** are deposits that have already been discovered and that can be mined economically and legally at the present time. **Resources,** in contrast, are the entire amount of a given material that may become available for use in the future. Resources include reserves, plus discovered but currently unprofitable deposits, plus undiscovered deposits that geologists think they may be able to find eventually (Figure 22.1). Often, resources that are too poor in quality or quantity to be worth mining now or are too difficult to retrieve become profitable when new technology is developed or prices rise.

Reserves are considered a dependable measure of supply as long as economic and technological conditions remain as they are now. As conditions change, some resources become reserves and vice versa. The conversion of oil resources in the North Sea to productive oil fields, for example, was accelerated by new technology and by price increases brought on by political instability in the oil-exporting nations of the Middle East. The assessment of resources is much less certain than the assessment of reserves. Any figure cited as the resources of a particular material represents only an educated guess of how much will be available in the future.

Most useful geological materials are considered *nonrenewable* because geologic processes produce them more slowly than civilization uses them up. Coal and oil, for example, will be exhausted faster

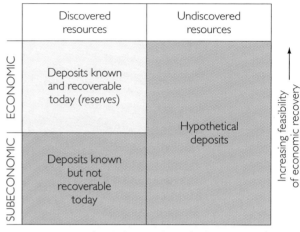

FIGURE 22.1 Categories that constitute total resources. Discovered resources consist of reserves, which are known deposits that are economically minable today, and deposits that are known but that are currently subeconomic. Undiscovered resources are hypothetical deposits that we may find. We can reasonably depend on reserves, but the areas that include known subeconomic deposits and estimates of undiscovered deposits represent only an educated guess of what might become available in the future.

than nature can replenish them. This fact makes it increasingly important to develop *renewable* resources such as solar energy, which is essentially infinite in supply, and fuels such as alcohol, which is derived from crops that can be replanted after they are harvested.

ENERGY

Energy is fundamental to everything. A crisis in the supply of energy can bring a modern society to a halt. Wars have been fought over access to supplies of fuel resources; economic recession and destructive currency inflation have resulted from gyrations in the price of oil.

Energy Use

As the world industrialized, the demand for energy increased and the types of energy used changed. The Industrial Revolution of the eighteenth and nineteenth centuries was powered by the energy from coal—in Britain from the coalfields of England and Wales, in continental Europe from the coal basins of

FIGURE **22.2**
Early oil wells. Signal Hill,
California. *(Chevron Corp.)*

western Germany and bordering countries, and in North America from the Appalachian coalfields of Pennsylvania and West Virginia. As industrialization expanded, so did the hunger for coal. Geological exploration for this fuel spread over much of the world, as coal use climbed at an ever-accelerating pace.

Half a century after the first oil well in America was drilled in 1859, oil and gas were beginning to displace coal as the fuels of choice (Figure 22.2). Not only did they burn more cleanly, producing no ash, but they could be transported by pipeline as well as by rail and ship.

The last quarter of the twentieth century saw the introduction of nuclear energy, with expectations that it would provide a large, new, low-cost, environmentally benign source of energy. These expectations were not realized, however, because safety concerns, the inability to dispose of nuclear wastes, and the escalating costs of stringent safety measures slowed the growth of nuclear energy.

The advanced nations depend primarily on oil, coal, natural gas, and some nuclear energy. In poorer countries, wood is an important source of fuel. The history of energy use in the United States is summarized in the graph in Figure 22.3. In 1850, coal

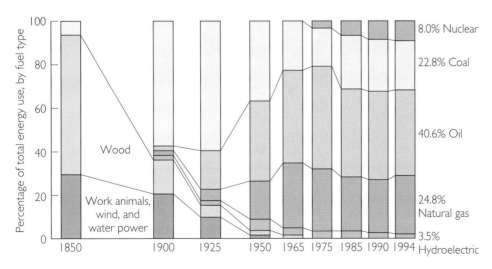

FIGURE **22.3**
Percentages of various types of energy used in the United States from 1850 to 1994. *(U.S. Energy Information Agency, 1995.)*

accounted for less than 5 percent of U.S. energy use. Today, oil, coal, and natural gas supply about 90 percent of the energy used in the United States. Consumption of this supply is distributed among residential and commercial uses (35.8 percent), industrial use (37.0 percent), and transportation (27.1 percent). Approximately half of the energy produced is lost in distribution and inefficient use.

Fossil Fuels

A century and a half ago, most of the energy used in the United States came from the burning of wood. A wood fire, in chemical terms, releases energy by the combustion of organic matter consisting of compounds of carbon with hydrogen attached to the carbon atoms. Organic matter is produced by plants or animals. The organic matter in this case is a tree produced from carbon dioxide and water, with energy for the transformation supplied by sun-

light—the biological process of photosynthesis (see Chapter 24). Thus, we can look upon a piece of wood or any piece of plant matter as a photosynthetic product that can be returned by burning or decay to the carbon dioxide and water from which it was made.

If we burn wood that was buried and transformed into the combustible rock known as coal 300 million years ago, we are using energy stored by photosynthesis from late Paleozoic sunlight. We are recovering "fossilized" energy. Crude oil and natural gas were also created by a process of burial and chemical transformation of dead organic matter to a combustible liquid and gas, respectively. We refer to all such resources derived from natural organic materials, from coal to oil and natural gas, as **fossil fuels** (Figure 22.4).

OIL AND NATURAL GAS

Crude oil (petroleum) and natural gas in minable form develop under special conditions in the environmental and geologic history of a region. Both are the organic debris of former life—plants, bacteria, algae and other microorganisms—that has been buried, transformed, and preserved in marine sediments.

How Oil and Gas Form

Oil and gas begin to form when more organic matter is produced than is destroyed by scavengers and decay. This condition exists in environments where the production of organic matter is high—as in the coastal waters of the sea, where large numbers of organisms thrive—and the supply of oxygen in bottom sediments is inadequate to decompose all organic matter. Many offshore sedimentary basins on continental shelves satisfy both of these conditions. In such environments, and to a lesser degree in some river deltas and inland seas, the rate of sedimentation is high and organic matter is buried and protected from decomposition.

During millions of years of burial, chemical reactions triggered by the elevated temperatures at depth slowly transform some of the organic material into liquid and gaseous compounds of hydrogen and carbon (hydrocarbons). The hydrocarbons are the combustible materials of oil and natural gas. Compaction of muddy organic sediment into **source beds** forces the hydrocarbon-containing fluids and gases into adjacent beds of permeable rock (such as

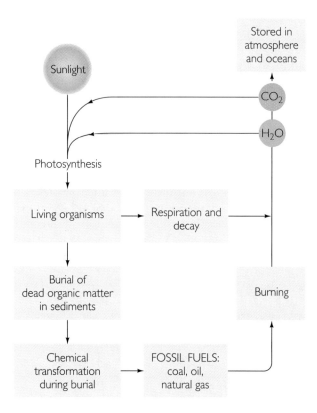

FIGURE **22.4** Photosynthesis produces organic matter from carbon dioxide (CO_2) and water (H_2O). If dead organic matter is buried and transformed into coal, oil, or natural gas, it becomes a fossilized product of photosynthesis—a fossil fuel. The burning of fossil fuels releases the carbon dioxide and water from which they were made.

sandstones or porous limestones), which we call **oil reservoirs.** The low density of oil and gas causes them to rise to the highest place they can reach, where they float atop the water that almost always occupies the pores of permeable formations.

OIL TRAPS Geological conditions that favor the large-scale accumulation of oil and natural gas are combinations of structure and rock types that create an impermeable barrier to upward migration—an **oil trap.** One type of oil trap is formed by an anticline in which an impermeable shale overlies a permeable bed of sandstone (Figure 22.5a). The oil and gas accumulate at the crest of the anticline—the gas highest, the oil next—both floating on the groundwater that saturates the sandstone. An oil trap caused by a structural deformation like the one that created the anticline is called a *structural trap.* Similarly, an angular unconformity or displacement at a fault may place a dipping permeable limestone bed opposite an impermeable shale, creating a structural trap for oil (Figure 22.5b). An oil trap created by the original pattern of sedimentation, as when a dipping permeable sandstone bed thins out against an impermeable shale (Figure 22.5c), is called a *stratigraphic trap.* Oil can also be trapped against an impermeable mass of salt, as in a salt dome (Figure 22.5d).

SOURCE BEDS In their search for oil, geologists have mapped thousands of structural and stratigraphic traps all over the world. Only a fraction of them have proved to contain any oil or gas, because the traps alone are not enough. A trap will contain oil only if source beds were present, if the necessary chemical reactions took place, and if the oil could migrate into the trap and stay there without being disturbed by subsequent severe heating or deformation. Although oil and gas are not rare, most of the easy-to-find deposits have already been located, and new fields are becoming more difficult to find.

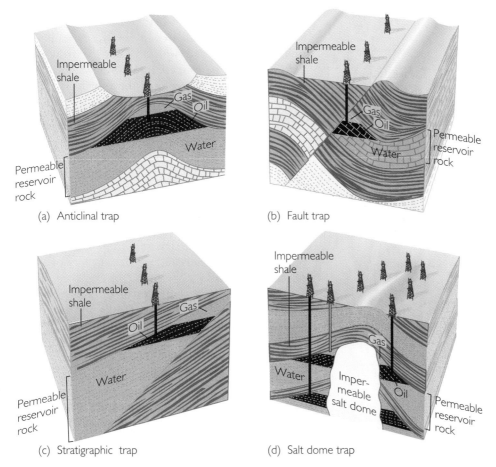

FIGURE 22.5 Oil traps: (a) anticlinal trap; (b) fault trap; (c) stratigraphic trap; (d) oil trapped by a salt dome. Natural gas (green) and oil (black) are trapped by an impermeable layer above the permeable oil-producing formation. Oil floats above the waterline.

The World Distribution of Oil and Natural Gas

If you were to visit the exploration and research offices of any large oil company, you would find maps and reports of all the regions containing geologic sections of sedimentary rock in which the company has operated, because where there is a thick section of sedimentary rocks, there may be oil. Thirty-one of the 50 states of the United States produce oil for the market, and small, noncommercial occurrences are known in most of the others. Many of Canada's provinces produce oil.

Two of the richest and most important oil-producing regions in the world are the Middle East and the area around the Gulf of Mexico and the Caribbean. The oil fields of the Middle East, including Iran, Kuwait, Saudi Arabia, Iraq, and the Baku region in Azerbaijan, contain about two-thirds of the world's known reserves. The highly productive Gulf coast—Caribbean area includes the Louisiana-Texas region, Mexico, Colombia, Venezuela, and Trinidad. Saudi Arabia holds the largest reserves. The United States ranks eighth. Figure 22.6 summarizes oil reserves in various parts of the world.

Why are oil and natural gas so unevenly distributed about the globe? A political scientist once jokingly asked, "Why is it that oil is always found in underdeveloped countries?" The distribution of oil, of course, is determined not by political boundaries but by the geologic history of a region. For example, continental shelves are good places to explore for oil. They easily meet the conditions required for rich oil deposits. Their sediments are old enough to accommodate the slow process of oil formation and young enough to be preserved in an environment that is little eroded, metamorphosed, or deformed. Yet most shelf deposits, even those of the U.S. continental shelf, are still untapped. One reason is concern about the environment. Another is the difficulty and expense of drilling in deep water. Many oil companies are waiting for oil prices to rise before they make the huge investments that would be required. Some companies willing to make the investment in offshore drilling have faced the opposition of a local population concerned with environmental damage, as in California.

Oil and the Environment

Pollution is the major problem of offshore drilling. The environment around Santa Barbara, California, suffered great damage in 1969 when oil was accidentally released from an offshore drilling platform. In 1979, a well that was being drilled in the Gulf of Mexico off the Yucatan coast "blew out," spilling as much as 100,000 barrels of oil a day for many weeks before it could be capped. In 1988, an explosion destroyed a drilling platform in the North Sea, killing many oil workers and marine animals. The grounding of the tanker *Exxon Valdez* off the coast of Alaska in 1989, with the release of 240,000 barrels of crude oil in pristine coastal waters, was covered widely by TV and newspapers and heightened public awareness of the severe ecological damage that can result from an oil spill (Figure 22.7). Despite such incidents and the difficulty of guaranteeing the safety of a well or a tanker, proponents of oil development believe that careful design of equipment and safety procedures can greatly reduce the chances of a serious accident.

There are large resources of oil and gas under the coastal plain of northern Alaska and under the submerged continental shelves of the United States. Many people argue that eventually they will have to be drilled to satisfy the world's growing energy needs. A skeptical public, however, is not convinced that such drilling and production can be done without serious threat to pristine environments. There is currently a strident political debate in Washington about whether or not to allow drilling for oil and natural gas in the Arctic National Wildlife Refuge. There is no doubt that these resources would be an

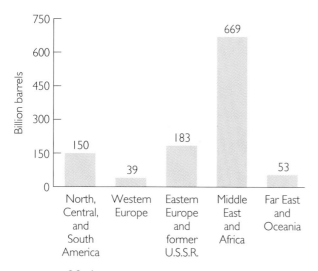

FIGURE **22.6** Estimated world reserves of crude oil, 1995. (*Data from World Oil.*)

FIGURE 22.7 The effect on wildlife of an oil spill from an oil tanker, the *Exxon Valdez*, in Prince William Sound, Alaska. *(UPI/Corbis-Bettmann.)*

economic boon to the nation. But oil and gas production involves building roads, pipelines, and housing in a very delicate ecological environment that is a particularly important breeding area for caribou, musk-oxen, snow geese, and other wildlife (Figure 22.8).

Oil: An Exhaustible Resource

Inconceivable as it may seem to the landowner who sees a gusher spurting oil from a derrick on his property, that oil well will eventually run dry. What the world wants to know is: How soon will *all* the wells run dry? Forty years ago, many oil producers were optimistic. Supplies seemed so immense, and so many scientific and engineering innovations held promise of tapping new sources, that no one saw a need to worry about the future. A more sober analysis of the total quantity of oil remaining on Earth, however, presents a picture of the steady dwindling of a fixed supply. World demand has accelerated rapidly. Twice as much oil was removed from the ground in the past 20 years as in the previous 100 years. It takes millions of years to create oil, and humankind is using it up in centuries. Natural processes cannot replenish the oil supply as quickly as we are using it.

FIGURE 22.8 The Alaska pipeline transports oil from northern to southern Alaska, where tankers pick up the oil and carry it to refineries. *(Thomas Kitchin/Tom Stack.)*

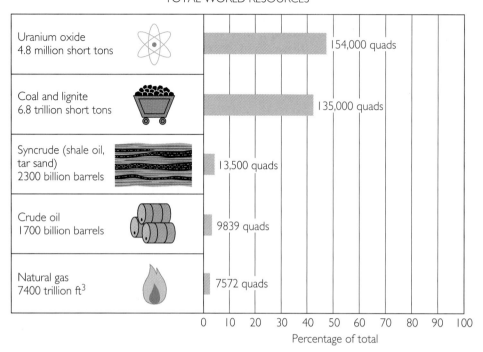

TOTAL WORLD RESOURCES

Uranium oxide 4.8 million short tons		154,000 quads
Coal and lignite 6.8 trillion short tons		135,000 quads
Syncrude (shale oil, tar sand) 2300 billion barrels		13,500 quads
Crude oil 1700 billion barrels		9839 quads
Natural gas 7400 trillion ft^3		7572 quads

Percentage of total

FIGURE 22.9 Remaining nonrenewable world energy resources amount to about 320,000 quads. Amounts are given in conventional units of weight (short tons), volume (barrels), and energy content (quads). A short ton is 2000 pounds or 907.20 kilograms; a barrel of oil is 42 gallons. Coal and lignite resources, for example, amount to 6.8 trillion short tons, equivalent to 135,000 quads, or 42 percent of total energy resources.

Figure 22.9 shows that the world's remaining energy resources of all types are estimated to be about 320,000 quads.[1] Of this total, nearly 10,000 quads are oil (the equivalent of 1700 billion barrels). How long this supply of oil will last depends on the rate at which we use oil and on our success in converting resources into reserves. At current rates of world use (20 billion barrels per year), oil production will start declining in 2015 or 2020, and most of the remaining oil will be depleted in about 85 years. To round off an uncertain number, say 100 years. The remaining oil supply will last longer if consumption declines, as it did because of conservation and the industrial slowdown that came with the recession of the early 1980s, or it may run out sooner if use picks up, as seems likely.

WHEN CONSUMPTION EXCEEDS PRODUCTION
Here are some thought-provoking facts. The United States consumes more oil than it produces. It produces only about 11 percent of the world's oil but consumes about 26 percent. This means that of the total amount of oil used in the United States—about 6 billion barrels per year—nearly half is imported. These annual imports cost anywhere from $50 billion to $70 billion, depending on the current price of a barrel of oil, and account for a large part of the U.S. trade deficit—the excess of imports over exports. Of the total energy consumed in the United States each year, oil accounts for about 40 percent. Most of the oil (about 60 percent) is used for transportation.

---

[1]Fuel resources are measured in units appropriate to the material, for example, barrels of oil, tons of coal. To make it easier to compare the energy available from different fuel resources, however, a common unit called the quad is used. A **quad** is a measure of the energy that can be extracted from a given amount of fuel. The quad is based on a standard measure of energy called the British thermal unit (Btu). One Btu is the amount of energy needed to raise the temperature of 1 pound of water by 1°F. One quad equals 10^{15} Btu. The United States uses about 84 quads of energy a year.

At current use rates, remaining U.S. oil reserves (60 billion barrels) amount to roughly a 20-year supply. Estimates of remaining reserves vary considerably. In 1989, reserves were thought to be about half that amount; current estimates are higher because new methods have been developed to recover more of the oil from existing fields. In any case, 20 years is not much time before the United States would become dependent on imports for most of its oil, at a cost of perhaps several hundred billion dollars annually.

New technology might be developed to find oil deposits that have been missed by current methods of exploration and to enhance the recovery of oil from existing domestic oil fields. About half of the oil in a field is now left in the ground because it is difficult to pump it out. Water can be pumped to the periphery of a field to force more of this oil out of a central well. To loosen up very viscous oils and allow them to flow more easily, steam or carbon dioxide can be injected into the oil reservoir under pressure. A large amount of oil may become economically recoverable in the United States by these enhancement techniques, and this could increase domestic reserves significantly. The amount recovered could be equivalent to a few decades of imports.

Many people in the United States think that the dependence on foreign sources of oil is potentially destabilizing to the economy. Many oil-exporting nations have not been reliable suppliers and have cut off exports because of political disagreements and wars. Other industrial nations that depend on oil imports, such as Japan, France, and Germany, have similar concerns.

REDUCING OIL CONSUMPTION   Oil-importing nations have options to reduce their vulnerability. They can change their pattern of oil use to reduce their need for imports. Automobile and airplane engines can be designed to use fuel much more efficiently. If the efficiency of the U.S. automobile fleet were to improve by 5 miles per gallon of gasoline, current oil imports could be reduced by almost half. Automobiles can run on natural gas, electric batteries, or alcohol produced from biomass (grain, sugarcane). These fuels also would be less polluting than gasoline. Brazil fuels almost all of its automobiles with alcohol made from sugarcane and so does not need to import oil for transportation. New technology is needed, however, to improve these alternative fuels, lower their costs, and arrange for efficient, large-scale production and distribution. Improvements are also needed in the engines that use these fuels.

With about 100 years' worth of oil remaining beneath the Earth at current rates of use before nature's oil legacy is gone, we cannot yet proclaim the end of the age of oil. A century may be barely enough time to plan and put into place an orderly transfer to new systems for transportation and for other energy uses in the home and in industry. We need to make greater use of mass transportation rather than automobiles; design more fuel-efficient automobiles and airplanes; exploit alternative sources of fuel; and place greater emphasis on energy conservation in industry, commerce, and the home. Without such planning for alternatives and conservation, many nations may face social and economic disruption when transportation systems grind to a halt, production lines stop, and homes grow cold in winter because oil has become so scarce that its price has risen out of reach.

Natural Gas

The resources of natural gas are comparable to those of crude oil (see Figure 22.9) and may exceed them in the decades ahead. Estimates of natural gas resources have been rising in recent years. Exploration for this relatively clean fuel has increased, and geologic traps have been identified in new geological settings, such as very deep formations, overthrust belts, coal beds, tight (that is, somewhat impermeable) sandstones, and shales. The world's resources of natural gas are less depleted than oil resources because natural gas is a relative newcomer on the energy scene. It has been used on a large scale only in the United States and the former Soviet Union.

The burning of natural gas releases less carbon dioxide per unit of energy than the combustion of coal or oil. It is less polluting (little ash or acid rain precursors are released), and it is easily transportable. Natural gas from fields in Siberia, for example, is piped to factories and homes in Germany. For these reasons, natural gas is a premium fuel. Natural gas accounts for about 25 percent of all fossil fuels consumed in the United States each year, most of it for industry and commerce (52 percent), followed by residential use (28 percent) and generation of electric power (18 percent). More than half of American homes and a great majority of commercial and industrial buildings are connected to a network of underground pipelines that draw gas from fields in the United States, Canada, and Mexico. American resources should last about 35 years at current rates of use, probably longer with the tapping of accumulations in the new settings we mentioned earlier.

COAL

The abundant plant fossils found in coal beds indicate that coal is formed from large accumulations of plant materials, of the sort that would occur in wetlands. As the luxuriant plant growth of a wetland dies, it falls to the waterlogged soil. Rapid burial by falling leaves and immersion in water protect the dead twigs, branches, and leaves from complete decay because the bacteria that decompose vegetative matter are cut off from the oxygen they need. The vegetation accumulates and gradually turns into peat, a porous brown mass of organic matter in which twigs, roots, and other plant parts can still be recognized (Figure 22.10). The accumulation of peat in an oxygen-poor environment can be seen in modern swamps and peat bogs. Peat burns readily when it is dried because it is 50 percent carbon.

Over time, with continued burial, the peat is compressed and heated. Chemical transformations increase the peat's already high carbon content, and it becomes *lignite,* a very soft brownish-black coal-like material containing about 70 percent carbon. The higher temperatures and structural deformation that accompany greater depths of burial may metamorphose the lignite into *subbituminous* and *bituminous coal,* or soft coal, and ultimately to *anthracite,* or hard coal. The greater the metamorphism, the harder and brighter the coal and the higher its carbon content, which increases its heat value. Anthracite is over 90 percent carbon.

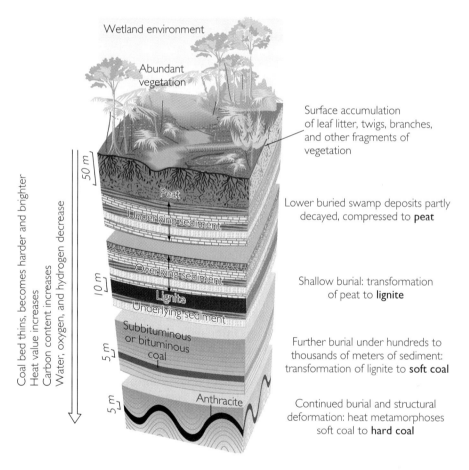

FIGURE 22.10 The process by which coal beds form begins with the deposition of vegetation. Protected from complete decay and oxidation in a wetland environment, the deposit is buried and compressed into peat. Subjected to further burial, peat undergoes mild metamorphism, which transforms it successively into lignite, subbituminous and bituminous (soft) coal, and anthracite (hard coal), as the deposit becomes more deeply buried, temperature rises, and structural deformation progresses.

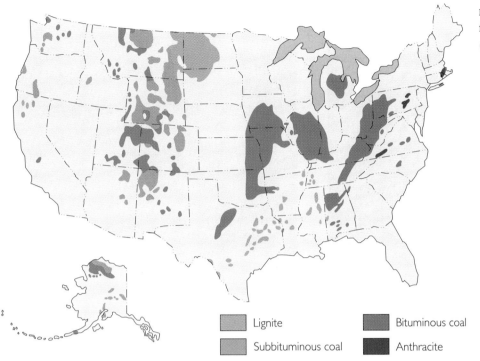

FIGURE 22.11 Coalfields of the United States. *(U.S. Bureau of Mines.)*

Lignite

Subbituminous coal

Bituminous coal

Anthracite

Coal Resources

According to some estimates (see Figure 22.9), about 6.8 trillion short tons of coal remain in the world. The leading producers are the United States (Figure 22.11), the former Soviet Union, and China, which together hold about 85 percent of the world's coal resources (former Soviet Union, 50 percent; China, 20 percent; United States, 15 percent). Domestic coal resources in the United States would last for 300 to 400 years at current rates of use—about a billion tons a year. Coal has supplied an increasing proportion of the energy needs of the United States since 1975, when the price of oil began to rise, and currently accounts for about 23 percent of the energy consumed (see Figure 22.3).

If oil supplies become scarce, coal could be converted to liquid or gaseous fuels comparable to those obtained from crude oil today. The cost of synthetic oil, called **syncrude,** that is derived from coal is higher than that of crude oil at today's prices, but with depletion of world oil reserves and political instability in the Middle East, the gap between the two could decrease in the years ahead. A few commercial-scale synthetic oil and gas plants were constructed in the United States in the 1970s, when oil-exporting nations in the Middle East cut back on oil production for political reasons. With today's low cost of imported oil, these plants are not eco-nomical and most have been shut down. If prices increase in the years ahead, however, as most observers expect, several million barrels of syncrude per day may well be produced in the United States in the next century.

Environmental Costs of Coal

There are serious problems with the recovery and use of coal—problems that make it less desirable than oil or gas, whether the coal is burned or converted to synthetic liquid fuel. Much coal contains appreciable amounts of sulfur, which vaporizes during combustion and liberates noxious sulfur oxides to the atmosphere. Acid rain, formed when these gases combine with rainwater, is becoming a serious problem in the northeastern United States, Canada, Scandinavia, and eastern Europe (see Chapter 24). Coal ash is the inorganic residue that remains after coal is burned. It contains metal impurities from the coal, some of which are toxic. Ash can amount to several tons for every 100 tons of coal burned and poses a significant disposal problem. It can escape from smokestacks, posing a health risk to people downwind. Strip mining, the removal of soil and surface sediments to expose coal beds, can ravage the countryside if the land is not restored

FIGURE 22.12 Coal strip mine near Shamokin, Pennsylvania. *Left:* Strip mining of anthracite. *Right:* After reclamation. *(Peter Kresan.)*

(Figure 22.12). Underground mining accidents take the lives of miners each year, and many more suffer from black lung, a debilitating inflammation of the lungs caused by the inhalation of coal particles. These human costs are as important to consider as the possible dangers of nuclear reactors.

Government regulations now require phasing in the use of technologies for the "clean" combustion or liquefaction of coal. The restoration of land ravaged by strip mining and the reduction of danger to miners are now mandated by law. But these measures are expensive and add to the cost of coal. This drawback, however, is unlikely to prevent the increased use of this fuel, which is so much more abundant than oil. Many countries have no other fuel resources, and other countries will not be able to afford to import oil, which will increase in price as the supply diminishes.

OIL SHALE AND TAR SANDS

Syncrude can also be extracted from oil shales and tar sands as well as coal. **Oil shales** are fine-grained sedimentary rocks containing a large proportion of solid organic matter that can be distilled by heating to yield oil. This organic matter has the same general origin as the other common organic components of fossil fuels, all of which ultimately derive from plant and animal matter. Some oil shales can produce up to 150 gallons of oil per ton, but most yield from 25 to 50 gallons per ton (Figure 22.13).

Tar sand is a sandy deposit of organic matter impregnated with a tarry substance made up almost entirely of hydrocarbons. Oil of a type similar to liquid petroleum can be recovered from tar sand by heating, and geological evidence suggests that its tarry components are a transformed product of a formerly liquid petroleum. Some deposits of tar sands are oil pools that have dried up and become tarry by loss of the more volatile hydrocarbons. One of the largest deposits, the Athabaska tar sands of Cretaceous age, is found in Alberta, Canada. Other major deposits are known to exist in Venezuela.

World syncrude resources derived from oil shale and tar sand are half again as great as all crude oil resources (see Figure 22.9). The Athabaska tar sands are already producing oil for the market. Nearly half of the world's syncrude resources are in the United States. This is a potentially significant future resource.

Like synthetic oil derived from coal, syncrude extracted from oil shale will be more expensive than crude oil in the near term, but prices could converge in the next century as oil supplies diminish. Disposal of the waste shale that remains after the oil has been extracted presents a serious problem. And the extraction process requires large amounts of water, a commodity in short supply in Wyoming, Utah, and Colorado, where oil shales are plentiful. The difficulty and expense of extraction and associated environmental problems make it unlikely that syncrude will displace much crude oil in the fossil fuel market in the next few decades.

FIGURE 22.13 Syncrude production in the principal oil-shale group of the Green River formation of Colorado, Utah, and Wyoming. Almost half of the world's supply of oil shale is found in this formation. At this facility, oil shale mined from the cliffs is heated to produce synthetic oil, which is subsequently refined into common petroleum products. The plant can produce 10,000 barrels of syncrude each day. *(Unocal Corp.)*

ALTERNATIVES TO FOSSIL FUELS

If crude oil and gas continue to be used as the major resources for satisfying the world's voracious appetite for energy, the great bulk of the world's supply will be exhausted within a century. Oil shale and tar sands will add some decades, but coal will eventually become the predominant fossil fuel in many countries. It may be reassuring to know that at modest annual energy growth rates, say, 3 percent per year, coal and other fossil fuels can meet the world's energy needs for about 100 years or longer. This security, however, may be false. Carbon dioxide released during the combustion of fossil fuels may trigger climatic changes that could force a shift away from these traditional fuels before they are depleted, to avoid a climatic crisis.

These estimates do not take into account the possibility that we may learn to meet much of our growing energy need in nontraditional ways: through increased efficiency in the use of fossil fuels and by the development and use of alternative energy sources such as nuclear energy, solar energy, geothermal energy, and energy derived from biomass. To the extent that these sources can be used, the pressure on our fossil fuel resources can be relaxed and their life extended.

Nuclear Energy Fueled by Uranium

Although the first use of uranium 235 (^{235}U) was in an atomic bomb in 1944, the nuclear physicists who first observed the vast energy released when its nucleus splits spontaneously (called *fission*) foresaw the possibility of peaceful applications of this new energy source. After World War II, their predictions were fulfilled as countries all over the world built nuclear reactors to produce **nuclear energy:** the fission of ^{235}U releases heat to make steam, which then drives turbines to create electricity. The fission of a piece of ^{235}U releases 3 million times more energy than a piece of coal of the same mass. In the United States, some 110 nuclear reactors now provide about 20 percent of the electric energy we use. France derives 75 percent of its electric energy from nuclear power. Today, more than 400 nuclear reactors are producing electricity in 25 countries. If its full potential is realized, nuclear energy can meet the world's needs for electric energy for hundreds of years.

URANIUM RESERVES One aspect of nuclear energy is most definitely in the province of geology: the question of reserves of uranium. Uranium is present in very small amounts in Earth's crust, constituting only 0.00016 percent of the average continental crustal rock. The isotope that fissions and releases energy, ^{235}U, constitutes only 1 of every 139 atoms of uranium mined. In terms of energy content, however, uranium is potentially our largest minable energy resource (see Figure 22.9). It is typically found as small quantities of the uranium oxide mineral uraninite (also called pitchblende) in veins in granites and other felsic igneous rocks. Uranium can also be found in sedimentary rocks. Under near-surface groundwater conditions, uranium in igneous rocks may become oxidized and dissolved, be transported in groundwater, and later be reprecipitated as uraninite in sedimentary rocks.

22.1   LIVING ON EARTH

Radioactive Waste Disposal

There are about 425 nuclear power plants in the world, generating less than 20 percent of the total electricity used, with about 110 of these in the United States. Each 1000-megawatt nuclear electric power plant produces about 30 metric tons of highly radioactive spent fuel each year. Regardless of one's views on the future of nuclear energy and the safety of nuclear power generation, the nation faces the unresolved problem of permanent and safe storage of wastes already generated, which will remain dangerous for hundreds of thousands of years.

Power plant wastes are now stored temporarily on the surface at each reactor site, and most sites will reach maximum capacity for such storage in a few years. We face the additional problem of finding long-term storage for the highly radioactive wastes generated by the military nuclear reactors that made the fissionable materials used in bombs. These are stored temporarily at sites in the states of South Carolina, Idaho, and Washington.

There is general agreement among geologists and other scientists that deep geological containment eventually will be the best long-term solution. Some thought is being given to storage on the bottom of the deep ocean (see Feature 17.3), but this option is still speculative. Geological containment involves construction of a facility deep underground in stable geological formations—that is, ones with no tectonic or volcanic activity, no faults or fractures penetrating to the surface, and very well understood hydrological conditions that demonstrate negligible flow of groundwater through the repository. Groundwater is of particular concern because it can corrode the containers and carry the radioactive materials away. Climatic change that will increase rainfall and inadvertent drilling into a waste site by a future society are considerations that are not negligible in designing a repository.

After treatment to reduce the volume of the radioactive materials and transform them to a nonsoluble, relatively inert substance such as glass, they would be placed in containers that might last for hundreds of years and stored in engineered tunnels or caverns deep underground. The idea is to design a system of multiple barriers that would decrease the chances of leakage into the environment. The geology and groundwater conditions would be selected to prevent or slow the movement of materials through rock formations, so that even if they escaped from their containers and came into contact with the rock they could not reach the surface until the radioactivity had decayed to harmless levels.

Federal regulations limit the release of radioactive materials to the environment during

Testing the storage capability for spent nuclear fuel in subterranean basalt caverns under the Hanford Reservation in Washington. (*U.S. Department of Energy/Mark Marten/Photo Researchers.*)

the first 10,000 years after containment: fewer than 1000 deaths must result from inhalation or ingestion of any materials that reach the environment for each 100,000 tons of stored material. This is equivalent to an average of less than 10 deaths per century.

Congress has designated Miocene volcanic ash-flow tuffs under Yucca Mountain in Nevada as the repository for civilian reactor wastes. Construction has not yet started, and the state of Nevada is fighting this decision.

The Waste Isolation Pilot Plant (WIPP) in Carlsbad, New Mexico, has been constructed in Permian salt beds 655 m below the surface as a test repository for military wastes generated in the nuclear weapons programs. Regulatory agencies have not yet granted approval to store wastes there. Facilities able to receive wastes from temporary storage sites will not be ready as planned.

Environmentalists and the state of Nevada support tough standards and argue that they have not been met at proposed waste sites. Others feel that the regulations require scientists to guarantee the behavior of rock formations for thousands of years into the future—something that realistically cannot be done. They call for more flexible standards like those in place in other countries such as Switzerland and Canada. For example, without relaxing overall safety requirements, waste repositories could be constructed and used with the expectation of learning from realistic conditions. If an unanticipated problem occurred, as almost always happens and cannot be guaranteed against, the wastes could be safely retrieved if necessary or the construction design modified.

There are risks associated with both points of view. On the one hand, compliance with overly conservative standards will cause long delays and may never be achieved, which poses the risks of relying on temporary storage at many sites on the surface. On the other hand, approving a flexible process that requires trust may not satisfy those who live near a repository. The debate continues.

The United States is known as a major source of uranium partly as a result of government support of exploration in the 1950s and 1960s to ensure a uranium supply during the nuclear arms race. The richest ores in the United States are found in the Triassic and Jurassic sedimentary rocks of the Colorado Plateau, in western Colorado and adjacent parts of Utah, Arizona, Wyoming, and New Mexico. Large resources are also found in Canada, Australia, South Africa, and Brazil, in that order. Little is known publicly about the uranium resources of the former Soviet Union and China. At one time, there was uncertainty about the adequacy of uranium resources. Currently, there is a glut of uranium and its price has plummeted. Not only has the demand for uranium slowed along with the lack of growth of nuclear power, but substantial new deposits have been discovered in Australia and elsewhere. The uranium and plutonium from demilitarized Russian nuclear weapons may be brought to the civilian market in competition with mined uranium.

NUCLEAR ENERGY HAZARDS The debate over the adequacy of uranium resources will undoubtedly be rekindled if enthusiasm for nuclear power revives, but such a revival may not occur because of the serious problems that have emerged with the use of nuclear energy. At one time, nuclear energy was promoted as the cheap, clean alternative to fossil fuels. The costs of building and maintaining reactors have since proved prohibitive. A more important obstacle to revival has been two accidents that created a public mind-set against nuclear energy.

The first nuclear accident was at the Three Mile Island reactor in Pennsylvania in 1979. A reactor was destroyed; radioactive debris was released but confined within the containment building. Although no one was harmed, most experts agree that it was a close call. Much more serious was the destruction of a nuclear reactor in the town of Chernobyl in Ukraine in 1986. The reactor went out of control because of poor design and human error and was destroyed. Radioactive debris spilled into the atmosphere and was carried by winds over Scandinavia and western Europe. Contamination of buildings and soil has made hundreds of square miles of land surrounding Chernobyl uninhabitable. Food supplies in many countries were contaminated by the fallout and had to be destroyed. Excess deaths from cancer caused by exposure to the fallout may be in the thousands over the next 40 years.

The uranium "used up" in nuclear reactors leaves behind dangerous radioactive wastes that must be disposed of (see Feature 22.1). A system of safe long-term waste disposal is not yet available, and reactor wastes are being held in temporary storage at reactor sites. In a few years, the limits of space available for temporary storage in the United States will be reached. Although many scientists believe that geological containment—the burial of nuclear wastes in deep, stable, impermeable rock formations—is a workable solution, there is not yet a generally approved plan for storage of the most dangerous wastes for the hundreds of thousands of years required before they cease to be radioactive. France and Sweden have built underground nuclear waste depositories, but the United States is still in the stage of research, development, and testing. It is also embroiled in litigation as some states battle the federal government to keep waste repositories from being built within their borders. No one wants to leave to future generations the combined legacy of depleted energy resources and a possibly unmanageable environmental hazard.

In the United States, these unresolved problems have essentially halted the installation of new plants, and new installations have been slowed in other countries as well. Concerns about the safety of nuclear reactors, their high cost, and the safe disposal of radioactive waste must be allayed before we can light up the world with nuclear energy.

Solar Energy

Since all of our fossil energy sources ultimately come from the Sun anyway, why not convert its rays to energy? In principle, the Sun can provide us with all the energy we need, in all the forms we use. Light from the Sun can be converted to heat and electricity. It can even be used to obtain hydrogen, which can be used as a gaseous fuel, from water. **Solar energy** is risk-free and nondepletable—the Sun will continue to shine for at least the next several billion years. Unfortunately, the technology currently available for large-scale conversion of solar energy to useful forms is inefficient and expensive, although the situation is improving.

In the near term, the only form of solar energy likely to be available at costs nearly competitive with those of other sources is heat for homes, water, and industrial and agricultural processes. Many homes and factories are beginning to use solar energy for these purposes, motivated in part by government tax credits and other incentives.

ELECTRICITY FROM SUNLIGHT Solar energy can be used to generate electricity in several ways.

One way is by photovoltaic conversion, in which solar cells of the kind that power space satellites collect and convert light to electricity directly (Figure 22.14). Another is by solar thermal conversion, in which sunlight is converted first to heat and then to electricity. For example, sunlight can be concentrated by means of mirrors and lenses on a tank of water to convert the water to steam, which drives an electric generator. These systems are being used for installations where costs are no impediment, such as demonstration projects, or in remote areas where alternatives are not available. But large-scale solar-generated electricity is not yet an alternative to conventional sources in general use. The efficiency of converting sunlight to electricity is improving but is still too low, and the costs of installing and maintaining the systems are still too high. Solar energy must await further technological advances to become competitive. Commercial-scale solar electric power plants can present significant environmental problems. A plant with 100-megawatt electric capacity (about 10 percent of the capacity of a nuclear power plant) located in a desert region would require at least a square mile of land and might alter the local climate by significantly changing the balance of solar radiation in the area.

INDIRECT SOURCES OF SOLAR ENERGY Solar energy is also stored in **biomass**—material of biological origin, such as wood, grain, sugar, and municipal wastes. Energy can be extracted by direct combustion or by processes that convert biomass to gaseous or liquid fuels, such as methane and alcohol. Converting the biomass in garbage also puts waste material to good use and eases the growing problem of finding sites for its disposal. The technology for some of these processes is well developed, but their price is not yet competitive without some form of government subsidy. The growing of crops that can be converted to energy also presents ecological problems: the need for water, fertilizer, and pesticides. An ethical issue also arises if food crops are displaced by "energy crops" when so many people in the world are going hungry.

In a sense, **hydroelectric energy** is a form of solar energy because it depends on rainfall, and the energy that drives weather comes from the Sun. Hydroelectric energy is derived from water that falls by the force of gravity and is made to drive electric turbines. Waterfalls or artificial reservoirs behind dams provide the water. Hydroelectric energy is clean, relatively riskless, and cheap. Hydroelectric energy delivers about 3 quads annually, or about 4 percent of the annual energy consumption in the

FIGURE 22.14 Solar cells convert sunlight, a renewable resource, to electric energy at this utility in a remote village in Nepal. (*Ned Gillette/The Stock Market.*)

United States. Significant expansion of the present capacity would be resisted in the United States, however, because it would involve the drowning of farmlands and wilderness areas under reservoirs behind dams. In addition, poorly sited and designed dams become expensive to maintain because silt deposits accumulate on the upstream side and have to be removed.

Wind power, or the use of a windmill to drive an electric generator, is also a form of solar energy (see the photograph at the beginning of this chapter). Its use is slowly growing in some places as designs improve and costs are brought down. Solar energy (exclusive of hydroelectric power) supplies only a few tenths of 1 percent of consumption.

BENEFITS OF SOLAR ENERGY How much solar energy can we depend on in the years ahead? Solar energy enthusiasts believe that some 20 quads per year (exclusive of hydroelectric power) might be supplied in the United States by the year 2010. This is equivalent to about half of the oil we now use. Others think it more realistic to figure on less than 10 quads per year. All agree that under either scenario, important social benefits would be realized: conservation of other energy resources, diversification of energy supply so that we are not overly dependent on a single source, and reduction of fuel imports. With adequate research and development,

solar energy can probably become economically competitive and a major source of energy in the next century.

Geothermal Energy

In Chapter 19, we saw that Earth's internal heat, fueled by radioactivity, provides the energy for plate tectonics and continental drift, mountain building, and earthquakes. The same internal heat can be harnessed to drive electric generators and heat homes. **Geothermal energy** is produced when underground heat is transferred by water that is heated as it passes through a subsurface region of hot rocks (a **heat reservoir**) that may be hundreds or thousands of feet deep. The water is brought to the surface as hot water or steam through boreholes drilled for the purpose. The water is usually naturally occurring groundwater that seeps down along fractures; less typically, the water is artificially introduced by being pumped down from the surface.

Eighteen countries now generate electricity using geothermal heat. By far the most abundant form of geothermal energy occurs at the relatively low temperatures of 80 to 180°C. Water-circulating heat reservoirs in this temperature range are able to extract enough heat to warm residential, commercial, and industrial spaces. More than 20,000 apartments in France are now heated by warm underground water drawn from a heat reservoir in a geologic structure near Paris called the Paris Basin. Iceland sits on the Mid-Atlantic Ridge, a volcanic structure that was discussed in Chapters 5 and 20. Reykjavík, the capital of Iceland, is entirely heated by geothermal energy derived from volcanic heat.

Geothermal reservoirs with temperatures above 180°C are useful for generating electricity. They occur primarily in regions of recent volcanism as hot, dry rock, natural hot water, or natural steam (Figure 22.15). The latter two sources are limited to those few areas where surface water seeps down through underground faults and fractures to reach deep rocks heated by recent magmatic activity. Naturally occurring water heated above the boiling point and naturally occurring steam are highly prized resources for which geologists are searching (Figure 22.16). The world's largest supply of natural steam occurs at The Geysers, 120 km north of San Francisco. Enough electricity (over 600 megawatts) to meet about half the needs of San Francisco is currently being generated there. This field is now in its third decade of production and is beginning to show signs of decline, perhaps because of overdevelopment. Some 70 geothermal electric generating plants are in operation in California, Utah, Nevada, and Hawaii, generating 2800 megawatts of power—enough to supply about a million people.

Extracting heat from very hot, dry rocks presents a more difficult problem: the rocks must be fractured at depth to permit the circulation of water, and the water must be provided artificially. The rocks are fractured by water pumped down at very high pressures. Experiments are under way to develop technologies for exploiting this resource.

Like most of the other energy sources we have looked at, geothermal energy presents some environmental problems. Regional subsidence can occur if hot groundwater is withdrawn without being replaced. In addition, geothermally heated waters can contain salts and toxic materials dissolved from the hot rock. These waters present a disposal problem if they are not reinjected.

The contribution of geothermal energy to the world's energy future is difficult to estimate. Only $\frac{1}{40}$ of a quad per year of geothermal energy is currently produced in the United States, and perhaps twice that amount in the entire world. The resource is in a sense nonrenewable, because in most places the heat is being drawn out of a reservoir more rapidly than it can be replenished by slow geological processes, and because heat flows very slowly through solid rock. In many places, however—California, Hawaii, the Philippines, Japan, Mexico, the rift valleys of Africa—the resource is potentially so large that its future will depend on the economics of production. At present, we know how to use only naturally occurring hot water or steam deposits. Although the potential is enormous, our guess is that in the near term geothermal energy can make important local contributions only where the proximity of the resource to the user and the economics are favorable, as they are in California, New Zealand, and Iceland. Geothermal energy probably will not make large-scale contributions to the world energy budget until well into the next century, if ever.

FIGURE **22.15** The Geysers, the world's largest supply of natural steam. The geothermal energy is converted into electricity for San Francisco, 120 km to the south. (*Pacific Gas and Electric.*)

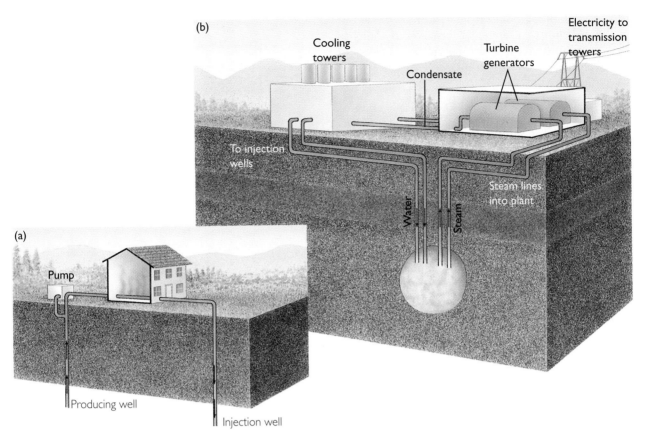

FIGURE 22.16 Applications of geothermal power. (a) Low-temperature hot water is used to heat a building and is then reinjected into the reservoir. (b) High-temperature steam drives turbines to generate electricity. Water condensed from steam is reinjected into the reservoir.

CONSERVATION

In a real sense, using energy more efficiently is like discovering a new source of fuel. It has been calculated that since the rise in oil prices in 1973 turned us all into conservationists, the world has saved more energy than it has gained from all new sources discovered in the same time. Some experts believe that conservation alone could cut the energy used by the industrialized nations in half. The savings in the United States could amount to some $200 billion a year, much of it spent on imported oil. Savings of this magnitude would reduce the cost of our products and make them more competitive in the world market, reduce our dependence on imported oil, and lower our trade deficit significantly. The kinds of practices that could lead to these savings require the application of mostly familiar technologies: changes from incandescent to fluorescent lighting; better home insulation; more efficient refrigerators, air conditioners, furnaces, and other appliances; more efficient motors, pumps, and other industrial devices; and better-performing automobile engines.

Political leadership and public education will be required to induce us to undertake these worthwhile changes today. In the years ahead, dwindling reserves and higher energy prices will force us to do so.

ENERGY POLICY

In view of the diversity and abundance of energy resources, you may well wonder why the world faces an energy crisis. There are several reasons:

1. The world demand for energy will increase, driven by the industrialization of China and other developing countries.

2. In the next few decades, world oil production will peak and begin to decline.

3. "Shortages"—the disruption of supply on political grounds by oil-producing nations—and the resulting escalation of prices may shock the world economic and political order severely.

4. Even without political interruptions in supply, global climatic change may restrict our ability to use fossil fuels, precipitating an energy crisis of even greater severity.

All nations aspire to grow economically and improve the quality of their people's lives. The challenge is to provide the energy to fuel this growth in the face of the depletion of oil resources and the possible need to cap or reduce the emission of carbon dioxide by reducing the use of fossil fuels (Figure 22.17). A partial solution, and the cheapest one, is to reduce the waste of energy by using it more efficiently. In the next few decades, however, we may also have to shift to a different mix of energy supply—drawn from natural gas, synthetic fuels for transportation, nuclear energy, renewable resources such as solar and biomass energy, and a gradually decreasing dependence on oil and coal. It is to be hoped that nuclear technology will improve in safety and regain public confidence and that advances in the technology for renewable energy sources will lower their costs.

Fossil fuels are too cheap in the United States. They are not taxed as much as in other advanced nations, so there is too little attention paid to conservation and the introduction of renewable resources. If the full social costs of fossil and nuclear fuels were included in the prices—the costs of cleaning up acid rain, oil spills, and other environmental damage; the costs of storing nuclear wastes; the cost of trade deficits; the costs of global warming; the military costs of defending oil supplies (for example, the defense of Kuwait against the Iraqi invasion)—renewable and geothermal energy resources could compete very well with fossil fuels. Some experts believe that even today the United States can meet 30 percent of its energy demand with renewable resources at competitive prices. Unfortunately, the political will to achieve these results does not yet exist.

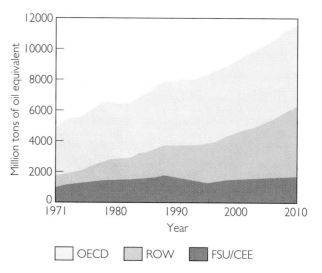

FIGURE 22.17 World energy demand 1971–2010. OECD, the Organization for Economic Cooperation and Development, consists primarily of the advanced industrialized nations. FSU/CEE is the former Soviet Union and Central and Eastern Europe. ROW, the rest of the world, refers primarily to China and other developing nations. The energy units are in terms of the energy of a million tons of oil. ROW is the largest factor in the growth of energy demand. (*World Energy Outlook,* 1994 edition, International Energy Agency, Paris, France.)

Nevertheless, we are in a race against time. We must develop the several options of renewable or indefinitely sustainable energy resources before remaining oil resources are depleted or before we are forced to shift away from fossil fuels because of their harmful effects on climate. Whether the transition to an era of energy security is smooth or rough depends on our determination, technological skill, and ability to solve the attendant complex social and political problems. In our view, the technological advances required are achievable if we start to develop them now. Whether the sociopolitical problems will be resolved in time is less certain.

SUMMARY

What is the origin of oil and natural gas? Oil and gas form from organic matter deposited in marine sediments, typically in the coastal waters of the sea. The organic materials are buried as the sedimentary layers grow in thickness. Under heat and high pressure, the organic matter is transformed into liquid and gaseous hydrocarbons, compounds of carbon

and hydrogen. Oil and gas accumulate in geologic traps that confine the fluids within impermeable barriers.

What environmental concerns are connected with the production of oil? Pollution in the production and transportation of oil is a major problem. Proponents of oil production believe that careful design of equipment and safety procedures can greatly reduce the chances of oil spills and satisfy the energy needs of our civilization. A significant body of public opinion is unconvinced, however, and exploration for new reserves is therefore restricted in many coastal regions and pristine wilderness areas, such as northern Alaska.

Why is there concern about the world's oil supply? Oil is a finite resource: it will be depleted faster than nature can replenish it. Therefore, as the supply is withdrawn from the oil reservoirs of the world, its availability diminishes, its price rises, and other types of energy sources will have to be found. At current rates of use, the remaining oil available for transportation, heating, and generation of electricity will be depleted in about 100 years.

Why is natural gas a premium fuel? Natural gas burns cleanly, releases less carbon dioxide per unit of energy than other fossil fuels, and is comparable in supply to oil resources.

What is the origin of coal and how big a resource is it? Coal is formed by the compaction and chemical alteration of wetland vegetation. It is present as huge resources in sedimentary rocks. We have used only about 2.5 percent of the world's minable coal.

What is the trade-off between risk and benefit in the use of coal? Coal mining and pollution caused by coal burning are risky to human life and to the environment. Coal combustion is a major source of carbon dioxide and the acid emissions that are the precursors to acid rain. Because of its abundance and low cost, however, the use of coal is likely to increase in the next decades for generation of electric power and for conversion to liquid and gaseous fuels.

Are there any other fossil fuels that can be drawn upon as oil reserves disappear? Reserves of oil shale and tar sands are great and may be extensively exploited in the next few decades as oil costs rise and technology improves. These fossil fuels will last well into the next century, when alternative energy sources should be available. Combustion of these fuels, however, like that of coal and oil, has environmental consequences of some concern.

Is nuclear energy a solution to the world's energy problem? Nuclear power from the fission of uranium 235 can be a major energy source, but only if its costs do not keep escalating and the public can be assured of its safety. It has the advantage that it does not release carbon dioxide and the disadvantage that safe repositories must be found to store radioactive wastes for hundreds of thousands of years. Known high-grade reserves of uranium 235 ore can support the projected use of conventional nuclear power plants for a few decades, longer if advanced reactors are installed early in the next century. The use of nuclear energy could extend our resources of fossil fuels.

What are the prospects for alternative energy sources? Alternative energy sources include hydroelectric power, solar energy, biomass energy, and geothermal power, none of which has any immediate prospect of being an adequate response to world energy needs. With advances in technology and reduction in cost, however, solar and biomass energy could become major energy sources in the next century.

What should be the goal of energy policy? The goal of energy policy should be to guide the nations of the world through the transition from oil to the more plentiful fossil fuels and to nuclear energy if its safety can be improved, thus buying time to develop nonpolluting and unlimited energy sources to replace fossil fuels in the next century. An important component of this policy must be the more efficient use of energy to conserve resources and reduce environmental damage. The resources and technology exist to achieve this goal peacefully and without economic dislocation if the sociopolitical problems can be worked out.

KEY TERMS AND CONCEPTS

sustainable development (p. 562)
reserves (p. 562)
resources (p. 562)
fossil fuels (p. 564)
source bed (p. 564)
oil reservoir (p. 565)

oil trap (p. 565)
quad (p. 568)
syncrude (p. 571)
oil shale (p. 572)
tar sand (p. 572)
nuclear energy (p. 573)
solar energy (p. 576)

biomass (p. 576)
hydroelectric energy (p. 576)
wind power (p. 577)
geothermal energy (p. 577)
heat reservoir (p. 577)

EXERCISES

1. What sedimentary environments favor the formation of sediments containing organic matter that might later be transformed into petroleum? Give some modern examples.

2. Do you think the continental slope and rise might be good places to drill for oil if we can invent the necessary technology? Why or why not?

3. Which of the following factors are most important in estimating the future supply of oil and gas: (a) rate of oil accumulation, (b) rate of natural seepage of oil, (c) rate of pumping of oil from known reserves, (d) rate of discovery of new reserves, (e) total amount of oil now present in the Earth?

4. Taking benefits and risks into consideration, rank according to relative importance all the forms of fossil fuels and explain how their rankings might differ at the end of the next century.

5. Name four regions of the world that are major sources of oil.

6. Which three countries have the largest coal reserves?

7. Contrast the risks and benefits of nuclear fission and coal combustion as energy sources.

8. How would you use knowledge of the distribution of plate boundaries to make a map showing the areas of Earth most likely to be sources of geothermal power?

9. In what important respects do uranium reserves differ from fossil fuel reserves?

10. Give three examples of energy derived from biomass.

11. What social costs should be included in evaluating the true costs of energy derived from fossil fuels, from nuclear fission, and from geothermal heat?

12. What do you think will be the major sources of energy in the year 2050? In the year 3000?

THOUGHT QUESTIONS

1. A tax of $1 per gallon on gasoline would generate $100 billion of income for the federal government and reduce the budget deficit substantially. It would force drivers to switch to more fuel-efficient automobiles, thereby reducing imports of oil. Even with this increase, the price of gasoline would still be lower in the United States than in many European countries. Despite all these advantages, such a tax is unlikely to be enacted. What are the arguments against it?

2. Should we stockpile oil against future shortages caused by political disruptions? What are the costs and benefits of doing so?

3. Taking into account the costs of oil imports to the economy, the risks of relying on foreign exporters, and the environmental consequences, do you think we should remove restrictions on the production of oil from the continental shelves and wildlife preserves?

4. Explain the statement that increased conservation is the cheapest new energy source.

5. Would you rather live near a nuclear reactor or a plant that generates electricity by burning coal?

6. You are the U.S. representative to the United Nations. You have just made a speech extolling the importance of sustainable development. How would you respond to the representative of a developing country that has to increase energy production to grow economically, who claims that her country does not have the capital to invest in renewable energy and criticizes the United States for being profligate in the use of resources?

LONG-TERM TEAM PROJECT

See Chapter 11.

SHORT-TERM TEAM PROJECT: ALTERNATIVES TO OIL

Regions of the world that produce oil often suffer conflict over access to and use of this nonrenewable resource. The Nigerian activist Ken Saro-Wiwa was executed by his own government, despite worldwide protest, for agitating against oil companies accused by knowledgeable people of polluting the Niger delta. Considering the turmoil that often surrounds the use of oil, you and a partner investigate and report on an alternative energy source that could be widely used because of global availability.

SUGGESTED READINGS

Abelson, P. H. 1987. Energy futures. *American Scientist* 75:584–593.

Department of Energy. 1991. *National Energy Strategy*. Washington, D.C.: U.S. Government Printing Office.

Energy Information Administration. 1994. *Annual Energy Review*. Washington, D.C.: U.S. Department of Energy.

Energy for Planet Earth. 1990. Special issue of *Scientific American* (September).

Gibbons, J. H., and P. D. Blair. 1991. U.S. energy transition: On getting from here to there. *Physics Today* (July):21–30.

Goldenberg, Jose. 1995. Energy needs in developing countries and sustainability. *Science* 269:1058–1059.

National Research Council. 1991. *Policy Implications of Greenhouse Warming*. Washington, D.C.: National Academy Press.

National Research Council. 1992. *Radioactive Waste Repository Licensing*. Washington, D.C.: National Academy Press.

National Research Council. 1993. *Solid-Earth Sciences and Society*. Washington, D.C.: National Academy Press.

Resource Reserve Definitions. 1980. Circular 831. Washington, D.C.: U.S. Geological Survey.

Stone, J. L. 1993. Photovoltaics: Unlimited electrical energy from the Sun. *Physics Today* (September):22–29.

Tenenbaum, David. 1995. Tapping the fire down below. *Technology Review* (January):38–47.

INTERNET SOURCES

Yucca Mountain Project (YMP)
🛈 **http://yucca_web.ymp.gov/**
The Yucca Mountain Project page is provided by the United States Office of Radioactive Waste Management. Features include a data base, news about the Yucca Mountain site, FAQ, maps, and images. The education program includes on-line lessons.

About Oil
🛈 **http://www.shellus.com/OilProducts/abtoil.html**
The Shell Oil Company provides this overview of petroleum exploration, production, transportation, and products.

A Matter of Degrees: A Primer on Global Warming
🛈 **http://www.ns.doc.ca/udo/warm-1.html**
The Atmospheric Environment Service of Environment Canada produced this extensive site, which is written largely as a series of questions and answers. A bibliography for each section, a glossary, and a series of images supplement the text.

23

Limestone blocks cut from limestone formation in an Indiana quarry. Limestone is an important source of building stone. Code numbers are used to match colors and textures. (*Jeffrey A. Wolin.*)

Mineral Resources from the Earth

From the Stone Age to the Bronze Age to the Iron Age, our evolving ability to use Earth's minerals is a measure of the progress of civilization. Today we mine mineral deposits deep in the Earth. We strip away the surface deposits on land and evaporate seawater to obtain dissolved metals. As we explore the seafloor, we are finding minerals deposited by hot springs and accumulations of manganese and other metals.

Like fossil fuels (see Chapter 22), minerals are vital to the functioning of a modern nation. Just about everything we use comes from the ground. Metallic mineral deposits are the source of the iron, aluminum, and all the other metals we use for everything from spacecraft to salt shakers. Nonmetallic mineral deposits provide salt, clay, gravel, building stone, the limestone from which cement is made, the sand from which glass and transistors are made, and many other important products.

Despite its benefits, mining has been called the original dirty industry. Mining strips more of the Earth's surface each year than the natural erosion of rivers. The waste products exceed those produced by the world accumulation of municipal garbage each year. Despite progress in developing clean and environmentally benign mining operations, mining and processing are major contributors to environmental degradation and poor health in many countries.

In this chapter, we will survey a broad range of Earth's mineral resources. In addition to their geological settings, we will discuss their economic and social context, including issues of distribution, depletion, environmental and health costs, recycling, substitution, and price.

MINERALS AS ECONOMIC RESOURCES

Although mining itself represents only a small part of the national income of the United States and Canada, these nations depend on materials extracted from the Earth. Without them, there would be no stone for buildings, phosphates for fertilizers, cement for construction, clays for ceramics, sand for silicon transistors and fiber-optic cables, and metals for just about everything. Annual consumption of minerals in the United States alone has been calculated to be 18,000 pounds per person (Figure 23.1).

Concentration of Minerals

The chemical elements of Earth's crust are widely distributed in many kinds of minerals, and those minerals are found in a great variety of rocks. In most places, any given element will be found homogenized with the other elements in amounts close to its average concentration in the crust. An ordinary granitic rock, for example, may contain a few percentage points of iron, close to the average concentration of iron in Earth's crust.

Elements that occur in higher concentrations have undergone some geologic process that has segregated much larger quantities of the element than normal. High concentrations of elements are found in a limited number of specific geological settings. These are the settings of economic interest, because the higher the concentration of a resource in a given deposit, the lower the cost to recover it.

ORE MINERALS   Rich deposits of minerals from which valuable metals can be recovered profitably are called **ores;** the minerals containing these metals are **ore minerals.** Ore minerals include sulfides (the main group), oxides, and silicates. Ore minerals in each of these groups are compounds of metallic elements with sulfur, oxygen, and silicon oxide, respectively. The copper ore mineral covelite, for example, is a copper sulfide (CuS). The iron ore mineral hematite (Fe_2O_3) is an iron oxide (Figure 23.2). The nickel ore mineral garnierite is a nickel silicate,

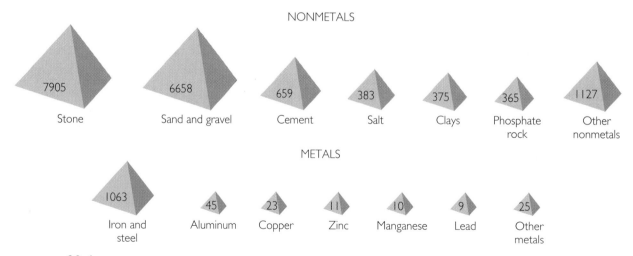

FIGURE 23.1 Average per capita consumption of nonfuel minerals in the United States (in pounds). During the 1980s, each person in the United States used more than 18,000 pounds of minerals each year. (*U.S. Bureau of Mines, 1992.*)

FIGURE 23.2 Iron ores (clockwise from left): magnetite, siderite, iron pyrite, hematite. *(Chip Clark.)*

$Ni_3Si_2O_5(OH)_4$. In addition, some metals, such as gold, are found in their native state—that is, uncombined with other elements (Figure 23.3).

CONCENTRATION FACTOR The **concentration factor** of an element in an ore body is the ratio of the element's abundance in the deposit to its average abundance in the crust. The *economical* concentration factor varies from element to element, depending on its average abundance (Table 23.1).

Iron, one of the most common elements of the crust, has an average abundance in crustal rocks of 5.8 percent. An *economical* iron ore—one that is profitable to mine under current costs of extraction, costs of transportation, and selling prices—must contain at least 50 percent iron, about 10 times the average crustal abundance. In other words, an iron ore becomes economical when its concentration factor is about 10. A less abundant metal, such as copper, which has a crustal abundance of 0.0058 percent, is concentrated by factors of at least 80 to 100 in its economical ores. Ore deposits of the rarer elements, such as mercury and gold, require concentration factors in the thousands to hundreds of thousands to be economical.

FIGURE 23.3 Gold occurring in the free state (native gold) on a quartz crystal. *(Chip Clark.)*

TABLE 23.1

ECONOMICAL CONCENTRATION FACTORS OF SOME COMMERCIALLY IMPORTANT ELEMENTS

ELEMENT	CRUSTAL ABUNDANCE (PERCENT BY WEIGHT)	ECONOMICAL CONCENTRATION FACTOR[1]
Aluminum	8.00	3–4
Iron	5.8	5–10
Copper	0.0058	80–100
Nickel	0.0072	150
Zinc	0.0082	300
Uranium	0.00016	1,200
Lead	0.00010	2,000
Gold	0.0000002	4,000
Mercury	0.000002	100,000

[1]Concentration factor = abundance in deposit divided by crustal abundance.

SOURCES: Data from B. J. Skinner, *Earth Resources,* Prentice Hall, 1969; D. A. Brobst and W. P. Pratt, *Mineral Resources of the U.S.,* USGS Professional Paper 820, 1973.

Supply of Minerals

Elements are so widely distributed in many common rocks that whether a particular deposit should be considered a resource or a reserve (see Chapter 22) depends on the costs of recovery and the selling price. Theoretically, with enough money and energy, we could extract both abundant and rare elements from any rock and never run out of minerals. Our practical concern, however, is the exhaustion of reserves, the identified mineral deposits that are profitable to mine and purify. We can reasonably expect that new discoveries will add to current reserves, but at an uncertain rate. Once the highest grade deposits are mined out, we will be forced to rely on deposits of lower grades. These could be more expensive to recover, although technological advances could compensate to some extent.

DECREASING DEMAND FOR MINERALS Despite predictions in the 1960s and 1970s that the world would run short of economically important minerals, current estimates indicate that they will be available and affordable for the next century or so. Rates of mineral consumption in the United States and other industrialized nations have slowed. These mature economies are shifting away from construction and manufacturing to service, technology, and other activities that require less raw materials. Demand for metals, in particular, has been reduced by **recycling** the metal content of discarded goods and by substitution of low-cost ceramics, composites, and plastics.

Recycling is a growth industry because it has become less expensive to extract certain elements from trash, such as aluminum cans and automobiles that have been discarded, than from deposits of ore minerals (Figure 23.4). About 45 percent of the

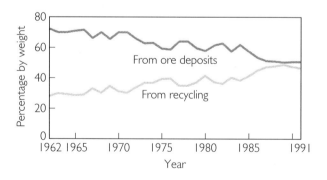

FIGURE 23.4 U.S. consumption of metals derived from ore deposits and from recycled materials. The consumption trends began to converge in the 1970s; by the 1990s, new and recycled metals were being consumed in nearly equal amounts. (Carroll Ann Hodges, *Science,* 268, 1995, p. 1307.)

gold, platinum, and aluminum now consumed is recycled material. Even larger fractions of lead (73 percent), copper (60 percent), and iron and steel (56 percent) are being recycled.

Recycling is a complex issue because it is not always economically or environmentally justified. The only rational way to decide whether or not to recycle is to compare the costs of mining, smelting, and transporting natural ores; disposing of wastes; and controlling pollution with the corresponding costs of recycling metals.

IMPORTS Mineral resources are ultimately finite, and it is wise to think about the long-term future. In the United States, for example, only a few deposits are available in quantities adequate to last for more than a few hundred years. Although the United States is the world's largest producer of raw minerals, it is also dependent on mineral imports, sometimes because imports are cheaper, sometimes because the country lacks reserves of a particular mineral.

Figure 23.5 shows that the United States satisfies more than 60 percent of its demand for 20 of the

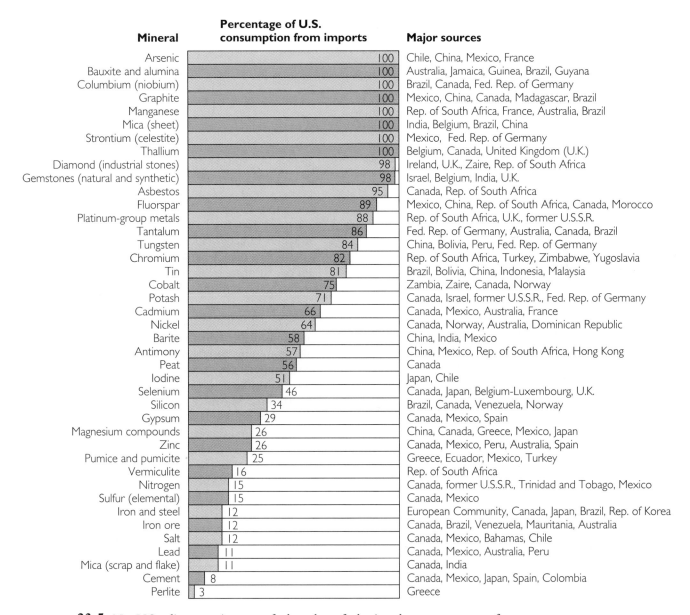

Mineral	Percentage of U.S. consumption from imports	Major sources
Arsenic	100	Chile, China, Mexico, France
Bauxite and alumina	100	Australia, Jamaica, Guinea, Brazil, Guyana
Columbium (niobium)	100	Brazil, Canada, Fed. Rep. of Germany
Graphite	100	Mexico, China, Canada, Madagascar, Brazil
Manganese	100	Rep. of South Africa, France, Australia, Brazil
Mica (sheet)	100	India, Belgium, Brazil, China
Strontium (celestite)	100	Mexico, Fed. Rep. of Germany
Thallium	100	Belgium, Canada, United Kingdom (U.K.)
Diamond (industrial stones)	98	Ireland, U.K., Zaire, Rep. of South Africa
Gemstones (natural and synthetic)	98	Israel, Belgium, India, U.K.
Asbestos	95	Canada, Rep. of South Africa
Fluorspar	89	Mexico, China, Rep. of South Africa, Canada, Morocco
Platinum-group metals	88	Rep. of South Africa, U.K., former U.S.S.R.
Tantalum	86	Fed. Rep. of Germany, Australia, Canada, Brazil
Tungsten	84	China, Bolivia, Peru, Fed. Rep. of Germany
Chromium	82	Rep. of South Africa, Turkey, Zimbabwe, Yugoslavia
Tin	81	Brazil, Bolivia, China, Indonesia, Malaysia
Cobalt	75	Zambia, Zaire, Canada, Norway
Potash	71	Canada, Israel, former U.S.S.R., Fed. Rep. of Germany
Cadmium	66	Canada, Mexico, Australia, France
Nickel	64	Canada, Norway, Australia, Dominican Republic
Barite	58	China, India, Mexico
Antimony	57	China, Mexico, Rep. of South Africa, Hong Kong
Peat	56	Canada
Iodine	51	Japan, Chile
Selenium	46	Canada, Japan, Belgium-Luxembourg, U.K.
Silicon	34	Brazil, Canada, Venezuela, Norway
Gypsum	29	Canada, Mexico, Spain
Magnesium compounds	26	China, Canada, Greece, Mexico, Japan
Zinc	26	Canada, Mexico, Peru, Australia, Spain
Pumice and pumicite	25	Greece, Ecuador, Mexico, Turkey
Vermiculite	16	Rep. of South Africa
Nitrogen	15	Canada, former U.S.S.R., Trinidad and Tobago, Mexico
Sulfur (elemental)	15	Canada, Mexico
Iron and steel	12	European Community, Canada, Japan, Brazil, Rep. of Korea
Iron ore	12	Canada, Brazil, Venezuela, Mauritania, Australia
Salt	12	Canada, Mexico, Bahamas, Chile
Lead	11	Canada, Mexico, Australia, Peru
Mica (scrap and flake)	11	Canada, India
Cement	8	Canada, Mexico, Japan, Spain, Colombia
Perlite	3	Greece

FIGURE 23.5 Net U.S. reliance on imports of selected nonfuel minerals as a percentage of consumption in 1993. *(U.S. Bureau of Mines, 1994.)*

42 listed minerals through imports. Chances are that the graphite in your pencil comes from Mexico or China. Those who are concerned with U.S. economic and defense security, and mining industry representatives worried about sales, are troubled by the country's heavy dependence on imports of several strategically critical metals. These are metals—such as cobalt, manganese, chromium, titanium, and the platinum group—without which an entire industry, such as the aircraft or chemicals industry, could collapse. Since 1939, therefore, the U.S. government has established stockpiles of minerals for use in an emergency, such as an economic boycott or a wartime cutoff of supply.

Some people think that the United States should increase domestic supply and decrease dependence

23.1   LIVING ON EARTH

Use of Federal Lands in the United States

The land owned and managed by the federal government amounts to about a quarter of the country and represents an important part of the political, social, and economic history of the United States. The stewards are primarily the Bureau of Land Management and the Forest Service, although the National Park Service and the Department of Defense also manage some of the land.

Over the years, these lands were bought at bargain prices, won in wars, or acquired by negotiations with other nations. The Louisiana Purchase from France in 1803 added huge land areas from the mouth of the Mississippi to what is now the state of Montana, at a cost of $15 million. Florida was bought from Spain in 1819 for $7 million. The treaty of 1848 following the war with Mexico added most of the Southwest to the public domain, including parts of Arizona and New Mexico and all of California, Nevada, and Utah. Great Britain gave us the lands that are now the states of Washington, Oregon, and Idaho. The last major acquisition was the purchase of Alaska from Russia for $7 million in 1867.

In addition to these lands, the continental shelf beyond a few miles offshore is owned and managed by the federal government. The continental shelf is valuable for its oil and gas resources, fisheries, and mineral resources.

Over U.S. history, some 60 percent of lands in the public domain were transferred to states, homesteaders, and various organizations to aid education (the land grant colleges), to build the privately owned railroads, to reward veterans, and for other purposes.

The federal lands have many uses, some in conflict with others. Their commercial value is very large: rents and royalties from private users amounted to $6.2 billion in 1992. About 80 percent of this federal income is from oil and gas producers, mostly offshore, followed by timber (less than 20 percent) and mineral-mining interests (about 1 percent). Cattle ranchers (grazing) and recreational users contribute less than 1 percent each. Recreational users constitute by far the largest group of individuals using the land.

The U.S. mining industry enjoys a huge subsidy legislated in the Mining Act of 1872, an unjustifiable relic of frontier days that still allows mining companies to buy federal land with mineral deposits for about $5.00 an acre or less. Mining was given the highest priority for land use in earlier times. Today, many people would assign greater value to wilderness protection and recreational use.

Federal ownership and management of these holdings is one of the most contentious political issues of the day, as evidenced by congressional debates, election campaigns, lobbying, and lawsuits brought against the federal government by states and private interests. Here are some of the policy issues that must be resolved:

- How much, if any, federal land should be sold off, to whom, and at what price?

Much of this discussion was drawn from Marion Clawson, *The Federal Lands Revisited*, Washington, D.C., Resources for the Future, 1983.

on imports by allowing greater access to its federal lands for exploitation of mineral resources. Others think that federal lands should be preserved for public recreational use (see Feature 23.1).

Economists point out that the United States has always been dependent on imports of many minerals, that alternative markets are always available, and that major interruptions of supply have not occurred. Imports may be advantageous as long as the supply is secure, the price is reasonable, and overall exports and imports are in rough balance. Moreover, even though the United States has consumed more minerals over the past five decades than were consumed in all previous time, most mineral reserves have increased as a result of advances in geologic knowledge and improvements in mining and metal extraction.

- If the government retains ownership, what kind of access should be given commercial interests, recreational users, and others, and on what terms?

- How much of the land should be set aside for wilderness preservation?

- To what degree should state and local governments participate in management of the lands?

A case can be made for federal retention of the lands. The federal government can look after national interests better than state and local governments. The federal government is best suited to organize and monitor the multiple diverse uses of the land. Public agencies can take a longer view of conservation and the needs of future generations than private interests driven by short-term profits.

There are also arguments for selling off much of the land. Private owners would manage the lands more efficiently than federal bureaucrats who lack management experience and are subject to political pressure. The greatest benefits to society come from the accumulated effect of individual entrepreneurs who work to achieve the greatest personal gain. There is not much evidence that government administration of the land for any purpose is more exemplary than that of the private sector. Jobs created by the private sector are more important than preserving some unimportant subspecies.

The time has come to try to achieve a national consensus on the uses of federal lands.

Suburbs encroaching on a salt pond next to San Diego Bay where waterfowl and shore birds nest and feed. Such wildlife refuges, if not protected, are unlikely to last. (*Joel Sartore/National Geographic Image Collection.*)

A major uncertainty is the future course of developing nations. World mineral consumption will continue to increase because of the growing demand for raw materials from developing countries where population growth defies control and there is a strong aspiration to industrialize and improve the standard of living. Can this demand be met from existing reserves? Are clean mining and environmental protection, together with recycling, substitution, and other conservation measures, economically feasible for these countries, which point out that the industrialized nations, now sensitive to the environment, have historically been among the biggest polluters during their own development?

THE GEOLOGY OF MINERAL DEPOSITS

Mineral deposits are created by various kinds of geologic processes, most of which have already been discussed in earlier chapters. In general, a mineral deposit forms when three conditions are satisfied:

1. A source of the minerals exists in a place where it is accessible to a natural transport mechanism.

2. A natural transport mechanism is available to move the minerals away from the source.

3. A site exists with a mechanism for the transport agent to deposit the minerals.

It is not luck that places ore deposits close to Earth's surface, where humans can reach them. Rocks near the surface contain cracks and open fractures (pressure closes them at greater depths), allowing easier transportation of ore-bearing fluids. Also, rocks near the surface are cooler, so that ore minerals precipitate from the hot fluids that carry them.

Hydrothermal Deposits

Many of the richest known ore deposits crystallized out of hot, aqueous solutions called *hydrothermal solutions*. These hot waters are the transport mechanism. They can emanate directly from the magma of an igneous intrusion (the source) and carry away in solution ore constituents derived from the soluble components of the magma (Figure 23.6). Hydrothermal solutions can also form when circulating groundwater contacts heated rock or a hot intrusion, reacts with it, and carries off ore constituents released by the reaction.

VEIN DEPOSITS   Ore constituents are often deposited in fractured rocks. The hot fluids flow easily through the fractures and joints, cooling rapidly in the process. Quick cooling causes fast precipitation of the ore constituents. The tabular or sheetlike deposits of precipitated minerals in the fractures and joints are called **vein deposits,** or simply **veins.** Some ores are found in veins; others are found in the rock adja-

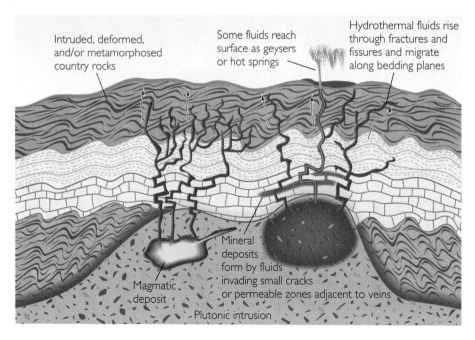

Intruded, deformed, and/or metamorphosed country rocks

Some fluids reach surface as geysers or hot springs

Hydrothermal fluids rise through fractures and fissures and migrate along bedding planes

Magmatic deposit

Mineral deposits form by fluids invading small cracks or permeable zones adjacent to veins

Plutonic intrusion

FIGURE 23.6 Many ore deposits are found in hydrothermal veins formed by hot solutions rising from magmatic intrusions.

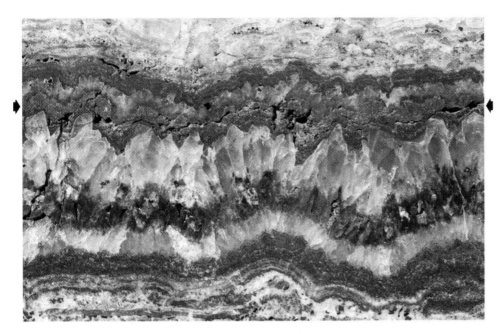

FIGURE 23.7 Quartz vein deposit (about 1 cm thick) containing gold and silver ores. Oatman, Arizona. *(Peter Kresan.)*

cent to the veins (country rock) that was altered when the hot solutions heated and infiltrated them. As the solutions react with surrounding rocks, they may precipitate ore minerals together with quartz, calcite, or other common vein-filling minerals. Vein deposits are a major source of gold (Figure 23.7).

Hydrothermal vein deposits are among the most important sources of metal ores. Typically, *metallic* ores occur as sulfides, such as iron sulfide (pyrite), lead sulfide (galena), zinc sulfide (sphalerite), mercury sulfide (cinnabar)—shown in Figure 23.8—and copper sulfide (covelite and chalcocite). When hydrothermal solutions reach the surface, they become hot springs and geysers, many of which precipitate mineral ores—including ores of lead, zinc, and mercury—as they cool.

FIGURE 23.8 Metal sulfide ores (from left to right): galena (lead sulfide), cinnabar (mercury sulfide), iron pyrite, sphalerite (zinc sulfide). Sulfides are the most common types of metallic ores. *(Chip Clark.)*

FIGURE 23.9 Copper ores (from left to right): chalcopyrite, malachite, chalcocite. Chalcopyrite and chalcocite are copper sulfide ores. Malachite is a carbonate of copper found in association with sulfides of copper. *(Chip Clark.)*

DISSEMINATED DEPOSITS Mineral deposits that are scattered through volumes of rock much larger than veins are called **disseminated deposits.** In igneous and sedimentary rocks alike, minerals are disseminated along abundant cracks and fractures. Among the economically important disseminated deposits are the porphyry-copper deposits of the southwestern United States and Chile. These deposits became mineralized when ore-forming minerals were introduced into a great number of tiny fractures in porphyritic felsic intrusives (granitic rocks containing large feldspar or quartz crystals in a finer-grained matrix) and country rocks surrounding the higher parts of the plutons. Some unknown process associated with the intrusion or its aftermath broke the rocks into millions of pieces. Hydrothermal solutions penetrated and recemented the rocks by precipitating ore minerals throughout the extensive network of tiny fractures. The most common copper mineral in porphyry is chalcopyrite, a copper sulfide (Figure 23.9). This widespread dispersal produced a low-grade but very large resource of many millions of tons of ore, which can be mined economically by large-scale methods (Figure 23.10).

FIGURE 23.10 Open-pit copper mine south of Tucson, Arizona. Open-pit mining is typical of the large-scale methods used to exploit widely disseminated ore deposits. *(Bob Lynn/Cyprus Minerals.)*

Extensive disseminated hydrothermal deposits may also occur in sedimentary rocks, such as in the lead-zinc province of the upper Mississippi River valley, which extends from southwestern Wisconsin to Kansas and Oklahoma. Because the ores in this province are not associated with a known magmatic intrusion that could have been a source of hydrothermal fluids, their origin is unknown. Some geologists speculate that the ores were deposited by groundwater. Groundwater may have penetrated hot crustal rocks at great depths and extracted soluble ore constituents, then moved upward into the overlying sedimentary rocks, where it precipitated its mineral load as fillings in cavities. In some cases, ore fluids appear to have dissolved some carbonates when they infiltrated limestone formations and then replaced them with equal volumes of new crystals of sulfide. The major minerals of the hydrothermal deposits in this province are lead sulfide (galena) and zinc sulfide (sphalerite).

Igneous Ore Deposits

The most important **igneous ore deposits**—deposits of ores in igneous rocks—are found as segregations of ore minerals near the bottom of intrusions. The deposits are formed when minerals crystallize from molten magma, settle, and accumulate on the floor of a magma chamber (see Chapter 4). Most of the chromium and platinum ores in the world, such as the deposits in South Africa and Montana, are found as layered accumulations of minerals that formed this way (Figure 23.11). One of the richest ore bodies ever found, at Sudbury, Ontario, is a large mafic intrusive formation containing great quantities of layered nickel, copper, and iron sulfides near its base. These sulfide deposits are believed to have formed from crystallization of a dense, sulfide-rich liquid that separated from the rest of the cooling magma and sank to the bottom of the chamber before it congealed.

PEGMATITES Pegmatites are extremely coarse-grained intrusive rocks of granitic composition that are usually found as veins, dikes, or lenses in granitic batholiths. Pegmatites form by fractional crystallization (Chapter 4) of a granitic magma. As the magma in a large granitic intrusion cools, the last melt to freeze solidifies as pegmatites, in which minerals that occurred only in trace amounts in the parent body are concentrated. For this reason, pegmatites may contain rare mineral deposits rich in such elements as boron, lithium, fluorine, niobium, and uranium and in gem minerals such as tourmaline.

FIGURE 23.11 Chromite (chrome ore, dark layer) occurring in a layered igneous intrusive. Bushveldt Complex, South Africa. *(Spence Titley.)*

23.2   INTERPRETING THE EARTH

The Great Canadian Diamond Rush

The announcement in 1991 of the discovery of a diamond-bearing kimberlite pipe at Lac de Gras in Canada's Northwest Territories triggered what may have been North America's largest land rush. Before 1991, sources of diamonds in economic amounts had been found on the Archean cratons of every continent except North America. The idea that the Canadian craton was a worthy prospect for diamonds became an obsession

Two geologists examining kimberlite core pipe samples in Canada's Northwest Territories, where Charles Fipke correctly predicted that diamonds would be found in economic amounts. *(Stephen Ferry/Gamma Liaison.)*

KIMBERLITES   One of the most valuable minerals, diamond, occurs chiefly in ultramafic igneous rocks called kimberlites, named for Kimberley, South Africa, where they are found in relative abundance. These rocks were forcefully intruded to the surface from deep in the crust and upper mantle in the form of long, narrow pipes. We know these diamond-bearing kimberlites originate at great depths because diamonds and other minerals found in them can be formed only under the conditions of extremely high pressure that exist in the upper mantle. Kimberlites erupt to the surface at high speed, propelled by pressurized volatiles such as H_2O and CO_2. No one has ever seen a kimberlite eruption. One geologist has likened it to a shotgun blast fired from the mantle through the lithosphere to the surface.

Rich accumulations of diamonds have also been found in alluvial deposits hundreds of kilometers from their kimberlite pipe source, transported there by rivers that picked up fragments eroded from a pipe and carried them downstream. Flowing ice sheets may also transport kimberlite fragments, as one venturesome geologist affirmed (Feature 23.2).

Sedimentary Mineral Deposits

Sedimentary mineral deposits include some of the world's most valuable mineral sources. Many economically important minerals segregate by chemical and physical means as an ordinary result of sedimentary processes (see Chapter 7).

NONMETALLIC SEDIMENTARY DEPOSITS   Limestones, for instance, chemically precipitated mainly by marine organisms, are used for cement, agricultural lime, and building stone. Pure quartz sands, left behind when mixed-mineral sands are abraded and winnowed by waves and currents so that all materials other than quartz are removed, are the raw materials

with Charles Fipke. This determined, independent Canadian geologist borrowed money and offered stock to raise funds for his exploration of a remote and inhospitable area some 200 km south of the Arctic Circle and 300 km northeast of Yellow Knife. In his fieldwork, he had to contend with blood-sucking mosquitoes, biting black flies, wolves and bears, and miserable climate.

Fipke was looking for trails of heavy minerals—such as ilmenite, diopside, and garnet—that are found in association with diamonds in kimberlite pipes. His idea was to follow the trails back to the kimberlite pipe at the source. He sampled eskers and other glacial drift deposited by the ice sheet that covered Canada until about 10,000 years ago. The eskers run roughly in the direction of ice movement. Fipke worked "up ice," that is, opposite to the direction of ice flow. Usually, a geologist would follow an upstream direction in searching for a source of minerals. Fipke reasoned that in the case of the Canadian Arctic, it

would have been the moving ice sheet rather than rivers that carried eroded mineral fragments of a kimberlite pipe away from its source.

Fipke's hunch paid off handsomely when he found kimberlite pipes containing diamonds in economic amounts. In the few short years since his discovery, some 250 companies have staked out claims over several hundred thousands of square kilometers of the Northwest Territories. Several hundred million dollars have been invested in geological exploration, including airplane surveys with remote sensors that could find kimberlite pipes. At least 130 kimberlite pipes have been discovered, of which 70 have been reported as worth mining for diamonds. Multi-billion-dollar mining operations are now being initiated.

A determined geologist, with a hypothesis that he believed in, transformed an area's economy, creating new wealth and thousands of jobs. Fipke has made Canada a significant player in the world market in diamonds.

for glassmaking and for the fiber-optic cables that are replacing copper wires in communication lines. Coarse sand and gravel, suitable for construction purposes, have been abundantly distributed in many areas of the northern United States and southern Canada by the Pleistocene glaciers; these materials are also widely distributed in channels and former channels of many rivers. Clays of high purity produced by prolonged weathering are used for ceramics in both home and industry. Evaporite deposits of gypsum, separated from seawater by precipitation, are used for plaster; sodium and potassium salts from evaporites have varied uses, from table salt to fertilizer. Phosphate rocks—marine shales and limestones enriched in phosphate by chemical reaction with deep seawater—are raw materials for the world's fertilizer industry.

METALLIC SEDIMENTARY DEPOSITS Sedimentary mineral deposits are also important sources of

copper, iron, and other metals. These deposits were chemically precipitated in sedimentary environments to which large quantities of metals were transported in solution. Some of the important sedimentary copper ores, such as those of the Permian Kupferschiefer (German for "copper slate") beds of Germany, may have precipitated from hot brines of hydrothermal origin, rich in metal sulfides, that interacted with sediments on the ocean bottom.

The major iron ores have been found in Precambrian sedimentary rocks. Earth's atmosphere was poor in oxygen early in its history (see Chapter 1), and it is now thought that the low availability of oxygen allowed an abundance of iron in its soluble, lower oxidation or ferrous form (Fe^{2+}) to be leached in great quantities from the land surface. The ferrous iron was transported in solution by groundwater to broad, shallow marine environments where it could be oxidized to its insoluble ferric form (Fe^{3+}) and precipitated. (Chapter 6 discusses oxidation states of

FIGURE 23.12 Precambrian banded iron beds. Rust-colored layers are limonite, interbedded with hematite and chert. Hammerslee, Australia. *(Spence Titley.)*

iron.) In many of these basins, the iron was deposited in thin layers alternating with layers of siliceous sediments (cherts). Such iron ores are called *banded iron formations* (Figure 23.12). The Lake Superior iron deposits, for many years the source of iron for the U.S. steel industry, are of this type.

PLACERS Possibly the most publicized (and romanticized) type of mineral prospecting is panning for gold: the gold seeker dredges up a flat pan of river sediment and sifts it in the hope of turning up the glint of a nugget. Many rich deposits of gold, diamonds, and other heavy minerals such as magnetite and chromite are found in **placers,** ore deposits that have been concentrated by the mechanical sorting action of river currents.

Because heavy minerals settle out of a current more quickly than lighter minerals such as quartz and feldspar, the heavy minerals tend to accumulate on river bottoms and sandbars, where the current is strong enough to keep the lighter minerals suspended and in transport but too weak to move the heavier material. In a similar manner, ocean waves preferentially deposit heavy minerals on the beach or on shallow offshore bars. The gold panner accomplishes the same thing: the shaking of a water-filled pan allows the lighter minerals to be washed away, leaving the heavier gold in the bottom of the pan (Figure 23.13). Because of the recent high prices of gold, old-fashioned gold panning is undergoing a revival.

Some placers can be traced upstream to the location of the original mineral deposit, usually of igneous origin, from which the minerals were eroded. Erosion of the Mother Lode, an extensive gold-bearing vein system lying along the western flanks of the Sierra Nevada batholith, produced the placers that were discovered in 1848 and led to the

FIGURE 23.13 Gold mining in the Sierra Nevada, California, 1852. Miners used running water to wash away valueless lighter minerals and concentrate the heavier gold. *(California State Archives.)*

California gold rush. The placers were discovered first, then their source. This was also the sequence of events that led to the discovery of the Kimberley diamond mines of South Africa two decades later.

ORE DEPOSITS AND PLATE TECTONICS

Plate-tectonics theory explains the various types of igneous activity in terms of the interactions of plates at boundaries where they separate or collide. Since igneous processes bring chemical elements and their mineral compounds from the interior to the surface, the theory of plate tectonics provides a foundation for understanding the origin of ore deposits. Such an understanding helps to explain existing ore deposits and can lead to the discovery of new ones.

In 1979, geologists exploring the seafloor at a plate-separation center (the East Pacific Rise) discovered hot springs, laden with dissolved minerals, venting on the seafloor (see Feature 17.3). These hot springs have their origin in seawater that circulates through fractures near the rift where the plates separate (Figure 23.14). The seawater is heated to temperatures of several hundred degrees Celsius when it comes in contact with magma or hot rocks deep in

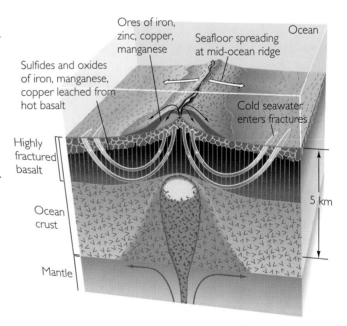

FIGURE 23.14 Cold seawater percolates through fractured volcanic rocks at mid-ocean ridges and is heated when it approaches the magma chamber below. The hot fluid leaches metals from the basaltic rock and rises to the seafloor. When the hot fluid with its dissolved metals vents into the cold ocean-bottom waters, the metals it is carrying in solution precipitate as rich sulfides of iron, zinc, copper, and other ores.

the crust. The heated seawater leaches minerals from the hot rocks and rises to the seafloor. When the hot waters, now loaded with dissolved minerals, reach the cooler upper crust and near-freezing ocean-bottom waters, the minerals precipitate. This is the origin of the "black smokers" shown in the photo in Feature 17.3. In this manner, enormous quantities of sulfide ores rich in zinc, copper, iron, and other metals are being deposited along mid-ocean spreading centers. High-grade deposits of native gold have been found in one submarine deposit on the Mid-Atlantic Ridge.

When current spreading centers were recognized as rich sources of mineral deposits, geologists began to look on land for the remains of ancient seafloor, which might also hold valuable resources. Some deposits were found in plate-collision zones where fragments of ancient oceanic lithosphere are occasionally emplaced on land. These deposits, known as ophiolites, are discussed in Chapter 20. The rich copper, lead, and zinc sulfide deposits in the ophiolites of Cyprus, the Philippines, the Apennines in Italy, and elsewhere probably owe their origin to the process of hydrothermal circulation along an-

cient mid-ocean rifts. The copper deposits of Cyprus were as important to the economy of ancient Greece as Middle East oil deposits are to the modern economy. Economically important deposits of chromite ores are occasionally found in deeper portions of ophiolites. They may have originated by fractional crystallization within magma chambers that underlie mid-ocean ridges.

Many other types of deposits of sulfide ores, of hydrothermal or igneous origin, are found at modern and ancient plate-collision boundaries, including those of the cordillera of North and South America, the eastern Mediterranean to Pakistan, the Philippine Islands, and Japan (see Figure 23.17). Figure 23.15 summarizes some of the associations between plate tectonics and mineral deposits. Deposits found in magmatic arcs are thought to result from the igneous activity that typically occurs in collision zones. One hypothesis proposes that some of these collision-boundary deposits represent the second stage in a two-stage ore-forming process. The first stage is the creation of mineral ores by hydrothermal activity at a mid-ocean spreading center. The second stage, separated in time and space from the first, is

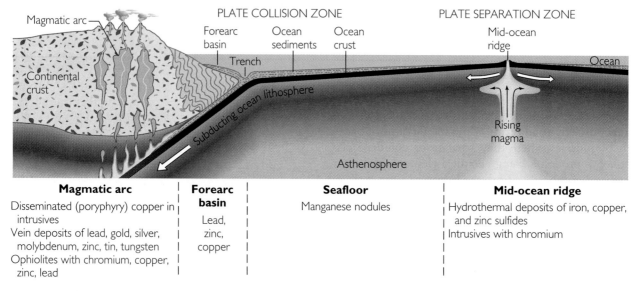

Magmatic arc	Forearc basin	Seafloor	Mid-ocean ridge
Disseminated (poryphyry) copper in intrusives Vein deposits of lead, gold, silver, molybdenum, zinc, tin, tungsten Ophiolites with chromium, copper, zinc, lead	Lead, zinc, copper	Manganese nodules	Hydrothermal deposits of iron, copper, and zinc sulfides Intrusives with chromium

FIGURE 23.15 The role of plate boundaries in the accumulation of mineral deposits. Ocean sediment and crust are enriched in metals by hydrothermal ore deposition along a mid-ocean ridge. Rising magma in the subduction zone is the source of ores that form the metal-bearing provinces of a magmatic belt such as the cordillera of North and South America. The melting of subducted sediment and crust may contribute ore constituents. Mineral-bearing oceanic fragments (ophiolites) accrete to the continent in the collision zone.

FIGURE 23.16 Manganese nodules are small concretions found on the deep seafloor that contain as much as 20 percent manganese and smaller amounts of iron, copper, and nickel. (This nodule is about 7.5 cm in diameter.) They have been dredged from the ocean floor to demonstrate the feasibility of mining the seabed. *(Chip Clark.)*

the subduction and partial melting at a collision zone of oceanic sediments and crust containing these previously concentrated minerals. As the plate descends into increasingly hot regions of the mantle, the metals "boil off"—that is, they melt and rise into the overriding plate along with magma. The iron, copper, molybdenum, lead, zinc, tin, and gold found along convergent plate boundaries could have been created by hydrothermal activity and reborn by igneous processes, all driven by plate-tectonic movements.

The seafloor away from plate boundaries, however, may be the first candidate for deep-sea mining because of the widespread occurrence of **manganese nodules,** potato-shaped aggregates of manganese, iron, copper, nickel, cobalt, and other metal oxides (Figure 23.16). The nodules vary in size, but most of them are a few centimeters in diameter. They are formed by the precipitation of these metal oxides from seawater, usually on a small nucleus such as a shark's tooth or a chip of rock. They are potentially valuable not only because of the gradual deple-

tion of high-grade deposits of manganese on land but also because they are rich in other metals. Deposits are estimated to be in the trillions of tons.

This brief summary of the geology of mineral deposits barely touches on the great diversity of geological settings in which various minerals of value are found. Some minerals or ores are found mainly or only in one kind of deposit; others are found in a variety of settings. Although there is probably an abundance of ore bodies on the deep seafloor, most known ore bodies are found on the continental crust. They either originated on the continent or occur as remnants of mineralized pieces of ocean crust thrust onto the continent in plate collisions. Figure 23.17 shows the locations of some of the major metallic ore deposits on the same map of continental deformation over geologic time that appears in Figure 21.2. Note that iron ores tend to be found in older parts of the crust and that ore deposits tend to be associated with orogenic belts. Table 23.2 shows the geologic occurrence and uses of some of the principal kinds of mineral deposits.

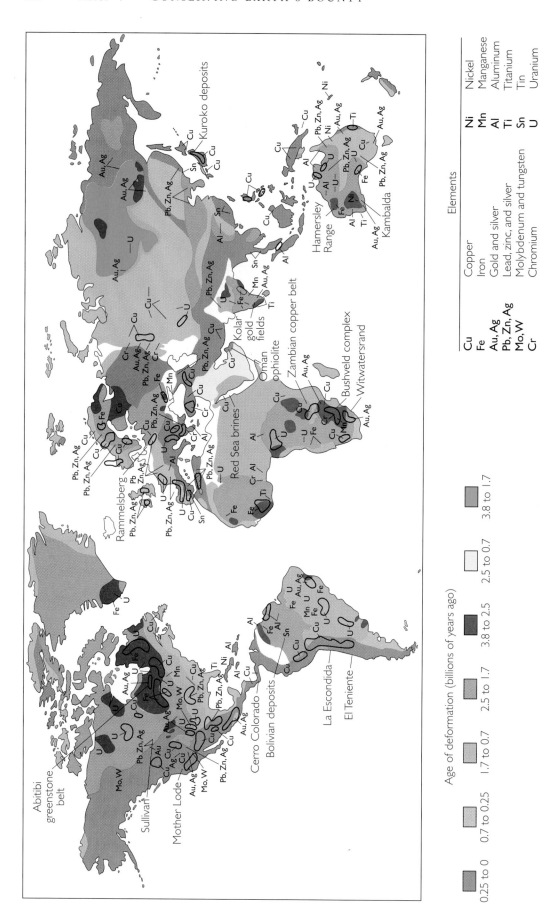

FIGURE 23.17 Locations of major metallic ore bodies on continents. Iron is concentrated primarily in Precambrian time. Note also the association of ore deposits with orogenic belts. Some famous ore bodies are identified by name. (After G. Brimhall, "The Genesis of Ores," *Scientific American*, May 1991, p. 84; based on a map by B. Clark Burchfiel.)

TABLE 23.2

PRINCIPAL TYPES OF ECONOMIC MINERAL DEPOSITS

MINERAL DEPOSIT	TYPICAL MINERALS	GEOLOGICAL OCCURRENCE	USES	MAJOR DEPOSITS / REMARKS
METALS PRESENT IN MAJOR AMOUNTS IN EARTH'S CRUST				
Iron	Hematite, Fe_2O_3 Magnetite, Fe_3O_4 Limonite, $FeO(OH)$	Sedimentary banded iron formation Contact metamorphic rocks Magmatic segregations Sedimentary bog iron ore	Manufactured materials, construction, etc.	*Mesabi, Minn.; Cornwall, Pa.; Kiruna, Sweden* Resources immense; economics determines exploitation
Aluminum	Gibbsite, $Al(OH_3)$ Diaspore, $AlO(OH)$	Bauxite: residual soils formed by deep chemical weathering	Lightweight manufactured materials	*Jamaica* Resources great, but expensive to smelt
Magnesium	Dolomite, $CaMg(CO_3)_2$ Magnesite, $MgCO_3$	Dissolved in seawater Hydrothermal veins, limestones	Lightweight alloy metal, insulators, chemical raw material	Most extracted from seawater; unlimited supply
Titanium	Ilmenite, $FeTiO_3$ Rutile, TiO_2	Magmatic segregations Placers	High-temperature alloys; paint pigment	*Allard Lake, Quebec; Kerala, India* Reserves large in relation to demand
Chromium	Chromite, $(Mg,Fe)_2CrO_4$	Magmatic segregations of mafic and ultramafic rocks	Steel alloys	*Bushveldt, South Africa* Extensive reserves in a number of large deposits
Manganese	Pyrolusite, MnO_2	Chemical sedimentary deposits, residual weathering deposits, seafloor nodules	Essential to steelmaking	*Ukraine* World's land resources moderate, but seafloor deposits immense
METALS PRESENT IN MINOR AMOUNTS IN EARTH'S CRUST				
Copper	Covelite, CuS Chalcocite, Cu_2S Digenite, Cu_9S_5 Chalcopyrite, $CuFeS_2$ Bornite, Cu_5FeS_4	Porphyry copper deposits Hydrothermal veins Contact metamorphic rocks Sedimentary deposits in shales (Kupferschiefer type)	Electrical wire and other products	*Bingham Canyon, Utah; Kupferschiefer, Germany; Poland*
Lead	Galena, PbS	Hydrothermal (replacement) deposits Contact metamorphic rocks Sedimentary deposits (Kupferschiefer type)	Storage batteries, gasoline additive (tetraethyl lead)	*Mississippi Valley; Broken Hill, Australia* Large resources; many lower grade deposits

(Continued)

TABLE **23.2** (*Continued*)

PRINCIPAL TYPES OF ECONOMIC MINERAL DEPOSITS

MINERAL DEPOSIT	TYPICAL MINERALS	GEOLOGICAL OCCURRENCE	USES	MAJOR DEPOSITS / REMARKS
METALS PRESENT IN MINOR AMOUNTS IN EARTH'S CRUST				
Zinc	Sphalerite, ZnS	Same as lead	Alloy metal	Same as lead
Nickel	Pentlandite, $(Ni, Fe)_9S_8$ Garnierite, $Ni_3Si_2O_5(OH)_4$	Magmatic segregations Residual weathering deposits	Alloy metal	*Sudbury, Ontario* High-grade ores limited; large resources of low-grade ores; also in seafloor manganese nodules
Silver	Argentite, Ag_2S In solid solution in copper, lead, and zinc sulfides	Hydrothermal veins with lead, zinc, and copper	Photographic chemicals; electrical equipment	Most produced as by-product of copper, lead, and zinc recovery
Mercury	Cinnabar, HgS	Hydrothermal veins	Electrical equipment, pharmaceuticals	*Almadén, Spain* Few high-grade deposits with limited reserves
Platinum	Native metal	Magmatic segregations (mafic rocks) Placers	Chemical and electrical industry; alloying metal	*Bushveldt, South Africa* Large reserves in relation to demand
Gold	Native metal	Hydrothermal veins Placers	Coinage; dentistry; jewelry	*Witwatersrand, South Africa* Reserves concentrated in a few larger deposits
NONMETALS				
Salt	Halite, NaCl	Evaporite deposits Salt domes	Food; chemicals	Resources unlimited; economics determines exploitation
Phosphate rock	Apatite, $Ca_5(PO_4)_3OH$	Marine phosphatic sedimentary rock Residual concentrations of nodules	Fertilizer	*Florida* High-grade deposits limited but extensive resources of low-grade deposits
Sulfur	Native sulfur Sulfide ore minerals	Caprock of salt domes (main source) Hydrothermal and sedimentary sulfides	Fertilizer manufacture; chemical industry	*Texas; Louisiana; Sicily* Native sulfur reserves limited but immense resources of sulfides
Potassium	Sylvite, KCl Carnallite, $KCl \cdot MgCl_2 \cdot 6H_2O$	Evaporite deposits	Fertilizer	*Carlsbad, New Mexico* Great resources of rich deposits
Diamond	Diamond, C	Kimberlite pipes Placers	Industrial abrasives	*Kimberley, South Africa* Synthetic diamond now commercially available

TABLE 23.2 *(Continued)*

PRINCIPAL TYPES OF ECONOMIC MINERAL DEPOSITS

MINERAL DEPOSIT	TYPICAL MINERALS	GEOLOGICAL OCCURRENCE	USES	MAJOR DEPOSITS REMARKS
		NONMETALS		
Gypsum	Gypsum, $CaSO_4 \cdot 2H_2O$ Anhydrite, $CaSO_4$	Evaporite deposits	Plaster	Immense resources widely distributed
Limestone	Calcite, $CaCO_3$ Dolomite, $CaMg(CO_3)_2$	Sedimentary carbonate rocks	Building stone, agricultural lime; cement	Widely distributed; transportation a major cost
Clay	Kaolinite $Al_2Si_2O_5(OH)_4$ Smectite[1] Illite[1]	Residual weathering deposits; sedimentary clays and shales	Ceramics: china, electrical; structural tile	Many large pure deposits; immense reserves of all grades
Asbestos	Chrysotile, $Mg_3Si_2O_5(OH)_4$	Ultramafic rocks altered and hydrated in near-surface crustal zones	Nonflammable fibers and products	*Southeastern Quebec* Limited high-grade reserves but great low-grade reserves

[1]Formula highly variable; a hydrous aluminum silicate with other cations, such as Na^+, K^+, Ca^{2+}, Mg^{2+}.

THE NEED TO FIND NEW MINERAL DEPOSITS

The total dollar value of all mineral resources, including fuels, produced in the United States has grown from less than $5 billion in 1952 to about $100 billion in recent years. This 20-fold growth in less than four decades (which includes some inflationary increases because of the lowering of the value of the dollar in that time period) took place in a highly industrialized society that had already built a huge technological capability and whose population grew by only about 60 percent in the same period. The recovery of many nations after World War II and the growth in their economies led to an enormous increase in consumption and production in the rest of the world, resulting in a relative decline in production and consumption of nonfuel minerals by the United States in comparison with the rest of the world.

Although the growth of mineral demand has now declined in the industrialized nations, unequal sharing of world mineral resources still remains and has far-reaching economic and political repercussions. For instance, North America, with less than 10 percent of the world's population, consumes almost 75 percent of the world's production of aluminum, whereas Asia and Africa, with about two-thirds of the world's population, together use only a little over 5 percent. The same extreme imbalance is associated with other materials as well. International relations are now deeply affected by struggles over the control of resources. One widespread cause of conflict is the demand by some mineral-rich developing nations for a greater degree of control over resources mined on their land by corporations based in North America or Europe. In recent years, several countries in Africa and South America have nationalized foreign-owned mining companies.

As the Earth's human population grows and people in less developed nations seek higher standards of living (more food, raw materials, energy, manufactured goods, construction of all types), the demand for mineral resources will be strong. To meet this demand, plate-tectonics theory and other new concepts may serve the geologist in seeking new high-grade ores and mineral deposits.

How important the unexplored deep sea will become is an open question. The answer depends in part on the development of efficient marine technology for deep-sea exploration and mining and the

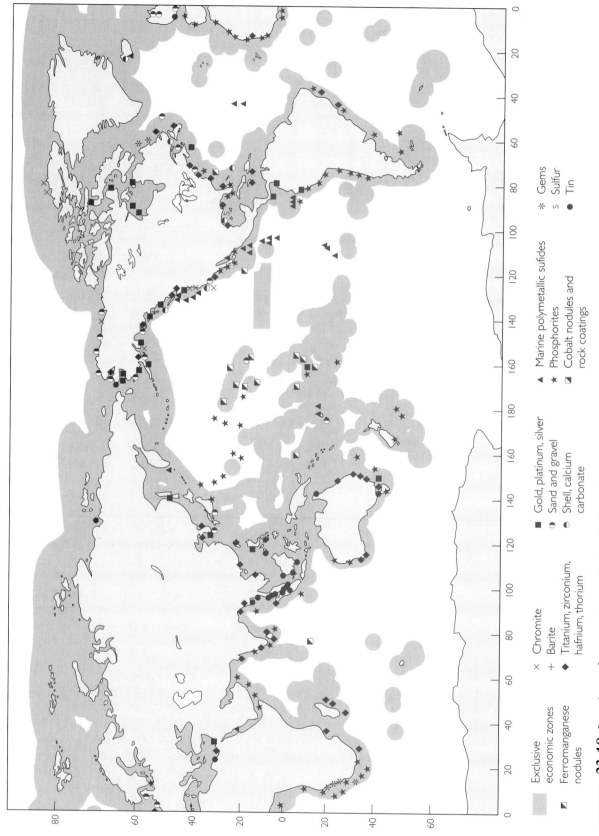

FIGURE 23.18 Locations of some major nonfuel seabed ore deposits. Note the concentrations at plate boundaries and continental shelves. The gray areas outline exclusive economic zones claimed by individual nations—the seas within 200 nautical miles of their territory. (After J. M. Broadus, "Sea Bed Materials," *Science,* vol. 235, 1987, pp. 853–859.)

resolution of legal issues regarding ownership of deep-sea deposits. It is generally agreed that a nation has exclusive rights to mineral deposits in the offshore area within 200 nautical miles of its coast—the so-called **exclusive economic zone** (Figure 23.18). Still in question is who owns the mineral deposits on the seafloor beyond this zone. Are the deposits owned by all nations in common? By the discoverer and developer? By both? Because the economies of many nations that export minerals are threatened by competition from new sources on the seafloor, these nations would like to limit such development and obtain a share of the profits.

These issues of international law have been debated for more than 20 years. In 1982, the United Nations adopted an agreement known as the Law of the Sea Treaty, providing a legal and regulatory system for the development of deep-sea resources, by a vote of 130 for, 4 against. The United States was one of the four nations opposed, because the treaty placed limits on seabed mineral production, and the United States feared that private companies that developed the technology and invested in production would be inadequately compensated.

A hard fact of life is that an equal per capita sharing of the world's available reserves would not be enough to bring everyone to a "satisfactory" level of consumption, certainly not to a level anywhere near that of an affluent country in Europe or North America. The problem is compounded by the continued rapid growth of the world's population (see Chapter 24).

The solution is probably a combination of conservation, recycling, and more efficient use of materials; increased discovery and the "clean" exploitation of mineral and energy resources; and development of low-cost, environmentally benign substitutes for minerals, such as fiber-optic cables made of glass instead of copper wire. In a sense, we are all dependent on design and manufacturing engineers, geologists, and scientists who develop new materials from plentiful resources. Ultimately, however, science will be of no avail unless population growth is stabilized.

The world clearly faces major readjustments in the decades ahead, and it is certainly not too soon to start working out equitable, humane, and lasting approaches to building a sustainable world. How well we use our planet will depend on how well we understand how it works and the extent to which the people of the world cooperate intelligently in the development and use of its resources—and do so in a manner that protects the environment. We have only one Earth. To continue to live on it, we must learn to use it wisely.

SUMMARY

What distinguishes an economical mineral deposit from one that is not economical? Mineral deposits of economic value are those in which an element occurs in much greater abundance than in the average crustal rock—great enough to make the deposit economically worthwhile to mine when the market price and the costs of mining, processing, and transportation are taken into account.

What is hydrothermal mineral deposition? Hydrothermal deposits, which are some of the most important ore deposits, are formed by hot water that emanates from igneous intrusions or by heated circulating groundwater or seawater. The heated water leaches soluble minerals in its path and transports them to cooler rocks, where they are precipitated in fractures, joints, and voids. These ores may occur in veins or in disseminated deposits, such as the copper-porphyry type.

What are some of the processes that lead to the formation of sedimentary mineral deposits? Ordinary chemical and mechanical sedimentary processes segregate such economically important minerals as limestone, sand and gravel, and evaporite salt deposits. Sedimentary ores of copper and iron have formed as precipitates in special sedimentary environments, the iron ores chiefly in Precambrian times. Placers are ore deposits, rich in gold or other heavy minerals, that were laid down by currents.

How do igneous ore deposits form? Igneous ore deposits typically occur when minerals crystallize from molten magma, settle, and accumulate on the floor of a magma chamber. They are often found as layered accumulations of minerals. The rich ore body at Sudbury, Ontario, for example, is a mafic intrusive formation that contains great quantities of layered nickel, copper, and iron sulfides near its base.

What insights into the formation of ores are provided by plate-tectonics theory? Whether it is mineral precipitation from hydrothermal fluids or crystal settling in a magmatic body, igneous activity is the ultimate source of many minerals. Igneous activity is a feature of plate spreading and subduction, so modern and ancient plate boundaries are places where many concentrations of minerals should be found.

What are some of the mineral policy issues faced by modern society? It is vital to discover new mineral deposits to support an increasingly industrialized world civilization. Prospects for finding new resources are good. The sea represents a largely untapped resource. Although technological advance can postpone the day of reckoning, stocks will become increasingly scarce. Therefore, conservation, recycling, and substitution of alternative materials will become increasingly important in the years ahead. These advances, however, will be unable to satisfy the demands of a world population that increases without limits.

KEY TERMS AND CONCEPTS

ores (p. 586)
ore minerals (p. 586)
concentration factor (p. 587)
recycling (p. 588)
hydrothermal vein deposits (veins)
 (p. 592)

disseminated deposits (p. 594)
igneous ore deposits (p. 595)
sedimentary mineral deposits
 (p. 596)
placers (p. 598)

manganese nodules (p. 601)
exclusive economic zone (p. 607)

EXERCISES

1. What are the characteristics of an economical ore deposit?

2. Describe the creation of an ore body by hydrothermal activity.

3. Give some examples of useful hydrothermally deposited minerals.

4. How do useful minerals become concentrated in an igneous ore deposit?

5. List several examples of minerals in igneous ore bodies.

6. What is the role of physical and chemical processes in the formation of sedimentary mineral deposits?

7. Identify some sites where sedimentary mineral deposits may be found and what the deposits consist of.

8. Contrast the process of ore formation in the sediment and crust of the deep sea with that in a convergence zone.

9. What useful minerals might be found in the exclusive economic zone of the United States? Of Canada? Of England?

10. Why are conservation and recycling of useful materials drawn from the Earth and the creation of substitutes important to the future of humankind?

THOUGHT QUESTIONS

1. Why would two advanced nations, such as the United States and Canada, take opposite positions with regard to the Law of the Sea Treaty?

2. Should the United States spend large sums to stockpile strategically important minerals (those essential for economic or defense needs) for use in an emergency in the event that foreign supply is cut off?

3. In 1992, many nations with interests in Antarctica signed a treaty agreeing not to mine for minerals on that continent. Why did they do so? Do you agree with their decision?

4. Will plate-tectonics theory contribute to the search for ore bodies? How?

5. Should the Mining Act of 1872 be modified? How?

6. Do you think people in developed countries will be willing to reduce consumption in order to make raw materials more readily available to people in less developed countries?

LONG-TERM TEAM PROJECT

See Chapter 11.

SUGGESTED READINGS

Broadus, J. M. 1987. Sea bed materials. *Science* 235:853–859.

Carr, D. D., and N. Herz (eds.). 1988. *Concise Encyclopedia of Mineral Resources.* Cambridge, Mass.: MIT Press.

Clark, J. P., and F. R. Field III. 1985. How critical are critical materials? *Technology Review* (August/September).

Dorr, Ann. 1987. *Minerals: Foundations of Society.* Alexandria, Va.: American Geological Institute.

Guilbert, J. M., and C. F. Park, Jr. 1986. *The Geology of Ore Deposits.* New York: W. H. Freeman.

Heaton, George, Robert Repetto, and Rodney Sobin. 1991. *Transforming Technology: An Agenda for Environmentally Sustainable Growth in the 21st Century.* Washington, D.C.: World Resources Institute.

Hodges, Carroll Ann. 1995. Mineral resources, environmental issues, and land use. *Science* 268:1305–1312.

Keary, P., and F. J. Vine. 1990. Plate tectonics and economic geology. Chap. 11 in *Global Tectonics.* London: Blackwell Scientific Publications.

National Research Council. 1990. *Competitiveness of the U.S. Minerals and Metals Industry.* Washington, D.C.: National Academy Press.

Sawkins, F. J. 1984. *Metal Deposits in Relation to Plate Tectonics.* New York: Springer-Verlag.

Science Summit on World Population. 1993. Report of conference of national academies of sciences of 80 countries, New Delhi, October 24–27.

U.S. Bureau of Mines. 1994. *Minerals Yearbook.* Washington, D.C.: U.S. Government Printing Office.

INTERNET SOURCES

Bureau of Land Management
ⓘ **http://www.blm.gov/**
The Bureau of Land Management oversees more than 270 million acres of federal land for multiple uses, including mining, recreation, and agriculture. This site includes a description of the agency, its responsibilities, breaking news, state-by-state information, a map of sites of interest, and related Web sites.

US Geological Survey Minerals Information
ⓘ **http://minerals.er.usgs.gov/minerals/**
This site provides links to "statistics and information on the worldwide supply, demand, and flow of minerals and materials essential to the U.S. economy, the national security, and protection of the environment."

24

Bora–Bora, French Polynesia. Air, water, land, and life interact with each other and with the Earth's interior. *(Sylvain Gandadam/ Photo Researchers.)*

610

Earth Systems and Cycles

W hen we try to pick up anything by itself, we find it entwined with everything else in the universe.
—John Muir, *My First Summer in the Sierras* (1911)

The most successful strategy for solving a complex problem is to reduce it to simpler ones. To understand a complex system such as Earth, we study each of its subsystems separately, as if it existed alone. To get a perspective on the whole Earth, however, we must learn how its separate systems interact.

Changes to one of Earth's systems can have unanticipated effects on the whole Earth. An intense volcanic eruption can change global climate for a few years. An ice age can lead to the extinction or migration of many species. Over the past five decades, water impounded in some 100 artificial reservoirs in the high latitudes has resulted in the speeding of Earth's rotation, much as a skater speeds up when she draws her arms inward; this increase in Earth's rotation speed has reduced the length of a day by about 10 microseconds.

Throughout this book, we have emphasized the interdependence between Earth's internal and external processes. In this epilogue, we will look at how all of Earth's major systems—the interior, the lithosphere, the atmosphere, the hydrosphere, and the biosphere—are linked. We will see why scientists attribute past changes on Earth to geological perturbations of its systems and cycles. And we will learn why many people fear that human-induced perturbances could cause global catastrophe.

THE FORMATION OF EARTH SYSTEMS

Our planet is not a self-contained system. It resides within a solar system, which in turn exists within a galaxy, a tiny part of the universe. As the scale of each of these systems increases, so does its complexity and our ignorance.

Scientists surmise that the universe began some 10 billion to 15 billion years ago with the Big Bang, which put the laws of physics into place. In a manner not completely understood, these rules prescribe the essential features of our existence—for example, how elementary particles make up atoms; how all elements heavier than hydrogen and helium are made in the cores of stars; how stars begin and end; how planets are formed; and, ultimately, how life evolves. Humankind is linked to all these systems. We literally descend from the stars.

The Interior and the Lithosphere

The laws of physics force Earth toward thermal and gravitational equilibrium—a condition in which the hot core and mantle cool down, light materials rise to the surface, and heavy materials settle into the interior. The cooling of a molten Earth about 4.4 billion years ago led to a chemically differentiated planet with an iron core and progressively lighter materials in the mantle, lithosphere and crust, ocean, and atmosphere. Once the lithosphere solidified, perhaps 4 billion years ago, plate tectonics became the successor mechanism for cooling the hot interior.

Powered by convection currents in the mantle, plate tectonics continuously shapes the surface of the Earth by driving the creation and destruction (by subduction) of crustal plates. Plate tectonics thus connects the mantle and lithosphere through repeated cycles of subduction, mixing, melting, and differentiation. In essence, plate tectonics is a system of continuous distillation in which matter repeatedly extracted from and returned to the mantle becomes the highly differentiated magma that solidifies as granitic rocks—the building blocks of continents. Magmatic intrusions, orogenies, and the accretion of microplate terranes at colliding boundaries are mechanisms for assembling continents over geologic time. Light enough to stand high above the seafloor and resist subduction, the continents persist as the stable, dry, mineral-rich platforms on which land animals evolved and advanced.

The Atmosphere and the Hydrosphere

Perhaps the most fundamental interaction between Earth's internal and external systems is that the former created the latter. The best-supported theory is that outgassing during differentiation and volcanism provided Earth with an atmosphere and a hydrosphere (see Figure 1.8). Geologists are led to this conclusion because volcanic eruptions, which have been active throughout geologic time, are accompanied by substantial emissions of water, carbon dioxide, and other gases, some of which can be shown to be of primordial origin. Many rock-forming minerals known to be present in the mantle contain water and gases in their chemical compounds that would be released by heating and brought to the surface during differentiation. Additionally, bubbles of carbon dioxide and other gases are found as inclusions in samples of olivine and other minerals of mantle origin found at the surface, evidence of the presence of these gases in the mantle. Calculations based on these observations show that outgassing from the interior could provide enough gases and water to account for the oceans and an early atmosphere with carbon dioxide as a major constituent but not much oxygen. Additional gases may have been added as constituents of comets and meteorites, but Earth's interior is the primary source of the atmosphere and hydrosphere.

Unable to escape Earth's gravitation, the carbon dioxide, water vapor, and other relatively heavy gases—including nitrogen and sulfur gases—enveloped the planet. This early atmosphere lacked the oxygen that makes up 21 percent of the atmosphere today. Oxygen did not enter the atmosphere until photosynthetic organisms evolved, as described in the next section of this chapter.

As Earth cooled and atmospheric temperatures dropped to the point where water vapor condenses, the great bulk of the water rained out and formed the oceans and other components of the hydrosphere. The amount of water in the hydrosphere is fairly constant over periods of thousands of years. The water is distributed among four major reservoirs:

- The oceans, 1370 million km^3
- Ice, 30 million km^3
- Surface waters, 8 million to 19 million km^3
- Atmospheric moisture, 0.01 million km^3

Sea level varies as the distribution among these reservoirs, particularly the first three, changes.

The Biosphere

The **biosphere** (from *bios*, Greek for "life") is the sum of all the materials and processes involving organisms. Without the atmosphere and hydrosphere, Earth's biosphere could not have evolved.

THE GREENHOUSE EFFECT A little over 4 billion years ago, Earth's early atmosphere and hydrosphere had already formed. Light gases such as hydrogen (which is quickly lost to space) had escaped, while heavier gases such as water vapor, carbon dioxide, and sulfur dioxide had been left behind. This early atmosphere allowed all the components of sunlight—including ultraviolet (UV) rays, which are damaging to life—to reach Earth's surface. At the same time, carbon dioxide and water vapor in the atmosphere kept the Earth warm by trapping heat radiating from Earth's surface—a phenomenon known as the **greenhouse effect.**

Here is how the greenhouse effect works. Much of the radiant energy from the Sun is absorbed by the Earth's surface and then reemitted as invisible infrared (IR) heat rays. Just as a hot pavement radiates heat when it is warmed by the Sun, the Earth's surface radiates heat back to the atmosphere. The atmosphere, however, is not transparent to these infrared rays, because carbon dioxide and water molecules strongly absorb the infrared instead of allowing it to escape into space. As a result, the atmosphere is heated and radiates heat back to the surface. The process is called the greenhouse effect by analogy to the warming of a greenhouse, whose glass lets in visible light but lets little heat escape (Figure 24.1).

The more carbon dioxide, the warmer the atmosphere; the less carbon dioxide, the colder. Without any greenhouse effect, Earth's surface temperature would be well below freezing and the oceans would be a solid mass of ice. Venus, in contrast, has a "runaway" greenhouse effect because carbon dioxide makes up 96.5 percent of the

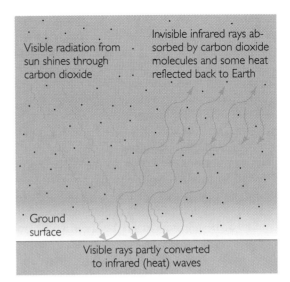

FIGURE 24.1 The greenhouse effect. Just as the glass of a greenhouse transmits light rays but holds in heat, the carbon dioxide of the atmosphere transmits visible radiation from the Sun but absorbs the infrared radiation from the ground surface and re-radiates it back to Earth.

planet's atmosphere. With a surface temperature of 475°C, all of its water boiled off, leaving a bone-dry, lifeless surface.

LIFE BEGINS In Earth's greenhouse, despite intense radiation and an atmosphere with little oxygen, life began. The evidence lies in microscopic fossils of bacteria, probably marine, at least 3.5 billion years old (Figure 24.2). There is a strong likelihood, however, that life had originated long

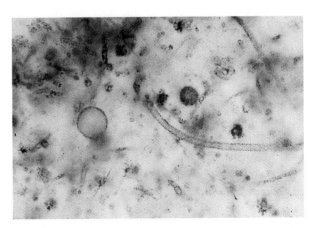

FIGURE 24.2 Fossilized microorganisms found in chert dated at 3.5 billion years, found in Ontario, Canada. (*M. Abbey/Photo Researchers.*)

before, perhaps 4 billion years ago, perhaps a few hundred million years later. The first step toward the evolution of these bacteria is thought to have involved the assembly of large organic molecules from gases such as methane and ammonia, with the energy for these transformations supplied by strong UV radiation. This step has been explored in many chemical experiments that show how various building blocks of life could have formed.

The next steps in the evolution of life occurred when, for reasons not yet understood, these organic molecules aggregated and formed a sort of general system for growth and metabolism. This kind of system could not be considered life because it was not self-reproducing, but it is thought of as protolife. Some scientists argue that these steps toward life took place at hydrothermal vents on mid-ocean ridges, which almost certainly existed at that time. Complex reactions among carbon-containing gases, seawater, magmatic fluids, and basalt go on at these superheated vents, and some aspects of the synthesis of life seem more likely to have taken place there than in the open surface waters of the ocean.

However and wherever these organic systems formed, the next step would have been the development of the first truly self-replicating molecule, RNA (ribonucleic acid). RNA, like DNA (deoxyribonucleic acid), is a nucleic acid basic to life. Both molecules are intimately involved in the process of self-replication. RNA is made of slightly different chemical components than DNA and is generally single-stranded rather than double-stranded. The RNA world was transitory and soon evolved to the more complex DNA world that has characterized the biosphere for most of geologic history.

Not all scientists subscribe to every one of these steps. A small number believe that cometary impacts brought to Earth not only the atmosphere's gases but also life itself. According to this view, life on Earth began when in-falling comets—snowballs of frozen gases—"colonized" the planet.

These early steps in the origin of life probably did not significantly affect the early atmosphere, which remained dominated by nitrogen as well as carbon dioxide. Even in the absence of oxygen, weathering and other surface processes would have gone on pretty much as they do today, although the lack of organisms and oxygen would have limited chemical weathering to some extent. Feldspar and carbonates would have weathered as they do now, but reduced metal minerals such as pyrite and magnetite would not have weathered by oxidation. The next evolutionary development, however, the arrival of photosynthesis, did profoundly affect the atmosphere, the hydrosphere, and the lithosphere.

Photosynthesis Changes the Earth

Photosynthesis is the process by which green organisms use chlorophyll (which colors them green) and the energy from sunlight to make carbohydrates—organic matter—out of carbon dioxide and water. The carbohydrates provide the energy needed to power the biosphere. In the chemical reaction below, the simple compound CH_2O represents all the many different carbohydrates that actually form:

$$\text{carbon dioxide} + \text{water} \xrightarrow{\substack{\text{in the presence of} \\ \text{sunlight and chlorophyll}}} \text{carbohydrate} + \text{oxygen}$$
$$CO_2 \quad + \quad H_2O \qquad\qquad CH_2O \quad + \quad O_2$$

For each 30 g of carbohydrate produced in this reaction, approximately 112,000 calories of energy from sunlight are converted into chemical energy and stored in plants. At the same time, for each atom of carbon taken from carbon dioxide (CO_2) and incorporated into organic matter, one molecule of oxygen (O_2) is formed.

Because animals cannot synthesize carbohydrates for themselves, they depend on food from plants or other animals for the energy they need to live. To retrieve that stored energy, the organism takes in oxygen and, in its cells, combines oxygen with carbohydrate. The chemical reaction that releases the energy stored by photosynthesis is called **respiration:**

$$\text{carbohydrate} + \text{oxygen} \xrightarrow{\substack{\text{release of} \\ \text{chemical energy}}} \text{carbon dioxide} + \text{water}$$
$$CH_2O \quad + \quad O_2 \qquad\qquad CO_2 \quad + \quad H_2O$$

In this reaction, one molecule of oxygen is used up for each atom of organic matter oxidized to carbon dioxide.

As you can see, photosynthesis and respiration are reciprocal. They link carbon dioxide, oxygen, and carbohydrate (or organic carbon, a term we use for all the great number of organic compounds biologically synthesized from the building blocks supplied by photosynthesis). Photosynthesis and respiration are the vehicles for cycling carbon dioxide and oxygen between the atmosphere, the oceans, the land surface, and the biosphere (Figure 24.3).

Bits and pieces of the organic matter produced by photosynthesis are incorporated into sediments, buried in the crust, and converted to organic carbon in sedimentary rocks (see Chapter 7). If photosynthesis and respiration were in perfect balance glob-

PHOTOSYNTHESIS

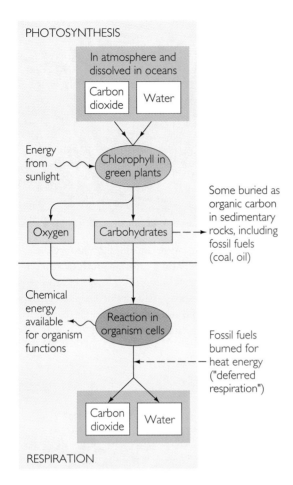

FIGURE 24.3 Photosynthesis and respiration cycle carbon dioxide and oxygen through the atmosphere, hydrosphere, land surface, and biosphere.

their own food: the organic matter manufactured by sunlight in combination with carbon dioxide and water. As organic matter was buried and altered diagenetically, atmospheric oxygen accumulated. As the oxygen level rose, oxidative chemical weathering began to affect the land surface as it does today. In addition, as oxygen increased, the ozone layer evolved.

Atmospheric oxygen molecules diffusing upward into the **stratosphere** (the upper atmosphere) were transformed by solar radiation into **ozone** (O_3). The stratospheric ozone layer absorbs certain portions of solar ultraviolet radiation before it reaches the surface, where it can damage and cause mutations in animal and plant cells. Without this protection from UV rays, life is unlikely to have flourished on land.

(a)

(b)

FIGURE 24.4 The lush canopy (a) and decaying organic matter on the floor (b) of the Hoh Rain Forest, a temperate-zone rain forest in Olympic National Park, Washington. The burial of organic matter over billions of years accounts for the level of oxygen in the Earth's atmosphere today. *(Kunio Owaki/The Stock Market.)*

ally, all organic matter would be used up in respiration, and the carbon dioxide and oxygen from the two processes would be in balance. The burial of organic matter creates an imbalance, an excess of photosynthesis (and the production of oxygen) over respiration (and the production of carbon dioxide). For each molecule of organic matter buried, there is a molecule of oxygen left over that cannot be used up by respiration. Thus, the oxygen in the atmosphere today is the legacy of organic matter buried over billions of years (Figure 24.4).

OXYGEN BECOMES A MAJOR GAS IN THE ATMOSPHERE The evolution of photosynthesis early in Earth's history had immense consequences. The biosphere changed as organisms evolved from being dependent on the relatively small amounts of organic matter produced by inorganic chemical processes and recycled from other organisms to photosynthetic organisms that could synthesize

EVIDENCE OF EARLY OXYGEN   When did oxygen become a significant part of the atmosphere? From fossil evidence, it seems likely that photosynthesis began a relatively short geologic time after the establishment of bacterial life and the evolution of cells with a nucleus. Some recent fossil bacteria discovered in rocks from Canada give an early date of 3.5 billion years ago (see Figure 24.2). A more conservative estimate based on other fossils is in the vicinity of 2.0 billion to 1.5 billion years ago.

The geological evidence given by abundant oxidized iron formations and studies of Precambrian paleosols (ancient soils, see Chapter 6) indicate that by 1.5 billion years ago, there was a significant amount of oxygen in the atmosphere. The explosion of life forms in the late Precambrian and Cambrian was probably stimulated at least in part by a rise in oxygen levels to levels comparable to today's (Figure 24.5).

Table 24.1 summarizes significant developments in the evolution of Earth and life.

FIGURE 24.5 A Precambrian soil 2.2 billion years old, at Daspoort Tunnel, near Pretoria, South Africa. The whiteness of the layer extending from lower left to upper right is due to a lack of iron oxide, which suggests that the early atmosphere did not contain enough oxygen to oxidize iron and precipitate it into the soil. (*Courtesy of H. D. Holland.*)

TABLE 24.1

A THEORY OF EARTH'S EVOLUTION

YEARS AGO[1]	EVENT
15 billion	The "Big Bang" brings the universe into being
4.6 billion	Our solar system takes shape
4.5–4 billion	—Earth accretes, melts, and differentiates
	—Early atmosphere forms primarily from heavier magmatic gases: water vapor, carbon dioxide, nitrogen, and gases of sulfur
	—With no oxygen, Earth's atmosphere cannot burn up invading bolides or deflect UV rays
	—With its heavy load of water vapor and carbon dioxide, Earth's atmosphere retains warmth
4–3.5 billion	—Organic molecules synthesize and stabilize in the seas
	—Organized aggregates of organic molecules form a system for metabolism and growth (protolife)
	—Life begins with the evolution of self-replicating molecules
3.5–1.5 billion	—Bacteria and photosynthetic unicellular organisms evolve
	—Oxygen waste from photosynthesis dramatically changes the atmosphere, hydrosphere, lithosphere, and biosphere, in ways that will allow life to form on land
1.5–0.5 billion	Multicellular marine organisms evolve
0.5 billion–present	Complex marine organisms evolve
0.45 billion–present	Land plants and animals evolve

[1]The timing of the events remains largely speculative.

GEOCHEMICAL CYCLES: TRACERS OF EARTH SYSTEMS

As we have seen, Earth changes because chemical elements such as carbon and oxygen circulate through it. Photosynthesis and respiration, for example, are part of the cycle that carries carbon through the atmosphere, biosphere, and hydrosphere and back again. Geologists have found that **geochemical cycles** such as the carbon cycle, much like the geological cycles of Hutton (see Chapter 3), maintain the Earth in more or less a steady state.

Earth's atmosphere, hydrosphere, and other systems can be viewed as groups of **reservoirs** for holding Earth's chemicals. These reservoirs are linked by transport processes between them. Geochemical cycles trace the flow, or **flux,** of Earth's chemicals from one reservoir to another. By quantifying the amounts of various materials that are stored and moved from one reservoir to another,

geochemical cycles give us new insights into the workings of the giant system of Earth. The **calcium cycle,** for example, traces the flux of calcium from one reservoir to another. The amount of calcium liberated by the weathering of limestone can be measured by the abundance of calcium in river waters, which transport calcium to the oceans.

The Calcium Cycle

The world ocean is a reservoir that we can describe by its mass and by the total amount of its dissolved solids and gases, which are constant from place to place and over geologic time. Among many other elements, the ocean contains about 5.6×10^{20} g of calcium, dissolved in a total ocean mass of about 1.4×10^{24} g. Calcium steadily enters this reservoir in large quantities through the rivers of the world, which transport dissolved calcium derived from the weathering of such minerals as calcite, gypsum, calcium feldspars, and other calcium silicates (Figure 24.6).

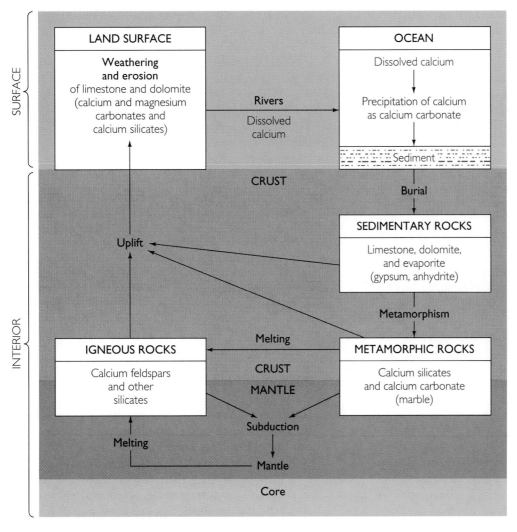

FIGURE 24.6 The geochemical cycle of calcium, showing the fluxes and reservoirs of calcium in the Earth's surface, crust, and mantle. There is relatively little calcium in the core.

If the ocean kept receiving this much calcium, it would quickly become supersaturated with respect to calcite and gypsum if there were no calcium flux from the ocean. The flux that removes the majority of excess calcium from the ocean is the *sedimentation of calcium carbonate* (see Chapter 7). A smaller amount of calcium is precipitated as gypsum in evaporite deposits. Like calcium, all elements transported by rivers to the ocean are precipitated as sediment from the ocean.

A generation ago, geochemists showed that the inflow of elements into the ocean is approximately equal to the outflow. This finding led to the conclusion that the oceans are close to a steady state—that is, not changing with time (Figure 24.7).

Residence Time

When the inflow of an element into a reservoir equals the outflow, the average length of time that an atom of the element spends in the reservoir before leaving is the element's **residence time.** Think of a crowded party where many more people have been invited than can fit in the room. As people come in, the room fills up, and then as more arrive, the crush gets to the point that people start to leave. At its most active, the party is in a steady state, with arrivals and departures balanced and the room filled, or saturated. Even though some guests come early and stay late while others leave after only a short time, there is an average length of time between each person's arrival and departure. This average is the residence time.

At the party, residence time is governed by the maximum number of people that can be jammed into the room (the *capacity* of the room) divided by the rate of arrivals (inflow) or departures (outflow). If the room's capacity is 30 people and a new person arrives every 2 minutes, we can calculate that the average residence time is 60 minutes.

RESIDENCE TIME IN THE OCEAN   We can visualize residence time in the ocean as the average time that elapses between an atom's entry into the ocean and its removal through sedimentation. Rivers carry to the ocean great quantities of calcium derived in large part from the weathering of limestone, which is both abundant at the surface and easily dissolved, so the inflow is high. There is so much calcium in the ocean that calcium's residence time in the ocean is long, about 850,000 years. Sodium has one of the longest residence times in the ocean—about 48 million years—because there is a huge amount of sodium in the reservoir and rivers contain relatively low amounts. Iron, in contrast, stays in the ocean only about 100 years, because the solubility of iron in seawater is extremely low, which limits the total amount in the ocean, and the inflow is large.

Knowing the residence time of an element in a given reservoir helps when predicting the behavior of toxic and radioactive elements. It is important, for example, to know how long crude oil spilled from tankers will remain in the ocean before decomposing or falling out as sediment onto the seafloor and beaches. Likewise, knowing the residence time of carbon dioxide in the ocean helps when predicting the effects of a rise in carbon dioxide in the oceans and in the atmosphere.

OCEAN–ATMOSPHERE INTERACTIONS   The atmosphere interacts with the oceans and the land along the thin layer of air immediately overlying Earth's surface (Figure 24.8). At the sea surface, water evaporates and rainfall returns water to the ocean. Also at the sea surface, gas molecules escape from their dissolved state in the water to enter the atmosphere. Their escape is balanced by the dissolution of gas molecules from the air above the water into the water at the surface. This gas exchange is accelerated by the evaporation of sea spray, which releases dissolved gases as well as dissolved salt in the form of tiny crystals.

RESIDENCE TIME IN THE ATMOSPHERE   Just as we can calculate residence times in the ocean, we can find the average length of time that a gas molecule spends in the atmosphere. Carbon dioxide, which makes up less than 5 percent of the atmosphere, has a residence time of 10 years. Oxygen, about 20 percent of the atmosphere, has a residence

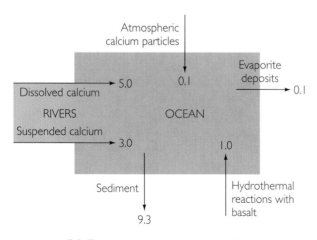

FIGURE 24.7 Quantitative estimate of calcium fluxes into and out of the ocean. Fluxes are in units of 10^{14} g per year. The inflow of calcium is approximately equal to the outflow, resulting in a steady state.

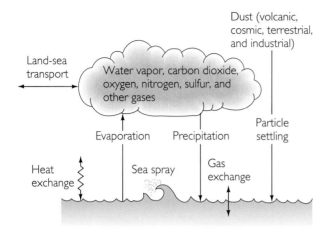

FIGURE 24.8 The oceans and land exchange gases with the atmosphere.

time of 6000 years. Sulfur dioxide, present in very small amounts in the atmosphere, has a residence time of hours to weeks. Nitrogen, more than 75 percent of the atmosphere, has a residence time of 400 million years. A molecule of nitrogen that went into

the atmosphere in the late Paleozoic era, about 300 million years ago, is likely to be there still.

Most residence times in the atmosphere are shorter than those in the ocean, a reflection of the smaller size of the reservoir. The atmosphere holds only one-sixtieth of the carbon dioxide that is held in the ocean in one form or another, as dissolved carbon dioxide and bicarbonate and carbonate ions.

The Carbon Cycle

We have already seen how two major aspects of the carbon cycle, photosynthesis and respiration, are related to the current mass of organic carbon, carbon dioxide, and oxygen at Earth's surface. Now, with our understanding of the relatively simple calcium cycle, we can add the calcium cycle and the sedimentation flux to get a more complete understanding of the **carbon cycle.**

Figure 24.9 depicts all the fluxes and reservoirs that govern the carbon cycle. Carbon enters the atmosphere in the form of carbon dioxide gas. Carbon dioxide is contributed to the atmosphere by three main routes:

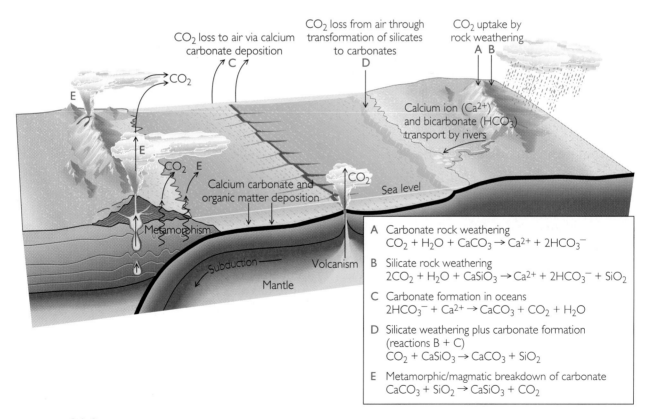

A Carbonate rock weathering
$CO_2 + H_2O + CaCO_3 \rightarrow Ca^{2+} + 2HCO_3^-$

B Silicate rock weathering
$2CO_2 + H_2O + CaSiO_3 \rightarrow Ca^{2+} + 2HCO_3^- + SiO_2$

C Carbonate formation in oceans
$2HCO_3^- + Ca^{2+} \rightarrow CaCO_3 + CO_2 + H_2O$

D Silicate weathering plus carbonate formation (reactions B + C)
$CO_2 + CaSiO_3 \rightarrow CaCO_3 + SiO_2$

E Metamorphic/magmatic breakdown of carbonate
$CaCO_3 + SiO_2 \rightarrow CaSiO_3 + CO_2$

FIGURE 24.9 The geochemical carbon cycle, showing reservoirs and fluxes. The fluxes of carbon dioxide into the atmosphere are balanced by fluxes out of the atmosphere. Similarly, the fluxes of carbonate and organic carbon into the Earth's interior are balanced by fluxes from the interior.

1. *Volcanism*—Outgassing from magmas formed in the mantle and crust injects carbon dioxide into the atmosphere.

2. *Sedimentation of calcium carbonate*—This reaction releases one molecule of carbon dioxide and one molecule of water for each molecule of calcium carbonate formed (see reaction C in Figure 24.9; see also the discussion of carbonate precipitation in Chapter 7).

3. *Metamorphism*—Carbon dioxide is liberated when carbonate minerals are replaced by silicate minerals, and some of this carbon dioxide eventually goes into the atmosphere (see reaction E in Figure 24.9; see also the discussion of contact metamorphism of limestones in Chapter 8).

These gains in atmospheric carbon dioxide are balanced by losses, particularly when carbon dioxide is transformed to dissolved bicarbonate ions by weathering (see reactions A and B in Figure 24.9; see also the discussion of weathering of silicates and carbonates in Chapter 6). In addition, carbon is buried in the crust in two forms: oxidized carbon in carbonate and reduced carbon in organic matter. As we have seen, burial of organic carbon releases oxygen to the atmosphere.

The carbon cycle keeps carbon in Earth's reservoirs balanced over the long term, but over the short term the burning of fossil fuels—coal and oil—is adding carbon dioxide to the atmosphere at a fast rate, threatening to affect global climate and the biosphere.

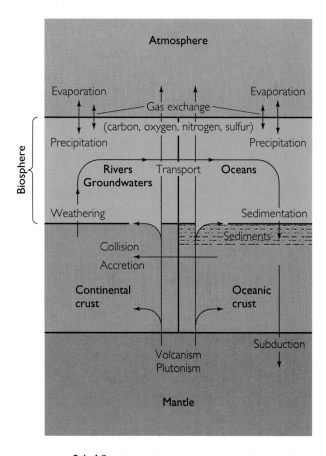

FIGURE 24.10 The major reservoirs and fluxes of the Earth system. In this steady-state model, all fluxes are balanced so that the amount of any substance in the mantle, lithosphere, hydrosphere, and atmosphere remains constant with time. The biosphere is spread out over the crust, hydrosphere, and lower atmosphere.

A Geochemical Model of Earth

A simplified geochemical model of the Earth is shown in Figure 24.10. The mantle is the reservoir that underlies both oceanic and continental crust. The flux between the mantle and the crust is volcanism (and plutonism). These are controlled by plate tectonics. The continental reservoir of lakes, rivers, and groundwater—collectively known, together with the oceans, as the hydrosphere—overlies and permeates the surface layers of continental crust. Into this reservoir drip the dissolved products of weathering, a flux from the continental crust. The biosphere consists of the total mass of living organisms and is spread through the hydrosphere, along the surface of the crust, and in the lower layers of the atmosphere. The biosphere gains and loses elements as organisms exchange carbon dioxide, oxygen, water, and other components with the hydrosphere, atmosphere, and land surface.

Sedimentation is the great flux that keeps the ocean in a steady state, primarily by counterbalancing the influx of river water. As sediments are buried, they become part of the oceanic crust. There they stay until they move into the mantle through subduction or become part of the continental crust through plate collisions. Plate collisions are an alternative flux that thrusts parts of the seafloor up onto volcanic island arcs and adjoining continental margins, transforming them to continental crust. Uplift of continental regions exposes crustal rocks to weathering and erosion, maintaining the balance of fluxes among all the reservoirs.

This balance sustains life as we know it on Earth. If the current balance were to change through some great perturbation, geological or human-induced, many of Earth's life forms would face devastation and perhaps extinction.

GEOLOGICAL PERTURBANCES AND CLIMATE CHANGE

Throughout geologic time since the early Precambrian, when life began, the course of biological evolution has both caused and been affected by changes in the atmosphere and hydrosphere—the systems primarily involved in weather and climate.

Earth's climate draws its energy from the solar radiation that warms the oceans, atmosphere, and land. The amount of radiation from the Sun reaching the top of the atmosphere measures a fairly consistent 1370 watts per square meter and is called the **solar constant.** Only since 1979 has the solar constant been measured accurately. Trace variations of about 0.1 percent have been detected and are being monitored for any effect on the environment. Life depends on the solar constant's not changing enough to significantly disrupt climate and weather.

The interacting systems of hydrosphere and atmosphere absorb solar energy and serve as reservoirs to store and transfer it. Ocean and air currents transport the stored heat, mitigating the extreme temperature differences that would otherwise occur between day and night, between summer and winter, and between equatorial and polar regions. Earth's climate is thus more equable than it would be otherwise and not so extreme as to create insurmountable hardships for life. From time to time, these interacting systems are perturbed, and Earth's climate changes as a result.

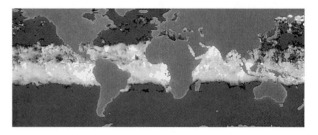

FIGURE 24.11 Global spread of sulfuric acid aerosols resulting from the June 1991 eruption of Mount Pinatubo. This false-color image from satellite data shows the concentration of aerosols as shades of yellow, from brown (lowest concentration) to white (highest concentration). The upper image, compiled within two weeks after the eruption, shows a slight increase over normal aerosol distribution. The lower image, compiled two months after the eruption, shows the buildup and spread of the aerosol layer over a vast area around the equator. *(Robert M. Carey, NOAA/Photo Researchers.)*

Internal Causes of Climate Change

Short-term global cooling can occur when sulfurous gases from large volcanic eruptions reach the stratosphere (Figure 24.11). There, sulfur gases combine with oxygen and water to form a layer of sulfuric acid aerosols (small droplets). This aerosol layer absorbs solar radiation and prevents it from reaching the lower atmosphere. As a result, the weather turns cold for a year or two until the aerosols fall out. In historical times, such events have caused crop losses, famine, and widespread suffering (see Chapter 5).

The geologic record suggests that long-term swings in global climate have occurred several times in Earth's evolution. To explain these climate changes, geologists are considering several hypotheses that link plate tectonics with the external environment. A new idea explains the climatic cooling and growth of ice sheets during the Cenozoic era by proposing a sequence of events that began with the collision of the Indian and Eurasian plates and the uplift of the Himalaya Mountains and the Tibetan Plateau (see Chapter 21). The Tibetan Plateau grew to a height of 5 km and an area half that of the United States to become a dominant feature of Earth's topography. The scientists behind this new idea[1] propose that the growth of the plateau would have had a profound effect on the monsoons of the region. Monsoons are major wind systems that reverse direction seasonally. They typically blow from cold to warm regions—from sea toward land in the summer because the summer sun heats the land, and from land toward sea in the winter. In southern Asia, the spring and summer heating of the Tibetan Plateau causes the air above to rise, creating a low-pressure region that draws in moisture-laden air from the adjacent oceans. The resulting winds rising over the southern slopes of the Himalayas bring heavy rains that swell the rivers.

The combination of heavy monsoon rains and high relief accelerates physical erosion and chemical

---

[1]M. E. Raymo and W. F. Ruddiman, 1992.

weathering of the newly exposed silicate rocks in the mountains. The carbonic acid in rainwater (formed by the solution of atmospheric carbon dioxide in water) accelerates the chemical breakdown of silicate rocks (see Chapter 6). The weathering products, which include minerals that lock up the carbon dioxide, are flushed away by rapidly moving streams, which constantly exposes fresh minerals to chemical attack. In simplified form, the reaction that removes atmospheric carbon dioxide is

$$CaSiO_3 + CO_2 \xrightarrow{\text{chemical weathering}} CaCO_3 + SiO_2$$

This reaction results in a drawdown of atmospheric carbon dioxide, a reduced greenhouse effect, and global cooling. Computer models of the growth of the Tibetan Plateau and climatic change seem to support this new hypothesis linking plate tectonics and the external environment.

Other mechanisms linking plate tectonics and global climate are also being researched. These include sea-level changes, the drift of continents over the poles, and tectonic movements that block ocean currents or open gateways for currents to flow through. The emergence of the Isthmus of Panama, for example, closed a passage connecting the Atlantic and Pacific oceans. If future geologic activity were to close the narrow channel between the Bahamas and Florida through which the Gulf Stream flows, temperatures in western Europe would drop drastically.

Climate Change and Mass Extinctions

Paleontologists have found dozens of mass extinctions in the fossil record. These show up as the rapid disappearance of large numbers of species and their replacement over time by new assortments of organisms. Global climate change is a favored explanation, and several causes of such change have been proposed, including superplumes (huge eruptions of flood basalts; see Chapter 5), impacts from bolides (asteroids and comets), and intense radiation from a nearby exploding star.

SUPERPLUME ERUPTIONS AND THE PERMIAN-TRIASSIC EXTINCTIONS The greatest mass extinction occurred 250 million years ago at the boundary between the Permian and Triassic periods, referred to as the P-T boundary (Figure 24.12), when some 90 percent of marine species, 70 percent of land vertebrates, and many families were wiped out. Some scientists think that this shock set the stage for the evolution of dinosaurs. The geologic

record shows a sea-level drop, an ice-sheet expansion, and an interval of acid rain at this time.

Geologists are searching for the causes. They have not yet found compelling evidence for a bolide, but they are impressed that the most voluminous known superplume eruption of continental flood basalt occurred in Siberia at about the time of the Permian-Triassic extinctions. Some argue that the close timing of these events could simply be coincidental and does not establish that one caused the other. Others argue that although a superplume and its drastic effects have never been witnessed, the Siberian flood volcanism should have disturbed Earth's environment profoundly. Estimates are that 2 million to 3 million km^3 of basaltic magma poured out over a period of a million years and covered an area of 2.5 million km^2, about one-fourth

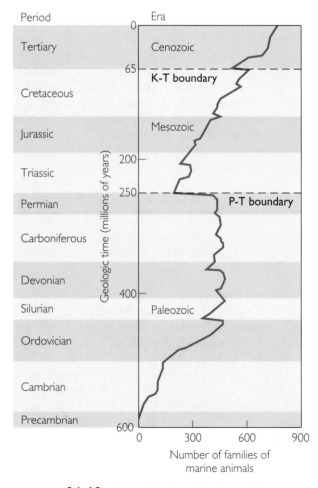

FIGURE 24.12 Marine life from the Precambrian to the Tertiary period. The fluctuation in the number of families of marine animals shows periods of growth and periods during which extinctions occurred. (After W. Alvarez and F. Asaro, 1990, *Scientific American*, October, 78–84.)

the area of the United States. About 20 percent of the Siberian superplume eruption was pyroclastic, and so volcanic dust would have reached the stratosphere and blanketed all of Earth. The superplume should also have released large amounts of sulfur dioxide (SO_2) gas, forming a haze of sulfuric acid aerosols. The dust and aerosols in the stratosphere would have blocked sunlight, chilled the climate, and devastated the food chain. The rainout of the aerosols (acid rain) could have been lethal to many ecosystems. Together, the chill and the acid rain would have pushed the environment beyond the limits of survival of many species. Some scientists believe that the 10 or 12 superplumes recorded by continental flood basalts over the past 250 million years may correlate with a similar number of mass extinctions.

Other scientists have proposed alternative mechanisms for the Permian-Triassic mass extinction. One hypothesis is that the drop in sea level evident at the time occurred because plate movements caused an increase in the volume of ocean basins. The drop in sea level would have exposed continental shelves and disrupted marine habitats. Oxidation of the exposed organic matter would have increased the amount of carbon dioxide in the atmosphere, warming and humidifying the climate and disrupting land habitats. Another hypothesis is that the oceans became deficient in oxygen because of a decline in ocean circulation. Some scientists propose that all of these mechanisms—superplume, drop in sea level, and oceanic oxygen deficiency—were at work, but spread over a few million years.

The Permian-Triassic mass extinction, dubbed "the mother of mass extinctions," is receiving a great deal of attention by scientists because it marked a profound change in the evolution of life.

BOLIDE IMPACT AND THE CRETACEOUS-TERTIARY EXTINCTIONS The second-greatest mass extinction occurred about 65 million years ago, at the boundary between the Cretaceous and Tertiary periods, referred to as the K-T boundary. Some scientists are convinced that volcanism was the cause and cite as evidence the major superplume eruption of flood basalts that occurred in India at that time. The most widely accepted hypothesis now, however, is that a giant bolide slammed into Earth, releasing 4 billion times more energy than the atomic bomb that destroyed Hiroshima. The impact would have blasted 1000 trillion tons of material out of the Earth and pulverized the bolide. The dust would have blanketed the Earth with an opaque cloud that blocked the Sun's light and heat. The

glowing ejecta returning to Earth would have ignited forest fires on a continental scale, sending a blanket of soot into the atmosphere and making it even more impenetrable to the Sun's energy. Earth passed through a freezing night that lasted for months. Photosynthesis stopped, the food chain collapsed, about half the planet's species died, and the 100-million-year reign of the dinosaurs ended. The course of evolution changed, perhaps luckily for humankind, because among the survivors were small mammals, our progenitors.

The **bolide impact hypothesis** has inspired much debate and research (see Feature 24.1). Evidence lies in a large crater on the Yucatan Peninsula of Mexico (Figure 24.13), together with a layer of breccia near the top of the Cretaceous strata containing shock-metamorphosed crystals of quartz and feldspar and rock fragments melted into glass. **Shock metamorphism** occurs when minerals are

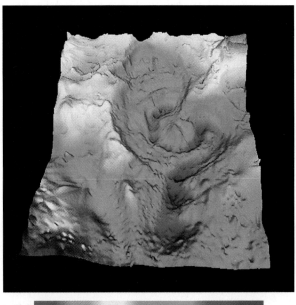

Highest gravity   Average gravity   Lowest gravity

FIGURE 24.13 Chicxulub impact crater in Yucatan Peninsula, Mexico. The crater is not visible at the surface but shows up in a three-dimensional map of gravity variations as a circular depression about 40 to 45 km in radius. The depression portrays a reduction of gravity over the crater because of the low densities of the impact debris and Tertiary sediments filling the crater compared to the higher density rock surrounding it. (*B. Sharpton, Lunar and Planetary Institute.*)

Mass Extinctions and the Bolide Impact Debate

About 65 million years ago, between the end of the Cretaceous period and the beginning of the Tertiary period (the K–T boundary), a mass extinction occurred in which half the life forms on Earth died out. Among the victims were the dinosaurs. Their disappearance allowed for the rapid evolutionary expansion in the Cenozoic era of mammals and, ultimately, humans.

Scientists have proposed several hypotheses to explain this mass extinction. None has inspired more debate, more research, and ultimately more confidence than the bolide impact hypothesis proposed in 1980 by Walter Alvarez.

Alvarez and a team of scientists from the University of California at Berkeley were measuring the iridium content of sedimentary layers near Gubbio, Italy (see photos). Iridium is rare in Earth's crust and upper mantle but relatively abundant in the rest of the solar system. The iridium content of sediments and sedimentary rock, therefore, is a clue to the rates at which cosmic dust settled on Earth over time.

Much to their surprise, the Alvarez team found concentrations of iridium in a 1-cm clay layer deposited exactly at the K–T boundary to be 30 times higher than in the sedimentary layers above and below it. Alvarez suspected that the unusual iridium level could somehow be connected to the K–T mass extinction. He and his team proposed that a large meteorite, perhaps 10 km in diameter, had collided with Earth at a speed of at least 75,000 km per hour. The force of the impact vaporized the meteorite and some 1000 trillion tons of Earth's crust into a globe-encircling dust cloud rich in iridium and other shock-metamorphosed debris. The environmental consequences of this catastrophic impact caused the extinctions that define the K–T boundary.

Like all good science, the Alvarez hypothesis was both creative and testable. It suggested new areas of research and generated lively debate among geologists and other scientists.

A 10-km meteorite should have produced an impact crater approximately 200 km in diameter. Where was the crater? Furthermore, the idea of such a catastrophic cause for the K–T extinctions was a direct challenge to the long-held view that all geologic change (including mass extinction) was primarily the result of gradual processes operating on a time scale of millions of years. Paleontologists pointed to the selective nature of the extinctions on land and in the sea. Why (or how) did some groups go extinct while others survived? Were the K–T extinctions truly instantaneous, or did they reflect a gradual dying out over the last several million years of the Cretaceous? Other scientists suggested that the iridium came not from a meteorite but from Earth's interior, brought to the surface 65 million years ago in an episode of widespread explosive activity that could also have been responsible for the K–T extinctions. In support of this hypothesis, they cited the Deccan Plateau basalts of northern India, dated at 65 million years and believed to have been formed by a vast superplume eruption of magma.

In the decade following Alvarez's publication of the bolide impact hypothesis, independent studies designed to test it found compelling support for it. A sampling of K–T boundary sediments at more than 100 locations around the world consistently revealed levels of iridium up to 1000 times higher than the average concentration in rocks of Earth's crust. This was powerful evidence for a bolide impact, but it did not refute the competing superplume hypothesis. Grains of quartz in sediments deposited at the K–T boundary were shown to have intersecting sets of microscopic parallel shock fractures. Shock-metamorphosed quartz crystals had previously been seen only at nuclear test sites and known impact craters. Those endorsing the superplume hypothesis suggested that the high pressures necessary to produce shock fractures could be generated in a magma chamber, but they were never able to provide convincing evidence. In some sections of the K–T boundary layer, scientists found spherical particles of glass believed to have been formed by rapid cooling of impact-generated droplets of molten rock. Chemical analysis of these glass particles offered convincing evidence that they could not have been produced by volcanism.

Despite all this evidence, the question of the impact location remained. Geologists acknowledged the crater could have been destroyed by subduction. Over the past 65 million years, approximately 30 percent of the ocean floor has been subducted. Still, the search for the elusive crater continued.

In the 1990s, researchers looking for oil-rich sedimentary basins discovered a sub-surface crater 200 km in diameter on the Yucatan Peninsula in Mexico, near the town of Chicxulub. The Chicxulub crater, as it is now known, contains impact-generated melt rock dated at 65 million years old and high levels of iridium characteristic of an extraterrestrial source. Careful study of sections of the K-T boundary layer around the Gulf of Mexico and the Caribbean Basin has also revealed layers of impact-generated glass particles up to 50 cm thick and chaotic sedimentary deposits that appear to have been churned up by impact-generated tsunamis.

After years of painstaking research, contentious debate, and the publication of more than 3000 professional papers, most scientists agree that an impact did occur. The results of such an impact, and especially its relation to the K-T extinctions, however, remain controversial. Paleontologists, in particular, continue to question why some species died out while others did not and to doubt that the extinctions occurred simultaneously and suddenly. Furthermore, little or no reliable evidence has been discovered that connects any other mass extinction with bolide impacts.

The story of the bolide impact hypothesis is an example of scientific methodology at its best. From careful analysis and observation, the Alvarez team proposed a creative and testable hypothesis. Testing has not only lent support to the original idea but has also generated new

(a) Strata at Gubbio, Italy, bearing millions of years of geologic history. *(Michael M. Follo.)*

(b) Close-up of strata at Gubbio, Italy. The head of the hammer marks the iridium-rich 1-cm clay layer deposited at the K-T boundary, 65 million years ago, when half the Earth's life forms, including the dinosaurs, died out. *(Michael M. Follo.)*

and important research on other aspects of impacts, mass extinctions, and environmental crises. The effects of an impact at the K-T boundary continue to be the subject of lively debate. And whatever the outcome of that debate, the most important and far-reaching legacy of the Alvarez hypothesis may ultimately be a greater appreciation of the role of catastrophic events in shaping the history of our planet and life on it.

subjected to the high pressures and temperatures of shock waves generated by impacts. Shock-metamorphosed minerals can be recognized by such effects as melting, deformed grains, parallel sets of fractures, conical fractures, and transformation to high-pressure forms. But the "smoking gun"—the most convincing evidence for the bolide impact hypothesis—is a high concentration of the element iridium in the K-T boundary layer. Iridium is the chemical fingerprint of a bolide impact because the element is abundant in bolides but rare in Earth's crust (Figure 24.14).

HUMAN ACTIVITY AND GLOBAL CHANGE

Given the likely role of a bolide in at least one and perhaps more mass extinctions, some prominent scientists are requesting that the U.S. government provide about $100 million to install a global network of telescopes to detect possible bolides months or years before they could reach Earth. What protective action to take if and when one is found will tax the skills of future generations of scientists and engineers. In the meantime, more immediate threats to life on Earth arise from human activity.

The term **global change** entered the world's vocabulary as evidence mounted that emissions from human activities could alter the chemistry of the atmosphere with disastrous worldwide consequences. These include

- Climate change due to an enhanced greenhouse effect
- Increased exposure to ultraviolet rays because of stratospheric ozone depletion
- Mass die-offs due to acid precipitation
- An overburdening of many Earth systems due to overpopulation

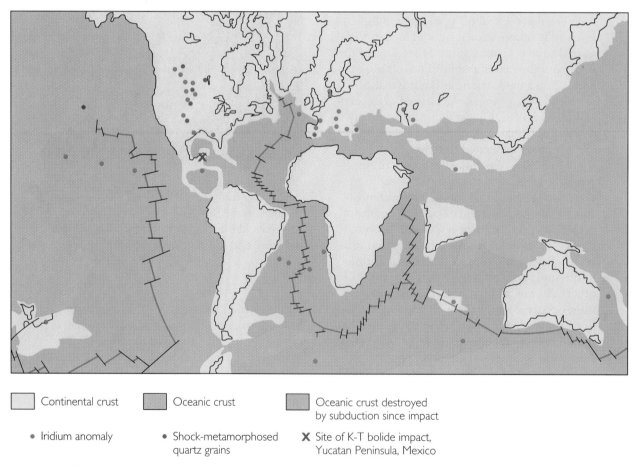

| Continental crust | Oceanic crust | Oceanic crust destroyed by subduction since impact |

- Iridium anomaly
- Shock-metamorphosed quartz grains
- **X** Site of K-T bolide impact, Yucatan Peninsula, Mexico

FIGURE 24.14 Worldwide distribution of iridium and shock-metamorphosed quartz at the K-T boundary. (After W. Alvarez and F. Asaro, *Scientific American*, October, 78–84.)

Greenhouse Gases and Global Warming

The Industrial Revolution began around 1850, and ever since then we have been burning carbon-based fuels—coal, oil, and natural gas—at an increasing rate. Without special treatment, every carbon atom burned turns into carbon dioxide in the atmosphere.

Not all of the increased load of carbon dioxide remains in the atmosphere. About half of it is absorbed by surface layers of the oceans and by life forms, mostly trees and other green plants. Growing plants take in CO_2. When they are burned or decay, they release CO_2. At the present rate at which we burn fossil fuels and destroy forests, we may expect the amount of carbon dioxide in the atmosphere to reach double the preindustrial level in the second half of the next century. Other greenhouse gases include methane, nitrous oxides, and the chlorofluorocarbons (CFCs) used as refrigerants, cleaning agents, and propellants. Under these conditions, the greenhouse effect that keeps Earth's surface warm could become dangerously enhanced. The combined effect of all these emissions could be a **global warming** of 1 to 3.5°C toward the end of the next century (Figure 24.15).

CONSEQUENCES OF GLOBAL WARMING   Global warming could cause changes in wind and rainfall patterns and soil moisture that could convert some of Earth's most productive agricultural regions into semiarid wastelands. The drought and unusually hot weather of 1988, 1991, and 1995 (the hottest year on record) and the record-breaking numbers of hurricanes in 1995 and 1996, although not necessarily related to the greenhouse effect, served to raise public consciousness of the seriousness of climatic change in the decades ahead. Because human-induced warming would be more rapid than any that occurs naturally, many plant and animal species would have difficulty adjusting or migrating. Those that could not cope with rapid warming would become extinct.

Along with changes in wind and rainfall, the oceans could warm and expand, raising the sea level as much as 60 cm, a serious problem in low-lying countries such as Bangladesh. If the continental ice sheets start to melt, the sea level could rise even higher; if the Antarctic and Greenland ice sheets melted entirely, the sea level would rise some 65 m. We agree, however, with most experts who study how continental glaciers form, grow, and shrink (see Chapter 15) that several hundred years would be required to achieve the necessary warming.

REDUCING GREENHOUSE GAS EMISSIONS With few exceptions, atmospheric scientists expect the climate to change because of the burning of fossil fuels. An international panel of scientists appointed by the United Nations reported in 1995 that part of the global warming of 0.5°C that has occurred in this century is due to human activity. No one can say at this time how much global warming will ultimately occur, or to what effect, but we must reckon with the possibility that a climatic crisis could prevent us from fully using our remaining resources of coal and other fossil fuels. Such a crisis could last a long time, because even if all human activities that generated carbon dioxide were to stop, it could take as much as 200 years for atmospheric carbon dioxide to return to its preindustrial level.

This uncertainty presents a problem for policymakers. How much should we spend to curb human-generated carbon dioxide emissions, and will the benefits justify the costs? On the one hand, too much spending could depress the economy and cause job losses. On the other hand, prevention might be less costly than coping with a disaster after it happens.

We could reduce the emissions of greenhouse gases by using fossil fuels more efficiently while making the transition to wind, water, and solar power; nuclear power redesigned for safety; fuels made from renewable resources such as wood and grain; and vehicles powered by electricity. The nations of the

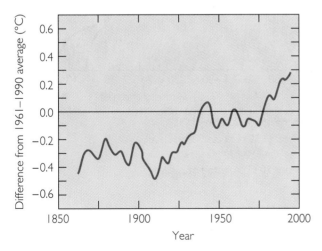

FIGURE 24.15 The trend of global warming. Although global temperatures fluctuate from year to year, the trend is clearly upward, reflecting a rise in carbon dioxide and other greenhouse gases. (*U.K. Meteorological Office, University of East Anglia.*)

world all agree to such ideas in principle, but many, including the United States, emphasize the costs of moving too quickly.

Policymakers must also face the issue of fairness in international politics. The rich, industrially advanced nations, which have been the greatest emitters of greenhouse gases, will be able to adjust to standards for reducing emissions more easily than the developing countries. China, for example, depends on its huge coal deposits for its economic growth. Developing nations argue for financial and technological support to help them cope with the demand to reduce emissions.

The authors of this book believe that the consequences of climate change could be so serious that it is not too soon for our leaders to start planning for reduced reliance on fossil fuels. The United States could reduce emissions of greenhouse gases by as much as 60 percent from 1990 levels at little cost by improving the efficiency of energy use—for example, insulating buildings; replacing incandescent lights with fluorescent lights; increasing the fuel efficiency of motor vehicles a few miles per gallon; and making greater use of natural gas, which emits less carbon dioxide when burned than coal. These modest steps constitute a low-cost insurance policy against global warming that also confers substantial benefits. These include reductions in expensive oil imports, lowered manufacturing costs, and increasingly healthful air quality.

CFCs and Ozone Depletion

Near Earth's surface, ozone gas (O_3) is a major constituent of smog. It undermines health, damages crops, and corrodes materials. Low-lying ozone forms when sunlight interacts with nitrogen oxides and other chemical wastes from industrial processes and automobile exhausts.

Ozone in Earth's stratosphere, 10 to 50 km above the surface, is another matter. There, solar radiation converts oxygen (O_2) to ozone (O_3), which forms a protective layer around the Earth that absorbs certain portions of cell-damaging UV radiation. Skin cancer, cataracts, impaired immune systems, and reduced crop yields are attributable to excessive UV exposure.

In 1996, the Nobel Prize in chemistry was awarded to Sherwood Rowland and Mario Molina for the hypothesis they advanced 22 years earlier that the protective ozone layer can be depleted by a reaction involving human-made compounds called chlorofluorocarbons (CFCs). CFCs, which are used

as refrigerants, propellants, and cleaning solvents, are stable and harmless except when they migrate to the stratosphere. High above Earth, the intense sunlight breaks down CFCs, releasing their chlorine. The chlorine reacts with the ozone molecules in the stratosphere and thins the protective ozone layer. Rowland and Molina's hypothesis was confirmed when a large hole in the ozone layer was discovered over Antarctica in 1985. Subsequently, **stratospheric ozone depletion** was found to be a global phenomenon. Elevated surface levels of UV radiation were observed to correspond to low ozone levels in 1992 and 1993.

In 1989, a group of nations entered into a global treaty to protect the ozone layer. This treaty, called the Montreal Protocol, now includes 140 countries. The main provisions of the Montreal Protocol are to phase out CFC production by 1996 and to set up a fund, paid for by developed nations, to help developing nations switch to ozone-safe chemicals (Figure 24.16). The phase-out of CFCs has been successful, and safer alternatives to CFCs are being manufactured.

The Montreal Protocol is a model of how scientists, industrial leaders, and government officials

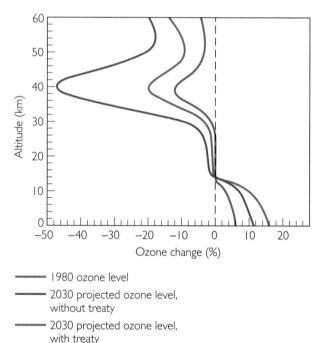

FIGURE 24.16 Projected ozone levels with and without the Montreal Protocol. The 140 nations that signed this international treaty have been successful at eliminating emissions of CFCs, which attack the ozone layer. Without such international cooperation, ozone concentrations would lessen dangerously. (After T. E. Graedel and P. J. Crutzen, 1993, *Atmospheric Change*, New York: W. H. Freeman.)

can work together to head off an environmental disaster. We can hope that it will serve as a role model for other situations requiring concerted response from the community of nations.

Acid Rain and Global Damage

In many industrialized areas of the world, the air is greatly polluted with sulfur-containing gases such as sulfur dioxide (SO_2). These gases are emitted from the smokestacks of power plants that burn coal containing large amounts of the mineral pyrite (iron sulfide, FeS_2), from smelters of sulfide ores, and from some factories. Coals mined in the eastern and midwestern regions of the United States contain more of these pollutants than coals from the western states. Although volcanoes and coastal marshes also add sulfur gases to the atmosphere, more than 90 percent of the sulfur emissions in eastern North America are of human origin.

Sulfur gases in the atmosphere react with oxygen and rainwater to form sulfuric acid, which is far stronger than the carbonic acid formed by carbon dioxide and rainwater. Some nitric acid is formed in the same way from nitrogen oxide gases (NO_x) emitted from smokestacks and automobile exhausts. Small amounts of sulfuric and nitric acids turn harmless rainwater into **acid rain.** Although it is much too weak to sting human skin, acid rain causes widespread damage to water and air, delicate organisms, and solid rock.

By acidifying sensitive lakes and streams, particularly those underlain by soils with limited ability to neutralize acidic compounds, acid rain has caused massive kills of fish in many lakes in Canada, the northeastern United States, and Scandinavia. A survey of more than 1000 lakes and thousands of miles of streams in the United States showed that 75 percent of the lakes and 50 percent of the streams were acidified by acid rain. In some acidified lakes and streams, fish species such as the brook trout have been completely eradicated. The salmon habitat of eastern Canada has been seriously affected. Acid rain also damages mountain forests, particularly those at higher elevations. Acid moisture in the air reduces visibility.

Acid rain causes noticeable damage to fabrics, paints, metals, and rocks, and it rapidly weathers stone monuments and outdoor sculptures (Figure 24.17). In Canada alone, acid rain causes about $1 billion in damage every year to buildings and monuments.

COAL BURNING   The relationship between acid rain and the burning of coal has been firmly established. Agencies of the U.S. and Canadian governments, and independent scientific panels, have tracked sulfur gases emitted by smokestacks to downwind locations where they are precipitated as acid rain. Careful tracing of pollutants from their sources to the sites of acid rain is necessary to demonstrate that it is coal-burning power plants that are responsible for much of the problem.

Most of the sulfurous emissions in North America come from coal-fired electric utility plants in the midwestern and eastern United States, and most of

FIGURE **24.17** A monument before and after deterioration caused by acid rain. (*Westfalisches Amt für Denkmalpflege.*)

the fallout occurs in these regions (Figure 24.18) and in eastern Canada. About 50 percent of the sulfate deposits in Canada originate in the United States—a source of friction between the two countries. The problem is equally critical in Europe and Asia, where scientists have noted damage to lakes and forests caused by acid rain. The damage is particularly heavy in China and in countries of the former Soviet bloc.

THE CLEAN AIR ACT   Concerned scientists have recommended restriction of sulfur emissions from power plants and smelters, but for a long time political resistance prevented effective control. Finally, after many years of wrangling, the U.S. Congress passed the Clean Air Act in 1991. The legislation requires coal-burning power plants to reduce their annual SO_2 emissions by 10 million tons and their annual NO_x emissions by 2 million tons by the year 2000. Deregulation of U.S. railroads has lowered transportation costs, making it less costly to ship cleaner western coal to midwestern and eastern power plants and thus making it easier to meet the new emission standards.

Unlimited Population Growth

The world population of 5.7 billion in 1995 is expected to double by 2050. About 95 percent of this growth will occur in developing countries, where 77 percent of the world's people live (Figure 24.19). The developed countries, with only 23 percent of the world's population, produce 85 percent of the world's economic output and withdraw the majority of the minerals and fossil fuels from Earth's finite resources. These resources fuel industry and provide the basis for food, shelter, transportation, recreation, and all the other aspects of a high standard of living. The use of these resources is also largely responsible for today's level of pollution. Uncontrolled use of resources is a serious and growing encroachment by humankind on the interacting systems of Earth's atmosphere, hydrosphere, and land surface. Warnings that Earth's resources are finite are already beginning to appear. In the decade between the mid-1980s and mid-1990s, production of food from both land and sea decreased relative to population growth. The availability of fresh water is down and, with it, the amount of land dedicated to agriculture.

The world faces several dilemmas. How do we reduce greenhouse gases and other pollutants without impeding economic growth? Even a developed nation such as the United States needs economic growth to generate the funds necessary to solve some of its social problems. How can we meet the legitimate aspirations of the developing countries for economic growth without adding to the planet's load of pollutants? If the developing countries were to achieve the high standard of living of the devel-

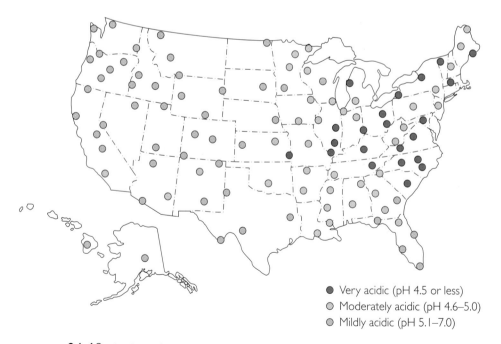

● Very acidic (pH 4.5 or less)
◐ Moderately acidic (pH 4.6–5.0)
○ Mildly acidic (pH 5.1–7.0)

FIGURE **24.18** Acidity of precipitation across the contiguous United States. The most acidic precipitation occurs in the east, mostly as a result of burning high-sulfur coal.

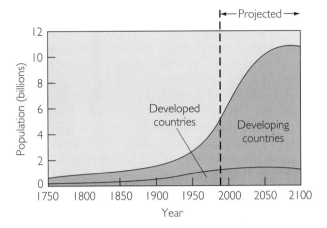

FIGURE **24.19** Population growth in developed and developing countries. *(From UN Population Division.)*

oped nations and consume Earth's resources in the same manner, the stress on the biosphere could not be sustained.

The population explosion in the developing countries compounds the problem of how to sustain both economic growth and a healthy environment. Unlimited population growth would lead to profound changes in the global environment (Figure 24.20). Human health would be threatened. The world would suffer an irreversible loss of biodiversity (the world array of wildlife and plants) with large-scale devastation of plant and animal habitats. There would be political unrest and mass migrations of people.

Some believe that the only alternative is for the developed nations to revert to the simpler lifestyles of earlier times, with a slower economic pace and much lower consumption of resources. In practice, this could mean smaller homes without air conditioning; fewer automobiles, televisions, and telephones; no packaged food—in short, fewer conveniences of all kinds. This philosophy has a small following, however, and is not likely to be adopted by leaders sensitive to electoral politics.

Is it possible to have **sustainable development** in both the developed and developing nations—that is, economic growth and improved standards of living that can last indefinitely and are environmentally benign? The stabilization of population growth and advances in science and technology might offer a solution.

According to a joint statement issued by 58 of the world's scientific academies,[2] high fertility rates can be correlated with poverty, high infant mortality, lowered life expectancy, low social status and educational levels of women, lack of access to reproductive health services, and inadequate availability and acceptance of contraceptives. Falling

[2]*Population Summit of the World's Scientific Academies, 1993.* National Academy Press, Washington, D.C.

FIGURE **24.20**
Unlimited population growth threatens to change the global environment in disastrous ways. *(Dilip Mehta/The Stock Market.)*

fertility rates go with increased incomes, improved standards of living, lowered infant mortality, increased life expectancy, increased adult literacy, and higher rates of female education and employment. Cultural influences also play an important role in fertility rates. These influences include the value placed on large families in developing countries, the lack of security for the elderly, and religious and political factors. The introduction of vaccines and antibiotics will continue to reduce death rates and increase population. It is projected that birth rates will not decline enough to reverse population growth for many decades.

A cooperative world effort could be undertaken by political leaders of both the developed and developing nations to address these root causes of population growth democratically and with sensitivity to human rights. The developed nations would have to contribute the necessary funds and personnel. Scientists could play an important role. They could invent more efficient ways to use fossil fuels, develop safe nuclear energy, and develop renewable energy sources such as solar and biomass energy. They could create more productive and disease-resistant crop species that would need fewer artificial fertilizers and chemical pesticides, thus reducing or eliminating chemical contamination of agricultural runoff water. They could develop new technology to reduce effluents from industrial processes. And they could encourage recycling and substitution, which would reduce the demand for materials and the enormous pollution caused by mining and smelting operations. To address the correlation of high infant mortality and low life expectancy with increased fertility rates, a global program would include improvements in sanitary engineering; the introduction and use of new vaccines and antibiotics; and the development and dissemination of more effective, simpler, and cheaper contraceptives.

Only by stabilizing population growth and using Earth's resources wisely can we avoid a calamity as serious as any humankind has ever faced.

SUMMARY

What is the origin of the atmosphere and hydrosphere? Outgassing during the early differentiation of Earth and continuing volcanic activity are the primary sources of the water and gases that make up the oceans and atmosphere, except for oxygen, which was added after photosynthetic organisms evolved.

What is the biosphere? The biosphere is the sum of all organisms living on and below the land surface, in the oceans, and in the atmosphere, including all the biological transformations that affect Earth's materials and processes.

What is the greenhouse effect? Much of the radiant energy from the Sun passes through the atmosphere and is absorbed by Earth's surface. The warmed surface radiates heat back to the atmosphere as infrared rays. Carbon dioxide and other trace gases absorb IR rays and prevent them from escaping. In this way, the atmosphere acts like the glass in a greenhouse, allowing solar radiant energy to pass through but trapping heat.

How did life begin? About 4 billion years ago, gases and liquids of the hydrosphere began to chemically react to form organic carbon under the energizing influence of strong UV radiation and other energy forms. These molecules increased in complexity through several stages, including a general growth and metabolism system; then a simple, biologically replicating RNA world; and finally the DNA world we now know as the biosphere. Soon after the DNA world evolved, the first primitive photosynthesizing organisms arose, profoundly changing the biosphere, the hydrosphere, and the atmosphere.

How did photosynthesis change the world? Photosynthesis is the chemical reaction by which carbon dioxide and water in the cells of organisms are chemically combined under the influence of sunlight to produce organic matter (carbohydrate) and oxygen. Photosynthesis provided an abundant source of biological energy and supplied oxygen to the atmosphere. Oxygen levels in the atmosphere became significant about 1.5 billion years ago. The change to an oxygen-rich atmosphere allowed for oxidative weathering, primarily of reduced iron minerals, and the evolution of oxygen-breathing animals.

What are geochemical cycles? Geochemical cycles trace the flux of Earth's elements from one reservoir to another, making it possible to quantify the amounts of elements that are stored in and moved between the oceans, atmosphere, land surface, crust, and mantle. The calcium cycle shows how calcium derived from weathering is transported by rivers to the oceans. Calcium's residence time in the oceans can be calculated as the total amount of the element in the reservoir divided by the inflow, when inflow balances outflow.

What is the carbon cycle? The carbon cycle includes photosynthesis and respiration; the inflow of carbon into the atmosphere from volcanism; the burning of organic materials; and the precipitation of calcium carbonate, the most abundant mineral containing carbon in its oxidized form. Outflows of carbon dioxide from the atmosphere are from the weathering of silicate and carbonate rocks and from the burial of carbon in the crust, which releases oxygen to the atmosphere.

How are extremes in temperature mitigated on Earth? The atmosphere and oceans store solar heat. Winds and ocean currents transfer the heat from place to place.

How can internal geologic processes cause climatic change? Over the short term of a few years, sulfuric acid aerosols emanating from large volcanic eruptions can absorb solar radiation before it reaches the lower atmosphere and thus lower global temperatures. Over the long term of millions of years, plate-tectonic movements can cause continents to drift over the poles, block or open gateways to ocean currents, and cause uplift, all of which alters weather systems and rates of chemical erosion that draw on atmospheric CO_2.

What is ozone depletion, and why should it concern us? CFCs are chemicals used as refrigerants and in industrial processes. When they reach the upper atmosphere, they react with ozone (O_3) and convert it to oxygen (O_2). Ozone absorbs harmful ultraviolet radiation streaming in from the Sun. Without the ozone shield in the upper atmosphere, many life forms would die off. An international agreement has been reached to stop the production of chemicals that destroy the ozone shield.

What is acid rain, and what is being done about it? Acid rain is rain that contains sulfuric acid and/or nitric acid, formed when sulfur and nitrogen gases react with rainwater. The sulfur gases are waste gases from power plants that burn coal containing pyrite and from smelters of sulfide ores. Nitric acid is a product of coal combustion and automobile exhausts. Acid rain can destroy fish stocks in lakes and rivers, damage mountain forests, and damage buildings and monuments. In the United States, the Clean Air Act was legislated to reduce acid emissions.

Why should we worry about unlimited population growth, and what can be done about it? With unlimited population growth, human activities would place great stress on the biosphere, cause an irreversible loss in biodiversity, and threaten human health and well-being. The root causes of unlimited population growth are known and to a large extent could be eliminated by international programs that improve the education, standard of living, economic and social status, and health of poor people in developing countries.

KEY TERMS AND CONCEPTS

biosphere (p. 613)
greenhouse effect (p. 613)
photosynthesis (p. 614)
respiration (p. 614)
stratosphere (p. 615)
ozone (p. 615)
geochemical cycle (p. 617)

reservoir (p. 617)
flux (p. 617)
calcium cycle (p. 617)
residence time (p. 618)
carbon cycle (p. 619)
solar constant (p. 621)
bolide impact hypothesis (p. 623)

shock metamorphism (p. 623)
global change (p. 626)
global warming (p. 627)
stratospheric ozone depletion
   (p. 628)
acid rain (p. 629)
sustainable development (p. 631)

EXERCISES

1. What is a greenhouse gas, and how does it affect Earth's climate?

2. What is residence time?

3. List two inflows of chemical components into the ocean and two outflows of chemical components from the ocean.

4. Define the biosphere.

5. How did oxygen arise in Earth's atmosphere?

6. List three causes of climate change that result from the interaction of Earth's internal and external systems.

7. What is meant by mass extinctions, and what kinds of events trigger them?

8. Make a list of human activities that have negative effects on the atmosphere and hydrosphere.

9. What are the important factors that determine rapid population growth? What are the consequences of unlimited population growth?

THOUGHT QUESTIONS

1. How would the calcium geochemical cycle be affected by a global increase in chemical weathering?

2. Devise a simple geochemical cycle for the element sodium, which is found in marine evaporites (halite) and clay minerals.

3. What would the carbon cycle have looked like after life had originated but before photosynthesis had evolved?

4. Assume that we keep pumping carbon dioxide into the atmosphere at a steadily increasing rate and Earth warms significantly in the next 100 years. How would this affect the global carbon cycle?

5. Can scientists offer credible advice about such politically and socially charged issues as population growth, global climatic change, and relations between the developed and developing nations? What should a government leader do when 8 out of 10 scientists say that human-induced global warming will occur unless something is done to head it off, while 2 out of 10 disagree?

6. An economist once wrote: "The predicted change in global temperature due to human activity is less than the difference in winter temperature between New York and Florida, so why worry?" Do you think the economist was right?

7. A high-level government official once suggested that ozone depletion is not worrisome because people could just use sunglasses, sunscreen lotion, and wide-brimmed hats to minimize the danger from UV rays. Do you think the official was right?

LONG-TERM TEAM PROJECT

See Chapter 11.

SHORT-TERM TEAM PROJECTS:

EARLY EARTH

Imagine that you have just bought a ticket on a time machine that will transport you and a classmate for a few hours to a time when the Earth was half its present age. Together, develop a list of six items that would and would not be useful to you to survive during that time, and explain why you could or could not use them. For example, an oxygen mask would be helpful, but a frying pan probably would not be.

SURFACE EFFECTS OF INTERIOR PROCESSES

Observation of unusual amounts of the element iridium, thought not to have a large terrestrial source, in a clay layer at the Cretaceous-Tertiary boundary led to the hypothesis that the impact of a bolide caused the extinction of about half the species on Earth, including dinosaurs. Competing with the bolide impact hypothesis was the idea that worldwide volcanism could have caused the extinctions and released the iridium. Three chemists from the University of Maryland reported in an issue of *Science* that airborne particles from the January 1983 eruption of Kilauea in Hawaii that had collected on air filters contained concentrations of iridium up to 17,000 times higher than in typical Hawaiian basalts. These authors wrote: "Since iridium enrichments have not previously been observed in volcanic emissions, the results for Kilauea suggest that it is part of an unusual volcanic system which may be fed by magma from the mantle."

For this short-term project, you and a partner should investigate and list the characteristics of volcanoes whose magma originates in the Earth's mantle and suggest reasons why they could not account for the iridium found at the Earth's surface.

SUGGESTED READINGS

Alvarez, W., and F. Asaro. 1990. What caused the mass extinction? An extraterrestrial impact. *Scientific American* (October):78–84.

Berner, Robert A., and A. C. Lasaga. 1989. Modeling the geochemical carbon cycle. *Scientific American* (March):74–81.

Cook, Elizabeth. 1996. *Marking a Milestone in Ozone Protection: Learning from the CFC Phase-Out.* Washington, D.C: World Resources Institute.

deDuve, C. 1995. The beginnings of life on Earth. *American Scientist* 83:428–437.

Erwin, D. H. 1996. The mother of mass extinctions. *Scientific American* (July):72–78.

Graedel, T. E., and P. J. Crutzen. 1993. *Atmospheric Change: An Earth System Perspective.* New York: W. H. Freeman.

National Academy of Sciences. 1992. *Policy Implications of Greenhouse Warming.* Washington, D.C.: National Academy Press.

National Research Council. 1993. *Solid Earth Sciences and Society.* Washington, D.C.: National Academy Press.

Office of Technology Assessment. 1991. *Changing by Degrees: Steps to Reduce Greenhouse Gases.* Washington, D.C.: U.S. Government Printing Office.

Raymo, M. E., and W. F. Ruddiman. 1992. Tectonic forcing of late Cenozoic climate. *Nature* 359:117–122.

Renne, P. R., and others. 1995. Synchrony and causal relations between Permian-Triassic boundary crises and Siberian flood volcanism. *Science* 269:1413–1416.

INTERNET SOURCES

Carbon Dioxide Information

❶ **http://cdiac.ESD.ORNL.GOV/cdiac/**

The Carbon Dioxide Information Analysis Center is operated by Oak Ridge National Laboratory and provides information to help "evaluate complex environmental issues, including potential climate change associated with elevated levels of carbon dioxide" and other greenhouse gases. The historic climate change data base is an outstanding resource.

Climate Processes over the Oceans

❶ **http://eos.atmos.washington.edu/**

Sponsored by the NASA Earth Observing System and located at the University of Washington, this site focuses on "the role of circulation, clouds, radiation, water vapor, and precipitation in climate change, and the role of ocean-atmosphere interactions in the energy and water cycles." The significance of midlatitude and tropical cloud systems, low-level clouds, and atmospheric mixing is described.

Asteroid and Comet Impact Hazards

❶ **http://ccf.arc.nasa.gov/sst/main.html**

This Web site at the NASA Ames Space Science Division is a gateway to a wealth of information about near-Earth objects, Earth-crossing asteroids, and current efforts to identify them. Links to a variety of related sites include The KT Event, Collisions with Earth, and The Collision of Comet Shoemaker-Levy 9 with Jupiter.

Climate Change

❶ **http://www.ucsusa.org/global/climate.html**

Sponsored by the Union of Concerned Scientists, this site includes a scientific overview of climate change, the policy arena, frequently asked questions, "Popular Myths Undone," and climate-change resources on the Web.

U.S. Census Bureau

❶ **http://www.census.gov/**

This is the gateway to the Census Bureau's information resources. The Population Clocks provide up-to-the-minute estimates of the U.S. and world populations and data on population growth. Links provide easy access to basic population data for every state and county.

Appendix 1 Conversion Factors

LENGTH

1 centimeter	0.3937 inch
1 inch	2.5400 centimeters
1 meter	3.2808 feet; 1.0936 yards
1 foot	0.3048 meter
1 yard	0.9144 meter
1 kilometer	0.6214 mile (statute); 3281 feet

LENGTH

1 mile (statute)	1.6093 kilometers
1 mile (nautical)	1.8531 kilometers
1 fathom	6 feet; 1.8288 meters
1 angstrom	10^{-8} centimeter
1 micrometer	0.0001 centimeter

VELOCITY

1 kilometer/hour	27.78 centimeters/second
1 mile/hour	17.60 inches/second

AREA

1 square centimeter	0.1550 square inch
1 square inch	6.452 square centimeters
1 square meter	10.764 square feet; 1.1960 square yards
1 square foot	0.0929 square meter
1 square kilometer	0.3861 square mile
1 square mile	2.590 square kilometers
1 acre (U.S.)	4840 square yards

VOLUME

1 cubic centimeter	0.0610 cubic inch
1 cubic inch	16.3872 cubic centimeters
1 cubic meter	35.314 cubic feet
1 cubic foot	0.02832 cubic meter
1 cubic meter	1.3079 cubic yards
1 cubic yard	0.7646 cubic meter
1 liter	1000 cubic centimeters; 1.0567 quarts (U.S. liquid)
1 gallon (U.S. liquid)	3.7853 liters

MASS

1 gram	0.03527 ounce
1 ounce	28.3495 grams
1 kilogram	2.20462 pounds
1 pound	0.45359 kilogram

PRESSURE

1 kilogram/square centimeter	0.96784 atmosphere; 0.98067 bar; 14.2233 pounds/square inch
1 bar	0.98692 atmosphere; 10^5 pascals

ENERGY

1 erg	2.39006×10^{-8} calorie (gram); 9.48451×10^{-11} Btu; 10^{-7} joule
1 quad	10^{15} Btu

POWER

1 watt	10^7 ergs/second; 0.001341 horsepower (U.S.); 0.05688 Btu/minute

Degrees

F	C
210	100
200	90
190	80
180	
170	70
160	
150	60
140	
130	50
120	
110	40
100	
90	30
80	
70	20
60	
50	10
40	
30	0
20	
10	10
0	
10	20

Appendix 2 Numerical Data Pertaining to Earth

Equatorial radius	6378 kilometers
Polar radius	6357 kilometers
Radius of sphere with Earth's volume	6371 kilometers
Volume	1.083×10^{27} cubic centimeters
Surface area	5.1×10^{18} square centimeters
Percent surface area of oceans	71
Percent surface area of land	29
Average elevation of land	623 meters
Average depth of oceans	3.8 kilometers
Mass	5.976×10^{27} grams
Density	5.517 grams/cubic centimeter
Gravity at equator	978.032 centimeters/second/second
Mass of atmosphere	5.1×10^{21} grams
Mass of ice	$25-30 \times 10^{21}$ grams
Mass of oceans	1.4×10^{24} grams
Mass of crust	2.5×10^{25} grams
Mass of mantle	4.05×10^{27} grams
Mass of core	1.90×10^{27} grams
Mean distance to Sun	1.496×10^{8} kilometers
Mean distance to Moon	3.844×10^{5} kilometers
Ratio: Mass of Sun/mass of Earth	3.329×10^{5}
Ratio: Mass of Earth/mass of Moon	81.303
Total geothermal energy reaching Earth's surface each year	10^{28} ergs; 80 quads
Earth's daily receipt of solar energy	1.49×10^{22} joules; 1.49×10^{29} ergs
U.S. energy consumption, 1994	88.5 quads

Appendix 3 Properties of the Most Common Minerals of Earth's Crust

Mineral or Group Name	Structure or Composition	Varieties and Chemical Composition	Form, Diagnostic Characteristics	Cleavage, Fracture	Color	Hardness
LIGHT-COLORED MINERALS, VERY ABUNDANT IN EARTH'S CRUST IN ALL MAJOR ROCK TYPES	FRAMEWORK SILICATES	*POTASSIUM FELDSPARS* KAlSi$_3$O$_8$ *Sanidine Orthoclase Microcline*	Cleavable coarsely crystalline or finely granular masses; isolated crystals or grains in rocks, most commonly not showing crystal faces	Two at right angles, one perfect and one good; pearly luster on perfect cleavage	White to gray, frequently pink or yellowish; some green	
FELDSPAR		*PLAGIOCLASE FELDSPARS* NaAlSi$_3$O$_8$ *Albite* CaAl$_2$Si$_2$O$_8$ *Anorthite*		Two at nearly right angles, one perfect and one good; fine parallel striations on perfect cleavage	White to gray, less commonly greenish or yellowish	6
QUARTZ		SiO$_2$	Single crystals or masses of 6-sided prismatic crystals; also formless crystals and grains or finely granular or massive	Very poor or nondetectable; conchoidal fracture	Colorless, usually transparent; also slightly colored smoky gray, pink, yellow	7
MICA	SHEET SILICATES	*MUSCOVITE* KAl$_3$Si$_3$O$_{10}$(OH)$_2$	Thin, disc-shaped crystals, some with hexagonal outlines; dispersed or aggregates	One perfect; splittable into very thin, flexible, transparent sheets	Colorless; slight gray or green to brown in thick pieces	2–2$\frac{1}{2}$
		BIOTITE K(Mg, Fe)$_3$AlSi$_3$O$_{10}$(OH)$_2$	Irregular, foliated masses; scaly aggregates	One perfect; splittable into thin, flexible sheets	Black to dark brown; translucent to opaque	2$\frac{1}{2}$–3
		CHLORITE (Mg,Fe)$_5$(Al,Fe)$_2$Si$_3$O$_{10}$(OH)$_8$	Foliated masses or aggregates of small scales	One perfect; thin sheets flexible but not elastic	Various shades of green	2–2$\frac{1}{2}$
DARK-COLORED MINERALS, ABUNDANT IN MANY KINDS OF IGNEOUS AND METAMORPHIC ROCKS	DOUBLE CHAINS	*TREMOLITE-ACTINOLITE* Ca$_2$(Mg,Fe)$_5$Si$_8$O$_{22}$(OH)$_2$	Long, prismatic crystals, usually 6-sided; commonly in fibrous masses or irregular aggregates	Two perfect cleavage directions at 56° and 124° angles	Pale to deep green Pure tremolite white	5–6
AMPHIBOLE		*HORNBLENDE* Complex Ca, Na, Mg, Fe, Al silicate				
PYROXENE	SINGLE CHAINS	*ENSTATITE-HYPERSTHENE* (Mg,Fe)$_2$Si$_2$O$_6$	Prismatic crystals, either 4- or 8-sided; granular masses and scattered grains	Two good cleavage directions at about 90°	Green and brown to grayish or greenish white	5–6
		DIOPSIDE (Ca,Mg)$_2$Si$_2$O$_6$			Light to dark green	
		AUGITE Complex Ca, Na, Mg, Fe, Al silicate			Very dark green to black	

(Continued)

Mineral or Group Name	Structure or Composition	Varieties and Chemical Composition	Form, Diagnostic Characteristics	Cleavage, Fracture	Color	Hardness
OLIVINE	ISOLATED TETRAHEDRA	$(Mg,Fe)_2SiO_4$	Granular masses and disseminated small grains	Conchoidal fracture	Olive to grayish green and brown	$6\frac{1}{2}$–7
GARNET		Ca, Mg, Fe, Al silicate	Isometric crystals, well formed or rounded; high specific gravity, 3.5–4.3	Conchoidal and irregular fracture	Red and brown, less commonly pale colors	$6\frac{1}{2}$–7
CALCITE	CARBONATES	$CaCO_3$	Coarsely to finely crystalline in beds, veins, and other aggregates; cleavage faces may show in coarser masses; calcite effervesces rapidly, dolomite slowly, only in powders	Three perfect cleavages, at oblique angles; splits to rhombohedral cleavage pieces	Colorless, transparent to translucent; variously colored by impurities	3
DOLOMITE		$CaMg(CO_3)_2$				$3\frac{1}{2}$–4
CLAY MINERALS	HYDROUS ALUMINO-SILICATES	*KAOLINITE* $Al_2Si_2O_5(OH)_4$ *ILLITE* Similar to muscovite +Mg,Fe *SMECTITE* Complex Ca, Na, Mg, Fe, Al silicate + H_2O	Earthy masses in soils; bedded; in association with other clays, iron oxides, or carbonates; plastic when wet; montmorillonite swells when wet	Earthy, irregular	White to light gray and buff; also gray to dark gray, greenish gray, and brownish depending on impurities and associated minerals	$1\frac{1}{2}$–$2\frac{1}{2}$
GYPSUM	SULFATES	$CaSO_4 \cdot 2H_2O$	Granular, earthy, or finely crystalline masses; tabular crystals	One perfect, splitting to fairly thin slabs or sheets; two other good cleavages	Colorless to white; transparent to translucent	2
ANHYDRITE		$CaSO_4$	Massive or crystalline aggregates in beds and veins	One perfect, one nearly perfect, one good; at right angles	Colorless, some tinged with blue	3–$3\frac{1}{2}$
HALITE	HALIDES	$NaCl$	Granular masses in beds; some cubic crystals; salty taste	Three perfect cleavages at right angles	Colorless, transparent to translucent	$2\frac{1}{2}$
OPAL-CHALCEDONY	SILICA	SiO_2 [Opal is an amorphous variety; chalcedony is a formless microcrystalline quartz.]	Beds in siliceous sediments and chert; in veins or banded aggregates	Conchoidal fracture	Colorless or white when pure, but tinged with various colors by impurities in bands, especially in agates	5–$6\frac{1}{2}$
MAGNETITE	IRON OXIDES	Fe_3O_4	Magnetic; disseminated grains, granular masses; occasional octahedral isometric crystals; high specific gravity, 5.2	Conchoidal or irregular fracture	Black, metallic luster	6

LIGHT-COLORED MINERALS, TYPICALLY AS ABUNDANT CONSTITUENTS OF SEDIMENTS AND SEDIMENTARY ROCKS

DARK-COLORED MINERALS, COMMON IN MANY ROCK TYPES

Mineral or Group Name	Structure or Composition	Varieties and Chemical Composition	Form, Diagnostic Characteristics	Cleavage, Fracture	Color	Hardness
HEMATITE	IRON OXIDES	Fe_2O_3	Earthy to dense masses, some with rounded forms, some granular or foliated; high specific gravity, 4.9–5.3	None; uneven, sometimes splintery fracture	Reddish brown to black	5–6
"LIMONITE"		GOETHITE [the major mineral of the mixture called "limonite," a field term] $HFeO_2$	Earthy masses, massive bodies or encrustations, irregular layers; high specific gravity, 3.3–4.3	One excellent in the rare crystals; usually an early fracture	Yellowish brown to dark brown and black	$5–5\frac{1}{2}$
KYANITE	ALUMINO-SILICATES	Al_2SiO_5	Long, bladed or tabular crystals or aggregates	One perfect and one poor, parallel to length of crystals	White to light-colored or pale blue	5 parallel to crystal length 7 across crystals
SILLIMANITE		Al_2SiO_5	Long, slender crystals or fibrous, felted masses	One perfect parallel to length, not usually seen	Colorless, gray to white	6–7
ANDALUSITE		Al_2SiO_5	Coarse, nearly square prismatic crystals, some with symmetrically arranged impurities	One distinct; irregular fracture	Red, reddish brown, olive-green	$7\frac{1}{2}$
FELDSPATHOIDS		NEPHELINE $(Na,K)AlSiO_4$	Compact masses or as embedded grains, rarely as small prismatic crystals	One distinct; irregular fracture	Colorless, white, light gray; gray-greenish in masses, with greasy luster	$5\frac{1}{2}–6$
		LEUCITE $KAlSi_2O_6$	Trapezohedral crystals embedded in volcanic rocks	One very imperfect	White to gray	$5\frac{1}{2}–6$
SERPENTINE		$Mg_6Si_4O_{10}(OH)_8$	Fibrous (asbestos) or platy masses	Splintery fracture	Green; some yellowish brownish, or gray; waxy or greasy luster in massive habit; silky luster in fibrous habit	4–6
TALC		$Mg_3Si_4O_{10}(OH)_2$ masses or aggregates	Foliated or compact masses or aggregates	One perfect, making thin flakes or scales; soapy feel	White to pale green; pearly or greasy luster	1
CORUNDUM		Al_2O_3	Some rounded, barrel-shaped crystals; most often as disseminated grains or granular masses (emery)	Irregular fracture	Usually brown, pink, or blue; emery black Gemstone varieties: ruby, sapphire	9

LIGHT-COLORED MINERALS, MAINLY IN IGNEOUS AND METAMORPHIC ROCKS AS COMMON OR MINOR CONSTITUENTS

(Continued)

	Mineral or Group Name	Structure or Composition	Varieties and Chemical Composition	Form, Diagnostic Characteristics	Cleavage, Fracture	Color	Hardness
DARK-COLORED MINERALS, COMMON IN METAMORPHIC ROCKS	EPIDOTE	SILICATES	$Ca_2(Al,Fe)Al_2Si_3O_{12}(OH)$	Aggregates of long prismatic crystals, granular or compact masses, embedded grains	One good, one poor at greater than right angles; conchoidal and irregular fracture	Green, yellow-green, gray, some varieties dark brown to black	6–7
	STAUROLITE		$Fe_2Al_9Si_4O_{22}(O,OH)_2$	Short prismatic crystals, some cross-shaped, usually coarser than matrix of rock	One poor	Brown, reddish, or dark brown to black	$7–7\frac{1}{2}$
METALLIC LUSTER, COMMON IN MANY ROCK TYPES, ABUNDANT IN VEINS	PYRITE	SULFIDES	FeS_2	Granular masses or well-formed cubic crystals in veins and beds or disseminated; high specific gravity, 4.9–5.2	Uneven fracture	Pale brass-yellow	$6–6\frac{1}{2}$
	GALENA		PbS	Granular masses in veins and disseminated; some cubic crystals; very high specific gravity, 7.3–7.6	Three perfect cleavages at mutual right angles, giving cubic cleavage fragments	Silver-gray	$2\frac{1}{2}$
	SPHALERITE		ZnS	Granular masses or compact crystalline aggregates; high specific gravity, 3.9–4.1	Six perfect cleavages at 60° to one another	White to green, brown, and black; submetallic luster	$3\frac{1}{2}–4$
	CHALCOPYRITE		$CuFeS_2$	Granular or compact masses; disseminated crystals; high specific gravity, 4.1–4.3	Uneven fracture	Brassy to golden-yellow	$3\frac{1}{2}–4$
	CHALCOCITE		Cu_2S	Fine-grained masses; high specific gravity, 5.5–5.8	Conchoidal fracture	Lead-gray to black; may tarnish green or blue	$2\frac{1}{2}–3$
MINERALS FOUND IN MINOR AMOUNTS IN A VARIETY OF ROCK TYPES AND IN VEINS OR PLACERS	RUTILE	TITANIUM OXIDES	TiO_2	Slender to prismatic crystals; granular masses; high specific gravity, 4.25	One distinct, one less distinct; conchoidal fracture	Reddish brown, some yellowish, violet, or black	$6–6\frac{1}{2}$
	ILMENITE		$FeTiO_3$	Compact masses, embedded grains, detrital grains in sand; high specific gravity, 4.79	Conchoidal fracture	Iron-black; metallic to submetallic luster	5–6
	ZEOLITES	SILICATES	Complex hydrous silicates; many varieties of minerals, including analcime, natrolite, phillipsite, heulandite, and chabazite	Well-formed radiating crystals in cavities in volcanics, veins, and hot springs; also as fine-grained and earthy bedded deposits	One perfect for most	Colorless, white, some pinkish	4–5

Appendix 4 Topographic and Geologic Maps

A map is a quantitative representation of the spatial distribution of some attribute or property of the Earth. It is a kind of graph in which the axes are lines of latitude and longitude and the positions of points on the surface (or beneath it) are plotted in relation to those axes or some other established reference. Geologists also want maps that describe the geological materials at or near the surface so that they can construct a three-dimensional mental picture of the geology from this two-dimensional graph. The practiced map reader can become proficient at deducing much of the geologic structure and history of an area.

The use of topographic maps (see Chapter 16) and geologic maps (see Chapter 10) has spread widely throughout our culture. To the more traditional users of such maps—such as geologists and surveyors—have been added city planners, industrial zoning commissions, and many members of the public seeking recreational areas for hiking, camping, fishing, and other activities. In 1993 the U.S. Geological Survey distributed nearly 7 million copies of its 70,000 published topographic maps and 18,000 copies of geologic and hydrologic maps from its open file reports. Maps are a necessity for all kinds of geological and mineral resource studies, as well as for studies of groundwater, flood control, soil management, and such environmental concerns as land-use planning, which involves the locations of highways, industrial areas, oil and gas pipelines, and recreational areas.

Scale

Because the size of the area covered, and thus the amount of detail that can be shown, is always important, maps are always drawn to an appropriate *scale*—that is, the relationship of a distance (or area) on the map to the true distance on the Earth. A map's scale is stated as a ratio, such as 1:24,000, which indicates that a distance of one unit on the map represents a distance of 24,000 such units on the Earth. It does not matter what the units are: a map of scale 1:24,000 is the same whether we use the metric or the English system. The scale can be thought of in any convenient units desired: 1 in. = 2000 ft, or 1 m = 24 km, or 10 cm = 2.4 km. For convenience, maps have a graphic scale, usually at the bottom margin, in which a distance such as 1 km or 1 mile, usually with subdivisions, is shown as it would appear on

the map. A common scale for detailed topographic and geologic maps is 1:24,000, used by the U.S. Geological Survey for most modern maps. The scale used for regional maps covering much larger areas is 1:250,000. Scales of 1:1,000,000 are used for aeronautical charts. In 1976 the U.S. Geological Survey introduced the first of a new series of 1:100,000 all-metric topographic maps on which graphic scales show both kilometers and miles.

Topographic Maps

A topographic map shows a region's landforms and elevations. On most maps, natural and constructed features of the surface are represented by conventional symbols. Those used by the U.S. Geological Survey are typical: rivers, lakes, and oceans are shown in blue; topography is shown in brown; constructed features are shown in black, with main highways and urban areas in red; green shaded areas show wooded land. Some special symbols may be shown on the explanation, or *legend,* of the map, which is usually displayed along the bottom margin. The most complex representations are the topographic elevations of the surface, usually shown on maps by contours (see Chapter 16). Special maps are sometimes prepared to show environmental variables, such as the distribution of slopes of various steepness.

Geologic Maps

Geologic maps are a representation of the distribution of rocks and other geologic materials of different lithologies and ages over the Earth's surface or below it. The geologist perceives the Earth not only in its surface expression of topography and patterns of land and water but also in terms of its pattern of subsurface structures, stratigraphic sequences, igneous intrusions, unconformities, and other geometric relationships of rocks. Just as an anatomist can visualize the muscles and bones beneath the skin, so can a geologist visualize details of the Earth's subsurface.

Detailed geologic maps are normally constructed on a topographic map base (Figure A4.1). This makes it easy to locate geologic structures with respect to surface features of the Earth. It is also important because topography is so often related to the nature of the underlying rocks and their structures.

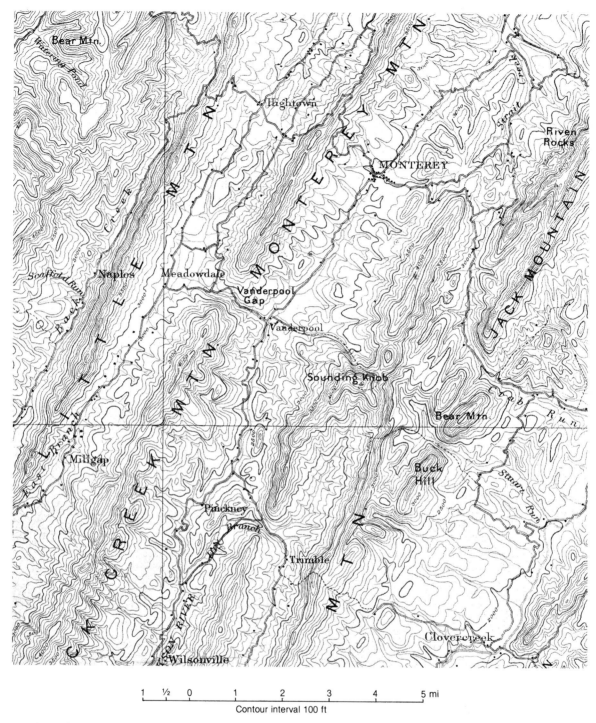

1 ½ 0 1 2 3 4 5 mi

Contour interval 100 ft

FIGURE A4.1 Topographic map *(above)* and geologic map with cross sections *(facing page)* of folded sedimentary rocks in the Valley and Ridge province of the Appalachian Mountains. Contours show the pronounced trends of valleys and ridges that reflect the parallel folds. The ridges have developed along the formations that are resistant to erosion, some at the crests of anticlines, such as Jack Mountain north of Crab Run, and others along the flanks of folds. The valleys are in the easily eroded formations, some in synclines, such as Jackson River, some

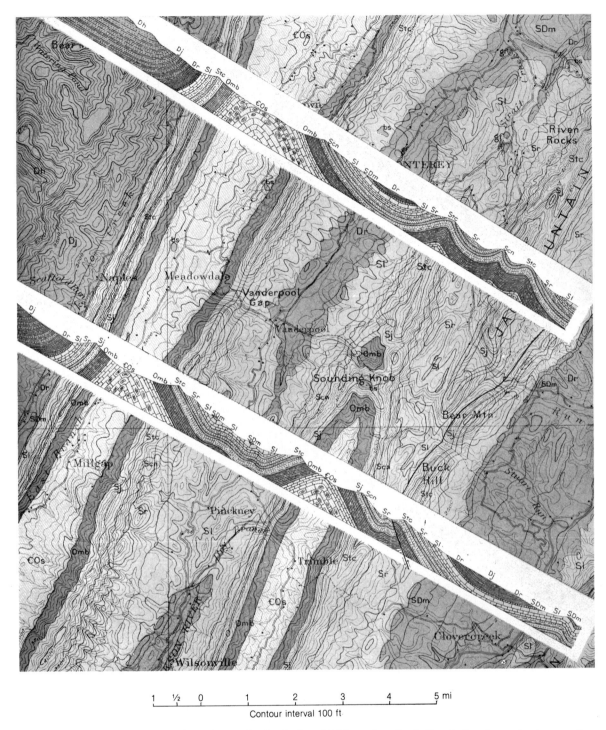

1 ½ 0   1   2   3   4   5 mi

Contour interval 100 ft

on anticlines, such as East Branch, and some in the flanks of folds, such as Back Creek. On the geologic map the pattern of anticlines and synclines can be read from the positions of formations of different age, such as at East Branch, where the oldest rocks, Cambrian and Ordovician formations (COs), are at the surface bordered on both sides by younger formations of Ordovician and Silurian age (Omb, Stc, Sj, and others). The cross sections make these relationships clearer and add some detail. *(U.S. Geological Survey.)*

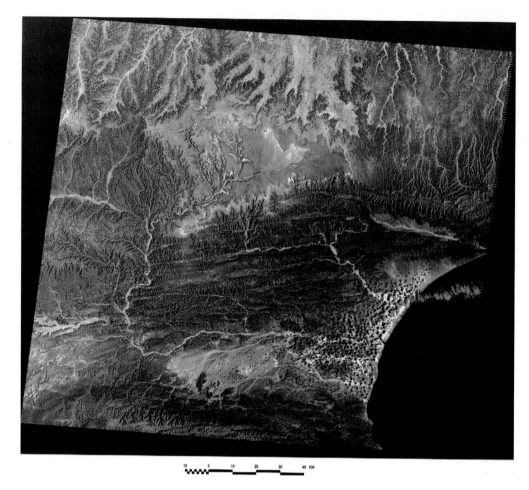

FIGURE A4.2 Satellite photo of stream drainage in the folded mountain belt of Al Ghaydah, Yemen. This image shows clear examples of trellis drainage *(center)* and dendritic drainage *(above)*.

Because it contains so much more information than a topographic map alone, a geologic map is the most valuable for many of the purposes for which maps are used.

Traditional geologic maps are made by a geologist who roams over the area and notes the kinds of rocks, sediments, and soils and their structural and stratigraphic relationships. Geologists cover the ground on foot to see most if not all of the outcrops; they are helped enormously by the automobile, sometimes by a helicopter, and in some places even by a horse or donkey.

Geologic mapping proceeds by the following steps:

1. *Principal observations.* Description of outcrop location, lithology, age, fossil content, and struc-tural attitude as measured by dip and strike, direction of fault movement, fold axes, and so forth (see Chapter 10). Plotting observations on work map.

2. *First integration.* Conceptualizing the spatial relationship of one outcrop to another by stratigraphic correlation of rocks of the same age, facies, degree of metamorphism and deformation. Grouping of mappable rock units into formations. Drawing of lines on the primitive geologic map of inferred connections where formations are hidden. Compilation of the complete or composite stratigraphic sequences and ages of deformational or igneous intrusive events.

3. *Synthesizing the map.* Visualizing the larger pattern of geologic relationships and constructing

the map, together with geologic cross sections made both to help geologists in their thinking and to illustrate more detail and inference from the map.

Analyses of rock composition, absolute age, and seismic, gravity, and magnetic data may be incorporated into the map. The geologist further draws on the geologic literature or personal experience of the geology of nearby and similar kinds of regions. The finished geologic map is a codified mass of information from which anyone familiar with geology can quickly read the nature of the Earth's crust in the area and a good deal of its geologic history.

In the last few decades much of traditional geologic mapping has been supplemented or even entirely supplanted by remote sensing by aerial and satellite photography or geophysical instruments. Inaccessible areas, such as those in some polar or desert regions, may be mapped almost entirely by this method, with the geologist ground-checking in scattered places. The mapping of the Moon is an extreme example of this approach. Mars is being mapped with no ground check at all except for the area immediately surrounding the *Viking* landing site.

Aerial images are now obtained by aircraft aerial photography, manned and unmanned satellites, and the space shuttle. These images may be in ordinary color, or they may be taken in infrared or other parts of the light spectrum, or they may be false-color radar images. Digital processing systems allow the information gathered by remote sensing devices to be processed into specialized maps that show geologic structures, erosional patterns, bodies of water, and the distribution and types of vegetation and urbanization. Figure A4.2, for example, is a satellite image that provides a map of stream drainage in a mountainous area of Yemen.

There are numerous environmental applications of remote mapping techniques. Aerial and satellite images can track changes in coastal zones, including alteration of sandy beaches, which can be a guide to shoreline engineering. Satellite mapping of the sea surface can be used for imaging the topography of the seafloor, and details of seafloor topography can be mapped by side-scanning radar (see Chapter 17). Satellites can also be used to monitor landslides, floods, volcanic eruptions, and strain changes that could signal an earthquake (see Chapter 18) and to assess the damages from these natural hazards.

There are many kinds of geologic maps. The most common shows the bedrock geology and gives a picture of what the land would look like if all soil were stripped away. Surficial geologic maps, on the other hand, emphasize the nature of soils, unconsolidated river sediment, sand dunes, and whatever other materials, including outcrops, appear at the surface. A special kind of surficial geologic map is used for environmental hazards. One kind shows areas of high-angle or unsupported slopes that are likely to slump or slide (see Chapter 11). Whatever the geologic purpose, there is a map that can be made to show the relevant data. There is no question that the map is both the best device for geological research into the origin of the distribution of important geologic characteristics over the Earth and the best way to illustrate the patterns discovered from such research.

Glossary

Words in *italic* have separate entries in the Glossary.
Specific minerals are defined in Appendix 3.

AA: A blocky and fragmented form of basaltic *lava* occurring in flows with fissured and angular surfaces.

ABLATION: The annual amount of ice and snow lost from a *glacier* by the processes of melting, *sublimation,* wind *erosion,* and *iceberg calving.*

ABSOLUTE AGE: The age in years of a particular geologic event or feature, generally obtained with *radiometric dating* techniques. (Compare *Relative age.*)

ABYSSAL PLAIN: A flat, sediment-covered province of the seafloor.

ACTIVE MARGIN: A *continental margin* characterized by *earthquakes,* igneous activity, and/or uplifted mountains resulting from convergent or transform *plate* motion.

A-HORIZON: The uppermost layer of a *soil,* containing organic material and leached *minerals.*

ALLUVIAL FAN: A low, cone-shaped deposit of *terrigenous sediment* formed where a *stream* undergoes an abrupt widening as it leaves a mountain front for an open valley.

AMPHIBOLITE: A mostly nonfoliated *metamorphic rock* consisting primarily of amphibole and plagioclase feldspar.

ANDESITE: A volcanic rock type intermediate in composition between *rhyolite* and *basalt;* the extrusive equivalent of *diorite.*

ANGLE OF REPOSE: The steepest slope angle at which a particular *sediment* will lie without cascading down.

ANGULAR UNCONFORMITY: An *unconformity* in which the *bedding* planes of the rocks above and below are not parallel.

ANION: Any negatively charged *ion;* the opposite of *cation.*

ANTECEDENT STREAM: A *stream* that existed before the present *topography* was created, thereby maintaining its original course despite changes in the structure of the underlying rocks and in topography.

ANTICLINE: A large upfold of strata, usually from 100 m to 300 km in width, whose *limbs* are lower than its center. (Compare *Syncline.*)

AQUICLUDE: A stratum with low *permeability* that acts as a barrier to the flow of *groundwater.* Also called "confining layer."

AQUIFER: A permeable *formation* that stores and transmits *groundwater* in sufficient quantity to supply wells.

ARETE: The sharp, jagged crest along the divide between glacial *cirques,* resulting from the headward *erosion* of the walls of adjoining cirques.

ARGILLITE: A low-grade *metamorphic rock* made from a shaly *sedimentary rock,* characterized by irregular fracture and lack of *foliation.*

ARTESIAN FLOW: Flow in a *confined aquifer,* in which the *groundwater* is at a greater pressure than in an *unconfined aquifer* at similar depths, thereby causing water in a well that penetrates a confined aquifer (an artesian well) to rise above the level of the *aquiclude.*

ASEISMIC RIDGE: A submarine ridge of volcanic origin, far from present plate boundaries and therefore characterized by the absence of seismic activity. (Compare *Mid-ocean ridge,* which is seismically active.)

ASTHENOSPHERE: The weak layer below the *lithosphere* that is marked by low *seismic wave* velocities and high seismic wave attenuation. Movement in the asthenosphere occurs by plastic deformation.

ASYMMETRICAL FOLD: A *fold* of strata in which the *dips* of the two *limbs* are unequal.

ATOLL: A continuous or broken circle of coral *reefs* and low coral islands surrounding a central lagoon.

ATOM: The smallest unit of an element that retains the element's physical and chemical properties.

ATOMIC NUMBER: The number of *protons* in the nucleus of an atom.

ATOMIC WEIGHT: The sum of the masses of the *protons* and *neutrons* in the atomic nucleus of an element.

AXIAL PLANE: In *folds,* the plane that most nearly separates two symmetrical *limbs.* In a simple *anticline,* it is vertical; in complex folding, it is perpendicular to the direction of compression.

648

BACKSHORE: The upper, generally dry, zone of the shore, extending landward from the upper limit of wave wash at high *tide* to the upper limit of shore-zone processes.

BACKWASH: The return flow of water down a beach after a wave has broken.

BADLAND: *Topography* characterized by intricate patterns of stream *erosion* developed on surfaces with little or no vegetative cover overlying *unconsolidated* or poorly cemented *clays,* silts, or sands.

BARCHAN: A crescent-shaped *eolian* sand *dune* that moves across a flat surface with its convex face upwind and its concave *slip face* downwind.

BARRIER ISLAND: A long, narrow island parallel to the shore, composed of sand and built by wave action.

BASAL SLIP: The sliding of a *glacier* along its base.

BASALT: A fine-grained, dark, *mafic igneous rock* composed largely of plagioclase feldspar and pyroxene: the extrusive equivalent of *gabbro.*

BASE LEVEL: The level below which a *stream* cannot erode: usually sea level, sometimes locally the level of a lake or resistant *formation.*

BASEMENT: The oldest rocks recognized in a given area; a complex of *metamorphic* and *igneous rocks* that underlies all the sedimentary *formations.* Usually Precambrian or Paleozoic in age.

BASIN (TECTONIC): A circular, synclinelike depression of strata that dips radically toward a central point.

BATHOLITH: A great, irregular mass of rock cutting across the country rock, with an exposed surface of at least 100 km²; usually an *intrusive igneous rock,* but sometimes derived from the country rock through very high temperature and pressure *metamorphism.* (See also *Discordant intrusion.*)

BAUXITE: A rock composed primarily of hydrous aluminum oxides and formed by intense *chemical weathering* in tropical areas with good drainage; a major ore of aluminum.

BEDDING: A characteristic of *sedimentary rocks* in which parallel planar surfaces separate layers of different grain sizes or compositions deposited at different times.

BEDDING SEQUENCE: A pattern of interbedding of different *sedimentary rock* types or sedimentary rocks with different *sedimentary structures* that is characteristic of a certain *sedimentary environment.*

BED LOAD: The *sediment* that a *stream* moves along the bottom of its *channel* by rolling and bouncing (*saltation*).

BEDROCK: The solid rock underlying *unconsolidated* surface materials, such as *soil.*

B-HORIZON: The intermediate layer in a *soil,* below the *A-horizon* and above the *C-horizon,* consisting of *clays* and oxide materials.

BIOCHEMICAL SEDIMENT, ROCK: A *sediment* or rock containing the mineral remains of organisms, such as shells, or *minerals* precipitated as a result of biological processes, such as in *iron formations.*

BIOMASS: Organic carbon-containing material of biological origin, including living and dead animals and plants.

BIOSPHERE: The parts of the atmosphere, hydrosphere, and lithosphere occupied by living organisms.

BIOTURBATION: The reworking of existing *sediments* by organisms.

BLOWOUT: (1) A parabola-shaped *eolian* sand *dune,* typically one blown back from a beach, that has its convex *slip face* oriented downwind. (2) A shallow circular or elliptical depression in sand or dry soil formed by wind erosion. (See also *Deflation.*)

BLUESCHIST: A *metamorphic rock* formed under conditions of high pressure (in excess of 5000 bars) and relatively low temperature, often containing the blue *minerals* glaucophane (an amphibole) and kyanite.

BOLIDE IMPACT HYPOTHESIS: The proposal that an extraterrestrial object slammed into Earth 65 million years ago, causing a global climate change that wiped out half of Earth's species, including the dinosaurs.

BOTTOMSET BED: A flat-lying bed of fine *sediment* deposited seaward of a *delta* and then buried by continued delta growth.

BOWEN REACTION SERIES: A simple schematic description of the order in which different *minerals* crystallize during the cooling and progressive *crystallization* of a *magma.*

BRAIDED STREAM: A *stream* so choked with *sediment* that it divides and recombines numerous times, forming many small and meandering *channels.*

BRECCIA: See *Sedimentary breccia; Volcanic breccia.*

BRITTLE MATERIAL: A material that breaks abruptly when its elastic limit is reached; the opposite of a *ductile material.*

BURIAL METAMORPHISM: A low-grade metamorphism in which buried *sedimentary rocks* are metamorphosed by the heat and pressure exerted by overlying *sediments* and sedimentary rocks; *bedding* and other *sedimentary structures* are preserved.

CALCIUM CYCLE: The set of processes that carry calcium through Earth's systems.

CALDERA: A large basin-shaped volcanic depression that can form after an eruption if the volcano collapses through the roof of the emptied *magma chamber*. Potentially catastrophic eruptions of a "resurgent caldera" can occur when fresh *magma* reenters the collapsed *magma chamber*.

CAPACITY (STREAM): The amount of *sediment* and detritus a *stream* can transport past any point in a given amount of time. (Compare *Competence*.)

CARBONATE COMPENSATION DEPTH: The ocean depth below which the solution rate of calcium carbonate ($CaCO_3$) becomes so great that no carbonate organisms or *sediments* are preserved on the seafloor.

CARBONATE PLATFORM: Broad, shallow area where both biological and nonbiological carbonates are deposited.

CARBONATE SEDIMENT, ROCK: A *sediment* or *sedimentary rock* formed from the accumulation of carbonate *minerals* precipitated organically or inorganically. Rocks are chiefly *limestone* and *dolostone*.

CATACLASTIC METAMORPHISM: High-pressure metamorphism occurring primarily by the crushing, shearing, and recrystallization of rock during tectonic movements and resulting in the formation of very fine grained rock.

CATION: Any positively charged *ion;* the opposite of *anion*.

CEMENTATION: A *lithification* process in which *minerals* are precipitated in the pore space of *sediments,* often binding the grains.

CENTRAL VENT: The largest vent of a volcano, situated at the center of its cone.

CHANNEL: The trough through which water flows in a stream *valley;* sometimes reserved for the deepest part of the streambed, in which the main current flows.

CHEMICAL SEDIMENT, ROCK: A *sediment* or *sedimentary rock* that is formed at or near its place of deposition by chemical precipitation, usually from seawater.

CHEMICAL WEATHERING: The set of all chemical reactions that can act on rock exposed to water and the atmosphere and so dissolve the *minerals* or change them to more stable forms.

CHERT: A *sedimentary rock* made up of chemical or biochemically precipitated silica.

C-HORIZON: The lowest layer of a *soil,* consisting of fragments of rock and their chemically weathered products.

CINDER CONE: A steep, conical hill built up about a volcanic vent and composed of coarse *pyroclastic rock* fragments expelled from the vent by escaping gases.

CIRQUE: The head of a glacial valley, usually with the form of one half of an inverted cone. The upper edges have the steepest slopes, approaching the vertical, and the base may be flat or hollowed out. The base is commonly occupied by a small lake or pond after deglaciation.

CLASTIC SEDIMENT, ROCK: A *sediment* or *sedimentary rock* formed from particles (clasts) derived from the *erosion* of preexisting rocks and mechanically transported.

CLAY: Any of a number of hydrous aluminosilicate *minerals* with sheetlike *crystal* structure, formed by the *weathering* and *hydration* of other silicates. Also, any mineral fragments smaller than 0.0039 mm.

CLEAVAGE (MINERAL): The tendency of a *crystal* to break along certain preferred planes in the crystal lattice; also, the geometric pattern of such a breakage.

CLEAVAGE (ROCK): The tendency of a rock to break along certain planes induced during deformation or metamorphism, usually in the direction of preferred orientation of the *minerals* in the rock.

COAL: The metamorphic product of stratified plant remains. It contains more than 50 percent carbon compounds and burns readily.

COAST: The strip of land adjacent to an ocean or sea and extending from low *tide* landward to the point of major change in landscape features.

COMPACTION: The decrease in volume and *porosity* of a *sediment* caused by burial beneath other sediments.

COMPETENCE (STREAM): A measure of the largest particle a *stream* is able to transport, not the total volume. (Compare *Capacity*.)

COMPOSITE VOLCANO: A volcanic cone containing layers of both *lava* flows and *pyroclastic rocks*. Synonym for *stratovolcano*.

COMPRESSIVE FORCES: Forces that squeeze together or shorten a body. Compressive forces dominate at *convergent plate boundaries*.

CONCENTRATION FACTOR: The ratio of the abundance of an element in a *mineral* deposit to its average abundance in the *crust*.

CONCORDANT INTRUSION: An *intrusive igneous rock* having contacts with the *country rock* that are parallel to *bedding* or *foliation* planes. (Compare *Discordant intrusion*.)

CONDUCTION: See *Heat conduction*.

CONFINED AQUIFER: An *aquifer* overlain by relatively impermeable strata (*aquicludes*), thereby causing the water to be contained under pressure. (Compare *Unconfined aquifer.*)

CONGLOMERATE: A *sedimentary rock,* a significant fraction of which is composed of rounded pebbles, cobbles, and boulders. The lithified equivalent of *gravel.*

CONSOLIDATED MATERIAL: *Sediment* that is lithified; that is, compacted and bound together by *mineral* cements.

CONTACT METAMORPHISM: Changes in the *mineralogy* and *texture* of rock resulting from the heat and pressure of an igneous intrusion in the near vicinity.

CONTINENTAL DRIFT: The horizontal displacement or rotation of continents relative to one another.

CONTINENTAL GLACIER: A continuous, thick *glacier* covering more than 50,000 km² and moving independently of minor topographic features. (Compare *Valley glacier.*)

CONTINENTAL MARGIN: The portion of the ocean floor extending from the *shoreline* to the landward edge of the *abyssal plain* and including the *continental shelf, slope,* and *rise.*

CONTINENTAL RISE: A broad and gently sloping ramp that rises from an *abyssal plain* to the *continental slope.*

CONTINENTAL SHELF: The gently sloping submerged edge of a continent, extending commonly to a depth of about 200 m to the edge of the *continental slope.*

CONTINENTAL SHELF DEPOSITS: *Sediments* laid down in a tectonically quiet *syncline* at a *passive continental margin.*

CONTINENTAL SLOPE: The region of steep slopes between the *continental shelf* and *continental rise.*

CONTINUOUS REACTION SERIES: A *reaction series* in which the same *mineral* crystallizes throughout the range of temperatures in question, but in which there is gradual change in the chemical composition of the mineral with changing temperature. (Compare *Discontinuous reaction series.*)

CONTOUR: A curve on a topographic map that connects points of equal *elevation.*

CONVECTION: A mechanism of heat transfer in a flowing material in which hot material from the bottom rises because of its lesser *density,* while cool surface material sinks. (Compare *Heat conduction.*)

CONVERGENT PLATE BOUNDARY: A boundary at which Earth's *plates* collide and area is lost either by shortening and crustal thickening or by *subduction* of one plate beneath the other. The site of volcanism, *earthquakes, trenches,* and mountain building. (See also *Subduction zone.*)

CORE: The central part of the Earth below a depth of 2900 km. It is thought to be composed of iron and nickel and to be molten on the outside with a central solid inner core.

COUNTRY ROCK: The rock into which an *igneous rock* intrudes or a *mineral* deposit is emplaced.

COVALENT BOND: A bond between atoms in which outer *electrons* are shared.

CRATER: A bowl-shaped pit at the summit of most volcanoes, around the central vent.

CRATON: A portion of a continent that has not been subjected to major deformation for a prolonged time, typically since Precambrian or early Paleozoic time.

CREEP: Slow, downhill *mass movement* of *soil* and *regolith* under gravitational force.

CREVASSE: Any large vertical crack in the surface of a *glacier* or snowfield.

CROSS–BEDDING: Inclined beds in a *sedimentary rock* that were formed at the time of deposition by currents of wind or water in the direction in which the bed slopes downward.

CRUST: The outermost layer of the *lithosphere,* consisting of relatively light, low-melting temperature materials. Continental crust consists largely of *granite* and *granodiorite.* Oceanic crust is mostly *basalt.*

CRYSTAL: A form of matter, characterized by flat surfaces, in which the *atoms, ions,* or molecules are arranged in all directions to form a regular, repeating network.

CRYSTAL HABIT: The general shape of a *crystal;* for example, cubic, prismatic, or fibrous.

CRYSTALLIZATION: The formation of crystalline solids from a gas or liquid, such as in the formation of crystalline *minerals* in *magma.*

CUESTA: An asymmetrical ridge with one steep and one gentle face formed where gently dipping beds of erosion-resistant rocks are undercut by *erosion* of a weaker bed underneath.

CYCLE OF EROSION: A proposed sequence of changes in a landscape that progresses from high, rugged, tectonically formed mountains to low, rounded hills and finally to worn-down, tectonically stable plains.

DACITE: Volcanic equivalent of *granodiorite.*

DEBRIS AVALANCHE: A fast downhill *mass movement* of *soil* and rock.

DEBRIS FLOW: A fluid *mass movement* of rock fragments supported by a muddy matrix. Debris flows differ from *earthflows* in that they generally contain coarser material and move faster than earthflows.

DEBRIS SLIDE: A *mass movement* of rock material and *soil* largely as one or more units along planes of weakness at the base of or within the rock material.

DEFLATION: The removal of *clay* and dust from dry *soil* by strong winds that gradually scoop out shallow depressions in the ground.

DELTA: A body of *sediment* deposited in an ocean or lake at the mouth of a *stream.*

DENDRITIC DRAINAGE: A *stream* system that branches irregularly, resembling a branching tree.

DENSITY: The mass per unit volume of a substance, commonly expressed in grams per cubic centimeter.

DEPOSITIONAL REMANENT MAGNETIZATION: A weak magnetization created in *sedimentary rocks* by the rotation of magnetic *crystals* into line with the ambient field during settling.

DESERT PAVEMENT: A residual deposit left when continued *deflation* removes the fine grains of a *soil* and leaves a surface covered with close-packed cobbles.

DESERT VARNISH: A dark coating commonly found on the surface of rock in the desert. It consists of *clays,* iron oxides, and magnesium oxides produced during *weathering.*

DIAGENESIS: The chemical and physical changes undergone by buried *sediments* during *lithification* and *compaction* into sedimentary rock.

DIATREME: A volcanic vent filled with *volcanic breccia* by the explosive escape of gases.

DIKE: A tabular *igneous* intrusion that cuts across structures of surrounding rock. (See also *Discordant intrusion.*)

DIORITE: A *plutonic* rock with composition intermediate between *granite* and *gabbro;* the intrusive equivalent of *andesite.*

DIP: The maximum angle by which a stratum or other planar feature deviates from the horizontal. The angle is measured in a plane perpendicular to the *strike.*

DISCHARGE: The rate of water movement through a *stream,* measured in units of volume per unit time; the exit of *groundwater* to the surface (opposite of *recharge*).

DISCONTINUOUS REACTION SERIES: A *reaction series* in which the end members have different *crystal* structures (are distinct *mineral* phases). (Compare *Continuous reaction series.*)

DISCORDANT INTRUSION: An *intrusive igneous rock* that has contacts with the *country rock* cutting across *bedding* or *foliation* planes. (Compare *Concordant intrusion.*)

DISTRIBUTARY: A smaller branch of a large *stream* that receives water from the main *channel;* the opposite of a *tributary.*

DIVERGENT PLATE BOUNDARY: A boundary at which Earth's *plates* move apart and new *lithosphere* is created; the site of *mid-ocean ridges,* shallow-focus *earthquakes,* and volcanism.

DIVIDE: A ridge of high ground separating two *drainage basins* emptied by different *streams.*

DOLOSTONE: A *sedimentary rock* composed primarily of dolomite, a carbonate *mineral* with the general formula $CaMg(CO_3)_2$.

DOME (TECTONIC): A round or elliptical anticlinal upwarp of strata in which strata dip away in all directions from the high point. (See also *Volcanic dome.*)

DRAAS: Extremely large (1 km or more long and over 100 m high) composite of sand *dunes* found in deserts.

DRAINAGE BASIN: A region of land surrounded by *divides* and crossed by *streams* that funnel all its water into the network of streams draining the area, usuually to converge eventually to one *river* or lake.

DRAINAGE NETWORK: The pattern of *tributaries,* large and small, of a *stream* system.

DRIFT (GLACIAL): A collective term for all the rock, sand, and *clay* that is deposited by a *glacier* either as *till* or as *outwash.*

DRUMLIN: A smooth, streamlined hill composed of *till* and, in many cases, *bedrock.*

DRY WASH: An intermittent streambed in a desert canyon that carries water only briefly after a rain; called "wadi" in the Near East.

DUCTILE MATERIAL: A material that can undergo considerable change in shape by plastic deformation before rupture occurs; the opposite of a *brittle material.*

DUNE: An elongated mound of sand formed by wind or water.

EARTHFLOW: A fluid *mass movement* of mainly fine-grained material, along with some broken rock, at slow or moderate speeds.

EARTHQUAKE: The violent motion of the ground caused by the passage of *seismic waves* radiating from a *fault* along which sudden movement has taken place.

EBB TIDE: The part of the *tide* cycle during which the water level is falling.

ECLOGITE: An extremely high-grade *metamorphic rock* containing the *minerals* garnet and pyroxene.

EFFLUENT STREAM: A *stream* or portion of a stream that receives some water from groundwater *discharge* because the stream's *elevation* is below the *groundwater table*. (Compare *Influent stream.*)

ELASTIC REBOUND THEORY: A theory of *fault* movement and *earthquake* generation holding that faults remain locked while strain energy accumulates in the rock formations on both sides, temporarily deforming them until a sudden slip along the fault releases the energy.

ELECTRON: A negatively charged atomic particle with a negligible mass (9.1×10^{-28} gram) and a charge of -1.6×10^{-19} coulomb, commonly expressed as -1. The position of an electron about an atomic nucleus is not fixed but is described as a region where an electron is most likely to be found.

ELEVATION: The vertical height of one point on the Earth above a given plane, usually sea level.

EOLIAN: Pertaining to or deposited by wind.

EON: The largest division of geologic time, embracing several *eras;* for example, the Phanerozoic eon, from 600 million years ago to the present.

EPEIROGENY: Large-scale, primarily vertical, movement of the *crust*, characteristically so gradual that rocks are little folded and faulted.

EPICENTER: The point on the Earth's surface directly above the *focus* of an *earthquake.*

EPOCH: A subdivision of a geologic *period,* often chosen to correspond to a *stratigraphic sequence.* Also used for a division of time corresponding to a *paleomagnetic* interval.

ERA: A division of geologic time including several *periods,* but smaller than an *eon.* Commonly recognized eras are Precambrian, Paleozoic, Mesozoic, and Cenozoic.

ERG: Extensive region, or "sea," of composite dunes formed by wind-transported sand and found in major deserts.

EROSION: The set of all processes by which *soil* and rock are loosened and moved downhill or downwind.

ERRATIC: Rock fragment (especially boulder-sized) carried by a *glacier* away from the *outcrop* from which it was derived, often into an area underlain by a rock type different from that of the rock fragment.

ESKER: A glacial deposit of sand and gravel in the form of a continuous, winding ridge, formed from the deposits of a *stream* flowing beneath the ice.

ESTUARY: A body of water along a coastline, open to the ocean but diluted by fresh water.

EXCLUSIVE ECONOMIC ZONE: The zone from a country's *coast* to 200 miles offshore in which the country has exclusive rights to *mineral* deposits.

EXFOLIATION: A *physical weathering* process in which sheets of rock are fractured and detached from an *outcrop.*

EXTRUSIVE IGNEOUS ROCK: An *igneous rock* formed from *lava* or from other products of volcanic material spewed out onto the surface of the Earth.

FAULT: A planar or gently curved fracture in the Earth's *crust* across which there has been relative displacement of the two blocks of rock parallel to the fracture.

FAULT-BLOCK MOUNTAINS: A mountain or range formed when the crust is broken into blocks of different elevations by normal *faulting*

FAULTING: The processes by which crustal forces cause a rock formation to break and slip along a *fault.*

FELSIC: An adjective used to describe a light-colored *igneous rock* that is poor in iron and magnesium and contains abundant feldspars and quartz.

FERRIC IRON: Iron with a $+3$ charge (Fe^{3+}).

FERROUS IRON: Iron with a $+2$ charge (Fe^{2+}).

FIRN: Old, dense, compacted snow.

FISSURE ERUPTION: A volcanic eruption emanating from an elongate fissure rather than a *central vent.*

FJORD: A former glacial valley with steep walls and a U-shaped profile, now occupied by the sea.

FLOOD BASALT: A basltic plateau extending many kilometers in flat, layered flows originating from *fissure eruptions.*

FLOODPLAIN: A level plain of stratified, *unconsolidated* sediment on either side of a *stream,* submerged during floods and built up by silt and sand carried out of the main *channel.*

FLOOD TIDE: The part of the *tide* cycle during which the water is rising or leveling off at high water.

FLUX: The flow of Earth's chemicals from one *reservoir* to another. (See also *Geochemical cycle.*)

FOCUS (EARTHQUAKE): The point along a *fault* at which the rupture occurs. Also called the "hypocenter." (See also *Epicenter.*)

FOLD: A bent or warped stratum or sequence of strata that was originally horizontal, or nearly so, and was subsequently deformed.

FOLD AXIS: Within each stratum involved in a *fold,* the axis connecting all the points in the center of the fold, from which both *limbs* bend.

FOLD BELT: Synonym for *Orogenic belt.*

FOLDING: The processes by which crustal forces deform an area of *crust* so that layers of rock are pushed into *folds.*

FOLIATION: A set of flat or wavy planes in a *metamorphic rock,* produced by structural deformation.

FORAMINIFER: A group of single-celled organisms whose secretions and calcite shells account for most of the ocean's carbonate sediments.

FORAMINIFERAL OOZE: A calcareous *pelagic sediment* composed of the shells of dead *foraminifera.*

FORESET BED: One of the inclined beds found in *cross-bedding;* also an inclined bed deposited on the outer front of a *delta.*

FORESHORE: The marine zone between the upper limit of wave wash at high *tide* and the low-tide mark.

FORMATION: The basic unit for the naming of rocks in *stratigraphy;* a set of rocks that are or once were horizontally continuous, that share some distinctive feature of lithology, and that are large enough to be mapped.

FOSSIL FUEL: A general term for combustible geologic deposits of hydrocarbons of biologic origin, including *coal,* oil, natural gas, *oil shales,* and *tar sands.*

FRACTIONAL CRYSTALLIZATION: The separation of a cooling *magma* into components by the successive formation and removal of *crystals* at progressively lower temperatures.

FRACTURE: The irregular breaking of a *crystal* along a surface not parallel to a crystal face; serves to identify *minerals.*

FRINGING REEF: A coral *reef* that is directly attached to a landmass not composed of coral.

FUMAROLE: Small volcanic vent that emits gas and steam from which minerals precipitate onto surrounding surfaces.

GABBRO: A black, coarse-grained, *intrusive igneous rock,* composed of calcic feldspars and pyroxene; the intrusive equivalent of *basalt.*

GEOCHEMICAL CYCLE: The set of processes that carry a particular chemical from reservoir to reservoir in Earth's systems.

GEOLOGIC TIME: The time from the formation of the Earth to the present, divided into periods of time during which known geological events have taken place.

GEOLOGIC TIME SCALE: The division of geologic history into *eras, periods,* and *epochs* accomplished through *stratigraphy* and *paleontology.*

GEOLOGY: The science of Earth—how it originated, how it evolved, how it works, and how human intervention can affect it.

GEOTHERM: A curve on a temperature-pressure or temperature-depth graph that describes how temperature in the Earth changes with depth. Different tectonic provinces are characterized by more or less rapid increases of temperature with depth.

GEOTHERMAL ENERGY: Energy generated by using the heat energy of the *crust,* especially in volcanic regions.

GLACIER: A mass of ice and surficial snow that persists throughout the year and flows downhill under its own weight. The size range is from 100 m to 10,000 km. (See also *Continental glacier; Valley glacier.*)

GLACIER SURGE: A period of unusually rapid movement of a glacier, sometimes lasting more than a year.

GLASS: A rock formed when *magma* or molten rock is cooled too rapidly to allow *crystal* growth.

GLASSY: Adjective indicating that a material does not have an orderly, repeating, three-dimensional array of atoms.

GLOBAL CHANGE: A change in climate that has worldwide effects on the environment, life, and other components of Earth.

GNEISS: A coarse-grained *regional metamorphic* rock that shows banding and parallel alignment of *minerals.*

GRADED BEDDING: A bed in which the coarsest particles are concentrated at the bottom and grade gradually upward into fine silt, the whole bed having been deposited by a waning current.

GRADED STREAM: A stream whose smooth *longitudinal profile* is unbroken by resistant ledges, lakes, or waterfalls and that exactly maintains the slope, velocity, and *discharge* required to carry its *sediment* load in equilibrium without *erosion* or sedimentation.

GRANITE: A felsic, coarse-grained, *intrusive igneous rock* composed of quartz, orthoclase feldspar, sodium-rich plagioclase feldspar, and micas. Also sometimes a metamorphic product. (Compare *Rhyolite.*)

GRANITIZATION: The formation of metamorphic *granite* from other rocks by recrystallization with or without complete melting.

GRANODIORITE: A *plutonic* rock similar to *granite* in composition, except that plagioclase feldspar is present in greater abundance than orthoclase feldspar; the intrusive equivalent of *dacite*.

GRANULITE: A nonfoliated *regional metamorphic* rock with coarse interlocking grains, generally formed under conditions of relatively high pressure and temperature.

GRAVEL: The coarsest *clastic sediment*, consisting mostly of particles larger than 2 mm and including cobbles and boulders.

GREENHOUSE EFFECT: A global warming effect in which carbon dioxide and water vapor absorb infrared radiation from Earth's surface and radiate it back to the surface.

GREENSCHIST: A *schist* containing chlorite and epidote (which are green) and formed by low-temperature, low-pressure metamorphism of mafic volcanic rocks.

GREENSTONE: A field term applied to any low-grade metamorphosed *mafic igneous rock* (for example, *basalt, gabbro,* or diabase). Chlorite accounts for their greenish cast.

GROUNDWATER: The mass of water in the ground (below the *unsaturated zone*) occupying the total pore space in the rock and moving slowly where *permeability* allows.

GROUNDWATER TABLE: The upper surface of the *saturated zone* of *groundwater*. Also called the "water table."

GUYOT: A flat-topped submarine mountain or *seamount*.

HALF-LIFE: The time required for half of a sample of a given radioactive *isotope* to decay.

HANGING VALLEY: The valley left by a melted glacial tributary that enters a larger glacial valley above its base, high up on the valley wall.

HARDNESS (MINERAL): A measure of the ease with which the surface of a *mineral* can be scratched.

HEAT CONDUCTION: The transfer of the vibrational energy of atoms and molecules, which constitutes heat energy, by the mechanism of atomic or molecular impact. (Compare *Convection*.)

HEAT RESERVOIR: A subsurface region containing enough heat to be used for *geothermal energy*.

HOGBACK: A *formation* similar to a *cuesta* in that it is a ridge formed by slower *erosion* of hard strata, but having two steep, equally inclined slopes.

HORNFELS: A high-temperature *metamorphic rock* of uniform grain size showing no *foliation*. Usually formed by *contact metamorphism*.

HOT SPOT: The volcanic surface expression of a *mantle plume*.

HUMUS: The decayed part of the organic matter in a *soil*.

HYDRATION: The absorption of water by a *mineral*, usually in *weathering*.

HYDROLOGIC CYCLE: The cyclical movement of water from the ocean to the atmosphere, through rain to the surface, through *runoff* and *groundwater* to *streams,* and back to the sea.

HYDROLOGY: The science of that part of the *hydrologic cycle* between rain and return to the sea; the study of the movement and characteristics of water on and within the land.

HYDROTHERMAL ACTIVITY: Any process involving high-temperature *groundwater,* especially the alteration and emplacement of *minerals* and the formation of hot springs and geysers.

HYDROTHERMAL METAMORPHISM: A form of *metamorphism,* frequently associated with *mid-ocean ridges,* in which hot fluids percolate through the crust and metamorphose the invaded rocks.

HYDROTHERMAL VEIN: A cluster of *minerals* precipitated by *hydrothermal activity* in a rock cavity.

ICEBERG CALVING: The breaking off of blocks of ice from a *glacier* when it moves to a shoreline, forming icebergs.

IGNEOUS ROCK: A rock formed by the solidification of a *magma*.

INFILTRATION: The movement of *groundwater* or hydrothermal water into rock or *soil* through pores and *joints*.

INFLUENT STREAM: A *stream* or portion of a stream that *recharges groundwater* through the stream bottom because its *elevation* is above the *groundwater table*. (Compare *Effluent stream*.)

INTRUSIVE IGNEOUS ROCK: *Igneous rock* that forced its way in a molten state into the *country rock*. Also called an "intrusion."

ION: An atom or group of atoms that has gained or lost *electrons* and so has a net electric charge.

IONIC BOND: A bond formed between atoms by electrostatic attraction between oppositely charged *ions*.

IRON FORMATION: A *sedimentary rock* containing much iron, usually more than 15 percent, as sulfide, oxide, hydroxide, or carbonate; a low-grade ore of iron.

ISLAND ARC: A linear or arc-shaped chain of volcanic islands formed at a *convergent plate boundary*. The island arc is formed in the overriding plate from rising melt derived from the subducted plate and from the *asthenosphere* above that plate.

ISOCHRON: A line connecting points of equal age.

ISOSTASY, PRINCIPLE OF: The mechanism whereby areas of the *crust* rise or subside until the mass of their *topography* is buoyantly supported or compensated by the thickness of crust below, which "floats" on the denser *mantle*. The theory of isostasy holds that continents and mountains are supported by low-density crustal "roots."

ISOTOPE: One of several forms of one element, all having the same number of *protons* in the nucleus, but differing in their number of *neutrons* and thus in their *atomic weight*.

JOINT: A large and relatively planar fracture in a rock across which there is no relative displacement of the two sides.

KAME: A ridgelike or hilly local glacial deposit of coarse *clastic sediment* formed as a *delta* at the *glacier* front by meltwater *streams*.

KARST TOPOGRAPHY: An irregular *topography* characterized by *sinkholes,* caverns, and lack of surface *streams;* formed in humid regions because an underlying carbonate *formation* has been riddled with underground drainage channels that capture the surface streams.

KETTLE: A hollow or depression formed in glacial deposits when *outwash* was deposited around a residual block of ice that later melted.

LAHAR: A *mudflow* of *unconsolidated volcanic ash,* dust, *breccia,* and boulders that occurs when *pyroclastic* or *lava* deposits mix with rain or the water of a lake, river, or melting glacier.

LAMINAR FLOW: A flow in which streamlines are straight or gently curved and parallel. (Compare *Turbulent flow.*)

LANDFORM: A characteristic landscape feature on the Earth's surface that attained its shape through the processes of *erosion* and *sedimentation;* for example, a hill or a valley.

LATERITE: A distinctive, deep-red *soil* formed in very humid regions, characterized by high alumina and iron oxide content, and produced by rapid *chemical weathering* of feldspar *minerals*.

LAVA: *Magma* that has reached the surface.

LEVEE: A ridge along a *stream* bank, formed by deposits left when floodwater slowed on leaving the *channel;* also an artificial barrier to floods built in the same form.

LIMB (FOLD):   The two relatively planar sides of a *fold*, one on either side of the *axial plane*.

LIMESTONE: A *sedimentary rock* composed mainly of calcium carbonate ($CaCO_3$), usually as the *mineral* calcite.

LINEAR DUNE: A long, narrow *eolian* sand *dune* that is aligned parallel to the direction of the prevailing wind.

LITHIFICATION: The chemical and physical diagenetic processes that bind and harden a *sediment* into a *sedimentary rock*. (See also *Compaction*.)

LITHOSPHERE: The outer, rigid shell of the Earth, situated above the *asthenosphere* and containing the *crust,* the uppermost part of the *mantle,* the continents, and the *plates*.

LOESS: An unstratified, wind-deposited, dusty *sediment* rich in *clay* minerals.

LONGITUDINAL PROFILE: A cross section of a *stream* from its head to its mouth, showing *elevation* versus distance to the mouth.

LONGSHORE CURRENT: A current that flows parallel to the *shoreline;* the summed longshore components of water motion of waves that break obliquely with respect to the shore.

LONGSHORE DRIFT: The movement of *sediment* along a beach by *swash* and *backwash* of waves that approach the shore obliquely.

LUSTER: The general quality of the shine of a *mineral* surface, described by such subjective terms as dull, glassy, or metallic.

MAFIC: Adjective describing dark-colored *minerals* rich in iron and magnesium and relatively poor in silica (for example, pyroxene, amphibole, or olivine); also, describing rocks rich in mafic minerals.

MAGMA: Molten rock material that forms *igneous rocks* upon cooling. Magma that reaches the surface is *lava*.

MAGMA CHAMBER: A magma-filled cavity within the *lithosphere*.

MAGNETIC STRATIGRAPHY: The study and correlation of polarity epochs and events in the history of the Earth's magnetic field as contained in magnetic rocks.

MAGNITUDE: A measure of *earthquake* size, determined by taking the common logarithm (base 10) of the largest ground motion observed during the arrival of a *P wave* or *seismic surface wave* and applying a standard correction for distance to the *epicenter*.

MANTLE: The main bulk of the solid Earth, between the *crust* and the *core,* ranging from depths of about 40 km to 2900 km. It is composed of dense, mafic silicate minerals and divided into concentric layers by phase changes that are caused by the increase in pressure with depth.

MANTLE PLUME: Rising jet of hot, partially molten material, thought to emanate from the deep *mantle* and responsible for intraplate *volcanism*.

MARBLE: The metamorphosed equivalent of *limestone* or other *carbonate rock*.

MARINE EVAPORITE SEDIMENT, ROCK: A *sediment* or *sedimentary rock* consisting of *minerals* precipitated by evaporating seawater. Includes salt and gypsum.

MASS MOVEMENT: A downhill movement of *soil* or fractured rock under the force of gravity.

MASS WASTING: The sum of all *mass movements* and related erosional phenomena.

MEANDER: A broad, semicircular curve in a *stream* that develops as the stream erodes the outer bank of a bend and deposits *sediment* (as *point bars*) against the inner bank.

MÉLANGE: A formation found at *convergent plate boundaries* consisting of a heterogeneous mixture of rock materials in which fragments of diverse composition, size, and *texture* were mixed and consolidated by tremendous deformational pressure.

MESA: A flat-topped, steep-sided upland topped by a resistant *formation*.

METALLIC BOND: A type of *covalent bond* in which freely mobile *electrons* are shared and dispersed among *ions* of metallic elements, which have the tendency to lose electrons and pack together as *cations*.

METAMORPHIC FACIES: Characteristic assemblages of *minerals* in *metamorphic rocks* that are indicative of the range of pressures and temperatures experienced during metamorphism.

METAMORPHIC ROCK: A rock whose original *mineralogy, texture,* or composition has been changed by the effects of pressure, temperature, or the gain or loss of chemical components.

METASOMATISM: A change in the bulk chemical composition of a rock by fluid transport of some chemical components into or out of the rock.

METEORIC WATER: Rainwater, snow, hail, and sleet.

MICROPLATE TERRANE: A block within an *orogenic belt* containing rock assemblages that contrast sharply with those in the surrounding areas, interpreted as small continents, *seamounts,* or *island arcs* that were accreted onto the larger continent at a *convergent plate boundary*.

MID-OCEAN RIDGE: A major elevated linear feature of the seafloor consisting of many small, slightly offset segments, with a total length of 200 to 20,000 km. A mid-ocean ridge occurs at a *divergent plate boundary,* a site where two *plates* are being pulled apart and new oceanic *lithosphere* is being created.

MIGMATITE: A rock with both igneous and metamorphic characteristics that shows large *crystal* and laminar flow structures. Probably formed metamorphically in the presence of water and without complete melting.

MINERAL: A naturally occurring, inorganic, crystalline solid with a specific chemical composition.

MINERALOGY: The study of *mineral* composition, structure, appearance, stability, occurrence, and associations. The mineralogy of a rock is the mineral assemblage contained within that rock.

MOHOROVIČIĆ DISCONTINUITY, MOHO: The boundary between *crust* and *mantle,* at a depth of 5 to 45 km, marked by a rapid increase in *seismic wave* velocity to more than 8 km per second.

MOHS SCALE OF HARDNESS: An empirical, ascending scale of *mineral* hardness. (See Table 2.3.)

MOMENT MAGNITUDE: A measure of *earthquake* size, determined by the slip of the *fault,* the area of the break, and the rigidity of rock. (See also *Richter magnitude*.)

MORAINE: A glacial deposit of *till* left at the margins of an ice sheet. Subdivided into ground moraine, lateral moraine, medial moraine, and end moraine.

MUDFLOW: A *mass movement* of material mostly finer than sand, along with some rock debris, lubricated with large amounts of water. The water tends to make mudflows move faster than *earthflows* or *debris flows*.

MUDSTONE: The lithified equivalent of mud; a fine-grained *sedimentary rock* similar to *shale* but less finely laminated.

MYLONITE: A very fine grained *metamorphic rock* commonly found in major thrust *faults* and produced by shearing and rolling during fault movement.

NATURAL LEVEE: See *Levee*.

NEAP TIDE: A *tide* cycle of unusually small amplitude that occurs twice monthly when the lunar and solar tides are opposed—that is, when the gravitational pull of the Sun is at right angles to that of the Moon. (Compare *Spring tide*.)

NEUTRON: An electrically neutral elementary particle in the atomic nucleus, having the mass of one *proton*.

NONFOLIATED METAMORPHIC ROCKS: *Metamorphic rocks* that have no preferred orientation of crystals, or very weak orientation, and thus show little or no slaty cleavage or schistosity.

OBSIDIAN: Dark volcanic *glass*, usually of *felsic* composition.

OFFSHORE: The marine zone extending from the breaker zone to the edge of the *continental shelf*.

OIL RESERVOIR: A bed of permeable and porous rock that contains commercially producible oil.

OIL SHALE: A dark-colored *shale* that contains organic material and that can be crushed and heated to liberate oil.

OIL TRAP: A tectonic or *sedimentary structure* that impedes upward movement of oil or gas and allows it to collect beneath the barrier.

OPHIOLITE SUITE: An assemblage of *mafic* and *ultramafic igneous rocks* with deep-sea *sediments* found on land, believed to be associated with *divergent plate boundaries* and the seafloor environment.

ORE DEPOSIT: A *sedimentary, igneous,* or *metamorphic rock* containing *minerals,* commonly metallic oxides or silicates, that can be commercially mined.

ORGANIC SEDIMENT, ROCK: A *sediment* or *sedimentary rock* consisting entirely or in part of organic carbon-rich deposits formed by the decay of once-living material after burial. Includes *coal* and organic carbon-rich *shales.*

ORIGINAL HORIZONTALITY, PRINCIPLE OF: The proposition that all sedimentary *bedding* is horizontal at the time of deposition.

OROGENIC BELT: A linear region that has been subjected to folding and other deformation in a mountain-building episode.

OROGENY: The tectonic process in which large areas are folded, thrust-faulted, metamorphosed, and subjected to *plutonism*. The cycle ends with uplift and the formation of mountains.

OUTCROP: A segment of *bedrock* exposed to the atmosphere.

OUTWASH: A *sediment* deposited by meltwater *streams* emanating from a *glacier.*

OVERTURNED FOLD: A *fold* in which a *limb* has tilted past vertical so that the older strata are uppermost.

OXBOW LAKE: A long, broad, crescent-shaped lake formed when a *stream* abandons a *meander* and takes a new course.

OXIDATION: A chemical reaction in which *electrons* are lost from an atom and its charge becomes more positive; a chemical combination of an element with oxygen.

OZONE: A molecule (O_3) that absorbs ultraviolet radiation in the *stratosphere* but creates smog when it forms near Earth's surface.

PAHOEHOE: A basaltic *lava* flow with a *glassy*, smooth, and ropy surface.

PALEOMAGNETIC STRATIGRAPHY: A branch of *stratigraphy* in which the remanent magnetization recorded in a rock is used to place the rock on the "magnetic" time scale constructed from known temporal variations in the Earth's magnetic field.

PALEOMAGNETISM: The remanent magnetization recorded in ancient rocks; allows the reconstruction of Earth's ancient magnetic field and the positions of the continents.

PALEONTOLOGY: The science of fossils of ancient life forms and their evolution.

PANGAEA: Supercontinent that coalesced in the latest Paleozoic era and comprised all present continents. The breakup of Pangaea began in Mesozoic time, as inferred from *paleomagnetic* and other data.

PARTIAL MELTING: A process in which heating melts some of the *minerals* in a mass of rock while the rest remain solid. Partial melting occurs because the minerals that compose a rock melt at different temperatures.

PASSIVE MARGIN: A *continental margin* characterized by thick, flat-lying, shallow-water *sediments* with only limited tectonism related to divergent *plate* motion.

PEAT: A marsh or swamp deposit of water-soaked plant remains containing more than 50 percent carbon.

PEDALFER: A common *soil* type in temperate regions; characterized by an abundance of iron oxides and *clay* minerals deposited in the *B-horizon* by leaching.

PEDIMENT: A planar, sloping rock surface forming a ramp up to the front of a retreating mountain range in an arid region. It may be covered locally by thin alluvial deposits.

PEDOCAL: A common *soil* type of arid regions, characterized by accumulation of calcium carbonate in the *B-horizon.*

PEGMATITE: A *vein* of extremely coarse-grained granite, often containing economic amounts of rare elements.

PELAGIC SEDIMENT: A deep-sea *sediment* composed of fine-grained detritus that slowly settles from surface waters. Common constituents are *clay, foraminiferal ooze,* and *silica ooze.*

PERCHED WATER TABLE: The upper surface of an isolated body of *groundwater* that is perched above and separated from the main body of groundwater by an *aquiclude*.

PERIDOTITE: A coarse-grained *ultramafic igneous rock* composed of olivine with small amounts of pyroxene and amphibole.

PERIOD (GEOLOGIC): The most commonly used unit of geologic time, representing one subdivision of an *era*.

PERIOD (WAVE): The time interval between the arrival of successive crests in a homogeneous wave train.

PERMAFROST: A permanently frozen aggregate of ice and *soil* occurring in very cold regions.

PERMEABILITY: The ability of a *formation* to transmit *groundwater* or other fluids through pores and cracks.

PHENOCRYST: A large *crystal* surrounded by a finer matrix in an *igneous rock*. An igneous rock that contains abundant phenocrysts is called a *porphyry*.

PHOSPHORITE: A *sedimentary rock* composed largely of calcium phosphate, usually as a variety of the mineral apatite and largely in the form of concretions and nodules. Primary ore of phosphate *minerals* and elemental phosphorus.

PHOTOSYNTHESIS: The process by which green organisms use chlorophyll and the energy from sunlight to make organic matter (carbohydrates) out of carbon dioxide and water. (See also *Respiration*.)

PHREATIC EXPLOSION: A volcanic eruption of steam, mud, and debris caused by the expansion of steam formed when *magma* comes in contact with *groundwater* or seawater.

PHYLLITE: A *metamorphic rock* that is intermediate in grade between slate and mica *schist*. Small *crystals* of micas give a silky sheen to the *cleavage* surfaces.

PHYSICAL WEATHERING: The set of all physical processes by which an *outcrop* is broken up into smaller particles.

PILLOW LAVA: A basaltic *lava* that forms under water when many small tongues of lava break through the chilled ocean floor and quickly solidify into a rock *formation* resembling a pile of sandbags.

PLACER: A *clastic* sedimentary deposit of a valuable *mineral* or native metal in unusually high concentration, usually segregated because of its greater *density*.

PLANETARY DIFFERENTIATION: The process by which heating, cooling, and gravitation sorted the materials of our planet so that it evolved into concentric layers that differ chemically and physically.

PLASTIC FLOW: Deformation of the shape or volume of a substance without fracturing.

PLATE: One of the dozen or more segments of the *lithosphere* that ride as distinct units over the *asthenosphere*.

PLATEAU: An extensive upland region at high *elevation* with respect to its surroundings.

PLATE TECTONICS: The theory and study of *plate* formation, movement, interactions, and destruction; the attempt to explain *seismicity, volcanism,* mountain-building, and evidence of *paleomagnetism* in terms of plate motions.

PLATFORM: A sediment-covered, tectonically stable, almost level region of a continent.

PLAYA, PLAYA LAKE: The flat floor of a closed basin in an arid region, usually rich in evaporate minerals. It may be occupied by an intermittent lake.

PLUNGING FOLD: A *fold* whose axis is not horizontal but dips.

PLUTON: A large igneous intrusion, at least 1 km^3, formed at depth in the *crust*.

PLUTONISM: Igneous activity at depth in the *crust*.

POINT BAR: A deposit of *sediment* on the inner bank of a *meander* that forms because the *stream* velocity is lower against the inner bank.

POLYMORPH: One of two or more alternative possible structures for a single chemical compound; for example, the minerals calcite and aragonite are polymorphs of calcium carbonate ($CaCO_3$).

POROSITY: The percentage of the total volume of a rock that is pore space (not occupied by *mineral* grains).

PORPHYROBLAST: A large *crystal* in a finer-grained matrix in a *metamorphic rock:* analogous to a *phenocryst* in an *igneous rock*.

PORPHYRY: An *igneous rock* containing abundant *phenocrysts* suspended in a finely crystalline matrix.

POTABLE WATER: Water that is agreeable to the taste and not dangerous to the health.

POTHOLE: A hemispherical hole in the *bedrock* of a streambed, formed by abrasion of small pebbles and cobbles in a strong current.

PRECIPITATE: To drop out of a saturated solution as crystals; the crystals that drop out of a saturated solution.

PRESSURE: A force distributed over a surface divided by the area of the surface. Confining pressure is uniform in all directions, while directed pressure is exerted in a particular direction.

PROTON: With neutrons, one of the two types of particles in the atomic nucleus that account for most of the atom's mass. The mass of a proton is valued at 1 unit and is equivalent to the mass of 1836 electrons. Each proton has a positive electrical charge of 1.6×10^{-19} coulombs, expressed as $+1$.

PUMICE: A form of volcanic *glass,* usually of *felsic* composition, so filled with holes from the escape of gas during quenching that it resembles a sponge and has very low *density.* (Compare *Obsidian.*)

P WAVE: The primary or fastest wave traveling away from a seismic event through the solid rock and consisting of a train of compressions and dilations of the material.

PYROCLAST: Fragment of volcanic material ejected during an eruption.

PYROCLASTIC FLOW: A glowing cloud of *volcanic ash,* fragments of volcanic rock, and gases that moves rapidly downhill away from the eruptive center during a volcanic eruption.

PYROXENE GRANULITE: A coarse-grained *regional metamorphic* rock containing pyroxene; formed at high temperatures and pressures deep in the *crust.*

QUARTZITE: A very hard, nonfoliated, white *metamorphic rock* formed from *sandstones* rich in quartz sand grains and quartz cement.

RADIAL DRAINAGE: A system of *streams* running in a radial pattern away from the center of a circular *elevation,* such as a volcano or *dome.*

RADIOACTIVITY: The emission of energetic particles and/or radiation during radioactive decay.

RADIOMETRIC DATING: The method of obtaining ages of geological materials by measuring the relative abundances of radioactive parent and daughter *isotopes* in them.

RAIN SHADOW: An area of low rainfall on the leeward slope of a mountain range.

REACTION SERIES: A series of chemical reactions occurring in a cooling *magma* by which a *mineral* formed at high temperature becomes unstable in the melt and reacts to form another mineral.

RECHARGE: In *hydrology,* the replenishment of *groundwater,* usually by infiltration of *meteoric water* through the *soil.*

RECTANGULAR DRAINAGE: A system of *streams* in which each straight segment of each stream takes one of two characteristic perpendicular directions, usually following sets of *joints.*

RECURRENCE INTERVAL: The average time interval between occurrences of a geologic event, such as a flood or *earthquake,* of a given or greater *magnitude.*

REEF: A mound or ridge-shaped organic structure that is built by calcareous organisms, is wave resistant, and stands in relief above the surrounding seafloor.

REGIONAL METAMORPHISM: *Metamorphism* caused by deep burial or strong tectonic forces that impose high temperatures and high pressures over large belts or regions of the crust.

REGOLITH: The layer of loose, heterogeneous material lying on top of *bedrock;* includes *soil,* unweathered fragments of parent rock, and rock fragments weathered from the bedrock.

REJUVENATION (OF MOUNTAINS): Renewed uplift in a mountain chain on the site of earlier uplifts, returning the area to a more youthful stage of the *cycle of erosion.*

RELATIVE AGE: The age of a geologic event or feature relative to other geologic events or features and expressed in terms of the *geologic time scale.* (Compare *Absolute age.*)

RELATIVE HUMIDITY: The amount of water vapor in the air, expressed as a percentage of the total amount of water vapor that the air could hold at that temperature if saturated.

RELIEF: The maximum regional difference in *elevation.*

RESERVES: Deposits of *minerals, coal,* or oil and gas that have been shown to be extractable profitably with existing technology. "Proven reserves" are those for which good estimates of the quantity and quality have been made. (See also *Resources.*)

RESERVOIR: A source or place of residence for elements in a chemical cycle or *hydrologic cycle.*

RESIDENCE TIME: The average length of time that an *atom* of a particular element spends in a *reservoir* before leaving.

RESOURCES: Discovered and undiscovered deposits of *minerals, coal,* or oil and gas that are or may become available for use in the future; includes *reserves,* plus discovered deposits not now commercially or technologically extractable, plus undiscovered deposits that may be inferred to exist. (See also *Reserves.*)

RESPIRATION: The process by which carbohydrates combine with oxygen to produce carbon dioxide and water and to release energy. (See also *Photosynthesis*.)

RETROGRADE METAMORPHISM: Metamorphism in which a rock that has been metamorphosed to a fairly high grade is later remetamorphosed at lower temperature and pressure to a lower grade.

RHYOLITE: The fine-grained volcanic or extrusive equivalent of *granite*, light-brown to gray and compact.

RICHTER MAGNITUDE: A measure of *earthquake* size, determined by taking the common logarithm (base 10) of the largest ground motion observed during the arrival of a *P wave* or *seismic surface wave* and applying a standard correction for distance to the *epicenter*. (See also *Moment magnitude*.)

RIFT VALLEY: A *fault* trough formed at a *divergent plate boundary* or other area of tension.

RIPPLE: A very small *dune* of sand or silt whose long dimension is formed at right angles to the current.

RIVER: A general term for a relatively large *stream*, or the main branches of a stream system.

ROCK AVALANCHE: The rapid, downhill-flowing *mass movement* of broken rock material, during which further breakage of the material may occur.

ROCK CYCLE: The set of geologic processes by which each of the three great groups of rocks is produced from the other two: *sedimentary rocks* are metamorphosed to *metamorphic rocks* or melted to create *igneous rocks*, and all rocks may be uplifted and eroded to make *sediments*, which lithify to sedimentary rocks.

ROCKFALL: The relatively free falling of a newly detached segment of *bedrock* from a cliff or other steep slope.

ROCK FLOUR: A glacial *sediment* of extremely fine (silt and clay-size) ground rock formed by abrasion of rocks at the base of the *glacier*.

ROCKSLIDE: The *mass movement* of large blocks of detached *bedrock* sliding more or less as a unit.

RUNOFF: The amount of rainwater that does not infiltrate the ground but leaves an area in surface drainage.

SALTATION: The movement of sand or fine *sediment* by short jumps above the ground or streambed under the influence of a current too weak to keep it permanently suspended.

SANDBLASTING: A *physical weathering* process in which rock is eroded by the impact of sand grains carried by the wind, frequently leading to *ventifact* formation of pebbles and cobbles.

SANDSTONE: A *clastic rock* composed of grains from 0.0625 to 2 mm in diameter, usually quartz, feldspar, and rock fragments, bound together by a cement of quartz, carbonate, or other *minerals,* or by a matrix of *clay* minerals.

SATURATED ZONE: The zone of *soil* and rock in which pores are completely filled with *groundwater*.

SCHIST: A *metamorphic rock* characterized by strong *foliation* or *schistosity*.

SCHISTOSITY: The parallel arrangement of sheety or prismatic *minerals* like micas and amphiboles resulting from metamorphism.

SCIENTIFIC METHOD: A general research strategy, based on creative analyses of verifiable data, by which scientists propose and test hypotheses that explain some aspect of how the physical realm works.

SEAFLOOR SPREADING: The mechanism by which new seafloor is created at ridges at *divergent plate boundaries* as adjacent *plates* move apart. This process may continue at a few centimeters per year through many geologic *periods*.

SEAMOUNT: An isolated, tall mountain on the seafloor that may extend more than 1 km from base to peak.

SEDIMENT: Any of a number of materials deposited at Earth's surface by physical agents (such as wind, water, and ice), chemical agents (precipitation from oceans, lakes, and rivers), or biological agents (organisms, living and dead).

SEDIMENTARY BASIN: A region of considerable extent (at least 10,000 km^2) that is the site of accumulation of a large thickness of *sediments*.

SEDIMENTARY BRECCIA: A *clastic rock* composed mainly of large angular fragments.

SEDIMENTARY ENVIRONMENT: A geographically limited area where *sediments* are preserved; characterized by its *landforms*, climate, relative energy of water and wind currents, biological activity, and the relative abundance of various chemical substances.

SEDIMENTARY ROCK: A rock formed by the accumulation and *cementation* of *mineral* grains by wind, water, or ice transportation to the site of deposition or by chemical precipitation at the site.

SEDIMENTARY STRUCTURE: Any structure of a sedimentary or weakly metamorphosed rock that was formed at the time of deposition; includes *bedding, crossbedding, graded bedding, ripples,* scour marks, mudcracks.

SEISMIC GAP METHOD: A predictive model for *earthquake* occurrences along active *fault* zones based on the study of segments that have experienced little or no movement and are thought to be under high *stress*.

SEISMICITY: The worldwide or local distribution of *earthquakes* in space and time; a general term for the number of earthquakes in a unit of time.

SEISMIC SURFACE WAVE: A *seismic wave* that follows the Earth's surface only, with a speed less than that of *S waves.*

SEISMIC WAVE: A general term for the elastic waves produced by *earthquakes* or explosions. (See also *P wave; S wave; Seismic surface wave.*)

SEISMOGRAPH: An instrument for magnifying and recording the motions of the Earth's surface that are caused by *seismic waves.*

SHADOW ZONE: A zone 105° to 142° from the *epicenter* of an *earthquake* in which there is no penetration of *seismic waves* through the Earth because of *wave refraction* or because the waves are not transmitted upon entering the liquid *core.*

SHALE: A very fine grained *clastic rock* composed of silt and *clay* that tends to part along *bedding* planes. (See also *Oil shale.*)

SHEARING FORCES: Forces that deform a body so that parts of the body on opposite sides of a plane slide past one another; that is, forces acting tangentially to the plane. Shearing forces dominate at *transform fault plate boundaries.*

SHIELD: A large region of stable, ancient *basement* rocks within a continent.

SHIELD VOLCANO: A large, broad volcanic cone with very gentle slopes built up by nonviscous basaltic *lavas.*

SHOCK METAMORPHISM: *Metamorphism* that occurs when minerals are subjected to the high pressures and temperatures of shock waves generated by impacts.

SHORELINE: The straight or sinuous, smooth or irregular interface between land and sea.

SILICA OOZE: A *pelagic sediment* consisting of the remains of tiny organisms that have shells made of amorphous silica.

SILICATE ROCK: An *igneous* or *metamorphic rock* made up largely of silicate minerals, such as feldspar, mica, or garnet.

SILICEOUS SEDIMENTARY ROCK: Rock containing abundant free silica of either organic or inorganic origin, formed by biochemical, chemical, or physical deposition of silica.

SILL: A horizontal, tabular igenous intrusion running between parallel layers of bedded country rock. (See also *Concordant intrusion.*)

SILTSTONE: A *clastic rock* that contains mostly silt-sized material, from 0.0039 to 0.062 mm.

SINKHOLE: A small, steep depression caused in *karst topography* by the dissolution and collapse of subterranean caverns in carbonate formations.

SLATE: The lowest grade of foliated *metamorphic rock,* easily split into thin sheets; produced primarily by the metamorphism of shale.

SLIP FACE: The steep downwind face of a *dune* on which sand is deposited in *cross-beds* at the *angle of repose.*

SLIP (FAULT):   The motion of one face of a *fault* relative to the other.

SLUMP: A slow *mass movement* of *unconsolidated materials* that slide as a unit.

SOIL: The surface accumulation of sand, *clay,* and humus that composes the *regolith,* but excluding the larger fragments of unweathered rock.

SOLAR CONSTANT: The amount of radiation from the Sun that reaches the top of the atmosphere; about 1370 watts per square meter.

SOLIFLUCTION: The *creep* of *soil* saturated with water and/or ice, caused by alternate freezing and thawing; most common in polar regions.

SOLUBILITY (MINERAL):   The extent to which a *mineral* can dissolve in water; the amount of the mineral dissolved in water when the solution reaches the saturation point.

SORTING: A measure of the homogeneity of the sizes of particles in a *sediment* or *sedimentary rock.*

SOURCE BED: *Organic sediment* or rock that liberates oil or gas when heated during burial. Usually source beds are organic-rich "black" *shales* or *limestones.*

SPECIFIC GRAVITY: The ratio of the *density* of a given substance to the density of water.

SPHEROIDAL WEATHERING: A *physical* and *chemical weathering* process in which curved layers split off from a rounded boulder, leaving a spherical inner core.

SPIT: A long range of sand deposited by *longshore currents* and *longshore drift* where the *coast* takes an abrupt inward turn. It is attached to land at the upstream end.

SPRING TIDE: A *tide* cycle of unusually large amplitude that occurs twice monthly when the lunar and solar tides are in phase. (Compare *Neap tide.*)

STACKS: Isolated rocky prominences or pinnacles left standing above a marine platform as erosional remnants.

STALACTITE: An icicle or toothlike deposit of calcite or aragonite hanging from the roof of a cave. It is deposited by evaporation and precipitation from solutions seeping through *limestone.*

STALAGMITE: An inverted icicle-shaped deposit that builds up on a cave floor beneath a *stalactite* and is formed by the same process as a stalactite.

STOCK (VOLCANIC): An intrusion with the characteristics of a *batholith* but less than 100 km^2 in area.

STRATIFICATION: The characteristic layering or *bedding* of *sedimentary rocks.*

STRATIGRAPHIC SEQUENCE: A set of deposited beds that reflects the changing conditions and *sedimentary environments* that define the geologic history of a region.

STRATIGRAPHY: The science of description, correlation, and classification of strata in *sedimentary rocks,* including the interpretation of the *sedimentary environments* of those strata.

STRATOSPHERE: The upper atmosphere, 10 to 50 km above the surface, where a protective *ozone* layer forms.

STRATOVOLCANO: See *Composite volcano.*

STREAK: The fine deposit of mineral dust left on an abrasive surface when a *mineral* is scraped across it; especially the characteristic color of the dust.

STREAM: A general term for any body of water, large or small, that moves under the force of gravity in a *channel.* (Compare *River.*)

STREAM PIRACY: The *erosion* of a *divide* between two *streams* by the more *competent* stream, leading to the capture of all or part of the drainage of the slower stream by the faster.

STRESS: The force exerted, in terms of force per unit area, when one body presses upon, pulls upon, or pushes tangentially against another body.

STRIATION (GLACIAL): Scratches left on *bedrock* and boulders by overriding ice, showing the direction of glacial motion.

STRIKE: The angle between true north and the horizontal line contained in any planar feature (inclined bed, *dike, fault* plane, and so forth); also the geographic direction of this horizontal line.

SUBDUCTION: The sinking of an oceanic *plate* beneath an overriding plate; occurs at *convergent plate boundaries.*

SUBDUCTION ZONE: The zone between a sinking oceanic *plate* and an overriding plate, descending away from a *trench* and characterized by high *seismicity.* (See also *Convergent plate boundary.*)

SUBLIMATION: A phase change in which a substance passes between the solid and gaseous states without passing through the liquid state. Glaciers can lose ice through sublimation.

SUBMARINE CANYON: An underwater canyon in the *continental shelf.*

SUBMARINE FAN: A *terrigenous* cone- or fan-shaped deposit located at the foot of a *continental slope,* usually seaward of large *rivers* and *submarine canyons.*

SUBSIDENCE: A gentle *epeirogenic* movement in which a broad area of the *crust* sinks without appreciable deformation.

SUPERPOSED STREAM: A *stream* that flows through resistant *formations* because its course was established at a higher level on uniform rocks before downcutting began.

SUPERPOSITION, PRINCIPLE OF: The principle that, except in extremely deformed strata, a bed that overlies another bed is always the younger.

SURF: The foamy, bubbly surface of water waves as they break close to shore.

SURFACE TENSION: The attractive force between molecules at a surface.

SURFACE WAVE: See *Seismic surface wave.*

SURF ZONE: An offshore belt along which the waves collapse into breakers as they approach the shore.

SUSPENDED LOAD: The fine *sediment* kept suspended in a *stream* because the *settling velocity* of the sediment is lower than the upward velocity of eddies.

SUSTAINABLE DEVELOPMENT: Growth in economies and in living standards that can last indefinitely without harming the environment.

SUTURE: A zone of intensely deformed rocks that marks the boundary where two continents collided.

SWASH: The landward rush of water from a breaking wave up the slope of the beach.

S WAVE: The secondary *seismic wave,* which travels more slowly than the *P wave* and consists of elastic vibrations transverse to the direction of travel. S waves cannot penetrate a liquid.

SWELL: An oceanic water wave with a *wavelength* on the order of 30 m or more and a *wave height* of approximately 2 m or less that may travel great distances from its source.

SYNCLINE: A large downfold, whose *limbs* are higher than its center. (Compare *Anticline.*)

SYNCRUDE: Synthetic oil produced from *coal.*

TALUS: A deposit of large angular fragments of physically weathered *bedrock,* usually at the base of a cliff or steep slope.

TAR SAND: A sandy deposit of organic matter impregnated with a tarry substance made up mostly of hydrocarbons, from which petroleum can be extracted.

TENSIONAL FORCES: Forces that stretch a body and pull it apart. Tensional forces dominate at *divergent plate boundaries.*

TERRACE (STREAM VALLEY): A flat, steplike surface above the *floodplain* in a stream *valley,* marking a former floodplain that existed at the higher level before regional uplift or an increase in *discharge* caused the *stream* to erode into the former floodplain.

TERRIGENOUS SEDIMENT: *Sediment* eroded from the land surface.

TEXTURE (ROCK): The rock characteristics of grain or *crystal* size, size variability, rounding or angularity, and preferred orientation.

THERMOREMENANT MAGNETIZATION: A permanent magnetization acquired by *minerals* in *igneous rocks* during *crystallization.*

TIDAL FLAT: A broad, flat region of muddy or sandy *sediment,* covered and uncovered in each *tide* cycle.

TIDAL SURGE: Waves that overrun a beach and batter sea cliffs when an intense storm passes near the shore during a *spring tide.*

TIDE: The rise and fall of the water level of the ocean that occurs twice a day and is caused by the gravitational attraction of the Moon and, to a lesser degree, the Sun, with greater force on the parts of the Earth facing and opposite the Moon (and Sun).

TILL: An unstratified and poorly sorted *sediment* containing all sizes of fragments from *clay* to boulders, deposited by glacial action.

TILLITE: The lithified equivalent of *till.*

TOPOGRAPHY: The shape of the Earth's surface, above and below sea level; the set of *landforms* in a region; the distribution of *elevations.*

TOPSET BED: A horizontal sedimentary bed formed at the top of a *delta* and overlying the *foreset beds.*

TRACE ELEMENT: An element that appears in a *mineral* in a concentration of less than 1 percent (often less than 0.001 percent).

TRANSFORM FAULT PLATE BOUNDARY: A boundary at which Earth's *plates* slide horizontally past each other, approximately at right angles to their *divergent plate boundaries.*

TRANSPIRATION: The release of water vapor by plants into the atmosphere.

TRANSVERSE DUNE: A *dune* that has its axis perpendicular (transverse) to the prevailing winds or to a current. The upwind or upcurrent side has a gentle slope, and the downwind or downcurrent side lies at the *angle of repose.*

TRELLIS DRAINAGE: A system of *streams* in which *tributaries* tend to lie in parallel valleys formed in steeply dipping beds in folded belts.

TRENCH: A long, narrow, deep trough in the seafloor; marks the line along which a *plate* bends down into a *subduction zone.*

TRIBUTARY: A *stream* that discharges water into a larger stream.

TSUNAMI: A large destructive wave caused by seafloor movements in an *earthquake.*

TURBIDITE: The sedimentary deposit of a *turbidity current,* typically showing *graded bedding.*

TURBIDITY CURRENT: A mass of mixed water and *sediment* that flows downhill along the bottom of an ocean or lake because it is denser than the surrounding water. It may reach high speeds and erode rapidly.

TURBULENT FLOW: A high-velocity flow in which streamlines are neither parallel nor straight but curled into small tight eddies. (Compare *Laminar flow.*)

ULTRAMAFIC ROCK: An *igneous rock* consisting mainly of *mafic* minerals and containing less than 10 percent feldspar. Includes *peridotite, amphibolite,* dunite, and pyroxenite.

UNCONFINED AQUIFER: An *aquifer* that is not overlain by an *aquiclude,* thereby causing the level of water in a well that penetrates the aquifer to be at the level of the surrounding *groundwater table.* (Compare *Confined aquifer.*)

UNCONFORMITY: A surface that separates two strata. It represents an interval of time in which deposition stopped, *erosion* removed some *sediments* and rock, and deposition resumed. (See also *Angular unconformity.*)

UNCONSOLIDATED MATERIAL: Unlithified *sediment* that has no *mineral* cement or matrix binding its grains.

UNIFORMITARIANISM, PRINCIPLE OF: The concept that the processes that have shaped Earth through geologic time are the same as those observable today.

UNSATURATED ZONE: The region in the ground between the surface and the *groundwater table* in which pores are not completely filled with water.

UPWARPED MOUNTAINS: Mountains elevated by uplift of broad regions without *faulting*.

U-SHAPED VALLEY: A deep valley with steep upper walls that grade into a flat floor; typical shape of a valley eroded by a *glacier*.

VALLEY (STREAM): The entire area between the top of the slopes on either side of a *stream*.

VALLEY GLACIER: A *glacier* that is smaller than a *continental glacier* or an icecap and that flows mainly along well-defined valleys in mountainous regions.

VARVE: A thin pair of sedimentary layers grading upward from coarse to fine and light to dark, found in a glacial lake and representing one year's deposition.

VEIN: A deposit of foreign *minerals* within a rock fracture or *joint*.

VENTIFACT: A rock that exhibits the effects of *sandblasting* or "snowblasting" on its surfaces, which become flat with sharp edges in between.

VISCOSITY: A measure of a liquid's resistance to flow.

VOLCANIC ASH: A volcanic *sediment* of rock fragments, usually *glass*, less than 2 mm in diameter, that is formed when escaping gases force out a fine spray of *magma*.

VOLCANIC ASH-FLOW DEPOSIT: A layer of *volcanic ash* and debris deposited during a *pyroclastic flow*.

VOLCANIC BRECCIA: A pyroclastic rock in which all fragments are more than 2 mm in diameter.

VOLCANIC DOME: A rounded accumulation around a volcanic vent of congealed *lava* too viscous to flow away quickly; hence usually *rhyolite* lava.

VOLCANIC TUFF: A consolidated rock composed of pyroclastic rock fragments and fine *volcanic ash* welded together by their own heat.

VOLCANISM: The processes that form *volcanoes*; the progress of *magma* as it rises up through the crust, emerges onto the surface as *lava*, and solidifies into volcanic rocks and landforms.

VOLCANO: A hill or mountain that forms from the accumulation of matter that erupts at the surface.

WAVE-CUT TERRACE: A level surface formed by wave *erosion* of coastal *bedrock* to the bottom of the turbulent breaker zone. May appear above sea level if uplifted or if sea level drops.

WAVE HEIGHT: The vertical distance from the trough to the crest of a wave.

WAVELENGTH: The distance between two successive peaks, or between troughs, of a wave.

WAVE REFRACTION: The bending of water waves as they encounter different depths and bottom conditions, or of other waves as they pass from one medium to another of different properties.

WEATHERING: The set of all processes that decay and break up rock, by a combination of physical fracturing and chemical decomposition.

WORLD OCEAN: The combination of all the individual oceans (Atlantic, Pacific, and so on) considered as a single interconnected body of water.

YARDANG: A streamlined, sharp-crested ridge aligned with the direction of the prevailing wind in arid regions. Yardangs appear to have been carved by wind *erosion* and abrasion by silt and dust carried by the wind.

ZEOLITE: A class of silicate *minerals* containing water in cavities within the *crystal* structure. Formed by alteration at low temperature and pressure of other silicates, often volcanic *glass*.

ZONED CRYSTAL: A single *crystal* of one *mineral* which has a different chemical composition in its inner and outer parts; formed in minerals that can have variation in abundance of some elements and caused by the changing concentration of elements in a cooling *magma*.

Index

Page numbers in **boldface** refer to definitions; page numbers in *italics* refer to illustrations and tables. The Appendixes and Glossary are not covered by this index.

THE GEOLOGIC TIME SCALE

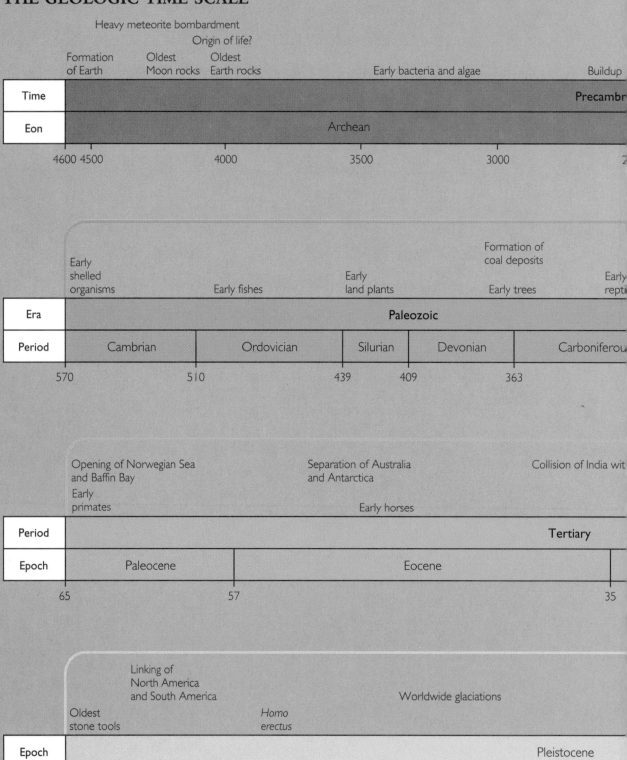

Heavy meteorite bombardment

Origin of life?

| Formation of Earth | Oldest Moon rocks | Oldest Earth rocks | | Early bacteria and algae | | Buildup |

Time								Precambr
Eon		Archean					

4600 4500   4000   3500   3000   2

				Formation of coal deposits	

| Early shelled organisms | | Early fishes | Early land plants | Early trees | Early repti |

Era			Paleozoic			
Period	Cambrian	Ordovician	Silurian	Devonian	Carboniferou

570   510   439   409   363

Opening of Norwegian Sea and Baffin Bay	Separation of Australia and Antarctica	Collision of India wit

Early primates

Early horses

Period			Tertiary
Epoch	Paleocene	Eocene	

65   57   35

Linking of North America and South America		Worldwide glaciations

| Oldest stone tools | Homo erectus | |

Epoch			Pleistocene

1.6   1.5   1.4   1.3   1.2   1.1   1   0.9

Time before present (millions of

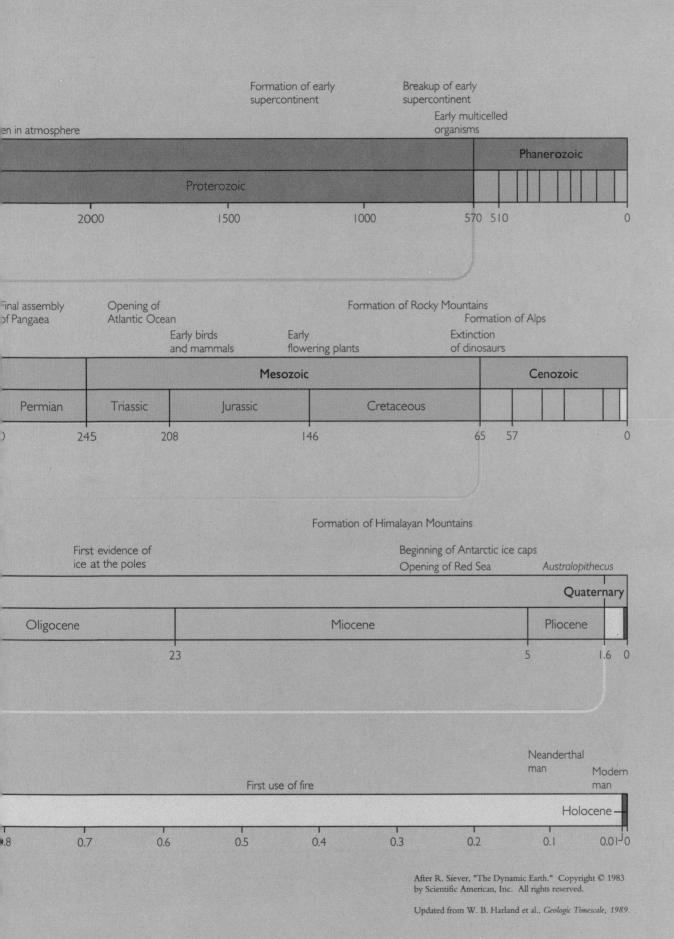

en in atmosphere

Formation of early
supercontinent

Breakup of early
supercontinent

Early multicelled
organisms

Phanerozoic

Proterozoic

| 2000 | 1500 | 1000 | 570 | 510 | 0 |

Final assembly
of Pangaea

Opening of
Atlantic Ocean

Early birds
and mammals

Early
flowering plants

Formation of Rocky Mountains

Formation of Alps

Extinction
of dinosaurs

Mesozoic

Cenozoic

| Permian | Triassic | Jurassic | Cretaceous | |
| 245 | 208 | 146 | 65 | 57 | 0 |

Formation of Himalayan Mountains

First evidence of
ice at the poles

Beginning of Antarctic ice caps

Opening of Red Sea

Australopithecus

Quaternary

| Oligocene | Miocene | Pliocene | |
| 23 | 5 | 1.6 | 0 |

Neanderthal
man

Modern
man

First use of fire

Holocene —

| .8 | 0.7 | 0.6 | 0.5 | 0.4 | 0.3 | 0.2 | 0.1 | 0.01 | 0 |